T0324212

Graduate Texts in Mathematics 136

Springer
New York
Berlin
Heidelberg
Barcelona
Hong Kong
London
Milan
Paris
Singapore
Tokyo

Graduate Texts in Mathematics

(continued after index)

William A. Adkins   Steven H. Weintraub

Algebra

An Approach via Module Theory

 Springer

William A. Adkins
Steven H. Weintraub
Department of Mathematics
Louisiana State University
Baton Rouge, LA 70803
USA

Mathematics Subject Classifications: 12-01, 13-01, 15-01, 16-01, 20-01

Library of Congress Cataloging-in-Publication Data
Adkins, William A.
    Algebra: an approach via module theory/William A. Adkins,
  Steven H. Weintraub.
      p.   cm. — (Graduate texts in mathematics; 136)
    Includes bibliographical references and indexes.
    ISBN 0-387-97839-9. — ISBN 3-540-97839-9
    1. Algebra.   2. Modules (Algebra)   I. Weintraub, Steven H.
  II. Title.   III. Series.
  QA154.A33   1992
  512′.4—dc20   92-11951

Printed on acid-free paper.

Production managed by Francine Sikorski; manufacturing supervised by Jacqui Ashri.
Photocomposed copy prepared using TeX.
Printed and bound by R.R. Donnelley and Sons, Harrisonburg, VA.
Printed in the United States of America.

9 8 7 6 5 4 3 2 (Corrected second printing, 1999)

ISBN 0-387-97839-9 Springer-Verlag New York Berlin Heidelberg
ISBN 3-540-97839-9 Springer-Verlag Berlin Heidelberg New York SPIN 10667846

Preface

This book is designed as a text for a first-year graduate algebra course. As necessary background we would consider a good undergraduate linear algebra course. An undergraduate abstract algebra course, while helpful, is not necessary (and so an adventurous undergraduate might learn some algebra from this book).

Perhaps the principal distinguishing feature of this book is its point of view. Many textbooks tend to be encyclopedic. We have tried to write one that is thematic, with a consistent point of view. The theme, as indicated by our title, is that of modules (though our intention has not been to write a textbook purely on module theory). We begin with some group and ring theory, to set the stage, and then, in the heart of the book, develop module theory. Having developed it, we present some of its applications: canonical forms for linear transformations, bilinear forms, and group representations.

Why modules? The answer is that they are a basic unifying concept in mathematics. The reader is probably already familiar with the basic role that vector spaces play in mathematics, and modules are a generalization of vector spaces. (To be precise, modules are to rings as vector spaces are to fields.) In particular, both abelian groups and vector spaces with a linear transformation are examples of modules, and we stress the analogy between the two—the basic structure theorems in each of these areas are special cases of the structure theorem of finitely generated modules over a principal ideal domain (PID). As well, our last chapter is devoted to the representation theory of a group G over a field $\mathbf{F}$, this being an important and beautiful topic, and we approach it from the point of view of such a representation being an $\mathbf{F}(G)$-module. On the one hand, this approach makes it very clear what is going on, and on the other hand, this application shows the power of the general theory we develop.

We have heard the joke that the typical theorem in mathematics states that something you do not understand is equal to something else you cannot compute. In that sense we have tried to make this book atypical. It has been our philosophy while writing this book to provide proofs with a maximum of insight and a minimum of computation, in order to promote understanding. However, since in practice it is necessary to be able to compute as well, we have included extensive material on computations. (For example, in our entire development in Chapter 4 of canonical forms for linear transformations we only have to compute one determinant, that of a companion matrix. But then Chapter 5 is almost entirely dedicated to computational methods for modules over a PID, showing how to find canonical forms and characteristic polynomials. As a second example, we derive the basic results about complex representations of finite groups in Section 8.3, without mentioning the word *character*, but then devote Section 8.4 to characters and how to use them.)

Here is a more detailed listing of the contents of the book, with emphasis on its novel features:

Chapter 1 is an introduction to (or review of) group theory, including a discussion of semidirect products.

Chapter 2 is an introduction to ring theory, covering a variety of standard topics.

In Chapter 3 we develop basic module theory. This chapter culminates in the structure theorem for finitely generated modules over a PID. (We then specialize to obtain the basic structure theorem for finitely generated Abelian groups.) We feel that our proof of this theorem is a particularly insightful one. (Note that in considering free modules we do not assume the corresponding results for vector spaces to be already known.) Noteworthy along the way is our introduction and use of the language of homological algebra and our discussion of free and projective modules.

We begin Chapter 4 with a treatment of basic topics in linear algebra. In principle, this should be a review, but we are careful to develop as much of the theory as possible over a commutative ring (usually a PID) rather than just restricting ourselves to a field. The matrix representation for module homomorphisms is even developed for modules over noncommutative rings, since this is needed for applications to Wedderburn's theorem in Chapter 7. This chapter culminates in the derivation of canonical forms (the rational canonical form, the (generalized) Jordan canonical form) for linear transformations. Here is one place where the module theory shows its worth. By regarding a vector space V over a field F, with a linear transformation T, as an $F[X]$-module (with X acting by T), these canonical forms are immediate consequences of the structure theorem for finitely generated torsion modules over a PID. We also derive the important special case of the real Jordan canonical form, and end the chapter by deriving the spectral theorem.

Chapter 5 is a computational chapter, showing how to obtain effectively (in so far as is possible) the canonical forms of Chapter 4 in concrete cases. Along the way, we introduce the Smith and Hermite canonical forms as well.

This chapter also has Dixon's proof of a criterion for similarity of matrices based solely on rank computations.

In Chapter 6 we discuss duality and investigate bilinear, sesquilinear, and quadratic forms, with the assistance of module theory, obtaining complete results in a number of important special cases. Among these are the cases of skew-symmetric forms over a PID, sesquilinear (Hermitian) forms over the complex numbers, and bilinear and quadratic forms over the real numbers, over finite fields of odd characteristic, and over the field with two elements (where the Arf invariant enters in the case of quadratic forms).

Chapter 7 has two sections. The first discusses semisimple rings and modules (deriving Wedderburn's theorem), and the second develops some multilinear algebra. Our results in both of these sections are crucial for Chapter 8.

Our final chapter, Chapter 8, is the capstone of the book, dealing with group representations mostly, though not entirely, in the semisimple case. Although perhaps not the most usual of topics in a first-year graduate course, it is a beautiful and important part of mathematics. We view a representation of a group G over a field $\mathbf{F}$ as an $\mathbf{F}(G)$-module, and so this chapter applies (or illustrates) much of the material we have developed in this book. Particularly noteworthy is our treatment of induced representations. Many authors define them more or less ad hoc, perhaps mentioning as an aside that they are tensor products. We define them as tensor products and stick to that point of view (though we provide a recognition principle not involving tensor products), so that, for example, Frobenius reciprocity merely becomes a special case of adjoint associativity of Hom and tensor product.

The interdependence of the chapters is as follows:

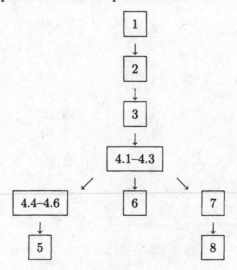

We should mention that there is one subject we do not treat. We do not discuss any field theory in this book. In fact, in writing this book we were careful to avoid requiring any knowledge of field theory or algebraic number theory as a prerequisite.

We use standard set theoretic notation. For the convenience of the reader, we have provided a very brief introduction to equivalence relations and Zorn's lemma in an appendix. In addition, we provide an index of notation, with a reference given of the first occurrence of the symbol.

We have used a conventional decimal numbering system. Thus a reference to Theorem 4.6.23 refers to item number 23 in Section 6 of Chapter 4, which happens to be a theorem. Within a given chapter, the chapter reference is deleted.

The symbol □ is used to denote the end of a proof; the end of proof symbol □ with a blank line is used to indicate that the proof is immediate from the preceding discussion or result.

The material presented in this book is for the most part quite standard. We have thus not attempted to provide references for most results. The bibliography at the end is a collection of standard works on algebra.

We would like to thank the editors of Springer-Verlag for allowing us the opportunity, during the process of preparing a second printing, to correct a number of errors which appeared in the first printing of this book. Moreover, we extend our thanks to our colleagues and those readers who have taken the initiative to inform us of the errors they have found. Michal Jastrzebski and Lyle Ramshaw, in particular, have been most helpful in pointing out mistakes and ambiguities.

Baton Rouge, Louisiana   William A. Adkins
                                                            Steven H. Weintraub

Contents

Chapter 1

Groups

In this chapter we introduce groups and prove some of the basic theorems in group theory. One of these, the structure theorem for finitely generated abelian groups, we do not prove here but instead derive it as a corollary of the more general structure theorem for finitely generated modules over a PID (see Theorem 3.7.22).

1.1 Definitions and Examples

(1.1) Definition. *A* **group** *is a set G together with a binary operation*

$$\cdot : G \times G \to G$$

satisfying the following three conditions:

(a) *$a \cdot (b \cdot c) = (a \cdot b) \cdot c$ for all a, b, $c \in G$. (Associativity)*
(b) *There exists an element $e \in G$ such that $a \cdot e = e \cdot a = a$ for all $a \in G$. (Existence of an identity element)*
(c) *For each $a \in G$ there exists a $b \in G$ such that $a \cdot b = b \cdot a = e$. (Existence of an inverse for each $a \in G$)*

It is customary in working with binary operations to write $a \cdot b$ rather than $\cdot(a, b)$. Moreover, when the binary operation defines a group structure on a set G then it is traditional to write the group operation as ab. One exception to this convention occurs when the group G is **abelian**, i.e., if $ab = ba$ for all a, $b \in G$. If the group G is abelian then the group operation is commonly written additively, i.e., one writes $a + b$ rather than ab. This convention is not rigidly followed; for example, one does not suddenly switch to additive notation when dealing with a group that is a subset of a group written multiplicatively. However, when dealing specifically with abelian groups the additive convention is common. Also, when dealing with abelian groups the identity is commonly written $e = 0$, in conformity with the additive notation. In this chapter, we will write e for the identity of general groups, i.e., those written multiplicatively, but when we study group representation theory in Chapter 8, we will switch to 1 as the identity for multiplicatively written groups.

To present some examples of groups we must give the set G and the operation $\cdot : G \times G \to G$ and then check that this operation satisfies (a), (b), and (c) of Definition 1.1. For most of the following examples, the fact that the operation satisfies (a), (b), and (c) follows from properties of the various number systems with which you should be quite familiar. Thus details of the verification of the axioms are generally left to the reader.

(1.2) Examples.

(1) The set $\mathbf{Z}$ of integers with the operation being ordinary addition of integers is a group with identity $e = 0$, and the inverse of $m \in \mathbf{Z}$ is $-m$. Similarly, we obtain the additive group $\mathbf{Q}$ of rational numbers, $\mathbf{R}$ of real numbers, and $\mathbf{C}$ of complex numbers.

(2) The set $\mathbf{Q}^*$ of nonzero rational numbers with the operation of ordinary multiplication is a group with identity $e = 1$, and the inverse of $a \in \mathbf{Q}^*$ is $1/a$. $\mathbf{Q}^*$ is abelian, but this is one example of an abelian group that is not normally written with additive notation. Similarly, there are the abelian groups $\mathbf{R}^*$ of nonzero real numbers and $\mathbf{C}^*$ of nonzero complex numbers.

(3) The set $\mathbf{Z}_n = \{0, 1, \ldots, n-1\}$ with the operation of addition modulo n is a group with identity 0, and the inverse of $x \in \mathbf{Z}_n$ is $n-x$. Recall that addition modulo n is defined as follows. If $x, y \in \mathbf{Z}_n$, take $x + y \in \mathbf{Z}$ and divide by n to get $x + y = qn + r$ where $0 \le r < n$. Then define $x + y \pmod{n}$ to be r.

(4) The set U_n of complex n^{th} roots of unity, i.e., $U_n = \{\exp((2k\pi i)/n) : 0 \le k \le n - 1\}$ with the operation of multiplication of complex numbers is a group with the identity $e = 1 = \exp(0)$, and the inverse of $\exp((2k\pi i)/n)$ is $\exp((2(n - k)\pi i)/n)$.

(5) Let $\mathbf{Z}_n^* = \{m : 1 \le m < n \text{ and } m \text{ is relatively prime to } n\}$. Under the operation of multiplication modulo n, $\mathbf{Z}_n^*$ is a group with identity 1. Details of the verification are left as an exercise.

(6) If X is a set let S_X be the set of all bijective functions $f : X \to X$. Recall that a function is bijective if it is one-to-one and onto. Functional composition gives a binary operation on S_X and with this operation it becomes a group. S_X is called the group of **permutations** of X or the **symmetric group** on X. If $X = \{1, 2, \ldots, n\}$ then the symmetric group on X is usually denoted S_n and an element α of S_n can be conveniently indicated by a $2 \times n$ matrix

$$\alpha = \begin{pmatrix} 1 & 2 & \cdots & n \\ \alpha(1) & \alpha(2) & \cdots & \alpha(n) \end{pmatrix}$$

where the entry in the second row under k is the image $\alpha(k)$ of k under the function α. To conform with the conventions of functional composition, the product $\alpha\beta$ will be read from right to left, i.e., first do β and then do α. For example,

$$\begin{pmatrix} 1 & 2 & 3 & 4 \\ 3 & 2 & 4 & 1 \end{pmatrix} \begin{pmatrix} 1 & 2 & 3 & 4 \\ 3 & 4 & 1 & 2 \end{pmatrix} = \begin{pmatrix} 1 & 2 & 3 & 4 \\ 4 & 1 & 3 & 2 \end{pmatrix}.$$

(7) Let $GL(n, \mathbf{R})$ denote the set of $n \times n$ invertible matrices with real entries. Then $GL(n, \mathbf{R})$ is a group under matrix multiplication. Let $SL(n, \mathbf{R}) = \{T \in GL(n, \mathbf{R}) : \det T = 1\}$. Then $SL(n, \mathbf{R})$ is a group under matrix multiplication. (In this example, we are assuming familiarity with basic properties of matrix multiplication and determinants. See Chapter 4 for details.) $GL(n, \mathbf{R})$ (respectively, $SL(n, \mathbf{R})$) is known as the **general linear group** (respectively, **special linear group**) of degree n over $\mathbf{R}$.

(8) If X is a set let $\mathcal{P}(X)$ denote the power set of X, i.e., $\mathcal{P}(X)$ is the set of all subsets of X. Define a product on $\mathcal{P}(X)$ by the formula $A \triangle B = (A \setminus B) \cup (B \setminus A)$. $A \triangle B$ is called the symmetric difference of A and B. It is a straightforward exercise to verify the associative law for the symmetric difference. Also note that $A \triangle A = \emptyset$ and $\emptyset \triangle A = A \triangle \emptyset = A$. Thus $\mathcal{P}(X)$ with the symmetric difference operation is a group with $\emptyset$ as identity and every element as its own inverse. Note that $\mathcal{P}(X)$ is an abelian group.

(9) Let $\mathcal{C}(\mathbf{R})$ be the set of continuous real-valued functions defined on $\mathbf{R}$ and let $\mathcal{D}(\mathbf{R})$ be the set of differentiable real-valued functions defined on $\mathbf{R}$. Then $\mathcal{C}(\mathbf{R})$ and $\mathcal{D}(\mathbf{R})$ are groups under the operation of function addition.

One way to explicitly describe a group with only finitely many elements is to give a table listing the multiplications. For example the group $\{1, -1\}$ has the multiplication table

$\cdot$	1	-1
1	1	-1
-1	-1	1

whereas the following table

·	e	a	b	c
e	e	a	b	c
a	a	e	c	b
b	b	c	e	a
c	c	b	a	e

is the table of a group called the **Klein 4-group**. Note that in these tables each entry of the group appears exactly once in each row and column. Also the multiplication is read from left to right; that is, the entry at the intersection of the row headed by α and the column headed by β is the product $\alpha\beta$. Such a table is called a **Cayley diagram** of the group. They are sometimes useful for an explicit listing of the multiplication in small groups.

The following result collects some elementary properties of a group:

(1.3) Proposition. *Let G be a group.*

(1) *The identity e of G is unique.*
(2) *The inverse b of $a \in G$ is unique. We denote it by a^{-1}.*
(3) *$(a^{-1})^{-1} = a$ for all $a \in G$ and $(ab)^{-1} = b^{-1}a^{-1}$ for all $a, b \in G$.*
(4) *If $a, b \in G$ the equations $ax = b$ and $ya = b$ each have unique solutions in G.*
(5) *If $a, b, c \in G$ then $ab = ac$ implies that $b = c$ and $ab = cb$ implies that $a = c$.*

Proof. (1) Suppose e' is also an identity. Then $e' = e'e = e$.

(2) Suppose $ab = ba = e$ and $ab' = b'a = e$. Then $b = eb = (b'a)b = b'(ab) = b'e = b'$, so inverses are unique.

(3) $a(a^{-1}) = (a^{-1})a = e$, so $(a^{-1})^{-1} = a$. Also $(ab)(b^{-1}a^{-1}) = a(bb^{-1})a^{-1} = aa^{-1} = e$ and similarly $(b^{-1}a^{-1})(ab) = e$. Thus $(ab)^{-1} = b^{-1}a^{-1}$.

(4) $x = a^{-1}b$ solves $ax = b$ and $y = ba^{-1}$ solves $ya = b$, and any solution must be the given one as one sees by multiplication on the left or right by a^{-1}.

(5) If $ab = ac$ then $b = a^{-1}(ab) = a^{-1}(ac) = c$. $\square$

The results in part (5) of Proposition 1.3 are known as the cancellation laws for a group.

The associative law for a group G shows that a product of the elements a, b, c of G can be written unambiguously as abc. Since the multiplication is binary, what this means is that any two ways of multiplying $a, b,$ and c (so that the order of occurrence in the product is the given order) produces the same element of G. With three elements there are only two choices for multiplication, that is, $(ab)c$ and $a(bc)$, and the law of associativity says that these are the same element of G. If there are n elements of G then the law of associativity combined with induction shows that we can write $a_1 a_2 \cdots a_n$ unambiguously, i.e., it is not necessary to include parentheses to indicate which sequence of binary multiplications occurred to arrive at an element of G involving all of the a_i. This is the content of the next proposition.

(1.4) Proposition. *Any two ways of multiplying the elements a_1, a_2, ..., a_n in a group G in the order given (i.e., removal of all parentheses produces the juxtaposition $a_1 a_2 \cdots a_n$) produces the same element of G.*

Proof. If $n = 3$ the result is clear from the associative law in G.

Let $n > 3$ and consider two elements g and h obtained as products of a_1, a_2, ..., a_n in the given order. Writing g and h in terms of the last multiplications used to obtain them gives

$$g = (a_1 \cdots a_i) \cdot (a_{i+1} \cdots a_n)$$

and

$$h = (a_1 \cdots a_j) \cdot (a_{j+1} \cdots a_n).$$

Since i and j are less than n, the induction hypothesis implies that the products $a_1 \cdots a_i$, $a_{i+1} \cdots a_n$, $a_1 \cdots a_j$, and $a_{j+1} \cdots a_n$ are unambiguously defined elements in G. Without loss of generality we may assume that $i \leq j$. If $i = j$ then $g = h$ and we are done. Thus assume that $i < j$. Then, by the induction hypothesis, parentheses can be rearranged so that

$$g = (a_1 \cdots a_i)((a_{i+1} \cdots a_j)(a_{j+1} \cdots a_n))$$

and

$$h = ((a_1 \cdots a_i)(a_{i+1} \cdots a_j))(a_{j+1} \cdots a_n).$$

Letting $A = (a_1 \cdots a_i)$, $B = (a_{i+1} \cdots a_j)$, and $C = (a_{j+1} \cdots a_n)$ the induction hypothesis implies that A, B, and C are unambiguously defined elements of G. Then

$$g = A(BC) = (AB)C = h$$

and the proposition follows by the principle of induction.   □

Since products of n elements of G are unambiguous once the order has been specified, we will write $a_1 a_2 \cdots a_n$ for such a product, without any specification of parentheses. Note that the only property of a group used in Proposition 1.4 is the associative property. Therefore, Proposition 1.4 is valid for *any* associative binary operation. We will use this fact to be able to write unambiguous multiplications of elements of a ring in later chapters. A convenient notation for $a_1 \cdots a_n$ is $\prod_{i=1}^{n} a_i$. If $a_i = a$ for all i then $\prod_{i=1}^{n} a$ is denoted a^n and called the nth power of a. Negative powers of a are defined by $a^{-n} = (a^{-1})^n$ where $n > 0$, and we set $a^0 = e$. With these notations the standard rules for exponents are valid.

(1.5) Proposition. *If G is a group and $a \in G$ then*

(1) $a^m a^n = a^{m+n}$, *and*
(2) $(a^m)^n = a^{mn}$ *for all integers m and n.*

Proof. Part (1) follows from Proposition 1.4 while part (2) is an easy exercise using induction.   □

1.2 Subgroups and Cosets

Let G be a group and let $H \subseteq G$ be a subset. H is called a **subgroup** of G if H together with the binary operation of G is a group. The first thing to note is that this requires that H be closed under the multiplication of G, that is, ab is in H whenever a and b are in H. This is no more than the statement that the multiplication on G is defined on H. Furthermore, if H is a subgroup of G then H has an identity e' and G has an identity e. Then $e'e = e'$ since e is the identity of G and $e'e' = e'$ since e' is the identity of H. Thus $e'e = e'e'$ and left cancellation of e' (in the group G) gives $e = e'$. Therefore, the identity of G is also the identity of any subgroup H of G. Also, if $a \in H$ then the inverse of a as an element of H is the same as the inverse of a as an element of G since the inverse of an element is the unique solution to the equations $ax = e = xa$.

(2.1) Proposition. *Let G be a group and let H be a nonempty subset of G. Then H is a subgroup if and only if the following two conditions are satisfied.*

(1) *If $a, b \in H$ then $ab \in H$.*
(2) *If $a \in H$ then $a^{-1} \in H$.*

Proof. If H is a subgroup then (1) and (2) are satisfied as was observed in the previous paragraph. If (1) and (2) are satisfied and $a \in H$ then $a^{-1} \in H$ by (2) and $e = aa^{-1} \in H$ by (1). Thus conditions (a), (b), and (c) in the definition of a group are satisfied for H, and hence H is a subgroup of G.   □

(2.2) Remarks. (1) Conditions (1) and (2) of Proposition 2.1 can be replaced by the following single condition.

(1)′ If $a, b \in H$ then $ab^{-1} \in H$.

Indeed, if $(1)'$ is satisfied then whenever $a \in H$ it follows that $e = aa^{-1} \in H$ and then $a^{-1} = ea^{-1} \in H$. Thus $a \in H$ implies that $a^{-1} \in H$. Also, if $a, b \in H$ then $b^{-1} \in H$ so that $ab = a(b^{-1})^{-1} \in H$. Therefore, $(1)'$ implies (1) and (2). The other implication is clear.

(2) If H is finite then only condition (1) of Proposition 2.1 is necessary to ensure that H is a subgroup of G. To see this suppose that H is a finite set and suppose that $a, b \in H$ implies that $ab \in H$. We need to show that $a^{-1} \in H$ for every $a \in H$. Thus let $a \in H$ and let $T_a : H \to H$ be defined by $T_a(b) = ab$. Our hypothesis implies that $T_a(H) \subseteq H$. If $T_a(b) = T_a(c)$ then $ab = ac$ and left cancellation in the group G (Proposition 1.3 (5)) shows that $b = c$. Hence T_a is an injective map and, since H is assumed to be finite, it follows that T_a is bijective, so the equation $ax = c$ is solvable in H for any choice of $c \in H$. Taking $c = a$ shows that $e \in H$ and then taking $c = e$ shows that $a^{-1} \in H$. Therefore, condition (2) of Proposition 2.1 is satisfied and H is a subgroup of G.

(3) If G is an abelian group with the additive notation, then $H \subseteq G$ is a subgroup if and only if $a - b \in H$ whenever $a, b \in H$.

(2.3) Proposition. *Let I be an index set and let H_i be a subgroup of G for each $i \in I$. Then $H = \bigcap_{i \in I} H_i$ is a subgroup of G.*

Proof. If $a, b \in H$ then $a, b \in H_i$ for all $i \in I$. Thus $ab^{-1} \in H_i$ for all $i \in I$. Hence $ab^{-1} \in H$ and H is a subgroup by Remark 2.2 (1). $\qquad \square$

(2.4) Definition. *Let G and H be groups and let $f : G \to H$ be a function. Then f is a **group homomorphism** if $f(ab) = f(a)f(b)$ for all $a, b \in G$. A **group isomorphism** is an invertible group homomorphism. If f is a group homomorphism, let*

$$\mathrm{Ker}(f) = \{a \in G : f(a) = e\}$$

and

$$\mathrm{Im}(f) = \{h \in H : h = f(a) \quad \text{for some } a \in G\}.$$

$\mathrm{Ker}(f)$ *is the **kernel** of the homomorphism f and $\mathrm{Im}(f)$ is the **image** of f.*

It is easy to check that f is invertible as a group homomorphism if and only if it is invertible as a function between sets, i.e., if and only if it is bijective.

(2.5) Proposition. *Let $f : G \to H$ be a group homomorphism. Then $\mathrm{Ker}(f)$ and $\mathrm{Im}(f)$ are subgroups of G and H respectively.*

Proof. First note that $f(e) = f(ee) = f(e)f(e)$, so by cancellation in H we conclude that $f(e) = e$. Then $e = f(e) = f(aa^{-1}) = f(a)f(a^{-1})$ for all $a \in G$. Thus $f(a^{-1}) = f(a)^{-1}$ for all $a \in G$. Now let $a, b \in \mathrm{Ker}(f)$. Then $f(ab^{-1}) = f(a)f(b^{-1}) = f(a)f(b)^{-1} = ee^{-1} = e$, so $ab^{-1} \in \mathrm{Ker}(f)$

and $\mathrm{Ker}(f)$ is a subgroup of G. Similarly, if $f(a)$, $f(b) \in \mathrm{Im}(f)$ then $f(a)f(b)^{-1} = f(ab^{-1}) \in \mathrm{Im}(f)$, so $\mathrm{Im}(f)$ is a subgroup of H. $\square$

(2.6) Definition. *Let S be a subset of a group G. Then $\langle S \rangle$ denotes the intersection of all subgroups of G that contain S. The subgroup $\langle S \rangle$ is called* **the subgroup generated by** S. *If S is finite and $G = \langle S \rangle$ we say that G is* **finitely generated.** *If $S = \{a\}$ has only one element and $G = \langle S \rangle$ then we say that G is a* **cyclic** *group.*

(2.7) Proposition. *Let S be a nonempty subset of a group G. Then*

$$\langle S \rangle = \{a_1 a_2 \cdots a_n : n \in \mathbf{N} \text{ and } a_i \text{ or } a_i^{-1} \in S \text{ for } 1 \le i \le n\}.$$

That is, $\langle S \rangle$ is the set of all finite products consisting of elements of S or inverses of elements of S.

Proof. Let H denote the set of elements of G obtained as a finite product of elements of S or $S^{-1} = \{a^{-1} : a \in S\}$. If a, $b \in H$ then ab^{-1} is also a finite product of elements from $S \cup S^{-1}$, so $ab^{-1} \in H$. Thus H is a subgroup of G that contains S. Any subgroup K of G that contains S must be closed under multiplication by elements of $S \cup S^{-1}$, so K must contain H. Therefore, $H = \langle S \rangle$. $\square$

(2.8) Examples. You should provide proofs (where needed) for the claims made in the following examples.

(1) The additive group $\mathbf{Z}$ is an infinite cyclic group generated by the number 1.
(2) The multiplicative group $\mathbf{Q}^*$ is generated by the set $S = \{1/p : p$ is a prime number$\} \cup \{-1\}$.
(3) The group $\mathbf{Z}_n$ is cyclic with generator 1.
(4) The group U_n is cyclic with generator $\exp(2\pi i/n)$.
(5) The even integers are a subgroup of $\mathbf{Z}$. More generally, all the multiples of a fixed integer n form a subgroup of $\mathbf{Z}$ and we will see shortly that these are all the subgroups of $\mathbf{Z}$.
(6) If $\alpha = \left(\begin{smallmatrix} 1 & 2 & 3 \\ 2 & 3 & 1 \end{smallmatrix}\right)$ then $H = \{e, \alpha, \alpha^2\}$ is a subgroup of the symmetric group S_3. Also, S_3 is generated by α and $\beta = \left(\begin{smallmatrix} 1 & 2 & 3 \\ 2 & 1 & 3 \end{smallmatrix}\right)$.
(7) If $\beta = \left(\begin{smallmatrix} 1 & 2 & 3 \\ 2 & 1 & 3 \end{smallmatrix}\right)$ and $\gamma = \left(\begin{smallmatrix} 1 & 2 & 3 \\ 3 & 2 & 1 \end{smallmatrix}\right)$ then $S_3 = \langle \beta, \gamma \rangle$.
(8) A matrix $A = [a_{ij}]$ is upper triangular if $a_{ij} = 0$ for $i > j$. The subset $T(n, \mathbf{R}) \subseteq \mathrm{GL}(n, \mathbf{R})$ of invertible upper triangular matrices is a subgroup of $\mathrm{GL}(n, \mathbf{R})$.
(9) If G is a group let $Z(G)$, called the **center of** G, be defined by

$$Z(G) = \{a \in G : ab = ba \quad \text{for all } b \in G\}.$$

Then $Z(G)$ is a subgroup of G.

(10) If G is a group and $x \in G$, then the **centralizer of** x is the subset $C(x)$ of G defined by

$$C(x) = \{a \in G : ax = xa\}.$$

$C(x)$ is a subgroup of G and $C(x) = G$ if and only if $x \in Z(G)$. Also note that $C(x)$ always contains the subgroup $\langle x \rangle$ generated by x.

(11) If G is a group and $a, b \in G$, then $[a, b] = a^{-1}b^{-1}ab$ is called the **commutator** of a and b. The subgroup G' generated by all the commutators of elements of G is called the **commutator subgroup** of G. Another common notation for the commutator subgroup is $[G, G]$. See Exercise 22 for some properties of the commutator subgroup.

(12) A convenient way to describe some groups is by giving generators and relations. Rather than giving formal definitions we shall be content to illustrate the method with two examples of groups commonly expressed by generators and relations. For the first, the **quaternion group** is a group with 8 elements. There are two generators a and b subject to the three relations (and no others):

$$a^4 = e; \qquad b^2 = a^2; \qquad b^{-1}ab = a^{-1}.$$

We leave it for the reader to check that

$$Q = \{e, a, a^2, a^3, b, ab, a^2b, a^3b\}.$$

For a concrete description of Q as a subgroup of $\mathrm{GL}(2, \mathbf{C})$, see Exercise 24.

(13) As our second example of a group expressed by generators and relations, the **dihedral group of order** $2n$, denoted D_{2n}, is a group generated by two elements x and y subject to the three relations (and no others):

$$x^n = e; \qquad y^2 = e; \qquad yxy^{-1} = x^{-1}.$$

Again, we leave it as an exercise to check that

$$D_{2n} = \{e, x, x^2, \ldots, x^{n-1}, y, yx, yx^2, \ldots, yx^{n-1}\}.$$

Thus, D_{2n} has $2n$ elements. The dihedral group will be presented as a group of symmetries in Section 1.6, and it will be studied in detail from the point of view of representation theory in Chapter 8.

(2.9) Definition. *The **order** of G, denoted $|G|$, is the cardinality of the set G. The **order of an element** $a \in G$, denoted $o(a)$ is the order of the subgroup generated by a. (In general, $|X|$ will denote the cardinality of the set X, with $|X| = \infty$ used to indicate an infinite set.)*

(2.10) Lemma. *Let G be a group and $a \in G$. Then*

(1) $o(a) = \infty$ *if and only if $a^n \neq e$ for any $n > 0$.*

(2) *If $o(a) < \infty$, then $o(a)$ is the smallest positive integer n such that $a^n = e$.*

(3) *$a^k = e$ if and only if $o(a) \mid k$.*

Proof. (1) If $a^n \neq e$ for any $n > 0$, then $a^r \neq a^s$ for any $r \neq s$ since $a^r = a^s$ implies $a^{r-s} = e = a^{s-r}$, and if $r \neq s$, then $r - s > 0$ or $s - r > 0$, which is excluded by our hypothesis. Thus, if $a^n \neq e$ for $n > 0$, then $|\langle a \rangle| = \infty$, so $o(a) = \infty$. If $a^n = e$ then let a^m be any element of $\langle a \rangle$. Writing $m = qn + r$ where $0 \leq r < n$ we see that $a^m = a^{nq+r} = a^{nq} a^r = (a^n)^q a^r = e^q a^r = a^r$. Thus $\langle a \rangle = \{e, a, a^2, \ldots, a^{n-1}\}$ and $o(a) \leq n < \infty$.

(2) By part (1), if $o(a) < \infty$ then there is an $n > 0$ such that $a^n = e$ and for each such n the argument in (1) shows that $\langle a \rangle = \{e, a, \ldots, a^{n-1}\}$. If we choose n as the smallest positive integer such that $a^n = e$ then we claim that the powers a^i are all distinct for $0 \leq i \leq n - 1$. Suppose that $a^i = a^j$ for $0 \leq i < j \leq n - 1$. Then $a^{j-i} = e$ and $0 < j - i < n$, contradicting the choice of n. Thus $o(a) = n = $ smallest positive integer such that $a^n = e$.

(3) Assume that $a^k = e$, let $n = o(a)$, and write $k = nq + r$ where $0 \leq r < n$. Then $e = a^k = a^{nq+r} = a^{nq} a^r = a^r$. Part (2) shows that we must have $r = 0$ so that $k = nq$. $\square$

We will now characterize all subgroups of cyclic groups. We start with the group **Z**.

(2.11) Theorem. *If H is a subgroup of **Z** then H consists of all the multiples of a fixed integer m, i.e., $H = \langle m \rangle$.*

Proof. If $H = \{0\}$ we are done. Otherwise H contains a positive integer since H contains both n and $-n$ whenever it contains n. Let m be the least positive integer in H. Then we claim that $H = \{km : k \in \mathbf{Z}\} = \langle m \rangle$. Indeed, let $n \in H$. Then write $n = qm + r$ where $0 \leq r < m$. Since $n \in H$ and $m \in H$, it follows that $r = n - qm \in H$ because H is a subgroup of **Z**. But $0 \leq r < m$ so the choice of m forces $r = 0$, otherwise r is a smaller positive integer in H than m. Hence $n = qm$ so that every element of H is a multiple of m, as required. $\square$

We now determine all subgroups of a cyclic group G. Assume that $G = \langle a \rangle$ and let H be a subgroup of G such that $H \neq \{e\}$. If H contains a power a^{-m} with a negative exponent then it also contains the inverse a^m, which is a positive power of a. Arguing as in Theorem 2.11, let m be the smallest positive integer such that $a^m \in H$. Let a^s be an arbitrary element of H and write $s = qm + r$ where $0 \leq r < m$. Then $a^r = a^{s-qm} = a^s (a^m)^{-q} \in H$ since a^s and a^m are in H. Thus we must have $r = 0$ since $r < m$ and m is the smallest positive integer with $a^m \in H$. Therefore, $s = qm$ and $a^s = (a^m)^q$ so that all elements of H are powers of a^m.

If a is of finite order n so that $a^n = e$ then n must be divisible by m because $e = a^n \in H$ so that $n = qm$ for some q. In this case, $H =$

$\{e, a^m, a^{2m}, \ldots, a^{(q-1)m}\}$. Therefore, $|H| = q = n/m$. However, if the order of a is infinite, then $H = \{e, a^{\pm m}, a^{\pm 2m}, \ldots\} = \langle a^m \rangle$ is also infinite cyclic. Thus we have proved the following result.

(2.12) Theorem. *Any subgroup H of a cyclic group $G = \langle a \rangle$ is cyclic. Moreover, either $H = \langle e \rangle$ or $H = \langle a^m \rangle$ where m is the smallest positive power of a that is in H. If G is infinite then m is arbitrary and H is infinite cyclic. If $|G| = n$ then $m \mid n$ and $|H| = n/m$. If m is any factor of n then there is exactly one subgroup H of G of order n/m, namely, $H = \langle a^m \rangle$.*

The above theorem gives a complete description of cyclic groups and their subgroups. From this description, it is easy to see that any two cyclic groups of order n are isomorphic, as well as any two infinite cyclic groups are isomorphic. Indeed, if $G = \langle a \rangle$ and $H = \langle b \rangle$ where $|G| = |H| = n$ then define $f : G \to H$ by $f(a^m) = b^m$ for all m. One checks that f is a group isomorphism. In particular, every cyclic group of order n is isomorphic to the additive group $\mathbf{Z}_n$ of integers modulo n (see Example 1.2 (3)), and any infinite cyclic group is isomorphic to the additive group $\mathbf{Z}$.

(2.13) Definition. *Let G be a group and H a subgroup. For a fixed element $a \in G$ we define two subsets of G:*

(1) *The **left coset** of H in G determined by a is the set $aH = \{ah : h \in H\}$. The element a is called a representative of the left coset aH.*
(2) *The **right coset** of H in G determined by a is the set $Ha = \{ha : h \in H\}$. The element a is called a representative of the right coset Ha.*

Remark. Unfortunately, there is no unanimity on this definition in the mathematical world. Some authors define left and right cosets as we do; others have the definitions reversed.

A given left or right coset of H can have many different representatives. The following lemma gives a criterion for two elements to represent the same coset.

(2.14) Lemma. *Let H be a subgroup of G and let $a, b \in G$. Then*

(1) $aH = bH$ *if and only if* $a^{-1}b \in H$, *and*
(2) $Ha = Hb$ *if and only if* $ab^{-1} \in H$.

Proof. We give the proof of (1). Suppose $a^{-1}b \in H$ and let $b = ah$ for some $h \in H$. Then $bh' = a(hh')$ for all $h' \in H$ and $ah_1 = (ah)(h^{-1}h_1) = b(h^{-1}h_1)$ for all $h_1 \in H$. Thus $aH = bH$. Conversely, suppose $aH = bH$. Then $b = be = ah$ for some $h \in H$. Therefore, $a^{-1}b = h \in H$. □

(2.15) Theorem. *Let H be a subgroup of G. Then the left cosets (right cosets) of H form a partition of G.*

Proof. Define a relation L on G by setting $a \sim_L b$ if and only if $a^{-1}b \in H$. Note that

(1) $a \sim_L a$,
(2) $a \sim_L b$ implies $b \sim_L a$ (since $a^{-1}b \in H$ implies that $b^{-1}a = (a^{-1}b)^{-1} \in H$), and
(3) $a \sim_L b$ and $b \sim_L c$ implies $a \sim_L c$.

Thus, L is an equivalence relation on G and the equivalence classes of L, denoted $[a]_L$, partition G. (See the appendix.) That is, the equivalence classes $[a]_L$ and $[b]_L$ are identical or they do not intersect. But

$$
\begin{aligned}
[a]_L &= \{b \in G : a \sim_L b\} \\
&= \{b \in G : a^{-1}b \in H\} \\
&= \{b \in G : b = ah \text{ for some } h \in H\} \\
&= aH.
\end{aligned}
$$

Thus, the left cosets of H partition G and similarly for the right cosets. □

The function $\phi_a : H \to aH$ defined by $\phi_a(h) = ah$ is bijective by the left cancellation property. Thus, every left coset of H has the same cardinality as H, i.e., $|aH| = |H|$ for every $a \in G$. Similarly, by the right cancellation law the function $\psi_a(h) = ha$ from H to Ha is bijective so that every right coset of H also has the same cardinality as H. In particular, all right and left cosets of H have the same cardinality, namely, that of H itself.

(2.16) Definition. *If H is a subgroup of G we define the **index of H in G**, denoted $[G : H]$, to be the number of left cosets of H in G. The left cosets of H in G are in one-to-one correspondence with the right cosets via the correspondence $aH \leftrightarrow Ha^{-1} = (aH)^{-1}$. Therefore, $[G : H]$ is also the number of right cosets of H in G.*

(2.17) Theorem. (Lagrange) *If H is a subgroup of a finite group G, then $[G : H] = |G|/|H|$, and in particular, $|H|$ divides $|G|$.*

Proof. The left cosets of H partition G into $[G : H]$ sets, each of which has exactly $|H|$ elements. Thus, $|G| = [G : H]|H|$. □

(2.18) Corollary. *If G is a finite group and $a \in G$ then $o(a) \mid |G|$.*

Proof. □

(2.19) Corollary. *If $|G| = n$, then $a^n = e$ for all $a \in G$.*

Proof.   □

(2.20) Corollary. *If $|G| = p$ where p is prime, then G is a cyclic group.*

Proof. Choose $a \in G$ with $a \neq e$ and consider the subgroup $H = \langle a \rangle$. Then $H \neq \{e\}$, and since $|H| \mid |G| = p$, it follows that $|H| = p$, so $H = G$.   □

(2.21) *Remark.* The converse of Theorem 2.17 is false in the sense that if m is an integer dividing $|G|$, then there need not exist a subgroup H of G with $|H| = m$. A counterexample is given in Exercise 31. It is true, however, when m is prime. This will be proved in Theorem 4.7.

(2.22) Definition. *If G is any group, then the **exponent** of G is the smallest natural number n such that $a^n = e$ for all $a \in G$. If no such n exists, we say that G has infinite exponent.*

If $|G| < \infty$, then Corollaries 2.18 and 2.19 show that the exponent of G divides the order of G.

There is a simple multiplication formula relating indices for a chain of subgroups $K \subseteq H \subseteq G$.

(2.23) Proposition. *Let G be a group and H, K subgroups with $K \subseteq H$. If $[G : K] < \infty$ then*

$$[G : K] = [G : H][H : K].$$

Proof. Choose one representative a_i $(1 \leq i \leq [G : H])$ for each left coset of H in G and one representative b_j $(1 \leq j \leq [H : K])$ for each left coset of K in H. Then we claim that the set

$$\{a_i b_j : 1 \leq i \leq [G : H], \; 1 \leq j \leq [H : K]\}$$

consists of exactly one representative from each left coset of K in G. To see this, let cK be a left coset of K in G. Then $c \in a_i H$ for a unique a_i so that $c = a_i h$. Then $h \in b_j K$ for a unique b_j so that $c = a_i b_j k$ for uniquely determined a_i, $b_j k$. Therefore, $cK = a_i b_j K$ for unique a_i, b_j, and we conclude that the number of left cosets of K in G is $[G : H][H : K]$.   □

(2.24) *Remark.* If $|G| < \infty$ then Proposition 2.23 follows immediately from Lagrange's theorem. Indeed, in this case $[G : K] = |G|/|K| = (|G|/|H|)(|H|/|K|) = [G : H][H : K]$.

(2.25) Examples.

(1) If $G = \mathbf{Z}$ and $H = 2\mathbf{Z}$ is the subgroup of even integers, then the cosets of H consist of the even integers and the odd integers. Thus,

$[\mathbf{Z} : 2\mathbf{Z}] = 2$. Since $\mathbf{Z}$ is abelian, it is not necessary to distinguish between left and right cosets.

(2) If $G = \mathbf{Z}$ and $H = n\mathbf{Z}$, then $[\mathbf{Z} : n\mathbf{Z}] = n$ where the coset $m+H$ consists of all integers that have the same remainder as m upon division by n.

(3) Let $G = S_3 = \{e, \alpha, \alpha^2, \beta, \alpha\beta, \alpha^2\beta\}$ where $\alpha = \left(\begin{smallmatrix} 1 & 2 & 3 \\ 2 & 3 & 1 \end{smallmatrix}\right)$ and $\beta = \left(\begin{smallmatrix} 1 & 2 & 3 \\ 2 & 1 & 3 \end{smallmatrix}\right)$. If $H = \langle \beta \rangle$, then the left cosets of H in G are

$$H = \{e, \beta\} \qquad \alpha H = \{\alpha, \alpha\beta\} \qquad \alpha^2 H = \{\alpha^2, \alpha^2\beta\},$$

while the right cosets are

$$H = \{e, \beta\} \qquad H\alpha = \{\alpha, \alpha^2\beta\} \qquad H\alpha^2 = \{\alpha^2, \alpha\beta\}.$$

Note that, in this example, left cosets are not the same as right cosets.

(4) Let $G = GL(2, \mathbf{R})$ and let $H = SL(2, \mathbf{R})$. Then A, $B \in GL(2, \mathbf{R})$ are in the same left coset of H if and only if $A^{-1}B \in H$, which means that $\det(A^{-1}B) = 1$. This happens if and only if $\det A = \det B$. Similarly, A and B are in the same right coset of H if and only if $\det A = \det B$. Thus in this example, left cosets of H are also right cosets of H. A set of coset representatives consists of the matrices

$$\left\{ \begin{bmatrix} a & 0 \\ 0 & 1 \end{bmatrix} : a \in \mathbf{R}^* \right\}.$$

Therefore, the set of cosets of H in G is in one-to-one correspondence with the set of nonzero real numbers.

(5) **Groups of order ≤ 5.** Let G be a group with $|G| \leq 5$. If $|G| = 1, 2, 3,$ or 5 then Corollary 2.20 shows that G is cyclic. Suppose now that $|G| = 4$. Then every element $a \neq e \in G$ has order 2 or 4. If G has an element a of order 4 then $G = \langle a \rangle$ and G is cyclic. If G does not have any element of order 4 then $G = \{e, a, b, c\}$ where $a^2 = b^2 = c^2 = e$ since each nonidentity element must have order 2. Now consider the product ab. If $ab = e$ then $ab = a^2$, so $b = a$ by cancellation. But a and b are distinct elements. Similarly, ab cannot be a or b, so we must have $ab = c$. A similar argument shows that $ba = c$, $ac = b = ca$, $bc = a = cb$. Thus, G has the Cayley diagram of the Klein 4-group. Therefore, we have shown that there are exactly two nonisomorphic groups of order 4, namely, the cyclic group of order 4 and the Klein 4-group.

The left cosets of a subgroup were seen (in the proof of Theorem 2.14) to be a partition of G by describing an explicit equivalence relation on G. There are other important equivalence relations that can be defined on a group G. We will conclude this section by describing one such equivalence relation.

(2.26) Definition. *Let G be a group and let a, $b \in G$. Then a is **conjugate** to b if there is a $g \in G$ such that $b = gag^{-1}$. It is easy to check that conjugacy* *is an equivalence relation on G. The equivalence classes are called* **conjugacy classes**. *Let $[a]_C$ denote the conjugacy class of the element $a \in G$.*

(2.27) Proposition. *Let G be a group and let $a \in G$. Then*

$$|[a]_C| = [G : C(a)]$$

where $C(a)$ is the centralizer of the element a.

Proof. Since

$$gag^{-1} = hah^{-1} \Leftrightarrow g^{-1}h \in C(a)$$
$$\Leftrightarrow gC(a) = hC(a),$$

there is a bijective function $\phi : [a]_C \to G/C(a) = $ the set of left cosets of $C(a)$, defined by $\phi(gag^{-1}) = gC(a)$, which gives the result. $\square$

(2.28) Corollary. (Class equation) *Let G be a finite group. Then*

$$|G| = |Z(G)| + \sum [G : C(a)]$$

where the sum is over a complete set of nonconjugate a not in $Z(G)$.

Proof. Since $|[a]_C| = 1$ if and only if $a \in Z(G)$, the above equation is nothing more than the partition of G into equivalence classes under conjugation, with the observation that all equivalence classes consisting of a single element have been grouped into $|Z(G)|$. $\square$

1.3 Normal Subgroups, Isomorphism Theorems, and Automorphism Groups

If G is a group, let $\mathcal{P}^*(G)$ denote the set of all nonempty subsets of G and define a multiplication on $\mathcal{P}^*(G)$ by the formula

$$ST = \{st : s \in S, \quad t \in T\}$$

where $S, T \in \mathcal{P}^*(G)$. Since the multiplication in G is associative it follows that the multiplication in $\mathcal{P}^*(G)$ is associative, so that parentheses are not necessary in multiplications such as $STUV$. If $S = \{s\}$ then we will write sT or Ts instead of $\{s\}T$ or $T\{s\}$. In particular, if H is a subgroup of G and $a \in G$ then the left coset aH is just the product in $\mathcal{P}^*(G)$ of the subsets $\{a\}$ and H of G and there is no ambiguity in the notation aH. The subset $\{e\} \in \mathcal{P}^*(G)$ satisfies $eS = Se = S$ for all $S \in \mathcal{P}^*(G)$. Thus $\mathcal{P}^*(G)$ has an identity element for its multiplication, namely, $\{e\}$, and hence $\mathcal{P}^*(G)$ forms what is called a **monoid** (a set with an associative multiplication with an identity element), but it is not a group except in the trivial case $G = \{e\}$ since an inverse will not exist (using the multiplication on $\mathcal{P}^*(G)$) for any subset S of G with $|S| > 1$. If $S \in \mathcal{P}^*(G)$ let $S^{-1} = \{s^{-1} : s \in S\}$. Note, however, that S^{-1} is not the inverse of S under the multiplication of $\mathcal{P}^*(G)$ except when S contains only one element. If H is a subgroup of G, then $HH = H$, and if $|H| < \infty$, then Remark 2.2 (2) implies that this equality is equivalent to H being a subgroup of G. If H is a subgroup of G then $H^{-1} = H$ since subgroups are closed under inverses.

Now consider the following question. Suppose H, $K \in \mathcal{P}^*(G)$ are subgroups of G. Then under what conditions is HK a subgroup of G? The following lemma gives one answer to this question; another answer will be provided later in this section after the concept of normal subgroup has been introduced.

(3.1) Lemma. *If H and K are subgroups of G then HK is a subgroup if and only if $HK = KH$.*

Proof. If HK is a subgroup, then HK contains all inverses of elements of HK. Thus, $HK = (HK)^{-1} = K^{-1}H^{-1} = KH$.

Conversely, suppose that $HK = KH$. Then HK is closed under inverses since $(HK)^{-1} = KH = HK$, and it is closed under products since $(HK)(HK) = HKHK = HHKK = HK$. Thus, HK is a subgroup by Proposition 2.1. $\square$

The equality $HK = KH$ is an equality of subsets of G; it should not be confused with element by element commutativity. In terms of elements, $HK = KH$ means that any product hk ($h \in H$, $k \in K$) can also be written $k_1 h_1$ for some $k_1 \in K$, $h_1 \in H$. If G is abelian this is of course automatic.

We now consider the question of when the subset of $\mathcal{P}^*(G)$ consisting of all the left cosets of a subgroup H is closed under the multiplication on $\mathcal{P}^*(G)$.

(3.2) Definition. *If H is a subgroup of G then $G/H \subseteq \mathcal{P}^*(G)$ will denote the set of all left cosets of H in G. It is called the* **coset space** *of H in G.*

Consider two left cosets of H, say aH and bH. If $(aH)(bH) = cH$, then $ab \in cH$, and hence $cH = abH$. Therefore, to ask if G/H is closed under multiplication is to ask if the equation $(aH)(bH) = abH$ is true for all $a, b \in G$.

(3.3) Lemma. *If H is a subgroup of G, then $(aH)(bH) = abH$ for all $a, b \in G$ if and only if $cHc^{-1} = H$ for all $c \in G$.*

Proof. Suppose $cHc^{-1} = H$ for all $c \in G$. Then $cH = Hc$ for all $c \in G$, so

$$(aH)(bH) = a(Hb)H = a(bH)H = abHH = abH.$$

Conversely, if $(aH)(bH) = abH$ for all $a, b \in G$, then

$$cHc^{-1} \subseteq cHc^{-1}H = cc^{-1}H = H$$

for all $c \in G$. Replacing c by c^{-1} (since $c^{-1} \in G$) gives an inclusion $c^{-1}Hc \subseteq H$ and multiplying on the left by c and the right by c^{-1} gives $H \subseteq cHc^{-1}$. Hence, $cHc^{-1} = H$ for all $c \in G$. $\square$

(3.4) Definition. *A subgroup N of G is said to be* **normal**, *denoted $N \lhd G$, if $aNa^{-1} = N$ for all $a \in G$.*

(3.5) *Remark.* The argument in Lemma 3.3 shows that N is normal in G if and only if $aNa^{-1} \subseteq N$ for all $a \in G$. This is frequently easier to check than the equality $aNa^{-1} = N$. Also note that Definition 3.4 is equivalent to $aN = Na$ for all $a \in G$.

(3.6) Proposition. *If $N \lhd G$, then the coset space $G/N \subseteq \mathcal{P}^*(G)$ forms a group under the multiplication inherited from $\mathcal{P}^*(G)$.*

Proof. By Lemma 3.3, G/N is closed under the multiplication on $\mathcal{P}^*(G)$. Since the multiplication on $\mathcal{P}^*(G)$ is already associative, it is only necessary to check the existence of an identity and inverses. But the coset $N = eN$ satisfies

$$(eN)(aN) = eaN = aN = aeN = (aN)(eN),$$

so N is an identity of G/N. Also

$$(aN)(a^{-1}N) = aa^{-1}N = eN = N = a^{-1}aN = (a^{-1}N)(aN)$$

so that $a^{-1}N$ is an inverse of aN. Therefore, the axioms for a group structure on G/N are satisfied. $\square$

(3.7) Definition. *If $N \lhd G$, then G/N is called the* **quotient group** *of G by N.*

(3.8) *Remark.* If $N \lhd G$ and $|G| < \infty$, then Lagrange's theorem (Theorem 2.17) shows that $|G/N| = [G : N] = |G|/|N|$.

(3.9) Examples.

(1) If G is abelian, then every subgroup of G is normal.
(2) $\mathrm{SL}(n, \mathbf{R})$ is a normal subgroup of $\mathrm{GL}(n, \mathbf{R})$. Indeed, if $A \in \mathrm{GL}(n, \mathbf{R})$ and $B \in \mathrm{SL}(n, \mathbf{R})$ then

$$\det(ABA^{-1}) = (\det A)(\det B)(\det A)^{-1} = 1$$

so that $ABA^{-1} \in \mathrm{SL}(n, \mathbf{R})$ for all $A \in \mathrm{GL}(n, \mathbf{R})$ and $B \in \mathrm{SL}(n, \mathbf{R})$. The quotient group $\mathrm{GL}(n, \mathbf{R})/\mathrm{SL}(n, \mathbf{R})$ is isomorphic to $\mathbf{R}^*$, the multiplicative group of nonzero real numbers. This will follow from Theorem 3.11 (to be proved shortly) by considering the homomorphism $\det : \mathrm{GL}(n, \mathbf{R}) \to \mathbf{R}^*$. The details are left as an exercise.

(3) The subgroup $T(n, \mathbf{R})$ of upper triangular matrices is *not* a normal subgroup of $\mathrm{GL}(n, \mathbf{R})$. For example, take $n = 2$ and let $A = \begin{bmatrix} 1 & 0 \\ 1 & 1 \end{bmatrix}$ and $B = \begin{bmatrix} 1 & 1 \\ 0 & 1 \end{bmatrix}$. Then $ABA^{-1} = \begin{bmatrix} 0 & 1 \\ -1 & 2 \end{bmatrix} \notin T(2, \mathbf{R})$. A similar example can be constructed for any $n > 1$. Thus the set of cosets $\mathrm{GL}(n, \mathbf{R})/T(n, \mathbf{R})$ does not form a group under the operation of coset multiplication.

(4) If $\alpha = \begin{pmatrix} 1 & 2 & 3 \\ 2 & 3 & 1 \end{pmatrix}$, then $H = \{e, \alpha, \alpha^2\}$ is a normal subgroup of the symmetric group S_3 (check it). If $\beta \notin H$ then the cosets are H and βH.

(5) Let $K = \langle \beta \rangle \subseteq S_3$ where $\beta = \begin{pmatrix} 1 & 2 & 3 \\ 2 & 1 & 3 \end{pmatrix}$. Then the left cosets of K in G are

$$K = \{e, \beta\} \qquad \alpha K = \{\alpha, \alpha\beta\} \qquad \alpha^2 K = \{\alpha^2, \alpha^2\beta\}$$

where α is the permutation defined in Example 3.9 (4). Then

$$K(\alpha K) = \{e, \alpha\}\{\alpha, \alpha\beta\} = \{\alpha, \alpha\beta, \alpha^2, \alpha^2\beta\} \neq \alpha K.$$

Therefore, the product of two cosets of K is not a coset of K, and in particular, K is not a normal subgroup of S_3. A straightforward calculation shows that $\alpha K \alpha^{-1} \neq K$.

(3.10) Proposition. *Let* $f : G \to H$ *be a group homomorphism. Then* $\mathrm{Ker}(f) \triangleleft G$.

Proof. Let $a \in G$ and $b \in \mathrm{Ker}(f)$. Then

$$f(aba^{-1}) = f(a)f(b)f(a^{-1}) = f(a)ef(a)^{-1} = e$$

so $aba^{-1} \in \mathrm{Ker}(f)$ for all $b \in \mathrm{Ker}(f)$, $a \in G$ and $\mathrm{Ker}(f)$ is normal by Remark 3.5. $\qquad\square$

In fact, Proposition 3.10 describes *all* possible normal subgroups of a group G. To see this let $N \triangleleft G$ and define a function $\pi : G \to G/N$ by the formula $\pi(a) = aN$. By the definition of multiplication on G/N we see that

$$\pi(ab) = abN = (aN)(bN) = \pi(a)\pi(b).$$

Thus, π is a group homomorphism (called the **natural projection** or simply **natural map**) from G to G/N. Note that $\mathrm{Ker}(\pi) = N$ and therefore N is the kernel of a group homomorphism. Since N was an arbitrary normal subgroup of G, it follows that *the normal subgroups of G are precisely the kernels of all possible group homomorphisms from G to some other group.*

We now present some general results, which are commonly called the **noether isomorphism theorems**. Similar results will also be seen in the theory of rings and the theory of modules.

(3.11) Theorem. (First isomorphism theorem) *Let* $f : G \to H$ *be a group homomorphism with kernel* K. *Then* $G/K \cong \mathrm{Im}(f)$ ($\cong$ *means is isomorphic to*).

Proof. Define a function $\overline{f} : G/K \to \mathrm{Im}(f)$ by the formula $\overline{f}(aK) = f(a)$. The first thing that needs to be checked is that this is a well-defined function since the coset aK may also be a coset bK. It is necessary to check that $f(a) = f(b)$ in this case. But $aK = bK$ if and only if $a^{-1}b \in K$, which means that $f(a^{-1}b) = e$ or $f(a) = f(b)$. Therefore, $\overline{f}$ is a well-defined function on G/K. Also

$$\overline{f}((aK)(bK)) = \overline{f}(abK) = f(ab) = f(a)f(b) = \overline{f}(aK)\overline{f}(bK)$$

so that $\overline{f}$ is a homomorphism. $\overline{f}$ is clearly surjective and $\mathrm{Ker}(\overline{f}) = K$ which is the identity of G/K. Hence $\overline{f}$ is an isomorphism. $\square$

Recall from Lemma 3.1 that the product HK of two subgroups H, K is a subgroup if and only if $HK = KH$. There is a simple criterion for this commutativity.

(3.12) Lemma. *Let* H, K *be subgroups of* G. *If either* H *or* K *is normal in* G, *then* HK *is a subgroup of* G.

Proof. Suppose $K \triangleleft G$. Then $aK = Ka$ for all $a \in G$. In particular, $HK = KH$, so HK is a subgroup. $\square$

(3.13) Theorem. (Second isomorphism theorem) *Let* H *and* N *be subgroups of* G *with* $N \triangleleft G$. *Then* $H/(H \cap N) \cong HN/N$.

Proof. Let $\pi : G \to G/N$ be the natural map and let π_0 be the restriction of π to H. Then π_0 is a homomorphism with $\mathrm{Ker}(\pi_0) = H \cap N$. Thus,

$$H/(H \cap N) = H/\mathrm{Ker}(\pi_0) \cong \mathrm{Im}(\pi_0).$$

But the image of π_0 is the set of all cosets of N having representatives in H. Therefore, $\mathrm{Im}(\pi_0) = HN/N$. $\square$

(3.14) Theorem. (Third isomorphism theorem) *Let* $N \triangleleft G$, $H \triangleleft G$ *and assume that* $N \subseteq H$. *Then*

$$G/H \cong (G/N)/(H/N).$$

Proof. Define a function $f : G/N \to G/H$ by the formula $f(aN) = aH$. It is easy to check (do it) that this is a well-defined group homomorphism. Then

$$\mathrm{Ker}(f) = \{aN : aH = H\} = \{aN : a \in H\} = H/N.$$

The result then follows from the first isomorphism theorem. $\square$

(3.15) Theorem. (Correspondence theorem) *Let $N \lhd G$ and let $\pi :$ $G \to G/N$ be the natural map. Then the function $H \mapsto H/N$ defines a one-to-one correspondence between the set of all subgroups of G containing N and the set of all subgroups of G/N. This correspondence satisfies the following properties.*

(1) $H_1 \subseteq H_2$ *if and only if* $H_1/N \subseteq H_2/N$, *and in this case*

$$[H_2 : H_1] = [H_2/N : H_1/N].$$

(2) $H \lhd G$ *if and only if* $H/N \lhd G/N$.

Proof. Letting

$$S_1 = \{H : H \text{ is a subgroup of } G \text{ containing } N\}$$

and

$$S_2 = \{\text{subgroups of } G/N\},$$

define $\alpha : S_1 \to S_2$ by $\alpha(H) = H/N = \mathrm{Im}(\pi|_H)$. Suppose $H_1/N = H_2/N$ where H_1, $H_2 \in S_1$. We claim that $H_1 = H_2$. Let $h_1 \in H_1$. Then $h_1 N \in H_2/N$, so $h_1 N = h_2 N$ where $h_2 \in H_2$. Therefore, $H_1 \subseteq H_2$ and a similar argument shows that $H_2 \subseteq H_1$ so that $H_1 = H_2$. Thus α is one-to-one. If $K \in S_2$ then $\pi^{-1}(K) \in S_1$ and $\alpha(\pi^{-1}(K)) = K$ so that α is surjective. We conclude that α is a $1-1$ correspondence between S_1 and S_2.

Now consider properties (1) and (2). The fact that $H_1 \subseteq H_2$ if and only if $H_1/N \subseteq H_2/N$ is clear. To show that $[H_2 : H_1] = [H_2/N : H_2/N]$ it is necessary to show that the set of cosets aH_1 (for $a \in H_2$) is in one-to-one correspondence with the set of cosets $\bar{a}H_1/N$ (for $\bar{a} \in H_2/N$). This is left as an exercise.

Suppose $H \lhd G$. Then $H/N \lhd G/N$ since

$$(aN)(H/N)(aN)^{-1} = (aHa^{-1})/N = H/N.$$

Conversely, let H/N be a normal subgroup of G/N. Then if $\pi_1 :$ $G/N \to (G/N)/(H/N)$ is the natural map we see that $\mathrm{Ker}(\pi_1 \circ \pi) = H$. Thus, $H \lhd G$. $\qquad \square$

The following result is a simple, but useful, criterion for normality of a subgroup:

(3.16) Proposition. *Let H be a subgroup of G with $[G : H] = 2$. Then $H \lhd G$.*

Proof. Let $a \in G$. If $a \in H$ then certainly $aHa^{-1} = H$. If $a \notin H$ then $G = H \cup aH$ (since $[G : H] = 2$), so the left coset of H containing a is $G \setminus H$. But also $G = H \cup Ha$ (since $[G : H] = 2$), so the right coset of H containing a is $G \setminus H$. Hence, $aH = Ha$ so that $aHa^{-1} = H$ for all $a \in G$ and $H \lhd G$. $\qquad \square$

(3.17) Definition. *If G is a group then an* **automorphism** *of G is a group isomorphism $\phi : G \rightarrow G$. Aut(G) will denote the set of all automorphisms of G. Under the operation of functional composition Aut(G) is a group; in fact, it is a subgroup of the symmetric group S_G on the set G (Example 1.2 (6)).*

(3.18) Examples.

(1) Aut$(\mathbf{Z}) \cong \mathbf{Z}_2$. To see this let $\phi \in$ Aut$(\mathbf{Z})$. Then if $\phi(1) = r$ it follows that $\phi(m) = mr$ so that $\mathbf{Z} = \text{Im}(\phi) = \langle r \rangle$. Therefore, r must be a generator of $\mathbf{Z}$, i.e., $r = \pm 1$. Hence $\phi(m) = m$ or $\phi(m) = -m$ for all $m \in \mathbf{Z}$.

(2) Let $G = \{(a, b) : a, b \in \mathbf{Z}\}$. Then Aut$(G)$ is not abelian. Indeed,

$$\text{Aut}(G) \cong \text{GL}(2, \mathbf{Z}) = \left\{ \begin{bmatrix} a & b \\ c & d \end{bmatrix} : a, b, c, d \in \mathbf{Z} \quad \text{and} \quad ad - bc = \pm 1 \right\}.$$

(3) Let V be the Klein 4-group. Then Aut$(V) \cong S_3$ (exercise).

(3.19) Definition. *If $a \in G$ define $I_a : G \rightarrow G$ by $I_a(b) = aba^{-1}$. Then $I_a \in$ Aut(G). An automorphism of G of the form I_a for some $a \in G$ is called an* **inner automorphism** *or* **conjugation** *of G. All other automorphisms are called* **outer automorphisms** *of G. Let Inn(G) denote the set of all inner automorphisms of G. Define a function $\Phi : G \rightarrow$ Aut(G) by $\Phi(a) = I_a$. Thus Im$(\Phi) =$ Inn(G).*

(3.20) Proposition. *Φ is a group homomorphism with Im$(\Phi) =$ Inn(G) and*

$$\text{Ker}(\Phi) = Z(G).$$

Recall (Example 2.8 (9)) that $Z(G)$ denotes the center of G, i.e.,

$$Z(G) = \{a \in G : ab = ba \quad \text{for all } b \in G\}.$$

Proof. $\Phi(ab)(c) = I_{ab}(c) = (ab)c(ab)^{-1} = a(bcb^{-1})a^{-1} = I_a(I_b(c)) = I_a \circ I_b(c)$. Thus Φ is a homomorphism, and the rest is clear. $\square$

(3.21) Corollary. *Inn$(G) \cong G/Z(G)$.*

Proof. $\square$

(3.22) Example.

(1) The group S_3 has $Z(S_3) = \{e\}$ (check this). Thus Inn$(S_3) \cong S_3$. Recall that $S_3 = \{e, \alpha, \alpha^2, \beta, \alpha\beta, \alpha^2\beta\}$ (see Example 2.8 (6)). Note that α and β satisfy $\alpha^3 = e = \beta^2$ and $\alpha\beta = \alpha^2\beta$. The elements α and α^2 have order 3 and $\beta, \alpha\beta$, and $\alpha^2\beta$ all have order 2. Thus if $\phi \in$ Aut(S_3)

then $\phi(\alpha) \in \{\alpha, \alpha^2\}$ and $\phi(\beta) \in \{\beta, \alpha\beta, \alpha^2\beta\}$. Since S_3 is generated by $\{\alpha, \beta\}$, the automorphism ϕ is completely determined once $\phi(\alpha)$ and $\phi(\beta)$ are specified. Thus $|\operatorname{Aut}(S_3)| \le 6$ and we conclude that

$$\operatorname{Aut}(S_3) = \operatorname{Inn}(S_3) \cong S_3.$$

(2) If G is abelian then every nontrivial automorphism of G is an outer automorphism.

In general it is difficult to compute $\operatorname{Aut}(G)$ for a given group G. There is, however, one important special case where the computation is possible.

(3.23) Proposition. $\operatorname{Aut}(\mathbf{Z}_n) \cong \mathbf{Z}_n^*$.

Proof. Recall that $\mathbf{Z}_n^* = \{m : 1 \le m < n \text{ and } (m, n) = 1\}$ with the operation of multiplication modulo n, and $\mathbf{Z}_n = \{m : 0 \le m < n\} = \langle 1 \rangle$ with the operation of addition modulo n. Let $\phi \in \operatorname{Aut}(\mathbf{Z}_n)$. Since 1 is a generator of $\mathbf{Z}_n$, ϕ is completely determined by $\phi(1) = m$. Since ϕ is an isomorphism and $o(1) = n$, we must have $o(m) = o(\phi(1)) = n$. Let $d = (m, n)$, the greatest common divisor of m and n. Then $n \mid (n/d)m$, so $(n/d)m = 0$ in $\mathbf{Z}_n$. Since n is the smallest multiple of m that gives $0 \in \mathbf{Z}_n$, we must have $d = 1$, i.e., $m \in \mathbf{Z}_n^*$.

Also, any $m \in \mathbf{Z}_n^*$ determines an element $\phi_m \in \operatorname{Aut}(\mathbf{Z}_n)$ by the formula $\phi_m(r) = rm$. To see this we need to check that ϕ_m is an automorphism of $\mathbf{Z}_n$. But if $\phi_m(r) = \phi_m(s)$ then $rm = sm$ in $\mathbf{Z}_n$, which implies that $(r - s)m = 0 \in \mathbf{Z}_n$. But $(m, n) = 1$ implies that $r - s$ is a multiple of n, i.e., $r = s$ in $\mathbf{Z}_n$.

Therefore, we have a one-to-one correspondence of sets

$$\operatorname{Aut}(\mathbf{Z}_n) \longleftrightarrow \mathbf{Z}_n^*$$

given by

$$\phi_m \longleftrightarrow m.$$

Furthermore, this is an isomorphism of groups since

$$\phi_{m_1}(\phi_{m_2}(r)) = \phi_{m_1}(m_2 r) = m_1 m_2 r = \phi_{m_1 m_2}(r).$$

$\square$

1.4 Permutation Representations and the Sylow Theorems

If X is any set, then the set $S_X = \{\text{one-to-one correspondences } f : X \to X\}$ is a group under functional composition. S_X is called the **symmetric group** on X or group of permutations of X. A **permutation group** is a subgroup of S_X for some set X. The following theorem, due to Cayley, shows that *all* groups can be considered as permutation groups if the set X is appropriately chosen:

(4.1) Theorem. (Cayley) *Any group G is isomorphic to a subgroup of the symmetric group S_G.*

Proof. Define $\Phi : G \to S_G$ by the formula $\Phi(a)(b) = ab$. That is, $\Phi(a)$ is the function on G that multiplies each $b \in G$ by a on the left. By Proposition 1.3 (4) and (5) it follows that each $\Phi(a)$ is a bijective function on G so that $\Phi(a) \in S_G$. Also Φ is a group homomorphism since

$$\Phi(ab)(c) = (ab)c = a(bc) = \Phi(a)(bc) = \Phi(a)(\Phi(b)(c)) = (\Phi(a) \circ \Phi(b))(c).$$

Now

$$\mathrm{Ker}(\Phi) = \{a \in G : ab = b \text{ for all } b \in G\} = \{e\}.$$

Thus, Φ is injective, so by the first isomorphism theorem $G \cong \mathrm{Im}(\Phi) \subseteq S_G$. $\square$

(4.2) *Remark.* The homomorphism Φ is called the **left regular representation** of G. If $|G| < \infty$ then Φ is an isomorphism only when $|G| \leq 2$ since if $|G| > 2$ then $|S_G| = |G|! > |G|$. This same observation shows that Theorem 4.1 is primarily of interest in showing that nothing is lost if one chooses to restrict consideration to permutation groups. As a practical matter, the size of S_G is so large compared to that of G that rarely is much insight gained with the use of the left regular representation of G in S_G. It does, however, suggest the possibility of looking for smaller permutation groups that might contain a copy of G. One possibility for this will be considered now.

By a **permutation representation** of G we mean any homomorphism $\phi : G \to S_X$ for some set X. The left regular representation is one such example with $X = G$. Another important example, where $|X|$ may be substantially smaller than $|G|$, is obtained by taking $X = G/H$ where H is a subgroup of G. We are not assuming that H is normal in G, so the coset space G/H is only a set, not necessarily a group. Define $\Phi_H : G \to S_{G/H}$ by the formula $\Phi_H(a)(bH) = abH$.

(4.3) Proposition. *If H is a subgroup of G then $\Phi_H : G \to S_{G/H}$ is a group homomorphism and $\mathrm{Ker}\,(\Phi_H)$ is the largest normal subgroup of G contained in H.*

Proof. If $abH = acH$, then $bH = cH$, so $\Phi_H(a)$ is a one-to-one function on G/H and it is surjective since $\Phi_H(a)(a^{-1}bH) = bH$. Thus, $\Phi_H(a) \in S_{G/H}$. The fact that Φ_H is a group homomorphism is the same calculation as that used to show that Φ was a group homomorphism in the proof of Cayley's theorem. Thus, $\text{Ker}(\Phi_H) \lhd G$ and if $a \in \text{Ker}(\Phi_H)$ then $\Phi_H(a)$ acts as the identity on G/H. Thus, $aH = \Phi_H(a)(H) = H$ so that $a \in H$. Therefore, $\text{Ker}(\Phi_H)$ is a normal subgroup of G contained in H. Now suppose that $N \lhd G$ and $N \subseteq H$. Let $a \in N$. Then $\Phi_H(a)(bH) = abH = ba'H = bH$ since $b^{-1}ab = a' \in N \subseteq H$. Therefore, $N \subseteq \text{Ker}(\Phi_H)$ and $\text{Ker}(\Phi_H)$ is the largest normal subgroup of G contained in H. $\qquad\qquad\qquad\qquad\qquad\qquad$ $\square$

As an example of the usefulness of Proposition 4.3, we will indicate how to use this result to prove the existence of normal subgroups of certain groups.

(4.4) Corollary. *Let H be a subgroup of the finite group G and assume that $|G|$ does not divide $[G : H]!$. Then there is a subgroup $N \subseteq H$ such that $N \neq \{e\}$ and $N \lhd G$.*

Proof. Let N be the kernel of the permutation representation Φ_H. By Proposition 4.3 N is the largest normal subgroup of G contained in H. To see that $N \neq \{e\}$, note that $G/N \cong \text{Im}(\Phi_H)$, which is a subgroup of $S_{G/H}$. Thus,

$$|G|/|N| = |\,\text{Im}(\Phi_H)|\,\big|\,|S_{G/H}| = [G : H]!.$$

Since $|G|$ does not divide $[G : H]!$, we must have that $|N| > 1$ so that $N \neq \{e\}$. $\qquad\qquad\qquad\qquad\qquad\qquad\qquad\qquad\qquad\qquad$ $\square$

(4.5) Corollary. *Let H be a subgroup of the finite group G such that*

$$\big(|H|, ([G : H] - 1)!\big) = 1.$$

Then $H \lhd G$.

Proof. Let $N = \text{Ker}(\Phi_H)$. Then $N \subseteq H$ and $G/N \cong \text{Im}(\Phi_H)$ so that

$$(|G|/|N|)\,\big|\,[G : H]! = (|G|/|H|)!.$$

Therefore,

$$(|G|/|H|) \cdot (|H|/|N|)\,\big|\,[G : H]!$$

so that $(|H|/|N|) \mid ([G : H] - 1)!$. But $|H|$ and $([G : H] - 1)!$ have no common factors so that $|H|/|N|$ must be 1, i.e., $H = N$. $\qquad\qquad\qquad$ $\square$

(4.6) Corollary. *Let p be the smallest prime dividing $|G|$. Then any subgroup of G of index p is normal.*

Proof. Let H be a subgroup of G with $[G : H] = p$ and let $r = |H| = |G|/p$. Then every prime divisor of r is $\geq p$ so that

$$\big(|H|, ([G : H] - 1)!\big) = (r, (p - 1)!) = 1.$$

By Corollary 4.5, $H \lhd G$. $\qquad\qquad\qquad\qquad\qquad\qquad\qquad\qquad\qquad$ $\square$

The following result is a partial converse of Lagrange's theorem:

(4.7) Theorem. (Cauchy) *Let G be a finite group and let p be a prime dividing $|G|$. Then G has a subgroup of order p.*

Proof. If we can find an element a of order p, then $\langle a \rangle$ is the desired subgroup. To do this consider the set

$$X = \{\bar{a} = (a_0, a_1, \dots, a_{p-1}) : a_i \in G \quad \text{and} \quad a_0 a_1 \cdots a_{p-1} = e\}.$$

Then we have a permutation representation of the group $\mathbf{Z}_p$ on X where the homomorphism $\phi : \mathbf{Z}_p \to S_X$ is given by

$$\phi(i)(\bar{a}) = \phi(i)(a_0, \dots, a_{p-1}) = (a_i, a_{i+1}, \dots, a_p, a_0, \dots, a_{i-1}).$$

Note that $(a_i \cdots a_p) = (a_0 \cdots a_{i-1})^{-1}$ so that $\phi(i)(\bar{a}) \in X$.

We may define an equivalence relation on X by $\bar{a} \sim \bar{b}$ if $\phi(i)(\bar{a}) = \bar{b}$ for some i. Then X is partitioned into equivalence classes, and it is easy to see that each equivalence class consists of either exactly one or exactly p elements of X. If n_1 and n_p denote the number of equivalence classes with 1 and p elements respectively, then

$$|X| = n_1 \cdot 1 + n_p \cdot p.$$

Now X has $|G|^{p-1}$ elements (since we may choose $a_0, \dots, a_{p-2}$ arbitrarily, and then $a_{p-1} = (a_0 \cdots a_{p-2})^{-1}$), and this number is a multiple of p. Thus we see that n_1 must be divisible by p as well. Now $n_1 \geq 1$ since there is an equivalence class $\{(e, \dots, e)\}$. Therefore, there must be other equivalence classes with exactly one element. All of these are of the form $\{(a, \dots, a)\}$ and by the definition of X, such an element of X gives $a \in G$ with $a^p = e$. $\square$

(4.8) *Remark.* Note that Corollary 4.6 is a generalization of Proposition 3.16. Proposition 4.3 and its corollaries are useful in beginning a study of the structural theory of finite groups. One use of permutation representations in the structure theory of finite groups is the proof of Cauchy's theorem presented above. The next is in proving the Sylow theorems, which are substantial generalizations of Cauchy's theorem. We begin our presentation of the Sylow theorems by indicating what we mean by an action of a group on a set.

(4.9) Definition. *Let G be a group and let X be a set. By an* **action of G on X** *we mean a permutation representation $\Phi : G \to S_X$. In general, we shall write gx for $\Phi(g)(x)$. The fact that Φ is a homomorphism means that $g(hx) = (gh)x$ for all $g, h \in G$ and $x \in X$, while $ex = x$ where $e \in G$ is* the identity. Associated to $x \in X$ there is a subset Gx of X and a subgroup $G(x)$ of G defined as follows:

(1) $Gx = \{gx : g \in G\}$ is called the **orbit** of x.
(2) $G(x) = \{g \in G : gx = x\}$ is called the **stabilizer** of x.

(4.10) Lemma. *Let the group G act on a finite set X. Then*

$$|Gx| = [G : G(x)]$$

for each $x \in G$.

Proof. Since

$$gx = hx \Leftrightarrow g^{-1}h \in G(x)$$
$$\Leftrightarrow gG(x) = hG(x),$$

there is a bijective function $\phi : Gx \to G/G(x)$ defined by $\phi(gx) = gG(x)$, which gives the result. $\square$

(4.11) Lemma. *Let the group G act on a finite set X. Then*

$$|X| = \sum [G : G(x)]$$

where the sum is over a set consisting of one representative of each orbit of G.

Proof. The orbits of G form a partition X, and hence $|X| = \sum |Gx|$ where the sum is over a set consisting of one representative of each orbit of G. The result then follows from Lemma 4.10. $\square$

(4.12) Remark. Note that Lemma 4.11 generalizes the class equation (Corollary 2.28), which is the special case of Lemma 4.11 when $X = G$ and G acts on X by conjugation.

(4.13) Definition. (1) *If p is a prime, a finite group G is a p-**group** if $|G| = p^n$ for some $n \geq 1$.*
(2) *H is a p-**subgroup** of a group G if H is a subgroup of G and H is a p-group.*
(3) *Let G be an arbitrary finite group, p a prime, and p^n the highest power of p dividing $|G|$ (i.e., p^n divides $|G|$, but p^{n+1} does not). H is a p-**Sylow subgroup** of G if H is a subgroup of G and $|H| = p^n$.*

The three parts of the following theorem are often known as the three Sylow theorems:

(4.14) Theorem. (Sylow) *Let G be a finite group and let p be a prime dividing $|G|$.*

(1) *G has a p-Sylow subgroup, and furthermore, every p-subgroup of G is contained in some p-Sylow subgroup.*
(2) *The p-Sylow subgroups of G are all mutually conjugate.*
(3) *The number of p-Sylow subgroups of G is congruent to 1 modulo p and divides $|G|$.*

Proof. Let $m = |G|$ and write $m = p^n k$ where k is not divisible by p and $n \geq 1$. We will first prove that G has a p-Sylow subgroup by induction on m. If $m = p$ then G itself is a p-Sylow subgroup. Thus, suppose that $m > p$ and consider the class equation of G (Corollary 2.28):

$$(4.1) \qquad |G| = |Z(G)| + \sum [G : C(a)]$$

where the sum is over a complete set of nonconjugate a not in $Z(G)$. There are two possibilities to consider:

(1) For some a, $[G : C(a)]$ is not divisible by p. In that case, $|C(a)| = |G|/[G : C(a)] = p^n k'$ for some k' dividing k. Then p divides $|C(a)|$ and $|C(a)| < |G|$, so by induction $C(a)$ has a subgroup H of order p^n, which is then also a p-Sylow subgroup of G.
(2) $[G : C(a)]$ is divisible by p for all $a \notin Z(G)$. Then, since $|G|$ is divisible by p, we see from Equation (4.1) that p divides $|Z(G)|$. By Cauchy's theorem (Theorem 4.7), there is an $x \in Z(G)$ of order p. Let $N = \langle x \rangle$. If $n = 1$ (i.e., p divides $|G|$, but p^2 does not) then N itself is a p-Sylow subgroup of G. Otherwise, note that since $N \subseteq Z(G)$, it follows that $N \lhd G$ (Exercise 21). Consider the projection map $\pi : G \to G/N$. Now $|G/N| = p^{n-1}k < |G|$, so by induction, G/N has a subgroup H with $|H| = p^{n-1}$, and then $\pi^{-1}(H)$ is a p-Sylow subgroup of G.

Thus, we have established that G has a p-Sylow subgroup P. Let X be the set of all subgroups of G conjugate to P. (Of course, any subgroup conjugate to P has the same order as P, so it is also a p-Sylow subgroup of G.) The group G acts on X by conjugation, and since all elements of X are conjugate to P, there is only one orbit. By Lemma 4.11, we have $|X| = [G : G(P)]$. But $P \subseteq G(P)$, so $[G : G(P)]$ divides k and, in particular, is not divisible by p. Thus, $|X|$ is relatively prime to p.

Now let H be an arbitrary p-subgroup of G, and consider the action of H on X by conjugation. Again by Lemma 4.11,

$$(4.2) \qquad |X| = \sum [H : H(x)].$$

Since $|X|$ is not divisible by p, some term on the right-hand side of Equation (4.2) must not be divisible by p; since H is a p-group, that can only happen if it is equal to one. Thus, there is some p-Sylow subgroup P' of G, conjugate to P, with $hP'h^{-1} = P'$ for all $h \in H$, i.e., with $HP' = P'H$. But then Lemma 3.1 implies that HP' is a subgroup of G. Since

$$|HP'| = |H||P'|/|H \cap P'|$$

(see Exercise 17), it follows that HP' is also a p-subgroup of G. Since P' is a p-Sylow subgroup, this can only happen if $HP' = P'$, i.e., if $H \subseteq P'$. Thus part (1) of the theorem is proved.

To see that (2) is true, let H itself be any p-Sylow subgroup of G. Then $H \subseteq P'$ for some conjugate P' of P, and since $|H| = |P'|$, we must have $H = P'$ so that H is conjugate to P. This gives that X consists of all the p-Sylow subgroups of G, and hence, $|X| = [G : G(P)]$ divides $|G|$. Now take $H = P$. Equation (4.2) becomes

(4.3) $$|X| = \sum [P : P(x)].$$

Then, for $x = P$, $[P : P(x)] = 1$, while if x is a representative of any other orbit, $[P : P(x)]$ is divisible by p, showing that $|X|$ is congruent to 1 modulo p. Thus part (3) is verified.   □

The Sylow theorems are a major tool in analyzing the structure of finite groups. In Section 1.7, as an application of these theorems, we will classify all finite groups of order ≤ 15.

1.5 The Symmetric Group and Symmetry Groups

Recall that if $X = \{1, 2, \ldots, n\}$ then we denote S_X by S_n and we can write a typical element $\alpha \in S_n$ as a two-rowed array

$$\alpha = \begin{pmatrix} 1 & 2 & \cdots & n \\ \alpha(1) & \alpha(2) & \cdots & \alpha(n) \end{pmatrix}.$$

This notation is somewhat cumbersome so we introduce a simpler notation which is frequently more useful.

(5.1) Definition. *An element $i \in X = \{1, 2, \ldots, n\}$ is fixed by $\alpha \in S_n$ if $\alpha(i) = i$. $\alpha \in S_n$ is an r-cycle or* **cycle of length** *r if there are r integers $i_1, i_2, \ldots, i_r \in X$ such that*

$$\alpha(i_1) = i_2, \qquad \alpha(i_2) = i_3, \qquad \ldots, \qquad \alpha(i_{r-1}) = i_r, \qquad \alpha(i_r) = i_1$$

and such that α fixes all other $i \in X$. The r-cycle α is denoted $(i_1 i_2 \cdots i_r)$. If α is an r-cycle, note that $o(\alpha) = r$. A 2-cycle is called a **transposition**. *Two cycles $\alpha = (i_1 \cdots i_r)$ and $\beta = (j_1 \cdots j_s)$ are* **disjoint** *if*

$$\{i_1, \ldots, i_r\} \cap \{j_1, \ldots, j_s\} = \emptyset.$$

That is, every element moved by α is fixed by β.

As an example of the increased clarity of the cycle notation over the 2-rowed notation, consider the following permutation in S_9.

$$\alpha = \begin{pmatrix} 1 & 2 & 3 & 4 & 5 & 6 & 7 & 8 & 9 \\ 3 & 9 & 7 & 4 & 2 & 1 & 6 & 8 & 5 \end{pmatrix}.$$

α is not a cycle, but it is a product of disjoint cycles, namely,

$$\alpha = (1\,3\,7\,6)(9\,5\,2)(4)(8).$$

Since 1-cycles represent the identity function, it is customary to omit them and write $\alpha = (1\,3\,7\,6)(9\,5\,2)$. This expression for α generally gives more information and is much cleaner than the 2-rowed notation. There are, however, two things worth pointing out concerning the cycle notation. First the cycle notation is not unique. For an r-cycle $(i_1 \cdots i_r)$ there are r different cycle notations for the same r-cycle:

$$(i_1 \cdots i_r) = (i_2\,i_3 \cdots i_r\,i_1) = \cdots = (i_r\,i_1 \cdots i_{r-1}).$$

The second observation is that the cycle notation does not make it clear which symmetric group S_n the cycle belongs to. For example, the transposition $(1\,2)$ has the same notation as an element of every S_n for $n \geq 2$.

In practice, this ambiguity is not a problem. We now prove a factorization theorem for permutations.

(5.2) Lemma. *Disjoint cycles commute.*

Proof. Suppose α and β are disjoint cycles in S_n, and let $i \in X = \{1, 2, \ldots, n\}$. If i is fixed by both α and β then $\alpha\beta(i) = i = \beta\alpha(i)$. If α moves i, then α also moves $\alpha(i)$, and thus, β fixes both of these elements. Therefore, $\alpha\beta(i) = \alpha(i) = \beta\alpha(i)$. Similarly, if β moves i then $\alpha\beta(i) = \beta(i) = \beta\alpha(i)$. □

(5.3) Theorem. *Every $\alpha \in S_n$ with $\alpha \neq e$ can be written uniquely (except for order) as a product of disjoint cycles of length ≥ 2.*

Proof. We first describe an algorithm for producing the factorization. Let k_1 be the smallest integer in $X = \{1, 2, \ldots, n\}$ that is not fixed by α (k_1 exists since $\alpha \neq e$) and then choose the smallest positive r_1 with $\alpha^{r_1}(k_1) = k_1$ (such an r_1 exists since $o(\alpha) < \infty$). Then let α_1 be the r_1-cycle

$$\alpha_1 = (k_1\,\alpha(k_1)\,\alpha^2(k_1) \cdots \alpha^{r_1-1}(k_1)).$$

Now let $X_1 = X \setminus \{k_1, \alpha(k_1), \ldots, \alpha^{r_1-1}\}$.

If every $k \in X_1$ is fixed by α then $\alpha = \alpha_1$ and we are finished. Otherwise let k_2 be the smallest integer in X_1 not fixed by α and then let r_2 be the smallest positive integer with $\alpha^{r_2}(k_2) = k_2$. Then let α_2 be the r_2-cycle

$$\alpha_2 = (k_2\,\alpha(k_2)\,\alpha^2(k_2) \cdots \alpha^{r_2-1}(k_2)).$$

It is clear from the construction that α_1 and α_2 are disjoint cycles. Continuing in this manner we eventually arrive at a factorization

$$\alpha = \alpha_1 \alpha_2 \cdots \alpha_s$$

of α into a product of disjoint cycles.

We now consider the question of uniqueness of the factorization. Suppose that

$$\alpha = \alpha_1 \alpha_2 \cdots \alpha_s = \beta_1 \beta_2 \cdots \beta_t$$

where each of these is a factorization of α into disjoint cycles of length ≥ 2. We must show that $s = t$ and $\alpha_i = \beta_{\phi(i)}$ for some $\phi \in S_s$. Let $m = \max\{s, t\}$. If $m = 1$ then $\alpha = \alpha_1 = \beta_1$ and uniqueness is clear. We proceed by induction on m. Suppose that $m > 1$ and let k be an element of X that is moved by α. Then some α_i and β_j must also move k. Since disjoint cycles commute, we can, without loss of generality, suppose that α_1 and β_1 move k. Since none of the other α_i or β_j move k, it follows that

$$\alpha_1^\ell(k) = \alpha^\ell(k) = \beta_1^\ell(k) \qquad \text{for all } \ell.$$

Thus, $o(\alpha_1) = o(\beta_1) = r = $ smallest r with $\alpha^r(k) = k$. Hence,

$$\alpha_1 = (k\, \alpha(k)\, \cdots\, \alpha^{r-1}(k)) = \beta_1.$$

Multiplying by α_1^{-1} gives a factorization

$$\alpha_1^{-1}\alpha = \alpha_2 \cdots \alpha_s = \beta_2 \cdots \beta_t,$$

and the proof is completed by induction on m. ☐

(5.4) Corollary. *Every $\alpha \in S_n$ is a product of transpositions.*

Proof. By Theorem 5.3, it is sufficient to factor any cycle as a product of transpositions. But

$$(i_1\, i_2\, \cdots\, i_r) = (i_1\, i_r)(i_1\, i_{r-1}) \cdots (i_1\, i_2)$$

is such a factorization. ☐

In contrast to the uniqueness of the factorization of a permutation into disjoint cycles, writing a permutation as a product of transpositions is not very well behaved. First, the transpositions may not commute. For example, $(1\,3)(1\,2) = (1\,2\,3) \neq (1\,3\,2) = (1\,2)(1\,3)$. Second, the factorization is not uniquely determined, e.g., $(1\,2\,3) = (1\,3)(1\,2) = (1\,3)(1\,2)(2\,3)(2\,3)$. There is, however, one observation that can be made concerning this factorization; namely, the number of transpositions occurring in both factorizations is even. While we have shown only one example, this is in fact a result that is true in general. Specifically, *the number of transpositions occurring in any factorization of a permutation as a product of transpositions is always odd or always even.* This will be verified now.

If $\alpha = (i_1 \cdots i_r)$ then $\alpha = (i_1 i_r) \cdots (i_1 i_2)$ so that an r-cycle α can be written as a product of $(o(\alpha) - 1)$ transpositions. Hence, if $\alpha \neq e \in S_n$ is written in its cycle decomposition $\alpha = \alpha_1 \cdots \alpha_s$ then α is the product of $f(\alpha) = \sum_{i=1}^{s}(o(\alpha_i) - 1)$ transpositions. We also set $f(e) = 0$. Now suppose that

$$\alpha = (a_1 \, b_1)(a_2 \, b_2) \cdots (a_t \, b_t)$$

is written as an arbitrary product of transpositions. We claim that $f(\alpha) - t$ is even. To see this note that

$$(a \, i_1 \, i_2 \cdots i_r \, b \, j_1 \cdots j_s)(a \, b) = (a \, j_1 \cdots j_s)(b \, i_1 \cdots i_r)$$

and (since $(a \, b)^2 = e$)

$$(a \, j_1 \cdots j_s)(b \, i_1 \cdots i_r)(a \, b) = (a \, i_1 \, i_2 \cdots i_r \, b \, j_1 \cdots j_s)$$

where it is possible that no i_k or j_k is present. Hence, if a and b both occur in the same cycle in the cycle decomposition of α it follows that $f(\alpha \cdot (a \, b)) = f(\alpha) - 1$, while if they occur in different cycles or are both not moved by α then $f(\alpha \cdot (a \, b)) = f(\alpha) + 1$. In any case

$$f(\alpha \cdot (a \, b)) - f(\alpha) \equiv 1 \pmod 2.$$

Continuing this process gives

$$0 = f(e) = f(\alpha \cdot (a_t b_t) \cdots (a_1 b_1)) \equiv f(\alpha) + t \pmod 2.$$

We conclude that any factorization of α into a product of t transpositions has both $f(\alpha)$ and t even or both odd, which is what we wished to verify. Because of this fact we can make the following definition.

(5.5) Definition. *A permutation $\alpha \in S_n$ is **even** if α can be written as a product of an even number of transpositions. α is **odd** if α can be written as a product of an odd number of transpositions. Define the **sign** of α, denoted $\mathrm{sgn}(\alpha)$, by*

$$\mathrm{sgn}(\alpha) = \begin{cases} 1 & \text{if } \alpha \text{ is even,} \\ -1 & \text{if } \alpha \text{ is odd.} \end{cases}$$

The argument in the previous paragraph shows that a permutation cannot be both even and odd. Thus $\mathrm{sgn} : S_n \to \{1, -1\}$ *is a well-defined function, and moreover, it is a group homomorphism. The kernel of* sgn, *i.e., the set of even permutations, is a normal subgroup of S_n called the **alternating group** and denoted A_n.*

(5.6) Remark. Note that the above argument gives a method for computing $\mathrm{sgn}(\alpha)$. Namely, decompose $\alpha = \alpha_1 \cdots \alpha_s$ into a product of cycles and compute $f(\alpha) = \sum_{i=1}^{s}(o(\alpha_i) - 1)$. Then $\mathrm{sgn}(\alpha) = 1$ if $f(\alpha)$ is even and $\mathrm{sgn}(\alpha) = -1$ if $f(\alpha)$ is odd.

There is an alternative method that does not require that α be first decomposed into a product of cycles. We have defined α as a bijection of $\{1, \ldots, n\}$. Let

$$\tilde{f}(\alpha) = |\{(i, j) : 1 \le i < j \le n \quad \text{and} \quad \alpha(j) < \alpha(i)\}|.$$

Then $\text{sgn}(\alpha) = 1$ if $\tilde{f}(\alpha)$ is even and $\text{sgn}(\alpha) = -1$ if $\tilde{f}(\alpha)$ is odd. We leave the proof of this as an exercise for the reader.

(5.7) Proposition. $|A_n| = n!/2$.

Proof. Since $\text{sgn} : S_n \to \{1, -1\}$ is a group homomorphism, the first isomorphism theorem gives

$$S_n/A_n \cong \text{Im}(\text{sgn}) = \{1, -1\}.$$

Thus, $n! = |S_n| = 2|A_n|$. $\qquad\qquad\qquad\qquad\qquad\qquad\qquad\qquad\qquad\qquad\square$

(5.8) Proposition. *If $n > 2$ then A_n is generated by all the 3-cycles in S_n.*

Proof. An element of A_n is a product of terms of the form $(i\,j)(k\,l)$ or $(i\,j)(i\,k)$ where i, j, k, l are distinct. (If $n = 3$, only the latter product is possible.) But

$$(i\,j)(i\,k) = (i\,k\,j)$$

and

$$(i\,j)(k\,l) = (i\,k\,j)(i\,k\,l)$$

so that every element of A_n is a product of 3-cycles. $\qquad\qquad\qquad\qquad\square$

If G is a group recall (Definition 2.26) that two elements $a, b \in G$ are conjugate if $b = cac^{-1}$ for some $c \in G$. In general, it is not easy to determine if two elements of G are conjugate, but for the group S_n there is a simple criterion for conjugacy based on the cycle decomposition (factorization) of $\alpha, \beta \in S_n$. We will say that α and β have the same cycle structure if their factorizations into disjoint cycles produce the same number of r-cycles for each r.

(5.9) Proposition. (1) *If $\alpha \in S_n$ and $\beta = (i_1 \cdots i_r)$ is an r-cycle, then $\alpha\beta\alpha^{-1}$ is the r-cycle $(\alpha(i_1) \cdots \alpha(i_r))$.*
    (2) *Any two r-cycles in S_n are conjugate.*

Proof. (1) If $j \notin \{\alpha(i_1), \ldots, \alpha(i_r)\}$ then $\alpha^{-1}(j)$ is fixed by β so that $\alpha\beta\alpha^{-1}(j) = j$. Also

$$\alpha\beta\alpha^{-1}(\alpha(i_1)) = \alpha(i_2)$$

$$\vdots$$

$$\alpha\beta\alpha^{-1}(\alpha(i_{r-1})) = \alpha(i_r)$$
$$\alpha\beta\alpha^{-1}(\alpha(i_r)) = \alpha(i_1)$$

so that $\alpha\beta\alpha^{-1} = (\alpha(i_1) \cdots \alpha(i_r))$.

(2) Let $\beta = (i_1 \cdots i_r)$ and $\gamma = (j_1 \cdots j_r)$ be any two r-cycles in S_n. Define $\alpha \in S_n$ by $\alpha(i_k) = j_k$ for $1 \leq k \leq r$ and extend α to a permutation in any manner. Then by part (1) $\alpha\beta\alpha^{-1} = \gamma$.   □

(5.10) Corollary. *Two permutations $\alpha, \beta \in S_n$ are conjugate if and only if they have the same cycle structure.*

Proof. Suppose that $\gamma\alpha\gamma^{-1} = \beta$. Then if $\alpha = \alpha_1 \cdots \alpha_s$ is the cycle decomposition of α, it follows from Proposition 5.9 (1) that

$$\beta = \gamma\alpha\gamma^{-1} = (\gamma\alpha_1\gamma^{-1})(\gamma\alpha_2\gamma^{-1}) \cdots (\gamma\alpha_s\gamma^{-1})$$

is the cycle decomposition of β. Thus, α and β have the same cycle structure.

The converse is analogous to the proof of Proposition 5.9 (2); it is left to the reader.   □

(5.11) Example. Let $H = \{e, (1\,2)(3\,4), (1\,3)(2\,4), (1\,4)(2\,3)\} \subseteq S_4$. Then H is a subgroup of S_4 isomorphic with the Klein 4-group, and since H consists of all permutations in S_4 with cycle type $(a\,b)(c\,d)$ (where a, b, c, d are all distinct), it follows from Corollary 5.10 that $H \triangleleft S_4$. Let $K = \{e, (1\,2)(3\,4)\}$. Then K is a normal subgroup of H (since H is abelian), but K is *not* normal in S_4 since any other permutation of cycle type $(a\,b)(c\,d)$ can be obtained from $(1\,2)(3\,4)$ by conjugation in S_4. Therefore, normality is not a transitive property on the set of all subgroups of a group G.

Let $X \subset \mathbf{R}^n$. By a **symmetry** of X we mean a function $f : \mathbf{R}^n \to \mathbf{R}^n$ such that $f(X) \subseteq X$ and $\|x - y\| = \|f(x) - f(y)\|$ for all $x, y \in \mathbf{R}^n$. The set of all symmetries of X under functional composition forms a group, called the symmetry group of X. If $X = P_n \subseteq \mathbf{R}^2$ is a regular polygon with n vertices then a symmetry is completely determined by the action on the vertices (since it is easy to see from the triangle inequality that lines must go to lines and adjacent vertices must go to adjacent vertices) so that we get a permutation representation of the symmetry group of P_n, denoted D_{2n}, as a subgroup of S_n. D_{2n} is called the **dihedral group of order** $2n$. If P_n is taken on the unit circle centered at $(0,0)$ with one vertex at $(1,0)$ then the symmetries of P_n are the rotations through an angle of $\theta_k = 2k\pi/n$ around $(0,0)$ for $0 \leq k < n$ and the reflections through the lines from each vertex and from the midpoint of each side to the center of the circle. (There are always n such distinct lines.) Thus $|D_{2n}| = 2n$ when there are n rotations and n reflections. If we let α be the rotation through the angle θ_1 and β the reflection through the x-axis, then

$$D_{2n} = \{\alpha^i\beta^j : 0 \leq i < n, j = 0, 1\}.$$

It is easy to check that $o(\alpha) = n$ and that $\beta\alpha\beta = \alpha^{-1}$. If the vertices of P_n are numbered $n, 1, 2, \ldots, n-1$ counterclockwise starting at $(1,0)$, then D_{2n} is identified as a subgroup of S_n by

$$\alpha \longleftrightarrow (1\,2\,\cdots\,n)$$

$$\beta \longleftrightarrow \begin{cases} (1\,n-1)(2\,n-2)\,\cdots\,((n-1)/2\ (n+1)/2) & \text{when } n \text{ is odd,} \\ (1\,n-1)(2\,n-2)\,\cdots\,(n/2-1\ (n/2)+1) & \text{when } n \text{ is even.} \end{cases}$$

Thus, we have arrived at a concrete representation of the dihedral group that was described by means of generators and relations in Example 2.8 (13).

(5.12) Examples.

(1) If X is the rectangle in $\mathbf{R}^2$ with vertices $(0, 1)$, $(0, 0)$, $(2, 0)$, and $(2, 1)$ labelled from 1 to 4 in the given order, then the symmetry group of X is the subgroup

$$H = \{e, (1\,3)(2\,4), (1\,2)(3\,4), (1\,4)(2\,3)\}$$

of S_4, which is isomorphic to the Klein 4-group.

(2) $D_6 \cong S_3$ since D_6 is generated as a subgroup of S_3 by the permutations $\alpha = (1\,2\,3)$ and $\beta = (2\,3)$.

(3) D_8 is a (nonnormal) subgroup of S_4 of order 8. If $\alpha = (1\,2\,3\,4)$ and $\beta = (1\,3)$ then

$$D_8 = \{e, \alpha, \alpha^2, \alpha^3, \beta, \alpha\beta, \alpha^2\beta, \alpha^3\beta\}.$$

There are two other subgroups of S_4 conjugate to D_8 (exercise).

1.6 Direct and Semidirect Products

(6.1) Definition. *If N and H are groups the* **(external) direct product** *of N and H, denoted $N \times H$, is the cartesian product set $N \times H$ with the multiplication defined componentwise, i.e.,*

$$(n, h)(n', h') = (nn', hh').$$

It is easy to check that $N \times H$ is a group with this multiplication. Associated to $N \times H$ there are some natural homomorphisms

$$\pi_N : N \times H \to N \quad ((n, h) \mapsto n)$$
$$\pi_H : N \times H \to H \quad ((n, h) \mapsto h)$$
$$\iota_N : N \to N \times H \quad (n \mapsto (n, e))$$
$$\iota_H : H \to N \times H \quad (h \mapsto (e, h)).$$

The homomorphisms π_N and π_H are called the **natural projections** while ι_N and ι_H are known as the **natural injections**. The word canonical is used interchangeably with natural when referring to projections or injections. Note the following relationships among these homomorphisms

$$\text{Ker}(\pi_H) = \text{Im}(\iota_N)$$
$$\text{Ker}(\pi_N) = \text{Im}(\iota_H)$$
$$\pi_H \circ \iota_H = 1_H$$
$$\pi_N \circ \iota_N = 1_N$$

(1_G refers to the identity homomorphism of the group G). In particular, $N \times H$ contains a normal subgroup

$$\tilde{N} = \text{Im}(\iota_N) = \text{Ker}(\pi_H) \cong N$$

and a normal subgroup

$$\tilde{H} = \text{Im}(\iota_H) = \text{Ker}(\pi_N) \cong H$$

such that $\tilde{N} \cap \tilde{H} = \{(e, e)\}$ is the identity in $N \times H$ and $N \times H = \tilde{N}\tilde{H}$. Having made this observation, we make the following definition.

(6.2) Definition. *Let G be a group with subgroups N and H such that*

$$N \cap H = \{e\} \quad and \quad NH = G.$$

(1) *If N and H are both normal, then we say that G is the **internal direct product** of N and H.*
(2) *If N is normal (but not necessarily H), then we say that G is the **semidirect product** of N and H.*

The relationship between internal and external direct products is given by the following result. We have already observed that $N \times H$ is the internal direct product of $\tilde{N}$ and $\tilde{H}$, which are subgroups of $N \times H$ isomorphic to N and H respectively.

(6.3) Proposition. *If G is the internal direct product of subgroups N and H, then $G \cong N \times H$.*

Proof. Let $a \in G$. Then $a = nh$ for some $n \in N$, $h \in H$. Suppose we may also write $a = n_1 h_1$ for some $n_1 \in N$, $h_1 \in H$. Then $nh = n_1 h_1$ so that $n^{-1} n_1 = hh_1^{-1} \in N \cap H = \{e\}$. Therefore, $n = n_1$ and $h = h_1$ so that the factorization $a = nh$ is unique.

Define $f : G \to N \times H$ by $f(a) = (n, h)$ where $a = nh$. This function is well defined by the previous paragraph, which also shows that f is a one-to-one correspondence. It remains to check that f is a group homomorphism.

Suppose that $a = nh$ and $b = n_1 h_1$. Then $ab = nhn_1 h_1$. We claim that $hn_1 = n_1 h$ for all $n_1 \in N$ and $h \in H$. Indeed, $(hn_1 h^{-1})n_1^{-1} \in N$ since N is normal in G and $h(n_1 h^{-1} n_1^{-1}) \in H$ since H is normal. But $N \cap H = \{e\}$, so $hn_1 h^{-1} n_1^{-1} \in N \cap H = \{e\}$, and thus, $hn_1 = n_1 h$. Therefore,

$$f(ab) = f(nhn_1 h_1) = f(nn_1 hh_1) = (nn_1,\ hh_1) = (n,\ h)(n_1,\ h_1) = f(a)f(b)$$

so that f is a group homomorphism, and, hence, a group isomorphism since it is a one-to-one correspondence. $\square$

(6.4) Examples.

(1) Recall that if G is a group then the **center** of G, denoted $Z(G)$, is the set of elements that commute with all elements of G. It is a normal subgroup of G. Now, if N and H are groups, then it is an easy exercise (do it) to show that $Z(N \times H) = Z(N) \times Z(H)$. As a consequence, one obtains the fact that *the product of abelian groups is abelian.*

(2) The group $\mathbf{Z}_2 \times \mathbf{Z}_2$ is isomorphic to the Klein 4-group. Therefore, the two nonisomorphic groups of order 4 are $\mathbf{Z}_4$ and $\mathbf{Z}_2 \times \mathbf{Z}_2$.

(3) All the hypotheses in the definition of internal direct product are necessary for the validity of Proposition 6.3. For example, let $G = S_3$, $N = A_3$, and $H = \langle (1\,2) \rangle$. Then $N \triangleleft G$ but H is not a normal subgroup of G. It is true that $G = NH$ and $N \cap H = \{e\}$, but $G \not\cong N \times H$ since G is not abelian, but $N \times H$ is abelian.

(4) In the previous example S_3 is the semidirect product of $N = A_3$ and $H = \langle (1\,2) \rangle$.

(6.5) Lemma. *If G is the semidirect product of N and H then every $a \in G$ can be written uniquely as $a = nh$ where $n \in N$ and $h \in H$.*

Proof. By hypothesis, $G = NH$, so existence of the factorization is clear. Suppose $a = n_1 h_1 = n_2 h_2$. Then $n_2^{-1} n_1 = h_2 h_1^{-1} \in N \cap H = \{e\}$. Therefore, $n_1 = n_2$ and $h_1 = h_2$. $\square$

According to this lemma, G is set theoretically the cartesian product set $N \times H$, but the group structures are different.

If G is the semidirect product of N and H, then the second isomorphism theorem (Theorem 3.12) shows that

$$H = H/(H \cap N) \cong (HN)/N = (NH)/N = G/N.$$

Thus, H is determined once we have N. A natural question is then, given groups N and H, identify all groups G such that G is the semidirect product of subgroups $\tilde{N}$ and $\tilde{H}$ where $\tilde{N} \cong N$ and $\tilde{H} \cong H$. As one answer to this problem, we will present a construction showing how to produce all semidirect products. We start with the following definition:

(6.6) Definition. *Let N and H be groups. An **extension of** N **by** H *is a group G such that*

(1) G *contains N as a normal subgroup.*
(2) $G/N \cong H$.

The first isomorphism theorem shows that for G to be an extension of N by H means that there is an **exact sequence** of groups and group homomorphisms

$$1 \longrightarrow N \overset{\theta}{\longrightarrow} G \overset{\pi}{\longrightarrow} H \longrightarrow 1.$$

In this sequence, $1 = \{e\}$ and exactness means that π is surjective, θ is injective, and $\operatorname{Ker}(\pi) = \operatorname{Im}(\theta)$.

The extension G of N by H is a **split extension** if there is a homomorphism $\alpha : H \to G$ such that $\pi \circ \alpha = 1_H$. In this case we say that the above sequence is a **split exact sequence**.

The relationship between semidirect products and extensions is given by the following result:

(6.7) Proposition. *G is a semidirect product of N and H if and only if G is a split extension of N by H.*

Proof. Suppose G is a semidirect product of N and H with $N \lhd G$. Define $\pi : G \to H$ by $\pi(a) = h$ where $a = nh$. Lemma 6.5 shows that π is well defined. To see that π is a homomorphism, note that $h_1 n_2 h_1^{-1} = n_2' \in N$ whenever $h_1, n_2 \in N$ (because $N \lhd G$). Thus,

$$\pi(n_1 h_1 n_2 h_2) = \pi(n_1 n_2' h_1 h_2) = h_1 h_2 = \pi(n_1 h_1)\pi(n_2 h_2),$$

so π is a homomorphism. It is clear that $\operatorname{Im}(\pi) = H$ and $\operatorname{Ker}(\pi) = N$. Let $\alpha : H \to G$ be the inclusion map, i.e., $\alpha(h) = h$. Then $\pi \circ \alpha(h) = h$ for all $h \in H$, so the extension determined by π is split.

Conversely, assume that G is a split extension of N by H with $\pi : G \to H$ and $\alpha : H \to G$ the homomorphisms given by the definition of split extension. Then $N = \operatorname{Ker}(\pi) \lhd G$ and $\widetilde{H} = \operatorname{Im}(\alpha)$ is a subgroup of G. Suppose that $a \in N \cap \widetilde{H}$. Then $\pi(a) = e$ and $a = \alpha(h)$ for some $h \in H$ so that $h = \pi(\alpha(h)) = \pi(a) = e$. Therefore, $a = \alpha(e) = e$, and we conclude that $N \cap \widetilde{H} = \{e\}$. Now let $a \in G$ and write

$$a = (a \cdot \alpha(\pi(a))^{-1}) \cdot \alpha(\pi(a)) = nh.$$

Clearly, $h \in \widetilde{H}$ and

$$\pi(n) = \pi(a \cdot \alpha(\pi(a))^{-1}) = \pi(a)\pi(\alpha(\pi(a))^{-1}) = \pi(a)\pi(a)^{-1} = e,$$

so $n \in N$. Therefore, G is a semidirect product of N and $\widetilde{H} \cong H$. $\qquad\square$

(6.8) *Remark.* Comparing the definitions of semidirect product and direct product, we see that if G is the semidirect product of N and H with H normal (in addition to N), then G is in fact the (internal) direct product of these subgroups. Of course, in an abelian group every subgroup is normal, so for abelian groups the notion of semidirect product reduces to that of direct product. In particular, we see from Proposition 6.7 that given a *split* exact sequence of *abelian* groups

$$1 \longrightarrow N \xrightarrow{\ \theta\ } G \xrightarrow{\ \pi\ } H \longrightarrow 1$$

we have that $G \cong N \times H$.

We now consider a way to construct split extensions of N by H, which according to Proposition 6.7 is equivalent to constructing semidirect products. Let N and H be groups and let $\phi : H \to \mathrm{Aut}(N)$ be a group homomorphism. We will write $\phi_h \in \mathrm{Aut}(N)$ instead of $\phi(h)$. Then define $G = N \rtimes_\phi H = N \rtimes H$ to be the set $N \times H$ with the multiplication defined by

$$(n_1, h_1)(n_2, h_2) = (n_1 \phi_{h_1}(n_2), h_1 h_2).$$

We identify N and H with the subsets $N \times \{e\}$ and $\{e\} \times H$, respectively.

(6.9) Theorem. *With the above notation,*

(1) $G = N \rtimes_\phi H$ *is a group,*
(2) H *is a subgroup of G and $N \triangleleft G$,*
(3) G *is a split extension of N by H, and*
(4) $hnh^{-1} = \phi_h(n)$ *for all $h \in H \subseteq G$ and $n \in N \subseteq G$.*

Proof. (1) (e, e) is easily seen to be the identity of G. For inverses, note that

$$(\phi_{h^{-1}}(n^{-1}), h^{-1})(n, h) = (\phi_{h^{-1}}(n^{-1}) \cdot \phi_{h^{-1}}(n), h^{-1}h)$$
$$= (\phi_{h^{-1}}(e), e) = (e, e)$$

and

$$(n, h)(\phi_{h^{-1}}(n^{-1}), h^{-1}) = (n\phi_h(\phi_{h^{-1}}(n^{-1})), hh^{-1})$$
$$= (n\phi_e(n^{-1}), e) = (nn^{-1}, e) = (e, e).$$

Thus, $(n, h)^{-1} = (\phi_{h^{-1}}(n^{-1}), h^{-1})$.

To check associativity, note that

$$((n_1, h_1)(n_2, h_2))(n_3, h_3) = (n_1 \phi_{h_1}(n_2), h_1 h_2)(n_3, h_3)$$
$$= (n_1 \phi_{h_1}(n_2)\phi_{h_1 h_2}(n_3), h_1 h_2 h_3)$$
$$= (n_1 \phi_{h_1}(n_2)\phi_{h_1}(\phi_{h_2}(n_3)), h_1 h_2 h_3)$$
$$= (n_1 \phi_{h_1}(n_2\phi_{h_2}(n_3)), h_1 h_2 h_3)$$
$$= (n_1, h_1)(n_2\phi_{h_2}(n_3), h_2 h_3)$$
$$= (n_1, h_1)((n_2, h_2)(n_3, h_3)).$$

(2) It is clear from the definition that N and H are subgroups of G. Let $\pi : N \rtimes_\phi H \to H$ be defined by $\phi(n, h) = h$. Then π is a group homomorphism since $\pi((n_1, h_1)(n_2, h_2)) = \pi(n_1 \phi_{h_1}(n_2), h_1 h_2) = h_1 h_2 = \pi(n_1, h_1)\pi(n_2, h_2)$ and $N = \text{Ker}(\pi)$, so $N \lhd G$.

(3) Let $\alpha : H \to G$ be defined by $\alpha(h) = (e, h)$. Then α is a homomorphism and $\pi \circ \alpha = 1_H$.

(4)

$$
\begin{aligned}
(e, h)(n, e)(e, h)^{-1} &= (e, h)(n, e)(e, h^{-1}) \\
&= (\phi_h(n), h)(e, h^{-1}) \\
&= (\phi_h(n)\phi_h(e), hh^{-1}) \\
&= (\phi_h(n), e).
\end{aligned}
$$

$\square$

(6.10) Examples.

(1) Let $\phi : H \to \text{Aut}(N)$ be defined by $\phi(h) = 1_N$ for all $h \in H$. Then $N \rtimes_\phi H$ is just the direct product of N and H.

(2) If $\phi : \mathbf{Z}_2 \to \text{Aut}(\mathbf{Z}_n)$ is defined by $1 \mapsto \phi_1(a) = -a$ where $\mathbf{Z}_2 = \{0, 1\}$, then $\mathbf{Z}_n \rtimes_\phi \mathbf{Z}_2 \cong D_{2n}$.

(3) The construction in Example (2) works for any abelian group A in place of $\mathbf{Z}_n$ and gives a group $A \rtimes_\phi \mathbf{Z}_2$. Note that $A \rtimes_\phi \mathbf{Z}_2 \not\cong A \times \mathbf{Z}_2$ unless $a^2 = e$ for all $a \in A$.

(4) $\mathbf{Z}_{p^2}$ is a nonsplit extension of $\mathbf{Z}_p$ by $\mathbf{Z}_p$. Indeed, define $\pi : \mathbf{Z}_{p^2} \to \mathbf{Z}_p$ by $\pi(r) = r \pmod{p}$. Then $\text{Ker}(\pi)$ is the unique subgroup of $\mathbf{Z}_{p^2}$ of order p, i.e., $\text{Ker}(\pi) = \langle p \rangle \subseteq \mathbf{Z}_{p^2}$. But then any nonzero homomorphism $\alpha : \mathbf{Z}_p \to \mathbf{Z}_{p^2}$ must have $|\text{Im}(\alpha)| = p$ and, since there is only one subgroup of $\mathbf{Z}_{p^2}$ of order p, it follows that $\text{Im}(\alpha) = \text{Ker}(\pi)$. Therefore, $\pi \circ \alpha = 0 \neq 1_{\mathbf{Z}_p}$ so that the extension is nonsplit.

(6.11) Remark. Note that *all* semidirect products arise via the construction of Theorem 6.9 as follows. Suppose $G = NH$ is a semidirect product. Define $\phi : H \to \text{Aut}(N)$ by $\phi_h(n) = hnh^{-1}$. Then the map $\Phi : G \to N \rtimes_\phi H$, defined by $\Phi(nh) = (n, h)$, is easily seen to be an isomorphism. Note that Φ is well defined by Lemma 6.5 and is a homomorphism by Theorem 6.9 (4).

1.7 Groups of Low Order

This section will illustrate the group theoretic techniques introduced in this chapter by producing a list (up to isomorphism) of all groups of order at most 15. The basic approach will be to consider the prime factorization of

$|G|$ and study groups with particularly simple prime factorizations for their order. First note that groups of prime order are cyclic (Corollary 2.20) so that every group of order 2, 3, 5, 7, 11, or 13 is cyclic. Next we consider groups of order p^2 and pq where p and q are distinct primes.

(7.1) Proposition. *If p is a prime and G is a group of order p^2, then $G \cong \mathbf{Z}_{p^2}$ or $G \cong \mathbf{Z}_p \times \mathbf{Z}_p$.*

Proof. If G has an element of order p^2, then $G \cong \mathbf{Z}_{p^2}$. Assume not. Let $e \neq a \in G$. Then $o(a) = p$. Set $N = \langle a \rangle$. Let $b \in G$ with $b \notin N$, and set $H = \langle b \rangle$. Then $N \cong \mathbf{Z}_p$ and $H \cong \mathbf{Z}_p$, and by Corollary 4.6, $N \lhd G$ and $H \lhd G$; so

$$G \cong N \times H \cong \mathbf{Z}_p \times \mathbf{Z}_p$$

by Proposition 6.3. $\square$

(7.2) Proposition. *Let p and q be primes such that $p > q$ and let G be a group of order pq.*

(1) *If q does not divide $p - 1$, then $G \cong \mathbf{Z}_{pq}$.*
(2) *If $q \mid p - 1$, then $G \cong \mathbf{Z}_{pq}$ or $G \cong \mathbf{Z}_p \rtimes_\phi \mathbf{Z}_q$ where*

$$\phi : \mathbf{Z}_q \to \mathrm{Aut}(\mathbf{Z}_p) \cong \mathbf{Z}_p^*$$

is a nontrivial homomorphism. All nontrivial homomorphisms produce isomorphic groups.

Proof. By Cauchy's theorem (Theorem 4.7) G has a subgroup N of order p and a subgroup H of order q, both of which are necessarily cyclic. Then $N \lhd G$ since $[G : N] = q$ and q is the smallest prime dividing $|G|$ (Corollary 4.6). Since it is clear that $N \cap H = \langle e \rangle$ and $NH = G$, it follows that G is the semidirect product of N and H.

The map $\phi : H \to \mathrm{Aut}(N)$ given by $\phi_h(n) = hnh^{-1}$ is a group homomorphism, so if q does not divide $|\mathrm{Aut}(N)| = |\mathrm{Aut}(\mathbf{Z}_p)| = |\mathbf{Z}_p^*| = p - 1$, then ϕ is the trivial homomorphism. Hence $\phi_h = 1_N$ for all $h \in H$, i.e., $nh = hn$ for all $h \in H$, $n \in N$. Hence $H \lhd G$ and $G \cong \mathbf{Z}_p \times \mathbf{Z}_q \cong \mathbf{Z}_{pq}$ (see Exercise 11). If $q \mid p - 1$ then there are nontrivial homomorphisms

$$\phi : \mathbf{Z}_q \to \mathrm{Aut}(N) \cong \mathbf{Z}_p^*$$

and for some homomorphism ϕ,

$$G \cong \mathbf{Z}_p \rtimes_\phi \mathbf{Z}_q.$$

Therefore, if $N = \langle a \rangle$ and $H = \langle b \rangle$, then $G = \langle a, b \rangle$, subject to the relations

$$a^p = e, \qquad b^q = e, \qquad b^{-1}ab = a^r$$

where $r^q \equiv 1 \pmod{p}$. If $r = 1$ then ϕ is trivial, H is normal, and $G \cong \mathbf{Z}_p \times \mathbf{Z}_q$. Otherwise, G is nonabelian. We leave it as an exercise to verify that all choices of $r \neq 1$ produce isomorphic groups. Thus, if $q \mid p - 1$, then there are exactly two nonisomorphic groups of order pq. $\square$

(7.3) Corollary. *If $|G| = 2p$, where p is an odd prime, then $G \cong \mathbf{Z}_{2p}$ or $G \cong D_{2p}$.*

Proof. The only nontrivial homomorphism $\phi : \mathbf{Z}_2 \to \mathrm{Aut}(\mathbf{Z}_p) = \mathbf{Z}_p^*$ is the homomorphism $1 \mapsto \phi_1$ with $\phi_1(a) = -a$. Apply Example 6.10 (2). $\square$

(7.4) *Remark.* The results obtained so far completely describe all groups of order ≤ 15, except for groups of order 8 and 12. We shall analyze each of these two cases separately.

Groups of Order 8

We will consider first the case of abelian groups of order 8.

(7.5) Proposition. *If G is an abelian group of order 8, then G is isomorphic to exactly one of the following groups:*

(1) $\mathbf{Z}_8$,
(2) $\mathbf{Z}_4 \times \mathbf{Z}_2$, *or*
(3) $\mathbf{Z}_2 \times \mathbf{Z}_2 \times \mathbf{Z}_2$.

Proof. Case 1: Suppose that G has an element of order 8. Then G is cyclic and, hence, isomorphic to $\mathbf{Z}_8$.

Case 2: Suppose every element of G has order 2. Let $\{a, b, c\} \subseteq G \backslash \{e\}$ with $c \neq ab$. Then $H = \langle a, b \rangle$ is a subgroup of G isomorphic to $\mathbf{Z}_2 \times \mathbf{Z}_2$. Furthermore, $H \cap \langle c \rangle = \langle e \rangle$ and $H \langle c \rangle = G$ so that

$$G \cong H \times \langle c \rangle \cong \mathbf{Z}_2 \times \mathbf{Z}_2 \times \mathbf{Z}_2.$$

Case 3: If G does not come under Case 1 or Case 2, then G is not cyclic and not every element has order 2. Therefore, G has an element a of order 4. We claim that there is an element $b \notin \langle a \rangle$ such that $b^2 = e$. To see this, let c be any element not in $\langle a \rangle$. If $c^2 = e$, take $b = c$. Otherwise, we must have $o(c) = 4$. Since $|G/\langle a \rangle| = 2$, it follows that $c^2 \in \langle a \rangle$. Since a^2 is the only element of $\langle a \rangle$ of order 2, it follows that $c^2 = a^2$. Let $b = ac$. Then

$$b^2 = a^2 c^2 = a^4 = e.$$

Proposition 6.3 then shows that

$$G \cong \langle a \rangle \times \langle b \rangle \cong \mathbf{Z}_4 \times \mathbf{Z}_2.$$

Since every abelian group of order 8 is covered by Case 1, Case 2, or Case 3, the proof is complete. $\square$

Now consider the case of nonabelian groups of order 8.

(7.6) Proposition. *If G is a nonabelian group of order 8, then G is isomorphic to exactly one of the following two groups:*

(1) $Q =$ *the quaternion group, or*
(2) $D_8 =$ *the dihedral group of order 8.*

Proof. Since G is not abelian, it is not cyclic so G does not have an element of order 8. Similarly, if $a^2 = e$ for all $a \in G$, then G is abelian (Exercise 8); therefore, there is an element $a \in G$ of order 4. Let b be an element of G not in $\langle a \rangle$. Since $[G : \langle a \rangle] = 2$, the subgroup $\langle a \rangle \triangleleft G$. But $|G/\langle a \rangle| = 2$ so that $b^2 \in \langle a \rangle$. Since $o(b)$ is 2 or 4, we must have $b^2 = e$ or $b^2 = a^2$. Since $\langle a \rangle \triangleleft G$, $b^{-1}ab$ is in $\langle a \rangle$ and has order 4. Since G is not abelian, it follows that $b^{-1}ab = a^3$. Therefore, G has two generators a and b subject to one of the following sets of relations:

(1) $a^4 = e$, $b^2 = e$, $b^{-1}ab = a^3$;
(2) $a^4 = e$, $b^2 = a^2$, $b^{-1}ab = a^3$.

In case (1), G is isomorphic to D_8, while in case (2) G is isomorphic to Q. We leave it as an exercise to check that Q and D_8 are not isomorphic. $\square$

(7.7) Remarks. (1) Propositions 7.5 and 7.6 together show that there are precisely 5 distinct isomorphism classes of groups of order 8; 3 are abelian and 2 are nonabelian.

(2) D_8 is a semidirect product of $\mathbf{Z}_4$ and $\mathbf{Z}_2$ as was observed in Example 6.10 (2). However, Q is a nonsplit extension of $\mathbf{Z}_4$ by $\mathbf{Z}_2$, or of $\mathbf{Z}_2$ by $\mathbf{Z}_2 \times \mathbf{Z}_2$. In fact Q is not a semidirect product of proper subgroups.

Groups of Order 12

To classify groups of order 12, we start with the following result.

(7.8) Proposition. *Let G be a group of order $p^2 q$ where p and q are distinct primes. Then G is the semidirect product of a p-Sylow subgroup H and a q-Sylow subgroup K.*

Proof. If $p > q$ then $H \triangleleft G$ by Corollary 4.6.

If $q > p$ then $1 + kq \mid p^2$ for some $k \geq 0$. Since $q > p$, this can only occur if $k = 0$ or $1 + kq = p^2$. The latter case forces q to divide $p^2 - 1 = (p+1)(p-1)$. Since $q > p$, we must have $q = p+1$. This can happen only if $p = 2$ and $q = 3$. Therefore, in the case $q > p$, the q-Sylow subgroup K is a normal subgroup of G, except possibly when $|G| = 2^2 \cdot 3 = 12$.

To analyze this case, let K be a 3-Sylow subgroup of a group G of order 12. If K is not normal in G, then the number of 3-Sylow subgroups of G is 4. Let these 3-Sylow subgroups be K_1, K_2, K_3, and K_4. Then $K_1 \cup K_2 \cup K_3 \cup K_4$ accounts for 9 distinct elements of G.

The remaining elements, together with the identity e, must form the 2-Sylow subgroup H of G. Hence, we must have $H \triangleleft G$.

Therefore, we have shown that at least one of H (a p-Sylow subgroup of G) or K (a q-Sylow subgroup of G) is normal in G. Since it is clear that $H \cap K = \langle e \rangle$ and $HK = G$, it follows that G is a semidirect product of H and K. $\qquad\square$

(7.9) Proposition. *A nonabelian group G of order 12 is isomorphic to exactly one of the following groups:*

(1) A_4,
(2) D_{12}, or
(3) $T = \mathbf{Z}_3 \rtimes_\phi \mathbf{Z}_4$ *where* $\phi : \mathbf{Z}_4 \to \mathrm{Aut}(\mathbf{Z}_3) \cong \mathbf{Z}_2$ *is the nontrivial homomorphism.*

Proof. Let H be a 2-Sylow subgroup and K a 3-Sylow subgroup of G. By Proposition 7.8 and the fact that G is nonabelian, exactly one of H and K is normal in G.

Case 1: Suppose $H \triangleleft G$. Then K is not normal in G. Since $[G : K] = 4$, there is a permutation representation $\Phi_K : G \to S_4$. By Proposition 4.3, $\mathrm{Ker}(\Phi_K)$ is the largest normal subgroup of G contained in K. Since K has prime order and is not normal, it follows that G is injective so that

$$G \cong \mathrm{Im}(\Phi_K) \subseteq S_4.$$

It is an easy exercise to show that A_4 is the only subgroup of S_4 of order 12; therefore, $G \cong A_4$ if the 2-Sylow subgroup is normal in G.

Case 2: Suppose $K \triangleleft G$ and $H \cong \mathbf{Z}_4$. In this case

$$G \cong \mathbf{Z}_3 \rtimes_\phi \mathbf{Z}_4$$

where $\phi : \mathbf{Z}_4 \to \mathrm{Aut}(K)$ is a nontrivial homomorphism, but the only nontrivial automorphism of $\mathbf{Z}_3$ is $a \mapsto a^{-1}$ where $K = \langle a \rangle$. In this case $G \cong T$.

Case 3: Suppose $K \triangleleft G$ and $H \cong \mathbf{Z}_2 \times \mathbf{Z}_2$. Let $K = \langle a \rangle$ and let

$$\phi : H \to \mathrm{Aut}(K) \cong \mathbf{Z}_2$$

be the conjugation homomorphism. Then $H \cong (\mathrm{Ker}(\phi)) \times \mathbf{Z}_2$, so let $\mathrm{Ker}(\phi) = \langle c \rangle$ and let $d \in H$ with $\phi(d) \neq 1_K$. Then $c^{-1}ac = a$ and $d^{-1}ad = a^{-1} = a^2$. Let $b = ac$. Then $o(b) = 6$, $d \notin \langle b \rangle$, and

$$d^{-1}bd = d^{-1}acd = d^{-1}adc = a^2c = (ac)^{-1} = b^{-1}.$$

Thus, $G \cong D_{12}$. $\qquad\square$

It remains to consider the case of abelian groups of order 12.

(7.10) Proposition. *If G is an abelian group of order 12, then G is isomorphic to exactly one of the following groups:*

(1) $\mathbf{Z}_{12}$, *or*
(2) $\mathbf{Z}_2 \times \mathbf{Z}_6$.

Proof. Exercise.   □

By combining the results of this section we arrive at the following table of distinct groups of order at most 15. That is, every group of order ≤ 15 is isomorphic to exactly one group in this table.

Table 7.1. Groups of order ≤ 15

Order	Abelian Groups	Nonabelian Groups	Total Number
1	$\{e\}$		1
2	$\mathbf{Z}_2$		1
3	$\mathbf{Z}_3$		1
4	$\mathbf{Z}_4$		2
	$\mathbf{Z}_2 \times \mathbf{Z}_2$		
5	$\mathbf{Z}_5$		1
6	$\mathbf{Z}_6$	S_3	2
7	$\mathbf{Z}_7$		1
8	$\mathbf{Z}_8$	Q	5
	$\mathbf{Z}_4 \times \mathbf{Z}_2$	D_8	
	$\mathbf{Z}_2 \times \mathbf{Z}_2 \times \mathbf{Z}_2$		
9	$\mathbf{Z}_9$		2
	$\mathbf{Z}_3 \times \mathbf{Z}_3$		
10	$\mathbf{Z}_{10}$	D_{10}	2
11	$\mathbf{Z}_{11}$		1
12	$\mathbf{Z}_{12}$	A_4	5
	$\mathbf{Z}_2 \times \mathbf{Z}_6$	D_{12}	
		$\mathbf{Z}_3 \rtimes_\phi \mathbf{Z}_4$	
13	$\mathbf{Z}_{13}$		1
14	$\mathbf{Z}_{14}$	D_{14}	2
15	$\mathbf{Z}_{15}$		1

1.8 Exercises

1. Prove that $\mathbf{Z}_n^*$ is a group. (See Example 1.2 (5).)
2. Prove that $\mathcal{P}(X)$ (Example 1.2 (8)) with the symmetric difference operation is a group.
3. Write the Cayley diagram for the group S_3.
4. Write the Cayley diagram for the group $\mathbf{Z}_{12}^*$.
5. Let G be a group, $g \in G$, and define a new multiplication $\cdot$ on G by the formula $a \cdot b = agb$ for all $a, b \in G$. Prove that G with the multiplication $\cdot$ is a group. What is the identity of G under $\cdot$? If $a \in G$ what is the inverse of a under $\cdot$?
6. Suppose that G is a set and $\cdot$ is an associative binary operation on G such that there is an element $e \in G$ with $e \cdot a = a$ for all $a \in G$ and such that for each $a \in G$ there is an element $b \in G$ with $b \cdot a = e$. Prove that $(G, \cdot)$ is a group. The point of this exercise is that it is sufficient to assume associativity, a left identity, and left inverses in order to have a group. Similarly, left can be replaced with right in the hypotheses.
7. Prove that $\mathbf{R}^* \times \mathbf{R}$ is a group under the multiplication defined by

$$(a, b)(c, d) = (ac, ad + b).$$

   Is this group abelian?
8. Prove that if $a^2 = e$ for all a in a group G, then G is abelian.
9. Let $V \subseteq \mathrm{GL}(2, \mathbf{R})$ be the set

$$V = \left\{ \begin{bmatrix} 1 & 0 \\ 0 & 1 \end{bmatrix}, \begin{bmatrix} -1 & 0 \\ 0 & 1 \end{bmatrix}, \begin{bmatrix} 1 & 0 \\ 0 & -1 \end{bmatrix}, \begin{bmatrix} -1 & 0 \\ 0 & -1 \end{bmatrix} \right\}.$$

   Prove that V is a subgroup of $\mathrm{GL}(2, \mathbf{R})$ that is isomorphic to the Klein 4-group.
10. For fixed positive integers b_0, m_0, and n_0 consider the subset $S \subset \mathrm{GL}(3, \mathbf{Z})$ defined by

$$S = \left\{ \begin{bmatrix} 1 & m & n \\ 0 & 1 & b \\ 0 & 0 & 1 \end{bmatrix} : m_0 \mid m,\ n_0 \mid n,\ b_0 \mid b \right\}.$$

    When is S a subgroup? The notation $a \mid b$ for integers a and b means that a divides b.
11. Let G be a group and let $a, b \in G$ be elements such that $ab = ba$.
    (a) Prove that $o(ab) \mid o(a)o(b)$.
    (b) If $ab = ba$ and $\langle a \rangle \cap \langle b \rangle = \langle e \rangle$, show that

$$o(ab) = \mathrm{lcm}\{o(a), o(b)\}.$$

    ($\mathrm{lcm}\{n, m\}$ refers to the least common multiple of the integers n and m.)
    (c) If $ab = ba$ and $o(a)$ and $o(b)$ are relatively prime, then $o(ab) = o(a)o(b)$.
    (d) Give a counterexample to show that these results are false if we do not assume commutativity of a and b.
12. If $\sigma : G \to H$ is a group homomorphism then $o(\sigma(a)) \mid o(a)$ for all $a \in G$ with $o(a) < \infty$. If σ is an isomorphism then $o(\sigma(a)) = o(a)$.
13. (a) A group G is abelian if and only if the function $f : G \to G$ defined by $f(a) = a^{-1}$ is a group homomorphism.

(b) A group G is abelian if and only if the function $g : G \to G$ defined by $g(a) = a^2$ is a group homomorphism.

14. Let G be the multiplicative group of positive real numbers and let H be the additive group of all reals. Prove that $G \cong H$. (Hint: Remember the properties of the logarithm function.)

15. Write all the subgroups of S_3.

16. Let G be a group and let H_1, H_2 be subgroups of G. Prove that $H_1 \cup H_2$ is a subgroup of G if and only if $H_1 \subseteq H_2$ or $H_2 \subseteq H_1$. Is the analogous result true for three subgroups H_1, H_2, H_3?

17. If G is a finite group and H and K are subgroups, prove that

$$|H||K| = |H \cap K||HK|.$$

18. Prove that the intersection of two subgroups of finite index is a subgroup of finite index. Prove that the intersection of finitely many subgroups of finite index is a subgroup of finite index.

19. Let X be a finite set and let $Y \subseteq X$. Let G be the symmetric group S_X and define H and K by

$$H = \{f \in G : f(y) = y \text{ for all } y \in Y\}$$
$$K = \{f \in G : f(y) \in Y \text{ for all } y \in Y\}.$$

If $|X| = n$ and $|Y| = m$ compute $[G : H]$, $[G : K]$, and $[K : H]$.

20. If G is a group let $Z(G) = \{a \in G : ab = ba \text{ for all } b \in G\}$. Then prove that $Z(G)$ is an abelian subgroup of G. $Z(G)$ is called the *center* of G. If $G = GL(n, \mathbf{R})$ show that

$$Z(G) = \{aI_n : a \in \mathbf{R}^*\}.$$

21. Let G be a group and let $H \subseteq Z(G)$ be a subgroup of the center of G. Prove that $H \lhd G$.

22. (a) If G is a group, prove that the commutator subgroup G' is a normal subgroup of G, and show that G/G' is abelian.
    (b) If H is any normal subgroup of G such that G/H is abelian, show that $G' \subseteq H$.

23. If G is a group of order $2n$ show that the number of elements of G of order 2 is odd.

24. Let Q be the multiplicative subgroup of $GL(2, \mathbf{C})$ generated by

$$A = \begin{bmatrix} 0 & i \\ i & 0 \end{bmatrix} \quad \text{and} \quad B = \begin{bmatrix} 0 & 1 \\ -1 & 0 \end{bmatrix}.$$

    (a) Show that A and B satisfy the relations $A^4 = I, A^2 = B^2, B^{-1}AB = A^{-1}$. (Thus, Q is a concrete representation of the quaternion group.)
    (b) Prove that $|Q| = 8$ and list all the elements of Q in terms of A and B.
    (c) Compute $Z(Q)$ and prove that $Q/Z(Q)$ is abelian.
    (d) Prove that every subgroup of Q is normal.

25. Let n be a fixed positive integer. Suppose a group G has exactly one subgroup H of order n. Prove that $H \lhd G$.

26. Let $H \lhd G$ and assume that G/H is abelian. Show that every subgroup $K \subseteq G$ containing H is normal.

27. Let G_n be the multiplicative subgroup of $GL(2, \mathbf{C})$ generated by

$$A = \begin{bmatrix} \zeta & 0 \\ 0 & \zeta^{-1} \end{bmatrix} \quad \text{and} \quad B = \begin{bmatrix} 0 & 1 \\ 1 & 0 \end{bmatrix}$$

where $\zeta = \exp(2\pi i/n)$. Verify that G_n is isomorphic to the dihedral group D_{2n}. (See Example 2.8 (13).)

28. Let G be a group of order n. If G is generated by two elements of order 2, show that $G \cong \mathbf{Z}_2 \times \mathbf{Z}_2$ if $n = 4$ and $G \cong D_n$ if $n > 4$.

29. Let G be a nonabelian group of order 6. Prove that $G \cong S_3$.

30. (a) If $H \triangleleft G$ and $[G : H] = n$, then show that $a^n \in H$ for all $a \in G$.
    (b) Show that the result in part (a) is false if H is not normal in G.

31. Show that the alternating group A_4 of order 12 does not have a subgroup of order 6. (Hint: Find at least 8 elements of A_4 that are squares, and apply Exercise 30.)

32. Recall (Definition 4.13) that a group G is called a p-group if $|G| = p^n$ for some integer $n \geq 1$.
    (a) If G is a p-group, show that $Z(G) \neq \langle e \rangle$. (Hint: Use the class equation (Corollary 2.28).)
    (b) If $|G| = p^n$, show that G has a subgroup of order p^m for every $0 \leq m \leq n$.

33. Let $G = \left\{ \begin{bmatrix} a & b \\ 0 & 1 \end{bmatrix} \in \mathrm{GL}(2, \mathbf{R}) \right\}$. Prove that G is a subgroup of $\mathrm{GL}(2, \mathbf{R})$ and that G is isomorphic to the group $\mathbf{R}^* \times \mathbf{R}$ with the multiplication defined in Exercise 7.

34. (a) Find all homomorphisms $\phi : \mathbf{Z} \to \mathbf{Z}_n$.
    (b) Find all homomorphisms $\phi : \mathbf{Z}_7 \to \mathbf{Z}_{16}$.
    (c) What is a condition on finite cyclic groups G and H that ensures there is a homomorphism $\phi : G \to H$ other than the zero homomorphism?

35. Let $\mathrm{Hom}(\mathbf{Z}_n, \mathbf{Z}_m)$ be the set of all group homomorphisms from $\mathbf{Z}_n$ to $\mathbf{Z}_m$. Let d be the greatest common divisor of m and n. Show that $|\mathrm{Hom}(\mathbf{Z}_n, \mathbf{Z}_m)| = d$.

36. If n is odd, show that $D_{4n} \cong D_{2n} \times \mathbf{Z}_2$.

37. Write the class equations (Corollary 2.28) for the quaternion group Q and the dihedral group D_8.

38. Verify that the alternating group A_5 has no nontrivial normal subgroups. (Hint: The class equation.) (The trivial subgroups of a group G are $\{e\}$ and G.) A group with no nontrivial normal subgroups is called **simple**. It is known that A_n is simple for all $n \neq 4$.

39. Suppose that G is an abelian group of order n. If $m \mid n$ show that G has a subgroup of order m. Compare this result with Exercise 31.

40. (a) Write each of the following permutations as a product of disjoint cycles:

$$\alpha = \begin{pmatrix} 1 & 2 & 3 & 4 & 5 & 6 \\ 6 & 5 & 4 & 1 & 2 & 3 \end{pmatrix}$$

$$\beta = \begin{pmatrix} 1 & 2 & 3 & 4 & 5 & 6 & 7 & 8 \\ 8 & 1 & 3 & 6 & 5 & 7 & 4 & 2 \end{pmatrix}$$

$$\gamma = \begin{pmatrix} 1 & 2 & 3 & 4 & 5 & 6 & 7 & 8 & 9 \\ 2 & 3 & 4 & 5 & 6 & 7 & 8 & 9 & 1 \end{pmatrix}$$

$$\delta = \begin{pmatrix} 1 & 2 & 3 & 4 & 5 & 6 & 7 & 8 & 9 \\ 5 & 8 & 9 & 2 & 1 & 4 & 3 & 6 & 7 \end{pmatrix}.$$

(b) Let $\sigma \in S_{10}$ be the permutation

$$\sigma = \begin{pmatrix} 1 & 2 & 3 & 4 & 5 & 6 & 7 & 8 & 9 & 10 \\ 3 & 5 & 4 & 1 & 7 & 10 & 2 & 6 & 9 & 8 \end{pmatrix}.$$

Compute $o(\sigma)$ and calculate σ^{100}.

41. Let $H \subseteq S_n$ be defined by $H = \{f \in S_n : f(1) = 1\}$. Prove that H is a subgroup of S_n that is isomorphic to S_{n-1}. Is $H \triangleleft S_n$?

42. (a) Prove that an r–cycle is even (odd) if and only if r is odd (even).

(b) Prove that a permutation σ is even if and only if there are an even number of even order cycles in the cycle decomposition of σ.

43. Show that if a subgroup G of S_n contains an odd permutation then G has a normal subgroup H with $[G : H] = 2$.

44. For $\alpha \in S_n$, let

$$\widetilde{f}(\alpha) = |\{(i, j) : 1 \leq i < j \quad \text{and} \quad \alpha(j) < \alpha(i)\}|.$$

(For example, if

$$\alpha = \begin{pmatrix} 1 & 2 & 3 & 4 & 5 \\ 2 & 5 & 1 & 4 & 3 \end{pmatrix} \in S_5,$$

then $\widetilde{f}(\alpha) = 5$.) Show that $\operatorname{sgn}(\alpha) = 1$ if $\widetilde{f}(\alpha)$ is even and $\operatorname{sgn}(\alpha) = -1$ if $\widetilde{f}(\alpha)$ is odd. Thus, $\widetilde{f}$ provides a method of determining if a permutation is even or odd without the factorization into disjoint cycles.

45. (a) Prove that S_n is generated by the transpositions $(1\,2)$, $(1\,3)$, ..., $(1\,n)$.
(b) Prove that S_n is generated by $(1\,2)$ and $(1\,2\cdots n)$.

46. In the group S_4 compute the number of permutations conjugate to each of the following permutations: $e = (1)$, $\alpha = (1\,2)$, $\beta = (1\,2\,3)$, $\gamma = (1\,2\,3\,4)$, and $\delta = (1\,2)(3\,4)$.

47. (a) Find all the subgroups of the dihedral group D_8.
(b) Show that D_8 is not isomorphic to the quaternion group Q. Note, however, that both groups are nonabelian groups of order 8. (Hint: Count the number of elements of order 2 in each group.)

48. Construct two nonisomorphic nonabelian groups of order p^3 where p is an odd prime.

49. Show that any group of order 312 has a nontrivial normal subgroup.

50. Show that any group of order 56 has a nontrivial normal subgroup.

51. Show $\operatorname{Aut}(\mathbf{Z}_2 \times \mathbf{Z}_2) \cong S_3$.

52. How many elements are there of order 7 in a simple group of order 168? (See Exercise 38 for the definition of simple.)

53. Classify all groups (up to isomorphism) of order 18.

54. Classify all groups (up to isomorphism) of order 20.

Chapter 2

Rings

2.1 Definitions and Examples

(1.1) Definition. *A* **ring** $(R, +, \cdot)$ *is a set R together with two binary operations $+ : R \times R \to R$ (addition) and $\cdot : R \times R \to R$ (multiplication) satisfying the following properties.*

(a) $(R, +)$ *is an abelian group. We write the identity element as 0.*
(b) $a \cdot (b \cdot c) = (a \cdot b) \cdot c$ *($\cdot$ is associative).*
(c) $a \cdot (b + c) = a \cdot b + a \cdot c$ *and* $(b + c) \cdot a = b \cdot a + c \cdot a$ *($\cdot$ is left and right distributive over $+$).*

As in the case of groups, it is conventional to write ab instead of $a \cdot b$. A ring will be denoted simply by writing the set R, with the multiplication and addition being implicit in most cases. If $R \neq \{0\}$ and multiplication on R has an identity element, i.e., there is element $1 \in R$ with $a1 = 1a = a$ for all $a \in R$, then R is said to be a **ring with identity**. In this case $1 \neq 0$ (see Proposition 1.2). We emphasize that the ring $R = \{0\}$ is **not** a ring with identity.

Convention. *In Sections 2.1 and 2.2, R will denote an arbitrary ring. In the rest of this book, the word "ring" will always mean "ring with identity."*

If the multiplication on R is commutative, i.e., $ab = ba$ for all $a, b \in R$, then R is called a **commutative ring**. The standard rules of sign manipulation, which are familiar from the real or complex numbers, are also true in a general ring R. The verification of the following rules are left as an exercise.

(1.2) Proposition. *Let R be a ring. Then if $a, b, c \in R$ the following rules are valid.*

(1) $a0 = 0a = 0$.

(2) $(-a)b = a(-b) = -(ab)$.
(3) $ab = (-a)(-b)$.
(4) $a(b - c) = ab - ac$ and $(a - b)c = ac - bc$.
(5) *If R has an identity then $(-1)a = -a$.*
(6) *If R has an identity then $1 \neq 0$.*

If $a \neq 0$ and $b \neq 0$ are elements of R such that $ab = 0$ then a and b are called **zero divisors** of the ring R. Note that 0 is *not* a zero divisor. If R has an identity, an element $a \in R$ is a **unit** if a has a multiplicative inverse, that is, if there is a $b \in R$ with $ab = 1 = ba$. R^* will denote the set of all units of R. R^* is a group, called the **group of units** of R.

(1.3) Lemma. *Let R be a ring and $a \neq 0$ an element of R that is not a zero divisor. If $b, c \in R$ satisfy $ab = ac$, then $b = c$. Similarly, if $ba = ca$, then $b = c$.*

Proof. If $ab = ac$, then

$$0 = ab - ac = a(b - c),$$

and since a is not a zero divisor and $a \neq 0$, this implies that $b - c = 0$, i.e., $b = c$. The other half is similar.   □

(1.4) Definition. (1) *A ring R is an **integral domain** if it is a commutative ring with identity such that R has no zero divisors.*

(2) *A ring R with identity is a **division ring** if $R^* = R \setminus \{0\}$, i.e., every nonzero element of R has a multiplicative inverse.*

(3) *A **field** is a commutative division ring.*

Thus, a commutative ring R with identity is an integral domain if and only if the equation $ab = 0$ (for $a, b \in R$) implies that $a = 0$ or $b = 0$. R is a division ring if and only if the equations $ax = b$ and $ya = b$ are solvable in R for every $b \in R$ and $a \neq 0 \in R$.

(1.5) Proposition. *A finite integral domain is a field.*

Proof. Let R be a finite integral domain and let $a \neq 0 \in R$. Define $\phi_a :$ $R \to R$ by $\phi_a(b) = ab$. Suppose that $\phi_a(b) = \phi_a(c)$. Then $ab = ac$, so $b = c$ by Lemma 1.3. Therefore, ϕ_a is an injective function on the finite set R so that $|R| = |\phi_a(R)|$, and hence, $\phi_a(R) = R$. In particular, the equation $ax = 1$ is solvable for every $a \neq 0$ and R is a field.   □

(1.6) Remark. The conclusion of Proposition 1.5 is valid under much weaker hypotheses. In fact, there is a theorem of Wedderburn that states that the commutativity follows from the finiteness of the ring. Specifically, Wedderburn proved that any finite division ring is automatically a field. This result requires more background than the elementary Proposition 1.5 and will not be presented here.

(1.7) Definition. *If R is a ring with identity, then the* **characteristic of** R, *denoted* char(R), *is the smallest natural number n such that $n \cdot 1 = 0$. If $n \cdot 1 \neq 0$ for all $n \in \mathbf{N}$, then we set* char(R) $= 0$.

(1.8) Proposition. *If R is an integral domain, then* char(R) $= 0$ *or* char(R) *is prime.*

Proof. Suppose that char(R) $= n \neq 0$. If n is composite, then we may write $n = rs$ where $1 < r < n$ and $1 < s < n$. Then, by the definition of characteristic, $r \cdot 1 \neq 0$ and $s \cdot 1 \neq 0$. But $0 = n \cdot 1 = (r \cdot 1)(s \cdot 1)$. Therefore, the ring R has zero divisors. This contradicts the fact that R is an integral domain, so n must be prime. $\qquad\square$

(1.9) Definition. *A subset S of a ring R is a* **subring** *if S, under the operations of multiplication and addition on R, is a ring. Thus S is a subring of R if and only if S is an additive subgroup of R that is closed under multiplication.*

We will now present a number of examples of rings. Many of the mathematical systems with which you are already familiar are rings. Thus the integers $\mathbf{Z}$ are an integral domain, while the rational numbers $\mathbf{Q}$, the real numbers $\mathbf{R}$, and the complex numbers $\mathbf{C}$ are fields.

(1.10) Examples.

(1) $2\mathbf{Z} = \{$even integers$\}$ is a subring of the ring $\mathbf{Z}$ of integers. $2\mathbf{Z}$ does not have an identity and thus it fails to be an integral domain, even though it has no zero divisors.

(2) $\mathbf{Z}_n$, the integers under addition and multiplication modulo n, is a ring with identity. $\mathbf{Z}_n$ has zero divisors if and only if n is composite. Indeed, if $n = rs$ for $1 < r < n$, $1 < s < n$ then $rs = 0$ in $\mathbf{Z}_n$ and $r \neq 0, s \neq 0$ in $\mathbf{Z}_n$, so $\mathbf{Z}_n$ has zero divisors. Conversely, if $\mathbf{Z}_n$ has zero divisors then there is an equation $ab = 0$ in $\mathbf{Z}_n$ with $a \neq 0, b \neq 0$ in $\mathbf{Z}_n$. By choosing representatives of a and b in $\mathbf{Z}$ we obtain an equation $ab = nk$ in $\mathbf{Z}$ where we may assume that $0 < a < n$ and $0 < b < n$. Therefore, every prime divisor of k divides either a or b so that after enough divisions we arrive at an equation $rs = n$ where $0 < r < n$ and $0 < s < n$, i.e., n is composite.

(3) Example (2) combined with Proposition 1.5 shows that $\mathbf{Z}_n$ is a field if and only if n is a prime number. In particular, we have identified some finite fields, namely, $\mathbf{Z}_p$ for p a prime.

(4) There are finite fields other than the fields $\mathbf{Z}_p$. We will show how to construct some of them after we develop the theory of polynomial rings.

Table 1.1. Multiplication and addition for a field with four elements

+	0	1	a	b
0	0	1	a	b
1	1	0	b	a
a	a	b	0	1
b	b	a	1	0

·	0	1	a	b
0	0	0	0	0
1	0	1	a	b
a	0	a	b	1
b	0	b	1	a

For now we can present a specific example via explicit addition and multiplication tables. Let $F = \{0, 1, a, b\}$ have addition and multiplication defined by Table 1.1. One can check directly that $(F, +, \cdot)$ is a field with 4 elements. Note that the additive group $(F, +) \cong \mathbf{Z}_2 \times \mathbf{Z}_2$ and that the multiplicative group $(F^*, \cdot) \cong \mathbf{Z}_3$.

(5) Let $\mathbf{Z}[i] = \{m + ni : m, n \in \mathbf{Z}\}$. Then $\mathbf{Z}[i]$ is a subring of the field of complex numbers called the ring of **gaussian integers**. As an exercise, check that the units of $\mathbf{Z}[i]$ are $\{\pm 1, \pm i\}$.

(6) Let $d \neq 0, 1 \in \mathbf{Z}$ be square-free (i.e., n^2 does not divide d for any $n > 1$) and let $\mathbf{Q}[\sqrt{d}] = \{a + b\sqrt{d} : a, b \in \mathbf{Q}\} \subseteq \mathbf{C}$. Then $\mathbf{Q}[\sqrt{d}]$ is a subfield of $\mathbf{C}$ called a **quadratic field**.

(7) Let X be a set and $\mathcal{P}(X)$ the power set of X. Then $(\mathcal{P}(X), \triangle, \cap)$ is a commutative ring with identity where addition in $\mathcal{P}(X)$ is the symmetric difference (see Example 1.2 (8)) and the product of A and B is $A \cap B$. For this ring, $0 = \emptyset \in \mathcal{P}(X)$ and $1 = X \in \mathcal{P}(X)$.

(8) Let R be a ring with identity and let $M_{m,n}(R)$ be the set of $m \times n$ matrices with entries in R. If $m = n$ we will write $M_n(R)$ in place of $M_{n,n}(R)$. If $A = [a_{ij}] \in M_{m,n}(R)$ we let $\mathrm{ent}_{ij}(A) = a_{ij}$ denote the entry of A in the ith row and j^{th} column for $1 \leq i \leq m$, $1 \leq j \leq n$. If $A, B \in M_{m,n}(R)$ then the sum is defined by the formula

$$\mathrm{ent}_{ij}(A + B) = \mathrm{ent}_{ij}(A) + \mathrm{ent}_{ij}(B),$$

while if $A \in M_{m,n}(R)$ and $B \in M_{n,p}(R)$ the product $AB \in M_{m,p}(R)$ is defined by the formula

$$\mathrm{ent}_{ij}(AB) = \sum_{k=1}^{n} \mathrm{ent}_{ik}(A)\,\mathrm{ent}_{kj}(B).$$

In particular, note that addition and multiplication are always defined for two matrices in $M_n(R)$, and with these definitions of addition and multiplication, $M_n(R)$ is a ring with identity, called a **matrix ring**. The identity of $M_n(R)$ is the matrix I_n defined by $\mathrm{ent}_{ij}(I_n) = \delta_{ij}$ where δ_{ij} is the kronecker delta

$$\delta_{ij} = \begin{cases} 1 & \text{if } i = j, \\ 0 & \text{if } i \neq j. \end{cases}$$

There are mn matrices E_{ij} $(1 \leq i \leq m, 1 \leq j \leq n)$ in $M_{m,n}$ that are particularly useful in many calculations concerning matrices. E_{ij} is defined by the formula $\text{ent}_{kl}(E_{ij}) = \delta_{ki}\delta_{lj}$, that is, E_{ij} has a 1 in the ij position and 0 elsewhere. Therefore, any $A = [a_{ij}] \in M_{m,n}(R)$ can be written as

$$(1.1) \qquad A = \sum_{i=1}^{m} \sum_{j=1}^{n} a_{ij} E_{ij}.$$

Note that the symbol E_{ij} does not contain notation indicating which $M_{m,n}(R)$ the matrix belongs to. This is determined from the context. There is the following matrix product rule for the matrices E_{ij} (when the matrix multiplications are defined):

$$(1.2) \qquad E_{ij} E_{kl} = \delta_{jk} E_{il}.$$

In case $m = n$, note that $E_{ii}^2 = E_{ii}$, and when $n > 1$, $E_{11} E_{12} = E_{12}$ while $E_{12} E_{11} = 0$. Therefore, if $n > 1$ then the ring $M_n(R)$ is not commutative and there are zero divisors in $M_n(R)$. The matrices $E_{ij} \in M_n(R)$ are called **matrix units**, but they are definitely not (except for $n = 1$) units in the ring $M_n(R)$. A unit in the ring $M_n(R)$ is an invertible matrix, so the group of units of $M_n(R)$ is called the **general linear group** $GL(n, R)$ of degree n over the ring R.

(9) There are a number of important subrings of $M_n(R)$. To mention a few, there is the ring of diagonal matrices

$$D_n(R) = \{A \in M_n(R) : \text{ent}_{ij}(A) = 0 \text{ if } i \neq j\},$$

the ring of upper triangular matrices

$$T^n(R) = \{A \in M_n(R) : \text{ent}_{ij}(A) = 0 \text{ if } i > j\},$$

and the ring of lower triangular matrices

$$T_n(R) = \{A \in M_n(R) : \text{ent}_{ij}(A) = 0 \text{ if } i < j\}.$$

All three of these subrings of $M_n(R)$ are rings with identity, namely the identity of $M_n(R)$. The subrings of strictly upper triangular matrices ST^n and strictly lower triangular matrices ST_n do not have an identity. A matrix is strictly upper triangular if all entries on and below the diagonal are 0 and strictly lower triangular means that all entries on and above the diagonal are 0.

(10) Let F be a subfield of the real numbers $\mathbf{R}$ and let $x, y \in F$ with $x > 0$ and $y > 0$. Define a subring $Q(-x, -y; F)$ of $M_2(\mathbf{C})$ by

$$Q(-x, -y; F) =$$

$$\left\{ \begin{bmatrix} a + b\sqrt{-x} & c\sqrt{-y} + d\sqrt{xy} \\ c\sqrt{-y} - d\sqrt{xy} & a - b\sqrt{-x} \end{bmatrix} : a, b, c, d \in F \right\}.$$

(In these formulas $\sqrt{-x}$ and $\sqrt{-y}$ denote the square roots with positive imaginary parts.) It is easy to check that $Q(-x, -y; F)$ is closed under matrix addition and matrix multiplication so that it is a subring of $M_2(\mathbf{C})$. Let

$$1 = \begin{bmatrix} 1 & 0 \\ 0 & 1 \end{bmatrix}, \qquad i = \begin{bmatrix} \sqrt{-x} & 0 \\ 0 & -\sqrt{-x} \end{bmatrix},$$

$$j = \begin{bmatrix} 0 & \sqrt{-y} \\ \sqrt{-y} & 0 \end{bmatrix}, \quad \text{and} \quad k = \begin{bmatrix} 0 & \sqrt{xy} \\ -\sqrt{xy} & 0 \end{bmatrix}.$$

Then $Q(-x, -y; F) = \{a\mathbf{1} + b\mathbf{i} + c\mathbf{j} + d\mathbf{k} : a, b, c, d \in F\}$. Note that

$$\begin{cases} \mathbf{i}^2 = -x\mathbf{1}, \ \mathbf{j}^2 = -y\mathbf{1}, \ \mathbf{k}^2 = -xy\mathbf{1} \\ \mathbf{ij} = -\mathbf{ji} = \mathbf{k} \\ \mathbf{ik} = -\mathbf{ki} = x\mathbf{j} \\ \mathbf{kj} = -\mathbf{jk} = y\mathbf{i}. \end{cases}$$

If $h = a\mathbf{1} + b\mathbf{i} + c\mathbf{j} + d\mathbf{k} \in \mathbf{H}$, let $\overline{h} = a\mathbf{1} - b\mathbf{i} - c\mathbf{j} - d\mathbf{k}$. Then

$$h\overline{h} = (a^2 + xb^2 + yc^2 + xyd^2)\mathbf{1}$$

so that if $h \neq 0$ then h is invertible, and in particular, $h^{-1} = \alpha\overline{h}$ where $\alpha = 1/(a^2 + xb^2 + yc^2 + xyd^2)$. Therefore, $Q(-x, -y; F)$ is a division ring, but it is not a field since it is not commutative. $Q(-x, -y; F)$ is called a **definite quaternion algebra**. In case $F = \mathbf{R}$, all these quaternion algebras are isomorphic (see Exercise 13 for some special cases of this fact) and $\mathbf{H} = Q(-1, -1; \mathbf{R})$ is called the **ring of quaternions**. (The notation is chosen to honor their discoverer, Hamilton.) Note that the subset

$$\{\pm\mathbf{1}, \pm\mathbf{i}, \pm\mathbf{j}, \pm\mathbf{k}\}$$

of $\mathbf{H}$ is a multiplicative group isomorphic to the quaternion group Q of Example 1.2.8 (12). (Also see Exercise 24 of Chapter 1.) In case F is a proper subfield of $\mathbf{R}$, the quaternion algebras $Q(-x, -y; F)$ are not all mutually isomorphic, i.e., the choice of x and y matters here.

(11) Let A be an abelian group. An **endomorphism** of A is a group homomorphism $f : A \to A$. Let $\text{End}(A)$ denote the set of all endomorphisms of A and define multiplication and addition on $\text{End}(A)$ by

$$(f + g)(a) = f(a) + g(a) \quad \text{(addition of functions)},$$
$$(fg)(a) = f(g(a)) \quad \text{(functional composition)}.$$

With these operations, $\text{End}(A)$ is a ring (exercise). In general, $\text{End}(A)$ is not commutative. The group of units of $\text{End}(A)$ is the automorphism group of A.

(12) Let $\mathbf{Z}^+$ denote the set of all nonnegative integers. If R is a ring with identity, let $R[X]$ denote the set of all functions $f : \mathbf{Z}^+ \to R$ such that

$f(n) = 0$ for all but a finite number of natural numbers n. Define a ring structure on the set $R[X]$ by the formulas

$$(f + g)(n) = f(n) + g(n)$$

$$(fg)(n) = \sum_{m=0}^{n} f(m)g(n - m).$$

It is easy to check that $R[X]$ is a ring with these operations (do it). $R[X]$ is called the **ring of polynomials in the indeterminate X with coefficients in R.** Notice that the indeterminate X is nowhere mentioned in our definition of $R[X]$. To show that our description of $R[X]$ agrees with that with which you are probably familiar, we define X as a function on $\mathbf{Z}^+$ as follows

$$X(n) = \begin{cases} 1 & \text{if } n = 1, \\ 0 & \text{if } n \neq 1. \end{cases}$$

Then the function X^n satisfies

$$X^n(m) = \begin{cases} 1 & \text{if } m = n, \\ 0 & \text{if } m \neq n. \end{cases}$$

Therefore, any $f \in R[X]$ can be written uniquely as

$$f = \sum_{n=0}^{\infty} f(n)X^n$$

where the summation is actually finite since only finitely many $f(n) \neq 0$. Note that X^0 means the identity of $R[X]$, which is the function $1 : \mathbf{Z}^+ \to R$ defined by

$$1(n) = \begin{cases} 1 & \text{if } n = 0, \\ 0 & \text{if } n > 0. \end{cases}$$

We do not need to assume that R is commutative in order to define the polynomial ring $R[X]$, but many of the theorems concerning polynomial rings will require this hypothesis, or even that R be a field. However, for some applications to linear algebra it is convenient to have polynomials over noncommutative rings.

(13) We have defined the polynomial ring $R[X]$ very precisely as functions from $\mathbf{Z}^+$ to R that are 0 except for at most finitely many nonnegative integers. We can similarly define the polynomials in several variables. Let $(\mathbf{Z}^+)^n$ be the set of all n-tuples of nonnegative integers and, if R is a commutative ring with identity, define $R[X_1, \ldots, X_n]$ to be the set of all functions $f : (\mathbf{Z}^+)^n \to R$ such that $f(\alpha) = 0$ for all but at most finitely many $\alpha \in (\mathbf{Z}^+)^n$. Define ring operations on $R[X_1, \ldots, X_n]$ by

$$\left\{\begin{array}{c} (f+g)(\alpha) = f(\alpha) + g(\alpha) \\ (fg)(\alpha) = \sum_{\beta+\gamma=\alpha} f(\beta)g(\gamma) \end{array}\right\} \quad \text{for all} \quad \alpha \in (\mathbf{Z}^+)^n.$$

Define the indeterminate X_i by

$$X_i(\alpha) = \begin{cases} 1 & \text{if } a = (\delta_{i1}, \ldots, \delta_{in}), \\ 0 & \text{otherwise.} \end{cases}$$

If $\alpha = (\alpha_1, \ldots, \alpha_n) \in (\mathbf{Z}^+)^n$ we write $X^\alpha = X_1^{\alpha_1} X_2^{\alpha_2} \cdots X_n^{\alpha_n}$ and we leave it as an exercise to check the following formula, which corresponds to our intuitive understanding of what a polynomial in several variables is. If $f \in R[X_1, \ldots, X_n]$ then we can write

$$f = \sum_{\alpha \in (\mathbf{Z}^+)^n} a_\alpha X^\alpha$$

where $a_\alpha = 0$ except for at most finitely many $\alpha \in (\mathbf{Z}^+)^n$. In fact, $a_\alpha = f(\alpha)$. Note that R is a subring of $R[X_1, \ldots, X_n]$ in a natural way, and, more generally, $R[X_1, \ldots, X_{n-1}]$ is a subring of $R[X_1, \ldots, X_n]$.

(14) If R is a ring with identity then the ring of **formal power series** $R[[X]]$ is defined similarly to the ring of polynomials. Specifically, $R[[X]]$ is the set of all functions $f : \mathbf{Z}^+ \to R$ with the same formulas for addition and multiplication as in $R[X]$. The only difference is that we do not assume that $f(n) = 0$ for all but a finite number of n. We generally write a formal power series as an expression

$$f(X) = \sum_{n=0}^{\infty} a_n X^n.$$

Since we cannot compute infinite sums (at least without a topology and a concept of limit), this expression is simply a convenient way to keep track of $f(n)$ for all n. In fact, $a_n = f(n)$ is the meaning of the above equation. With this convention, the multiplication and addition of formal power series proceeds by the rules you learned in calculus for manipulating power series. A useful exercise to become familiar with algebra in the ring of formal power series is to verify that $f \in R[[X]]$ is a unit if and only if $f(0)$ is a unit in R.

(15) Let G be a group and let R be a ring with identity. Let $R(G)$ be the set of all functions $f : G \to R$ such that $f(a) \neq 0$ for at most a finite number of $a \in G$. Define multiplication and addition on $R(G)$ by the formulas

$$(f+g)(a) = f(a) + g(a)$$
$$(fg)(a) = \sum_{b \in G} f(b)g(b^{-1}a).$$

Note that the summation used in the definition of product in $R(G)$ is a finite sum since $f(b) \neq 0$ for at most finitely many $b \in G$. The ring

$R(G)$ is called the **group ring** of G with coefficients from R. This is a ring that is used in the representation theory of groups. The product in $R(G)$ is called the **convolution product**. If S is only a semigroup (a set with an associative binary operation) then one can form a similar ring called the semigroup ring $R(S)$. If S is a monoid, $R(S)$ is a ring with identity. We leave the details to the reader, but we point out that the semigroup ring $R(\mathbf{Z}^+)$ is nothing more than the polynomial ring $R[X]$.

This list of examples should be referred to whenever new concepts for rings are introduced to see what the new concepts mean for some specific rings.

We conclude this introductory section by commenting that the generalized associative laws proved for groups in Proposition 1.1.4 is also valid for multiplication in a ring since the proof of the group theoretic result used only the associative law for groups; inverses and the group identity were not used. In particular, if R is a ring and $a_1, \ldots, a_n$ are elements of R, then the product $\prod_{i=1}^{n} a_i$ is well defined so that we can define a^n $(n \geq 1)$, and if R has an identity we can also define $a^0 = 1$. Since a ring has two operations related by the distributive laws, there should be some form of generalized distributive law valid for rings; this is the content of the following result:

(1.11) Proposition. *Let R be a ring and let $a_1, \ldots, a_m, b_1, \ldots, b_n \in R$. Then*

$$(a_1 + \cdots + a_m)(b_1 + \cdots + b_n) = \sum_{i=1}^{m} \sum_{j=1}^{n} a_i b_j.$$

Proof. For $m = 1$ the proof is by induction on n using the left distributive law. Then proceed by induction on m. $\qquad\square$

Recall that the binomial coefficients are given by $\binom{n}{r} = n!/(r!(n-r)!)$. The binomial theorem is proved by induction on n, exactly the same as the proof for real numbers; we leave the proof as an exercise.

(1.12) Proposition. (Binomial theorem) *Let R be a ring with identity and let $a, b \in R$ with $ab = ba$. Then for any $n \in \mathbf{N}$*

$$(a+b)^n = \sum_{k=0}^{n} \binom{n}{k} a^k b^{n-k}.$$

Proof. Exercise. $\qquad\square$

2.2 Ideals, Quotient Rings, and Isomorphism Theorems

A function $f : R \to S$, where R and S are rings, is a **ring homomorphism** if

$$f(a+b) = f(a) + f(b)$$

and

$$f(ab) = f(a)f(b) \qquad \text{for all} \quad a, b \in R.$$

If f is invertible (i.e., there exists a ring homomorphism $g : S \to R$ such that $f \circ g = 1_S$ and $g \circ f = 1_R$), then we say that f is a **ring isomorphism**. As with group homomorphisms, f is invertible as a ring homomorphism if it is a bijective function. If f is a ring homomorphism, then we let $\mathrm{Ker}(f) = \{a \in R : f(a) = 0\}$ and $\mathrm{Im}(f) = \{b \in S : b = f(a) \text{ for some } a \in R\}$. Thus $\mathrm{Ker}(f)$ and $\mathrm{Im}(f)$ are the kernel and image, respectively, of f when viewed as a group homomorphism between the abelian group structures on R and S so that $\mathrm{Ker}(f)$ and $\mathrm{Im}(f)$ are abelian subgroups of R and S respectively. Moreover, since f also preserves multiplication, it follows that $\mathrm{Ker}(f)$ and $\mathrm{Im}(f)$ are subrings of R and S respectively. From our study of groups we know that not every subgroup of a group can be the kernel of a group homomorphism—the subgroup must be normal. In the case of a ring homomorphism $f : R \to S$, $\mathrm{Ker}(f)$ is automatically normal since R is an abelian group under addition, but the multiplicative structure imposes a restriction on the subring $\mathrm{Ker}(f)$. Specifically, note that if $a \in \mathrm{Ker}(f)$ and $r \in R$, then $f(ar) = f(a)f(r) = 0f(r) = 0$ and $f(ra) = f(r)f(a) = f(r)0 = 0$ so that $ar \in \mathrm{Ker}(f)$ and $ra \in \mathrm{Ker}(f)$ whenever $a \in \mathrm{Ker}(f)$ and $r \in R$. This is a stronger condition than being a subring, so we make the following definition.

(2.1) Definition. *Let R be a ring and let $I \subseteq R$. We say that I is an* **ideal** *of R if and only if*

    (1) *I is an additive subgroup of R,*
    (2) *$rI \subseteq I$ for all $r \in R$, and*
    (3) *$Ir \subseteq I$ for all $r \in R$.*

A subset $I \subseteq R$ satisfying (1) and (2) is called a **left ideal** of R, while if I satisfies (1) and (3), then I is called a **right ideal** of R. Thus an ideal of R is both a left and a right ideal of R. Naturally, if R is commutative then the concepts of left ideal, right ideal, and ideal are identical, but for noncommutative rings they are generally distinct concepts.

We have already observed that the following result is true.

(2.2) Lemma. *If $f : R \to S$ is a ring homomorphism, then $\mathrm{Ker}(f)$ is an ideal of R.*

Every ring R has at least two ideals, namely, $\{0\}$ and R are both ideals of R. For division rings, these are the only ideals, as we see from the following observation.

(2.3) Lemma. *If R is a division ring, then the only ideals of R are R and $\{0\}$.*

Proof. Let $I \subseteq R$ be an ideal such that $I \neq \{0\}$. Let $a \neq 0 \in I$ and let $b \in R$. Then the equation $ax = b$ is solvable in R, so $b \in I$. Therefore, $I = R$. □

(2.4) Corollary. *If R is a division ring and $f : R \to S$ is a ring homomorphism then f is injective or $f \equiv 0$.*

Proof. If $\mathrm{Ker}(f) = \{0\}$ then f is injective; if $\mathrm{Ker}(f) = R$ then $f \equiv 0$. □

(2.5) Remarks. (1) In fact, the converse of Lemma 2.2 is also true; that is, every ideal is the kernel of some ring homomorphism. The proof of this requires the construction of the quotient ring, which we will take up next.

(2) The converse of Lemma 2.3 is false. See Remark 2.28.

If $I \subseteq R$ is an ideal then the quotient group R/I is well defined since I is a subgroup (and hence a normal subgroup) of the additive abelian group R. Let $\pi : R \to R/I$, defined by $\pi(r) = r + I$, be the natural projection map. We will make the abelian group R/I into a ring by defining a multiplication on R/I. First recall that coset addition in R/I is defined by $(r + I) + (s + I) = (r + s) + I$. Now define coset multiplication by the formula $(r + I)(s + I) = rs + I$. All that needs to be checked is that this definition is independent of the choice of coset representatives. To see this, suppose $r + I = r' + I$ and $s + I = s' + I$. Then $r' = r + a$ and $s' = s + b$ where $a, b \in I$. Thus,

$$
\begin{aligned}
r's' &= (r + a)(s + b) \\
&= rs + as + rb + ab \\
&= rs + c
\end{aligned}
$$

where $c = as + rb + ab \in I$ because I is an ideal. Therefore, $rs + I = r's' + I$ and multiplication of cosets is well defined. By construction the natural map $\pi : R \to R/I$ is a ring homomorphism so that $I = \mathrm{Ker}(\pi)$ is the kernel of a ring homomorphism. Note that if R is commutative then R/I is commutative for any I, and if R has an identity and I is a proper ideal then $1 + I$ is the identity of the quotient ring R/I. If $a, b \in R$ we will use the notation $a \equiv b \pmod{I}$ to mean $a - b \in I$, i.e., $a + I = b + I$. This is a generalization of the concept of congruence of integers modulo an integer n.

The noether isomorphism theorems for groups, Theorems 1.3.11 to 1.3.15, have direct analogues for rings.

(2.6) Theorem. (First isomorphism theorem) *Let* $f : R \to S$ *be a ring homomorphism. Then* $R/\operatorname{Ker}(f) \cong \operatorname{Im}(f)$. *($\cong$ means ring isomorphism.)*

Proof. Let $K = \operatorname{Ker}(f)$. From Theorem 1.3.11 we know that $\overline{f} : R/K \to \operatorname{Im}(f)$, defined by $\overline{f}(a + K) = f(a)$, is a well-defined isomorphism of groups. It only remains to check that multiplication is preserved. But

$$\overline{f}((a + K)(b + K)) = \overline{f}(ab + K) = f(ab) = f(a)f(b) = \overline{f}(a + K)\overline{f}(b + K),$$

so $\overline{f}$ is a ring homomorphism and hence an isomorphism. $\qquad\square$

(2.7) Theorem. (Second isomorphism theorem) *Let* R *be a ring,* $I \subseteq R$ *an ideal, and* $S \subseteq R$ *a subring. Then* $S + I$ *is a subring of* R, I *is an ideal of* $S + I$, $S \cap I$ *is an ideal of* S, *and there is an isomorphism of rings*

$$(S + I)/I \cong S/(S \cap I).$$

Proof. Suppose that $s, s' \in S$, $a, a' \in I$. Then

$$(s + a)(s' + a') = ss' + (as' + sa' + aa') \in S + I,$$

so $S + I$ is closed under multiplication and hence is a subring of R. (It is already an additive subgroup from the theory of groups.) The fact that I is an ideal of $S + I$ and $S \cap I$ is an ideal of S is clear. Let $\pi : R \to R/I$ be the natural homomorphism and let π_0 be the restriction of π to S. Then π_0 is a ring homomorphism with $\operatorname{Ker}(\pi_0) = S \cap I$ and the first isomorphism theorem gives $S/(S \cap I) = S/\operatorname{Ker}(\pi_0) \cong \operatorname{Im}(\pi_0)$. But $\operatorname{Im}(\pi_0)$ is the set of all cosets of I with representatives in S. Therefore, $\operatorname{Im}(\pi_0) = (S + I)/I$. $\quad\square$

(2.8) Theorem. (Third isomorphism theorem) *Let* R *be a ring and let* I *and* J *be ideals of* R *with* $I \subseteq J$. *Then* J/I *is an ideal of* R/I *and*

$$R/J \cong (R/I)/(J/I).$$

Proof. Define a function $f : R/I \to R/J$ by $f(a + I) = a + J$. It is easy to check that this is a well-defined ring homomorphism. Then

$$\operatorname{Ker}(f) = \{a + I : a + J = J\} = \{a + I : a \in J\} = J/I.$$

The result then follows from the first isomorphism theorem. $\qquad\square$

(2.9) Theorem. (Correspondence theorem) *Let* R *be a ring,* $I \subseteq R$ *an ideal of* R, *and* $\pi : R \to R/I$ *the natural map. Then the function* $S \mapsto S/I$ *defines a one-to-one correspondence between the set of all subrings of* R *containing* I *and the set of all subrings of* R/I. *Under this correspondence, ideals of* R *containing* I *correspond to ideals of* R/I.

Proof. According to the correspondence theorem for groups (Theorem 1.3.15) there is a one-to-one correspondence between additive subgroups of R/I and additive subgroups of R containing I. It is only necessary to check that under this correspondence (which is $H \mapsto H/I$) subrings correspond to subrings and ideals correspond to ideals. We leave this to check as an exercise. $\square$

(2.10) Lemma. *Let R be a ring and let $\{S_\alpha\}_{\alpha \in A}$ be a family of subrings (resp., ideals) of R. Then $S = \bigcap_{\alpha \in A} S_\alpha$ is a subring (resp., ideal) of R.*

Proof. Suppose $a, b \in S$. Then $a, b \in S_\alpha$ for all $\alpha \in A$ so that $a - b$ and $ab \in S_\alpha$ for all $\alpha \in A$. Thus, $a - b$ and $ab \in S$, so S is a subring of R. If each S_α is an ideal and $r \in R$ then ar and $ra \in S_\alpha$ for all $\alpha \in A$ so $ar, ra \in S$ and S is an ideal. $\square$

(2.11) Definition. *If $X \subseteq R$ is a subset then the **subring generated by** X is the smallest subring of R containing X and the **ideal generated by** X is the smallest ideal of R containing X. By Lemma 2.10 this is just the intersection of all subrings (resp., ideals) containing X. We will use the notation $\langle X \rangle$ to denote the **ideal** generated by X.*

(2.12) Lemma. *Let $X \subseteq R$ be a nonempty subset of a ring R.*

(1) *The subring of R generated by X is the sum or difference of all finite products of elements of X.*

(2) *If R is a ring with identity, then the ideal of R generated by X is the set*

$$RXR = \left\{ \sum_{i=1}^{n} r_i x_i s_i : r_i, s_i \in R, \ x_i \in X, \ n \geq 1 \right\}.$$

(3) *If R is a commutative ring with identity, then the ideal of R generated by X is the set*

$$RX = \left\{ \sum_{i=1}^{n} r_i x_i : r_i \in R, \ x_i \in X, \ n \geq 1 \right\}.$$

Proof. (2) Every ideal containing X certainly must contain RXR. It is only necessary to observe that RXR is indeed an ideal of R, and hence it is the smallest ideal of R containing X. Parts (1) and (3) are similar and are left for the reader. $\square$

(2.13) *Remarks.* (1) The description given in Lemma 2.12 (2) of the ideal generated by X is valid for rings with identity. There is a similar, but more complicated description for rings without an identity. We do not present it since we shall not have occasion to use such a description.

(2) If $X = \{a\}$ then the ideal generated by X in a commutative ring R with identity is the set $Ra = \{ra : r \in R\}$. Such an ideal is said to be **principal**. An integral domain R in which every ideal of R is principal is

called a **principal ideal domain** (PID). The integers $\mathbf{Z}$ are an example of a PID. Another major example of a PID is the polynomial ring $F[X]$ where F is a field. This will be verified in Section 2.4.

There is another useful construction concerning ideals. If R is a ring and $I_1, \ldots, I_n$ are ideals of R then we define the **sum of the ideals** $I_1, \ldots, I_n$ by

$$\sum_{i=1}^{n} I_i = \{a_1 + \cdots + a_n : a_i \in I_i \text{ for } 1 \le i \le n\}.$$

We also define the **product of the ideals** $I_1, \ldots, I_n$ by

$$\prod_{i=1}^{n} I_i = \left\{ \sum_{j=1}^{m} a_{1j} \cdots a_{nj} : a_{ij} \in I_i \quad (1 \le i \le n) \quad \text{and } m \text{ is arbitrary} \right\}.$$

(2.14) Lemma. *If $I_1, \ldots, I_n$ are ideals of R then $\sum_{i=1}^{n} I_i$ is an ideal of R. In fact, $\sum_{i=1}^{n} I_i$ is the ideal generated by the union $\cup_{i=1}^{n} I_i$.*

Proof. Exercise. □

(2.15) Definition. *An ideal M in a ring R is called **maximal** if $M \ne R$ and M is such that if I is an ideal with $M \subseteq I \subseteq R$ then $I = M$ or $I = R$.*

(2.16) Theorem. *Let R be a ring with identity and let $I \ne R$ be an ideal of R. Then there is a maximal ideal of R containing I.*

Proof. Let S be the set of all ideals J of R that contain I and are not equal to R. Then $S \ne \emptyset$ since $I \in S$. Partially order S by inclusion and let $C = \{J_\alpha\}_{\alpha \in A}$ be a chain in S. Let $J = \cup_{\alpha \in A} J_\alpha$. Then J is an ideal of R since if $a, b \in J$, $r \in R$, it follows that $a, b \in J_\alpha$ for some α because C is a chain (i.e., $J_\alpha \subseteq J_\beta$ or $J_\beta \subseteq J_\alpha$ for every $\alpha, \beta \in A$). Thus $a - b, ab, ar, ra \in J_\alpha \subseteq J$, so J is an ideal. Furthermore, $J \ne R$ since $1 \notin J_\alpha$ for any α. Thus J is an upper bound for the chain C so Zorn's lemma implies that S has a maximal element M and a maximal element of S is clearly a maximal ideal of R. □

(2.17) Corollary. *In a ring with identity there are always maximal ideals.*

Proof. □

(2.18) Theorem. *Let R be a commutative ring with identity. Then an ideal M of R is maximal if and only if R/M is a field.*

Proof. Suppose R/M is a field. Then $0 \ne 1 \in R/M$, so $M \ne R$. But the only ideals in a field are $\{0\}$ and the whole field, so the correspondence theorem shows that there are no ideals of R properly between M and R. Thus, M is maximal.

Conversely, suppose that M is a maximal ideal. It is necessary to show that every $\bar{a} = a + M \in R/M$ has an inverse if $\bar{a} \neq 0 \in R/M$, i.e., if $a \notin M$. Consider the ideal $Ra + M$ of R. Since $a \in Ra + M$ and $a \notin M$, it follows that $Ra + M = R$ since M is assumed to be maximal. In particular, $ra + m = 1$ for some $r \in R, m \in M$. Let $\bar{r} = r + M$. Then

$$\bar{r}\,\bar{a} = (r + M)(a + M) = ra + M = (1 - m) + M = 1 + M = \bar{1} \in R/M.$$

Thus, $\bar{a}^{-1} = \bar{r}$ and R/M is a field. □

(2.19) Definition. *An ideal P in a commutative ring R is said to be **prime** if $P \neq R$ and P is such that if $ab \in P$ then $a \in P$ or $b \in P$. An element $p \in R$ is **prime** if the ideal $Rp = \langle p \rangle$ is a prime ideal.*

(2.20) Theorem. *Let R be a commutative ring with identity. Then an ideal P of R is prime if and only if R/P is an integral domain.*

Proof. Suppose P is a prime ideal. Then $P \neq R$, so $R/P \neq \{0\}$. Hence R/P is a ring with identity $1 + P$. Given $\bar{a} = a + P, \bar{b} = b + P \in R/P$, suppose $\bar{a}\bar{b} = \bar{0}$. Then $(a + P)(b + P) = P$, so $ab + P = P$ and $ab \in P$. Since P is a prime ideal, this implies that $a \in P$ or $b \in P$, i.e., $\bar{a} = \bar{0}$ or $\bar{b} = \bar{0}$. Therefore, R/P is an integral domain.

Conversely, suppose R/P is an integral domain and suppose that $ab \in P$. Then $\bar{a}\bar{b} = \bar{0} \in R/P$. Therefore, $\bar{a} = \bar{0}$ or $\bar{b} = \bar{0}$, i.e., $a \in P$ or $b \in P$. Thus P is a prime ideal. □

(2.21) Theorem. *Let R be a commutative ring with identity and let I be an ideal. If I is maximal, then I is prime.*

Proof. If I is maximal then R/I is a field and hence an integral domain. Therefore, I is prime by Theorem 2.20.

More directly, suppose that $I \subseteq R$ is not prime and let $a, b \in R$ with $a \notin I, b \notin I$, but $ab \in I$. Let

$$J = \{x \in R : ax \in I\}.$$

Then $I \subseteq J$ since I is an ideal, and J is clearly an ideal. Also, $J \neq I$ since $b \in J$ but $b \notin I$, and $J \neq R$ since $1 \notin J$. Therefore, I is not maximal. The theorem follows by contraposition. □

(2.22) Corollary. *Let $f : R \to S$ be a surjective homomorphism of commutative rings with identity.*

(1) *If S is a field then $\mathrm{Ker}(f)$ is a maximal ideal of R.*
(2) *If S is an integral domain then $\mathrm{Ker}(f)$ is a prime ideal of R.*

Proof. $R/\mathrm{Ker}(f) \cong \mathrm{Im}(f) = S$. Now apply Theorems 2.20 and 2.18. □

(2.23) Examples.

(1) We compute all the ideals of the ring of integers $\mathbf{Z}$. We already know that all subgroups of $\mathbf{Z}$ are of the form $n\mathbf{Z} = \{nr : r \in \mathbf{Z}\}$. But if $s \in \mathbf{Z}$ and $nr \in n\mathbf{Z}$ then $s(nr) = nrs = (nr)s$ so that $n\mathbf{Z}$ is an ideal of $\mathbf{Z}$. Therefore, the ideals of $\mathbf{Z}$ are the subsets $n\mathbf{Z}$ of multiples of a fixed integer n. The quotient ring $\mathbf{Z}/n\mathbf{Z}$ is the ring $\mathbf{Z}_n$ of integers modulo n. It was observed in the last section that $\mathbf{Z}_n$ is a field if and only if n is a prime number.

(2) Define $\phi : \mathbf{Z}[X] \to \mathbf{Z}$ by $\phi(a_0 + a_1 X + \cdots + a_n X^n) = a_0$. This is a surjective ring homomorphism, and hence, $\mathrm{Ker}(\phi)$ is a prime ideal. In fact, $\mathrm{Ker}(\phi) = \langle X \rangle = $ ideal generated by X.

Now define $\psi : \mathbf{Z}[X] \to \mathbf{Z}_2[X]$ by

$$\psi(a_0 + a_1 X + \cdots + a_n X^n) = \bar{a}_0 + \bar{a}_1 X + \cdots + \bar{a}_n X^n$$

where $\bar{a}_i = \pi(a_i)$ if $\pi : \mathbf{Z} \to \mathbf{Z}_2$ is the natural projection map. $\mathbf{Z}_2[X]$ is an integral domain (see Section 2.4), so $\mathrm{Ker}(\psi)$ is a prime ideal. In fact,

$$\mathrm{Ker}(\psi) = \{a_0 + a_1 X + \cdots + a_n X^n : a_i \text{ is even for all } i\}$$
$$= \langle 2 \rangle = \text{ ideal generated by } 2.$$

Next consider the map $\psi' : \mathbf{Z}[X] \to \mathbf{Z}_2$ defined by

$$\psi'(a_0 + a_1 X + \cdots + a_n X^n) = \bar{a}_0 = \pi(a_0).$$

Then $\mathrm{Ker}(\psi')$ is a maximal ideal since ψ' is a surjective ring homomorphism to the field $\mathbf{Z}_2$. In fact,

$$\mathrm{Ker}(\psi') = \langle 2, X \rangle = \text{ ideal generated by } \{2, X\}.$$

Note that $\langle X \rangle \subsetneq \langle 2, X \rangle$ and $\langle 2 \rangle \subsetneq \langle 2, X \rangle$. Thus, we have some examples of prime ideals that are not maximal.

(3) Let G be a group and R a ring. Then there is a map

$$\mathrm{aug} : R(G) \to R$$

called the **augmentation map** defined by

$$\mathrm{aug}\left(\sum_{g \in G} r_g \cdot g \right) = \sum_{g \in G} r_g$$

where we have denoted elements of $R(G)$ by the formal sum notation $\sum_{g \in G} r_g \cdot g$ rather than the formally equivalent functional notation $f : G \to R$ with $f(g) = r_g$. We leave it as an exercise to check that aug is a ring homomorphism. The ideal $I = \mathrm{Ker}(\mathrm{aug}) \subseteq R(G)$ is called the

augmentation ideal. If R is an integral domain, then I is prime, and if R is a field, then I is maximal.

(4) Let F be a field and let $T^n(F)$ be the ring of $n \times n$ upper triangular matrices with entries in F, and let $ST^n(F)$ be the subring of strictly upper triangular matrices. Then $ST^n(F)$ is an ideal in the ring $T^n(F)$ and the quotient

$$T^n(F)/ST^n(F) \cong D_n(F)$$

where $D_n(F)$ is the ring of $n \times n$ diagonal matrices with entries in F. To see this define a ring homomorphism $\phi : T^n(F) \to D_n(F)$ by

$$\phi\left(\begin{bmatrix} a_{11} & a_{12} & \cdots & a_{1n} \\ 0 & a_{22} & \cdots & a_{2n} \\ \vdots & \vdots & \ddots & \vdots \\ 0 & 0 & \cdots & a_{nn} \end{bmatrix}\right) = \begin{bmatrix} a_{11} & 0 & \cdots & 0 \\ 0 & a_{22} & \cdots & 0 \\ \vdots & \vdots & \ddots & \vdots \\ 0 & 0 & \cdots & a_{nn} \end{bmatrix}.$$

It is easy to check (do it) that ϕ is a ring homomorphism and that $\operatorname{Ker}(\phi) = ST^n(F)$ and $\operatorname{Im}(\phi) = D_n(F)$, so the result follows from the first isomorphism theorem.

The next result is an extension to general commutative rings of the classical Chinese remainder theorem concerning simultaneous solution of congruences in integers. For example, the Chinese remainder theorem is concerned with solving congruences such as

$$x \equiv -1 \pmod{15}$$
$$x \equiv 3 \pmod{11}$$
$$x \equiv 6 \pmod{8}.$$

(2.24) Theorem. (Chinese remainder theorem) *Let R be a commutative ring with identity and let $I_1, \ldots, I_n$ be ideals of R such that $I_i + I_j = R$ for all $i \neq j$. (A collection of such ideals is said to be **coprime** or **relatively prime**.) Given elements $a_1, \ldots, a_n \in R$ there exists $a \in R$ such that*

$$a \equiv a_i \pmod{I_i} \quad \text{for } 1 \leq i \leq n.$$

Also, $b \in R$ is a solution of the simultaneous congruence

$$x \equiv a_i \pmod{I_i} \quad (1 \leq i \leq n)$$

if and only if

$$b \equiv a \pmod{I_1 \cap \cdots \cap I_n}.$$

Proof. We will first do the special case of the theorem where $a_1 = 1$ and $a_j = 0$ for $j > 1$. For each $j > 1$, since $I_1 + I_j = R$, we can find $b_j \in I_1$ and $c_j \in I_j$ with $b_j + c_j = 1$. Then $\prod_{j=2}^n (b_j + c_j) = 1$, and since each $b_j \in I_1$, it follows that $1 = \prod_{j=2}^n (b_j + c_j) \in I_1 + \prod_{j=2}^n I_j$. Therefore, there is $\alpha_1 \in I_1$ and $\beta_1 \in \prod_{j=2}^n I_j$ such that $\alpha_1 + \beta_1 = 1$. Observe that β_1 solves the required congruences in the special case under consideration. That is,

$$\beta_1 \equiv 1 \pmod{I_1}$$
$$\beta_1 \equiv 0 \pmod{I_j} \quad \text{for} \quad j \neq 1$$

since $\beta_1 - 1 \in I_1$ and $\beta_1 \in \prod_{j=2}^n I_j \subseteq I_j$ for $j \neq 1$.

By a similar construction we are able to find $\beta_2, \ldots, \beta_n$ such that

$$\beta_i \equiv 1 \pmod{I_i}$$
$$\beta_i \equiv 0 \pmod{I_j} \quad \text{for} \quad j \neq i.$$

Now let $a = a_1\beta_1 + \cdots + a_n\beta_n$. Then

$$a \equiv a_i \pmod{I_i} \quad (1 \leq i \leq n).$$

Now suppose also that $b \equiv a_i \pmod{I_i}$ for all i. Then $b - a \equiv 0 \pmod{I_i}$ so that $b - a \in I_i$ for all i, i.e.,

$$b - a \in \bigcap_{i=1}^n I_i.$$

The converse is clear.   □

There is another version of the Chinese remainder theorem. In order to state it we will need to introduce the concept of cartesian, or direct, product of rings. We will only be concerned with finite cartesian products. Thus let $R_1, \ldots, R_n$ be finitely many rings and let $\prod_{i=1}^n R_i = R_1 \times \cdots \times R_n$ denote the cartesian product set. On the set $\prod_{i=1}^n R_i$ we may define addition and multiplication componentwise, i.e.,

$$(a_1, \ldots, a_n) + (b_1, \ldots, b_n) = (a_1 + b_1, \ldots, a_n + b_n)$$
$$(a_1, \ldots, a_n)(b_1, \ldots, b_n) = (a_1 b_1, \ldots, a_n b_n),$$

to make $\prod_{i=1}^n R_i$ into a ring called the **direct product** of $R_1, \ldots, R_n$. Given $1 \leq i \leq n$ there is a natural projection homomorphism $\pi_i : \prod_{j=1}^n R_j \to R_i$ defined by $\pi_i(a_1, \ldots, a_n) = a_i$.

(2.25) Corollary. *Let R be a commutative ring with identity and let $I_1, \ldots, I_n$ be ideals of R such that $I_i + I_j = R$ if $i \neq j$. Define*

$$f : R \to \prod_{i=1}^n R/I_i$$

by $f(a) = (a+I_1, \ldots, a+I_n)$. Then f is surjective and $\text{Ker}(f) = I_1 \cap \cdots \cap I_n$. Thus

$$R/(I_1 \cap \cdots \cap I_n) \cong \prod_{i=1}^n R/I_i.$$

Proof. Surjectivity follows from Theorem 2.24, and it is clear from the definition of f that $\mathrm{Ker}(f) = I_1 \cap \cdots \cap I_n$. $\qquad\square$

Suppose that we take $R = \mathbf{Z}$ in Corollary 2.25. Let $m \in \mathbf{Z}$ and let $m = \prod_{i=1}^{k} p_i^{r_i}$ be the factorization of m into distinct prime powers. Then if $I_i = \langle p_i^{r_i} \rangle$ it follows that $I_i + I_j = \mathbf{Z}$ if $i \neq j$. Moreover, $\langle m \rangle = I_1 \cap \cdots \cap I_k$ so that Corollary 2.25 applies to give an isomorphism of rings

$$\mathbf{Z}_m \cong \prod_{i=1}^{k} \mathbf{Z}_{p_i^{r_i}}.$$

A practical method of applying the Chinese remainder theorem in Euclidean domains will be presented in Section 2.5. (This class of rings, defined there, includes the integers $\mathbf{Z}$.)

We conclude this section by computing all the ideals of the full matrix ring $M_n(R)$ where R is a commutative ring with identity. To do this we will make use of the matrix units E_{ij} introduced in Example 1.10 (8). Recall that E_{ij} has a 1 in the ij position and 0 elsewhere.

(2.26) Theorem. *Let R be a ring with identity and let I be an ideal of $M_n(R)$. Then there is an ideal $J \subseteq R$ such that $I = M_n(J)$.*

Proof. First note that if J is any ideal of R then $M_n(J)$ is an ideal of $M_n(R)$. This follows immediately from the definition of multiplication and addition in $M_n(R)$. Now suppose that I is an ideal in $M_n(R)$, and define

$$J = \{a \in R : \text{ there exists } A \in I \text{ with } \mathrm{ent}_{11}(A) = a\}.$$

Observe that J is an ideal. Indeed, if $a, b \in J$, $r \in R$ then $a = \mathrm{ent}_{11}(A)$, $b = \mathrm{ent}_{11}(B)$ for some $A, B \in I$. Then $a - b = \mathrm{ent}_{11}(A - B)$, $ra = \mathrm{ent}_{11}((rE_{11}A))$, and $ar = \mathrm{ent}_{11}(A(rE_{11}))$. But $A - B$, $(rE_{11})A$, and $A(rE_{11})$ are all in I since I is an ideal. Therefore, J is an ideal of R.

Now we show that $I = M_n(J)$. First observe that

$$E_{1k} \left(\sum_{i,j=1}^{n} a_{ij} E_{ij} \right) E_{\ell 1} = a_{k\ell} E_{11} \in I.$$

Thus, whenever $A \in I$ every entry of A is in J since the above calculation shows that every entry of A can be moved to the 1,1 position of some matrix in I. Hence, $I \subseteq M_n(J)$. But $M_n(J)$ is generated by elements aE_{ij} where $a = \mathrm{ent}_{11}(A)$ for some $A \in I$. Since $aE_{ij} = E_{i1}AE_{1j} \in I$ because I is an ideal, it follows that each generator of $M_n(J)$ is in I, so $M_n(J) \subseteq I$, and we conclude that $M_n(J) = I$. $\qquad\square$

(2.27) Corollary. *If D is a division ring then $M_n(D)$ has no nontrivial proper ideals, i.e., the only ideals are $\{0\}$ and $M_n(D)$.*

Proof. The only ideals of D are $\{0\}$ and D so that the only ideals of $M_n(D)$ are $\{0\} = M_n(\{0\})$ and $M_n(D)$. $\square$

(2.28) *Remark.* Note that Corollary 2.27 shows that commutativity is a crucial hypothesis in Theorem 2.18. One might conjecture that an ideal M of any ring R (with identity) is maximal if and only if R/M is a division ring. But this is false since $\langle 0 \rangle$ is a maximal ideal of $M_n(D)$ (by Corollary 2.27), but clearly, $M_n(D)$ is not a division ring.

2.3 Quotient Fields and Localization

If F is a field and R is a subring of F then R is an integral domain and the smallest subfield of F containing R is the subset

$$Q(R) = \{a/b \in F : a, b \in R,\ b \neq 0\}.$$

Of course, if $c \neq 0 \in R$ then $a/b = (ac)/(bc) \in F$ since a/b just means ab^{-1}. Thus $Q(R)$ is obtained from R in the same manner in which the rational numbers $\mathbf{Q}$ are obtained from the integers $\mathbf{Z}$. We will now consider the converse situation. Suppose we are given an integral domain R. Can we find a field F that contains an isomorphic copy of R as a subring? The answer is yes, and in fact we will work more generally by starting with a commutative ring R and a subset S of R of elements that we wish to be able to invert. Thus we are looking for a ring R' that contains R as a subring and such that $S \subseteq (R')^*$, i.e., every element of S is a unit of R'. In the case of an integral domain R we may take $S = R \setminus \{0\}$ to obtain $Q(R)$.

(3.1) Definition. *If R is a commutative ring, a subset S of R is said to be* **multiplicatively closed** *if the product of any two elements of S is an element of S.*

Note that if S' is any nonempty subset of R, then the set S consisting of all finite products of elements of S' is multiplicatively closed. If S' contains no zero divisors, then neither does S.

(3.2) Definition. *Let R be a commutative ring and let $\emptyset \neq S \subseteq R$ be a multiplicatively closed subset of R containing no zero divisors. The* **localization of R away from** *S is a commutative ring R_S with identity, and an injective ring homomorphism $\phi : R \to R_S$ such that for all $a \in R_S$ there are $b \in R$ and $c \in S$ such that $\phi(c)$ is a unit in R_S and $a = \phi(b)\phi(c)^{-1}$. If R is an integral domain and $S = R \setminus \{0\}$ then we call R_S the* **quotient field** *of R and we denote it $Q(R)$.*

(3.3) *Remark.* If $S \subseteq R^*$, then R together with the identity map $1_R : R \to R$ is the localization R_S. In particular, if R is a field then any localization of R just reproduces R itself.

(3.4) **Definition.** *Let $S \subseteq R \setminus \{0\}$ be a nonempty multiplicatively closed subset of R containing no zero divisors, and let (R_S, ϕ) and (R'_S, ϕ') be localizations of R away from S. R_S and R'_S are said to be **equivalent** if there is an isomorphism $\beta : R_S \to R'_S$ such that $\beta \circ \phi = \phi'$, i.e., the following diagram of rings and ring homomorphisms commutes:*

$$
\begin{array}{ccc}
 & R & \\
{\scriptstyle \phi}\swarrow & & \searrow{\scriptstyle \phi'} \\
R_S & \xrightarrow{\ \beta\ } & R'_S
\end{array}
$$

(3.5) **Theorem.** *Let R be a commutative ring. If $S \subseteq R \setminus \{0\}$ is a multiplicatively closed subset containing no zero divisors, then there exists a localization R_S away from S, and R_S is unique up to equivalence.*

Proof. Define a relation $\sim$ on $R \times S$ by setting $(a, b) \sim (c, d)$ if and only if $ad = bc$. Observe that $\sim$ is an equivalence relation. Symmetry and reflexivity are clear. To check transitivity suppose that $(a, b) \sim (c, d)$ and $(c, d) \sim (e, f)$. Then $ad = bc$ and $cf = de$. Therefore, $adf = bcf$ and $bcf = bde$ so that $adf = bde$. But d is not a zero divisor and $d \neq 0$ since $d \in S$ so by Lemma 3.1, $af = be$. Therefore, $(a, b) \sim (e, f)$ and $\sim$ is transitive and, hence, is an equivalence relation.

Let $R_S = R \times S/ \sim$ be the set of equivalence classes of the equivalence relation $\sim$. We will denote the equivalence class of (a, b) by the suggestive symbol a/b. Note that $(a, b) \sim (ac, bc)$ whenever $c \in S$ (S is multiplicatively closed, so $bc \in S$ whenever $b, c \in S$). Thus $a/b = (ac)/(bc)$ for every $c \in S$. Define ring operations on R_S by the formulas

$$
\frac{a}{b} \cdot \frac{c}{d} = \frac{ac}{bd}
$$
$$
\frac{a}{b} + \frac{c}{d} = \frac{ad + bc}{bd}.
$$

Note that $bd \in S$ since S is multiplicatively closed. Since the symbol a/b denotes an equivalence class, it is necessary to check that these operations do not depend upon the choice of representative of the equivalence class used in their definition. To check this, suppose that $a/b = a'/b'$ and $c/d = c'/d'$. Then $ab' = ba'$ and $cd' = dc'$ so that $acb'd' = bda'c'$ and thus $(ac)/(bd) = (a'c')/(b'd')$ and the multiplication formula is well defined. Similarly,

$$
\begin{aligned}
(ad + bc)b'd' &= ab'dd' + bb'cd' \\
&= ba'dd' + bb'dc' \\
&= (a'd' + b'c')bd,
\end{aligned}
$$

so addition is also well defined.

We leave to the reader the routine check that these operations make R_S into a commutative ring with identity. Observe that $0/s = 0/s'$, $s/s = s'/s'$ for any $s, s' \in S$ and $0/s$ is the additive identity of R_S, while s/s is the multiplicative identity. If $s, t \in S$ then $(s/t)(t/s) = (st)/(st) = 1 \in R_S$ so that $(s/t)^{-1} = t/s$.

Now we define $\phi : R \to R_S$. Choose $s \in S$ and define $\phi_s : R \to R_S$ by $\phi_s(a) = (as)/s$. We claim that if $s' \in S$ then $\phi_{s'} = \phi_s$. Indeed, $\phi_s(a) = (as)/s$ and $\phi_{s'}(a) = (as')/s'$, but $(as)/s = (as')/s'$ since $ass' = as's$. Thus we may define ϕ to be ϕ_s for *any* $s \in S$.

Claim. ϕ *is a ring homomorphism.*

Indeed, if $a, b \in R$ and $s \in S$ then

$$\phi(ab) = \frac{abs}{s}$$
$$= \frac{abs^2}{s^2}$$
$$= \frac{as}{s} \cdot \frac{bs}{s}$$
$$= \phi(a)\phi(b)$$

and

$$\phi(a+b) = \frac{(a+b)s}{s} = \frac{as^2 + bs^2}{s^2}$$
$$= \frac{as}{s} + \frac{bs}{s}$$
$$= \phi(a) + \phi(b).$$

Note that $\phi(1) = s/s = 1_{R_S}$ and

$$\mathrm{Ker}(\phi) = \{a \in R : \frac{a}{s} = \frac{0}{s'}\}$$
$$= \{a \in R : as' = 0 \text{ for some } s' \in S\}$$
$$= \{0\}$$

since S contains no zero divisors. Therefore, ϕ is an injective ring homomorphism. Suppose that $a/b \in R_S$ where $a \in R$ and $b \in S$. Then

$$\frac{a}{b} = \frac{as}{s} \cdot \frac{s}{bs} = \phi(a)(\phi(b))^{-1}.$$

Therefore, we have constructed a localization of R away from S. It remains only to check uniqueness up to equivalence.

Suppose that $\phi : R \to R_S$ and $\phi' : R \to R'_S$ are two localizations of R away from S. Define $\beta : R_S \to R'_S$ as follows. Let $a \in R_S$. Then we may write $a = \phi(b)(\phi(c))^{-1}$ where $b \in R$ and $c \in S$, and we define $\beta(a) = \phi'(b)(\phi'(c))^{-1}$.

Claim. β *is well defined.*

If $\phi(b)(\phi(c))^{-1} = a = \phi(b')(\phi(c'))^{-1}$ then $\phi(b)\phi(c') = \phi(b')\phi(c)$ and hence $\phi(bc') = \phi(b'c)$. But ϕ is injective so $bc' = b'c$. Now apply the homomorphism ϕ' to conclude $\phi'(b)(\phi'(c))^{-1} = \beta(a) = \phi'(b')(\phi'(c'))^{-1}$ so β is well defined.

Claim. β *is bijective.*

Define $\gamma : R'_S \to R_S$ by $\gamma(a') = \phi(b')(\phi(c'))^{-1}$ if $a' = \phi'(b')(\phi'(c'))^{-1}$ for some $b' \in R, c' \in S$. Exactly as for β, one checks that γ is well defined. Then if $a \in R_S$, write $a = \phi(b)(\phi(c))^{-1}$ and compute

$$\gamma(\beta(a)) = \gamma(\phi'(b)(\phi'(c))^{-1})$$
$$= \phi(b)(\phi(c))^{-1}$$
$$= a.$$

Similarly, one shows that $\beta(\gamma(a')) = a'$ so that β is bijective.

It remains to check that β is a homomorphism. We will check that β preserves multiplication and leave the similar calculation for addition as an exercise. Let $a_1 = \phi(b_1)(\phi(c_1))^{-1}$ and $a_2 = \phi(b_2)(\phi(c_2))^{-1}$. Since ϕ is a ring homomorphism, $a_1 a_2 = \phi(b_1 b_2)(\phi(c_1 c_2))^{-1}$ so that

$$\beta(a_1 a_2) = \phi'(b_1 b_2)(\phi'(c_1 c_2))^{-1}$$
$$= (\phi'(b_1)(\phi'(c_1))^{-1})(\phi'(b_2)(\phi'(c_2))^{-1})$$
$$= \beta(a_1)\beta(a_2).$$

This completes the proof of Theorem 3.5. $\square$

(3.6) Examples.

(1) $Q(\mathbf{Z}) = \mathbf{Q}$.
(2) Let $p \in \mathbf{Z}$ be a prime and let $S_p = \{1, p, p^2, \ldots\} \subseteq \mathbf{Z}$. Then

$$\mathbf{Z}_{S_p} = \left\{ \frac{a}{b} \in \mathbf{Q} : b \text{ is a power of } p \right\}.$$

(3) Let R be an integral domain, let $P \subseteq R$ be a prime ideal and let $S = R \setminus P$. Then the definition of prime ideal shows that S is a multiplicatively closed subset of R and it certainly contains no zero divisors since R is an integral domain. Then R_S is isomorphic to a subring of the quotient field $Q(R)$. In fact

$$R_S = \left\{ \frac{a}{b} \in Q(R) : b \notin P \right\}.$$

R_S is said to be R **localized at the prime ideal** P. It is an example of a **local ring**, that is, a ring with a unique maximal ideal. For R_S the maximal ideal is $\phi(P) \subseteq R_S$.

(4) Let $d \neq 0, 1 \in \mathbf{Z}$ be an integer that is square-free, i.e., whose prime factorization contains no squares, and then define $\mathbf{Z}[\sqrt{d}] = \{a + b\sqrt{d} : a, b \in \mathbf{Z}\}$. It is easy to check that $\mathbf{Z}[\sqrt{d}]$ is a subring of the complex numbers $\mathbf{C}$ and thus the quotient field $Q(\mathbf{Z}[\sqrt{d}])$ can be identified with a subfield of the complex numbers. In fact,

$$Q(\mathbf{Z}[\sqrt{d}]) = \mathbf{Q}[\sqrt{d}] = \{a + b\sqrt{d} : a, b \in \mathbf{Q}\}.$$

(5) Let F be a field and let $F[X]$ be the ring of polynomials in the indeterminate X (see Example 1.10 (12)). Then $F[X]$ is an integral domain (see Section 2.4 for a proof) and its quotient field is denoted $F(X)$. It consists formally of quotients of polynomials in one variable, called **rational functions** over F.

2.4 Polynomial Rings

Let R be a commutative ring with identity. The polynomial ring $R[X]$ was defined in Example 1.10 (12). Recall that an element $f \in R[X]$ is a function $f : \mathbf{Z}^+ \to R$ such that $f(n) \neq 0$ for at most finitely many nonnegative integers n. If $X \in R[X]$ is defined by

$$X(n) = \begin{cases} 1 & \text{if } n = 1, \\ 0 & \text{if } n \neq 1, \end{cases}$$

then every $f \in R[X]$ can be written uniquely in the form

$$f = \sum_{n=0}^{\infty} a_n X^n$$

where the sum is finite since only finitely many $a_n = f(n) \in R$ are not 0. With this notation the multiplication formula becomes

$$\left(\sum_{n=0}^{\infty} a_n X^n\right)\left(\sum_{n=0}^{\infty} b_n X^n\right) = \sum_{n=0}^{\infty} c_n X^n$$

where

$$c_n = \sum_{m=0}^{n} a_m b_{n-m}.$$

It is traditional to denote elements of $R[X]$ by a symbol $f(X)$ but it is important to recognize that $f(X)$ does not mean a function f on the set

R evaluated at X; in fact, X is not an element of R. However, there is a concept of polynomial function, which we now describe.

Let S be another commutative ring with identity, let $\phi : R \to S$ be a ring homomorphism such that $\phi(1) = 1$, and let $u \in S$. Define a map $\phi_u : R[X] \to S$ by the formula

$$\phi_u(a_0 + a_1 X + \cdots + a_n X^n) = \phi(a_0) + \phi(a_1)u + \cdots + \phi(a_n)u^n.$$

We leave it as an exercise to check that ϕ_u is a ring homomorphism from $R[X]$ to S. It is called the **substitution homorphism determined by** $\phi :$ $R \to S$ **and** $u \in S$. If R is a subring of S and $\phi : R \to S$ is the inclusion homomorphism (i.e., $\phi(r) = r$), then ϕ_u just substitutes the element $u \in S$ for the indeterminate X in each polynomial $f(X) \in R[X]$. If $\mathrm{Ker}(\phi_u) = \{0\}$ then $R[X]$ can be identified with a subring of S, and this is a way we frequently think of polynomials. However, it may not be the case that $\mathrm{Ker}(\phi_u) = \{0\}$. For example, if $R = \mathbf{R}, S = \mathbf{C}$ and $u = i$, then $\mathrm{Ker}(\phi_u) = \langle X^2 + 1 \rangle$. In general, we let $R[u] = \mathrm{Im}(\phi_u)$.

The above discussion is formalized in the following result, the details of which are left to the reader. Note that R can be viewed as a subring of $R[X]$ via the identification $a \mapsto a \cdot 1$.

(4.1) Theorem. (Polynomial substitution theorem) *Let R and S be commutative rings with identity, let $\phi : R \to S$ be a ring homomorphism with $\phi(1) = 1$, and let $u \in S$. Then there is a unique ring homomorphism $\phi_u : R[X] \to S$ such that $\phi_u(X) = u$ and $\phi_u|_R = \phi$.*

Proof. Exercise. $\square$

If $f(X) = \sum_{m=0}^{\infty} a_m X^m \neq 0 \in R[X]$ we define the **degree of** f, denoted $\deg(f(X))$, by

$$\deg(f(X)) = \max\{m : a_m \neq 0\}.$$

Thus if $n = \deg(f(X))$ then we may write $f(X) = \sum_{m=0}^{n} a_m X^m$. We define $\deg(0) = -\infty$, and for convenience in manipulating degree formulas, we set $-\infty < n$ and $-\infty + n = -\infty$ for any $n \in \mathbf{Z}^+$. The coefficient a_n is called the **leading coefficient** of $f(X)$, while a_0 is called the **constant term**. If the leading coefficient of $f(X)$ is 1, then $f(X)$ is a **monic polynomial**.

(4.2) Lemma. *Let R be a commutative ring and let $f(X), g(X) \in R[X]$. Then*

(1) $\deg(f(X) + g(X)) \leq \max\{\deg(f(X)), \deg(g(X))\}$;
(2) $\deg(f(X)g(X)) \leq \deg(f(X)) + \deg(g(X))$; *and*
(3) *equality holds in* (2) *if the leading coefficient of $f(X)$ or $g(X)$ is not a zero divisor. In particular, equality holds in* (2) *if R is an integral domain.*

Proof. (1) is clear from the addition formula for polynomials.

For (2) and (3), suppose $\deg(f(X)) = n \geq 0$ and $\deg(g(X)) = m \geq 0$. Then

$$f(X) = a_0 + a_1 X + \cdots + a_n X^n \qquad \text{with } a_n \neq 0$$
$$g(X) = b_0 + b_1 X + \cdots + b_m X^m \qquad \text{with } b_m \neq 0.$$

Therefore,

$$f(X)g(X) = a_0 b_0 + (a_0 b_1 + a_1 b_0)X + \cdots + a_n b_m X^{m+n}$$

so that $\deg(f(X)g(X)) \leq n + m$ with equality if and only if $a_n b_m \neq 0$. If a_n or b_n is not a zero divisor, then it is certainly true that $a_n b_m \neq 0$.

The case for $f(X) = 0$ or $g(X) = 0$ is handled separately and is left to the reader. $\qquad \square$

(4.3) Corollary. *If R is an integral domain, then*

(1) *$R[X]$ is an integral domain, and*
(2) *the units of $R[X]$ are the units of R.*

Proof. (1) If $f(X) \neq 0, g(X) \neq 0$, then

$$\deg(f(X)g(X)) = \deg(f(X)) + \deg(g(X)) \geq 0 > -\infty,$$

and thus, $f(X)g(X) \neq 0$.

(2) If $f(X)g(X) = 1$ then $\deg(f(X)) + \deg(g(X)) = \deg(1) = 0$. Thus, $f(X)$ and $g(X)$ are both polynomials of degree 0, i.e., elements of R. Therefore, they are units not only in $R[X]$ but in R also. $\qquad \square$

We now consider the division algorithm for $R[X]$ where R is a commutative ring. Let $f(X) = a_0 + a_1 X + \cdots + a_n X^n \in R[X]$ and let $g(X) = b_0 + b_1 X + \cdots + b_{m-1} X^{m-1} + X^m$ be a monic polynomial in $R[X]$ of degree $m \geq 1$. If $n \geq m$, let $q_1(X) = a_n X^{n-m}$. Then $f_1(X) = f(X) - g(X)q_1(X)$ has degree $\leq n - 1$. If $\deg(f_1(X)) \geq m$ then repeat the process with $f(X)$ replaced by $f_1(X)$. After a finite number of steps we will arrive at a polynomial $f_s(X)$ of degree $< m$. Letting $q(X) = q_1(X) + \cdots + q_s(X)$ and $r(X) = f(X) - g(X)q(X)$ we obtain an equation $f(X) = g(X)q(X) + r(X)$ where $\deg(r(X)) < \deg(g(X))$. What we have described is the familiar long division process for polynomials.

(4.4) Theorem. (Division algorithm) *Let R be a commutative ring, let $f(X) \in R[X]$, and let $g(X) \in R[X]$ be a monic polynomial. Then there are unique polynomials $q(X)$ and $r(X)$ in $R[X]$ with $\deg(r(X)) < \deg(g(X))$ such that*

$$f(X) = g(X)q(X) + r(X).$$

Proof. Existence follows from the algorithm described in the previous paragraph. Now consider uniqueness. Suppose there are two such decompositions $f(X) = g(X)q(X) + r(X)$ and $f(X) = g(X)q_1(X) + r_1(X)$ with $\deg(r(X)) < \deg(g(X))$ and $\deg(r_1(X)) < \deg(g(X))$. Then

$$g(X)(q_1(X) - q(X)) = r(X) - r_1(X).$$

Since $g(X)$ is a monic polynomial, taking degrees of both sides gives

$$\deg(g(X)) + \deg(q_1(X) - q(X)) = \deg(r(X) - r_1(X)) < \deg(g(X)).$$

This forces $\deg(q_1(X) - q(X)) = -\infty$, i.e., $q_1(X) = q(X)$. It then follows that $r_1(X) = r(X)$, so uniqueness is established. $\square$

(4.5) Corollary. (Remainder theorem) *Let R be a commutative ring and let $a \in R$. Then for any $f(X) \in R[X]$*

$$f(X) = (X - a)q(X) + f(a)$$

for some $q(X) \in R[X]$.

Proof. By the division algorithm we may write $f(X) = (X-a)q(X)+r(X)$ where $\deg(r(X)) \le 0$. Therefore, $r(X) = r \in R$. Apply the substitution homomorphism $X \mapsto a$ to get $f(a) = (a-a)q(a) + r$ so that $r = f(a)$. $\square$

(4.6) Corollary. *Let R be a commutative ring, $f(X) \in R[X]$, and let $a \in R$. Then $f(a) = 0$ if and only if $X - a$ divides $f(X)$.*

Proof. $\square$

(4.7) Corollary. *Let R be an integral domain and let $f(X) \neq 0 \in R[X]$ be a polynomial of degree n. Then there are at most n roots of $f(X)$ in R, i.e., there are at most n elements $a \in R$ with $f(a) = 0$.*

Proof. If $n = 0$ the result is certainly true since $f(X) = a_0 \neq 0$ and thus $f(a) = a_0 \neq 0$ for every $a \in R$, i.e., $f(X)$ has no roots.

Now let $n > 0$ and suppose, by induction, that the result is true for polynomials of degree $< n$. If there are no roots of $f(X)$ in R, then there is nothing to prove. Thus suppose that there is at least 1 root $a \in R$. Then by Corollary 4.6, $X - a$ divides $f(X)$, so we may write $f(X) = (X - a)q(X)$ where $\deg(q(X)) = n - 1$. By the induction hypothesis, there are at most $n - 1$ roots of $q(X)$ in R. Now let b be any root of $f(X)$ so that $0 = f(b) = (b-a)q(a)$. Hence $b = a$ or b is a root of $q(X)$. We conclude that $f(X)$ has at most $(n - 1) + 1 = n$ roots in R. $\square$

(4.8) Remark. This result is false if R is not an integral domain. For example if $R = \mathbf{Z}_2 \times \mathbf{Z}_2$ then all four elements of R are roots of the quadratic polynomial $X^2 - X \in R[X]$.

(4.9) Corollary. *Let R be an integral domain and let $f(X)$, $g(X) \in R[X]$ be polynomials of degree $\le n$. If $f(a) = g(a)$ for $n + 1$ distinct $a \in R$, then $f(X) = g(X)$.*

Proof. The polynomial $h(X) = f(X) - g(X)$ is of degree $\leq n$ and has greater than n roots. Thus $h(X) = 0$ by Corollary 4.7.   □

In the case R is a field, there is the following complement to Corollary 4.9.

(4.10) Proposition. (Lagrange Interpolation) *Let F be a field and let a_0, a_1, ..., a_n be $n + 1$ distinct elements of F. Let c_0, c_1, ..., c_n be arbitrary (not necessarily distinct) elements of F. Then there exists a unique polynomial $f(X) \in F[X]$ of degree $\leq n$ such that $f(a_i) = c_i$ for $0 \leq i \leq n$.*

Proof. Uniqueness follows from Corollary 4.9, so it is only necessary to demonstrate existence. To see existence, for $0 \leq i \leq n$, let $P_i(X) \in F[X]$ be defined by

$$(4.1) \qquad P_i(X) = \prod_{j \neq i} \frac{(X - a_j)}{(a_i - a_j)}.$$

Note that $\deg P_i(X) = n$ and $P_i(a_j) = \delta_{ij}$. Thus, we may take

$$(4.2) \qquad f(X) = \sum_{i=0}^{n} c_i P_i(X)$$

to conclude the proof.   □

(4.11) *Remark.* The polynomial given in Equation (4.2) is known as **Lagrange's form of the interpolation polynomial**. Of course, the interpolation polynomial is unique, but there is more than one way to express the interpolation polynomial; in the language of vector spaces (see Chapter 3), this is simply the observation that there is more than one choice of a basis of the vector space $\mathcal{P}_n(F)$ of polynomials in $F[X]$ of degree at most n. The set $\{P_i(X) : 0 \leq i \leq n\}$ is one such basis. Another basis of $\mathcal{P}_n(F)$ is the set of polynomials $\widetilde{P}_0(X) = 1$ and

$$\widetilde{P}_i(X) = (X - a_0) \cdots (X - a_{i-1})$$

for $1 \leq i \leq n$. Any polynomial $f(X) \in F[X]$ of degree at most n can be written uniquely as

$$(4.3) \qquad f(X) = \sum_{i=0}^{n} \alpha_i \widetilde{P}_i(X).$$

The coefficients α_i can be computed from the values $f(a_i)$ for $0 \leq i \leq n$. The details are left as an exercise. The expression in Equation (4.3) is known as **Newton's form of interpolation**; it is of particular importance in numerical computations.

(4.12) Theorem. *Let F be a field. Then $F[X]$ is a principal ideal domain* (PID).

Proof. Let I be an ideal of $F[X]$. If $I = \{0\}$ then I is a principal ideal, so suppose that $I \neq \{0\}$. Choose $g(X) \in I$ such that $g(X) \neq 0$ and such that $\deg(g(X)) \leq \deg(f(X))$ for all $f(X) \in I \setminus \{0\}$. We claim that $I = (g(X))$. Since F is a field, we may multiply $g(X)$ by an element of F to get a monic polynomial, which will also be in I. Thus we may suppose that $g(X)$ is a monic polynomial. Let $f(X) \in I$. Then by the division algorithm we may write $f(X) = g(X)q(X) + r(X)$ where $\deg(r(X)) < \deg(g(X))$. But $r(X) = f(X) - g(X)q(X) \in I$, so $r(X)$ must be 0 since $g(X)$ was chosen to have minimal degree among all nonzero elements of I. Thus $f(X) = g(X)q(X)$ and we conclude that $I = (g(X))$. Since I was an arbitrary ideal of $F[X]$, it follows that $F[X]$ is a PID. $\qquad\square$

(4.13) *Remarks.* (1) If I is a nonzero ideal of $F[X]$, then there is a unique monic polynomial of minimal degree in I. This is an immediate consequence of the proof of Theorem 4.12.

(2) Both the statement and proof of Theorem 4.12 will be generalized to the case of Euclidean domains in Section 2.5 (Theorem 5.19).

If R is a PID that is not a field, then it is not true that the polynomial ring $R[X]$ is a PID. In fact, if $I = \langle p \rangle$ is a nonzero proper ideal of R, then $J = \langle p, X \rangle$ is not a principal ideal (see Example 2.23 (2)). It is, however, true that every ideal in the ring $R[X]$ has a finite number of generators. This is the content of the following result.

(4.14) Theorem. (Hilbert basis theorem) *Let R be a commutative ring in which every ideal is finitely generated. Then every ideal of the polynomial ring $R[X]$ is also finitely generated.*

Proof. The ideal $\langle 0 \rangle \subseteq R[X]$ is certainly finitely generated, so let $I \subseteq R[X]$ be a nonzero ideal. If $f(X) \in R[X]$ is not zero, we will let $\mathrm{lc}(f(X))$ denote the leading coefficient of $f(X)$. For $n = 0, 1, 2, \ldots$, let

$$I_n = \{a \in R : \mathrm{lc}(f(X)) = a \quad \text{for some } f(X) \in I \text{ of degree } n\} \cup \{0\}.$$

Note that I_n is an ideal of R for all n. Since $\mathrm{lc}(f(X)) = \mathrm{lc}(Xf(X))$, it follows that $I_n \subseteq I_{n+1}$ for all n. Let $J = \cup_{n=0}^{\infty} I_n$. Then J is an ideal of R and hence is finitely generated, and since the sequence of ideals I_n is increasing, it is easy to see that $J = I_n$ for some n. Also, I_m is finitely generated for any m. For $0 \leq m \leq n$, let

$$\{a_{m1}, \ldots, a_{mk_m}\}$$

generate I_m and choose $f_{mj}(X) \in I$ such that

$$\deg(f_{mj}(X)) = m \qquad \text{and} \qquad \mathrm{lc}(f_{mj}(X)) = a_{mj}$$

for all $0 \leq m \leq n$ and $1 \leq j \leq k_m$. Let

$$\widehat{I} = \langle f_{mj}(X) : 0 \leq m \leq n, 1 \leq j \leq k_m \rangle.$$

Claim. $I = \widehat{I}$.

Suppose that $f(X) \in I$. If $f(X) = 0$ or $\deg(f(X)) = 0$ then $f(X) \in \widehat{I}$ is clear, so assume that $\deg(f(X)) = r > 0$ and proceed by induction on r. Let $a = \text{lc}(f(X))$. If $r \leq n$, then $a \in I_r$, so we may write

$$a = c_1 a_{r1} + \cdots + c_{k_r} a_{rk_r}.$$

Then $\text{lc}(\sum_{i=1}^{k_r} c_i f_{ri}(X)) = a$ so that

$$\deg\left(f(X) - \sum_{i=1}^{k_r} c_i f_{ri}(X)\right) < r$$

and $f(X) - \sum_{i=1}^{k_r} c_i f_{ri}(X) \in \widehat{I}$ by induction. Thus, $f(X) \in \widehat{I}$ in case $r \leq n$. If $r > n$ then $a \in I_r = I_n$ so that

$$a = c_1 a_{n1} + \cdots + c_{k_n} a_{nk_n}.$$

Then, as above,

$$\deg\left(f(X) - \sum_{i=1}^{k_n} c_i X^{r-n} f_{ni}(X)\right) < r,$$

and hence $f(X) \in \widehat{I}$ by induction on r.

Thus, $I \subseteq \widehat{I}$ and the other inclusion is clear. Hence I is finitely generated. □

(4.15) Corollary. *Let R be a commutative ring in which every ideal is finitely generated. Then every ideal of $R[X_1, \ldots, X_n]$ is finitely generated.*

Proof. This follows from Theorem 4.14 by induction on n since it is an easy exercise to verify that

$$R[X_1, \ldots, X_n] \cong (R[X_1, \ldots, X_{n-1}])[X_n].$$

□

(4.16) Corollary. *Let F be a field. Then every ideal in the polynomial ring*

$$F[X_1, \ldots, X_n]$$

is finitely generated.

Proof. □

The theory of principal ideal domains will be studied in Section 2.5. At the present time we have two major examples of PIDs, namely, $\mathbf{Z}$ and $F[X]$ for F a field, to which the theory will apply. Further examples will be given in Section 2.5. We will conclude this section with the following concept, which is defined by means of polynomials.

(4.17) Definition. *Let F be a field. F is said to be **algebraically closed** if every nonconstant polynomial $f(X) \in F[X]$ has a root in F.*

(4.18) *Remarks.*

(1) According to Corollary 4.6, F is algebraically closed if and only if the nonconstant irreducible polynomials in $F[X]$ are precisely the polynomials of degree 1.

(2) The complex numbers $\mathbf{C}$ are algebraically closed. This fact, known as the **fundamental theorem of algebra**, will be assumed known at a number of points in the text. We shall not, however, present a proof.

(3) If F is an arbitrary field, then there is a field K such that $F \subseteq K$ and K is algebraically closed. One can even guarantee that every element of K is **algebraic** over F, i.e., if $a \in K$ then there is a polynomial $f(X) \in F[X]$ such that $f(a) = 0$. Again, this is a fact that we shall not prove since it involves a few subtleties of set theory that we do not wish to address.

2.5 Principal Ideal Domains and Euclidean Domains

The fundamental theorem of arithmetic concerns the factorization of an integer into a unique product of prime numbers. In this section we will show that the fundmental theorem of arithmetic is also valid in an arbitrary principal ideal domain. At present we have only two major examples of PIDs, namely, $\mathbf{Z}$ and $F[X]$ for F a field. Some examples will be presented of other PIDs. We will start by defining precisely the concepts of factorization needed to state and prove the extension of the fundamental theorem of arithmetic.

(5.1) Definition. *Let R be an integral domain and let $a,\ b \in R \setminus \{0\}$.*

(1) *a and b are **associates** if $a = ub$ for some unit $u \in R$. We can define an equivalence relation on R by setting $a \sim b$ if and only if a and b are associates in R.*

(2) *a **divides** b (written $a \mid b$) if $b = ac$ for some $c \in R$.*

(3) *A nonunit a is **irreducible** if $a = bc$ implies that b or c is a unit.*

(4) *A nonunit a is **prime** if $a \mid bc$ implies that $a \mid b$ or $a \mid c$.*

(5.2) *Remark.* Let R be an integral domain and let $a, b \in R \setminus \{0\}$.

(1) a and b are associates if and only if $a \mid b$ and $b \mid a$.

(2) Recall that $\langle a \rangle$ denotes the ideal of R generated by a. Then $a \mid b$ if and only if $\langle b \rangle \subseteq \langle a \rangle$ and a and b are associates if and only if $\langle a \rangle = \langle b \rangle$.

(3) a is a prime element of R if and only if $\langle a \rangle$ is a prime ideal.

(4) If $a \mid b$ then $au \mid bv$ for any units u, v.

(5) If $a \mid b$ and a is not a unit, then b is not a unit. Indeed, if b is a unit then $bc = 1$ for some $c \in R$ and $b = ad$ since $a \mid b$. Then $(ad)c = 1 = a(dc)$ so, a is a unit also.

(6) If p is a prime in R and $p \mid a_1 \cdots a_n$ then $p \mid a_i$ for some i (exercise).

(5.3) Lemma. *Let R be an integral domain. If $a \in R$ is prime, then a is irreducible.*

Proof. Let $a \in R$ be prime and suppose that $a = bc$. Then $a \mid b$ or $a \mid c$. To be specific, suppose that $a \mid b$ so that $b = ad$ for some $d \in R$. Then $a = bc = adc$, and since R is an integral domain, Lemma 1.3 shows that $dc = 1$ so that c is a unit. Thus a is irreducible. □

If $R = \mathbf{Z}$ then the concepts of prime and irreducible are the same, so that the converse of Lemma 5.3 is also valid. In fact, we shall show that the converse is valid in any PID, but it is not valid for every integral domain as the following examples show.

(5.4) Examples.

(1) Let F be a field and let

$$R = F[X^2, X^3]$$
$$= \{p(X) \in F[X] : p(X) = a_0 + a_2 X^2 + a_3 X^3 + \cdots + a_n X^n\}.$$

Then X^2 and X^3 are irreducible in R, but they are not prime since $X^2 \mid (X^3)^2 = X^6$, but X^2 does not divide X^3, and $X^3 \mid X^4 X^2 = X^6$, but X^3 does not divide either X^4 or X^2. All of these statements are easily seen by comparing degrees.

(2) *Let $\mathbf{Z}[\sqrt{-3}] = \{a + b\sqrt{-3} : a, b \in \mathbf{Z}\}$. In $\mathbf{Z}[\sqrt{-3}]$ the element 2 is irreducible but not prime.*

Proof. Suppose that $2 = (a + b\sqrt{-3})(c + d\sqrt{-3})$ with $a, b, c, d \in \mathbf{Z}$. Then taking complex conjugates gives $2 = (a - b\sqrt{-3})(c - d\sqrt{-3})$, so multiplying these two equations gives $4 = (a^2 + 3b^2)(c^2 + 3d^2)$. Since the equation in integers $\alpha^2 + 3\beta^2 = 2$ has no solutions, it follows that we must have $a^2 + 3b^2 = 1$ or $c^2 + 3d^2 = 1$ and this forces $a = \pm 1$, $b = 0$ or $c = \pm 1$, $d = 0$. Therefore, 2 is irreducible in $\mathbf{Z}[\sqrt{-3}]$. Note that 2 is not a unit in $\mathbf{Z}[\sqrt{-3}]$ since the equation $2(a + b\sqrt{-3}) = 1$ has no solution with $a, b \in \mathbf{Z}$.

To see that 2 is not prime, note that 2 divides $4 = (1 + \sqrt{-3})(1 - \sqrt{-3})$ but 2 does not divide either of the factors $1 + \sqrt{-3}$ or $1 - \sqrt{-3}$ in $\mathbf{Z}[\sqrt{-3}]$. We conclude that 2 is not a prime in the ring $\mathbf{Z}[\sqrt{-3}]$.   □

(5.5) Definition. *Let R be an integral domain and let A be a subset of R containing at least one nonzero element. We say that $d \in R$ is a* **greatest common divisor** *(gcd) of A if*

(1) *$d \mid a$ for all $a \in A$, and*
(2) *if $e \in R$ and $e \mid a$ for all $a \in A$, then $e \mid d$.*

If 1 is a gcd of A, then we say that the set A is **relatively prime**. *We say that $m \neq 0 \in R$ is a* **least common multiple** *(lcm) of A if $0 \notin A$, if*

(1) *$a \mid m$ for all $a \in A$, and*
(2) *if $e \in R$ and $a \mid e$ for all $a \in R$, then $m \mid e$.*

Note that any two gcds of A are associates. Thus the gcd (if it exists) is well defined up to multiplication by a unit. The following result shows that in a PID there exists a gcd of any nonempty subset of nonzero elements.

(5.6) Theorem. *Let R be a PID and let A be a subset of R containing at least one nonzero element.*

(1) *An element $d \in R$ is a gcd of A if and only if d is a generator for the ideal $\langle A \rangle$ generated by A.*
(2) *If $A = \{a_1, \ldots, a_n\}$ is finite and $a_i \neq 0$ for $1 \leq i \leq n$, then an element $m \in R$ is a lcm of A if and only if m is a generator of the ideal*

$$\langle a_1 \rangle \cap \cdots \cap \langle a_n \rangle.$$

Proof. (1) Suppose that d is a generator of the ideal $\langle A \rangle$. Certainly $d \mid a$ for each $a \in A$ since $a \in \langle A \rangle = \langle d \rangle$. Also, since $d \in \langle A \rangle$, it follows that $d = \sum_{i=1}^{n} r_i a_i$ for $r_1, \ldots, r_n \in R$ and $a_1, \ldots, a_n \in A$. Therefore, if $e \mid a$ for all $a \in A$, then $e \mid d$ so that d is a gcd of A.

Conversely, suppose that d is a gcd of the set A and let $\langle A \rangle = \langle c \rangle$. Then $d \mid a$ for all $a \in A$ so that $a \in \langle d \rangle$. Hence,

$$\langle c \rangle = \langle A \rangle \subseteq \langle d \rangle.$$

But, for each $a \in A$, $a \in \langle c \rangle$ so that $c \mid a$. Since d is a gcd of A, it follows that $c \mid d$, i.e., $\langle d \rangle \subseteq \langle c \rangle$. Hence $\langle c \rangle = \langle d \rangle$ and d is a generator of the ideal $\langle A \rangle$.

(2) Exercise.   □

(5.7) Corollary. *Let R be a PID and $a \in R \setminus \{0\}$. Then a is prime if and only if a is irreducible.*

Proof. Lemma 5.3 shows that if a is prime, then a is irreducible. Now assume that a is irreducible and suppose that $a \mid bc$. Let $d = \gcd\{a, b\}$. Thus $a = de$ and $b = df$. Since $a = de$ and a is irreducible, either d is a unit or e is a unit. If e is a unit, then $a \mid b$ because a is an associate of d and $d \mid b$. If d is a unit, then $d = ar' + bs'$ for some r', $s' \in R$ (since $d \in \langle a, b \rangle$). Therefore, $1 = ar + bs$ for some $r, s \in R$, and hence, $c = arc + bsc$. But $a \mid arc$ and $a \mid bsc$ (since $a \mid bc$ by assumption), so $a \mid c$ as required. $\square$

(5.8) Corollary. *Let R be a PID and let $I \subseteq R$ be a nonzero ideal. Then I is a prime ideal if and only if I is a maximal ideal.*

Proof. By Theorem 2.21, if I is maximal then I is prime. Conversely, suppose that I is prime. Then since R is a PID we have that $I = \langle p \rangle$ where p is a prime element of R. If $I \subseteq J = \langle a \rangle$ then $p = ra$ for some $r \in R$. But p is prime and hence irreducible, so either r or a is a unit. If r is a unit then $\langle p \rangle = \langle a \rangle$, i.e., $I = J$. If a is a unit then $J = R$. Thus I is maximal. $\square$

(5.9) Definition. *Let R be a ring. We say that R satisfies the **ascending chain condition** (ACC) on ideals if for any chain*

$$I_1 \subseteq I_2 \subseteq I_3 \subseteq \cdots$$

*of ideals of R there is an n such that $I_k = I_n$ for all $k \geq n$, i.e., the chain is eventually constant. A ring that satisfies the ascending chain condition is said to be **Noetherian**.*

The following characterization of Noetherian rings uses the concept of maximal element in a partially ordered set. Recall what this means (see the appendix). If X is a partially ordered set (e.g., $X \subseteq \mathcal{P}(Y)$ where the partial order is set inclusion), then a maximal element of X is an element $m \in X$ such that if $m \leq x$ for some $x \in X$, then $m = x$. That is, a element $m \in X$ is maximal if there is no element strictly larger than m in the partial order of X. For example, if

$$X = \{\langle 2 \rangle, \langle 3 \rangle, \langle 12 \rangle\}$$

is a set consisting of the given ideals of the ring $\mathbf{Z}$ with the partial order being inclusion of sets, then both $\langle 2 \rangle$ and $\langle 3 \rangle$ are maximal elements of X.

(5.10) Proposition. *Let R be a ring. The following conditions on R are equivalent.*

(1) *R is Noetherian.*
(2) *Every ideal of R is finitely generated.*
(3) *Every nonempty subset of ideals of R has a maximal element.*

In particular, a PID is Noetherian.

Proof. (1) $\Rightarrow$ (3) Suppose that $\mathcal{S} = \{I_\alpha\}_{\alpha \in A}$ is a nonempty set of ideals of R that does not have a maximal element.

Then choose $I_1 \in \mathcal{S}$. Since $\mathcal{S}$ does not have a maximal element, there is an element $I_2 \in \mathcal{S}$ such that $I_1 \subsetneq I_2$. Similarly, I_2 is not a maximal element so there is $I_3 \in \mathcal{S}$ such that $I_2 \subsetneq I_3$. In this manner we can construct a strictly increasing chain of ideal in R, which contradicts the assumption that R satisfies the ACC. Therefore, $\mathcal{S}$ must contain a maximal element if R satisfies the ACC.

$(3) \Rightarrow (2)$ Let I be an ideal of R and consider the family $\mathcal{S}$ of all finitely generated ideals of R that are contained in I. By hypothesis, there is a maximal element $J \in \mathcal{S}$. Let $a \in I$. Then the ideal $J + \langle a \rangle \in \mathcal{S}$ and it contains J. Since J is maximal in $\mathcal{S}$, it follows that $J = J + \langle a \rangle$, i.e., $a \in J$. Therefore, $I = J$ and I is finitely generated.

$(2) \Rightarrow (1)$ Suppose that every ideal of R is finitely generated. Let

$$I_1 \subseteq I_2 \subseteq I_3 \subseteq \cdots$$

be a chain of ideals of R and let $I = \bigcup_{n=1}^{\infty} I_n$. Then I is an ideal of R so that $I = \langle a_1, \ldots, a_m \rangle$ for some $a_i \in R$. But $a_i \in I = \bigcup_{n=1}^{\infty} I_n$ for $1 \le i \le m$ so $a_i \in I_{n_i}$ for some n_i. Since we have a chain of ideals, it follows that there is an n such that $a_i \in I_n$ for all i. Thus, for any $k \ge n$ there is an inclusion

$$\langle a_1, \ldots, a_m \rangle \subseteq I_k \subseteq I = \langle a_1, \ldots, a_m \rangle$$

so that $I_k = I = I_n$ for all $k \ge n$ and R satisfies the ACC on ideals. $\qquad \square$

(5.11) *Remark.* In light of Definition 5.9 and Proposition 5.10, the Hilbert basis theorem (Theorem 4.14) and its corollary (Corollary 4.15) are often stated:

> *If R is a commutative Noetherian ring, then the polynomial ring $R[X_1, \ldots, X_n]$ is also a commutative Noetherian ring.*

(5.12) Theorem. (Fundamental theorem of arithmetic) *Let R be a PID. Then any nonzero $a \in R$ can be written as $a = up_1 \cdots p_n$ where u is a unit and each p_i is a prime. Moreover, this factorization is essentially unique. That is, if $a = vq_1 \cdots q_m$ where v is a unit and each q_i is a prime, then $m = n$ and for some permutation $\sigma \in S_n$, q_i is an associate of $p_{\sigma(i)}$ for $1 \le i \le n$.*

Proof. We first prove existence of the factorization. Let $a \ne 0 \in R$. Then if a is a unit we are done and if a is a prime we are done. Otherwise write $a = a_1 b_1$ where neither a_1 nor b_1 is a unit. (Recall that in a PID, prime and irreducible elements are the same.) Thus $\langle a \rangle \subsetneq \langle b_1 \rangle$. If b_1 is a prime, stop. Otherwise, write $b_1 = a_2 b_2$ with neither a_2 nor b_2 a unit. Continue in this way to get a chain

$$\langle a \rangle \subsetneq \langle b_1 \rangle \subsetneq \langle b_2 \rangle \subsetneq \cdots$$

By the ascending chain condition, this must stop at some $\langle b_n \rangle$. Therefore b_n is a prime and we conclude that every $a \neq 0 \in R$ that is not a unit is divisible by some prime.

Therefore, if $a \neq 0$ is not a unit, write $a = p_1 c_1$ where p_1 is a prime. Thus $\langle a \rangle \subsetneq \langle c_1 \rangle$. If c_1 is a unit, stop. Otherwise, write $c_1 = p_2 c_2$ with p_2 a prime so that $\langle c_1 \rangle \subsetneq \langle c_2 \rangle$. Continue in this fashion to obtain a chain of ideals

$$\langle a \rangle \subsetneq \langle c_1 \rangle \subsetneq \langle c_2 \rangle \subsetneq \cdots .$$

By the ACC this must stop at some c_n, and by the construction it follows that this $c_n = u$ is a unit. Therefore,

$$a = p_1 c_1 = p_1 p_2 c_2 = \cdots = p_1 p_2 \cdots p_n u$$

so that a is factored into a finite product of primes times a unit.

Now consider uniqueness of the factorization. Suppose that

$$a = p_0 p_1 \cdots p_n = q_0 q_1 \cdots q_m$$

where $p_1, \ldots, p_n, q_1, \ldots, q_m$ are primes while p_0 and q_0 are units of R. We will use induction on $k = \min\{m, n\}$. If $n = 0$ then a is a unit, so $a = q_0$ and hence $m = 0$. Also $m = 0$ implies $n = 0$. Thus the result is true for $k = 0$. Suppose that $k > 0$ and suppose that the result is true for all elements $b \in R$ that have a factorization with fewer than k prime elements. Then $p_n \mid q_0 q_1 \cdots q_m$, so p_n divides some q_i since p_n is a prime element. After reordering, if necessary, we can assume that $p_n \mid q_m$. But q_m is prime, so $q_m = p_n c$ implies that c is a unit. Thus, p_n and q_m are associates. Let

$$a' = a/p_n = p_0 p_1 \cdots p_{n-1} = (q_0 c) q_1 \cdots q_{m-1}.$$

Then a' has a factorization with fewer than k prime factors, so the induction hypothesis implies that $n - 1 = m - 1$ and q_i is an associate of $p_{\sigma(i)}$ for some $\sigma \in S_{n-1}$, and the argument is complete. $\square$

(5.13) Corollary. *Let R be a PID and let $a \neq 0 \in R$. Then*

$$a = u p_1^{n_1} \cdots p_k^{n_k}$$

where $p_1, \ldots, p_k$ are distinct primes and u is a unit. The factorization is essentially unique.

Proof. $\square$

(5.14) *Remark.* Note that the proof of Theorem 5.12 actually shows that if R is any commutative Noetherian ring, then every nonzero element $a \in R$ has a factorization into irreducible elements, i.e., any $a \in R$ can be factored as $a = u p_1 \cdots p_n$ where u is a unit of R and $p_1, \ldots, p_n$ are irreducible (not *prime*) elements, but this factorization is not necessarily unique; however, this is not a particularly useful result.

For non-Noetherian rings we do not even have the factorization into irreducible elements. Examples 5.15 (4) and (5) are examples.

(5.15) Examples.

(1) In $F[X^2, X^3]$ there are two different factorizations of X^6 into irreducible elements, namely, $(X^2)^3 = X^6 = (X^3)^2$.

(2) In $Z[\sqrt{-3}]$ there is a factorization

$$4 = 2 \cdot 2 = (1 + \sqrt{-3})(1 - \sqrt{-3})$$

into two essentially different products of irreducibles.

(3) If F is a field and $p(X) \in F[X]$ is an irreducible polynomial, then $p(X)$ is prime since $F[X]$ is a PID. Thus the ideal $\langle p(X) \rangle$ is maximal according to Corollary 5.8. Hence the quotient ring $F[X]/\langle p(X) \rangle$ is a field. If $p \in Z$ is a prime number let $\mathbf{F}_p$ denote the field Z_p. Then $\mathbf{F}_2[X]/\langle X^2 + X + 1 \rangle$ is a field with 4 elements, $\mathbf{F}_3[X]/\langle X^2 + 1 \rangle$ is a field with 9 elements, while $\mathbf{F}_2[X]/\langle X^3 + X + 1 \rangle$ is a field with 8 elements. In fact, one can construct for any prime $p \in Z$ and $n \geq 1$ a field $\mathbf{F}_q$ with $q = p^n$ elements by producing an irreducible polynomial of degree n in the polynomial ring $\mathbf{F}_p[X]$. (It turns out that $\mathbf{F}_q$ is unique up to isomorphism, but the proof of this requires Galois theory, which we do not treat here.)

(4) Let $H(\mathbf{C})$ be the ring of complex analytic functions on the entire complex plane. (Consult any textbook on complex analysis for verification of the basic properties of the ring $H(\mathbf{C})$.) The units of $H(\mathbf{C})$ are precisely the complex analytic functions $f : \mathbf{C} \to \mathbf{C}$ such that $f(z) \neq 0$ for all $z \in \mathbf{C}$. Furthermore, if $a \in \mathbf{C}$ then the function $z - a$ divides $f(z) \in H(\mathbf{C})$ if and only if $f(a) = 0$. From this it is easy to see (exercise) that the irreducible elements of $H(\mathbf{C})$ are precisely the functions $(z - a)f(z)$ where $f(z) \neq 0$ for all $z \in \mathbf{C}$. Thus, a complex analytic function $g(z)$ can be written as a finite product of irreducible elements if and only if g has only finitely many zeros. Therefore, the complex analytic function $\sin(z)$ cannot be written as a finite product of irreducible elements in the ring $H(\mathbf{C})$. (Incidentally, according to Remark 5.14, this shows that the ring $H(\mathbf{C})$ is not Noetherian.)

(5) Let R be the subring of $\mathbf{Q}[X]$ consisting of all polynomials whose constant term is an integer, i.e.,

$$R = \{f(X) = a_0 + a_1 X + \cdots + a_n X^n \in \mathbf{Q}[X] : a_0 \in \mathbf{Z}\}.$$

The units of R are the constant polynomials ± 1. Note that for any nonzero integer k, there is a factorization $X = k(X/k) \in R$, and neither factor is a unit of R. This readily implies that X does not have a factorization into irreducibles. (Again, Remark 5.14 implies that R is not Noetherian, but this is easy to see directly:

$$\langle X \rangle \subsetneq \langle X/2 \rangle \subsetneq \langle X/4 \rangle \subsetneq \langle X/8 \rangle \subsetneq \cdots$$

is an infinite ascending chain.)

Both of the examples of PIDs that we currently have available, namely, $\mathbf{Z}$ and $F[X]$ for F a field, actually have more structure than just that of a PID. Specifically, both $\mathbf{Z}$ and $F[X]$ have a measure of the size of elements, $|n|$ for $n \in \mathbf{Z}$ and $\deg(p(X))$ for $p(X) \in F[X]$, together with a division algorithm, which allows one to divide one element by another with a remainder "smaller" than the divisor. In fact the division algorithm was precisely what was needed to prove that $\mathbf{Z}$ and $F[X]$ are principal ideal domains. We formalize this property of $\mathbf{Z}$ and $F[X]$ with the following definition.

(5.16) Definition. *An integral domain R is a* **Euclidean domain** *if there is a function $v : R \setminus \{0\} \rightarrow \mathbf{Z}^{+} = \mathbf{N} \cup \{0\}$ such that*

(1) *for all $a, b \in R \setminus \{0\}$, $v(a) \leq v(ab)$; and*
(2) *given $a, b \in R$ with $a \neq 0$, there are $q, r \in R$ with $b = aq + r$ such that $r = 0$ or $v(r) < v(a)$.*

(5.17) Examples.

(1) $\mathbf{Z}$ together with $v(n) = |n|$ is a Euclidean domain.
(2) If F is a field, then $F[X]$ together with $v(p(X)) = \deg(p(X))$ is a Euclidean domain (Theorem 4.4).

(5.18) Lemma. *If R is a Euclidean domain and $a \in R \setminus \{0\}$, then $v(1) \leq v(a)$. Furthermore, $v(1) = v(a)$ if and only if a is a unit.*

Proof. First note that $v(1) \leq v(1 \cdot a) = v(a)$. If a is a unit, let $ab = 1$. Then $v(a) \leq v(ab) = v(1)$, so $v(1) = v(a)$. Conversely, suppose that $v(1) = v(a)$ and divide 1 by a. Thus, $1 = aq + r$ where $r = 0$ or $v(r) < v(a) = v(1)$. But $v(1) \leq v(r)$ for any $r \neq 0$, so the latter possibility cannot occur. Therefore, $r = 0$ so $1 = aq$ and a is a unit. $\square$

(5.19) Theorem. *Let R be a Euclidean domain. Then R is a PID.*

Proof. Let $I \subseteq R$ be a nonzero ideal and let

$$S = \{v(a) : a \in I \setminus \{0\}\} \subseteq \mathbf{Z}^{+}.$$

This set has a smallest element n_0. Choose $a \in I$ with $v(a) = n_0$. We claim that $I = Ra$. Since $a \in I$, it is certainly true that $Ra \subseteq I$. Now let $b \in I$. Then $b = aq + r$ for $q, r \in R$ with $r = 0$ or $v(r) < v(a)$. But $r = b - aq \in I$, so $v(a) \leq v(r)$ if $r \neq 0$. Therefore, we must have $r = 0$ so that $b = aq \in Ra$. Hence, $I = Ra$ is principal. $\square$

In a Euclidean domain R the classical Euclidean algorithm for finding the gcd of two integers works for finding the gcd of two elements of R. This algorithm is the following. Given $a_1, a_2 \in R \setminus \{0\}$ write

$$
\begin{aligned}
a_1 &= a_2 q_1 + a_3 & \text{with} \quad a_3 &= 0 \text{ or } v(a_3) < v(a_2) \\
a_2 &= a_3 q_2 + a_4 & \text{with} \quad a_4 &= 0 \text{ or } v(a_4) < v(a_3) \\
a_3 &= a_4 q_3 + a_5 & \text{with} \quad a_5 &= 0 \text{ or } v(a_5) < v(a_4)
\end{aligned}
$$
$$
\vdots
$$

Since $v(a_2) > v(a_3) > v(a_4) > \cdots$, this process must terminate after a finite number of steps, i.e., $a_{n+1} = 0$ for some n. For this n we have

$$a_{n-1} = a_n q_{n-1} + 0.$$

Claim. $a_n = \gcd\{a_1, a_2\}$.

Proof. If $a, b \in R$ then denote the gcd of $\{a, b\}$ by the symbol (a, b). Theorem 5.5 shows that the gcd of $\{a, b\}$ is a generator of the ideal generated by $\{a, b\}$.

Now we claim that $(a_i, a_{i+1}) = (a_{i+1}, a_{i+2})$ for $1 \le i \le n - 1$. Since $a_i = a_{i+1} q_i + a_{i+2}$, it follows that

$$
\begin{aligned}
x a_i + y a_{i+1} &= x(a_{i+1} q_i + a_{i+2}) + y a_{i+1} \\
&= (x q_i + y) a_{i+1} + x a_{i+2}.
\end{aligned}
$$

Thus, $\langle a_i, a_{i+1} \rangle \subseteq \langle a_{i+1}, a_{i+2} \rangle$, and similarly

$$
\begin{aligned}
r a_{i+1} + s a_{i+2} &= r a_{i+1} + s(a_i - a_{i+1} q_i) \\
&= s a_i + (r - q_i) a_{i+1}
\end{aligned}
$$

so that $\langle a_{i+1}, a_{i+2} \rangle \subseteq \langle a_i, a_{i+1} \rangle$. Hence, $(a_i, a_{i+1}) = (a_{i+1}, a_{i+2})$, and we conclude

$$(a_1, a_2) = (a_2, a_3) = \cdots = (a_{n-1}, a_n).$$

But $(a_{n-1}, a_n) = a_n$ since $a_n \mid a_{n-1}$, and the claim is proved. $\qquad \square$

This result gives an algorithmic procedure for computing the gcd of two elements in a Euclidean domain. By reversing the sequence of steps used to compute $d = (a, b)$ one can arrive at an explicit expression $d = ra + sb$ for the gcd of a and b. We illustrate with some examples.

(5.20) Examples.

(1) We use the Euclidean algorithm to compute $(1254, 1110)$ and write it as $r1254 + s1110$. Using successive divisions we get

$$1254 = 1110 \cdot 1 + 144$$
$$1110 = 144 \cdot 7 + 102$$
$$144 = 102 \cdot 1 + 42$$
$$102 = 42 \cdot 2 + 18$$
$$42 = 18 \cdot 2 + 6$$
$$18 = 6 \cdot 3 + 0.$$

Thus, $(1254, 1110) = 6$. Working backward by substituting into successive remainders, we obtain

$$
\begin{aligned}
6 &= 42 - 18 \cdot 2 \\
&= 42 - (102 - 42 \cdot 2) \cdot 2 = 5 \cdot 42 - 2 \cdot 102 \\
&= 5 \cdot (144 - 102 \cdot 1) - 2 \cdot 102 = 5 \cdot 144 - 7 \cdot 102 \\
&= 5 \cdot 144 - 7 \cdot (1110 - 7 \cdot 144) = 54 \cdot 144 - 7 \cdot 1110 \\
&= 54 \cdot (1254 - 1110 \cdot 1) - 7 \cdot 1110 \\
&= 54 \cdot 1254 - 61 \cdot 1110.
\end{aligned}
$$

(2) Let $f(X) = X^2 - X + 1$ and $g(X) = X^3 + 2X^2 + 2 \in \mathbf{Z}_3[X]$. Then

$$g(X) = Xf(X) + 2X + 2$$

and

$$f(X) = (2X + 2)^2.$$

Thus, $(f(X), g(X)) = (2X + 2) = (X + 1)$ and $X + 1 = 2g(X) - 2Xf(X)$.

(3) We now give an example of how to solve a system of congruences in a Euclidean domain by using the Euclidean algorithm. The reader should refer back to the proof of the Chinese remainder theorem (Theorem 2.24) for the logic of our argument.

Consider the following system of congruences (in $\mathbf{Z}$):

$$
\begin{aligned}
x &\equiv -1 \pmod{15} \\
x &\equiv 3 \pmod{11} \\
x &\equiv 6 \pmod{8}.
\end{aligned}
$$

We apply the Euclidean algorithm to the pair $(88, 15)$:

$$
\begin{aligned}
88 &= 15 \cdot 5 + 13 \\
15 &= 13 \cdot 1 + 2 \\
13 &= 2 \cdot 6 + 1 \\
2 &= 1 \cdot 2.
\end{aligned}
$$

Now we substitute backward to obtain

$$1 = 13 - 2 \cdot 6$$
$$= 13 - (15 - 13 \cdot 1) \cdot 6 = -15 \cdot 6 + 13 \cdot 7$$
$$= -15 \cdot 6 + (88 - 15 \cdot 5) \cdot 7 = 88 \cdot 7 - 15 \cdot 41$$
$$= 616 - 15 \cdot 41,$$

so $616 \equiv 1 \pmod{15}$ and $616 \equiv 0 \pmod{88}$. Similarly, by applying the Euclidean algorithm to the pair $(120, 11)$, we obtain that $-120 \equiv 1 \pmod{11}$ and $-120 \equiv 0 \pmod{120}$, and by applying it to the pair $(165, 8)$, we obtain $-495 \equiv 1 \pmod 8$ and $-495 \equiv 0 \pmod{165}$. Then our solution is

$$x \equiv -1 \cdot (616) + 3 \cdot (-120) + 6 \cdot (-495) = -3946 \quad (\text{mod } 1320)$$

or, more simply,

$$x \equiv 14 \quad (\text{mod } 1320).$$

We will now give some examples of Euclidean domains other than $\mathbf{Z}$ and $F[X]$. If $d \ne 0, 1 \in \mathbf{Z}$ is square-free, we let $\mathbf{Z}[\sqrt{d}] = \{a + b\sqrt{d} : a, b \in \mathbf{Z}\}$. Then $\mathbf{Z}[\sqrt{d}]$ is a subring of the field of complex numbers $\mathbf{C}$ and the quotient field of $\mathbf{Z}[\sqrt{d}]$ is the quadratic field

$$\mathbf{Q}[\sqrt{d}] = \{a + b\sqrt{d} : a, b \in \mathbf{Q}\}.$$

(5.21) Proposition. *If $d \in \{-2, -1, 2, 3\}$ then the ring $\mathbf{Z}[\sqrt{d}]$ is a Euclidean domain with $v(a + b\sqrt{d}) = |a^2 - db^2|$.*

Proof. Note that $v(a + b\sqrt{d}) = |a^2 - db^2| \ge 1$ unless $a = b = 0$. It is a straightforward calculation to check that $v(\alpha\beta) = v(\alpha)v(\beta)$ for every α, $\beta \in \mathbf{Z}[\sqrt{d}]$. Then

$$v(\alpha\beta) = v(\alpha)v(\beta) \ge v(\alpha),$$

so part (1) of Definition 5.16 is satisfied.

Now suppose that $\alpha, \beta \in \mathbf{Z}[\sqrt{d}]$ with $\beta \ne 0$. Then in the quotient field of $\mathbf{Z}[\sqrt{d}]$ we may write $\alpha/\beta = x + y\sqrt{d}$ where $x, y \in \mathbf{Q}$. Since any rational number is between two consecutive integers and within $1/2$ of the nearest integer, it follows that there are integers $r, s \in \mathbf{Z}$ such that $|x - r| \le 1/2$ and $|y - s| \le 1/2$. Let $\gamma = r + s\sqrt{d}$ and $\delta = \beta((x - r) + (y - s)\sqrt{d})$. Then

$$\alpha = \beta(x + y\sqrt{d}) = \beta\gamma + \delta.$$

Since $r, s \in \mathbf{Z}$, it follows that $\gamma \in \mathbf{Z}[\sqrt{d}]$ and $\delta = \alpha - \beta\gamma \in \mathbf{Z}[\sqrt{d}]$ also. Then

$$v(\delta) = v(\beta)v((x - r) + (y - s)\sqrt{d})$$
$$= v(\beta)|(x - r)^2 - d(y - s)^2|.$$

But

$$|(x-r)^2 - d(y-s)^2| \leq |x-r|^2 + |d||y-s|^2$$
$$\leq (1/2)^2 + 3(1/2)^2 = 1.$$

The only possibility for equality is when $|x - r| = |y - s| = 1/2$ and $d = 3$. But in this case

$$|(x-r)^2 - d(y-s)^2| = |1/4 - 3/4| = 1/2 < 1.$$

Therefore, we always have $|(x-r)^2 - d(y-s)^2| < 1$ and we conclude that $v(\delta) < v(\beta)$. Hence $\mathbf{Z}[\sqrt{d}]$ is a Euclidean domain.   □

Complementary to Proposition 5.21, we have the following result:

(5.22) Proposition. *If $d < 0$ then $\mathbf{Z}[\sqrt{d}]$ is a PID if and only if $d = -1$ or $d = -2$.*

Proof. If $d = -1$ or $d = -2$ then by Proposition 5.21, $\mathbf{Z}[\sqrt{d}]$ is a Euclidean domain and hence a PID. For the converse we need the following lemma.

(5.23) Lemma. *2 is not a prime in $\mathbf{Z}[\sqrt{d}]$.*

Proof. Either d or $d - 1$ is even so that $2 \mid d(d - 1)$. But

$$d(d-1) = d^2 - d = (d + \sqrt{d})(d - \sqrt{d}),$$

so $2 \mid (d+\sqrt{d})(d-\sqrt{d})$ but neither $(d+\sqrt{d})/2$ nor $(d-\sqrt{d})/2$ are in $\mathbf{Z}[\sqrt{d}]$. Thus 2 divides the product of two numbers, but it divides neither of the numbers individually, so 2 is not a prime element in the ring $\mathbf{Z}[\sqrt{d}]$.   □

We now return to the proof of Proposition 5.22. We will show that if $d \leq -3$ then 2 is an irreducible element of $\mathbf{Z}[\sqrt{d}]$. Since in a PID, irreducible and prime are the same concept (Corollary 5.7), it will follow that $\mathbf{Z}[\sqrt{d}]$ cannot be a PID because Lemma 5.23 shows that 2 is not a prime in $\mathbf{Z}[\sqrt{d}]$.

Suppose $2 = \alpha\beta$ for $\alpha, \beta \in \mathbf{Z}[\sqrt{d}]$ with α and β not units. Then we must have $v(\alpha) > 1$ and $v(\beta) > 1$. Therefore,

$$4 = v(2) = v(\alpha)v(\beta),$$

and since $v(\alpha)$, $v(\beta) \in \mathbf{N}$ it follows that $v(\alpha) = v(\beta) = 2$. Thus if 2 is not irreducible in $\mathbf{Z}[\sqrt{d}]$ there is a number $\alpha = a + b\sqrt{d} \in \mathbf{Z}[\sqrt{d}]$ such that

$$v(\alpha) = a^2 - db^2 = \pm 2.$$

But if $d \leq -3$ and $b \neq 0$ then

$$a^2 - db^2 = a^2 + (-d)b^2 \geq 0 + 3 \cdot 1 > \pm 2,$$

while if $b = 0$ then

$$a^2 - db^2 = a^2 \neq \pm 2$$

when $a \in \mathbf{Z}$. Hence, if $d \leq -3$ there is no number $\alpha \in \mathbf{Z}[\sqrt{d}]$ with $v(\alpha) = 2$, and we conclude that 2 is irreducible in $\mathbf{Z}[\sqrt{d}]$. Therefore, Proposition 5.22 is proved. □

(5.24) Remarks.

(1) A complex number is **algebraic** if it is a root of a polynomial with integer coefficients, and a subfield $F \subseteq \mathbf{C}$ is said to be **algebraic** if every element of F is algebraic. If F is algebraic, the **integers** of F are those elements of F that are roots of a monic polynomial with integer coefficients. In the quadratic field $F = \mathbf{Q}[\sqrt{-3}]$, every element of the ring $\mathbf{Z}[\sqrt{-3}]$ is an integer of F, but it is not true, as one might expect, that $\mathbf{Z}[\sqrt{-3}]$ is all of the integers of F. In fact, the following can be shown: Let $d \neq 0, 1$ be a square-free integer. Then the ring of integers of the field $\mathbf{Q}[\sqrt{d}]$ is

$$\begin{cases} \mathbf{Z}[\sqrt{d}] & \text{if} \quad d \equiv 2, 3 \pmod{4}, \\ \mathbf{Z}\left[\frac{1+\sqrt{d}}{2}\right] & \text{if} \quad d \equiv 1 \pmod{4}. \end{cases}$$

In particular, the ring of integers of the field $\mathbf{Q}[\sqrt{-3}]$ is the ring

$$R = \left\{ a + b\left(\frac{-1+\sqrt{-3}}{2}\right) : a, b \in \mathbf{Z} \right\}.$$

We leave it as an exercise for the reader to prove that R is in fact a Euclidean domain. (Compare with Proposition 5.22.)

(2) So far all the examples we have seen of principal ideal domains have been Euclidean domains. Let

$$R = \left\{ a + b\left(\frac{1+\sqrt{-19}}{2}\right) : a, b \in \mathbf{Z} \right\}$$

be the ring of integers of the quadratic field $\mathbf{Q}[\sqrt{-19}]$. Then it can be shown that R is a principal ideal domain but R is not a Euclidean domain. The details of the verification are tedious but not particularly difficult. The interested reader is referred to the article *A principal ideal ring that is not a Euclidean ring* by J.C. Wilson, *Math. Magazine* (1973), pp. 34–38. For more on factorization in the rings of integers of quadratic number fields, see chapter XV of *Theory of Numbers* by G. H. Hardy and E. M. Wright (Oxford University Press, 1960).

2.6 Unique Factorization Domains

We have seen in Section 2.5 that the fundamental theorem of arithmetic holds for any PID. There are, however, rings that are not PIDs for which the fundamental theorem of arithmetic holds. In this section we will give a result that allows for the construction of such rings.

(6.1) Definition. *An integral domain R is a* **unique factorization domain** *(UFD) if every nonzero element a of R can be written essentially uniquely as $a = u p_1 \cdots p_r$ where u is a unit and each p_i is an irreducible element of R. Essentially uniquely means that if $a = v q_1 \cdots q_s$ where v is a unit and each q_j is irreducible, then $r = s$ and, after reordering (if necessary), q_i is an associate of p_i. By collecting associate primes together, we may write (essentially uniquely)*

$$a = u_1 p_1^{m_1} \cdots p_t^{m_t}$$

where p_i is not an associate of p_j if $i \neq j$. This is called the **prime factorization of** *a and the primes p_i are called the* **prime factors** *or* **prime divisors** *of a.*

(6.2) Lemma. *Let R be a UFD.*

(1) *An element $a \in R$ is irreducible if and only if it is prime.*
(2) *Any nonempty set of nonzero elments of R has a greatest common divisor.*

Proof. (1) Suppose $a \in R$ is irreducible and $a \mid bc$. Thus $ad = bc$ for some $d \in R$. Writing b, c, and d as a product of units and irreducible elements gives

$$a u_1 d_1 \cdots d_r = u_2 b_1 \cdots b_s u_3 c_1 \cdots c_t$$

where each b_i, c_j, and d_k is irreducible and each u_i is a unit. By uniqueness of factorization of bc, it follows that the irreducible element a is an associate of some b_i or c_j and, hence, $a \mid b$ or $a \mid c$.

(2) We prove this in the case that the set in question is $\{a, b\}$ consisting of two elements and leave the general case for the reader. Let $p_1, \ldots, p_r$ denote all the primes that are a prime factor of either a or b. Then we may write

$$a = p_1^{m_1} \cdots p_r^{m_r}$$

and

$$b = p_1^{n_1} \cdots p_r^{n_r}$$

where $0 \leq m_i$ and $0 \leq n_i$ for each i. Let $k_i = \min\{m_i, n_i\}$ for $1 \leq i \leq r$ and let

$$d = p_1^{k_1} \cdots p_r^{k_r}.$$

We claim that d is a gcd of a and b. It is clear that $d \mid a$ and $d \mid b$, so let e be any other common divisor of a and b. Since $e \mid a$, we may write $a = ec$

for some $c \in R$. Taking prime factorizations of e, a, and c and applying the unique factorization assumption, we conclude that any prime factor of e must also be a prime factor of a and the power of the prime that divides e can be no more than the power that divides a. Thus, since e also divides b, we may write

$$e = p_1^{\ell_1} \cdots p_r^{\ell_r}$$

where $\ell_i \leq \min\{m_i, n_i\} = k_i$. Therefore, every prime factor of e is also a prime factor of d and the power of the prime factor dividing e is at most that which divides d. Thus $e \mid d$ and d is a gcd of a and b. $\square$

Our goal is to prove that if R is a UFD then the polynomial ring $R[X]$ is also a UFD. This will require some preliminaries.

(6.3) Definition. *Let R be a UFD and let $f(X) \neq 0 \in R[X]$. A gcd of the coefficients of $f(X)$ is called the* **content** *of $f(X)$ and is denoted $\mathrm{cont}(f(X))$. The polynomial $f(X)$ is said to be* **primitive** *if $\mathrm{cont}(f(X)) = 1$.*

Note that the content of $f(X)$ is only uniquely defined up to multiplication by a unit of R. If $f(X)$ is a nonzero polynomial then we can write $f(X) = cf_1(X)$ where $f_1(X)$ is primitive and c is the content of $f(X)$.

(6.4) Lemma. (Gauss's lemma) *Let R be a UFD and let $f(X)$, $g(X)$ be nonzero polynomials in $R[X]$. Then*

$$\mathrm{cont}(f(X)g(X)) = \mathrm{cont}(f(X))\,\mathrm{cont}(g(X)).$$

In particular, if $f(X)$ and $g(X)$ are primitive, then the product $f(X)g(X)$ is primitive.

Proof. Write $f(X) = \mathrm{cont}(f(X))f_1(X)$ and $g(X) = \mathrm{cont}(g(X))g_1(X)$ where $f_1(X)$ and $g_1(X)$ are primitive polynomials. Then

$$f(X)g(X) = \mathrm{cont}(f(X))\,\mathrm{cont}(g(X))f_1(X)g_1(X),$$

so it is sufficient to check that $f_1(X)g_1(X)$ is primitive. Now let

$$f_1(X) = a_0 + a_1 X + \cdots + a_m X^m$$

and

$$g_1(X) = b_0 + b_1 X + \cdots + b_n X^n,$$

and suppose that the coefficients of $f_1(X)g_1(X)$ have a common divisor d other than a unit. Let p be a prime divisor of d. Then p must divide all of the coefficients of $f_1(X)g_1(X)$, but since $f_1(X)$ and $g_1(X)$ are primitive, p does not divide all the coefficients of $f_1(X)$ nor all of the coefficients of $g_1(X)$. Let a_r be the first coefficient of $f_1(X)$ not divisible by p and let b_s be the first coefficient of $g_1(X)$ not divisible by p. Consider the coefficient of X^{r+s} in $f_1(X)g_1(X)$. This coefficient is of the form

$$a_r b_s \neq (a_{r+1}b_{s-1} + a_{r+2}b_{s-2} + \cdots) + (a_{r-1}b_{s+1} + a_{r-2}b_{s+2} + \cdots).$$

By hypothesis p divides this sum and all the terms in the first parenthesis are divisible by p (because $p \mid b_j$ for $j < s$) and all the terms in the second parenthesis are divisible by p (because $p \mid a_i$ for $i < r$). Hence $p \mid a_r b_s$ and since p is prime we must have $p \mid a_r$ or $p \mid b_s$, contrary to our choice of a_r and b_s. This contradiction shows that no prime divides all the coefficients of $f_1(X)g_1(X)$, and hence, $f_1(X)g_1(X)$ is primitive. $\qquad\square$

(6.5) Lemma. *Let R be a UFD with quotient field F. If $f(X) \neq 0 \in F[X]$, then $f(X) = \alpha f_1(X)$ where $\alpha \in F$ and $f_1(X)$ is a primitive polynomial in $R[X]$. This factorization is unique up to multiplication by a unit of R.*

Proof. By extracting a common denominator d from the nonzero coefficients of $f(X)$, we may write $f(X) = (1/d)\widetilde{f}(X)$ where $\widetilde{f}(X) \in R[X]$. Then let $\alpha = \text{cont}(\widetilde{f}(X))/d = c/d \in F$. It follows that $f(X) = \alpha f_1(X)$ where $f_1(X)$ is a primitive polynomial. Now consider uniqueness. Suppose that we can also write $f(X) = \beta f_2(X)$ where $f_2(X)$ is a primitive polynomial in $R[X]$ and $\beta = a/b \in F$. Then we conclude that

$$(6.1) \qquad\qquad a d f_2(X) = c b f_1(X).$$

The content of the left side is ad and the content of the right side is cb, so $ad = ucb$ where $u \in R$ is a unit. Substituting this in Equation (6.1) and dividing by cb gives

$$u f_2(X) = f_1(X).$$

Thus the two polynomials differ by multiplication by the unit $u \in R$ and the coefficients satisfy the same relationship $\beta = a/b = u(c/d) = u\alpha$. $\qquad\square$

(6.6) Lemma. *Suppose R is a UFD with quotient field F. If $f(X) \in R[X]$ has positive degree and is irreducible in $R[X]$, then $f(X)$ is irreducible in $F[X]$.*

Proof. If $f(X) \in R[X]$ has positive degree and is irreducible in $R[X]$ then $f(X)$ is primitive since $\text{cont}(f(X)) \mid f(X)$ in $R[X]$. Suppose that $f(X)$ is reducible in $F[X]$. Thus $f(X) = g_1(X)g_2(X)$ where $g_i(X) \in F[X]$ and $\deg g_i(X) > 0$ for $i = 1$ and 2. Then $g_i(X) = \alpha_i f_i(X)$ where $\alpha_i \in F$ and $f_i(X) \in R[X]$ is primitive. Thus,

$$f(X) = \alpha_1 \alpha_2 f_1(X) f_2(X),$$

and the product $f_1(X)f_2(X)$ is primitive by Gauss's lemma. Thus, by Lemma 6.5, $f(X)$ and $f_1(X)f_2(X)$ differ by multiplication by a unit of R, which contradicts the irreducibility of $f(X)$ in $R[X]$. Thus, we conclude that $f(X)$ is irreducible in $F[X]$. $\qquad\square$

(6.7) Corollary. *Let R be a UFD with quotient field F. Then the irreducible elements of $R[X]$ are the irreducible elements of R and the primitive polynomials $f(X) \in R[X]$ which are irreducible in $F[X]$.*

Proof.   □

(6.8) Theorem. *If R is a UFD, then $R[X]$ is also a UFD.*

Proof. Let F be the quotient field of R and let $f(X) \neq 0 \in R[X]$. Since $F[X]$ is a UFD, we can write

$$f(X) = p_1(X) \cdots p_r(X)$$

where $p_i(X) \in F[X]$ is an irreducible polynomial for $1 \leq i \leq r$. By Lemma 6.5, $p_i(X) = \alpha_i q_i(X)$ where $\alpha_i \in F$ and $q_i(X) \in R[X]$ is a primitive polynomial. Thus, we have

$$f(X) = c q_1(X) \cdots q_r(X)$$

where $c = \alpha_1 \cdots \alpha_r \in F$. Write $c = a/b$ where $a, b \in R$. Then taking contents we get

$$\text{cont}(bf(X)) = \text{cont}(a q_1(X) \cdots q_r(X)) = a$$

by Gauss's lemma. Thus, $b\,\text{cont}(f(X)) = a$, so $b \mid a$ and $\text{cont}(f(X)) = c \in R$. Each $q_i(X)$ is irreducible in $F[X]$, and hence, it is irreducible in $R[X]$. Since R is a UFD, write $c = u d_1 \cdots d_s$ where each d_i is prime in R and $u \in R$ is a unit. Thus we have a factorization

$$f(X) = u d_1 \cdots d_s q_1(X) \cdots q_r(X)$$

of $f(X)$ into a product of irreducible elements of $R[X]$.

It remains to check uniqueness. Thus suppose we also have a factorization

$$f(X) = v b_1 \cdots b_t q_1'(X) \cdots q_k'(X)$$

where each $q_i'(X)$ is a primitive polynomial in $R[X]$ and each b_i is an irreducible element of R. By Corollary 6.7, this is what any factorization into irreducible elements of $R[X]$ must look like. Since this also gives a factorization in $F[X]$ and factorization there is unique, it follows that $r = k$ and $q_i(X)$ is an associate of $q_i'(X)$ in $F[X]$ (after reordering if necessary). But if primitive polynomials are associates in $F[X]$, then they are associates in $R[X]$ (Lemma 6.5). Furthermore,

$$\text{cont}(f(X)) = v b_1 \cdots b_t = u d_1 \cdots d_s,$$

so $s = t$ and b_i is an associate of d_i (after reordering if necessary) since R is a UFD. This completes the proof.   □

(6.9) Corollary. *Let R be a UFD. Then $R[X_1, \ldots, X_n]$ is also a UFD. In particular, $F[X_1, \ldots, X_n]$ is a UFD for any field F.*

Proof. Exercise.   □

(6.10) Example. We have seen some examples in Section 2.5 of rings that are not UFDs, namely, $F[X^2, X^3]$ and some of the quadratic rings (see Proposition 5.22 and Lemma 5.23). We wish to present one more example of a Noetherian function ring that is not a UFD. Let

$$S^1 = \{(x, y) \in \mathbf{R}^2 : x^2 + y^2 = 1\}$$

be the unit circle in $\mathbf{R}^2$ and let $I \subseteq \mathbf{R}[X, Y]$ be the set of all polynomials such that $f(x, y) = 0$ for all $(x, y) \in S^1$. Then I is a prime ideal of $\mathbf{R}[X, Y]$ and $\mathbf{R}[X, Y]/I$ can be viewed as a ring of functions on S^1 by means of

$$f(X, Y) + I \mapsto \overline{f}$$

where $\overline{f}(x, y) = f(x, y)$. We leave it for the reader to check that this is well defined. Let T be the set of all $f(X, Y) + I \in \mathbf{R}[X, Y]/I$ such that $\overline{f}(x, y) \neq 0$ for all $(x, y) \in S^1$. Then let the ring R be defined by localizing at the multiplicatively closed set T, i.e.,

$$R = (\mathbf{R}[X, Y]/I)_T.$$

Thus, R is a ring of functions on the unit circle, and a function in R is a unit if and only if the function never vanishes on S^1.

Claim. *R is not a UFD.*

Proof. Let $g(x, y) = x^2 + (y - 1)^2$ and $h(x, y) = x^2 + (y + 1)^2 \in R$. Then

$$
\begin{aligned}
gh(x, y) &= (x^2 + (y - 1)^2)(x^2 + (y + 1)^2) \\
&= x^4 + x^2((y - 1)^2 + (y + 1)^2) + (y^2 - 1)^2 \\
&= x^4 + x^2((y - 1)^2 + (y + 1)^2) + x^4
\end{aligned}
$$

so that x divides gh in R, but clearly x does not divide either g or h (since neither g nor h is zero at both $(0, 1)$ and $(0, -1) \in S^1$, but x is). Therefore the ideal $\langle x \rangle \subseteq R$ is not prime. The proof is completed by showing that x is an irreducible element of the ring R. To see this suppose that $x = f_1 f_2$ where neither is a unit. Then we must have $V(f_1) = \{(0, 1)\}$ and $V(f_2) = \{(0, -1)\}$ (or vice versa), where $V(g)$ means the set of zeros of the function g. Since f_1 and f_2 are continuous functions, it follows that f_1 does not change sign on $S^1 \setminus \{(0, 1)\}$ and f_2 does not change sign on $S^1 \setminus \{(0, -1)\}$. Therefore, $x = f_1 f_2$ will not change sign on the set $S^1 \setminus \{(0, 1), (0, -1)\}$, and this contradiction shows that x must be an irreducible element of R.   □

In general, the problem of explicitly determining if a polynomial in $R[X]$ is irreducible is difficult. For polynomials in $\mathbf{Z}[X]$ (and hence in $\mathbf{Q}[X]$) there is a procedure due to Kronecker that in principle can determine the factors of an integral polynomial in a finite number of steps. For this method see, for example, *Modern Algebra* by B.L. van der Waerden, Vol. I, p. 77. We shall content ourselves with the following simple criterion for irreducibility, which is sometimes of use. An example of its use will be in Corollary 6.12.

(6.11) Theorem. (Eisenstein's criterion) *Let R be a UFD with quotient field F. Let $f(X) = a_0 + a_1 X + \cdots + a_n X^n$ $(a_n \neq 0)$ be in $R[X]$ and suppose that $p \in R$ is a prime such that*

$$p \text{ does not divide } a_n,$$
$$p \mid a_i \text{ for } 0 \leq i \leq n - 1,$$
$$p^2 \text{ does not divide } a_0.$$

Then $f(X)$ is irreducible in $F[X]$.

Proof. If we write $f(X) = \text{cont}(f(X))f_1(X)$ then $f_1(X)$ is a primitive polynomial that also satisfies the hypotheses of the theorem. Thus without loss of generality we may suppose that $f(X)$ is a primitive polynomial. If there exists a factorization of $f(X)$ into factors of degree ≥ 1 in $F[X]$ then by Lemma 6.6 there is also a factorization in $R[X]$. Thus, suppose that we can write $f(X) = g(X)h(X)$ where $g(X), h(X) \in R[X]$. From Gauss's lemma we must have that $g(X)$ and $h(X)$ are primitive polynomials. Suppose

$$g(X) = b_0 + b_1 X + \cdots + b_\ell X^\ell$$
$$h(X) = c_0 + c_1 X + \cdots + c_m X^m$$

with $\ell, m \geq 1, b_\ell c_m \neq 0$, and $\ell + m = n$. Since $p \mid a_0 = b_0 c_0$ but p^2 does not divide a_0, it follows that $p \mid b_0$ or $p \mid c_0$ but not both. To be specific, suppose that $p \mid b_0$ but that p does not divide c_0. Not all the coefficients of $g(X)$ are divisible by p since $g(X)$ is primitive. Let b_i be the first coefficient of $g(X)$ that is not divisible by p so that $0 < i \leq \ell < n$. Then we have

$$a_i = b_i c_0 + b_{i-1} c_1 + \cdots + b_0 c_i.$$

But $p \mid a_i$ and $p \mid b_j$ for $j < i$, so $p \mid b_i c_0$. But p does not divide b_i and p does not divide c_0, so we have a contradiction. Therefore, we may not write $f(X) = g(X)h(X)$ with $\deg g(X) \geq 1, \deg h(X) \geq 1$, i.e., $f(X)$ is irreducible in $F[X]$. $\square$

The following is a useful consequence of Eisenstein's criterion:

(6.12) Corollary. *Let p be a prime number and let $f_p(X) \in \mathbf{Q}[X]$ be the polynomial*

$$f_p(X) = X^{p-1} + X^{p-2} + \cdots + X + 1 = \frac{X^p - 1}{X - 1}.$$

Then $f_p(X)$ is irreducible.

Proof. Since the map $g(X) \mapsto g(X+1)$ is a ring isomorphism of $\mathbf{Q}[X]$, it is sufficient to verify that $f_p(X+1)$ is irreducible. But p (by Exercise 1) divides every binomial coefficient $\binom{p}{k}$ for $1 \le k < p$, and hence,

$$f_p(X+1) = \frac{(X+1)^p - 1}{(X+1) - 1} = X^{p-1} + pX^{p-2} + \cdots + p.$$

Thus, $f_p(X+1)$ satisfies Eisenstein's criterion, and the proof is complete. □

2.7 Exercises

1. Prove the binomial theorem (Proposition 1.12). Give a counterexample if a and b do not commute. If p is a prime number, prove that the binomial coefficient $\binom{p}{k}$ is divisible by p for $0 < k < p$. Give an example to show that this result need not be true if p is not prime.

2. Let R be a ring with identity and let $a \in R$. The element a is said to be *nilpotent* if $a^n = 0$ for some $n \in \mathbf{N}$. It is said to be *idempotent* if $a^2 = a$. Prove the following assertions.
   (a) If R has no zero divisors, then the only nilpotent element of R is 0 and the only idempotent elements of R are 0 and 1.
   (b) No unit of R is nilpotent. The only idempotent unit of R is 1.
   (c) If a is nilpotent, then $1 - a$ is a unit. (Hint: Geometric series.) If a is idempotent, then $1 - a$ is idempotent.
   (d) If R is commutative and $N = \{a \in R : a \text{ is nilpotent}\}$, show that N is an ideal of R.
   (e) Provide a counterexample to part (d) if R is not commutative.
   (Note that in parts (b)–(e), the ring R is allowed to have zero divisors.)

3. For the ring $\mathcal{P}(X)$ of Example 1.10 (7), show that every $A \in \mathcal{P}(X)$ satisfies the equation $A^2 = A$. If $\mathcal{P}(X)$ is an integral domain, show that $|X| = 1$.

4. Continuing with the ring $\mathcal{P}(X)$, let $a \in X$ and define $I_a = \{A \in \mathcal{P}(X) : a \notin A\}$. Prove that I_a is a maximal ideal of $\mathcal{P}(X)$. What is $\mathcal{P}(X)/I_a$? For a finite set X determine all of the ideals of $\mathcal{P}(X)$ and show that every maximal ideal is an ideal I_a for some $a \in X$.

5. Prove Lemma 2.12 (1) and (3).

6. Let R be a ring with identity. Show that R is a division ring if and only if it has no left or right ideals other than $\{0\}$ and R.

7. (a) Solve the equation $6x = 7$ in the ring $\mathbf{Z}_{19}$, if possible.
   (b) Solve the equation $6x = 7$ in the ring $\mathbf{Z}_{20}$, if possible.

8. If R and S are rings with identities, prove that $(R \times S)^* = R^* \times S^*$. (Recall that R^* denotes the group of units of the ring R.)

9. Compute all the ideals, prime ideals, and maximal ideals of the ring $\mathbf{Z}_{60}$. What are the nilpotent elements of $\mathbf{Z}_{60}$?

10. Let R be a ring and let R^{op} ("op" for opposite) be the abelian group R, together with a new multiplication $a \cdot b$ defined by $a \cdot b = ba$, where ba denotes the given multiplication on R. Verify that R^{op} is a ring and that the identity function $1_R : R \to R^{\mathrm{op}}$ is a ring homomorphism if and only if R is a commutative ring.

11. (a) Let A be an abelian group. Show that $\mathrm{End}(A)$ is a ring. ($\mathrm{End}(A)$ is defined in Example 1.10 (11).)
    (b) Let F be a field and V an F-vector space. Show that $\mathrm{End}_F(V)$ is a ring. Here, $\mathrm{End}_F(V)$ denotes the set of all F-linear endomorphisms of V, i.e.,

$$\mathrm{End}_F(V) = \{h \in \mathrm{End}(V) : h(\alpha v) = \alpha h(v) \text{ for all } v \in V, \alpha \in F\}.$$

    In this definition, $\mathrm{End}(V)$ means the *abelian group* endomorphisms of V, and the ring structure is the same as that of $\mathrm{End}(V)$ in part (a).

12. (a) Let R be a ring without zero divisors. Show that if $ab = 1$ then $ba = 1$ as well. Thus, a and b are units of R.
    (b) Show that if a, b, $c \in R$ with $ab = 1$ and $ca = 1$, then $b = c$, and thus a (and b) are units. Conclude that if $ab = 1$ but $ba \neq 1$, then neither a nor b are units.
    (c) Let F be a field and $F[X]$ the polynomial ring with coefficients from F. $F[X]$ is an F-vector space in a natural way, so by Exercise 11, $R = \mathrm{End}_F(F[X])$ is a ring. Give an example of a, $b \in R$ with $ab = 1$ but $ba \neq 1$.

13. (a) Let x and y be arbitrary positive real numbers. Show that the quaternion algebras $Q(-x, -y; \mathbf{R})$ and $Q(-1, -1; \mathbf{R})$ are isomorphic.
    (b) Show that the quaternion algebras $Q(-1, -3; \mathbf{Q})$, $Q(-1, -7; \mathbf{Q})$, and $Q(-1, -11; \mathbf{Q})$ are all distinct.
    (c) Analogously to Example 1.10 (10), we may define indefinite quaternion algebras by allowing x or y to be negative. Show that, for any nonzero real number x and any subfield F of $\mathbf{R}$, $Q(1, x; F)$ and $Q(1, -x; F)$ are isomorphic.
    (d) Show that for any nonzero real number x, $Q(1, x; \mathbf{R})$ and $Q(1, 1; \mathbf{R})$ are isomorphic.
    (e) Show that for any subfield F of $\mathbf{R}$, $Q(1, 1; F)$ is isomorphic to $M_2(F)$, the ring of 2×2 matrices with coefficients in F. (Thus, $Q(1, 1; F)$ is not a division ring.)

14. Verify that $\mathbf{Z}[i]/\langle 3 + i \rangle \cong \mathbf{Z}_{10}$.

15. Let F be a field and let $R \subseteq F[X] \times F[Y]$ be the subring consisting of all pairs $(f(X), g(Y))$ such that $f(0) = g(0)$. Verify that $F[X, Y]/\langle XY \rangle \cong R$.

16. Let R be a ring with identity and let I be an ideal of R. Prove that $M_n(R/I) \cong M_n(R)/M_n(I)$.

17. Let F be a field and let $R = \left\{ \left[\begin{smallmatrix} a & b \\ 0 & 0 \end{smallmatrix} \right] \in M_2(F) \right\}$. Verify that R is a ring. Does R have an identity? Prove that the set $I = \left\{ \left[\begin{smallmatrix} 0 & b \\ 0 & 0 \end{smallmatrix} \right] \in R \right\}$ is a maximal ideal of R.

18. (a) Given the complex number $z = 1 + i$, let $\phi : \mathbf{R}[X] \to \mathbf{C}$ be the substitution homomorphism determined by z. Compute $\mathrm{Ker}(\phi)$.
    (b) Give an explicit isomorphism between the complex numbers $\mathbf{C}$ and the quotient ring $\mathbf{R}[X]/\langle X^2 - 2X + 2 \rangle$.

19. Let $R = \mathcal{C}([0, 1])$ be the ring of continuous real-valued functions on the interval $[0, 1]$. Let $T \subseteq [0, 1]$, and let

$$I(T) = \{f \in R : f(x) = 0 \text{ for all } x \in T\}.$$

    (a) Prove that $I(T)$ is an ideal of R.
    (b) If $x \in [0, 1]$ and $M_x = I(\{x\})$, prove that M_x is a maximal ideal of R and $R/M_x \cong \mathbf{R}$.

(c) If $S \subseteq R$ let

$$V(S) = \{x \in I : f(s) = 0 \text{ for all } f \in S\}.$$

Prove that S is a closed subset of $[0,1]$. (You may quote appropriate theorems on continuous functions.)

(d) If $I \subsetneq R$ is any ideal, then prove that $V(I) \neq \emptyset$. (Hint: Suppose that $V(I) = \emptyset$ and construct a function $f \in I$ such that $f(x) \neq 0$ for all $x \in [0,1]$. Conclude that $I = R$. At some point in your argument you will need to use the compactness of $[0,1]$.)

(e) Prove that any maximal ideal M of R is M_x for some $x \in [0,1]$.

20. Let $R \subseteq S$ be commutative rings with identity and let $d \in R$ be an element such that the equation $a^2 = d$ is solvable in S but not in R, and let $\sqrt{d}$ denote a solution to the equation in S. Define a set $R[\sqrt{d}]$ by

$$R[\sqrt{d}] = \{a + b\sqrt{d} : a, b \in R\} \subseteq S.$$

(a) Prove that $R[\sqrt{d}]$ is a commutative ring with identity.

(b) Prove that $\mathbf{Z}[\sqrt{d}]$ is an integral domain.

(c) If F is a field, prove that $F[\sqrt{d}]$ is a field.

21. (a) If $R = \mathbf{Z}$ or $R = \mathbf{Q}$ and d is not a square in R, show that $R[\sqrt{d}] \cong R[X]/\langle X^2 - d \rangle$ where $\langle X^2 - d \rangle$ is the principal ideal of $R[X]$ generated by $X^2 - d$.

(b) If $R = \mathbf{Z}$ or $R = \mathbf{Q}$ and d_1, d_2, and d_1/d_2 are not squares in $R \setminus \{0\}$, show that $R[\sqrt{d_1}]$ and $R[\sqrt{d_2}]$ are not isomorphic.
(The most desirable proof of these assertions is one which works for both $R = \mathbf{Z}$ and $R = \mathbf{Q}$, but separate proofs for the two cases are acceptable.)

(c) Let $R_1 = \mathbf{Z}_p[X]/\langle X^2 - 2 \rangle$ and $R_2 = \mathbf{Z}_p[X]/\langle X^2 - 3 \rangle$. Determine if $R_1 \cong R_2$ in case $p = 2$, $p = 5$, or $p = 11$.

22. Recall that R^* denotes the group of units of the ring R.

(a) Show that $(\mathbf{Z}[\sqrt{-1}])^* = \{\pm 1, \pm\sqrt{-1}\}$.

(b) If $d < -1$ show that $(\mathbf{Z}[\sqrt{d}])^* = \{\pm 1\}$.

(c) Show that

$$\mathbf{Z}\left[\frac{(1 + \sqrt{-3})}{2}\right]^* = \left\{\pm 1, \pm\frac{1 + \sqrt{-3}}{2}, \pm\frac{-1 + \sqrt{-3}}{2}\right\}.$$

(d) Let $d > 0 \in \mathbf{Z}$ not be a perfect square. Show that if $\mathbf{Z}[\sqrt{d}]$ has one unit other than ± 1, it has infinitely many.

(e) It is known that the hypothesis in part (d) is always satisfied. Find a unit in $\mathbf{Z}[\sqrt{d}]$ other than ± 1 for $2 \leq d \leq 15$, $d \neq 4, 9$.

23. Let $F \supseteq \mathbf{Q}$ be a field. An element $a \in F$ is said to be an algebraic integer if for some monic polynomial $p(X) \in \mathbf{Z}[X]$, we have $p(a) = 0$. Let $d \in \mathbf{Z}$ be a nonsquare.

(a) Show that if $a \in F$ is an algebraic integer, then a is a root of an irreducible monic polynomial $p(X) \in \mathbf{Z}[X]$. (Hint: Gauss's lemma.)

(b) Verify that $a + b\sqrt{d} \in \mathbf{Q}[\sqrt{d}]$ is a root of the quadratic polynomial $p(X) = X^2 - 2aX + (a^2 - b^2 d) \in \mathbf{Q}[X]$.

(c) Determine the set of algebraic integers in the fields $\mathbf{Q}[\sqrt{3}]$ and $\mathbf{Q}[\sqrt{5}]$. (See Remark 5.24 (1).)

24. If R is a ring with identity, then $\text{Aut}(R)$, called the *automorphism group of R*, denotes the set of all ring isomorphisms $\phi : R \to R$.

(a) Compute $\text{Aut}(\mathbf{Z})$ and $\text{Aut}(\mathbf{Q})$.

(b) Compute $\text{Aut}(\mathbf{Z}[\sqrt{d}])$ and $\text{Aut}(\mathbf{Q}[\sqrt{d}])$ if d is not a square in $\mathbf{Z}$ or $\mathbf{Q}$.

(c) If $a \neq 0 \in \mathbf{Q}, b \in \mathbf{Q}$ let $\phi_{a,b} : \mathbf{Q}[X] \to \mathbf{Q}[X]$ be the substitution homomorphism determined by $X \mapsto aX + b$. Prove that $\phi_{a,b} \in \text{Aut}(\mathbf{Q}[X])$. What is $\phi_{a,b}^{-1}$?

(d) If $\phi \in \text{Aut}(\mathbf{Q}[X])$ prove that there are $a \neq 0, b \in \mathbf{Q}$ such that $\phi = \phi_{a,b}$.

25. Let $\omega = \exp(2\pi i/n) \in \mathbf{C}$ and let $R = \mathbf{Q}[\omega]$. Show that $\text{Aut}(R) \cong \mathbf{Z}_n^*$.

26. Let R be a commutative ring with identity. Prove that $M_n(R[X]) \cong M_n(R)[X]$.

27. Let R be a commutative ring with identity and let $f = a_0 + a_1 X + \cdots + a_n X^n \in R[X]$ where $a_n \neq 0$. We know that if R is an integral domain then the units of $R[X]$ are the units of R (Corollary 4.3). This exercise will investigate when f is a unit, a zero divisor, or is nilpotent when R is *not* assumed to be an integral domain.

(a) Prove that f is a unit in $R[X]$ if and only if a_0 is a unit in R and $a_1, \ldots, a_n$ are nilpotent. (Hint: If $b_0 + b_1 X + \cdots + b_m X^m$ is the inverse of f, prove by induction on r that $a_n^{r+1} b_{m-r} = 0$. Conclude that a_n is nilpotent and apply Exercise 2 (c).)

(b) Prove that f is nilpotent if and only if $a_0, a_1, \ldots, a_n$ are nilpotent.

(c) Prove that f is a zero divisor in $R[X]$ if and only if there is a nonzero $a \in R$ such that $af = 0$. (Hint: Choose a nonzero polynomial $g = b_0 + b_1 X + \cdots + b_m X^m$ of least degree m such that $fg = 0$. If $m = 0$ we are done. Otherwise, $a_n b_m = 0$ and hence $a_n g = 0$ (because $(a_n g)f = a_n(gf) = 0$ and $\deg(a_n g) < \deg g$). Show by induction that $a_{n-r} g = 0$ for $(0 \leq r \leq n)$. Conclude that $b_m f = 0$.)

28. Factor the polynomial $X^2 + 3$ into irreducible polynomials in each of the following rings.

(a) $\mathbf{Z}_5[X]$.

(b) $\mathbf{Z}_7[X]$.

29. Let $F = \mathbf{Z}_5$ and consider the following factorization in $F[X]$:

$$(*) \qquad \begin{aligned} 3X^3 + 4X^2 + 3 &= (X+2)^2(3X+2) \\ &= (X+2)(X+4)(3X+1). \end{aligned}$$

Explain why $(*)$ does not contradict the fact that $F[X]$ is a UFD.

30. For what fields $\mathbf{Z}_p$ is $X^3 + 2X^2 + 2X + 4$ divisible by $X^2 + X + 1$?

31. In what fields $\mathbf{Z}_p$ is $X^2 + 1$ a factor of $X^3 + X^2 + 22X + 15$?

32. Find the gcd of each pair of elements in the given Euclidean domain and express the gcd in the form $ra + sb$.

(a) 189 and 301 in $\mathbf{Z}$.

(b) 1261 and 1649 in $\mathbf{Z}$.

(c) $X^4 - X^3 + 4X^2 - X + 3$ and $X^3 - 2X^2 + X - 2$ in $\mathbf{Q}[X]$.

(d) $X^4 + 4$ and $2X^3 + X^2 - 2X - 6$ in $\mathbf{Z}_3[X]$.

(e) $2 + 11i$ and $1 + 3i$ in $\mathbf{Z}[i]$.

(f) $-4 + 7i$ and $1 + 7i$ in $\mathbf{Z}[i]$.

33. Express $X^4 - X^2 - 2$ as a product of irreducible polynomials in each of the fields $\mathbf{Q}, \mathbf{R}, \mathbf{C}$, and $\mathbf{Z}_5$.

34. Let F be a field and suppose that the polynomial $X^m - 1$ has m distinct roots in F. If $k \mid m$, prove that $X^k - 1$ has k distinct roots in F.

35. If R is a commutative ring, let $F(R, R)$ be the set of all functions $f : R \to R$ and make $F(R, R)$ into a ring by means of addition and multiplication of functions, i.e., $(fg)(r) = f(r)g(r)$ and $(f + g)(r) = f(r) + g(r)$. Define a function $\Phi : R[X] \to F(R, R)$ by

$$\Phi(f(X))(r) = \phi_r(f(X)) = f(r)$$

for all $r \in R$.
(a) Show that Φ is injective if R is an infinite integral domain.
(b) Show that Φ is not injective if R is a finite field.

36. Let $f(X) \in \mathbf{Z}_p[X]$. Show that $f(X^p) = (f(X))^p$ and that the map $\Phi(f(X)) = (f(X))^p$ is a ring homomorphism.

37 Let F be a field and consider the substitution homomorphism

$$\phi : F[X, Y] \to F[T]$$

such that $\phi(X) = T^2$ and $\phi(Y) = T^3$. Show that $\mathrm{Ker}(\phi)$ is the principal ideal generated by $Y^2 - X^3$. What is $\mathrm{Im}(\phi)$?

38. Prove Proposition 4.10 (Lagrange interpolation) as a corollary of the Chinese remainder theorem.

39. Let F be a field and let $a_0, a_1, \ldots, a_n$ be $n + 1$ distinct elements of F. If $f : F \to F$ is a function, define the successive divided differences of f, denoted $f[a_0, \ldots, a_i]$, by means of the inductive formula:

$$f[a_0] = f(a_0)$$

$$f[a_0, a_1] = \frac{f[a_1] - f[a_0]}{a_1 - a_0}$$

$$\vdots$$

$$f[a_0, \ldots, a_n] = \frac{f[a_0, \ldots, a_{n-2}, a_n] - f[a_0, \ldots, a_{n-2}, a_{n-1}]}{a_n - a_{n-1}}.$$

Prove that the coefficient of $\widetilde{P}_i(X)$ in Equation (4.3) (Newton's interpolation formula) is the divided difference $f[a_0, \ldots, a_i]$.

40. (a) If $f(X) \in \mathbf{R}[X]$ and $z \in \mathbf{C}$ satisfies $f(z) = 0$, show that $f(\bar{z}) = 0$ (where $\bar{z}$ is the complex conjugate of z).
    (b) Let $r_0, r_1, \ldots, r_k$ be distinct real numbers, let $z_1, \ldots, z_m$ be distinct complex numbers with $z_j \neq \bar{z}_\ell$ for $1 \leq j, \ell \leq m$, and let $s_0, \ldots, s_k \in \mathbf{R}$, $w_1, \ldots, w_m \in \mathbf{C}$. Prove that there is a polynomial $f(X) \in \mathbf{R}[X]$ such that $f(r_i) = s_i$ for $0 \leq i \leq k$, while $f(z_j) = w_j$ for $1 \leq j \leq m$. What degree can we require for $f(X)$ in order to have a uniqueness statement as in Proposition 4.10?

41. Let F be a field and let $f, g \in F[X]$ with $\deg g \geq 1$. Show that there are unique $f_0, f_1, \ldots, f_d \in F[X]$ such that $\deg f_i < \deg g$ and

$$f = f_0 + f_1 g + \cdots + f_d g^d.$$

42. Let K and L be fields with $K \subseteq L$. Suppose that $f(X), g(X) \in K[X]$.
    (a) If $f(X)$ divides $g(X)$ in $L[X]$, prove that $f(X)$ divides $g(X)$ in $K[X]$.
    (b) Prove that the greatest common divisor of $f(X)$ and $g(X)$ in $K[X]$ is the same as the greatest common divisor of $f(X)$ and $g(X)$ in $L[X]$. (We will always choose the *monic* generator of the ideal $\langle f(X), g(X) \rangle$ as the greatest common divisor in a polynomial ring over a field.)

43. (a) Suppose that R is a Noetherian ring and $I \subseteq R$ is an ideal. Show that R/I is Noetherian.
    (b) If R is Noetherian and S is a subring of R, is S Noetherian?
    (c) Suppose that R is a commutative Noetherian ring and S is a nonempty multiplicatively closed subset of R containing no zero divisors. Prove that R_S (the localization of R away from S) is also Noetherian.

44. If R is a ring (not necessarily commutative) and $f(X) = a_0 + a_1 X + \cdots + a_n X^n \in R[X]$, then we say that $f(X)$ is *regular* of degree n if a_n is a unit of R. Note, in particular, that monic polynomials are regular and if R is a field then all nonzero polynomials in $R[X]$ are regular. Prove the following version of the division algorithm:

    Let $f(X) \in R[X]$ and let $g(X) \in R[X]$ be a regular polynomial of degree n. Then there are unique polynomials $q_1(X)$, $r_1(X)$, $q_2(X)$, and $r_2(X) \in R[X]$ such that $\deg r_1(X) < n$, $\deg r_2(X) < n$,

$$f(X) = q_1(X)g(X) + r_1(X)$$

    and

$$f(X) = g(X)q_2(X) + r_2(X).$$

    The two equations represent the left and right divisions of $f(X)$ by $g(X)$. In the special case that $g(X) = X - a$ for $a \in R$, prove the following version of these equations (noncommutative remainder theorem):

    Let $f(X) = a_1 + a_1 X + \cdots + a_n X^n \in R[X]$ and let $a \in R$. Then there are unique $q_{\mathcal{L}}(X)$ and $q_{\mathcal{R}}(X) \in R[X]$ such that

$$f(X) = q_{\mathcal{R}}(X)(X - a) + f_{\mathcal{R}}(a)$$

    and

$$f(X) = (X - a)q_{\mathcal{L}}(X) + f_{\mathcal{L}}(a)$$

    where

$$f_{\mathcal{R}}(a) = \sum_{k=0}^{n} a_k a^k \quad and \quad f_{\mathcal{L}}(a) = \sum_{k=0}^{n} a^k a_k$$

    are, respectively, the right and left evaluations of $f(X)$ at $a \in R$. (Hint: Use the formula

$$X^k - a^k = (X^{k-1} + X^{k-2}a + \cdots + Xa^{k-2} + a^{k-1})(X - a)$$
$$= (X - a)(X^{k-1} + aX^{k-2} + \cdots + a^{k-2}X + a^{k-1}).$$

    Then multiply on either the left or the right by a_k and sum over k to get the division formulas and the remainders.)

45. Let R be a UFD and let a and b be nonzero elements of R. Show that $ab = [a, b](a, b)$ where $[a, b] = \text{lcm}\{a, b\}$ and $(a, b) = \gcd\{a, b\}$.

46. Let R be a UFD. Show that d is a gcd of a and b $(a, b \in R \setminus \{0\})$ if and only if d divides both a and b and there is no prime p dividing both a/d and b/d. (In particular, a and b are relatively prime if and only if there is no prime p dividing both a and b.)

47. Let R be a UFD and let $\{r_i\}_{i=1}^{n}$ be a finite set of pairwise relatively prime nonzero elements of R (i.e., r_i and r_j are relatively prime whenever $i \neq j$). Let $a = \prod_{i=1}^{n} r_i$ and let $a_i = a/r_i$. Show that the set $\{a_i\}_{i=1}^{n}$ is relatively prime.

48. Let R be a UFD and let F be the quotient field of R. Show that $d \in R$ is a square in R if and only if it is a square in F (i.e., if the equation $a^2 = d$ has a solution with $a \in F$ then, in fact, $a \in R$). Give a counterexample if R is not a UFD.

49. Let x, y, z be integers with $\gcd\{x, y, z\} = 1$. Show that there is an integer a such that $\gcd\{x + ay, z\} = 1$. (Hint: The Chinese remainder theorem may be helpful.)

50. According to the Chinese remainder theorem there is an isomorphism of rings $\phi : \mathbf{Z}_{60} \to \mathbf{Z}_3 \times \mathbf{Z}_4 \times \mathbf{Z}_5$. Compute $\phi(26)$, $\phi(35)$, $\phi^{-1}(2, 3, 4)$, and $\phi^{-1}(1, 2, 2)$.

51. Solve the system of simultaneous congruences:

$$\begin{aligned} x &\equiv -3 \pmod{13} \\ x &\equiv 16 \pmod{18} \\ x &\equiv -2 \pmod{25} \\ x &\equiv 0 \pmod{29}. \end{aligned}$$

52. Solve the system of simultaneous congruences:

$$\begin{aligned} x &\equiv 6 \pmod{21} \\ x &\equiv 9 \pmod{33} \\ x &\equiv 2 \pmod{37}. \end{aligned}$$

53. For what values of $a \pmod{77}$ does the following system of simultaneous congruences have a solution?

$$\begin{aligned} x &\equiv 6 \pmod{21} \\ x &\equiv 9 \pmod{33} \\ x &\equiv a \pmod{77}. \end{aligned}$$

54. (a) Solve the following system of simultaneous congruences in $\mathbf{Q}[X]$:

$$\begin{aligned} f(X) &\equiv -3 \pmod{X+1} \\ f(X) &\equiv 12X \pmod{X^2 - 2} \\ f(X) &\equiv -4X \pmod{X^3}. \end{aligned}$$

    (b) Solve this system in $\mathbf{Z}_5[X]$.
    (c) Solve this system in $\mathbf{Z}_3[X]$.
    (d) Solve this system in $\mathbf{Z}_2[X]$.

55. Suppose that $m_1, m_2 \in \mathbf{Z}$ are not relatively prime. Then prove that there are integers a_1, a_2 for which there is no solution to the system of congruences:

$$\begin{aligned} x &\equiv a_1 \pmod{m_1} \\ x &\equiv a_2 \pmod{m_2}. \end{aligned}$$

56. Let R be a UFD. Prove that

$$f(X, Y) = X^4 + 2Y^2 X^3 + 3Y^3 X^2 + 4YX + 5Y + 6Y^2$$

    is irreducible in the polynomial ring $R[X, Y]$.

57. Prove that if R is a UFD and if $f(X)$ is a monic polynomial with a root in the quotient field of R, then that root is in R. (This result is usually called the *rational root theorem*.)

58. Suppose that R is a UFD and $S \subseteq R \setminus \{0\}$ is a multiplicatively closed subset. Prove that R_S is a UFD.

59. Let F be a field and let $F[[X]]$ be the ring of formal power series with coefficients in F. If $f = \sum_{n=0}^{\infty} a_n X^n \neq 0 \in F$, let $o(f) = \min\{n : a_n \neq 0\}$ and define $o(0) = \infty$. $o(f)$ is usually called the *order* of the power series f. Prove the following facts:
    (a) $o(fg) = o(f) + o(g)$.

(b) $o(f + g) \geq \min\{o(f), o(g)\}$.

(c) $f \mid g$ if and only if $o(f) \leq o(g)$.

(d) f is a unit if and only if $o(f) = 0$.

(e) If $f \neq 0$ then f is an associate of $X^{o(f)}$. Conclude that X is the only irreducible (up to multiplication by a unit) element of $F[[X]]$.

(f) $F[[X]]$ is a PID. In fact, every ideal is generated by X^k for some k.

(g) Is $F[[X]]$ a Euclidean domain?

60. If F is a field, let $F((X))$ denote the set of all formal Laurent series with coeffients in F, i.e.,

$$F((X)) = \left\{ \sum_{n=m}^{\infty} a_n X^n : a_n \in F, m \in \mathbf{Z} \text{ is arbitrary} \right\}$$

where the ring operations are defined as for $F[[X]]$. Prove that $F((X))$ is isomorphic to the quotient field of the integral domain $F[[X]]$.

61. Let R be a commutative ring and let $S = R[T_1, \ldots, T_n]$. Define $f(X) \in S[X]$ by

$$f(X) = \prod_{i=1}^{n}(X - T_i)$$

$$= \sum_{r=0}^{n}(-1)^r \sigma_r(T_1, \ldots, T_n) X^{n-r}.$$

(a) Show that

$$\sigma_r(T_1, \ldots, T_n) = \sum_{1 \leq i_1 < \cdots < i_r \leq n} T_{i_1} \cdots T_{i_r}.$$

Thus, $\sigma_1 = T_1 + \cdots + T_n$ and $\sigma_n = T_1 \cdots T_n$. σ_r is called the rth *elementary symmetric function*. Therefore, the coefficients of a polynomial are obtained by evaluating the elementary symmetric functions on the roots of the polynomial.

(b) If $g(X) = \prod_{i=1}^{n}(1 - T_i X)$, show that the coefficient of X^r is $(-1)^r \sigma_r$, i.e., the same as the coefficient of X^{n-r} in $f(X)$.

(c) Define $s_r(T_1, \ldots, T_n) = T_1^r + \cdots + T_n^r$ for $r \geq 1$, and let $s_0 = n$. Verify the following identities (known as *Newton's identities*) relating the power sums s_r and the elementary symmetric functions σ_k.

$$\sum_{k=0}^{r-1}(-1)^k \sigma_k s_{r-k} + (-1)^r r \sigma_r = 0 \qquad (1 \leq r \leq n)$$

$$\sum_{k=0}^{n}(-1)^k \sigma_k s_{r-k} = 0 \qquad (r > n).$$

(Hint: Do the following calculation in $S[[X]]$, where $'$ means derivative with respect to X:

$$\frac{g'(X)}{g(X)} = \sum_{i=1}^{n} \frac{-T_i}{1 - T_i X} = -\sum_{i=1}^{n}\sum_{j=0}^{\infty} T_i^{j+1} X^j = \sum_{j=0}^{\infty} s_{j+1} X^j.$$

Now multiply by $g(X)$ and compare coefficients of X^{r-1}.)

62. Let $p \in \mathbf{Z}$ be a prime and define

$$R = \{a = (a_1, a_2, a_3, \dots) :$$
$$a_k \in (\mathbf{Z}/p^k\mathbf{Z}), \ a_{k+1} \equiv a_k \pmod{p^k} \quad \text{for all} \quad k \geq 1\}.$$

(a) Show that R is a ring under the operations of componentwise addition and multiplication.

(b) Show that R is an integral domain. (Note that R contains $\mathbf{Z}$ as a subring so that $\text{char}(R) = 0$.)

(c) Let

$$P = \{a \in R : a_1 = 0 \in \mathbf{Z}/p\mathbf{Z}\}.$$

Show that every element of $R \setminus P$ is invertible. Show that P is a proper ideal of R. (Thus, P is the unique maximal ideal in R and so R is a local ring.)

(d) For $a \in R \setminus \{0\}$ define $v(a)$ to be p^{n-1} if n is the smallest value of k such that $a_k \neq 0 \in \mathbf{Z}/p^k\mathbf{Z}$. Show that $v : R \setminus \{0\} \to \mathbf{Z}^+$ makes R into a Euclidean domain.

Remark. The ring R plays an important role in mathematics; it is known as the *ring of p-adic integers* and its quotient field is known as the *field of p-adic numbers*.

Chapter 3

Modules and Vector Spaces

3.1 Definitions and Examples

Modules are a generalization of the vector spaces of linear algebra in which the "scalars" are allowed to be from an arbitrary ring, rather than a field. This rather modest weakening of the axioms is quite far reaching, including, for example, the theory of rings and ideals and the theory of abelian groups as special cases.

(1.1) Definition. *Let R be an arbitrary ring with identity (not necessarily commutative).*

(1) *A* **left R-module** *(or* **left module over R**) *is an abelian group M together with a scalar multiplication map*

$$\cdot : R \times M \to M$$

that satisfy the following axioms (as is customary we will write am in place of $\cdot(a, m)$ for the scalar multiplication of $m \in M$ by $a \in R$). In these axioms, a, b are arbitrary elements of R and m, n are arbitrary elements of M.

$(a_l) a(m + n) = am + an.$
$(b_l) (a + b)m = am + bm.$
$(c_l) (ab)m = a(bm).$
$(d_l) 1m = m.$

(2) *A* **right R-module** *(or* **right module over R**) *is an abelian group M together with a scalar multiplication map*

$$\cdot : M \times R \to M$$

that satisfy the following axioms (again a, b are arbitrary elements of R and m, n are arbitrary elements of M).

$(a_r) (m + n)a = ma + na.$
$(b_r) m(a + b) = ma + mb.$

$(c_r) m(ab) = (ma)b.$
$(d_r) m1 = m.$

(1.2) *Remarks.*

(1) If R is a commutative ring then any left R-module also has the structure of a right R-module by defining $mr = rm$. The only axiom that requires a check is axiom (c_r). But

$$m(ab) = (ab)m = (ba)m = b(am) = b(ma) = (ma)b.$$

(2) More generally, if the ring R has an antiautomorphism (that is, an additive homomorphism $\phi : R \to R$ such that $\phi(ab) = \phi(b)\phi(a)$) then any left R-module has the structure of a right R-module by defining $ma = \phi(a)m$. Again, the only axiom that needs checking is axiom (c_r):

$$
\begin{aligned}
(ma)b &= \phi(b)(ma) \\
&= \phi(b)(\phi(a)m) \\
&= (\phi(b)\phi(a))m \\
&= \phi(ab)m \\
&= m(ab).
\end{aligned}
$$

An example of this situation occurs for the group ring $R(G)$ where R is a ring with identity and G is a group (see Example 2.1.10 (15)). In this case the antiautomorphism is given by

$$\phi\left(\sum_{g \in G} a_g g\right) = \sum_{g \in G} a_g g^{-1}.$$

We leave it as an exercise to check that $\phi : R(G) \to R(G)$ is an antiautomorphism. Thus any left $R(G)$-module M is automatically a right $R(G)$-module.

(3) Let R be an arbitrary ring and let R^{op} ("op" for opposite) be the ring whose elements are the elements of R, whose addition agrees with that of R, but whose multiplication $\cdot$ is given by $a \cdot b = ba$ (where the multiplication on the right-hand side of this equation is that of R). Then any left R-module is naturally a right R^{op}-module (and viceversa). In fact, if M is a left R-module, define a right multiplication of elements of R^{op} (which are the same as elements of R) on M by $m \cdot a = am$. As in Remark 1.2 (1), the only axiom that requires checking is axiom (c_r). But

$$m \cdot (a \cdot b) = (a \cdot b)m = (ba)m = b(am) = b(m \cdot a) = (m \cdot a) \cdot b.$$

The theories of left R-modules and right R-modules are entirely parallel, and so, to avoid doing everything twice, we must choose to work on one side or the other. Thus, we shall work primarily with left R-modules unless explicitly indicated otherwise and we will define an **R-module** (or module over R) to be a left R-module. (Of course, if R is commutative, Remark 1.2 (1) shows there is no difference between left and right R-modules.) Applications of module theory to the theory of group representations will, however, necessitate the use of both left and right modules over noncommutative rings. Before presenting a collection of examples some more notation will be introduced.

(1.3) Definition. *Let R be a ring and let M, N be R-modules. A function $f : M \to N$ is an R-**module homomorphism** if*

(1) $f(m_1 + m_2) = f(m_1) + f(m_2)$ *for all m_1, $m_2 \in M$, and*
(2) $f(am) = af(m)$ *for all $a \in R$ and $m \in M$.*

The set of all R-module homomorphisms from M to N will be denoted $\mathrm{Hom}_R(M, N)$. In case $M = N$ we will usually write $\mathrm{End}_R(M)$ rather than $\mathrm{Hom}_R(M, M)$; elements of $\mathrm{End}_R(M)$ are called **endomorphisms**. If $f \in \mathrm{End}_R(M)$ is invertible, then it is called an **automorphism** of M. The group of all R-module automorphisms of M is denoted $\mathrm{Aut}_R(M)$ ($\mathrm{Aut}(M)$ if R is implicit). If $f \in \mathrm{Hom}_R(M, N)$ then we define $\mathrm{Ker}(f) \subseteq M$ and $\mathrm{Im}(f) \subseteq N$ to be the kernel and image of f considered as an abelian group homomorphism.

(1.4) Definition.

(1) *Let F be a field. Then an F-module V is called a **vector space** over F.*
(2) *If V and W are vector spaces over the field F then a **linear transformation** from V to W is an F-module homomorphism from V to W.*

(1.5) Examples.

(1) Let G be any abelian group and let $g \in G$. If $n \in \mathbf{Z}$ then define the scalar multiplication ng by

$$ng = \begin{cases} g + \cdots + g & (n \text{ terms}) \text{ if } n > 0, \\ 0 & \text{if } n = 0, \\ (-g) + \cdots + (-g) & (-n \text{ terms}) \text{ if } n < 0. \end{cases}$$

Using this scalar multiplication G is a $\mathbf{Z}$-module. Furthermore, if G and H are abelian groups and $f : G \to H$ is a group homomorphism, then f is also a $\mathbf{Z}$-module homomorphism since (if $n > 0$)

$$f(ng) = f(g + \cdots + g) = f(g) + \cdots + f(g) = nf(g)$$

and $f(-g) = -f(g)$.

(2) Let R be an arbitrary ring. Then R^n is both a left and a right R-module via the scalar multiplications

$$a(b_1, \ldots, b_n) = (ab_1, \ldots, ab_n)$$

and

$$(b_1, \ldots, b_n)a = (b_1 a, \ldots, b_n a).$$

(3) Let R be an arbitrary ring. Then the set of matrices $M_{m,n}(R)$ is both a left and a right R-module via left and right scalar multiplication of matrices, i.e.,

$$\text{ent}_{ij}(aA) = a \, \text{ent}_{ij}(A)$$

and

$$\text{ent}_{ij}(Aa) = (\text{ent}_{ij}(A))a.$$

(4) As a generalization of the above example, the matrix multiplication maps

$$M_m(R) \times M_{m,n}(R) \longrightarrow M_{m,n}(R)$$
$$(A, B) \longmapsto AB$$

and

$$M_{m,n}(R) \times M_n(R) \longrightarrow M_{m,n}(R)$$
$$(A, B) \longmapsto AB$$

make $M_{m,n}(R)$ into a left $M_m(R)$-module and a right $M_n(R)$-module.

(5) If R is a ring then a left ideal $I \subseteq R$ is a left R-module, while a right ideal $J \subseteq R$ is a right R-module. In both cases the scalar multiplication is just the multiplication of the ring R.

(6) If R is a ring and $I \subseteq R$ is an ideal then the quotient ring R/I is both a left R-module and a right R-module via the multiplication maps

$$R \times R/I \longrightarrow R/I$$
$$(a, b + I) \longmapsto ab + I$$

and

$$R/I \times R \longrightarrow R/I$$
$$(a + I, b) \longmapsto ab + I.$$

(7) M is defined to be an R-**algebra** if M is both an R-module and a ring, with the ring addition being the same as the module addition, and the multiplication on M and the scalar multiplication by R satisfying the following identity: For every $r \in R$, $m_1, m_2 \in M$,

(1.1) $$r(m_1 m_2) = (rm_1)m_2 = m_1(rm_2).$$

For example, every ring is a $\mathbf{Z}$-algebra, and if R is a commutative ring, then R is an R-algebra. Let R and S be rings and let $\phi : R \to S$ be a ring homomorphism with $\mathrm{Im}(\phi) \subseteq C(S) = \{a \in S : ab = ba$ for all $b \in S\}$, the center of S. If M is an S-module, then M is also an R-module using the scalar multiplication $am = (\phi(a))m$ for all $a \in R$ and $m \in M$. Since S itself is an S-module, it follows that S is an R-module, and moreover, since $\mathrm{Im}(\phi) \subseteq C(S)$, we conclude that S is an R-algebra. As particular cases of this construction, if R is a commutative ring, then the polynomial ring $R[X]$ and the matrix ring $M_n(R)$ are both R-algebras.

(8) If M and N are R-modules then $\mathrm{Hom}_R(M, N)$ is an abelian group via the operation $(f + g)(m) = f(m) + g(m)$. However, if we try to make $\mathrm{Hom}_R(M, N)$ into an R-module in the natural way by defining af by the formula $(af)(m) = a(f(m))$ we find that the function af need not be an R-module homomorphism unless R is a commutative ring. To see this, note that

$$(af)(rm) = a(f(rm)) = a(r(f(m))) = arf(m).$$

This last expression is equal to $r(af)(m) = raf(m)$ if R is a commutative ring, but not necessarily otherwise. Thus, if R is a commutative ring, then we may consider $\mathrm{Hom}_R(M, N)$ as an R-module for all M, N, while if R is not commutative then $\mathrm{Hom}_R(M, N)$ is only an abelian group. Since $\mathrm{End}_R(M)$ is also a ring using composition of R-module homomorphisms as the multiplication, and since there is a ring homomorphism $\phi : R \to \mathrm{End}_R(M)$ defined by $\phi(a) = a\,1_M$ where 1_M denotes the identity homomorphism of M, it follows from Example 1.5 (7) that $\mathrm{End}_R(M)$ is an R-algebra if R is a commutative ring.

(9) If G is an abelian group, then $\mathrm{Hom}_{\mathbf{Z}}(\mathbf{Z}, G) \cong G$. To see this, define $\Phi : \mathrm{Hom}_{\mathbf{Z}}(\mathbf{Z}, G) \to G$ by $\Phi(f) = f(1)$. We leave it as an exercise to check that Φ is an isomorphism of $\mathbf{Z}$-modules.

(10) Generalizing Example 1.5 (9), if M is an R-module then

$$\mathrm{Hom}_R(R, M) \cong M$$

as $\mathbf{Z}$-modules via the map $\Phi : \mathrm{Hom}_R(R, M) \to M$ where $\Phi(f) = f(1)$.

(11) Let R be a commutative ring, let M be an R-module, and let $S \subset \mathrm{End}_R(M)$ be a subring. (Recall from Example 1.5 (8) that $\mathrm{End}_R(M)$ is a ring, in fact, an R algebra.) Then M is an S-module by means of the scalar multiplication map $S \times M \to M$ defined by $(f, m) \mapsto f(m)$.

(12) As an important special case of Example 1.5 (11), let $T \in \mathrm{End}_R(M)$ and define a ring homomorphism $\phi : R[X] \to \mathrm{End}_R(M)$ by sending X to T and $a \in R$ to $a1_M$. (See the polynomial substitution theorem (Theorem 2.4.1).) Thus, if

$$f(X) = a_0 + a_1 X + \cdots + a_n X^n$$

then
$$\phi(f(X)) = a_0 1_M + a_1 T + \cdots + a_n T^n.$$

We will denote $\phi(f(X))$ by the symbol $f(T)$ and we let $\text{Im}(\phi) = R[T]$. That is, $R[T]$ is the subring of $\text{End}_R(M)$ consisting of "polynomials" in T. Then M is an $R[T]$ module by means of the multiplication

$$f(T)m = f(T)(m).$$

Using the homomorphism $\phi : R[X] \to R[T]$ we see that M is an $R[X]$-module using the scalar multiplication

$$f(X)m = f(T)(m).$$

This example is an extremely important one. It provides the basis for applying the theory of modules over principal ideal domains to the study of linear transformations; it will be developed fully in Section 4.4.

(13) We will present a concrete example of the situation presented in Example 1.5 (12). Let F be a field and define a linear transformation $T : F^2 \to F^2$ by $T(u_1, u_2) = (u_2, 0)$. Then $T^2 = 0$, so if $f(X) = a_0 + a_1 X + \cdots + a_m X^m \in F[X]$, it follows that $f(T) = a_0 1_{F^2} + a_1 T$. Therefore the scalar multiplication $f(X)u$ for $u \in F^2$ is given by

$$\begin{aligned}
f(X) \cdot (u_1, u_2) &= f(T)(u_1, u_2) \\
&= (a_0 1_{F^2} + a_1 T)(u_1, u_2) \\
&= (a_0 u_1 + a_1 u_2, a_0 u_2).
\end{aligned}$$

3.2 Submodules and Quotient Modules

Let R be a ring and M an R-module. A subset $N \subseteq M$ is said to be a submodule (or R-submodule) of M if N is a subgroup of the additive group of M that is also an R-module using the scalar multiplication on M. What this means, of course, is that N is a submodule of M if it is a subgroup of M that is closed under scalar multiplication. These conditions can be expressed as follows.

(2.1) Lemma. *If M is an R-module and N is a nonempty subset of M, then N is an R-submodule of M if and only if $am_1 + bm_2 \in N$ for all $m_1, m_2 \in N$ and $a, b \in R$.*

Proof. Exercise. $\qquad\square$

If F is a field and V is a vector space over F, then an F-submodule of V is called a **linear subspace** of V.

(2.2) Examples.

(1) If R is any ring then the R-submodules of the R-module R are precisely the left ideals of the ring R.

(2) If G is any abelian group then G is a $\mathbf{Z}$-module and the $\mathbf{Z}$-submodules of G are just the subgroups of G.

(3) Let $f : M \to N$ be an R-module homomorphism. Then $\mathrm{Ker}(f) \subseteq M$ and $\mathrm{Im}(f) \subseteq N$ are R-submodules (exercise).

(4) Continuing with Example 1.5 (12), let V be a vector space over a field F and let $T \in \mathrm{End}_F(V)$ be a fixed linear transformation. Let V_T denote V with the $F[X]$-module structure determined by the linear transformation T. Then a subset $W \subseteq V$ is an $F[X]$-submodule of the module V_T if and only if W is a linear subspace of V and $T(W) \subseteq W$, i.e., W must be a T-**invariant subspace** of V. To see this, note that $X \cdot w = T(w)$, and if $a \in F$, then $a \cdot w = aw$—that is to say, the action of the constant polynomial $a \in F[X]$ on V is just ordinary scalar multiplication, while the action of the polynomial X on V is the action of T on V. Thus, an $F[X]$-submodule of V_T must be a T-invariant subspace of V. Conversely, if W is a linear subspace of V such that $T(W) \subseteq W$ then $T^m(W) \subseteq W$ for all $m \geq 1$. Hence, if $f(X) \in F[X]$ and $w \in W$ then $f(X) \cdot w = f(T)(w) \in W$ so that W is closed under scalar multiplication and thus W is an $F[X]$-submodule of V.

(2.3) Lemma. *Let M be an R-module and let $\{N_\alpha\}_{\alpha \in A}$ be a family of submodules of M. Then $N = \bigcap_{\alpha \in A} N_\alpha$ is a submodule of M.*

Proof. Exercise. $\square$

We now consider quotient modules and the noether isomorphism theorems. Let M be an R-module and let $N \subseteq M$ be a submodule. Then N is a subgroup of the abelian group M, so we can form the quotient group M/N. Define a scalar multiplication map on the abelian group M/N by $a(m + N) = am + N$ for all $a \in R$, $m + N \in M/N$. Since N is an R-submodule of M, this map is well defined. Indeed, if $m + N = m' + N$ then $m - m' \in N$ so that $am - am' = a(m - m') \in N$ so that $am + N = am' + N$. The resulting R-module M/N is called the quotient module of M with respect to the submodule N. The noether isomorphism theorems, which we have seen previously for groups and rings, then have direct analogues for R-modules.

(2.4) Theorem. (First isomorphism theorem) *Let M and N be modules over the ring R and let $f : M \to N$ be an R-module homomorphism. Then $M/\mathrm{Ker}(f) \cong \mathrm{Im}(f)$.*

Proof. Let $K = \text{Ker}(f)$. From Theorem 1.3.10 we know that $\overline{f} : M/K \to \text{Im}(f)$ defined by $\overline{f}(m+K) = f(m)$ is a well-defined isomorphism of abelian groups. It only remains to check that $\overline{f}$ is an R-module homomorphism. But $\overline{f}(a(m+K)) = \overline{f}(am+K) = f(am) = af(m) = a\overline{f}(m+K)$ for all $m \in M$ and $a \in R$, so we are done. $\square$

(2.5) Theorem. (Second isomorphism theorem) *Let M be an R-module and let N and P be submodules. Then there is an isomorphism of R-modules*

$$(N + P)/P \cong N/(N \cap P).$$

Proof. Let $\pi : M \to M/P$ be the natural projection map and let π_0 be the restriction of π to N. Then π_0 is an R-module homomorphism with $\text{Ker}(\pi_0) = N \cap P$ and $\text{Im}(\pi_0) = (N + P)/P$. The result then follows from the first isomorphism theorem. $\square$

(2.6) Theorem. (Third isomorphism theorem) *Let M be an R-module and let N and P be submodules of M with $P \subseteq N$. Then*

$$M/N \cong (M/P)/(N/P).$$

Proof. Define $f : M/P \to M/N$ by $f(m+P) = m+N$. This is a well-defined R-module homomorphism and

$$\text{Ker}(f) = \{m + P : m + N = N\} = \{m + P : m \in N\} = N/P.$$

The result then follows from the first isomorphism theorem (Theorem 2.4). $\square$

(2.7) Theorem. (Correspondence theorem) *Let M be an R-module, N a submodule, and $\pi : M \to M/N$ the natural projection map. Then the function $P \mapsto P/N$ defines a one-to-one correspondence between the set of all submodules of M that contain N and the set of all submodules of M/N.*

Proof. Exercise. $\square$

(2.8) Definition. *If S is a subset of an R-module M then $\langle S \rangle$ will denote the intersection of all the submodules of M that contain S. This is called the* **submodule of M generated by** *S, while the elements of S are called* **generators** *of $\langle S \rangle$.*

Thus, $\langle S \rangle$ is a submodule of M that contains S and it is contained in every submodule of M that contains S, i.e., $\langle S \rangle$ is the smallest submodule of M containing S. If $S = \{x_1, \ldots, x_n\}$ we will usually write $\langle x_1, \ldots, x_n \rangle$

rather than $\langle\{x_1, \ldots, x_n\}\rangle$ for the submodule generated by S. There is the following simple description of $\langle S \rangle$.

(2.9) Lemma. *Let M be an R-module and let $S \subseteq M$. If $S = \emptyset$ then $\langle S \rangle = \{0\}$, while $\langle S \rangle = \{\sum_{i=1}^{n} a_i s_i : n \in \mathbf{N}, a_i \in R, s_i \in S, 1 \le i \le n\}$ if $S \ne \emptyset$.*

Proof. Exercise.   □

(2.10) Definition. *We say that the R-module M is **finitely generated** if $M = \langle S \rangle$ for some finite subset S of M. M is **cyclic** if $M = \langle m \rangle$ for some element $m \in M$. If M is finitely generated then let $\mu(M)$ denote the minimal number of generators of M. If M is not finitely generated, then let $\mu(M) = \infty$. We will call $\mu(M)$ the **rank** of M.*

(2.11) Remarks.

(1) We have $\mu(\{0\}) = 0$ by Lemma 2.9 (1), and $M \ne \{0\}$ is cyclic if and only if $\mu(M) = 1$.
(2) The concept of cyclic R-module generalizes the concept of cyclic group. Thus an abelian group G is cyclic (as an abelian group) if and only if it is a cyclic $\mathbf{Z}$-module.
(3) If R is a PID, then any R-submodule M of R is an ideal, so $\mu(M) = 1$.
(4) For a general ring R, it is not necessarily the case that if N is a submodule of the R-module M, then $\mu(N) \le \mu(M)$. For example, if R is a polynomial ring over a field F in k variables, $M = R$, and $N \subseteq M$ is the submodule consisting of polynomials whose constant term is 0, then $\mu(M) = 1$ but $\mu(N) = k$. Note that this holds even if $k = \infty$. We shall prove in Corollary 6.4 that this phenomenon cannot occur if R is a PID. Also see Remark 6.5.

If M is a finitely generated R-module and N is any submodule, then M/N is clearly finitely generated, and in fact, $\mu(M/N) \le \mu(M)$ since the image in M/N of any generating set of M is a generating set of M/N. There is also the following result, which is frequently useful for constructing arguments using induction on $\mu(M)$.

(2.12) Proposition. *Suppose M is an R-module and N is a submodule. If N and M/N are finitely generated, then so is M and*

$$\mu(M) \le \mu(N) + \mu(M/N).$$

Proof. Let $S = \{x_1, \ldots, x_k\} \subseteq N$ be a minimal generating set for N and if $\pi : M \to M/N$ is the natural projection map, choose $T = \{y_1, \ldots, y_\ell\} \subseteq M$ so that $\{\pi(y_1), \ldots, \pi(y_\ell)\}$ is a minimal generating set for M/N. We claim that $S \cup T$ generates M so that $\mu(M) \le k + \ell = \mu(N) + \mu(M/N)$. To see this suppose that $x \in M$. Then $\pi(x) = a_1 \pi(y_1) + \cdots + a_\ell \pi(y_\ell)$. Let $y =$

$a_1 y_1 + \cdots + a_\ell y_\ell \in \langle T \rangle$. Then $\pi(x-y) = 0$ so that $x - y \in \text{Ker}(\pi) = N = \langle S \rangle$. It follows that $x = (x - y) + y \in \langle S \cup T \rangle$, and the proof is complete.   □

(2.13) Definition. *If $\{N_\alpha\}_{\alpha \in A}$ is a family of R-submodules of M, then the* **submodule generated by** $\{N_\alpha\}_{\alpha \in A}$ *is $\langle \bigcup_{\alpha \in A} N_\alpha \rangle$. This is just the set of all sums $n_{\alpha_1} + \cdots + n_{\alpha_k}$ where $n_{\alpha_i} \in N_{\alpha_i}$. Instead of $\langle \bigcup_{\alpha \in A} N_\alpha \rangle$, we will use the notation $\sum_{\alpha \in A} N_\alpha$; if the index set A is finite, e.g., $A = \{1, \dots, m\}$, we will write $N_1 + \cdots + N_m$ for the submodule generated by $N_1, \dots, N_m$.*

(2.14) Definition. *If R is a ring, M is an R-module, and X is a subset of M, then the* **annihilator** *of X, denoted $\text{Ann}(X)$, is defined by*

$$\text{Ann}(X) = \{a \in R : ax = 0 \text{ for all } x \in X\}.$$

It is easy to check that $\text{Ann}(X)$ is a left ideal of R, and furthermore, if $X = N$ is a submodule of M, then $\text{Ann}(N)$ is an ideal of R. If R is commutative and $N = \langle x \rangle$ is a cyclic submodule of M with generator x, then

$$\text{Ann}(N) = \{a \in R : ax = 0\}.$$

This fact is not true if the ring R is not commutative. As an example, let $R = M_n(\mathbf{R}) = M$ and let $x = E_{11}$ be the matrix with a 1 in the 1 1 position and 0 elsewhere. It is a simple exercise to check that $\text{Ann}(E_{11})$ consists of all matrices with first column 0, while $\text{Ann}(\langle E_{11} \rangle) = \langle 0 \rangle$.

If R is commutative and N is cyclic with generator x then we will usually write $\text{Ann}(x)$ rather than $\text{Ann}(\langle x \rangle)$. In this situation, the ideal $\text{Ann}(x)$ is frequently called the **order ideal** of x. To see why, consider the example of an abelian group G and an element $g \in G$. Then G is a $\mathbf{Z}$-module and

$$\text{Ann}(g) = \{n \in \mathbf{Z} : ng = 0\}$$
$$= \langle p \rangle$$

where $p = o(g)$ if $o(g) < \infty$ and $p = 0$ if $\langle g \rangle$ is infinite cyclic.

Example. Let F be a field, V a vector space, $T \in \text{End}_F(V)$ a linear transformation, and let V_T be the $F[X]$ module determined by T (Example 1.5 (12)). If $v \in V$ then

$$\text{Ann}(v) = \{f(X) \in F[X] : f(T)(v) = 0\}.$$

Note that this is a principal ideal $\langle g(X) \rangle$ since $F[X]$ is a PID.

(2.15) Proposition. *Let R be a ring and let $M = \langle m \rangle$ be a cyclic R-module. Then $M \cong R/\text{Ann}(m)$.*

Proof. The function $f : R \to M$ defined by $f(a) = am$ is a surjective R-module homomorphism with $\text{Ker}(f) = \text{Ann}(m)$. The result follows by the first isomorphism theorem.   □

(2.16) Corollary. *If F is a field and M is a nonzero cyclic F-module then $M \cong F$.*

Proof. A field has only the ideals $\{0\}$ and F, and $1 \cdot m = m \neq 0$ if $m \neq 0$ is a generator for M. Thus, $\mathrm{Ann}(m) \neq F$, so it must be $\{0\}$.   □

If M is an R-module and $I \subseteq R$ is an ideal then we can define the product of I and M by

$$IM = \left\{ \sum_{i=1}^{n} a_i m_i : n \in \mathbf{N}, \ a_i \in I, \ m_i \in M \right\}.$$

The set IM is easily checked to be a submodule of M. The product IM is a generalization of the concept of product of ideals. If R is commutative and $I \subseteq \mathrm{Ann}(M)$ then there is a map

$$R/I \times M \to M$$

defined by $(a + I)m = am$. To see that this map is well defined, suppose that $a + I = b + I$. Then $a - b \in I \subseteq \mathrm{Ann}(M)$ so that $(a - b)m = 0$, i.e., $am = bm$. Therefore, whenever an ideal $I \subseteq \mathrm{Ann}(M)$, M is also an R/I module. A particular case where this occurs is if $N = M/IM$ where I is any ideal of R. Then certainly $I \subseteq \mathrm{Ann}(N)$ so that M/IM is an R/I-module.

(2.17) Definition. *Let R be an integral domain and let M be an R-module. We say that an element $x \in M$ is a* **torsion element** *if $\mathrm{Ann}(x) \neq \{0\}$. Thus an element $x \in M$ is torsion if and only if there is an $a \neq 0 \in R$ such that $ax = 0$. Let M_τ be the set of torsion elements of M. M is said to be* **torsion-free** *if $M_\tau = \{0\}$, and M is a* **torsion module** *if $M = M_\tau$.*

(2.18) Proposition. *Let R be an integral domain and let M be an R-module.*

(1) *M_τ is a submodule of M, called the* **torsion submodule**.
(2) *M/M_τ is torsion-free.*

Proof. (1) Let $x, y \in M_\tau$ and let $c, d \in R$. There are $a \neq 0$, $b \neq 0 \in R$ such that $ax = 0$ and $by = 0$. Since R is an integral domain, $ab \neq 0$. Therefore, $ab(cx + dy) = bc(ax) + ad(by) = 0$ so that $cx + dy \in M_\tau$.

(2) Suppose that $a \neq 0 \in R$ and $a(x + M_\tau) = 0 \in (M/M_\tau)_\tau$. Then $ax \in M_\tau$, so there is a $b \neq 0 \in R$ with $(ba)x = b(ax) = 0$. Since $ba \neq 0$, it follows that $x \in M_\tau$, i.e., $x + M_\tau = 0 \in M/M_\tau$.   □

(2.19) Examples.

(1) If G is an abelian group then the torsion $\mathbf{Z}$-submodule of G is the set of all elements of G of finite order. Thus, $G = G_\tau$ means that every element of G is of finite order. In particular, any finite abelian

group is torsion. The converse is not true. For a concrete example, take $G = \mathbf{Q}/\mathbf{Z}$. Then $|G| = \infty$, but every element of $\mathbf{Q}/\mathbf{Z}$ has finite order since $q(p/q + \mathbf{Z}) = p + \mathbf{Z} = 0 \in \mathbf{Q}/\mathbf{Z}$. Thus $(\mathbf{Q}/\mathbf{Z})_\tau = \mathbf{Q}/\mathbf{Z}$.

(2) An abelian group is torsion-free if it has no elements of finite order other than 0. As an example, take $G = \mathbf{Z}^n$ for any natural number n. Another useful example to keep in mind is the additive group $\mathbf{Q}$.

(3) Let $V = F^2$ and consider the linear transformation $T : F^2 \to F^2$ defined by $T(u_1, u_2) = (u_2, 0)$. See Example 1.5 (13). Then the $F[X]$ module V_T determined by T is a torsion module. In fact $\text{Ann}(V_T) = \langle X^2 \rangle$. To see this, note that $T^2 = 0$, so $X^2 \cdot u = 0$ for all $u \in V$. Thus, $\langle X^2 \rangle \subseteq \text{Ann}(V_T)$. The only ideals of $F[X]$ properly containing $\langle X^2 \rangle$ are $\langle X \rangle$ and the whole ring $F[X]$, but $X \notin \text{Ann}(V_T)$ since $X \cdot (0, 1) = (1, 0) \neq (0, 0)$. Therefore, $\text{Ann}(V_T) = \langle X^2 \rangle$.

The following two observations are frequently useful; the proofs are left as exercises:

(2.20) Proposition. *Let R be an integral domain and let M be a finitely generated torsion R-module. Then $\text{Ann}(M) \neq (0)$. In fact, if $M = \langle x_1, \dots, x_n \rangle$ then*

$$\text{Ann}(M) = \text{Ann}(x_1) \cap \dots \cap \text{Ann}(x_n) \neq (0).$$

Proof. Exercise.   □

(2.21) Proposition. *Let F be a field and let V be a vector space over F, i.e., an F-module. Then V is torsion-free.*

Proof. Exercise.   □

3.3 Direct Sums, Exact Sequences, and Hom

Let $M_1, \dots, M_n$ be a finite collection of R-modules. Then the cartesian product set $M_1 \times \dots \times M_n$ can be made into an R-module by the operations

$$(x_1, \dots, x_n) + (y_1, \dots, y_n) = (x_1 + y_1, \dots, x_n + y_n)$$
$$a(x_1, \dots, x_n) = (ax_1, \dots, ax_n)$$

where the 0 element is, of course, $(0, \dots, 0)$. The R-module thus constructed is called the **direct sum** of $M_1, \dots, M_n$ and is denoted

$$M_1 \oplus \dots \oplus M_n \quad \left(\text{or} \quad \bigoplus_{i=1}^{n} M_i\right).$$

The direct sum has an important homomorphism property, which, in fact, can be used to characterize direct sums. To describe this, suppose that $f_i : M_i \to N$ are R-module homomorphisms. Then there is a map

$$f : M_1 \oplus \cdots \oplus M_n \to N$$

defined by

$$f(x_1, \ldots , x_n) = \sum_{i=1}^{n} f_i(x_i).$$

We leave it as an exercise to check that f is an R-module homomorphism.

Now consider the question of when a module M is isomorphic to the direct sum of finitely many submodules. This result should be compared with Proposition 1.6.3 concerning internal direct products of groups.

(3.1) Theorem. *Let M be an R-module and let $M_1, \ldots , M_n$ be submodules of M such that*

(1) $M = M_1 + \cdots + M_n$, *and*
(2) *for $1 \le i \le n$,*

$$M_i \cap (M_1 + \cdots + M_{i-1} + M_{i+1} + \cdots + M_n) = 0.$$

Then

$$M \cong M_1 \oplus \cdots \oplus M_n.$$

Proof. Let $f_i : M_i \to M$ be the inclusion map, that is, $f_i(x) = x$ for all $x \in M_i$ and define

$$f : M_1 \oplus \cdots \oplus M_n \to M$$

by

$$f(x_1, \ldots , x_n) = x_1 + \cdots + x_n.$$

f is an R-module homomorphism and it follows from condition (1) that f is surjective. Now suppose that $(x_1, \ldots , x_n) \in \mathrm{Ker}(f)$. Then $x_1 + \cdots + x_n = 0$ so that for $1 \le i \le n$ we have

$$x_i = -(x_1 + \cdots + x_{i-1} + x_{i+1} + \cdots + x_n).$$

Therefore,

$$x_i \in M_i \cap (M_1 + \cdots + M_{i-1} + M_{i+1} + \cdots + M_n) = 0$$

so that $(x_1, \ldots , x_n) = 0$ and f is an isomorphism. $\square$

Our primary emphasis will be on the finite direct sums of modules just constructed, but for the purpose of allowing for potentially infinite rank free modules, it is convenient to have available the concept of an arbitrary direct sum of R-modules. This is described as follows. Let $\{M_j\}_{j \in J}$ be a family of R-modules indexed by the (possibly infinite) set J. Then the cartesian product set $\prod_{j \in J} M_j$ is the set of all the indexed sets of elements $(x_j)_{j \in J}$ where x_j is chosen from M_j. This set is made into an R-module by the coordinate-wise addition and scalar multiplication of elements. More precisely, we define

$$(x_j)_{j \in J} + (y_j)_{j \in J} = (x_j + y_j)_{j \in J}$$
$$a(x_j)_{j \in J} = (ax_j)_{j \in J}.$$

For each $k \in J$ there is an R-module homomorphism $\pi_k : \prod_{j \in J} M_j \to M_k$ defined by $\pi_k((x_j)_{j \in J}) = x_j$, that is, π_k picks out the element of the indexed set $(x_j)_{j \in J}$ that is indexed by k. We define the **direct sum** of the indexed family $\{M_j\}_{j \in J}$ of R-modules to be the following submodule $\bigoplus_{j \in J} M_j$ of $\prod_{j \in J} M_j$:

$$\bigoplus_{j \in J} M_j = \{(x_j)_{j \in J} : x_j = 0 \text{ except for finitely many indices } j \in J\}.$$

It is easy to check that $\oplus_{j \in J} M_j$ is a submodule of $\prod_{j \in J} M_j$.

To get a feeling for the difference between direct sums and direct products when the index set is infinite, note that the polynomial ring $R[X]$, as an R-module (ignoring the multiplicative structure), is just a countable direct sum of copies of R, in fact, the nth copy of R is indexed by the monomial X^n. However, the formal power series ring $R[[X]]$, as an R-module, is just a countable direct product of copies of R. Again, the nth copy of R is indexed by the monomial X^n. Each element of the polynomial ring has only finitely many monomials with nonzero coefficients, while an element of the formal power series ring may have all coefficients nonzero.

The homomorphism property of the finite direct sum of R-modules extends in a natural way to arbitrary direct sums. That is, suppose that N is an arbitrary R-module and that for each $j \in J$ there is an R-module homomorphism $f_j : M_j \to N$. Then there is a map $f : \oplus_{j \in J} M_j \to N$ defined by $f((x_j)_{j \in J}) = \sum_{j \in J} f_j(x_j)$. Note that this sum can be considered as a well-defined finite sum since $x_j = 0$ except for finitely many indices $j \in J$. (Note that this construction does not work for infinite direct products.) We leave it as an exercise to check that f is an R-module homomorphism.

The characterization of when an R-module M is isomorphic to the direct sum of submodules is essentially the same as the characterization provided in Theorem 3.1. We state the result, but the verification is left as an exercise.

(3.2) Theorem. *Let M be an R-module and let $\{M_j\}_{j \in J}$ be a family of submodules such that*

(1) $M = \sum_{j \in J} M_j = \langle \bigcup_{j \in J} M_j \rangle$, *and*
(2) $M_k \cap \sum_{j \in J \setminus \{k\}} M_j = \{0\}$ *for every $k \in J$.*

Then
$$M \cong \bigoplus_{j \in J} M_j.$$

Proof. Exercise.   □

(3.3) Definition. *If M is an R-module and $M_1 \subseteq M$ is a submodule, we say that M_1 is a* **direct summand** *of M, or is* **complemented** *in M, if there is a submodule $M_2 \subseteq M$ such that $M \cong M_1 \oplus M_2$.*

(3.4) Example. Let $R = \mathbf{Z}$ and $M = \mathbf{Z}_{p^2}$. If $M_1 = \langle p \rangle$ then M_1 is not complemented since M_1 is the only subgroup of M of order p, so condition (2) of Theorem 3.1 is impossible to satisfy.

The concept of exact sequences of R-modules and R-module homomorphisms and their relation to direct summands is a useful tool to have available in the study of modules. We start by defining exact sequences of R-modules.

(3.5) Definition. *Let R be a ring. A sequence of R-modules and R-module homomorphisms*

$$\cdots \longrightarrow M_{i-1} \xrightarrow{f_i} M_i \xrightarrow{f_{i+1}} M_{i+1} \longrightarrow \cdots$$

is said to be **exact at M_i** *if $\operatorname{Im}(f_i) = \operatorname{Ker}(f_{i+1})$. The sequence is said to be* **exact** *if it is exact at each M_i.*

As particular cases of this definition note that

(1) $0 \longrightarrow M_1 \xrightarrow{f} M$ is exact if and only if f is injective,
(2) $M \xrightarrow{g} M_2 \longrightarrow 0$ is exact if and only if g is surjective, and
(3) the sequence

(3.1) $$0 \longrightarrow M_1 \xrightarrow{f} M \xrightarrow{g} M_2 \longrightarrow 0$$

is exact if and only if f is injective, g is surjective, and $\operatorname{Im}(f) = \operatorname{Ker}(g)$. Note that the first isomorphism theorem (Theorem 2.4) then shows that $M_2 \cong M/\operatorname{Im}(f)$. $M/\operatorname{Im}(f)$ is called the **cokernel** of f and it is denoted $\operatorname{Coker}(f)$.

(3.6) Definition.

(1) *The sequence (3.1), if exact, is said to be a* **short exact sequence**.
(2) *The sequence (3.1) is said to be a* **split exact sequence** *(or just* **split***) if it is exact and if $\operatorname{Im}(f) = \operatorname{Ker}(g)$ is a direct summand of M.*

In the language of exact sequences, Proposition 2.12 can be stated as follows:

(3.7) Proposition. *Let* $0 \longrightarrow M_1 \longrightarrow M \longrightarrow M_2 \longrightarrow 0$ *be a short exact sequence of R-modules. If M_1 and M_2 are finitely generated, then so is M, and moreover,*

$$\mu(M) \le \mu(M_1) + \mu(M_2).$$

Proof.   □

(3.8) Example. Let p and q be distinct primes. Then we have short exact sequences

$$(3.2) \qquad\qquad 0 \longrightarrow \mathbf{Z}_p \xrightarrow{\phi} \mathbf{Z}_{pq} \xrightarrow{\psi} \mathbf{Z}_q \longrightarrow 0$$

and

$$(3.3) \qquad\qquad 0 \longrightarrow \mathbf{Z}_p \xrightarrow{f} \mathbf{Z}_{p^2} \xrightarrow{g} \mathbf{Z}_p \longrightarrow 0$$

where $\phi(m) = qm \in \mathbf{Z}_{pq}$, $f(m) = pm \in \mathbf{Z}_{p^2}$, and ψ and g are the canonical projection maps. Exact sequence (3.2) is split exact while exact sequence (3.3) is not split exact. Both of these observations are easy consequences of the material on cyclic groups from Chapter 1; details are left as an exercise.

There is the following useful criterion for a short exact sequence to be split exact.

(3.9) Theorem. *If*

$$(3.4) \qquad\qquad 0 \longrightarrow M_1 \xrightarrow{f} M \xrightarrow{g} M_2 \longrightarrow 0$$

is a short exact sequence of R-modules, then the following are equivalent:

(1) *There exists a homomorphism $\alpha : M \to M_1$ such that $\alpha \circ f = 1_{M_1}$.*
(2) *There exists a homomorphism $\beta : M_2 \to M$ such that $g \circ \beta = 1_{M_2}$.*
(3) *The sequence (3.4) is split exact.*
      If these equivalent conditions hold then

$$M \cong \mathrm{Im}(f) \oplus \mathrm{Ker}(\alpha)$$
$$\cong \mathrm{Ker}(g) \oplus \mathrm{Im}(\beta)$$
$$\cong M_1 \oplus M_2.$$

*The homomorphisms α and β are said to **split** the exact sequence (3.4) or be a **splitting**.*

Proof. Suppose that (1) is satisfied and let $x \in M$. Then

$$\alpha(x - f(\alpha(x))) = \alpha(x) - (\alpha \circ f)(\alpha(x)) = 0$$

since $\alpha \circ f = 1_{M_1}$. Therefore, $x - f(\alpha(x)) \in \mathrm{Ker}(\alpha)$ so that

$$M = \mathrm{Ker}(\alpha) + \mathrm{Im}(f).$$

Now suppose that $f(y) = x \in \mathrm{Ker}(\alpha) \cap \mathrm{Im}(f)$. Then

$$0 = \alpha(x) = \alpha(f(y)) = y,$$

and therefore, $x = f(y) = 0$. Theorem 3.1 then shows that

$$M \cong \mathrm{Im}(f) \oplus \mathrm{Ker}(\alpha).$$

Define $\beta : M_2 \to M$ by

(3.5)
$$\beta(u) = v - f(\alpha(v))$$

where $g(v) = u$. Since g is surjective, there is such a $v \in M$, but it may be possible to write $u = g(v)$ for more than one choice of v. Therefore, we must verify that β is well defined. Suppose that $g(v) = u = g(v')$. Then $v - v' \in \mathrm{Ker}(g) = \mathrm{Im}(f)$ so that

$$(v - f(\alpha(v))) - (v' - f(\alpha(v'))) = (v - v') + (f(\alpha(v')) - f(\alpha(v)))$$
$$\in \mathrm{Im}(f) \cap \mathrm{Ker}(\alpha)$$
$$= \{0\}.$$

We conclude that β is well defined. Since it is clear from the construction of β that $g \circ \beta = 1_{M_2}$, we have verified that (1) implies (2) and that $M \cong \mathrm{Im}(f) \oplus \mathrm{Ker}(\alpha)$, i.e., that (3) holds.

The proof that (2) implies (1) and (3) is similar and is left as an exercise.

Suppose that (3) holds, that is, $M \cong M' \oplus M''$ where $M' = \mathrm{Ker}(g) = \mathrm{Im}(f)$. Let $\pi_1 : M \to M'$ and $\pi_2 : M \to M''$ be the projections, and $\iota : M'' \to M$ be the inclusion. Note that $\pi_1 \circ f : M_1 \to M'$ and $g \circ \iota : M'' \to M_2$ are isomorphisms. Define $\alpha : M \to M_1$ by $\alpha = (\pi_1 \circ f)^{-1} \circ \pi_1$ and $\beta : M_2 \to M$ by $\beta = \iota \circ (g \circ \iota)^{-1}$. Then $\alpha \circ f = 1_{M_1}$ and $g \circ \beta = 1_{M_2}$, so (1) and (2) hold. $\qquad \square$

If M and N are R-modules, then the set $\mathrm{Hom}_R(M, N)$ of all R-module homomorphisms $f : M \to N$ is an abelian group under function addition. According to Example 1.5 (8), $\mathrm{Hom}_R(M, N)$ is also an R-module provided that R is a commutative ring. Recall that $\mathrm{End}_R(M) = \mathrm{Hom}_R(M)$ denotes the endomorphism ring of the R-module M, and the ring multiplication is composition of homomorphisms. Example 1.5 (8) shows that $\mathrm{End}_R(M)$ is an R-algebra if the ring R is commutative. Example 1.5 (10) shows that $\mathrm{Hom}_R(R, M) \cong M$ for any R-module M.

Now consider R-modules M, M_1, N, and N_1, and let $\phi : N \to N_1$, $\psi : M \to M_1$ be R-module homomorphisms. Then there are functions

$$\phi_* : \mathrm{Hom}_R(M, N) \to \mathrm{Hom}_R(M, N_1)$$

and

$$\psi^* : \mathrm{Hom}_R(M_1, N) \to \mathrm{Hom}_R(M, N)$$

defined by

$$\phi_*(f) = \phi \circ f \qquad \text{for all} \quad f \in \text{Hom}_R(M, N)$$

and

$$\psi^*(g) = g \circ \psi \qquad \text{for all} \quad g \in \text{Hom}_R(M_1, N).$$

It is straightforward to check that $\phi_*(f+g) = \phi_*(f) + \phi_*(g)$ and $\psi^*(f+g) = \psi^*(f) + \psi^*(g)$ for appropriate f and g. That is, ϕ_* and ψ^* are homomorphisms of abelian groups, and if R is commutative, then they are also R-module homomorphisms.

Given a sequence of R-modules and R-module homomorphisms

(3.6) $$\cdots \longrightarrow M_{i-1} \xrightarrow{\phi_i} M_i \xrightarrow{\phi_{i+1}} M_{i+1} \longrightarrow \cdots$$

and an R-module N, then $\text{Hom}_R(\ , N)$ and $\text{Hom}_R(N, \ )$ produce two sequences of abelian groups (R-modules if R is commutative):

(3.7) $$\cdots \longrightarrow \text{Hom}_R(N, M_{i-1}) \xrightarrow{(\phi_i)_*} \text{Hom}_R(N, M_i)$$
$$\xrightarrow{(\phi_{i+1})_*} \text{Hom}_R(N, M_{i+1}) \longrightarrow \cdots$$

and

(3.8) $$\cdots \longleftarrow \text{Hom}_R(M_{i-1}, N) \xleftarrow{(\phi_i)^*} \text{Hom}_R(M_i, N)$$
$$\xleftarrow{(\phi_{i+1})^*} \text{Hom}_R(M_{i+1}, N) \longleftarrow \cdots.$$

A natural question is to what extent does exactness of sequence (3.6) imply exactness of sequences (3.7) and (3.8). One result along these lines is the following.

(3.10) Theorem. *Let*

(3.9) $$0 \longrightarrow M_1 \xrightarrow{\phi} M \xrightarrow{\psi} M_2$$

be a sequence of R-modules and R-module homomorphisms. Then the sequence (3.9) is exact if and only if the sequence

(3.10) $$0 \longrightarrow \text{Hom}_R(N, M_1) \xrightarrow{\phi_*} \text{Hom}_R(N, M) \xrightarrow{\psi_*} \text{Hom}_R(N, M_2)$$

is an exact sequence of $\mathbf{Z}$-modules for all R-modules N.
    If

(3.11) $$M_1 \xrightarrow{\phi} M \xrightarrow{\psi} M_2 \longrightarrow 0$$

is a sequence of R-modules and R-module homomorphisms, then the sequence (3.11) is exact if and only if the sequence

(3.12) $$0 \longrightarrow \text{Hom}_R(M_2, N) \xrightarrow{\psi^*} \text{Hom}_R(M, N) \xrightarrow{\phi^*} \text{Hom}_R(M_1, N)$$

is an exact sequence of $\mathbf{Z}$-modules for all R-modules N.

Proof. Assume that sequence (3.9) is exact and let N be an arbitrary R-module. Suppose that $f \in \text{Hom}_R(N, M)$ and $\phi_*(f) = 0$. Then

$$0 = \phi \circ f(x) = \phi(f(x))$$

for all $x \in N$. But ϕ is injective, so $f(x) = 0$ for all $x \in N$. That is, $f = 0$, and hence, ϕ_* is injective.

Since $\psi \circ \phi = 0$ (because sequence (3.9) is exact at M), it follows that

$$\psi_*(\phi_*(f)) = \psi \circ \phi_*(f) = \psi \circ \phi \circ f = 0$$

for all $f \in \text{Hom}_R(N, M)$. Thus $\text{Im}(\phi_*) \subseteq \text{Ker}(\psi_*)$. It remains to check the other inclusion. Suppose that $g \in \text{Hom}_R(N, M)$ with $\psi_*(g) = 0$, i.e., $\psi(g(x)) = 0$ for all $x \in N$. Since $\text{Ker}(\psi) = \text{Im}(\phi)$, for each $x \in N$, we may write $g(x) = \phi(y)$ with $y \in M_1$. Since ϕ is injective, y is uniquely determined by the equation $g(x) = \phi(y)$. Thus it is possible to define a function $f : N \to M_1$ by $f(x) = y$ whenever $g(x) = \phi(y)$. We leave it as an exercise to check that f is an R-module homomorphism. Since $\phi_*(f) = g$, we conclude that $\text{Ker}(\psi_*) = \text{Im}(\phi_*)$ so that sequence (3.10) is exact.

Exactness of sequence (3.12) is a similar argument, which is left as an exercise.

Conversely, assume that sequence (3.10) is exact for all R-modules N. Then ϕ_* is injective for all R-modules N. Then letting $N = \text{Ker}(\phi)$ and $\iota : N \to M_1$ be the inclusion, we see that $\phi_*(\iota) = \phi \circ \iota = 0$. Since $\phi_* : \text{Hom}_R(N, M_1) \to \text{Hom}_R(N, M)$ is injective, we see that $\iota = 0$, i.e., $N = \langle 0 \rangle$. Thus, ϕ is injective.

Now letting $N = M_1$ we see that

$$0 = (\psi_* \circ \phi_*)(1_{M_1}) = \psi \circ \phi.$$

Thus $\text{Im}(\phi) \subseteq \text{Ker}(\psi)$. Now let $N = \text{Ker}(\psi)$ and let $\iota : N \to M$ be the inclusion. Since $\psi_*(\iota) = \psi \circ \iota = 0$, exactness of Equation (3.10) implies that $\iota = \phi_*(\alpha)$ for some $\alpha \in \text{Hom}_R(N, M_1)$. Thus,

$$\text{Im}(\phi) \supseteq \text{Im}(\iota) = N = \text{Ker}(\psi),$$

and we conclude that sequence (3.9) is exact.

Again, exactness of sequence (3.11) is left as an exercise. □

Note that, even if

$$0 \longrightarrow M_1 \xrightarrow{\phi} M \xrightarrow{\psi} M_2 \longrightarrow 0$$

is a short exact sequence, the sequences (3.10) and (3.12) need not be short exact, i.e., neither ψ_* or ϕ^* need be surjective. Following are some examples to illustrate this.

(3.11) Example. Consider the following short exact sequence of **Z**-modules:

$$(3.13) \qquad 0 \longrightarrow \mathbf{Z} \xrightarrow{\phi} \mathbf{Z} \xrightarrow{\psi} \mathbf{Z}_m \longrightarrow 0$$

where $\phi(i) = mi$ and ψ is the canonical projection map. If $N = \mathbf{Z}_n$ then sequence (3.12) becomes

$$0 \longrightarrow \mathrm{Hom}_{\mathbf{Z}}(\mathbf{Z}_m, \mathbf{Z}_n) \longrightarrow \mathrm{Hom}_{\mathbf{Z}}(\mathbf{Z}, \mathbf{Z}_n) \xrightarrow{\phi^*} \mathrm{Hom}_{\mathbf{Z}}(\mathbf{Z}, \mathbf{Z}_n),$$

which, by Example 1.5 (10), becomes

$$0 \longrightarrow \mathrm{Hom}_{\mathbf{Z}}(\mathbf{Z}_m, \mathbf{Z}_n) \longrightarrow \mathbf{Z}_n \xrightarrow{\phi^*} \mathbf{Z}_n$$

so that

$$\mathrm{Hom}_{\mathbf{Z}}(\mathbf{Z}_m, \mathbf{Z}_n) = \mathrm{Ker}(\phi^*).$$

Let $d = \gcd(m, n)$, and write $m = m'd$, $n = n'd$. Let $f \in \mathrm{Hom}_{\mathbf{Z}}(\mathbf{Z}, \mathbf{Z}_n)$. Then, clearly, $\phi^*(f) = 0$ if and only if $\phi^*(f)(1) = 0$. But

$$\phi^*(f)(1) = f(m \cdot 1) = mf(1) = m'df(1).$$

Since m' is relatively prime to n, we have $m'df(1) = 0$ if and only if $df(1) = 0$, and this is true if and only if $f(1) \in n'\mathbf{Z}_n$. Hence, $\mathrm{Ker}(\phi^*) = n'\mathbf{Z}_n \cong \mathbf{Z}_d$, i.e.,

$$(3.14) \qquad \mathrm{Hom}_{\mathbf{Z}}(\mathbf{Z}_m, \mathbf{Z}_n) \cong \mathbf{Z}_d.$$

This example also shows that even if

$$0 \longrightarrow M_1 \longrightarrow M \longrightarrow M_2 \longrightarrow 0$$

is exact, the sequences (3.10) and (3.12) are not, in general, part of short exact sequences. For simplicity, take $m = n$. Then sequence (3.12) becomes

$$(3.15) \qquad 0 \longrightarrow \mathbf{Z}_n \longrightarrow \mathbf{Z}_n \xrightarrow{\phi^*} \mathbf{Z}_n$$

with $\phi^* = 0$ so that ϕ^* is not surjective, while sequence (3.10) becomes

$$(3.16) \qquad 0 \longrightarrow \mathrm{Hom}_{\mathbf{Z}}(\mathbf{Z}_n, \mathbf{Z}) \longrightarrow \mathrm{Hom}_{\mathbf{Z}}(\mathbf{Z}_n, \mathbf{Z}) \xrightarrow{\psi_*} \mathrm{Hom}_{\mathbf{Z}}(\mathbf{Z}_n, \mathbf{Z}_n).$$

Since $\mathrm{Hom}_{\mathbf{Z}}(\mathbf{Z}_n, \mathbf{Z}) = 0$ and $\mathrm{Hom}_{\mathbf{Z}}(\mathbf{Z}_n, \mathbf{Z}_n) \cong \mathbf{Z}_n$, sequence (3.16) becomes

$$0 \longrightarrow 0 \longrightarrow 0 \xrightarrow{\psi_*} \mathbf{Z}_n$$

and ψ_* is certainly not surjective.

These examples show that Theorem 3.10 is the best statement that can be made in complete generality concerning preservation of exactness under application of Hom_R. There is, however, the following criterion for the preservation of short exact sequences under Hom:

(3.12) Theorem. *Let N be an arbitrary R-module. If*

$$(3.17) \qquad 0 \longrightarrow M_1 \xrightarrow{\phi} M \xrightarrow{\psi} M_2 \longrightarrow 0$$

is a split short exact sequence of R-modules, then

$$(3.18) \quad 0 \longrightarrow \operatorname{Hom}_R(N, M_1) \xrightarrow{\phi_*} \operatorname{Hom}_R(N, M) \xrightarrow{\psi_*} \operatorname{Hom}_R(N, M_2) \longrightarrow 0$$

and

$$(3.19) \quad 0 \longrightarrow \operatorname{Hom}_R(M_2, N) \xrightarrow{\psi^*} \operatorname{Hom}_R(M, N) \xrightarrow{\phi^*} \operatorname{Hom}_R(M_1, N) \longrightarrow 0$$

are split short exact sequences of abelian groups (R-modules if R is commutative).

Proof. We will prove the split exactness of sequence (3.18); (3.19) is similar and it is left as an exercise. Given Theorem 3.10, it is only necessary to show that ψ_* is surjective and that there is a splitting for sequence (3.18). Let $\beta : M_2 \to M$ split the exact sequence (3.17) and let $f \in \operatorname{Hom}_R(N, M_2)$. Then

$$\begin{aligned}
\psi_* \circ \beta_*(f) &= \psi_*(\beta \circ f) \\
&= (\psi \circ \beta) \circ f \\
&= (1_{M_2}) \circ f \\
&= \left(1_{\operatorname{Hom}_R(N, M_2)}\right)(f).
\end{aligned}$$

Thus, $\psi_* \circ \beta_* = 1_{\operatorname{Hom}_R(N, M_2)}$ so that ψ_* is surjective and β_* is a splitting of exact sequence (3.18). $\qquad\square$

(3.13) Corollary. *Let M_1, M_2, and N be R-modules. Then*

$$(3.20) \qquad \operatorname{Hom}_R(N, M_1 \oplus M_2) \cong \operatorname{Hom}_R(N, M_1) \oplus \operatorname{Hom}_R(N, M_2)$$

and

$$(3.21) \qquad \operatorname{Hom}_R(M_1 \oplus M_2, N) \cong \operatorname{Hom}_R(M_1, N) \oplus \operatorname{Hom}_R(M_2, N).$$

The isomorphisms are $\mathbf{Z}$-module isomorphisms (R-module isomorphisms if R is commutative).

Proof. Both isomorphisms follow by applying Theorems 3.12 and 3.9 to the split exact sequence

$$0 \longrightarrow M_1 \xrightarrow{\iota} M_1 \oplus M_2 \xrightarrow{\pi} M_2 \longrightarrow 0$$

where $\iota(m) = (m, 0)$ is the canonical injection and $\pi(m_1, m_2) = m_2$ is the canonical projection. $\qquad\square$

(3.14) *Remarks.*

(1) Notice that isomorphism (3.20) is given explicitly by

$$\Phi(f) = (\pi_1 \circ f, \, \pi_2 \circ f)$$

where $f \in \operatorname{Hom}_R(N, M_1 \oplus M_2)$ and $\pi_i(m_1, m_2) = m_i$ (for $i = 1, 2$); while isomorphism (3.21) is given explicitly by

$$\Psi(f) = (f \circ \iota_1, \, f \circ \iota_2)$$

where $f \in \operatorname{Hom}_R(M_1 \oplus M_2, N)$, $\iota_1 : M_1 \to M_1 \oplus M_2$ is given by $\iota_1(m) = (m, 0)$ and $\iota_2 : M_2 \to M_1 \oplus M_2$ is given by $\iota_2(m) = (0, m)$.

(2) Corollary 3.13 actually has a natural extension to arbitrary (not necessarily finite) direct sums. We conclude this section by stating this extension. The proof is left as an exercise for the reader.

(3.15) Proposition. *Let $\{M_i\}_{i \in I}$ and $\{N_j\}_{j \in J}$ be indexed families (not necessarily finite) of R-modules, and let $M = \oplus_{i \in I} M_i$, $N = \oplus_{j \in J} N_j$. Then*

$$\operatorname{Hom}_R(M, N) \cong \prod_{i \in I} \left(\bigoplus_{j \in J} \operatorname{Hom}_R(M_i, N_j) \right).$$

Proof. Exercise. □

3.4 Free Modules

(4.1) Definition. *Let R be a ring and let M be an R-module. A subset $S \subseteq M$ is said to be R-**linearly dependent** if there exist distinct $x_1, \ldots, x_n$ in S and elements $a_1, \ldots, a_n$ of R, not all of which are 0, such that*

$$a_1 x_1 + \cdots + a_n x_n = 0.$$

*A set that is not R-linearly dependent is said to be R-**linearly independent**.*

When the ring R is implicit from the context, we will sometimes write linearly dependent (or just dependent) and linearly independent (or just independent) in place of the more cumbersome R-linearly dependent or R-linearly independent. In case S contains only finitely many elements $x_1, x_2, \ldots, x_n$, we will sometimes say that $x_1, x_2, \ldots, x_n$ are R-linearly dependent or R-linearly independent instead of saying that $S = \{x_1, \ldots, x_n\}$ is R-linearly dependent or R-linearly independent.

(4.2) *Remarks.*

(1) To say that $S \subseteq M$ is R-linearly independent means that whenever there is an equation

$$a_1 x_1 + \cdots + a_n x_n = 0$$

where $x_1, \ldots, x_n$ are distinct elements of S and $a_1, \ldots, a_n$ are in R, then

$$a_1 = \cdots = a_n = 0.$$

(2) Any set S that contains a linearly dependent set is linearly dependent.

(3) Any subset of a linearly independent set S is linearly independent.

(4) Any set that contains 0 is linearly dependent since $1 \cdot 0 = 0$.

(5) A set $S \subseteq M$ is linearly independent if and only if every finite subset of S is linearly independent.

(4.3) Definition. *Let M be an R-module. A subset S of M is a* **basis** *of M if S generates M as an R-module and if S is R-linearly independent. That is, $S \subseteq M$ is a basis if and only if $M = \{0\}$, in which case $S = \emptyset$ is a basis, or $M \neq \{0\}$ and*

(1) *every $x \in M$ can be written as*

$$x = a_1 x_1 + \cdots + a_n x_n$$

for some $x_1, \ldots, x_n \in S$ and $a_1, \ldots, a_n \in R$, and

(2) *whenever there is an equation*

$$a_1 x_1 + \cdots + a_n x_n = 0$$

where $x_1, \ldots, x_n$ are distinct elements of S and $a_1, \ldots, a_n$ are in R, then

$$a_1 = \cdots = a_n = 0.$$

It is clear that conditions (1) and (2) in the definition of basis can be replaced by the single condition:

(1′) $S \subseteq M$ is a basis of $M \neq \{0\}$ if and only if every $x \in M$ can be written *uniquely* as

$$x = a_1 x_1 + \cdots + a_n x_n$$

for $a_1, \ldots, a_n \in R$ and $x_1, \ldots, x_n \in S$.

(4.4) Definition. *An R-module M is a* **free R-module** *if it has a basis.*

(4.5) *Remark.* According to Theorem 3.2, to say that $S = \{x_j\}_{j \in J}$ is a basis of M is equivalent to M being the direct sum of the family $\{Rx_j\}_{j \in J}$

of submodules of M, where $\text{Ann}(x_j) = \{0\}$ for all $j \in J$. Moreover, if J is any index set, then $N = \oplus_{j \in J} R_j$, where $R_j = R$ for all $j \in J$, is a free R-module with basis $S = \{e_j\}_{j \in J}$, where $e_j \in N$ is defined by $e_j = (\delta_{jk})_{k \in J}$. Here, δ_{jk} is the kronecker delta function, i.e., $\delta_{jk} = 1 \in R$ whenever $j = k$ and $\delta_{jk} = 0 \in R$ otherwise. N is said to be **free on the index set** J.

(4.6) Examples.

(1) If R is a field then R-linear independence and R-linear dependence in a vector space V over R are the same concepts used in linear algebra.

(2) R^n is a free module with basis $S = \{e_1, \ldots, e_n\}$ where

$$e_i = (0, \ldots, 0, 1, 0, \ldots, 0)$$

with a 1 in the ith position.

(3) $M_{m,n}(R)$ is a free R-module with basis

$$S = \{E_{ij} : 1 \le i \le m, 1 \le j \le n\}.$$

(4) The ring $R[X]$ is a free R-module with basis $\{X^n : n \in \mathbf{Z}^+\}$. As in Example 4.6 (2), $R[X]$ is also a free $R[X]$-module with basis $\{1\}$.

(5) If G is a finite abelian group then G is a $\mathbf{Z}$-module, but no nonempty subset of G is $\mathbf{Z}$-linearly independent. Indeed, if $g \in G$ then $|G| \cdot g = 0$ but $|G| \ne 0$. Therefore, finite abelian groups can never be free $\mathbf{Z}$-modules, except in the trivial case $G = \{0\}$ when $\emptyset$ is a basis.

(6) If R is a commutative ring and $I \subseteq R$ is an ideal, then I is an R-module. However, if I is not a principal ideal, then I is not free as an R-module. Indeed, no generating set of I can be linearly independent since the equation $(-a_2)a_1 + a_1 a_2 = 0$ is valid for any a_1, $a_2 \in R$.

(7) If M_1 and M_2 are free R-modules with bases S_1 and S_2 respectively, then $M_1 \oplus M_2$ is a free R-module with basis $S_1' \cup S_2'$, where

$$S_1' = \{(x, 0) : x \in S_1\} \quad \text{and} \quad S_2' = \{(0, y) : y \in S_2\}.$$

(8) More generally, if $\{M_j\}_{j \in J}$ is a family of free R-modules and $S_j \subseteq M_j$ is a basis of M_j for each $j \in J$, then $M = \oplus_{j \in J} M_j$ is a free R-module and $S = \cup_{j \in J} S_j'$ is a basis of M, where $S_j' \subseteq M$ is defined by

$$S_j' = \{s_{j\alpha}' = (\delta_{jk} s_{j\alpha})_{k \in J} : s_{j\alpha} \in S_j\}.$$

Informally, S_j' consists of all elements of M that contain an element of S_j in the jth component and 0 in all other components. This example incorporates both Example 4.6 (7) and Example 4.6 (2).

Example 4.6 (5) can be generalized to the following fact.

(4.7) Lemma. *Let M be an R-module where R is a commutative ring. Then an element $x \in M$ is R-independent if and only if $\mathrm{Ann}(x) = \{0\}$. In particular, an element $a \in R$ is an R-independent subset of the R-module R if and only if a is not a zero divisor.*

Proof. Exercise.   □

(4.8) Proposition. *Let R be an integral domain and let M be a free R-module. Then M is torsion-free.*

Proof. Let M have a basis $S = \{x_j\}_{j \in J}$ and let $x \in M_\tau$. Then $ax = 0$ for some $a \neq 0 \in R$. Write $x = \sum_{j \in J} a_j x_j$. Then

$$0 = ax = \sum_{j \in J}(aa_j)x_j.$$

Since S is a basis of M, it follows that $aa_j = 0$ for all $j \in J$, and since $a \neq 0$ and R is an integral domain, we conclude that $a_j = 0$ for all $j \in J$. Therefore, $x = 0$, and hence, $M_\tau = \langle 0 \rangle$ so that M is torsion-free.   □

The existence of a basis for an R-module M greatly facilitates the construction of R-module homomorphisms from M to another R-module N. In fact, there is the following important observation.

(4.9) Proposition. *Let M be a free R-module with basis S, let N be any R-module, and let $h : S \to N$ be any function. Then there is a unique $f \in \mathrm{Hom}_R(M, N)$ such that $f|_S = h$.*

Proof. Let $S = \{x_j\}_{j \in J}$. Then any $x \in M$ can be written uniquely as $x = \sum_{j \in J} a_j x_j$ where at most finitely many a_j are not 0. Define $f : M \to N$ by

$$f(x) = \sum_{j \in J} a_j h(x_j).$$

It is straightforward to check that $f \in \mathrm{Hom}_R(M, N)$ and that $f|_S = h$.   □

Remark. The content of Proposition 4.9 is usually expressed as saying that the value of a homomorphism can be arbitrarily assigned on a basis.

(4.10) Corollary. *Suppose that M is a free R-module with basis $S = \{x_j\}_{j \in J}$. Then*

$$\mathrm{Hom}_R(M, N) \cong \prod_{j \in J} N_j$$

where $N_j = N$ for all $j \in J$.

Proof. Define $\Phi : \mathrm{Hom}_R(M, N) \to \prod_{j \in J} N_j$ by $\Phi(f) = (f(x_j))_{j \in J}$. Then Φ is an isomorphism of abelian groups (R-modules if R is commutative).   □

(4.11) Theorem. *Let R be a commutative ring and let M and N be finitely generated free R-modules. Then $\mathrm{Hom}_R(M, N)$ is a finitely generated free R-module.*

Proof. Let $\mathcal{B} = \{v_1, \dots, v_m\}$ be a basis of M and $\mathcal{C} = \{w_1, \dots, w_n\}$ a basis of N. Define $f_{ij} \in \mathrm{Hom}_R(M, N)$ for $1 \le i \le m$ and $1 \le j \le n$ by

$$f_{ij}(v_k) = \begin{cases} w_j & \text{if } k = i, \\ 0 & \text{if } k \ne i. \end{cases}$$

f_{ij} is a uniquely defined element of $\mathrm{Hom}_R(M, N)$ by Proposition 4.9.

We claim that $\{f_{ij} : 1 \le i \le m; \ 1 \le j \le n\}$ is a basis of $\mathrm{Hom}_R(M, N)$. To see this suppose that $f \in \mathrm{Hom}_R(M, N)$ and for $1 \le i \le m$ write

$$f(v_i) = a_{i1}w_1 + \cdots + a_{in}w_n.$$

Let

$$g = \sum_{i=1}^m \sum_{j=1}^n a_{ij} f_{ij}.$$

Then

$$g(v_k) = a_{k1}w_1 + \cdots + a_{kn}w_n = f(v_k)$$

for $1 \le k \le m$, so $g = f$ since the two homomorphisms agree on a basis of M. Thus, $\{f_{ij} : 1 \le i \le m; \ 1 \le j \le n\}$ generates $\mathrm{Hom}_R(M, N)$, and we leave it as an exercise to check that this set is linearly independent and, hence, a basis. $\square$

(4.12) Remarks.

(1) A second (essentially equivalent) way to see the same thing is to write $M \cong \oplus_{i=1}^m R$ and $N \cong \oplus_{j=1}^n R$. Then, Corollary 3.13 shows that

$$\mathrm{Hom}_R(M, N) \cong \bigoplus_{i=1}^m \bigoplus_{j=1}^n \mathrm{Hom}_R(R, R).$$

But any $f \in \mathrm{Hom}_R(R, R)$ can be written as $f = f(1) \cdot 1_R$. Thus $\mathrm{Hom}_R(R, R) \cong R$ so that

$$\mathrm{Hom}_R(M, N) \cong \bigoplus_{i=1}^m \bigoplus_{j=1}^n R.$$

(2) The hypothesis of finite generation of M and N is crucial for the validity of Theorem 4.11. For example, if $R = \mathbf{Z}$ and $M = \oplus_1^\infty \mathbf{Z}$ is the free $\mathbf{Z}$-module on the index set $\mathbf{N}$, then Corollary 4.10 shows that

$$\mathrm{Hom}_{\mathbf{Z}}(M, \mathbf{Z}) \cong \prod_1^\infty \mathbf{Z}.$$

But the $\mathbf{Z}$-module $\prod_1^\infty \mathbf{Z}$ is not a free $\mathbf{Z}$-module. (For a proof of this fact (which uses cardinality arguments), see I. Kaplansky, *Infinite Abelian Groups*, University of Michigan Press, (1968) p. 48.)

(4.13) Proposition. *Let M be a free R-module with basis $S = \{x_j\}_{j \in J}$. If I is an ideal of R, then IM is a submodule of M and the quotient module M/IM is an R/I-module. Let $\pi : M \to M/IM$ be the projection map. Then M/IM is a free R/I-module with basis $\pi(S) = \{\pi(x_j)\}_{j \in J}$.*

Proof. Exercise.   □

(4.14) Proposition. *Every R-module M is the quotient of a free module and if M is finitely generated, then M is the quotient of a finitely generated free R-module. In fact, we may take $\mu(F) = \mu(M)$.*

Proof. Let $S = \{x_j\}_{j \in J}$ be a generating set for the R-module M and let $F = \oplus_{j \in J} R_j$ where $R_j = R$ be the free R-module on the index set J. Define the homomorphism $\psi : F \to M$ by

$$\psi((a_j)_{j \in J}) = \sum_{j \in J} a_j x_j.$$

Since S is a generating set for M, ψ is surjective and hence $M \cong F/\mathrm{Ker}(\psi)$. Note that if $|S| < \infty$ then F is finitely generated. (Note that every module has a generating set S since we may take $S = M$.) Since M is a quotient of F, we have $\mu(M) \leq \mu(F)$. But F is free on the index set J (Remark 4.5), so $\mu(F) \leq |J|$, and since J indexes a generating set of M, it follows that $\mu(F) \leq \mu(M)$ if S is a minimal generating set of M. Hence we may take F with $\mu(F) = \mu(M)$.   □

(4.15) Definition. *If M is an R-module then a short exact sequence*

$$0 \longrightarrow K \longrightarrow F \longrightarrow M \longrightarrow 0$$

where F is a free R-module is called a **free presentation** *of M.*

Thus, Proposition 4.14 states that every module has a free presentation.

(4.16) Proposition. *If F is a free R-module then every short exact sequence*

$$0 \longrightarrow M_1 \longrightarrow M \overset{f}{\longrightarrow} F \longrightarrow 0$$

of R-modules is split exact.

Proof. Let $S = \{x_j\}_{j \in J}$ be a basis of the free module F. Since f is surjective, for each $j \in J$ there is an element $y_j \in M$ such that $f(y_j) = x_j$. Define $h : S \to M$ by $h(x_j) = y_j$. By Proposition 4.9, there is a unique $\beta \in$

$\mathrm{Hom}_R(F, M)$ such that $\beta|_S = h$. Since $f \circ \beta(x_j) = x_j = 1_F(x_j)$ for all $j \in J$, it follows that $f \circ \beta = 1_F$, and the result follows from Theorem 3.9. $\qquad\square$

(4.17) Corollary.

(1) *Let M be an R-module and $N \subseteq M$ a submodule with M/N free. Then $M \cong N \oplus (M/N)$.*

(2) *If M is an R-module and F is a free R-module, then $M \cong \mathrm{Ker}(f) \oplus F$ for every surjective homomorphism $f : M \to F$.*

Proof. (1) Since M/N is free, the short exact sequence

$$0 \longrightarrow N \longrightarrow M \longrightarrow M/N \longrightarrow 0$$

is split exact by Proposition 4.16 Therefore, $M \cong N \oplus (M/N)$ by Theorem 3.9.

(2) Take $N = \mathrm{Ker}(f)$ in part (1). $\qquad\square$

(4.18) Corollary. *Let N be an arbitrary R-module and F a free R-module. If*

(4.1) $$0 \longrightarrow M_1 \overset{\phi}{\longrightarrow} M \overset{\psi}{\longrightarrow} F \longrightarrow 0$$

is a short exact sequence of R-modules, then

$$0 \longrightarrow \mathrm{Hom}_R(N, M_1) \overset{\phi_*}{\longrightarrow} \mathrm{Hom}_R(N, M) \overset{\psi_*}{\longrightarrow} \mathrm{Hom}_R(N, F) \longrightarrow 0$$

is a (split) short exact sequence of abelian groups (R-modules if R is commutative).

Proof. By Proposition 4.16, the sequence (4.1) is split exact, so the corollary follows immediately from Theorem 3.12. $\qquad\square$

(4.19) Remark. It is a theorem that any two bases of a free module over a commutative ring R have the same cardinality. This result is proved for finite-dimensional vector spaces by showing that any set of vectors of cardinality larger than that of a basis must be linearly dependent. The same procedure works for free modules over any commutative ring R, but it does require the theory of solvability of homogeneous linear equations over a commutative ring. However, the result can be proved for R a PID without the theory of solvability of homogeneous linear equations over R; we prove this result in Section 3.6. The result for general commutative rings then follows by an application of Proposition 4.13.

The question of existence of a basis of a module, that is, to ask if a given module is free, is a delicate question for a general commutative ring R. We have seen examples of **Z**-modules, namely, finite abelian groups, which are not free. We will conclude this section with the fact that all modules over division rings, in particular, vector spaces, are free modules. In Section 3.6 we will study in detail the theory of free modules over a PID.

(4.20) Theorem. *Let D be a division ring and let V be a D-module. Then V is a free D-module. In particular, every vector space V has a basis.*

Proof. The proof is an application of Zorn's lemma.

Let S be a generating set for V and let $B_0 \subseteq S$ be *any* linearly independent subset of S (we allow $B_0 = \emptyset$). Let $\mathcal{T}$ be the set of all linearly independent subsets of S containing B_0 and partially order $\mathcal{T}$ by inclusion. If $\{B_i\}$ is a chain in $\mathcal{T}$, then $\cup B_i$ is a linearly independent subset of S that contains B_0; thus, every chain in $\mathcal{T}$ has an upper bound. By Zorn's lemma, there is a maximal element in $\mathcal{T}$, so let B be a maximal linearly independent subset of S containing B_0. We claim that $S \subseteq \langle B \rangle$ so that $V = \langle S \rangle \subseteq \langle B \rangle$. Let $v \in S$. Then the maximality of B implies that $V \cup \{v\}$ is linearly dependent so that there is an equation

$$\sum_{i=1}^{m} a_i v_i + bv = 0$$

where $v_1, \ldots, v_m$ are distinct elements of B and $a_1, \ldots, a_m, b \in D$ are not all 0. If $b = 0$ it would follow that $\sum_{i=1}^{m} a_i v_i = 0$ with not all the scalars $a_i = 0$. But this contradicts the linear independence of B. Therefore, $b \neq 0$ and we conclude

$$v = b^{-1}(bv) = \sum_{i=1}^{m} (-b^{-1} a_i) v_i \in \langle B \rangle.$$

Therefore, $S \subseteq \langle B \rangle$, and as observed above, this implies that B is a basis of V. $\square$

The proof of Theorem 4.20 actually proved more than the existence of a basis of V. Specifically, the following more precise result was proved.

(4.21) Theorem. *Let D be a division ring and let V be a D-module. If S spans V and $B_0 \subseteq S$ is a linearly independent subset, then there is a basis B of V such that $B_0 \subseteq B \subseteq S$.*

Proof. $\square$

(4.22) Corollary. *Let D be a division ring, and let V be a D-module.*

(1) *Any linearly independent subset of V can be extended to a basis of V.*
(2) *A maximal linearly independent subset of V is a basis.*
(3) *A minimal generating set of V is a basis.*

Proof. Exercise. $\square$

Notice that the above proof used the existence of inverses in the division ring D in a crucial way. We will return in Section 3.6 to study criteria that ensure that a module is free if the ring R is assumed to be a PID. Even when R is a PID, e.g., $R = \mathbf{Z}$, we have seen examples of R modules that are not free, so we will still be required to put restrictions on the module M to ensure that it is free.

3.5 Projective Modules

The property of free modules given in Proposition 4.16 is a very useful one, and it is worth investigating the class of those modules that satisfy this condition. Such modules are characterized in the following theorem.

(5.1) Theorem. *The following conditions on an R-module P are equivalent.*

(1) *Every short exact sequence of R-modules*

$$0 \longrightarrow M_1 \longrightarrow M \longrightarrow P \longrightarrow 0$$

splits.

(2) *There is an R-module P' such that $P \oplus P'$ is a free R-module.*

(3) *For any R-module N and any surjective R-module homomorphism $\psi :$ $M \to P$, the homomorphism*

$$\psi_* : \operatorname{Hom}_R(N,\, M) \to \operatorname{Hom}_R(N,\, P)$$

is surjective.

(4) *For any surjective R-module homomorphism $\phi : M \to N$, the homomorphism*

$$\phi_* : \operatorname{Hom}_R(P,\, M) \to \operatorname{Hom}_R(P,\, N)$$

is surjective.

Proof. (1) $\Rightarrow$ (2). Let $0 \longrightarrow K \longrightarrow F \longrightarrow P \longrightarrow 0$ be a free presentation of P. Then this short exact sequence splits so that $F \cong P \oplus K$ by Theorem 3.9.

(2) $\Rightarrow$ (3). Suppose that $F = P \oplus P'$ is free. Given a surjective R-module homomorphism $\psi : M \to P$, let $\psi' = \psi \oplus 1_{P'} : M \oplus P' \to P \oplus P' = F$; this is also a surjective homomorphism, so there is an exact sequence

$$0 \longrightarrow \operatorname{Ker}(\psi') \longrightarrow M \oplus P' \xrightarrow{\psi'} F \longrightarrow 0.$$

Since F is free, Proposition 4.16 implies that this sequence is split exact; Theorem 3.12 then shows that

$$\psi'_* : \operatorname{Hom}_R(N,\, M \oplus P') \to \operatorname{Hom}_R(N,\, P \oplus P')$$

is a surjective homomorphism. Now let $f \in \mathrm{Hom}_R(N, P)$ be arbitrary and let $f' = \iota \circ f$, where $\iota : P \to P \oplus P'$ is the inclusion map. Then there is an $\tilde{f} \in \mathrm{Hom}_R(N, M \oplus P')$ with $\psi'_*(\tilde{f}) = f'$. Let $\pi : M \oplus P' \to M$ and $\pi' : P \oplus P' \to P$ be the projection maps. Note that $\pi' \circ \iota = 1_P$ and $\psi \circ \pi = \pi' \circ \psi'$. Then

$$\psi_*(\pi \circ \tilde{f}) = \psi \circ (\pi \circ \tilde{f})$$
$$= \pi' \circ \psi' \circ \tilde{f}$$
$$= \pi' \circ f'$$
$$= (\pi' \circ \iota) \circ f$$
$$= f.$$

Therefore, ψ_* is surjective.

(3) $\Rightarrow$ (4). Let $0 \longrightarrow K \longrightarrow F \overset{\psi}{\longrightarrow} P \longrightarrow 0$ be a free presentation of P. By property (3), there is a $\beta \in \mathrm{Hom}_R(P, F)$ such that $\psi_*(\beta) = 1_P$, i.e., $\psi \circ \beta = 1_P$. Let $\phi : M \to N$ be any surjective R-module homomorphism and let $f \in \mathrm{Hom}_R(P, N)$. Then there is a commutative diagram of R-module homomorphisms

$$F \overset{\psi}{\longrightarrow} P \longrightarrow 0$$
$$\downarrow f$$
$$M \overset{\phi}{\longrightarrow} N \longrightarrow 0$$

with exact rows. Let $S = \{x_j\}_{j \in J}$ be a basis of F. Since ϕ is surjective, we may choose $y_j \in M$ such that $\phi(y_j) = f \circ \psi(x_j)$ for all $j \in J$. By Proposition 4.9, there is an R-module homomorphism $g : F \to M$ such that $g(x_j) = y_j$ for all $j \in J$. Since $\phi \circ g(x_j) = \phi(y_j) = f \circ \psi(x_j)$, it follows that $\phi \circ g = f \circ \psi$. Define $\tilde{f} \in \mathrm{Hom}_R(P, M)$ by $\tilde{f} = g \circ \beta$ and observe that

$$\phi_*(\tilde{f}) = \phi \circ (g \circ \beta)$$
$$= f \circ \psi \circ \beta$$
$$= f \circ 1_P$$
$$= f.$$

Hence, $\phi_* : \mathrm{Hom}_R(P, M) \to \mathrm{Hom}_R(P, N)$ is surjective.

(4) $\Rightarrow$ (1). A short exact sequence

$$0 \longrightarrow M_1 \longrightarrow M \overset{\psi}{\longrightarrow} P \longrightarrow 0,$$

in particular, includes a surjection $\psi : M \to P$. Now take $N = P$ in part (4). Thus,

$$\psi_* : \mathrm{Hom}_R(P, M) \to \mathrm{Hom}_R(P, P)$$

is surjective. Choose $\beta : P \to M$ with $\psi_*(\beta) = 1_P$. Then β splits the short exact sequence and the result is proved.   □

(5.2) Definition. *An R-module P satisfying any of the equivalent conditions of Theorem 5.1 is called* **projective**.

As noted before Theorem 5.1, projective modules are introduced as the class of modules possessing the property that free modules were shown to possess in Proposition 4.16. Therefore, we have the following fact:

(5.3) Proposition. *Free R-modules are projective.*

Proof.   □

(5.4) Corollary. *Let R be an integral domain. If P is a projective R-module, then P is torsion-free.*

Proof. By Theorem 5.1 (2), P is a submodule of a free module F over R. According to Proposition 4.8, every free module over an integral domain is torsion-free, and every submodule of a torsion-free module is torsion-free.

                                                                                            □

(5.5) Corollary. *An R-module P is a finitely generated projective R-module if and only if P is a direct summand of a finitely generated free R-module.*

Proof. Suppose that P is finitely generated and projective. By Proposition 4.14, there is a free presentation

$$0 \longrightarrow K \longrightarrow F \longrightarrow P \longrightarrow 0$$

such that F is free and $\mu(F) = \mu(P) < \infty$. By Theorem 5.1, P is a direct summand of F.

Conversely, assume that P is a direct summand of a finitely generated free R-module F. Then P is projective, and moreover, if $P \oplus P' \cong F$ then $F/P' \cong P$ so that P is finitely generated.   □

(5.6) Examples.

(1) Every free module is projective.

(2) Suppose that m and n are relatively prime natural numbers. Then as abelian groups $\mathbf{Z}_{mn} \cong \mathbf{Z}_m \oplus \mathbf{Z}_n$. It is easy to check that this isomorphism is also an isomorphism of $\mathbf{Z}_{mn}$-modules. Therefore, $\mathbf{Z}_m$ is a direct summand of a free $\mathbf{Z}_{mn}$-module, and hence it is a projective $\mathbf{Z}_{mn}$-module. However, $\mathbf{Z}_m$ is not a free $\mathbf{Z}_{mn}$ module since it has fewer than mn elements.

(3) Example 5.6 (2) shows that projective modules need not be free. We will present another example of this phenomenon in which the ring R is an integral domain so that simple cardinality arguments do not suffice. Let $R = \mathbf{Z}[\sqrt{-5}]$ and let I be the ideal $I = \langle 2, 1 + \sqrt{-5} \rangle = \langle a_1, a_2 \rangle$. It is easily shown that I is not a principal ideal, and hence by Example 4.6 (6), we see that I cannot be free as an R-module. We claim that I

is a projective R-module. To see this, let $b = 1 - \sqrt{-5} \in R$, let F be a free R-module with basis $\{s_1, s_2\}$, and let $\phi : F \to I$ be the R-module homomorphism defined by

$$\phi(r_1 s_1 + r_2 s_2) = r_1 a_1 + r_2 a_2.$$

Now define an R-module homomorphism $\alpha : I \to F$ by

$$\alpha(a) = -as_1 + ((ab)/2)s_2.$$

Note that this makes sense because 2 divides ab for every $a \in I$. Now for $a \in I$,

$$\begin{aligned}
\phi \circ \alpha(a) &= \phi(-as_1 + ((ab)/2)s_2) \\
&= -aa_1 + ((ab)/2)a_2 \\
&= -aa_1 + aa_2 b/2 \\
&= -2a + 3a \\
&= a
\end{aligned}$$

so that α is a splitting of the surjective map ϕ. Hence, $F \cong \operatorname{Ker}(\phi) \oplus I$ and by Theorem 5.1, I is a projective R-module.

Concerning the construction of new projective modules from old ones, there are the following two simple facts:

(5.7) Proposition. *Let $\{P_j\}_{j \in J}$ be a family of R-modules, and let $P = \oplus_{j \in J} P_j$. Then P is projective if and only if P_j is projective for each $j \in J$.*

Proof. Suppose that P is projective. Then by Theorem 5.1, there is an R-module P' such that $P \oplus P' = F$ is a free R-module. Then

$$F = P \oplus P' = \left(\bigoplus_{j \in J} P_j\right) \oplus P',$$

and hence, each P_j is also a direct summand of the free R-module F. Thus, P_j is projective.

Conversely, suppose that P_j is projective for every $j \in J$ and let P_j' be an R-module such that $P_j \oplus P_j' = F_j$ is free. Then

$$\begin{aligned}
P \oplus \left(\bigoplus_{j \in J} P_j'\right) &\cong \bigoplus_{j \in J} (P_j \oplus P_j') \\
&\cong \bigoplus_{j \in J} F_j,
\end{aligned}$$

Since the direct sum of free modules is free (Example 4.6 (8)), it follows that P is a direct summand of a free module, and hence P is projective. $\square$

(5.8) Proposition. *Let R be a commutative ring and let P and Q be finitely generated projective R-modules. Then $\mathrm{Hom}_R(P, Q)$ is a finitely generated projective R-module:*

Proof. Since P and Q are finitely generated projective R-modules, there are R-modules P' and Q' such that $P \oplus P'$ and $Q \oplus Q'$ are finitely generated free modules. Therefore, by Theorem 4.11, $\mathrm{Hom}_R(P \oplus P', Q \oplus Q')$ is a finitely generated free R-module. But

$$\mathrm{Hom}_R(P \oplus P', Q \oplus Q') \cong \mathrm{Hom}_R(P, Q) \oplus \mathrm{Hom}_R(P, Q')$$
$$\oplus \mathrm{Hom}_R(P', Q) \oplus \mathrm{Hom}_R(P', Q')$$

so that $\mathrm{Hom}_R(P, Q)$ is a direct summand of a finitely generated free R-module, and therefore, it is projective and finitely generated by Corollary 5.5. □

Example 5.6 (3) was an example of an ideal in a ring R that was projective as an R-module, but not free. According to Example 4.6 (6), an ideal I in a ring R is free as an R-module if and only if the ideal is principal. It is a natural question to ask which ideals in a ring R are projective as R-modules. Since this turns out to be an important question in number theory, we will conclude our brief introduction to the theory of projective modules by answering this question for integral domains R.

(5.9) Definition. *Let R be an integral domain and let K be the quotient field of R. An ideal $I \subseteq R$ is said to be **invertible** if there are elements $a_1, \ldots, a_n \in I$ and $b_1, \ldots, b_n \in K$ such that*

(5.1) $b_i I \subseteq R$ for $1 \le i \le n$, and

(5.2) $a_1 b_1 + \cdots + a_n b_n = 1$.

(5.10) Examples.

(1) If $I \subseteq R$ is the principal ideal $I = \langle a \rangle$ where $a \ne 0$, then I is an invertible ideal. Indeed, let $b = 1/a \in K$. Then any $x \in I$ is divisible by a in R so that $bx = (1/a)x \in R$, while $a(1/a) = 1$.

(2) Let $R = \mathbf{Z}[\sqrt{-5}]$ and let $I = \langle 2, 1 + \sqrt{-5} \rangle$. Then it is easily checked that I is not principal, but I is an invertible ideal. To see this, let $a_1 = 2$, $a_2 = 1 + \sqrt{-5}$, $b_1 = -1$, and $b_2 = (1 - \sqrt{-5})/2$. Then

$$a_1 b_1 + a_2 b_2 = -2 + 3 = 1.$$

Furthermore, $a_1 b_2$ and $a_2 b_2$ are in R, so it follows that $b_2 I \subseteq R$, and we conclude that I is an invertible ideal.

The following result characterizes which ideals in an integral domain R are projective modules. Note that the theorem is a generalization of Example 5.6 (3):

(5.11) Theorem. *Let R be an integral domain and let $I \subseteq R$ be an ideal. Then I is a projective R-module if and only if I is an invertible ideal.*

Proof. Suppose that I is invertible and choose $a_1, \ldots, a_n \in I$ and $b_1, \ldots, b_n$ in the quotient field K of R so that Equations (5.1) and (5.2) are satisfied. Let $\phi : R^n \to I$ be defined by

$$\phi(x_1, \ldots, x_n) = a_1 x_1 + \cdots + a_n x_n,$$

and define $\beta : I \to R^n$ by

$$\beta(a) = (ab_1, \ldots, ab_n).$$

Note that $ab_i \in R$ for all i by Equation (5.1). Equation (5.2) shows that

$$\phi \circ \beta(a) = \sum_{i=1}^{n} a_i(ab_i) = a \left(\sum_{i=1}^{n} a_i b_i \right) = a$$

for every $a \in I$. Therefore $\phi \circ \beta = 1_P$ and Theorem 3.9 implies that I is a direct summand of the free R-module R^n, so I is a projective R-module.

Conversely, assume that the ideal $I \subseteq R$ is projective as an R-module. Then I is a direct summand of a free R-module F, so there are R-module homomorphisms $\phi : F \to I$ and $\beta : I \to F$ such that $\phi \circ \beta = 1_I$. Let $S = \{x_j\}_{j \in J}$ be a basis of F. Given $x \in I$, $\beta(x) \in F$ can be written uniquely as

$$(5.3) \qquad \beta(x) = \sum_{j \in J} c_j x_j.$$

For each $j \in J$, let $\psi_j(x) = c_j$. This gives a function $\psi_j : I \to R$, which is easily checked to be an R-module homomorphism. If $a_j = \phi(x_j) \in I$, note that

(5.4) for each $x \in I$, $\psi_j(x) = 0$ except for at most finitely many $j \in J$;
(5.5) for each $x \in I$, Equation (5.3) shows that

$$x = \phi(\beta(x)) = \sum_{j \in J} \psi_j(x) a_j.$$

Given $x \neq 0 \in I$ and $j \in J$, define $b_j \in K$ (K is the quotient field of R) by

$$(5.6) \qquad b_j = \frac{\psi_j(x)}{x}.$$

The element $b_j \in K$ depends on $j \in J$ but not on the element $x \neq 0 \in I$. To see this, suppose that $x' \neq 0 \in I$ is another element of I. Then

$$x' \psi_j(x) = \psi_j(x'x) = \psi_j(xx') = x \psi_j(x')$$

so that $\psi_j(x)/x = \psi_j(x')/x'$. Therefore, for each $j \in J$ we get a uniquely defined $b_j \in K$. By property (5.4), at most finitely many of the b_j are not 0. Label the nonzero b_j by $b_1, \ldots, b_n$. By property (5.5), if $x \neq 0 \in I$ then

$$x = \sum_{j=1}^{n} \psi_j(x) a_j = \sum_{j=1}^{n} (b_j x) a_j = x \left(\sum_{j=1}^{n} b_j a_j \right).$$

Cancelling $x \neq 0$ from this equation gives

$$a_1 b_1 + \cdots + a_n b_n = 1$$

where $a_1, \ldots, a_n \in I$ and $b_1 \cdots, b_n \in K$. It remains to check that $b_j I \subseteq R$ for $1 \leq j \leq n$. But if $x \neq 0 \in I$ then $b_j = \psi_j(x)/x$ so that $b_j x = \psi_j(x) \in R$. Therefore, I is an invertible ideal and the theorem is proved.   □

(5.12) *Remark.* Integral domains in which every ideal is invertible are known as **Dedekind domains**, and they are important in number theory. For example, the ring of integers in any algebraic number field is a Dedekind domain.

3.6 Free Modules over a PID

In this section we will continue the study of free modules started in Section 3.4, with special emphasis upon theorems relating to conditions which ensure that a module over a PID R is free. As examples of the types of theorems to be considered, we will prove that all submodules of a free R-module are free and all finitely generated torsion-free R-modules are free, provided that the ring R is a PID. Both of these results are false without the assumption that R is a PID, as one can see very easily by considering an integral domain R that is not a PID, e.g., $R = \mathbf{Z}[X]$, and an ideal $I \subseteq R$ that is not principal, e.g., $\langle 2, X \rangle \subseteq \mathbf{Z}[X]$. Then I is a torsion-free submodule of R that is not free (see Example 4.6 (6)).

Our analysis of free modules over PIDs will also include an analysis of which elements in a free module M can be included in a basis and a criterion for when a linearly independent subset can be included in a basis. Again, these are basic results in the theory of finite-dimensional vector spaces, but the case of free modules over a PID provides extra subtleties that must be carefully analyzed.

We will conclude our treatment of free modules over PIDs with a fundamental result known as the invariant factor theorem for finite rank submodules of free modules over a PID R. This result is a far-reaching generalization of the freeness of submodules of free modules, and it is the basis for the fundamental structure theorem for finitely generated modules over PIDs which will be developed in Section 3.7.

We start with the following definition:

(6.1) Definition. Let M be a *free R-module*. Then the **free rank** of M, denoted free-rank$_R(M)$, is the minimal cardinality of a basis of M.

Since we will not be concerned with the fine points of cardinal arithmetic, we shall not distinguish among infinite cardinals so that

$$\text{free-rank}_R(M) \in \mathbf{Z}^+ \cup \{\infty\}.$$

Since a basis is a generating set of M, we have the inequality $\mu(M) \leq$ free-rank$_R(M)$. We will see in Corollary 6.18 that for an arbitrary commutative ring R and for every free R-module, free-rank$_R(M) = \mu(M)$ and all bases of M have this cardinality.

(6.2) Theorem. Let R be a PID, and let M be a free R-module. If $N \subseteq M$ is a submodule, then N is a free R-module, and

$$\text{free-rank}_R(N) \leq \text{free-rank}_R(M).$$

Proof. We will first present a proof for the case where free-rank$_R(M) < \infty$. This case will then be used in the proof of the general case. For those who are only interested in the case of finitely generated modules, the proof of the second case can be safely omitted.

Case 1. free-rank$_R(M) < \infty$.

We will argue by induction on $k = $ free-rank$_R(M)$. If $k = 0$ then $M = \langle 0 \rangle$ so $N = \langle 0 \rangle$ is free of free-rank 0. If $k = 1$, then M is cyclic so $M = \langle x \rangle$ for some nonzero $x \in M$. If $N = \langle 0 \rangle$ we are done. Otherwise, let $I = \{a \in R : ax \in N\}$. Since I is an ideal of R and R is a PID, $I = \langle d \rangle$; since $N \neq \langle 0 \rangle$, $d \neq 0$. If $y \in N$ then $y = ax = rdx \in \langle dx \rangle$ so that $N = \langle dx \rangle$ is a free cyclic R-module. Thus free-rank$_R(N) = 1$ and the result is true for $k = 1$.

Assume by induction that the result is true for all M with free-rank k, and let M be a module with free-rank$_R(M) = k+1$. Let $S = \{x_1, \ldots, x_{k+1}\}$ be a basis of M and let $M_k = \langle x_1, \ldots, x_k \rangle$. If $N \subseteq M_k$ we are done by induction. Otherwise $N \cap M_k$ is a submodule of M_k which, by induction, is free of free-rank $\ell \leq k$. Let $\{y_1, \ldots, y_\ell\}$ be a basis of $N \cap M_k$. By Theorem 2.5

$$N/(N \cap M_k) \cong (N + M_k)/M_k \subseteq M/M_k = \langle x_{k+1} + M_k \rangle.$$

By the $k = 1$ case of the theorem, $(N + M_k)/M_k$ is a free cyclic submodule of M/M_k with basis $dx_{k+1} + M_k$ where $d \neq 0$. Choose $y_{\ell+1} \in N$ so that $y_{\ell+1} = dx_{k+1} + x'$ for some $x' \in M_k$. Then $(N + M_k)/M_k = \langle y_{\ell+1} + M_k \rangle$.

We claim that $S' = \{y_1, \ldots, y_\ell, y_{\ell+1}\}$ is a basis of N. To see this, let $y \in N$. Then $y + M_k = a_{\ell+1}(y_{\ell+1} + M_k)$ so that $y - a_{\ell+1}y_{\ell+1} \in N \cap M_k$, which implies that $y - a_{\ell+1}y_{\ell+1} = a_1 y_1 + \cdots a_\ell y_\ell$. Thus S' generates N. Suppose that $a_1 y_1 + \cdots + a_{\ell+1} y_{\ell+1} = 0$. Then $a_{\ell+1}(dx_{k+1} + x') + a_1 y_1 + \cdots + a_\ell y_\ell = 0$ so that $a_{\ell+1} dx_{k+1} \in M_k$. But S is a basis of M so we must have $a_{\ell+1}d = 0$; since $d \neq 0$ this forces $a_{\ell+1} = 0$. Thus $a_1 y_1 + \cdots + a_\ell y_\ell = 0$ which implies that $a_1 = \cdots = a_\ell = 0$ since $\{y_1, \cdots, y_\ell\}$ is linearly independent. Therefore S' is linearly independent and hence a basis of N, so that N is free with free-rank$_R(N) \leq \ell + 1 \leq k + 1$. This proves the theorem in Case 1.

Case 2. free-rank$_R(M) = \infty$.

Since $\langle 0 \rangle$ is free with basis $\emptyset$, we may assume that $N \neq \langle 0 \rangle$. Let $S = \{x_j\}_{j \in J}$ be a basis of M. For any subset $K \subseteq J$ let $M_K = \langle \{x_k\}_{k \in K} \rangle$ and let $N_K = N \cap M_K$. Let $\mathcal{T}$ be the set of all triples (K, K', f) where $K' \subseteq K \subseteq J$ and $f : K' \to N_K$ is a function such that $\{f(k)\}_{k \in K'}$ is a basis of N_K. We claim that $\mathcal{T} \neq \emptyset$.

Since $N \neq \langle 0 \rangle$ there is an $x \neq 0 \in N$, so we may write $x = a_1 x_{j_1} + \cdots + a_k x_{j_k}$. Hence $x \in N_K$ where $K = \{j_1, \ldots, j_k\}$. But M_K is a free R-module with free-rank$_R(M_K) \leq k < \infty$ and N_K is a nonzero submodule. By Case 1, N_K is free with free-rank$_R(N_K) = \ell \leq k$. Let $\{y_1, \ldots, y_\ell\}$ be a basis of N_K, and let $K' = \{j_1, \ldots, j_\ell\}$, and define $f : K' \to N_K$ by $f(j_i) = y_i$ for $1 \leq i \leq \ell$. Then $(K, K', f) \in \mathcal{T}$ so that $\mathcal{T} \neq \emptyset$, as claimed.

Now define a partial order on $\mathcal{T}$ by setting $(K, K', f) \leq (L, L', g)$ if $K \subseteq L$, $K' \subseteq L'$, and $g|_{K'} = f$. If $\{(K_\alpha, K'_\alpha, f_\alpha)\}_{\alpha \in A} \subseteq \mathcal{T}$ is a chain, then $\left(\bigcup_{\alpha \in A} K_\alpha, \bigcup_{\alpha \in A} K'_\alpha, F \right)$ where $F|_{K'_\alpha} = f_\alpha$ is an upper bound in $\mathcal{T}$ for the chain. Therefore, Zorn's lemma applies and there is a maximal element (K, K', f) of $\mathcal{T}$.

Claim. $K = J$.

Assuming the claim is true, it follows that $M_K = M$, $N_K = N \cap M_K = N$, and $\{f(k)\}_{k \in K'}$ is a basis of N. Thus, N is a free module (since it has a basis), and since S was an arbitrary basis of M, we conclude that N has a basis of cardinality $\leq$ free-rank$_R(M)$, which is what we wished to prove.

It remains to verify the claim. Suppose that $K \neq J$ and choose $j \in J \setminus K$. Let $L = K \cup \{j\}$. If $N_K = N_L$ then $(K, K', f) \lneqq (L, K', f)$, contradicting the maximality of (K, K', f) in $\mathcal{T}$. If $N_K \neq N_L$, then

$$N_L/(N_L \cap M_K) \cong (N_L + M_K)/M_K \subseteq M_L/M_K = \langle x_j + M_K \rangle.$$

By Case 1, $(N_L + M_K)/M_K$ is a free cyclic submodule with basis $dx_j + M_K$ where $d \neq 0$. Choose $z \in N_L$ so that $z = dx_j + w$ for some $w \in M_K$. Then $(N_L + M_K)/M_K = \langle z + M_K \rangle$. Now let $L' = K' \cup \{j\}$ and define $f' : L' \to N_L$ by

$$f'(k) = \begin{cases} f(k) & \text{if } k \in K', \\ z & \text{if } k = j. \end{cases}$$

We need to show that $\{f'(k)\}_{k \in L'}$ is a basis of N_L. But if $x \in N_L$ then $x + M_K = cz + M_K$ for some $c \in R$. Thus $x - cz \in M_K \cap N = N_K$ so that

$$x - cz = \sum_{k \in K'} b_k f(k)$$

where $b_k \in R$. Therefore, $\{f(k)\}_{k \in L'}$ generates N_L.

Now suppose $\sum_{k \in L'} b_k f'(k) = 0$. Then

$$b_j z + \sum_{k \in K'} b_k f(k) = 0$$

so that

$$db_j x_j + b_j w + \sum_{k \in K'} b_k f(k) = 0.$$

That is, $db_j x_j \in M_K \cap \langle x_j \rangle = \langle 0 \rangle$, and since $S = \{x_\ell\}_{\ell \in J}$ is a basis of M, we must have $db_j = 0$. But $d \neq 0$, so $b_j = 0$. This implies that $\sum_{k \in K'} b_k f(k) = 0$. But $\{f(k)\}_{k \in K'}$ is a basis of N_K, so we must have $b_k = 0$ for all $k \in K'$. Thus $\{f'(k)\}_{k \in L'}$ is a basis of N_L. We conclude that $(K, K', f) \lneq (L, L', f')$, which contradicts the maximality of (K, K', f). Therefore, the claim is verified, and the proof of the theorem is complete. $\square$

(6.3) Corollary. *Let R be a PID and let P be a projective R-module. Then P is free.*

Proof. By Proposition 4.14, P has a free presentation

$$0 \longrightarrow K \longrightarrow F \longrightarrow P \longrightarrow 0.$$

Since P is projective, this exact sequence splits and hence $F \cong P \oplus K$. Therefore, P is isomorphic to a submodule of F, and Theorem 6.2 then shows that P is free. $\square$

(6.4) Corollary. *Let M be a finitely generated module over the PID R and let $N \subseteq M$ be a submodule. Then N is finitely generated and*

$$\mu(N) \leq \mu(M).$$

Proof. Let

$$0 \longrightarrow K \longrightarrow F \overset{\phi}{\longrightarrow} M \longrightarrow 0$$

be a free presentation of M such that free-rank$(F) = \mu(M) < \infty$, and let $N_1 = \phi^{-1}(N)$. By Theorem 6.2, N_1 is free with

$$\mu(N_1) \leq \text{free-rank}(N_1) \leq \text{free-rank}(F) = \mu(M).$$

Since $N = \phi(N_1)$, we have $\mu(N) \leq \mu(N_1)$, and the result is proved. $\square$

(6.5) *Remark.* The hypothesis that R be a PID in Theorem 6.2 and Corollaries 6.3 and 6.4 is crucial. For example, consider the ring $R = \mathbf{Z}[X]$ and let $M = R$ and $N = \langle 2, X \rangle$. Then M is a free R-module and N is a submodule of M that is not free (Example 4.6 (6)). Moreover, $R = \mathbf{Z}[\sqrt{-5}]$, $P = \langle 2, 1 + \sqrt{-5} \rangle$ gives an example of a projective R-module P that is not free (Example 5.6 (3)). Also note that $2 = \mu(N) > \mu(M) = 1$ and $2 = \mu(P) > 1 = \mu(R)$.

Recall that if M is a free module over an integral domain R, then M is torsion-free (Proposition 4.8). The converse of this statement is false even under the restriction that R be a PID. As an example, consider the $\mathbf{Z}$-module $\mathbf{Q}$. It is clear that $\mathbf{Q}$ is a torsion-free $\mathbf{Z}$-module, and it is a simple exercise to show that it is not free. There is, however, a converse if the module is assumed to be finitely generated (and the ring R is a PID).

(6.6) Theorem. *If R is a PID and M is a finitely generated torsion-free R-module, then M is free and*

$$\text{free-rank}_R(M) = \mu(M).$$

Proof. The proof is by induction on $\mu(M)$. If $\mu(M) = 1$ then M is cyclic with generator $\{x\}$. Since M is torsion-free, $\text{Ann}(x) = \{0\}$, so the set $\{x\}$ is linearly independent and, hence, is a basis of M.

Now suppose that $\mu(M) = k > 0$ and assume that the result is true for all finitely generated torsion-free R-modules M' with $\mu(M') < k$. Let $\{x_1, \ldots, x_k\}$ be a minimal generating set for M, and let

$$M_1 = \{x \in M : ax \in \langle x_1 \rangle \quad \text{for some } a \neq 0 \in R\}.$$

Then M/M_1 is generated by $\{x_2 + M_1, \ldots, x_k + M_1\}$ so that $\mu(M/M_1) = j \leq k - 1$. If $ax \in M_1$ for some $a \neq 0 \in R$, then from the definition of M_1, $b(ax) \in \langle x_1 \rangle$ for some $b \neq 0$. Hence $x \in M_1$ and we conclude that M/M_1 is torsion-free. By the induction hypothesis, M/M_1 is free of free-rank j. Then Corollary 4.17 shows that $M \cong M_1 \oplus (M/M_1)$. We will show that M_1 is free of free-rank 1. It will then follow that

$$k = \mu(M) \leq \mu(M_1) + \mu(M/M_1) = 1 + j,$$

and since $j \leq k - 1$, it will follow that $j = k - 1$ and M is free of free-rank $= k$.

It remains to show that M_1 is free of rank 1. Note that if R is a field then $M_1 = R \cdot x_1$ and we are done. In the general case, M_1 is a submodule of M, so it is finitely generated by $\ell \leq k$ elements. Let $\{y_1, \ldots, y_\ell\}$ be a generating set for M_1 and suppose that $a_i y_i = b_i x_1$ with $a_i \neq 0$ for $1 \leq i \leq \ell$. Let $q_0 = a_1 \cdots a_\ell$.

Claim. *If $ax = bx_1$ with $a \neq 0$ then $a \mid bq_0$.*

To see this note that $x = \sum_{i=1}^{\ell} c_i y_i$ so that

$$
\begin{aligned}
q_0 x &= \sum_{i=1}^{\ell} c_i q_0 y_i \\
&= \sum_{i=1}^{\ell} c_i (q_0/a_i) a_i y_i \\
&= \sum_{i=1}^{\ell} c_i (q_0/a_i) b_i x_1 \\
&= \left(\sum_{i=1}^{\ell} c_i (q_0/a_i) b_i \right) x_1.
\end{aligned}
$$

Therefore.

$$
b q_0 x_1 = a q_0 x = a \left(\sum_{i=1}^{\ell} c_i (q_0/a_i) b_i \right) x_1.
$$

Since M_1 is torsion-free, it follows that

$$
b q_0 = a \left(\sum_{i=1}^{\ell} c_i (q_0/a_i) b_i \right),
$$

and the claim is proved.

Using this claim we can define a function $\phi : M_1 \to R$ by $\phi(x) = (b q_0)/a$ whenever $ax = b x_1$ for $a \neq 0$. We must show that ϕ is well defined. That is, if $ax = b x_1$ and $a'x = b'x$, then $(b q_0)/a = (b' q_0)/a'$. But $ax = b x_1$ and $a'x = b'x_1$ implies that $a'b x_1 = a'ax = ab'x_1$ so that $a'b = ab'$ because M is torsion-free. Thus $a'b q_0 = ab' q_0$ so that $(b q_0)/a = (b' q_0)/a'$ and ϕ is well defined. Furthermore, it is easy to see that ϕ is an R-module homomorphism so that $\mathrm{Im}(\phi)$ is an R-submodule of R, i.e., an ideal. Suppose that $\phi(x) = 0$. Then $ax = b x_1$ with $a \neq 0$ and $\phi(x) = (b q_0)/a = 0 \in R$. Since R is an integral domain, it follows that $b = 0$ and hence $ax = 0$. Since M is torsion-free we conclude that $x = 0$. Therefore, $\mathrm{Ker}(\phi) = \{0\}$ and

$$
M_1 \cong \mathrm{Im}(\phi) = Rc.
$$

Hence, M_1 is free of rank 1, and the proof is complete. $\square$

(6.7) Corollary. *If M is a finitely generated module over a field F, then M is free.*

Proof. Every module over a field is torsion-free (Proposition 2.20). $\square$

(6.8) Remark. We have already given an independent proof (based on Zorn's lemma) for Corollary 6.7, even without the finitely generated assumption (Theorem 4.20). We have included Corollary 6.7 here as an observation that it follows as a special case of the general theory developed for torsion-free finitely generated modules over a PID.

(6.9) Corollary. *If M is a finitely generated module over a PID R, then $M \cong M_\tau \oplus (M/M_\tau)$.*

Proof. There is an exact sequence of R-modules

$$0 \longrightarrow M_\tau \longrightarrow M \longrightarrow M/M_\tau \longrightarrow 0.$$

Hence, M/M_τ is finitely generated and by Proposition 2.18, it is torsion-free, so Theorem 6.6 shows that M/M_τ is free. Then Corollary 4.17 shows that $M \cong M_\tau \oplus (M/M_\tau)$. □

The main point of Corollary 6.9 is that any finitely generated module over a PID can be written as a direct sum of its torsion submodule and a free submodule. Thus an analysis of these modules is reduced to studying the torsion submodule, once we have completed our analysis of free modules. We will now continue the analysis of free modules over a PID R by studying when an element in a free module can be included in a basis. As a corollary of this result we will be able to show that any two bases of a finitely generated free R-module (R a PID) have the same number of elements.

(6.10) Example. Let R be a PID and view R as an R-module. Then an element $a \in R$ forms a basis of R if and only if a is a unit. Thus if R is a field, then every nonzero element is a basis of the R-module R, while if $R = \mathbf{Z}$ then the only elements of $\mathbf{Z}$ that form a basis of $\mathbf{Z}$ are 1 and -1. As a somewhat more substantial example, consider the $\mathbf{Z}$-module $\mathbf{Z}^2$. Then the element $u = (2, 0) \in \mathbf{Z}^2$ cannot be extended to a basis of $\mathbf{Z}^2$ since if v is any element of $\mathbf{Z}^2$ with $\{u, v\}$ linearly independent, the equation

$$\alpha u + \beta v = (1, 0)$$

is easily seen to have no solution $\alpha, \beta \in \mathbf{Z}$. Therefore, some restriction on elements of an R-module that can be included in a basis is necessary. The above examples suggest the following definition.

(6.11) Definition. *Let M be an R-module. A torsion-free element $x \neq 0 \in M$ is said to be **primitive** if $x = ay$ for some $y \in M$ and $a \in R$ implies that a is a unit of R.*

(6.12) Remarks.

(1) If R is a field, then *every* nonzero $x \in M$ is primitive.
(2) The element $x \in R$ is a primitive element of the R-module R if and only if x is a unit.

(3) The element $(2, 0) \in \mathbf{Z}^2$ is not primitive since $(2, 0) = 2 \cdot (1, 0)$.
(4) If $R = \mathbf{Z}$ and $M = \mathbf{Q}$, then *no* element of M is primitive.

(6.13) Lemma. *Let R be a PID and let M be a free R-module with basis $S = \{x_j\}_{j \in J}$. If $x = \sum_{j \in J} a_j x_j \in M$, then x is primitive if and only if $\gcd(\{a_j\}_{j \in J}) = 1$.*

Proof. Let $d = \gcd(\{a_j\}_{j \in J})$. Then $x = d(\sum_{j \in J}(a_j/d)x_j)$, so if d is not a unit then x is not primitive. Conversely, if $d = 1$ and $x = ay$ then

$$\sum_{j \in J} a_j x_j = x$$
$$= ay$$
$$= a\Big(\sum_{j \in J} b_j x_j\Big)$$
$$= \sum_{j \in J} a b_j x_j.$$

Since $S = \{x_j\}_{j \in J}$ is a basis, it follows that $a_j = a b_j$ for all $j \in J$. That is, a is a common divisor of the set $\{a_j\}_{j \in J}$ so that $a \mid d = 1$. Hence a is a unit and x is primitive. $\square$

(6.14) Lemma. *Let R be a PID and let M be a finitely generated R-module. If $x \in M$ has $\operatorname{Ann}(x) = \langle 0 \rangle$, then we may write $x = ax'$ where $a \in R$ and x' is primitive. (In particular, if M is not a torsion module, then M has a primitive element.)*

Proof. Let $x_0 = x$. If x_0 is primitive we are done. Otherwise, write $x_0 = a_1 x_1$ where $a_1 \in R$ is not a unit. Then $\langle x_0 \rangle \subsetneq \langle x_1 \rangle$. To see this, it is certainly true that $\langle x_0 \rangle \subseteq \langle x_1 \rangle$. If the two submodules are equal then we may write $x_1 = b x_0$ so that $x_0 = a_1 x_1 = a_1 b x_0$, i.e., $(1 - a_1 b) \in \operatorname{Ann}(x_0) = \langle 0 \rangle$. Therefore, $1 = a_1 b$ and a_1 is a unit, which contradicts the choice of a_1.

Now consider x_1. If x_1 is primitive, we are done. Otherwise, $x_1 = a_2 x_2$ where a_2 is not a unit, and as above we conclude that $\langle x_1 \rangle \subsetneq \langle x_2 \rangle$. Continuing in this way we obtain a chain of submodules

$$(6.1) \qquad \langle x_0 \rangle \subsetneq \langle x_1 \rangle \subsetneq \langle x_2 \rangle \subsetneq \cdots.$$

Either this chain stops at some i, which means that x_i is primitive, or (6.1) is an infinite properly ascending chain of submodules of M. We claim that the latter possibility cannot occur. To see this, let $N = \bigcup_{i=1}^{\infty} \langle x_i \rangle$. Then N is a submodule of the finitely generated module M over the PID R so that N is also finitely generated by $\{y_1, \ldots, y_k\}$ (Corollary 6.4). Since $\langle x_0 \rangle \subseteq \langle x_1 \rangle \subseteq \cdots$, there is an i such that $\{y_1, \ldots, y_k\} \subseteq \langle x_i \rangle$. Thus $N = \langle x_i \rangle$ and hence $\langle x_i \rangle = \langle x_{i+1} \rangle = \cdots$, which contradicts having an infinite properly ascending chain. Therefore, x_i is primitive for some i, and if we let $x' = x_i$ we conclude that $x = ax'$ where $a = a_1 a_2 \cdots a_i$. $\square$

(6.15) Remark. Suppose that M is a free R-module, where R is a PID, and $x \in M$. Then $\text{Ann}(x) = \langle 0 \rangle$, so $x = ax'$ where x' is a primitive element of M. If $S = \{x_j\}_{j \in J}$ is a basis of M, then we may write $x' = \sum_{j \in J} b_j x_j$ so that

$$x = ax' = \sum_{j \in J} ab_j x_j = \sum_{j \in J} c_j x_j.$$

Since $\gcd(\{b_j\}_{j \in J}) = 1$ (by Lemma 6.13) we see that $a = \gcd(\{c_j\}_{j \in J})$. The element $a \in R$, which is uniquely determined by x up to multiplication by a unit of R, is called the **content** of $x \in M$ and is denoted $c(x)$. (Compare with the concept of content of polynomials (Definition 2.6.3).) Thus, any $x \in M$ can be written

(6.2) $$x = c(x) \cdot x'$$

where x' is primitive.

(6.16) Theorem. *Let R be a PID and let M be a free R-module with*

$$\text{rank}(M) = k = \mu(M) = \text{free-rank}(M).$$

If $x \in M$ is primitive, then M has a basis of k elements containing x.

Proof. Assume first that $k < \infty$ and proceed by induction on k. Suppose $k = 1$ and let M have a basis $\{x_1\}$. Then $x = ax_1$ for some $a \in R$. Since x is primitive, it follows that a is a unit so that $\langle x \rangle = \langle x_1 \rangle = M$, hence $\{x\}$ is a basis of M.

The case $k = 2$ will be needed in the general induction step, so we present it separately. Thus suppose that M has a basis $\{x_1, x_2\}$ and let $x = rx_1 + sx_2$ where $r, s \in R$. Since x is primitive, $\gcd\{r, s\} = 1$, so we may write $ru + sv = 1$. Let $x_2' = -vx_1 + ux_2$. Then

$$x_1 = ux - sx_2'$$

and

$$x_2 = vx + rx_2'.$$

Hence, $\langle x, x_2' \rangle = M$. It remains to show that $\{x, x_2'\}$ is linearly independent. Suppose that $ax + bx_2' = 0$. Then

$$a(rx_1 + sx_2) + b(-vx_1 + ux_2) = 0.$$

Since $\{x_1, x_2\}$ is a basis of M, it follows that

$$ar - bv = 0$$

and

$$as + bu = 0.$$

Multiplying the first equation by u, multiplying the second by v, and adding shows that $a = 0$, while multiplying the first by $-s$, multiplying the second by r, and adding shows that $b = 0$. Hence, $\{x, x_2'\}$ is linearly independent and, therefore, a basis of M.

Now suppose that $\mu(M) = k > 2$ and that the result is true for all free R-modules of rank $< k$. By Theorem 6.6 there is a basis $\{x_1, \ldots, x_k\}$ of M. Let $x = \sum_{i=1}^{k} a_i x_i$. If $a_k = 0$ then $x \in M_1 = \langle x_1, \ldots, x_{k-1} \rangle$, so by induction there is a basis $\{x, x_2', \ldots, x_{k-1}'\}$ of M_1. Then $\{x, x_2', \ldots, x_{k-1}', x_k\}$ is a basis of M containing x. Now suppose that $a_k \neq 0$ and let $y = \sum_{i=1}^{k-1} a_i x_i$. If $y = 0$ then $x = a_k x_k$, and since x is primitive, it follows that a_k is a unit of R and $\{x_1, \ldots, x_{k-1}, x\}$ is a basis of M containing x in this case. If $y \neq 0$ then there is a primitive y' such that $y = by'$ for some $b \in R$. In particular, $y' \in M_1$ so that M_1 has a basis $\{y', x_2', \ldots, x_{k-1}'\}$ and hence M has a basis $\{y', x_2, \ldots, x_{k-1}', x_k\}$. But $x = a_k x_k + y = a_k x_k + by'$ and $\gcd(a_k, b) = 1$ since x is primitive. By the previous case $(k = 2)$ we conclude that the submodule $\langle x_k, y' \rangle$ has a basis $\{x, y''\}$. Therefore, M has a basis $\{x, x_2', \ldots, x_{k-1}', y''\}$ and the argument is complete when $k = \mu(M) < \infty$.

If $k = \infty$ let $\{x_j\}_{j \in J}$ be a basis of M and let $x = \sum_{i=1}^{n} a_i x_{j_i}$ for some finite subset $I = \{j_1, \ldots, j_n\} \subseteq J$. If $N = \langle x_{j_1}, \ldots, x_{j_n} \rangle$ then x is a primitive element in the finitely generated module N, so the previous argument applies to show that there is a basis $\{x, x_2', \ldots, x_n'\}$ of N. Then $\{x, x_2', \ldots, x_n'\} \cup \{x_j\}_{j \in J \setminus I}$ is a basis of M containing x. $\qquad \square$

(6.17) Corollary. *If M is a free module over a PID R, then* **every** *basis of M contains $\mu(M)$ elements.*

Proof. In case $\mu(M) < \infty$, the proof is by induction on $\mu(M)$. If $\mu(M) = 1$ then $M = \langle x \rangle$. If $\{x_1, x_2\} \subseteq M$ then $x_1 = a_1 x$ and and $x_2 = a_2 x$ so that $a_2 x_1 - a_1 x_2 = 0$, and we conclude that no subset of M with more than one element is linearly independent.

Now suppose that $\mu(M) = k > 1$ and assume the result is true for all free R-modules N with $\mu(N) < k$. Let $S = \{x_j\}_{j \in J} \subseteq M$ be any basis of M and choose $x \in S$. Since x is primitive (being an element of a basis), Theorem 6.16 applies to give a basis $\{x, y_2, \ldots, y_k\}$ of M with precisely $\mu(M) = k$ elements. Let $N = M/\langle x \rangle$ and let $\pi : M \to N$ be the projection map. It is clear that N is a free R-module with basis $\pi(S) \setminus \{\pi(x)\}$. By Proposition 2.12 it follows that $\mu(N) \geq k - 1$, and since $\{\pi(y_2), \ldots, \pi(y_k)\}$ generates N, we conclude that $\mu(N) = k - 1$. By induction, it follows that $|S| - 1 < \infty$ and $|S| - 1 = k - 1$, i.e., $|S| = k$, and the proof is complete in case $\mu(M) < \infty$.

In case $\mu(M) = \infty$, we are claiming that no basis of M can contain a finite number $k \in \mathbf{Z}^+$ of elements. This is proved by induction on k, the proof being similar to the case $\mu(M)$ finite, which we have just done. We leave the details to the reader. $\qquad \square$

(6.18) Corollary. *Let R be any commutative ring with identity and let M be a free R-module. Then* **every** *basis of M contains $\mu(M)$ elements.*

Proof. Let I be any maximal ideal of R (recall that maximal ideals exist by Theorem 2.2.16). Since R is commutative, the quotient ring $R/I = K$ is a field (Theorem 2.2.18), and hence it is a PID. By Proposition 4.13, the quotient module M/IM is a finitely generated free K-module so that Corollary 6.17 applies to show that every basis of M/IM has $\mu(M/IM)$ elements. Let $S = \{x_j\}_{j \in J}$ be an arbitrary basis of the free R-module M and let $\pi : M \to M/IM$ be the projection map. According to Proposition 4.13, the set $\pi(S) = \{\pi(x_j)\}_{j \in J}$ is a basis of M/IM over K, and therefore,

$$\mu(M) \leq |J| = \mu(M/IM) \leq \mu(M).$$

Thus, $\mu(M) = |J|$, and the corollary is proved. $\square$

(6.19) Remarks.

(1) If M is a free R-module over a commutative ring R, then we have proved that free-rank$(M) = \mu(M) =$ the number of elements in *any* basis of M. This common number we shall refer to simply as the **rank** of M, denoted $\text{rank}_R(M)$ or $\text{rank}(M)$ if the ring R is implicit. If R is a field we shall sometimes write $\dim_R(M)$ (the **dimension** of M over R) in place of $\text{rank}_R(M)$. Thus, a vector space M (over R) is finite dimensional if and only if $\dim_R(M) = \text{rank}_R(M) < \infty$.

(2) Corollary 6.18 is the invariance of rank theorem for finitely generated free modules over an arbitrary commutative ring R. The invariance of rank theorem is not valid for an arbitrary (possibly noncommutative) ring R. As an example, consider the **Z**-module $M = \oplus_{n \in \mathbf{N}} \mathbf{Z}$, which is the direct sum of countably many copies of **Z**. It is simple to check that $M \cong M \oplus M$. Thus, if we define $R = \text{End}_{\mathbf{Z}}(M)$, then R is a noncommutative ring, and Corollary 3.13 shows that

$$\begin{aligned}
R &= \text{End}_{\mathbf{Z}}(M) \\
&= \text{Hom}_{\mathbf{Z}}(M, M) \\
&\cong \text{Hom}_{\mathbf{Z}}(M, M \oplus M) \\
&\cong \text{Hom}_{\mathbf{Z}}(M, M) \oplus \text{Hom}_{\mathbf{Z}}(M, M) \\
&\cong R \oplus R.
\end{aligned}$$

The isomorphisms are isomorphisms of **Z**-modules. We leave it as an exercise to check that the isomorphisms are also isomorphisms of R-modules, so that $R \cong R^2$, and hence, the invariance of rank does not hold for the ring R. There is, however, one important class of noncommutative rings for which the invariance of rank theorem holds, namely, division rings. This will be proved in Proposition 7.1.14.

(6.20) Corollary. *If M and N are free modules over a PID R, at least one of which is finitely generated, then $M \cong N$ if and only if $\mathrm{rank}(M) = \mathrm{rank}(N)$.*

Proof. If M and N are isomorphic, then $\mu(M) = \mu(N)$ so that $\mathrm{rank}(M) = \mathrm{rank}(N)$. Conversely, if $\mathrm{rank}(M) = \mathrm{rank}(N)$, then Proposition 4.9 gives a homomorphism $f : M \to N$, which takes a basis of M to a basis of N. It is easy to see that f must be an isomorphism. $\qquad\square$

(6.21) *Remark.* One of the standard results concerning bases of finite-dimensional vector spaces is the statement that a subset $S = \{x_1, \ldots, x_n\}$ of a vector space V of dimension n is a basis provided that S is *either* a spanning set or linearly independent. Half of this result is valid in the current context of finitely generated free modules over a PID. The set $\{2\} \subseteq \mathbf{Z}$ is linearly independent, but it is not a basis of the rank 1 $\mathbf{Z}$-module $\mathbf{Z}$. There is, however, the following result.

(6.22) Proposition. *Let M be a finitely generated free R-module of $\mathrm{rank} = k$ where R is a PID. If $S = \{x_1, \ldots, x_k\}$ generates M, then S is a basis.*

Proof. Let $T = \{e_j\}_{j=1}^{k}$ be the standard basis of R^k. Then there is a homomorphism $\phi : R^k \to M$ determined by $\phi(e_j) = x_j$. Since $\langle S \rangle = M$, there is a short exact sequence

$$0 \longrightarrow K \longrightarrow R^k \overset{\phi}{\longrightarrow} M \longrightarrow 0$$

where $K = \mathrm{Ker}(\phi)$. Since M is free, Corollary 4.16 gives $R^k \cong M \oplus K$, and according to Theorem 6.2, K is also free of finite rank. Therefore,

$$k = \mathrm{rank}(M) + \mathrm{rank}(K) = k + \mathrm{rank}(K)$$

and we conclude that $\mathrm{rank}(K) = 0$. Hence ϕ is an isomorphism and S is a basis. $\qquad\square$

We will conclude this section with a substantial generalization of Theorem 6.2. This result is the crucial result needed for the structure theorem for finitely generated modules over a PID.

(6.23) Theorem. (Invariant factor theorem for submodules) *Let R be a PID, let M be a free R-module, and let $N \subseteq M$ be a submodule (which is automatically free by Theorem 6.2) of rank $n < \infty$. Then there is a basis S of M, a subset $\{x_1, \ldots, x_n\} \subseteq S$, and nonzero elements $s_1, \ldots, s_n \in R$ such that*

$$(6.3) \qquad\qquad \{s_1 x_1, \ldots, s_n x_n\} \qquad \text{is a basis of } N$$

and

$$(6.4) \qquad\qquad s_i \mid s_{i+1} \qquad \text{for} \quad 1 \leq i \leq n - 1.$$

Proof. If $N = \langle 0 \rangle$, there is nothing to prove, so we may assume that $N \neq \langle 0 \rangle$ and proceed by induction on $n = \text{rank}(N)$. If $n = 1$, then $N = \langle y \rangle$ and $\{y\}$ is a basis of N. By Lemma 6.14, we may write $y = c(y)x$ where $x \in M$ is a primitive element and $c(y) \in R$ is the content of y. By Theorem 6.16, there is a basis S of M containing the primitive element x. If we let $x_1 = x$ and $s_1 = c(y)$, then $s_1 x_1 = y$ is a basis of N, so condition (6.3) is satisfied; (6.4) is vacuous for $n = 1$. Therefore, the theorem is proved for $n = 1$.

Now assume that $n > 1$. By Lemma 6.14, each $y \in N$ can be written as $y = c(y) \cdot y'$ where $c(y) \in R$ is the content of y (Remark 6.15) and $y' \in M$ is primitive. Let

$$S = \{\langle c(y)\rangle : y \in N\}.$$

This is a nonempty collection of ideals of R. Since R is Noetherian, Proposition 2.5.10 implies that there is a maximal element of S. Let $\langle c(y) \rangle$ be such a maximal element. Thus, $y \in N$ and $y = c(y) \cdot x$, where $x \in M$ is primitive. Let $s_1 = c(y)$. Choose any basis T of M that contains x. This is possible by Theorem 6.16 since $x \in M$ is primitive. Let $x_1 = x$ and write $T = \{x_1\} \cup T' = \{x_1\} \cup \{x'_j\}_{j \in J'}$. Let $M_1 = \langle \{x'_j\}_{j \in J'} \rangle$ and let $N_1 = M_1 \cap N$.

Claim. $N = \langle s_1 x_1 \rangle \oplus N_1$.

To see this, note that $\langle s_1 x_1 \rangle \cap N_1 \subseteq \langle x_1 \rangle \cap M_1 = \langle 0 \rangle$ because T is a basis of M. Let $z \in N$. Then, with respect to the basis T, we may write

$$(6.5) \qquad\qquad z = a_1 x_1 + \sum_{j \in J'} b_j x'_j.$$

Let $d = (s_1, a_1) = \gcd\{s_1, a_1\}$. Then we may write $d = us_1 + va_1$ where $u, v \in R$. If $w = uy + vz$, then Equation (6.5) shows that

$$\begin{aligned}
w &= uy + vz \\
&= (us_1 + va_1)x_1 + \sum_{j \in J'} vb_j x'_j \\
&= dx_1 + \sum_{j \in J'} vb_j x'_j.
\end{aligned}$$

Writing $w = c(w) \cdot w'$ where $c(w)$ is the content of w and $w' \in M$ is primitive, it follows from Lemma 6.13 that $c(w) \mid d$ (because $c(w)$ is the greatest common divisor of all coefficients of w when expressed as a linear combination of any basis of M). Thus we have a chain of ideals

$$\langle s_1 \rangle \subseteq \langle d \rangle \subseteq \langle c(w) \rangle,$$

and the maximality of $\langle s_1 \rangle$ in S shows that $\langle s_1 \rangle = \langle c(w) \rangle = \langle d \rangle$. In particular, $\langle s_1 \rangle = \langle d \rangle$ so that $s_1 \mid a_1$, and we conclude that

$$z = b_1(s_1 x_1) + \sum_{j \in J'} b_j x'_j.$$

That is, $z \in \langle s_1 x_1 \rangle + N_1$. Theorem 3.1 then shows that

$$N \cong \langle s_1 x_1 \rangle \oplus N_1,$$

and the claim is proved.

By Theorem 6.2, N_1 is a free R-module since it is a submodule of the free R-module M. Furthermore, by the claim we see that

$$\operatorname{rank}(N_1) = \operatorname{rank}(N) - 1 = n - 1.$$

Applying the induction hypothesis to the pair $N_1 \subseteq M_1$, we conclude that there is a basis S' of M_1 and a subset $\{x_2, \dots, x_n\}$ of S', together with nonzero elements $s_2, \dots, s_n$ of R, such that

(6.6) $\{s_2 x_2, \dots, s_n x_n\}$ is a basis of N_1

and

(6.7) $s_i \mid s_{i+1}$ for $2 \le i \le n - 1$.

Let $S = S' \cup \{x_1\}$. Then the theorem is proved once we have shown that $s_1 \mid s_2$.

To verify that $s_1 \mid s_2$, consider the element $s_2 x_2 \in N_1 \subseteq N$ and let $z = s_1 x_1 + s_2 x_2 \in N$. When we write $z = c(z) \cdot z'$ where $z' \in M$ is primitive and $c(z) \in R$ is the content of z, Remark 6.15 shows that $c(z) = (s_1, s_2)$. Thus, $\langle s_1 \rangle \subseteq \langle c(z) \rangle$ and the maximality of $\langle s_1 \rangle$ in S shows that $\langle c(z) \rangle = \langle s_1 \rangle$, i.e., $s_1 \mid s_2$, and the proof of the theorem is complete. $\square$

(6.24) Example. Let $N \subseteq \mathbf{Z}^2$ be the submodule generated by $y_1 = (2, 4)$, $y_2 = (2, -2)$, and $y_3 = (2, 10)$. Then $c(y_1) = c(y_2) = c(y_3) = 2$. Furthermore, 2 divides every component of any linear combination of y_1, y_2, and y_3, so $s_1 = 2$ in the notation of Theorem 6.23. Let $v_1 = (1, 2)$. Then $y_1 = 2 v_1$. Extend v_1 to a basis of $\mathbf{Z}^2$ by taking $v_2 = (0, 1)$. Then

(6.8) $$N_1 = N \cap \langle (0, 1) \rangle = \langle (0, 6) \rangle.$$

To see this note that every $z \in N_1$ can be written as

$$z = a_1 y_1 + a_2 y_2 + a_3 y_3$$

where $a_1, a_3, a_3 \in \mathbf{Z}$ satisfy the equation

$$2 a_1 + 2 a_2 + 2 a_3 = 0.$$

Thus, $4 a_1 = -4 a_2 - 4 a_3$, and considering the second coordinate of z, we see that $z = (z_1, z_2)$ where

$$z_2 = 4 a_1 - 2 a_2 + 10 a_3 = -6 a_2 + 6 a_3 = 6(a_3 - a_2).$$

Therefore, $\{v_1, v_2\}$ is a basis of $\mathbf{Z}^2$, while $\{2 v_1, 6 v_2\}$ is a basis of N. To check, note that $y_1 = 2 v_1$, $y_2 = 2 v_1 - 6 v_2$, and $y_3 = 2 v_1 + 6 v_2$.

(6.25) *Remark.* In Section 3.7, we will prove that the elements $\{s_1, \ldots, s_n\}$ are determined just by the rank n submodule N and not by the particular choice of a basis S of M. These elements are called the **invariant factors** of the submodule N in the free module M.

3.7 Finitely Generated Modules over PIDs

The invariant factor theorem for submodules (Theorem 6.23) gives a complete description of a submodule N of a finitely generated free R-module M over a PID R. Specifically, it states that a basis of M can be chosen so that the first $n = \text{rank}(N)$ elements of the basis, multiplied by elements of R, provide a basis of N. Note that this result is a substantial generalization of the result from vector space theory, which states that any basis of a subspace of a vector space can be extended to a basis of the ambient space. We will now complete the analysis of finitely generated R-modules (R a PID) by considering modules that need not be free. If the module M is not free, then, of course, it is not possible to find a basis, but we will still be able to express M as a finite direct sum of cyclic submodules; the cyclic submodules may, however, have nontrivial annihilator. The following result constitutes the fundamental structure theorem for finitely generated modules over principal ideal domains.

(7.1) Theorem. *Let $M \neq 0$ be a finitely generated module over the PID R. If $\mu(M) = n$, then M is isomorphic to a direct sum of cyclic submodules*

$$M \cong Rw_1 \oplus \cdots \oplus Rw_n$$

such that

(7.1) $R \neq \text{Ann}(w_1) \supseteq \text{Ann}(w_2) \supseteq \cdots \supseteq \text{Ann}(w_n) = \text{Ann}(M).$

Moreover, for $1 \leq i < n$

(7.2) $\text{Ann}(w_i) = \text{Ann}\left(M/(Rw_{i+1} + \cdots + Rw_n)\right).$

Proof. Since $\mu(M) = n$, let $\{v_1, \ldots, v_n\}$ be a generating set of M and define an R-module homomorphism $\phi : R^n \to M$ by

$$\phi(a_1, \ldots, a_n) = \sum_{i=1}^{n} a_i v_i.$$

Let $K = \text{Ker}(\phi)$. Since K is a submodule of R^n, it follows from Theorem 6.2 that K is a free R-module of rank $m \leq n$. By Theorem 6.23, there is a basis $\{y_1, \ldots, y_n\}$ of R^n and nonzero elements $s_1, \ldots, s_m \in R$ such that

(7.3) $\{s_1y_1, \ldots, s_my_m\}$ is a basis for K

and

(7.4) $s_i \mid s_{i+1}$ for $1 \leq i \leq m-1$.

Let $w_i = \phi(y_i) \in M$ for $1 \leq i \leq n$. Then $\{w_1, \ldots, w_n\}$ generates M since ϕ is surjective and $\{y_1, \ldots, y_n\}$ is a basis of R^n. We claim that

$$M \cong Rw_1 \oplus \cdots \oplus Rw_n.$$

By the characterization of direct sum modules (Theorem 3.1), it is sufficient to check that if

(7.5) $a_1w_1 + \cdots + a_nw_n = 0$

where $a_i \in R$, then $a_iw_i = 0$ for all i. Thus suppose that Equation (7.5) is satisfied. Then

$$\begin{aligned} 0 &= a_1w_1 + \cdots + a_nw_n \\ &= a_1\phi(y_1) + \cdots + a_n\phi(y_n) \\ &= \phi(a_1y_1 + \cdots + a_ny_n) \end{aligned}$$

so that

$$a_1y_1 + \cdots + a_ny_n \in \mathrm{Ker}(\phi) = K = \langle s_1y_1, \ldots, s_my_m \rangle.$$

Therefore,

$$a_1y_1 + \cdots + a_ny_n = b_1s_1y_1 + \cdots + b_ms_my_m$$

for some $b_1, \ldots, b_m \in R$. But $\{y_1, \ldots, y_n\}$ is a basis of R^n, so we conclude that $a_i = b_is_i$ for $1 \leq i \leq m$ while $a_i = 0$ for $m+1 \leq i \leq n$. Thus,

$$a_iw_i = b_is_i\phi(y_i) = b_i\phi(s_iy_i) = 0$$

for $1 \leq i \leq m$ because $s_iy_i \in K = \mathrm{Ker}(\phi)$, while $a_iw_i = 0$ for $m+1 \leq i \leq n$ since $a_i = 0$ in this case. Hence

$$M \cong Rw_1 \oplus \cdots \oplus Rw_n.$$

Note that $\mathrm{Ann}(w_i) = \langle s_i \rangle$ for $1 \leq i \leq m$, and since $s_i \mid s_{i+1}$, it follows that

$$\mathrm{Ann}(w_1) \supseteq \mathrm{Ann}(w_2) \supseteq \cdots \supseteq \mathrm{Ann}(w_m),$$

while for $i > m$, since $\langle y_i \rangle \cap \mathrm{Ker}(\phi) = \langle 0 \rangle$, it follows that $\mathrm{Ann}(w_i) = \langle 0 \rangle$. Since $s_i \mid s_n$ for all i and since $\mathrm{Ann}(w_i) = \langle s_i \rangle$, we conclude that $s_nM = 0$. Hence, $\mathrm{Ann}(w_n) = \langle s_n \rangle = \mathrm{Ann}(M)$ and Equation (7.1) is satisfied. Since

$$M/(Rw_{i+1} + \cdots + Rw_n) \cong Rw_1 \oplus \cdots \oplus Rw_i,$$

Equation (7.2) follows from Equation (7.1). The proof is now completed by observing that $\mathrm{Ann}(w_i) \neq R$ for any i since, if $\mathrm{Ann}(w_i) = R$, then

$Rw_i = \langle 0 \rangle$, and hence, M could be generated by fewer than n elements. But $n = \mu(M)$, so this is impossible because $\mu(M)$ is the minimal number of generators of M. $\square$

A natural question to ask is to what extent is the cyclic decomposition provided by Theorem 7.1 unique. Certainly, the factors themselves are not unique as one can see from the example

$$\mathbf{Z}^2 \cong \mathbf{Z} \cdot (1,0) \oplus \mathbf{Z} \cdot (0,1)$$
$$\cong \mathbf{Z} \cdot (1,0) \oplus \mathbf{Z} \cdot (1,1).$$

More generally, if M is a free R-module of rank n, then any choice of basis $\{v_1, \ldots, v_n\}$ provides a cyclic decomposition

$$M \cong Rv_1 \oplus \cdots \oplus Rv_n$$

with $\mathrm{Ann}(v_i) = 0$ for all i. Therefore, there is no hope that the cyclic factors themselves are uniquely determined. What does turn out to be unique, however, is the chain of annihilator ideals

$$\mathrm{Ann}(w_1) \supseteq \cdots \supseteq \mathrm{Ann}(w_n)$$

where we require that $\mathrm{Ann}(w_i) \neq R$, which simply means that we do not allow copies of $\langle 0 \rangle$ in our direct sums of cyclic submodules. We reduce the uniqueness of the annihilator ideals to the case of finitely generated torsion R-modules by means of the following result. If M is an R-module, recall that the torsion submodule M_τ of M is defined by

$$M_\tau = \{x \in M : \mathrm{Ann}(x) \neq \langle 0 \rangle\}.$$

(7.2) Proposition. *If M and N are finitely generated modules over a PID R, then $M \cong N$ if and only if $M_\tau \cong N_\tau$ and* rank $M/M_\tau =$ rank N/N_τ.

Proof. Let $\phi : M \to N$ be an isomorphism. Then if $x \in M_\tau$, there is an $a \neq 0 \in R$ with $ax = 0$. Then $a\phi(x) = \phi(ax) = \phi(0) = 0$ so that $\phi(x) \in N_\tau$. Therefore, $\phi(M_\tau) \subseteq N_\tau$. Applying the same observation to ϕ^{-1} shows that $\phi(M_\tau) = N_\tau$. Thus, $\phi|_{M_\tau} : M_\tau \to N_\tau$ is an isomorphism; if $\pi : N \to N/N_\tau$ is the natural projection, it follows that $\mathrm{Ker}(\pi \circ \phi) = M_\tau$. The first isomorphism theorem then gives an isomorphism $M/M_\tau \cong N/N_\tau$. Since M/M_τ and N/N_τ are free R-modules of finite rank, they are isomorphic if and only if they have the same rank.

The converse follows from Corollary 6.20. $\square$

Therefore, our analysis of finitely generated R-modules over a PID R is reduced to studying finitely generated torsion modules M; the uniqueness of the cyclic submodule decomposition of finitely generated torsion modules is the following result.

(7.3) Theorem. *Let M be a finitely generated torsion module over a PID R, and suppose that there are cyclic submodule decompositions*

$$(7.6) \qquad M \cong Rw_1 \oplus \cdots \oplus Rw_k$$

and

$$(7.7) \qquad M \cong Rz_1 \oplus \cdots \oplus Rz_r$$

where

$$(7.8) \qquad \text{Ann}(w_1) \supseteq \cdots \supseteq \text{Ann}(w_k) \neq \langle 0 \rangle \qquad \text{with} \quad \text{Ann}(w_1) \neq R$$

and

$$(7.9) \qquad \text{Ann}(z_1) \supseteq \cdots \supseteq \text{Ann}(z_r) \neq \langle 0 \rangle \qquad \text{with} \quad \text{Ann}(z_1) \neq R.$$

Then $k = r$ and $\text{Ann}(w_i) = \text{Ann}(z_i)$ for $1 \leq i \leq k$.

Proof. Note that $\text{Ann}(M) = \text{Ann}(w_k) = \text{Ann}(z_r)$. Indeed,

$$\begin{aligned} \text{Ann}(M) &= \text{Ann}(Rw_1 + \cdots + Rw_k) \\ &= \text{Ann}(w_1) \cap \cdots \cap \text{Ann}(w_k) \\ &= \text{Ann}(w_k) \end{aligned}$$

since $\text{Ann}(w_1) \supseteq \cdots \supseteq \text{Ann}(w_k)$. The equality $\text{Ann}(M) = \text{Ann}(z_r)$ is the same argument.

We will first show that $k = r$. Suppose without loss of generality that $k \geq r$. Choose a prime $p \in R$ such that $\langle p \rangle \supseteq \text{Ann}(w_1)$, i.e., p divides the generator of $\text{Ann}(w_1)$. Then $\langle p \rangle \supseteq \text{Ann}(w_i)$ for all i. Since $p \in \text{Ann}(M/pM)$, it follows that M/pM is an R/pR-module and Equations (7.6) and (7.7) imply

$$(7.10) \qquad M/pM \cong Rw_1/(pRw_1) \oplus \cdots \oplus Rw_k/(pRw_k)$$

and

$$(7.11) \qquad M/pM \cong Rz_1/(pRz_1) \oplus \cdots \oplus Rz_r/(pRz_r).$$

Suppose that $pRw_i = Rw_i$. Then we can write $apw_i = w_i$ for some $a \in R$. Hence, $ap - 1 \in \text{Ann}(w_i) \subseteq \langle p \rangle$ by our choice of p, so $1 \in \langle p \rangle$, which contradicts the fact that p is a prime. Therefore, $pRw_i \neq Rw_i$ for all i and Equation (7.10) expresses the R/pR-module M/pM as a direct sum of cyclic R/pR-modules, none of which is $\langle 0 \rangle$. Since R/pR is a field (in a PID prime ideals are maximal), all R/pR-modules are free, so we conclude that M/pM is free of rank k. Moreover, Equation (7.11) expresses M/pM as a direct sum of r cyclic submodules, so it follows that $k = \mu(M/pM) \leq r$. Thus, $r = k$, and in particular, $Rz_i/(pRz_i) \neq 0$ since, otherwise, M/pM could be generated by fewer than k elements. Thus, $\langle p \rangle \supseteq \text{Ann}(z_i)$ for all i; if not, then $\langle p \rangle + \text{Ann}(z_i) = R$, so there are $a \in R$ and $c \in \text{Ann}(z_i)$ such that $ap + c = 1$. Then $z_i = apz_i + cz_i = apz_i \in pRz_i$, so $Rz_i/(pRz_i) = 0$, and we just observed that $Rz_i/(pRz_i) \neq 0$.

We are now ready to complete the proof. We will work by induction on $\ell(\text{Ann}(M))$ where, if $I = \langle a \rangle$ is an ideal of R, then $\ell(I)$ is the number of elements (counted with multiplicity) in a prime factorization of a. This number is well defined by the fundamental theorem of arithmetic for PIDs. Suppose that $\ell(\text{Ann}(M)) = 1$. Then $\text{Ann}(M) = \langle p \rangle$ where $p \in R$ is prime. Since $\text{Ann}(M) = \text{Ann}(w_k) = \text{Ann}(z_k) = \langle p \rangle$ and since $\langle p \rangle$ is a maximal ideal, Equations (7.8) and (7.9) imply that $\text{Ann}(w_i) = \langle p \rangle = \text{Ann}(z_i)$ for all i, and the theorem is proved in the case $\ell(\text{Ann}(M)) = 1$.

Now suppose the theorem is true for all finitely generated torsion R-modules N with $\ell(\text{Ann}(N)) < \ell(\text{Ann}(M))$, and consider the isomorphisms

$$(7.12) \qquad pM \cong pRw_1 \oplus \cdots \oplus pRw_k \cong pRw_{s+1} \oplus \cdots \oplus pRw_k$$

and

$$(7.13) \qquad pM \cong pRz_1 \oplus \cdots \oplus pRz_k \cong pRz_{t+1} \oplus \cdots \oplus pRz_k$$

where $\text{Ann}(w_1) = \cdots = \text{Ann}(w_s) = \text{Ann}(z_1) = \cdots = \text{Ann}(z_t) = \langle p \rangle$ and $\text{Ann}(w_{s+1}) \neq \langle p \rangle$, $\text{Ann}(z_{t+1}) \neq \langle p \rangle$ (s and t may be 0). Then $\text{Ann}(pM) = \langle a/p \rangle$ where $\text{Ann}(M) = \langle a \rangle$, so $\ell(\text{Ann}(pM)) = \ell(\text{Ann}(M)) - 1$. By induction we conclude that $k - s = k - t$, i.e., $s = t$, and $\text{Ann}(pw_i) = \text{Ann}(pz_i)$ for $s < i \leq k$. But $\text{Ann}(pw_i) = \langle a_i/p \rangle$ where $\text{Ann}(w_i) = \langle a_i \rangle$. Thus $\text{Ann}(w_i) = \text{Ann}(z_i)$ for all i and we are done. $\square$

Since $Rw_i \cong R/\text{Ann}(w_i)$ and since R/I and R/J are isomorphic R-modules if and only if $I = J$ (Exercise 10), we may rephrase our results as follows.

(7.4) Corollary. *Finitely generated modules over a PID R are in one-to-one correspondence with finite nonincreasing chains of ideals*

$$R \neq I_1 \supseteq I_2 \supseteq \cdots \supseteq I_n.$$

Such a chain of ideals corresponds to the module

$$M = R/I_1 \oplus \cdots \oplus R/I_n.$$

Note that $\mu(M) = n$ and if $I_{k+1} = \cdots = I_n = \langle 0 \rangle$ but $I_k \neq \langle 0 \rangle$, then

$$M \cong R/I_1 \oplus \cdots \oplus R/I_k \oplus R^{n-k}.$$

We will use the convention that the empty sequence of ideals ($n = 0$) corresponds to $M = \langle 0 \rangle$.

Proof. $\square$

(7.5) Definition. *If M is a finitely generated torsion module over a PID R and $M \cong Rw_1 \oplus \cdots \oplus Rw_n$ with $\text{Ann}(w_i) \supseteq \text{Ann}(w_{i+1})$ ($1 \leq i \leq n - 1$) and $\text{Ann}(w_i) \neq R$, then the chain of ideals $I_i = \text{Ann}(w_i)$ is called the **chain of invariant ideals** of M.*

Using this language, we can express our results as follows:

(7.6) Corollary. *Two finitely generated torsion modules over a* PID *are isomorphic if and only if they have the same chain of invariant ideals.*

Proof.   □

(7.7) *Remark.* In some cases the principal ideals $\text{Ann}(w_j)$ have a preferred generator a_j. In this case the generators $\{a_j\}_{j=1}^n$ are called the **invariant factors** of M.

The common examples are $R = \mathbf{Z}$, in which case we choose $a_j > 0$ so that $a_j = |\mathbf{Z}/\text{Ann}(w_j)|$, and $R = F[X]$, where we take monic polynomials as the preferred generators of ideals.

(7.8) Definition. *Let R be a* PID, *and let M be a finitely generated torsion R-module with chain of invariant ideals*

$$\langle s_1 \rangle \supseteq \langle s_2 \rangle \supseteq \cdots \supseteq \langle s_n \rangle.$$

We define $\text{me}(M) = s_n$ *and* $\text{co}(M) = s_1 \cdots s_n$.

Note that $\text{me}(M)$ and $\text{co}(M)$ are only defined up to multiplication by a unit, but in some cases ($R = \mathbf{Z}$ or $R = F[X]$) we have a preferred choice of generators of ideals. In these cases $\text{me}(M)$ and $\text{co}(M)$ are uniquely defined. Concerning the invariants $\text{me}(M)$ and $\text{co}(M)$, there is the following trivial but useful corollary of our structure theorems.

(7.9) Corollary. *Let M be a finitely generated torsion module over a* PID *R.*

(1) *If $a \in R$ with $aM = 0$, then $\text{me}(M) \mid a$.*
(2) *$\text{me}(M)$ divides $\text{co}(M)$.*
(3) *If $p \in R$ is a prime dividing $\text{co}(M)$, then p divides $\text{me}(M)$.*

Proof. (1) Since $\text{Ann}(M) = \langle s_n \rangle = \langle \text{me}(M) \rangle$ by Theorem 7.1 and the defintion of $\text{me}(M)$, it follows that if $aM = 0$, i.e., $a \in \text{Ann}(M)$, then $\text{me}(M) \mid a$.
    (2) Clearly s_n divides $s_1 \cdots s_n$.
    (3) Suppose that $p \mid s_1 \cdots s_n = \text{co}(M)$. Then p divides some s_i, but $\langle s_i \rangle \supseteq \langle s_n \rangle$, so $s_i \mid s_n$. Hence, $p \mid s_n = \text{me}(M)$.   □

(7.10) *Remark.* There are, unfortunately, no standard names for these invariants. The notation we have chosen reflects the common terminology in the two cases $R = \mathbf{Z}$ and $R = F[X]$. In the case $R = \mathbf{Z}$, $\text{me}(M)$ is the exponent and $\text{co}(M)$ is the order of the finitely generated torsion $\mathbf{Z}$-module

(= finite abelian group) M. In the case $R = F[X]$ of applications to linear algebra to be considered in Chapter 4, $\text{me}(V_T)$ will be the minimal polynomial and $\text{co}(V_T)$ will be the characteristic polynomial of the linear transformation $T \in \text{Hom}_F(V)$ where V is a finite-dimensional vector space over the field F and V_T is the $F[X]$-module determined by T (see Example 1.5 (12)).

There is another decomposition of a torsion R-module M into a direct sum of cyclic submodules which takes advantage of the prime factorization of any generator of $\text{Ann}(M)$. To describe this decomposition we need the following definition.

(7.11) Definition. *Let M be a module over the PID R and let $p \in R$ be a prime. Define the p-**component** M_p of M by*

$$M_p = \{x \in M : \text{Ann}(x) = \langle p^n \rangle \quad \text{for some} \quad n \in \mathbf{Z}^+\}.$$

*If $M = M_p$, then M is said to be p-**primary**, and M is **primary** if it is p-primary for some prime $p \in R$.*

It is a simple exercise to check that submodules, quotient modules, and direct sums of p-primary modules are p-primary (Exercise 54).

(7.12) Theorem. *If M is a finitely generated torsion module over a PID R, then M is a direct sum of primary submodules.*

Proof. Since M is a direct sum of cyclic submodules by Theorem 7.1, it is sufficient to assume that M is cyclic. Thus suppose that $M = \langle x \rangle$ and suppose that

$$\text{Ann}(x) = \langle a \rangle = \langle p_1^{r_1} \cdots p_n^{r_n} \rangle$$

where $p_1, \ldots, p_n$ are the distinct prime divisors of a. Let $q_i = a/p_i^{r_i}$. Then $1 = (q_1, \ldots, q_n) = \gcd\{q_1, \ldots, q_n\}$, so there are $b_1, \ldots, b_n \in R$ such that

(7.14) $1 = b_1 q_1 + \cdots + b_n q_n.$

Let $x_i = b_i q_i x$. Then Equation (7.14) implies that

$$x = x_1 + \cdots + x_n$$

so that

$$M = \langle x \rangle = \langle x_1 \rangle + \cdots + \langle x_n \rangle.$$

Suppose that $y \in \langle x_1 \rangle \cap (\langle x_2 \rangle + \cdots + \langle x_n \rangle)$. Then

$$y = c_1 x_1 = c_2 x_2 + \cdots + c_n x_n$$

and hence, $p_1^{r_1} y = c_1 b_1 p_1^{r_1} q_1 x = c_1 b_1 a x = 0$ and

$$q_1 y = c_2 \tilde{q}_2 p_2^{r_2} x_2 + \cdots + c_n \tilde{q}_n p_n^{r_n} x_n = 0,$$

where $\tilde{q}_j = q_1/p_j^{r_j}$. Therefore, $\{p_1^{r_1}, q_1\} \subseteq \mathrm{Ann}(y)$, but $(p_1^{r_1}, q_1) = 1$ so that $\mathrm{Ann}(y) = R$. Therefore, $y = 0$. A similar calculation shows that

$$\langle x_i \rangle \cap \left(\langle x_1 \rangle + \cdots + \widehat{\langle x_i \rangle} + \cdots + \langle x_n \rangle \right) = \langle 0 \rangle,$$

so by Theorem 3.1, $M \cong \langle x_1 \rangle \oplus \cdots \oplus \langle x_n \rangle$. $\square$

Combining Theorems 7.1 and 7.12, we obtain the following result:

(7.13) Theorem. *Any finitely generated torsion module M over a PID R is a direct sum of primary cyclic submodules.*

Proof. Suppose $M \cong Rw_1 \oplus \cdots \oplus Rw_n$ as in Theorem 7.1. Then if $\mathrm{Ann}(w_i) = \langle s_i \rangle$, we have $s_i \mid s_{i+1}$ for $1 \le i \le n-1$ with $s_1 \ne 1$ and $s_n \ne 0$ (since M is torsion). Let $p_1, \ldots, p_k$ be the set of distinct nonassociate primes that occur as a prime divisor of some invariant factor of M. Then

$$s_1 = u_1 p_1^{e_{11}} \cdots p_k^{e_{1k}}$$
$$\vdots$$
$$s_n = u_n p_1^{e_{n1}} \cdots p_k^{e_{nk}}$$

where the divisibility conditions imply that

$$0 \le e_{1j} \le e_{2j} \le \cdots \le e_{nj} \qquad \text{for} \quad 1 \le j \le k.$$

Then the proof of Theorem 7.12 shows that M is the direct sum of cyclic submodules with annihilators $\{p_j^{e_{ij}} : e_{ij} > 0\}$, and the theorem is proved. $\square$

(7.14) Definition. *The prime powers $\{p_j^{e_{ij}} : e_{ij} > 0, \ 1 \le j \le k\}$ are called the **elementary divisors** of M.*

(7.15) Theorem. *If M and N are finitely generated torsion modules over a PID R, then $M \cong N$ if and only if M and N have the same elementary divisors.*

Proof. Since M is uniquely determined up to isomorphism from the invariant factors, it is sufficient to show that the invariant factors of M can be recovered from a knowledge of the elementary divisors. Thus suppose that

$$\langle s_1 \rangle \supseteq \langle s_2 \rangle \supseteq \cdots \supseteq \langle s_n \rangle$$

is the chain of invariant ideals of the finitely generated torsion module M. This means that $s_i \mid s_{i+1}$ for $1 \le i < n$. Let $p_1, \ldots, p_k$ be the set of distinct nonassociate primes that occur as a prime divisor of some invariant factor of M. Then

$$s_1 = u_1 p_1^{e_{11}} \cdots p_k^{e_{1k}}$$

(7.15)
$$\vdots$$

$$s_n = u_n p_1^{e_{n1}} \cdots p_k^{e_{nk}}$$

where the divisibility conditions imply that

(7.16) $$0 \le e_{1j} \le e_{2j} \le \cdots \le e_{nj} \qquad \text{for} \quad 1 \le j \le k.$$

Thus, the elementary divisors of M are

(7.17) $$\{p_j^{e_{ij}} : e_{ij} > 0\}.$$

We show that the set of invariant factors (Equation (7.15)) can be reconstructed from the set of prime powers in Equation (7.17). Indeed, if

$$e_j = \max_{1 \le i \le n} e_{ij}, \qquad 1 \le j \le k,$$

then the inequalities (7.16) imply that s_n is an associate of $p_1^{e_1} \cdots p_k^{e_k}$. Delete

$$\{p_1^{e_1}, \ldots, p_k^{e_k}\}$$

from the set of prime powers in set (7.17), and repeat the process with the set of remaining elementary divisors to obtain s_{n-1}. Continue until all prime powers have been used. At this point, all invariant factors have been recovered. Notice that the number n of invariant factors is easily recovered from the set of elementary divisors of M. Since s_1 divides every s_i, it follows that every prime dividing s_1 must also be a prime divisor of every s_i. Therefore, in the set of elementary divisors, n is the maximum number of occurrences of $p^{e_{ij}}$ for a single prime p. $\square$

(7.16) Example. Suppose that M is the **Z**-module

$$M = \mathbf{Z}_{2^2} \times \mathbf{Z}_{2^2} \times \mathbf{Z}_3 \times \mathbf{Z}_{3^2} \times \mathbf{Z}_5 \times \mathbf{Z}_7 \times \mathbf{Z}_{7^2}.$$

Then the elementary divisors of M are 2^2, 2^2, 3, 3^2, 5, 7, 7^2. Using the algorithm from Theorem 7.15, we can recover the invariant factor description of M as follows. The largest invariant factor is the product of the highest power of each prime occurring in the set of elementary divisors, i.e., the least common multiple of the set of elementary divisors. That is, $s_2 = 7^2 \cdot 5 \cdot 3^2 \cdot 2^2 = 8820$. Note that the number of invariant factors of M is 2 since powers of the primes 2, 3, and 7 occur twice in the set of elementary divisors, while no prime has three powers among this set. Deleting 7^2, 5, 3^2, 2^2 from the set of elementary divisors, we obtain $s_1 = 7 \cdot 3 \cdot 2^2 = 84$. This uses all the elementary divisors, so we obtain

$$M \cong \mathbf{Z}_{84} \times \mathbf{Z}_{8820}.$$

We now present some useful observations concerning the invariants $me(M)$ and $co(M)$ where M is a torsion R-module (R a PID). See Definition 7.8 for the definition of these invariants. The verification of the results that we wish to prove require some preliminary results on torsion R-modules, which are of interest in their own right. We start with the following lemmas.

(7.17) Lemma. *Let M be a module over a PID R and suppose that $x \in M_\tau$. If $\text{Ann}(x) = \langle r \rangle$ and $a \in R$ with $(a, r) = d$ (recall that $(a, r) = \gcd\{a, r\}$), then $\text{Ann}(ax) = \langle r/d \rangle$.*

Proof. Since $(r/d)(ax) = (a/d)(rx) = 0$, it follows that $\langle r/d \rangle \subseteq \text{Ann}(ax)$. If $b(ax) = 0$, then $r \mid (ba)$, so $ba = rc$ for some $c \in R$. But $(a, r) = d$, so there are $s, t \in R$ with $rs + at = d$. Then $rct = bat = b(d - rs)$ and we see that $bd = r(ct + bs)$. Therefore, $b \in \langle r/d \rangle$ and hence $\text{Ann}(ax) = \langle r/d \rangle$. $\square$

(7.18) Lemma. *Let M be a module over a PID R, and let $x_1, \ldots, x_n \in M_\tau$ with $\text{Ann}(x_i) = \langle r_i \rangle$ for $1 \le i \le n$. If $\{r_1, \ldots, r_n\}$ is a pairwise relatively prime subset of R and $x = x_1 + \cdots + x_n$, then $\text{Ann}(x) = \langle a \rangle = \langle \prod_{i=1}^n r_i \rangle$. Conversely, if $y \in M_\tau$ is an element such that $\text{Ann}(y) = \langle b \rangle = \langle \prod_{i=1}^n s_i \rangle$ where $\{s_1, \ldots, s_n\}$ is a pairwise relatively prime subset of R, then we may write $y = y_1 + \cdots + y_n$ where $\text{Ann}(y_i) = \langle s_i \rangle$ for all i.*

Proof. Let $x = x_1 + \cdots + x_n$. Then $a = \prod_{i=1}^n r_i \in \text{Ann}(x)$ so that $\langle a \rangle \subseteq \text{Ann}(x)$. It remains to check that $\text{Ann}(x) \subseteq \langle a \rangle$. Thus, suppose that $bx = 0$. By the Chinese remainder theorem (Theorem 2.2.24), there are $c_1, \ldots, c_n \in R$ such that

$$c_i \equiv \begin{cases} 1 & (\text{mod } \langle r_i \rangle), \\ 0 & (\text{mod } \langle r_j \rangle), \quad \text{if } j \ne i. \end{cases}$$

Then, since $\langle r_j \rangle = \text{Ann}(x_j)$, we conclude that $c_i x_j = 0$ if $i \ne j$, so for each i with $1 \le i \le n$

$$0 = c_i bx = c_i b(x_1 + \cdots + x_n) = bc_i x_i.$$

Therefore, $bc_i \in \text{Ann}(x_i) = \langle r_i \rangle$, and since $c_i \equiv 1 \pmod{\langle r_i \rangle}$, it follows that $r_i \mid b$ for $1 \le i \le n$. But $\{r_1, \ldots, r_n\}$ is pairwise relatively prime and thus a is the least common multiple of the set $\{r_1, \ldots, r_n\}$. We conclude that $a \mid b$, and hence, $\text{Ann}(x) = \langle a \rangle$.

Conversely, suppose that $y \in M$ satisfies $\text{Ann}(y) = \langle b \rangle = \langle \prod_{i=1}^n s_i \rangle$ where the set $\{s_1, \ldots, s_n\}$ is pairwise relatively prime. As in the above paragraph, apply the Chinese remainder theorem to get $c_1, \ldots, c_n \in R$ such that

$$c_i \equiv \begin{cases} 1 & (\text{mod } \langle s_i \rangle), \\ 0 & (\text{mod } \langle s_j \rangle), \quad \text{if } j \ne i. \end{cases}$$

Since b is the least common multiple of $\{s_1, \ldots, s_n\}$, it follows that

$$1 \equiv c_1 + \cdots + c_n \pmod{\langle b \rangle},$$

and hence, if we set $y_i = c_i y$, we have

$$y_1 + \cdots + y_n = (c_1 + \cdots + c_n)y = y.$$

Since $\langle b, c_i \rangle = \left\langle \prod_{j \neq i} s_j \right\rangle$, Lemma 7.17 shows that $\mathrm{Ann}(y_i) = \mathrm{Ann}(c_i y) = \langle s_i \rangle$. $\square$

(7.19) Proposition. *Let R be a PID and suppose that M is a torsion R-module such that*

$$M \cong Rw_1 \oplus \cdots Rw_n$$

with $\mathrm{Ann}(w_i) = \langle t_i \rangle$. Then the prime power factors of the t_i $(1 \leq i \leq n)$ are the elementary divisors of M.

Proof. Let $p_1, \ldots, p_k$ be the set of distinct nonassociate primes that occur as a prime divisor of some t_i. Then we may write

$$t_1 = u_1 p_1^{e_{11}} \cdots p_k^{e_{1k}}$$

(7.18)
$$\vdots$$

$$t_n = u_n p_1^{e_{n1}} \cdots p_k^{e_{nk}}$$

where $u_1, \ldots, u_n$ are units in R and some of the exponents e_{ij} may be 0. The proof of Theorem 7.12 shows that

$$Rw_i \cong Rz_{i1} \oplus \cdots Rz_{ik}$$

where $\mathrm{Ann}(z_{ij}) = \langle p_j^{e_{ij}} \rangle$. For notational convenience we are allowing $z_{ij} = 0$ for those (i, j) with $e_{ij} = 0$. Therefore,

(7.19)
$$M \cong \bigoplus_{i,j} Rz_{ij}$$

where $\mathrm{Ann}(z_{ij}) = \langle p_j^{e_{ij}} \rangle$. Let $S = \{p_j^{e_{ij}}\}$ where we allow multiple occurrences of a prime power p^e, and let

$$\widetilde{S} = \{z_{ij}\}.$$

Let m be the maximum number of occurrences of positive powers of a single prime in S. If

(7.20)
$$f_{mj} = \max_{1 \leq i \leq n} e_{ij} \qquad \text{for} \quad 1 \leq j \leq k,$$

we define

(7.21)
$$s_m = p_1^{f_{m1}} \cdots p_k^{f_{mk}}.$$

Note that $f_{mj} > 0$ for $1 \leq j \leq k$.

Delete $\{p_1^{f_{m1}}, \ldots, p_k^{f_{mk}}\}$ from the set S and repeat the above process with the remaining prime powers until no further positive prime powers are available. Since a prime power for a particular prime p is used only once at each step, this will produce elements $s_1, \ldots, s_m \in R$. From the inductive description of the construction of s_i, it is clear that every prime dividing s_i also divides s_{i+1} to at least as high a power (because of Equation (7.21)). Thus,

$$s_i \mid s_{i+1} \quad \text{for} \quad 1 \le i < m.$$

Therefore, we may write

(7.22)
$$s_1 = u_1 p_1^{f_{11}} \cdots p_k^{f_{1k}}$$
$$\vdots$$
$$s_m = u_m p_1^{f_{m1}} \cdots p_k^{f_{mk}}$$

where

(7.23)
$$\{p_j^{f_{ij}} : f_{ij} > 0\} = \{p_\beta^{e_{\alpha\beta}} : e_{\alpha\beta} > 0\}$$

where repetitions of prime powers are allowed and where

(7.24)
$$0 \le f_{1j} \le f_{2j} \le \cdots \le f_{mj} \quad \text{for} \quad 1 \le j \le k$$

by Equation (7.20).

For each $p_j^{f_{ij}}$ $(1 \le i \le m)$, choose $w_{ij} \in \tilde{S}$ with $\text{Ann}(w_{ij}) = \langle p_j^{f_{ij}} \rangle$ and let $x_i = w_{i1} + \cdots + w_{ik}$. Lemma 7.18 shows that $\text{Ann}(x_i) = \langle s_i \rangle$ for $1 \le i \le m$, and thus,

$$Rx_i \cong R/\langle s_i \rangle \cong \bigoplus_{j=1}^{k} R/\langle p_j^{f_{ij}} \rangle \cong \bigoplus_{j=1}^{k} Rw_{ij}.$$

Equation (7.19) then shows that

$$M \cong \bigoplus_{\alpha,\beta} Rz_{\alpha\beta}$$
$$\cong \bigoplus_{i=1}^{m} \left(\bigoplus_{j=1}^{k} Rw_{ij} \right)$$
$$\cong Rx_1 \oplus \cdots \oplus Rx_m$$

where $\text{Ann}(x_i) = \langle s_i \rangle$. Since $s_i \mid s_{i+1}$ for $1 \le i < m$, it follows that $\{s_1, \ldots, s_m\}$ are the invariant factors of M, and since the set of prime power factors of $\{s_1, \ldots, s_m\}$ (counting multiplicities) is the same as the set of prime power factors of $\{t_1, \ldots, t_n\}$ (see Equation (7.23)), the proof is complete. □

(7.20) Corollary. *Let R be a PID, let $M_1, \ldots, M_k$ be finitely generated torsion R-modules, and let $M = \oplus_{i=1}^{k} M_i$. If $\{d_{i1}, \ldots, d_{i\ell_i}\}$ is the set of elementary divisors of M_i, then*

$$S = \{d_{ij} : 1 \leq i \leq k; 1 \leq j \leq \ell_i\}$$

is the set of elementary divisors of M.

Proof. By Theorem 7.1,

$$M_i \cong Rw_{i1} \oplus \cdots \oplus Rw_{ir_i}$$

where $\mathrm{Ann}(w_{ij}) = \langle s_{ij} \rangle$ and $s_{ij} \mid s_{i,j+1}$ for $1 \leq j \leq r_i$. The elementary divisors of M_i are the prime power factors of $\{s_{i1}, \ldots, s_{ir_i}\}$. Then

$$M = \bigoplus_{i=1}^{k} M_i \cong \bigoplus_{i,j} Rw_{ij}$$

where $\mathrm{Ann}(w_{ij}) = \langle s_{ij} \rangle$. The result now follows from Proposition 7.19. □

(7.21) Proposition. *Let R be a PID, let $M_1, \ldots, M_k$ be finitely generated torsion R-modules, and let $M = \oplus_{i=1}^{k} M_i$. Then*

$$(7.25) \qquad \mathrm{me}(M) = \mathrm{lcm}\{\mathrm{me}(M_1), \ldots, \mathrm{me}(M_k)\}$$

$$(7.26) \qquad \mathrm{co}(M) = \prod_{i=1}^{k} \mathrm{co}(M_i).$$

Proof. Since $\mathrm{Ann}(M) = \bigcap_{i=1}^{k} \mathrm{Ann}(M_i)$, Equation (7.25) follows since $\langle \mathrm{me}(M_i) \rangle = \mathrm{Ann}(M_i)$. Since $\mathrm{co}(M)$ is the product of all invariant factors of M, which is also the product of all the elementary divisors of M, Equation (7.26) follows from Corollary 7.20. □

The special case $R = \mathbf{Z}$ is important enough to emphasize what the results mean in this case. Suppose that M is an abelian group, i.e., a $\mathbf{Z}$-module. Then an element $x \in M$ is torsion if and only if $nx = 0$ for some $n > 0$. That is to say, $x \in M_\tau$ if and only if $o(x) < \infty$. Moreover, $\mathrm{Ann}(x) = \langle n \rangle$ means that $o(x) = n$. Thus the torsion submodule of M consists of the set of elements of finite order. Furthermore, M is finitely generated and torsion if and only if M is a finite abelian group. Indeed, if $M = \langle x_1, \ldots, x_k \rangle$ then any $x \in M$ can be written $x = n_1 x_1 + \cdots + n_k x_k$ where $0 \leq n_i \leq o(x_i) < \infty$ for $1 \leq i \leq k$. Therefore, $|M| \leq \prod_{i=1}^{k} o(x_i)$. Hence, the fundamental structure theorem for finitely generated abelian groups takes the following form.

(7.22) Theorem. *Any finitely generated abelian group M is isomorphic to $\mathbf{Z}^r \oplus M_1$ where $|M_1| < \infty$. The integer r is an invariant of M. Any finite abelian group is a direct sum of cyclic groups of prime power order and these prime power orders, counted with multiplicity, completely characterize* *the finite abelian group up to isomorpism. Also any finite abelian group is uniquely isomorphic to a group*

$$\mathbf{Z}_{s_1} \times \cdots \times \mathbf{Z}_{s_k}$$

where $s_i \mid s_{i+1}$ for all i.

Proof.   □

Given a natural number n it is possible to give a complete list of all abelian groups of order n, up to isomorphism, by writing $n = p_1^{r_1} \cdots p_k^{r_k}$ where $p_1, \ldots, p_k$ are the distinct prime divisors of n. Let M be an abelian group of order n. Then we may write M as a direct sum of its primary components

$$M \cong M_{p_1} \oplus \cdots \oplus M_{p_k}$$

where $|M_{p_i}| = p_i^{r_i}$. Then each primary component M_{p_i} can be written as a direct sum

$$M_{p_i} \cong \mathbf{Z}_{p_i^{e_{i1}}} \oplus \cdots \oplus \mathbf{Z}_{p_i^{e_{i\ell}}}$$

where

$$1 \leq e_{i1} \leq \cdots \leq e_{i\ell} \leq r_i$$

and

$$e_{i1} + \cdots + e_{i\ell} = r_i.$$

Furthermore, the main structure theorems state that M is determined up to isomorphism by the primes $p_1, \ldots, p_k$ and the partitions $e_{i1}, \ldots, e_{i\ell}$ of the exponents r_i. This is simply the statement that M is determined up to isomorphism by its elementary divisors. Therefore, to identify all abelian groups of order n, it is sufficient to identify all partitions of r_i, i.e., all ways to write $r_i = e_{i1} + \cdots + e_{i\ell}$ as a sum of natural numbers.

(7.23) Example. We will carry out the above procedure for $n = 600 = 2^3 \cdot 3 \cdot 5^2$. There are three primes, namely, 2, 3, and 5. The exponent of 2 is 3 and we can write $3 = 1 + 1 + 1$, $3 = 1 + 2$, and $3 = 3$. Thus there are three partitions of 3. The exponent of 3 is 1, so there is only one partition, while the exponent of 5 is 2, which has two partitions, namely, $2 = 1 + 1$ and $2 = 2$. Thus there are $3 \cdot 1 \cdot 2 = 6$ distinct, abelian groups of order 600. They are

$$\mathbf{Z}_2 \times \mathbf{Z}_2 \times \mathbf{Z}_2 \times \mathbf{Z}_3 \times \mathbf{Z}_5 \times \mathbf{Z}_5 \cong \mathbf{Z}_2 \times \mathbf{Z}_{10} \times \mathbf{Z}_{30}$$

$$\mathbf{Z}_2 \times \mathbf{Z}_2 \times \mathbf{Z}_2 \times \mathbf{Z}_3 \times \mathbf{Z}_{25} \cong \mathbf{Z}_2 \times \mathbf{Z}_2 \times \mathbf{Z}_{150}$$

$$\mathbf{Z}_2 \times \mathbf{Z}_4 \times \mathbf{Z}_3 \times \mathbf{Z}_5 \times \mathbf{Z}_5 \cong \mathbf{Z}_{10} \times \mathbf{Z}_{60}$$

$$\mathbf{Z}_2 \times \mathbf{Z}_4 \times \mathbf{Z}_3 \times \mathbf{Z}_{25} \cong \mathbf{Z}_2 \times \mathbf{Z}_{300}$$

$$\mathbf{Z}_8 \times \mathbf{Z}_3 \times \mathbf{Z}_5 \times \mathbf{Z}_5 \cong \mathbf{Z}_5 \times \mathbf{Z}_{120}$$

$$\mathbf{Z}_8 \times \mathbf{Z}_3 \times \mathbf{Z}_{25} \cong \mathbf{Z}_{600}$$

where the groups on the right are expressed in invariant factor form and those on the left are decomposed following the elementary divisors.

We will conclude this section with the following result concerning the structure of finite subgroups of the multiplicative group of a field. This is an important result, which combines the structure theorem for finite abelian groups with a bound on the number of roots of a polynomial with coefficients in a field.

(7.24) Theorem. *Let F be a field and let $G \subseteq F^* = F \setminus \{0\}$ be a finite subgroup of the multiplicative group F^*. Then G is a cyclic group.*

Proof. According to Theorem 7.1, G is isomorphic to a direct sum

$$G \cong \langle z_1 \rangle \oplus \cdots \oplus \langle z_n \rangle$$

where, if we let $k_i = o(z_i) = $ order of z_i, then $k_i \mid k_{i+1}$ for $1 \le i \le n-1$ and

$$\text{Ann}(G) = \text{Ann}(z_n) = (k_n)\mathbf{Z}.$$

In the language of Definition 7.8, $\text{me}(G) = k_n$. This means that $z^{k_n} = 1$ for all $z \in G$. Now consider the polynomial

(7.27) $$P(X) = X^{k_n} - 1.$$

Since F is a field, the polynomial $P(X)$ has at most k_n roots, because degree $P(X) = k_n$ (Corollary 2.4.7). But, as we have observed, every element of G is a root of $P(X)$, and

$$|G| = k_1 k_2 \cdots k_n.$$

Thus, we must have $n = 1$ and $G \cong \langle z_1 \rangle$ is cyclic.   □

(7.25) Corollary. *Suppose that F is a finite field with q elements. Then F^* is a cyclic group with $q - 1$ elements, and every element of F is a root of the polynomial $X^q - X$.*

Proof. Exercise.   □

(7.26) Corollary. *Let*

$$G_n = \{e^{2\pi i(k/n)} : 0 \le k \le n-1\} \subseteq \mathbf{C}^*.$$

Then G_n is the only subgroup of $\mathbf{C}^*$ of order n.

Proof. Let H be a finite subgroup of $\mathbf{C}^*$ with $|H| = n$. Then every element z of H has the property that $z^n = 1$. In other words, z is a root of the equation $X^n = 1$. Since this equation has at most n roots in $\mathbf{C}$ and since every element of G_n is a root of this equation, we have $z \in G_n$. Thus, we conclude that $H \subseteq G_n$ and hence $H = G_n$ because $n = |H| = |G_n|$. $\qquad\square$

3.8 Complemented Submodules

We will now consider the problem of extending a linearly independent subset of a free R-module to a basis. The example $\{2\} \subseteq \mathbf{Z}$ shows that some restrictions on the subset are needed, while Theorem 6.16 shows that any primitive element of a finitely generated free R-module (R a PID) can be extended to a basis.

(8.1) Definition. *Let M be an R-module and $S \subseteq M$ a submodule. Then S is said to be* **complemented** *if there exists a submodule $T \subseteq M$ with $M \cong S \oplus T$.*

Let M be a finitely generated free R-module with basis $\{v_1, \ldots, v_n\}$ and let $S = \langle v_1, \ldots, v_s \rangle$. Then S is complemented by $T = \langle v_{s+1}, \ldots, v_n \rangle$. This example shows that if $W = \{w_1, \ldots, w_k\}$ is a linearly independent subset of M, then a necessary condition for W to extend to a basis of M is that the submodule $\langle W \rangle$ be complemented. If R is a PID, then the converse is also true. Indeed, let T be a complement of $\langle W \rangle$ in M. Since R is a PID, T is free, so let $\{x_1, \ldots, x_r\}$ be a basis of T. Then it is easy to check that $\{w_1, \ldots, w_k, x_1, \ldots, x_r\}$ is a basis of M.

(8.2) Proposition. *Let R be a PID, let M be a free R-module, and let S be a submodule. Consider the following conditions on S.*

(1) *S is complemented.*
(2) *M/S is free.*
(3) *If $x \in S$ and $x = ay$ for some $y \in M$, $a \neq 0 \in R$, then $y \in S$.*

Then $(1) \Rightarrow (2)$ and $(2) \Rightarrow (3)$, while if M is finitely generated, then $(3) \Rightarrow (1)$.

Proof. $(1) \Rightarrow (2)$. If S is complemented, then there exists $T \subseteq M$ such that $S \oplus T \cong M$. Thus, $M/S \cong T$. But T is a submodule of a free module over a PID R, so T is free (Theorem 6.2).

$(2) \Rightarrow (3)$. Suppose M/S is free. If $x \in S$ satisfies $x = ay$ for some $y \in M$, $a \neq 0 \in R$, then $a(y + S) = S$ in M/S. Since free modules are torsion-free, it follows that $y + S = S$, i.e., $y \in S$.

(3) $\Rightarrow$ (1). Let M be a finite rank free R-module and let $S \subseteq M$ be a submodule satisfying condition (3). Then there is a short exact sequence

(8.1) $$0 \longrightarrow S \longrightarrow M \overset{\pi}{\longrightarrow} M/S \longrightarrow 0$$

where π is the projection map. Condition (3) is equivalent to the statement that M/S is torsion-free, so M/S is free by Theorem 6.6. But free modules are projective, so sequence (8.1) has a splitting $\alpha : M/S \to M$ and Theorem 3.9 shows that $M \cong S \oplus \mathrm{Im}(\alpha)$, i.e., S is complemented. □

(8.3) Remarks.

(1) A submodule S of M that satisfies condition (3) of Proposition 8.2 is called a **pure submodule** of M. Thus, a submodule of a finitely generated module over a PID is pure if and only if it is complemented.

(2) If R is a field, then *every* subspace $S \subseteq M$ satisfies condition (3) so that every subspace of a finite-dimensional vector space is complemented. Actually, this is true without the finite dimensionality assumption, but our argument has only been presented in the more restricted case. The fact that arbitrary subspaces of vector spaces are complemented follows from Corollary 4.21.

(3) The implication (3) $\Rightarrow$ (1) is false without the hypothesis that M be finitely generated. As an example, consider a free presentation of the **Z**-module **Q**:
$$0 \longrightarrow S \longrightarrow M \longrightarrow \mathbf{Q} \longrightarrow 0.$$

Since $M/S \cong \mathbf{Q}$ and **Q** is torsion-free, it follows that S satisfies condition (3) of Proposition 8.1. However, if S is complemented, then a complement $T \cong \mathbf{Q}$; so **Q** is a submodule of a free **Z**-module M, and hence **Q** would be free, but **Q** is not a free **Z**-module.

(8.4) Corollary. *If S is a complemented submodule of a finitely generated R-module (R a PID), then any basis for S extends to a basis for M.*

Proof. This was observed prior to Proposition 8.2. □

(8.5) Corollary. *If S is a complemented submodule of M, then $\mathrm{rank}\, S = \mathrm{rank}\, M$ if and only if $S = M$.*

Proof. A basis $\{v_1, \ldots, v_m\}$ of S extends to a basis $\{v_1, \ldots, v_n\}$ of M. But $n = m$, so $S = \langle v_1, \ldots, v_n \rangle = M$. □

If $M = \mathbf{Z}$ and $S = \langle 2 \rangle$, then $\mathrm{rank}\, S = \mathrm{rank}\, M$ but $M \neq S$. Of course, S is not complemented.

(8.6) Corollary. *If S is a complemented submodule of M, then*

$$\mathrm{rank}\, M = \mathrm{rank}\, S + \mathrm{rank}(M/S).$$

Proof. Let $S = \langle v_1, \ldots, v_m \rangle$ where $m = \operatorname{rank} S$. Extend this to a basis $\{v_1, \ldots, v_n\}$ of M. Then $T = \langle v_{m+1}, \ldots, v_n \rangle$ is a complement of S in M and $T \cong M/S$. Thus,

$$\operatorname{rank} M = n = m + (n - m) = \operatorname{rank} S + \operatorname{rank}(M/S). \qquad \square$$

(8.7) Proposition. *Let R be a PID and let $f : M \rightarrow N$ be an R-module homomorphism of finite-rank free R-modules. Then*

(1) $\operatorname{Ker}(f)$ *is a pure submodule, but*
(2) $\operatorname{Im}(f)$ *need not be pure.*

Proof. (1) Suppose $x \in \operatorname{Ker}(f)$, $a \neq 0 \in R$, and $y \in M$ with $x = ay$. Then $0 = f(x) = f(ay) = af(y)$. But N is free and, hence, torsion-free so that $f(y) = 0$. Hence, condition (3) of Proposition 8.2 is satisfied, so $\operatorname{Ker}(f)$ is complemented.

(2) If $f : \mathbf{Z} \rightarrow \mathbf{Z}$ is defined by $f(x) = 2x$, then $\operatorname{Im}(f) = 2\mathbf{Z}$ is not a complemented submodule of $\mathbf{Z}$. $\qquad \square$

(8.8) Proposition. *Let R be a PID and let $f : M \rightarrow N$ be an R-module homomorphism of finite-rank free R-modules. Then*

$$\operatorname{rank} M = \operatorname{rank}(\operatorname{Ker}(f)) + \operatorname{rank}(\operatorname{Im}(f)).$$

Proof. By the first isomorphism theorem, $\operatorname{Im}(f) \cong M/\operatorname{Ker}(f)$. But $\operatorname{Ker}(f)$ is a complemented submodule of M, so the result follows from Corollary 8.6. $\qquad \square$

(8.9) Corollary. *Let R be a PID and let M and N be finite-rank free R-modules with $\operatorname{rank}(M) = \operatorname{rank}(N)$. Let $f \in \operatorname{Hom}_R(M, N)$.*

(1) *If f is a surjection, then f is an isomorphism.*
(2) *If f is an injection and $\operatorname{Im}(f)$ is complemented, then f is an isomorphism.*

Proof. (1) By Proposition 8.8, $\operatorname{rank}(\operatorname{Ker}(f)) = 0$, i.e., $\operatorname{Ker}(f) = \langle 0 \rangle$, so f is an injection.

(2) By Proposition 8.8, $\operatorname{rank} N = \operatorname{rank} M = \operatorname{rank}(\operatorname{Im}(f))$. Since $\operatorname{Im}(f)$ is complemented by hypothesis, f is a surjection by Corollary 8.5. $\qquad \square$

(8.10) Proposition. *Let R be a field and let M and N be R-modules with $\operatorname{rank}(M) = \operatorname{rank}(N)$ finite. Let $f \in \operatorname{Hom}_R(M, N)$. Then the following are equivalent.*

(1) *f is an isomorphism.*

(8.11) Proposition. *Let M be a finite-rank free R-module (R a PID). If S and T are pure submodules, then*

(2) f is an injection.

(3) f is a surjection.

Proof. Since R is a field, $\text{Im}(f)$ is complemented (by Remark 8.3 (2)), so this is an immediate consequence of Corollary 8.9. $\square$

(8.11) Proposition. *Let M be a finite-rank free R-module (R a PID). If S and T are pure submodules, then*

$$\text{rank}(S + T) + \text{rank}(S \cap T) = \text{rank}\,S + \text{rank}\,T.$$

Proof. Note that if S and T are pure submodules of M, then $S \cap T$ is also pure. Indeed, if $ay \in S \cap T$ with $a \neq 0 \in R$, then $y \in S$ and $y \in T$ since these submodules are pure. Thus, $y \in S \cap T$, so $S \cap T$ is complemented by Proposition 8.2. Then

$$(S + T)/T \cong S/(S \cap T).$$

By Corollary 8.6, we conclude

$$\text{rank}(S + T) - \text{rank}(T) = \text{rank}(S) - \text{rank}(S \cap T).$$

$\square$

(8.12) Remark. It need not be true that $S + T$ is pure, even if S and T are both pure. For example, let $S = \langle (1, 1) \rangle \subseteq \mathbf{Z}^2$ and let $T = \langle (1, -1) \rangle \subseteq \mathbf{Z}^2$. Then S and T are both pure, but $S + T \neq \mathbf{Z}^2$, so it cannot be pure. In fact, $2 \cdot (1, 0) = (2, 0) = (1, 1) + (1, -1) \in S + T$, but $(1, 0) \notin S + T$.

3.9 Exercises

1. If M is an abelian group, then $\text{End}_{\mathbf{Z}}(M)$, the set of abelian group endomorphisms of M, is a ring under addition and composition of functions.
   (a) If M is a left R-module, show that the function $\phi : R \rightarrow \text{End}_{\mathbf{Z}}(M)$ defined by $\phi(r)(m) = rm$ is a ring homomorphism. Conversely, show that any ring homomorphism $\phi : R \rightarrow \text{End}_{\mathbf{Z}}(M)$ determines a left R-module structure on M.
   (b) Show that giving a right R-module structure on M is the same as giving a ring homomorphism $\phi : R^{\text{op}} \rightarrow \text{End}_{\mathbf{Z}}(M)$.

2. Show that an abelian group G admits the structure of a $\mathbf{Z}_n$-module if and only if $nG = \langle 0 \rangle$.

3. Show that the subring $\mathbf{Z}[\frac{p}{q}]$ of $\mathbf{Q}$ is not finitely generated as a $\mathbf{Z}$-module if $\frac{p}{q} \notin \mathbf{Z}$.

4. Let M be an S-module and suppose that $R \subseteq S$ is a subring. Then M is also an R-module by Example 1.5 (10). Suppose that $N \subseteq M$ is an R-submodule. Let $SN = \{sn : s \in S, n \in N\}$.
   (a) If $S = \mathbf{Q}$ and $R = \mathbf{Z}$, show that SN is the S-submodule of M generated by N.
   (b) Show that the conclusion of part (a) need not hold if $S = \mathbf{R}$ and $R = \mathbf{Q}$.

5. Let M be an R-module and let A, B, and C be submodules. If $C \subseteq A$, prove that
$$A \cap (B + C) = (A \cap B) + C.$$

This equality is known as the *modular law*. Show, by example, that this formula need not hold if C is not contained in A.

6. Let R be a commutative ring and let $S \subseteq R \setminus \{0\}$ be a multiplicatively closed subset of R containing no zero divisors. Let M be an R-module. Mimicking the construction of R_S (Theorem 2.3.5), we define M_S as follows. Define a relation $\sim$ on $M \times S$ by setting $(x, s) \sim (y, t)$ if and only if $utx = usy$ for some $u \in S$. Verify that this is an equivalence relation (see the proof of Theorem 2.3.5). We will denote the equivalence class of (x, s) by the suggestive symbol x/s.
   (a) Prove that M_S is an R_S-module via the operation $(a/s)(x/t) = (ax)/(st)$.
   (b) If $f : M \to N$ is an R-module homomorhism, show that $f_S : M_S \to N_S$ defined by $f_S(x/s) = f(x)/s$ is an R_S-module homomorphism.
   (c) If $x \in M$, show that $x/1 = 0$ in M_S if and only if $\operatorname{Ann}(x) \cap S \neq \emptyset$.

7. Let $R \subseteq F[X]$ be the subring
$$R = \{f(X) \in F[X] : f(X) = a_0 + a_2 X^2 + \cdots + a_n X^n\}.$$

Thus, $f(X) \in R$ if and only if the coefficient of X is 0. Show that $F[X]$ is a finitely generated R-module that is torsion-free but not free.

8. Show that $\mathbf{Q}$ is a torsion-free $\mathbf{Z}$-module that is not free.

9. (a) Let R be an integral domain, let M be a torsion R-module, and let N be a torsion-free R-module. Show that $\operatorname{Hom}_R(M, N) = \langle 0 \rangle$.
   (b) According to part (a), $\operatorname{Hom}_{\mathbf{Z}}(\mathbf{Z}_m, \mathbf{Z}) = \langle 0 \rangle$. If $n = km$, then $\mathbf{Z}_m$ is a $\mathbf{Z}_n$-module. Show that
$$\operatorname{Hom}_{\mathbf{Z}_n}(\mathbf{Z}_m, \mathbf{Z}_n) \cong \mathbf{Z}_m.$$

10. Let R be a commutative ring with 1 and let I and J be ideals of R. Prove that $R/I \cong R/J$ as R-modules if and only if $I = J$. Suppose we only ask that R/I and R/J be isomorphic rings. Is the same conclusion valid? (Hint: Consider $F[X]/(X - a)$ where $a \in F$ and show that $F[X]/(X - a) \cong F$ as rings.)

11. Prove Theorem 2.7.

12. Prove Lemma 2.9.

13. Let M be an R-module and let $f \in \operatorname{End}_R(M)$ be an idempotent endomorphism of M, i.e., $f \circ f = f$. (That is, f is an idempotent element of the ring $\operatorname{End}_R(M)$.) Show that
$$M \cong (\operatorname{Ker}(f)) \oplus (\operatorname{Im}(f)).$$

14. Prove the remaining cases in Theorem 3.10.

15. Let R be a PID and let a and $b \in R$ be nonzero elements. Then show that $\operatorname{Hom}_R(R/Ra, R/Rb) \cong R/Rd$ where $d = (a, b)$ is the greatest common divisor of a and b.

16. Compute $\operatorname{Hom}_{\mathbf{Z}}(\mathbf{Q}, \mathbf{Z})$.

17. Give examples of short exact sequences of R-modules
$$0 \longrightarrow M_1 \xrightarrow{\phi} M \xrightarrow{\phi'} M_2 \longrightarrow 0$$

and
$$0 \longrightarrow N_1 \overset{\psi}{\longrightarrow} N \overset{\psi'}{\longrightarrow} N_2 \longrightarrow 0$$
such that
(a) $M_1 \cong N_1$, $M \cong N$, $M_2 \not\cong N_2$;
(b) $M_1 \cong N_1$, $M \not\cong N$, $M_2 \cong N_2$;
(c) $M_1 \not\cong N_1$, $M \cong N$, $M_2 \cong N_2$.

18. Show that there is a split exact sequence
$$0 \longrightarrow m\mathbf{Z}_{mn} \longrightarrow \mathbf{Z}_{mn} \longrightarrow n\mathbf{Z}_{mn} \longrightarrow 0$$
of $\mathbf{Z}_{mn}$-modules if and only if $(m, n) = 1$.

19. Let N_1 and N_2 be submodules of an R-module M. Show that there is an exact sequence
$$0 \longrightarrow N_1 \cap N_2 \overset{\psi}{\longrightarrow} N_1 \oplus N_2 \overset{\phi}{\longrightarrow} N_1 + N_2 \longrightarrow 0$$
where $\psi(x) = (x,\, x)$ and $\phi(x,\, y) = x - y$.

20. Let R be an integral domain and let a and b be nonzero elements of R. Let $M = R/R(ab)$ and let $N = Ra/R(ab)$. Then M is an R-module and N is a submodule. Show that N is a complemented submodule in M if and only if there are $u,\, v \in R$ such that $ua + vb = 1$.

21. Let R be a ring, M a finitely generated R-module, and $\phi : M \to R^n$ a surjective R-module homomorphism. Show that $\mathrm{Ker}(\phi)$ is finitely generated. (Note that this is valid even when M has submodules that are not finitely generated.) (Hint: Consider the short exact sequence:
$$0 \longrightarrow K \longrightarrow M \overset{\phi}{\longrightarrow} R^n \longrightarrow 0.)$$

22. Suppose that
$$
\begin{array}{ccccccccc}
0 & \longrightarrow & M_1 & \overset{\phi}{\longrightarrow} & M & \overset{\phi'}{\longrightarrow} & M_2 & \longrightarrow & 0 \\
 & & \downarrow{\scriptstyle f} & & \downarrow{\scriptstyle g} & & \downarrow{\scriptstyle h} & & \\
0 & \longrightarrow & N_1 & \overset{\psi}{\longrightarrow} & N & \overset{\psi'}{\longrightarrow} & N_2 & \longrightarrow & 0
\end{array}
$$

is a commutative diagram of R-modules and R-module homomorphisms. Assume that the rows are exact and that f and h are isomorphisms. Then prove that g is an isomorphism.

23. Let R be a commutative ring and S a multiplicatively closed subset of R containing no zero divisors. If M is an R-module, then M_S was defined in Exercise 6. Prove that the operation of forming quotients with elements of S is exact. Precisely:

(a) Suppose that $M' \overset{f}{\to} M \overset{g}{\to} M''$ is a sequence of R-modules and homomorphisms which is exact at M. Show that the sequence
$$M'_S \overset{f_S}{\longrightarrow} M_S \overset{g_S}{\longrightarrow} M''_S$$
is an exact sequence of R_S-modules and homomorphisms.

(b) As a consequence of part (a), show that if M' is a submodule of M, then M'_S can be identified with an R_S-submodule of M_S.

(c) If N and P are R-submodules of M, show (under the identification of part (b)) that $(N + P)_S = N_S + P_S$ and $(N \cap P)_S = N_S \cap P_S$. (That is, formation of fractions commutes with finite sums and finite intersections.)

(d) If N is a submodule of M show that

$$(M/N)_S \cong (M_S)/(\bar{N}_S).$$

(That is, formation of fractions commutes with quotients.)

24. Let F be a field and let $\{f_i(X)\}_{i=0}^\infty$ be any subset of $F[X]$ such that $\deg f_i(X) = i$ for each i. Show that $\{f_i(X)\}_{i=0}^\infty$ is a basis of $F[X]$ as an F-module.

25. Let R be a commutative ring and consider $M = R[X]$ as an R-module. Then $N = R[X^2]$ is an R submodule. Show that M/N is isomorphic to $R[X]$ as an R-module.

26. Let G be a group and H a subgroup. If $\mathbf{F}$ is a field, then we may form the group ring $\mathbf{F}(G)$ (Example 2.1.9 (15)). Since $\mathbf{F}(G)$ is a ring and $\mathbf{F}(H)$ is a subring, we may consider $\mathbf{F}(G)$ as either a left $\mathbf{F}(H)$-module or a right $\mathbf{F}(H)$-module. As either a left or right $\mathbf{F}(H)$-module, show that $\mathbf{F}(G)$ is free of rank $[G : H]$. (Use a complete set $\{g_i\}$ of coset representatives of H as a basis.)

27. Let R and S be integral domains and let $\phi_1, \ldots, \phi_n$ be n distinct ring homomorphisms from R to S. Show that $\phi_1, \ldots, \phi_n$ are S-linearly independent in the S-module $F(R, S)$ of all functions from R to S. (Hint: Argue by induction on n, using the property $\phi_i(ax) = \phi_i(a)\phi_i(x)$, to reduce from a dependence relation with n entries to one with $n - 1$ entries.)

28. Let G be a group, let F be a field, and let $\phi_i : G \to F^*$ for $1 \le i \le n$ be n distinct group homomorphisms from G into the multiplicative group F^* of F. Show that $\phi_1, \ldots, \phi_n$ are linearly independent over F (viewed as elements of the F-module of all functions from G to F). (Hint: Argue by induction on n, as in Exercise 27.)

29. Let $R = \mathbf{Z}_{30}$ and let $A \in M_{2,3}(R)$ be the matrix

$$A = \begin{bmatrix} 1 & 1 & -1 \\ 0 & 2 & 3 \end{bmatrix}.$$

Show that the two rows of A are linearly independent over R, but that any two of the three columns are linearly dependent over R.

30. Let V be a finite-dimensional complex vector space. Then V is also a vector space over $\mathbf{R}$. Show that $\dim_\mathbf{R} V = 2 \dim_\mathbf{C} V$. (Hint: If

$$\mathcal{B} = \{v_1, \ldots, v_n\}$$

is a basis of V over $\mathbf{C}$, show that

$$\mathcal{B}' = \{v_1, \ldots, v_n, iv_1, \ldots, iv_n\}$$

is a basis of V over $\mathbf{R}$.)

31. Extend Exercise 30 as follows. Let L be a field and let K be a subfield of L. If V is a vector space over L, then it is also a vector space over K. Prove that

$$\dim_K V = [L : K] \dim_L V$$

where $[L : K] = \dim_K L$ is the dimension of L as a vector space over K. (Note that we are *not* assuming that $\dim_K L < \infty$.)

32. Let $K \subseteq L$ be fields and let V be a vector space over L. Suppose that $\mathcal{B} = \{u_\alpha\}_{\alpha \in \Gamma}$ is a basis of V as an L-module, and let W be the K-submodule of V generated by $\mathcal{B}$. Let $U \subseteq W$ be any K-submodule, and let U_L be the L-submodule of V generated by U. Prove that

$$U_L \cap W = U.$$

That is, taking L-linear combinations of elements of U does not produce any new elements of W.

That is, taking L-linear combinations of elements of U does not produce any new elements of W.

33. Let $K \subseteq L$ be fields and let $A \in M_n(K)$, $b \in M_{n,1}(K)$. Show that the matrix equation $AX = b$ has a solution $X \in M_{n,1}(K)$ if and only if it has a solution $X \in M_{n,1}(L)$.

34. Prove that the Lagrange interpolation polynomials (Proposition 2.4.10) and the Newton interpolation polynomials (Remark 2.4.11) each form a basis of the vector space $\mathcal{P}_n(F)$ of polynomials of degree $\leq n$ with coefficients from F.

35. Let $\mathcal{F}$ denote the set of all functions from $\mathbf{Z}^+$ to $\mathbf{Z}^+$, and let M be the free $\mathbf{Q}$-module with basis $\mathcal{F}$. Define a multiplication on M by the formula $(fg)(n) = f(n) + g(n)$ for all $f, g \in \mathcal{F}$ and extend this multiplication by linearity to all of M. Let f_m be the function $f_m(n) = \delta_{m,n}$ for all $m, n \geq 0$. Show that each f_m is irreducible (in fact, prime) as an element of the ring M. Now consider the function $f(n) = 1$ for all $n \geq 0$. Show that f does not have a factorization into irreducible elements in M. (Hint: It may help to think of f as the "infinite monomial"

$$X_0^{f(0)} X_1^{f(1)} \cdots X_m^{f(m)} \cdots.)$$

(Compare this exercise with Example 2.5.15.)

36. Let F be a field, and let

$$\mathcal{I} = \{p_\alpha(X) : p_\alpha(X) \text{ is an irreducible monic polynomial in } F[X]\}.$$

We will say that a rational function $h(X) = f(X)/g(X) \in F(X)$ is *proper* if $\deg(f(X)) < \deg(g(X))$. Let $F(X)$pr denote the set of all proper rational functions in $F[X]$.
(a) Prove that $F(X) \cong F[X] \oplus F(X)$pr as F-modules.
(b) Prove that

$$\mathcal{B} = \left\{ \frac{X^j}{(p_\alpha(X))^k} : p_\alpha(X) \in \mathcal{I};\ 0 \leq j < \deg(p_\alpha(X)),\ k \geq 1 \right\}$$

is a basis of $F(X)$pr as an F-module. The expansion of a proper rational function with respect to the basis $\mathcal{B}$ is known as the *partial fraction expansion*; it should be familiar from elementary calculus.

37. Prove that $\mathbf{Q}$ is not a projective $\mathbf{Z}$-module.

38. Let

$$R = \{f : [0, 1] \to \mathbf{R} :\quad f \text{ is continuous and } f(0) = f(1)\}$$

and let

$$M = \{f : [0, 1] \to \mathbf{R} :\quad f \text{ is continuous and } f(0) = -f(1)\}.$$

Then R is a ring under addition and multiplication of functions, and M is an R-module. Show that M is a projective R-module that is not free. (Hint: Show that $M \oplus M \cong R \oplus R$.)

39. Show that submodules of projective modules need not be projective. (Hint: Consider $p\mathbf{Z}_{p^2} \subseteq \mathbf{Z}_{p^2}$ as $\mathbf{Z}_{p^2}$-modules.) Over a PID, show that submodules of projective modules are projective.

40. (a) If R is a Dedekind domain, prove that R is Noetherian.
(b) If R is an integral domain that is a local ring (i.e., R has a unique maximal ideal), show that any invertible ideal I of R is principal.
(c) Let R be an integral domain and $S \subseteq R \setminus \{0\}$ a multiplicatively closed subset. If I is an invertible ideal of R, show that I_S is an invertible ideal of R_S.

(d) Show that in a Dedekind domain R, every nonzero prime ideal is maximal. (Hint: Let M be a maximal ideal of R containing a prime ideal P, and let $S = R \setminus M$. Apply parts (b) and (c).)

41. Show that $\mathbf{Z}[\sqrt{-3}]$ is not a Dedekind domain.

42. Show that $\mathbf{Z}[X]$ is not a Dedekind domain. More generally, let R be any integral domain that is not a field. Show that $R[X]$ is not a Dedekind domain.

43. Suppose R is a PID and $M = R\langle x \rangle$ is a cyclic R-module with $\operatorname{Ann} M = \langle a \rangle \neq \langle 0 \rangle$. Show that if N is a submodule of M, then N is cyclic with $\operatorname{Ann} N = \langle b \rangle$ where b is a divisor of a. Conversely, show that M has a unique submodule N with annihilator $\langle b \rangle$ for each divisor b of a.

44. Let R be a PID, M an R-module, $x \in M$ with $\operatorname{Ann}(x) = \langle a \rangle \neq \langle 0 \rangle$. Factor $a = u p_1^{n_1} \cdots p_k^{n_k}$ with u a unit and $p_1, \ldots, p_k$ distinct primes. Let $y \in M$ with $\operatorname{Ann}(y) = \langle b \rangle \neq \langle 0 \rangle$, where $b = u' p_1^{m_1} \cdots p_k^{m_k}$ with $0 \leq m_i < n_i$ for $1 \leq i \leq k$. Show that $\operatorname{Ann}(x + y) = \langle a \rangle$.

45. Let R be a PID, let M be a free R-module of finite rank, and let $N \subseteq M$ be a submodule. If M/N is a torsion R-module, prove that $\operatorname{rank}(M) = \operatorname{rank}(N)$.

46. Let R be a PID and let M and N be free R-modules of the same finite rank. Then an R-module homomorphism $f : M \to N$ is an injection if and only if $N/\operatorname{Im}(f)$ is a torsion R-module.

47. Let $u = (a, b) \in \mathbf{Z}^2$.
   (a) Show that there is a basis of $\mathbf{Z}^2$ containing u if and only if a and b are relatively prime.
   (b) Suppose that $u = (5, 12)$. Find a $v \in \mathbf{Z}^2$ such that $\{u, v\}$ is a basis of $\mathbf{Z}^2$.

48. Let M be a torsion module over a PID R and assume $\operatorname{Ann}(M) = \langle a \rangle \neq \langle 0 \rangle$. If $a = p_1^{r_1} \cdots p_k^{r_k}$ where $p_1, \ldots, p_k$ are the distinct prime factors of a, then show that $M_{p_i} = q_i M$ where $q_i = a/p_i^{r_i}$. Recall that if $p \in R$ is a prime, then M_p denotes the p-primary component of M.

49. Let M be a torsion-free R-module over a PID R, and assume that $x \in M$ is a primitive element. If $px = qx'$ show that $q \mid p$.

50. Find a basis and the invariant factors for the submodule of $\mathbf{Z}^3$ generated by $x_1 = (1, 0, -1)$, $x_2 = (4, 3, -1)$, $x_3 = (0, 9, 3)$, and $x_4 = (3, 12, 3)$.

51. Find a basis for the submodule of $\mathbf{Q}[X]^3$ generated by

$$f_1 = (2X - 1, X, X^2 + 3), \qquad f_2 = (X, X, X^2), \qquad f_3 = (X + 1, 2X, 2X^2 - 3).$$

52. Determine the structure of $\mathbf{Z}^3/K$ where K is generated by $x_1 = (2, 1, -3)$ and $x_2 = (1, -1, 2)$.

53. Let $R = \mathbf{R}[X]$ and suppose that M is a direct sum of cyclic R-modules with annihilators $(X - 1)^3$, $(X^2 + 1)^2$, $(X - 1)(X^2 + 1)^4$, and $(X + 2)(X^2 + 1)^2$. Determine the elementary divisors and invariant factors of M.

54. Let R be a PID and let $p \in R$ be a prime. Show that submodules, quotient modules, and direct sums of p-primary modules are p-primary.

55. An R-module M is said to be *irreducible* if $\langle 0 \rangle$ and M are the only submodules of M. Show that a torsion module M over a PID R is irreducible if and only if $M = R\langle x \rangle$ where $\operatorname{Ann}(x) = \langle p \rangle$ where p is prime. Show that if M is finitely generated, then M is *indecomposable* in the sense that M is not a direct sum of two nonzero submodules if and only if $M = R\langle x \rangle$ where $\operatorname{Ann}(x) = \langle 0 \rangle$ or $\operatorname{Ann}(z) = \langle p^e \rangle$ where p is a prime.

56. Let M be an R-module where R is a PID. We say that M is *divisible* if for each nonzero $a \in R$, $aM = M$.
   (a) Show that $\mathbf{Q}$ is a divisible $\mathbf{Z}$-module.

(b) Show that any quotient of a divisible R-module is divisible. It follows for example that $\mathbf{Q}/\mathbf{Z}$ is a divisible $\mathbf{Z}$-module.

(c) If R is not a field, show that no finitely generated R-module is divisible.

57. Determine all nonisomorphic abelian groups of order 360.

58. Use elementary divisors to describe all abelian groups of order 144 and 168.

59. Use invariant factors to describe all abelian groups of orders 144 and 168.

60. If p and q are distinct primes, use invariant factors to describe all abelian groups of order
    (a) $p^2 q^2$,
    (b) $p^4 q$,
    (c) p^5.

61. If p and q are distinct primes, use elementary divisors to describe all abelian groups of order $p^3 q^2$.

62. Let G, H, and K be finitely generated abelian groups. If $G \times K \cong H \times K$, show that $G \cong H$. Show by example that this need not be true if we do not assume that the groups are finitely generated.

63. Determine all integers for which there exists a unique abelian group of order n.

64. Show that two finite abelian groups are isomorphic if and only if they have the same number of elements of each order.

65. Let p be a prime and assume that a finite abelian group G has exactly k elements of order p. Find all possible values of k

66. Find a generator for the cyclic group F^* where F is each of the following fields (see Example 2.5.15 (3)):
    (a) $\mathbf{F}_2[X]/\langle X^2 + X + 1 \rangle$.
    (b) $\mathbf{F}_3[X]/\langle X^2 + 1 \rangle$.

67. Let
$$0 \longrightarrow M_1 \xrightarrow{f_1} M_2 \xrightarrow{f_2} \cdots \xrightarrow{f_n} M_{n+1} \longrightarrow 0$$

be an exact sequence of finite rank free modules and homomorphisms over a PID R. That is, f_1 is injective, f_n is surjective, and $\text{Im}(f_i) = \text{Ker}(f_{i+1})$ for $1 \leq i \leq n - 1$. Show that
$$\sum_{i=1}^{n+1} (-1)^{i+1} \text{rank}(M_i) = 0.$$

68. If $f(X_1, \ldots, X_n) \in R[X_1, \ldots, X_n]$, the degree of f is the highest degree of a monomial in f with nonzero coefficient, where
$$\deg(X_1^{i_1} \cdots X_n^{i_n}) = i_1 + \cdots + i_n.$$

Let F be a field. Given any five points $\{v_1, \ldots, v_5\} \subseteq F^2$, show that there is a quadratic polynomial $f(X_1, X_2) \in F[X_1, X_2]$ such that $f(v_i) = 0$ for $1 \leq i \leq 5$.

69. Let M and N be finite-rank free R-modules over a PID R and let $f \in \text{Hom}_R(M, N)$. If $S \subseteq N$ is a complemented submodule of N, show that $f^{-1}(S)$ is a complemented submodule of M.

70. Let R be a PID, and let $f : M \to N$ be an R-module homomorphism of finite rank free R-modules. If $S \subseteq N$ is a submodule, prove that
$$\text{rank}(f^{-1}(S)) = \text{rank}(S \cap \text{Im}(f)) + \text{rank}(\text{Ker}(f)).$$

71. Let $M_1 \xrightarrow{f} M \xrightarrow{g} M_2$ be a sequence of finite-rank R-modules and R-module homomorphisms, where R is a PID, and assume that $\text{Im}(f)$ is a complemented submodule of M.
    (a) Show that

$$\text{rank}(\text{Im}(g \circ f)) = \text{rank}(\text{Im}(f)) - \text{rank}(\text{Im}(f) \cap \text{Ker}(g)).$$

    (b) Show that

$$\text{rank}(\text{Im}(g \circ f)) = \text{rank}(\text{Im}(f) + \text{Ker}(g)) - \text{rank}(\text{Ker}(g)).$$

    If R is a field, then all submodules of R-modules are complemented, so these formulas are always valid in the case of vector spaces and linear transformations. Show, by example, that they need not be valid if $\text{Im}(f)$ is not complemented.

72. Let R be a PID, and let M, N, and P be finite rank free R-modules. Let $f : M \rightarrow N$ and $g : M \rightarrow P$ be homomorphisms. Suppose that $\text{Ker}(f) \subseteq \text{Ker}(g)$ and $\text{Im}(f)$ is a complemented submodule of N. Then show that there is a homomorphism $h : N \rightarrow P$ such that $g = h \circ f$.

73. Let F be a field and let V be a vector space over F. Suppose that $f, g \in V^* = \text{Hom}_F(V, F)$ such that $\text{Ker}(f) \subseteq \text{Ker}(g)$. Show that there is $a \in F$ such that $g = af$. Is this same result true if F is replaced by a PID?

74. Let R be a PID and let M be a finite rank free R-module. Let $\mathcal{C}_k(M)$ denote the set of complemented submodules of M of rank k. Let G be the group of units of the ring $\text{End}_R(M)$.
    (a) Show that $(\phi, N) \mapsto \phi(N)$ determines an action of the group G on the set $\mathcal{C}_k(M)$.
    (b) Show that the action defined in part (a) is transitive, i.e., given N_1, $N_2 \in \mathcal{C}_k(M)$ there is $\phi \in G$ that sends N_1 to N_2.

Chapter 4

Linear Algebra

The theme of the present chapter will be the application of the structure theorem for finitely generated modules over a PID (Theorem 3.7.1) to canonical form theory for a linear transformation from a vector space to itself. The fundamental results will be presented in Section 4.4. We will start with a rather detailed introduction to the elementary aspects of matrix algebra, including the theory of determinants and matrix representation of linear transformations. Most of this general theory will be developed over an arbitrary (in most instances, commutative) ring, and we will only specialize to the case of fields when we arrive at the detailed applications in Section 4.4.

4.1 Matrix Algebra

We have frequently used matrix rings as a source of examples, essentially using an assumed knowledge of matrix algebra (particularly matrix multiplication). The present section will give the primary formulas of the more elementary aspects of matrix algebra; many of the proofs will be left as exercises. *In this section, we do not assume that a ring R is commutative, except when explicitly stated.*

Let R be a ring. By an $m \times n$ matrix over R we mean a rectangular array

$$
A = \begin{bmatrix}
a_{11} & a_{12} & \cdots & a_{1n} \\
a_{21} & a_{22} & \cdots & a_{2n} \\
\vdots & \vdots & \ddots & \vdots \\
a_{m1} & a_{m2} & \cdots & a_{mm}
\end{bmatrix} = [a_{ij}]
$$

where $a_{ij} \in R$ for $1 \leq i \leq m$, $1 \leq j \leq n$. One can treat a matrix more formally as a function $f : I \times J \to R$ where $I = \{1, 2, \ldots, m\}$, $J = \{1, 2, \ldots, n\}$, and $f(i, j) = a_{ij}$, but we shall be content to think of a matrix in the traditional manner described above as a rectangular array consisting of m **rows** and n **columns**. Let $M_{m,n}(R)$ denote the set of all $m \times n$ matrices over R. When $m = n$ we will write $M_n(R)$ instead of $M_{n,n}(R)$, which is consistent with the notation introduced in Example 2.1.10 (8). If $A = [a_{ij}] \in M_{m,n}(R)$, then a_{ij} is called the ij^{th} entry of A; there are times when it will be convenient to denote this by $\text{ent}_{ij}(A) = a_{ij}$. The index i is referred to as the **row index**, and by the ith row of A we mean the $1 \times n$ matrix

$$\text{row}_i(A) = [\, a_{i1} \ \cdots \ a_{in} \,],$$

while the index j is the **column index** and the jth column of A is the $m \times 1$ matrix

$$\text{col}_j(A) = \begin{bmatrix} a_{1j} \\ \vdots \\ a_{mj} \end{bmatrix}.$$

If $A, B \in M_{m,n}(R)$ then we say that $A = B$ if and only if $\text{ent}_{ij}(A) = \text{ent}_{ij}(B)$ for all i, j with $1 \leq i \leq m$, $1 \leq j \leq n$.

The first order of business is to define algebraic operations on $M_{m,n}(R)$. Define an addition on $M_{m,n}(R)$ by

$$\text{ent}_{ij}(A + B) = \text{ent}_{ij}(A) + \text{ent}_{ij}(B)$$

whenever $A, B \in M_{m,n}(R)$. It is easy to see that this operation makes $M_{m,n}(R)$ into an abelian group with identity $0_{m,n}$ (the $m \times n$ matrix with all entries equal to $0 \in R$). We shall generally write 0 rather than $0_{m,n}$ for the zero matrix in $M_{m,n}(R)$. The additive inverse of $A \in M_{m,n}(R)$ is the matrix $-A$ defined by $\text{ent}_{ij}(-A) = -\text{ent}_{ij}(A)$. Now let $a \in R$ and $A \in M_{m,n}(R)$, and define the **scalar product** of a and A, denoted $aA \in M_{m,n}(R)$, by $\text{ent}_{ij}(aA) = a\,\text{ent}_{ij}(A)$. Similarly, $Aa \in M_{m,n}(R)$ is defined by $\text{ent}_{ij}(Aa) = \text{ent}_{ij}(A)\,a$. In the language to be introduced in Chapter 7, these scalar multiplications make $M_{m,n}(R)$ into an (R,R)-bimodule (Definition 7.2.1), i.e., $M_{m,n}(R)$ is both a left R-module and a right R-module under these scalar multiplications, and $a(Ab) = (aA)b$ for all $a, b \in R$ and $A \in M_{m,n}(R)$.

The addition and scalar multiplication of matrices arise naturally from thinking of matrices as functions $f : I \times J \to R$, namely, they correspond to addition and scalar multiplication of functions. However, multiplication of matrices is motivated by the relationship of matrices to linear transformations, which will be considered in Section 4.3. For now we simply present matrix multiplication via an explicit formula. We can multiply a matrix $A \in M_{m,n}(R)$ and a matrix $B \in M_{n,p}(R)$ and obtain a matrix AB (note the order) in $M_{m,p}(R)$ where AB is defined by the formula

$$\text{ent}_{ij}(AB) = \sum_{k=1}^{n} \text{ent}_{ik}(A)\,\text{ent}_{kj}(B).$$

Thus to multiply a matrix A by B (in the order AB) it is necessary that the number of columns of A is the same as the number of rows of B. Furthermore, the formula for $\text{ent}_{ij}(AB)$ involves only the ith row of A and the jth column of B. The multiplication formula can also be expressed as

$$\text{ent}_{ij}(AB) = [\text{row}_i(A)][\text{col}_j(B)]$$

where the formula for multiplying a row matrix by a column matrix is

$$[a_1 \quad \cdots \quad a_n] \begin{bmatrix} b_1 \\ \vdots \\ b_n \end{bmatrix} = \left[\sum_{k=1}^{n} a_k b_k \right].$$

Let $I_n \in M_n(R)$ be defined by $\text{ent}_{ij}(I_n) = \delta_{ij}$ where δ_{ij} is the kronecker δ function, i.e., $\delta_{ii} = 1$ and $\delta_{ij} = 0$ if $i \neq j$. The following lemma contains some basic properties of matrix multiplication. In part (c), the concept of center of a ring is needed. If R is a ring, then the **center of** R, denoted $C(R)$, is defined by

$$C(R) = \{a \in R : ab = ba \quad \text{for all } b \in R\}.$$

Note that $C(R)$ is a subring of R and R is commutative if and only if $R = C(R)$.

(1.1) Lemma. *The product map $M_{m,n}(R) \times M_{n,p}(R) \rightarrow M_{m,p}(R)$ satisfies the following properties (where $A, B,$ and C are matrices of appropriate sizes and $a \in R$):*

(1) $A(B + C) = AB + AC$.
(2) $(A + B)C = AC + BC$.
(3) $a(AB) = (aA)B$ and both equal $A(aB)$ when $a \in C(R)$.
(4) $I_m A = A$ and $AI_n = A$.
(5) $\text{row}_i(AB) = [\text{row}_i(A)]B$.
(6) $\text{col}_j(AB) = A[\text{col}_j(B)]$.
(7) *The product map $M_{m,n}(R) \times M_{n,p}(R) \times M_{p,q}(R) \rightarrow M_{m,q}(R)$ satisfies the associative law $A(BC) = (AB)C$.*

Proof. Exercise.   □

Remark. The content of the algebraic properties in Lemma 1.1 is that $M_{m,n}(R)$ is a left $M_m(R)$-module and a right $M_n(R)$-module. Note that $M_n(R)$ is a ring by (1), (2), and (7). The verification of (7) is an unenlightening computation, but we shall see in Remark 3.7 that this associativity is a consequence of the fact that composition of functions is associative. (Indeed, the basic reason for defining matrix multiplication as we have done is to make Proposition 3.6 true.) Also note that $M_n(R)$ is an algebra over R by Lemma 1.1 (3) if R is commutative, but not otherwise.

Recall (from Example 2.1 (8)) that the matrix $E_{ij} \in M_n(R)$ is defined to be the matrix with 1 in the ij^{th} position and 0 elsewhere. E_{ij} is called a **matrix unit**, but it should not be confused with a unit in the matrix ring

$M_n(R)$; unless $n = 1$ the matrix unit E_{ij} is *never* a unit in $M_n(R)$. We recall the basic properties of the matrices E_{ij} in the following lemma.

(1.2) Lemma. *Let* $\{E_{ij} : 1 \le i, j \le n\}$ *be the set of matrix units in* $M_n(R)$ *and let* $A = [a_{ij}] \in M_n(R)$. *Then*

(1) $E_{ij}E_{kl} = \delta_{jk}E_{il}$.
(2) $\sum_{i=1}^n E_{ii} = I_n$.
(3) $A = \sum_{i,j=1}^n a_{ij}E_{ij} = \sum_{i,j=1}^n E_{ij}a_{ij}$.
(4) $E_{ij}AE_{kl} = a_{jk}E_{il}$.

Proof. Exercise. ☐

Remark. When speaking of matrix units, we will generally mean the matrices $E_{ij} \in M_n(R)$. However, for every m, n there is a set $\{E_{ij}\}_{i=1\,j=1}^{m\,\,\,n} \subseteq M_{m,n}(R)$ where E_{ij} has 1 in the (i,j) position and 0 elsewhere. Then, with appropriately adjusted indices, items (3) and (4) in Lemma 1.2 are valid. Moreover, $\{E_{ij}\}_{i=1\,j=1}^{m\,\,\,n}$ is a basis of $M_{m,n}(R)$ as both a left R-module and a right R-module. Hence, if R is commutative so that rank makes sense (Remark 3.6.19), then it follows that $\mathrm{rank}_R(M_{m,n}(R)) = mn$.

If $a \in R$ the matrix $aI_n \in M_n(R)$ is called a **scalar matrix**. The set of scalar matrices $\{aI_n : a \in R\} \subseteq M_n(R)$ is a subring of $M_n(R)$ isomorphic to R, via the isomorphism $a \mapsto aI_n$. Note that $(aI_n)A = A(aI_n)$ for all $a \in C(R)$, $A \in M_n(R)$, where $C(R)$ is the center of R. Let $\nu : R \to M_n(R)$ be the ring homomorphism $\nu(a) = aI_n$. Then there is the following result.

(1.3) Lemma. *If* R *is a ring, then*

$$C(M_n(R)) = \nu(C(R)).$$

That is, the center of $M_n(R)$ *is the set of scalar matrices where the scalar is chosen from the center of* R.

Proof. Clearly $\nu(C(R)) \subseteq C(M_n(R))$. We show the converse. Let $A \in C(M_n(R))$ and let $1 \le j \le n$. Then $AE_{1j} = E_{1j}A$ and Lemma 1.1 (5) and (6) show that

$$\mathrm{col}_k(AE_{1j}) = A[\mathrm{col}_k(E_{1j})] = \delta_{jk}\,\mathrm{col}_1(A)$$

and

$$\mathrm{row}_k(E_{1j}A) = [\mathrm{row}_k(E_{1j})]A = \delta_{1k}\,\mathrm{row}_j(A).$$

Comparing entries in these n pairs of matrices shows that $a_{js} = 0$ if $s \ne j$ and $a_{11} = a_{jj}$. Since j is arbitrary, this shows that $A = a_{11}I_n$ is a scalar matrix. Since A must commute with all scalar matrices, it follows that $a_{11} \in C(R)$. ☐

A unit in the ring $M_n(R)$ is a matrix A such that there is some matrix B with $AB = BA = I_n$. Such matrices are said to be **invertible** or **unimodular**. The set of invertible matrices in $M_n(R)$ is denoted $\mathrm{GL}(n, R)$ and is called the **general linear group over R of dimension** n. Note that $\mathrm{GL}(n, R)$ is, in fact, a group since it is the group of units of a ring.

If $A = [a_{ij}] \in M_{m,n}(R)$, then the **transpose** of A, denoted $A^t \in M_{n,m}(R)$, is defined by

$$\mathrm{ent}_{ij}(A^t) = a_{ji} = \mathrm{ent}_{ji}(A).$$

The following formulas for transpose are straightforward and are left as an exercise.

(1.4) Lemma. *Let R be a ring.*

(1) *If $A, B \in M_{m,n}(R)$, then $(A + B)^t = A^t + B^t$.*
(2) *If R is commutative, if $A \in M_{m,n}(R)$ and $B \in M_{n,p}(R)$, then $(AB)^t = B^t A^t$.*
(3) *If R is commutative and $A \in \mathrm{GL}(n, R)$, then $(A^t)^{-1} = (A^{-1})^t$.*

Proof. $\square$

There is a particularly important R-module homomorphism from $M_n(R)$ to R defined as follows:

(1.5) Definition. *Define the R-module homomorphism* $\mathrm{Tr} : M_n(R) \to R$ *by*

$$\mathrm{Tr}(A) = \sum_{i=1}^{n} \mathrm{ent}_{ii}(A) = \sum_{i=1}^{n} a_{ii}.$$

The element $\mathrm{Tr}(A) \in R$ is called the **trace** *of the matrix $A \in M_n(R)$.*

The following is a simple but important result:

(1.6) Lemma. *Let R be a commutative ring.*

(1) *If $A \in M_{m,n}(R)$ and $B \in M_{n,m}(R)$, then*

$$\mathrm{Tr}(AB) = \mathrm{Tr}(BA).$$

(2) *If $A \in M_n(R)$ and $S \in \mathrm{GL}(n, R)$, then*

$$\mathrm{Tr}(S^{-1}AS) = \mathrm{Tr}(A).$$

Proof. (1)

$$\text{Tr}(AB) = \sum_{i=1}^{m} \text{ent}_{ii}(AB)$$

$$= \sum_{i=1}^{m} \sum_{k=1}^{n} \text{ent}_{ik}(A)\,\text{ent}_{ki}(B)$$

$$= \sum_{k=1}^{n} \sum_{i=1}^{m} \text{ent}_{ki}(B)\,\text{ent}_{ik}(A)$$

$$= \sum_{k=1}^{n} \text{ent}_{kk}(B)$$

$$= \text{Tr}(BA).$$

(2) $\text{Tr}(S^{-1}AS) = \text{Tr}(SS^{-1}A) = \text{Tr}(A)$.   □

(1.7) Definition. *If R is a ring and A, $B \in M_n(R)$, then we say that A and B are* **similar** *if there is a matrix $S \in \text{GL}(n, R)$ such that $S^{-1}AS = B$.*

Similarity determines an equivalence relation on $M_n(R)$, the significance of which will become clear in Section 4.3. For now, we simply note that Lemma 1.6 (2) states that, if R is commutative, then similar matrices have the same trace.

(1.8) Definition. *Let R be a ring and let $\{E_{ij}\}_{i,j=1}^{n} \subseteq M_n(R)$ be the set of matrix units. Then we define a number of particularly useful matrices in $M_n(R)$.*

(1) *For $\alpha \in R$ and $i \neq j$ define*

$$T_{ij}(\alpha) = I_n + \alpha E_{ij}.$$

The matrix $T_{ij}(\alpha)$ is called an **elementary transvection**. *$T_{ij}(\alpha)$ differs from I_n only in the ij^{th} position, where $T_{ij}(\alpha)$ has an α.*

(2) *If $\alpha \in R^*$ is a unit of R and $1 \leq i \leq n$, then*

$$D_i(\alpha) = I_n - E_{ii} + \alpha E_{ii}$$

is called an **elementary dilation**. *$D_i(\alpha)$ agrees with the identity matrix I_n except that it has an α (rather than a 1) in the ith diagonal position.*

(3) *The matrix*

$$P_{ij} = I_n - E_{ii} - E_{jj} + E_{ij} + E_{ji}$$

is called an **elementary permutation matrix**. *Note that P_{ij} is obtained from the identity I_n by interchanging rows i and j (or columns i and j).*

(1.9) Definition. *If R is a ring, then matrices of the form $D_i(\beta)$, $T_{ij}(\alpha)$, and P_{ij} are called **elementary matrices over** R. The integer n is not included in the notation for the elementary matrices, but is determined from the context.*

Examples. Suppose $n = 3$ and $\alpha \in R^*$. Then

$$T_{13}(\alpha) = \begin{bmatrix} 1 & 0 & \alpha \\ 0 & 1 & 0 \\ 0 & 0 & 1 \end{bmatrix} \qquad D_2(\alpha) = \begin{bmatrix} 1 & 0 & 0 \\ 0 & \alpha & 0 \\ 0 & 0 & 1 \end{bmatrix} \qquad P_{13} = \begin{bmatrix} 0 & 0 & 1 \\ 0 & 1 & 0 \\ 1 & 0 & 0 \end{bmatrix}.$$

Another useful class of matrices for which it is convenient to have an explicit notation is the set of diagonal matrices.

(1.10) Definition. *A matrix $A \in M_n(R)$ is **diagonal** if $\text{ent}_{ij}(A) = 0$ whenever $i \neq j$. Thus, a diagonal matrix A has the form $A = \sum_{i=1}^{n} a_i E_{ii}$ in terms of the matrix units E_{ij}. We shall use the notation $A = \text{diag}(a_1, \ldots, a_n)$ to denote the diagonal matrix $A = \sum_{i=1}^{n} a_i E_{ii}$. In particular, the elementary matrix $D_i(\beta) = \text{diag}(1, \ldots, 1, \beta, 1, \ldots, 1)$ where the β occurs in the ith diagonal position.*

Note the following formula for multiplication of diagonal matrices:

$$(1.1) \qquad \text{diag}(a_1, \ldots, a_n) \, \text{diag}(b_1, \ldots, b_n) = \text{diag}(a_1 b_1, \ldots, a_n b_n).$$

This observation will simplify the calculation in part (3) of the following collection of basic properties of elementary matrices.

(1.11) Lemma. *Let R be a ring.*

(1) *If α, $\beta \in R$ and $i \neq j$, then $T_{ij}(\alpha)T_{ij}(\beta) = T_{ij}(\alpha + \beta)$.*
(2) *If $\alpha \in R$ and $i \neq j$, then $T_{ij}(\alpha)$ is invertible and $[T_{ij}(\alpha)]^{-1} = T_{ij}(-\alpha)$.*
(3) *If $\beta \in R^*$ and $1 \leq i \leq n$, then $D_i(\beta)$ is invertible and $D_i(\beta)^{-1} = D_i(\beta^{-1})$.*
(4) *$P_{ij}^2 = I_n$, so P_{ij} is invertible and is its own inverse.*

Proof. (1)

$$T_{ij}(\alpha)T_{ij}(\beta) = (I + \alpha E_{ij})(I + \beta E_{ij})$$
$$= I + (\alpha + \beta)E_{ij}$$
$$= T_{ij}(\alpha + \beta)$$

since $E_{ij}^2 = 0$ if $i \neq j$.
  (2) $T_{ij}(\alpha)T_{ij}(-\alpha) = T_{ij}(0) = I_n$.
  (3) $D_i(\beta)D_i(\beta^{-1}) = \text{diag}(1, \ldots, \beta, \ldots, 1) \, \text{diag}(1, \ldots, \beta^{-1}, \ldots, 1) = I_n$ by Equation (1.1).
  (4) Exercise. $\square$

The matrices $T_{ij}(\alpha)$, $D_i(\alpha)$, and P_{ij} are generalizations to arbitrary rings of the elementary matrices used in linear algebra. This fact is formalized in the next result. To simplify the statement we introduce the following notation. Suppose R is a ring and $A \in M_m(R)$. The **left multiplication by** A is a function

$$\mathcal{L}_A : M_{m,n}(R) \to M_{m,n}(R)$$

defined by $\mathcal{L}_A(B) = AB$. Similarly, if $C \in M_n(R)$, the **right multiplication by** C is the function

$$\mathcal{R}_C : M_{m,n}(R) \to M_{m,n}(R)$$

defined by $\mathcal{R}_C(B) = BC$. Note that $\mathcal{L}_A$ is a right $M_n(R)$-module homomorphism, while $\mathcal{R}_C$ is a left $M_m(R)$-module homomorphism.

(1.12) Proposition. *Let R be a ring, let $\alpha \in R$, $\beta \in R^*$, and m, $n \in \mathbf{N}$ be given, and let $A \in M_{m,n}(R)$.*

(1) $\mathcal{L}_{T_{ij}(\alpha)}(A) = T_{ij}(\alpha)A$ *is obtained from A by replacing $\mathrm{row}_i(A)$ by $\alpha[\mathrm{row}_j(A)] + \mathrm{row}_i(A)$ and leaving the other rows intact.*

(2) $\mathcal{L}_{D_i(\beta)}(A) = D_i(\beta)A$ *is obtained from A by multiplying $\mathrm{row}_i(A)$ by β on the left and leaving the other rows intact.*

(3) $\mathcal{L}_{P_{ij}}(A) = P_{ij}A$ *is obtained from A by interchanging $\mathrm{row}_i(A)$ and $\mathrm{row}_j(A)$ and leaving the other rows intact.*

Proof. (1) $T_{ij}(\alpha)A = (I + \alpha E_{ij})A = A + \alpha E_{ij}A$. By Lemma 1.1 (5), $\mathrm{row}_k(E_{ij}A) = 0$ if $k \neq i$ and $\mathrm{row}_i(E_{ij}A) = [\mathrm{row}_i(E_{ij})]A = \mathrm{row}_j(A)$ since

$$\mathrm{row}_i(E_{ij}) = [0 \ \cdots \ 0 \ 1 \ 0 \ \cdots \ 0]$$

where the 1 is in the jth position. Thus, $\mathrm{row}_k(T_{ij}(\alpha)A) = \mathrm{row}_k(A)$ if $k \neq i$ while $\mathrm{row}_i(T_{ij}(\alpha)A) = \mathrm{row}_i(A) + \alpha[\mathrm{row}_j(A)]$.

(2) $D_i(\beta)A = (I - E_{ii} + \beta E_{ii})A = A + (\beta - 1)E_{ii}A$. The same calculation as in part (1) shows that $\mathrm{row}_k(D_i(\beta)A) = \mathrm{row}_k(A)$ if $k \neq i$ while

$$\mathrm{row}_i(D_i(\beta)A) = \mathrm{row}_i(A) + (\beta - 1)[\mathrm{row}_i(A)] = \beta[\mathrm{row}_i(A)].$$

(3) This is a similar calculation, which is left as an exercise. □

There is a corresponding result that relates right multiplication by elementary matrices to operations on the columns of a matrix. We state the result and leave the verification (using Lemma 1.1 (6)) to the reader.

(1.13) Proposition. *Let R be a ring, let $\alpha \in R$, $\beta \in R^*$, and m, $n \in \mathbf{N}$ be given, and let $A \in M_{m,n}(R)$.*

(1) $\mathcal{R}_{T_{ij}(\alpha)}(A) = AT_{ij}(\alpha)$ is obtained from A by replacing $\text{col}_j(A)$ by $[\text{col}_i(A)]\alpha + \text{col}_j(A)$ and leaving the other columns intact.

(2) $\mathcal{R}_{D_i(\beta)}(A) = AD_i(\beta)$ is obtained from A by multiplying $\text{col}_i(A)$ by β on the right and leaving the other columns intact.

(3) $\mathcal{R}_{P_{ij}}(A) = AP_{ij}$ is obtained from A by interchanging $\text{col}_i(A)$ and $\text{col}_j(A)$ and leaving the other columns intact.

Proof. $\square$

We conclude this introductory section by introducing the notation of partitioned matrices. Partitioning matrices into smaller submatrices is a technique that is frequently useful for verifying properties that may not be as readily apparent if the entire matrix is viewed as a whole. Thus, suppose that $A \in M_{m,n}(R)$. If $m = m_1 + \cdots + m_r$ and $n = n_1 + \cdots + n_s$, then we may think of A as an $r \times s$ block matrix

$$A = \begin{bmatrix} A_{11} & \cdots & A_{1s} \\ \vdots & \ddots & \vdots \\ A_{r1} & \cdots & A_{rs} \end{bmatrix}$$

where each block A_{ij} is a matrix of size $m_i \times n_j$ with entries in R. Two particularly important partitions of A are the partition by rows

$$A = \begin{bmatrix} A_1 \\ \vdots \\ A_m \end{bmatrix} \qquad \text{where} \qquad A_i = \text{row}_i(A) \in M_{1,n}(R)$$

and the partition by columns

$$A = \begin{bmatrix} B_1 & \cdots & B_n \end{bmatrix} \qquad \text{where} \qquad B_j = \text{col}_j(A) \in M_{m,1}(R).$$

Consider the problem of multiplying two partitioned matrices. Thus, suppose that

$$A = \begin{bmatrix} A_{11} & \cdots & A_{1s} \\ \vdots & \ddots & \vdots \\ A_{r1} & \cdots & A_{rs} \end{bmatrix}$$

and

$$B = \begin{bmatrix} B_{11} & \cdots & B_{1t} \\ \vdots & \ddots & \vdots \\ B_{s1} & \cdots & B_{st} \end{bmatrix}$$

are partitioned matrices. Can the product $C = AB$ be computed as a partitioned matrix $C = [C_{ij}]$ where $C_{ij} = \sum_{k=1}^{s} A_{ik}B_{kj}$? The answer is yes provided all of the required multiplications make sense. In fact, parts (5) and (6) of Lemma 1.1 are special cases of this type of multiplication for the partitions that come from rows and columns. Specifically, the equation

$\mathrm{row}_i(AB) = [\mathrm{row}_i(A)]B$ of Lemma 1.1 (5) translates into a product of partitioned matrices

$$AB = \begin{bmatrix} A_1 \\ \vdots \\ A_m \end{bmatrix} B = \begin{bmatrix} A_1 B \\ \vdots \\ A_m B \end{bmatrix}$$

where $A_i = \mathrm{row}_i(A)$, while the equation $\mathrm{col}_j(AB) = A[\mathrm{col}_j(B)]$ of Lemma 1.1 (6) translates into another product of partitioned matrices

$$AB = A\begin{bmatrix} B_1 & \cdots & B_p \end{bmatrix} = \begin{bmatrix} AB_1 & \cdots & AB_p \end{bmatrix}$$

where $B_j = \mathrm{col}_j(B)$.

For the product of general partitioned matrices, there is the following result.

(1.14) Proposition. *Let R be a ring, and let $A \in M_{m,n}(R)$, $B \in M_{n,p}(R)$. Suppose that $m = m_1 + \cdots + m_s$, $n = n_1 + \cdots + n_t$, and $p = p_1 + \cdots + p_u$, and assume that $A = [A_{ij}]$ and $B = [B_{ij}]$ are partitioned so that $A_{ij} \in M_{m_i,n_j}(R)$ while $B_{ij} \in M_{n_i,p_j}(R)$. Then the matrix $C = AB$ has a partition $C = [C_{ij}]$ where $C_{ij} \in M_{m_i,p_j}(R)$ and*

$$C_{ij} = \sum_{k=1}^{t} A_{ik} B_{kj}.$$

Proof. Suppose $1 \le \alpha \le m$ and $1 \le \beta \le p$. Then

$$(1.2) \qquad \mathrm{ent}_{\alpha\beta}(C) = \sum_{\gamma=1}^{n} a_{\alpha\gamma} b_{\gamma\beta}.$$

In the partition of C given by $m = m_1 + \cdots + m_s$ and $p = p_1 + \cdots + p_u$, we have that $\mathrm{ent}_{\alpha\beta}(C)$ is in a submatrix $C_{ij} \in M_{m_i,p_j}(R)$ so that $\mathrm{ent}_{\alpha\beta}(C) = \mathrm{ent}_{\omega\tau}(C_{ij})$ where $1 \le \omega \le m_i$ and $1 \le \tau \le p_j$. Thus we have a partition of $\mathrm{row}_\alpha(A)$ and $\mathrm{col}_\beta(B)$ as

$$\mathrm{row}_\alpha(A) = \begin{bmatrix} \mathrm{row}_\omega(A_{i1}) & \cdots & \mathrm{row}_\omega(A_{it}) \end{bmatrix}$$

and

$$\mathrm{col}_\beta(B) = \begin{bmatrix} \mathrm{col}_\tau(B_{1j}) \\ \vdots \\ \mathrm{col}_\tau(B_{tj}) \end{bmatrix}.$$

From Equation (1.2) we conclude that

$$\text{ent}_{\omega\tau}(C_{ij}) = \text{ent}_{\alpha\beta}(C)$$

$$= \sum_{\gamma=1}^{n} a_{\alpha\gamma} b_{\gamma\beta}$$

$$= \sum_{\gamma=1}^{n_1} a_{\alpha\gamma} b_{\alpha\gamma} + \sum_{\gamma=n_1+1}^{n_2} a_{\alpha\gamma} b_{\alpha\gamma} + \cdots + \sum_{\gamma=n_{t-1}+1}^{n_t} a_{\alpha\gamma} b_{\alpha\gamma}$$

$$= \text{ent}_{\omega\tau}(A_{i1}B_{1j}) + \text{ent}_{\omega\tau}(A_{i2}B_{2j}) + \cdots + \text{ent}_{\omega\tau}(A_{it}B_{tj}).$$

and the result is proved.   □

A particularly useful collection of partitioned matrices is the set of **block diagonal** matrices. A partitioned matrix

$$A = \begin{bmatrix} A_{11} & \cdots & A_{1s} \\ \vdots & \ddots & \vdots \\ A_{r1} & \cdots & A_{rs} \end{bmatrix}$$

is said to be a **block diagonal** matrix if $r = s$ and if $A_{ij} = 0$ whenever $i \neq j$. The submatrices A_{ii} are the diagonal blocks, but note that the blocks A_{ii} can be of any size. Generally, we will denote the diagonal blocks with the single subscript A_i. If

$$A = \begin{bmatrix} A_1 & 0 & \cdots & 0 \\ 0 & A_2 & \cdots & 0 \\ \vdots & \vdots & \ddots & \vdots \\ 0 & 0 & \cdots & A_r \end{bmatrix}$$

is a block diagonal matrix, then we say that A is the **direct sum** of the matrices $A_1, \dots, A_r$ and we denote this direct sum by

$$A = A_1 \oplus \cdots \oplus A_r.$$

Thus, if $A_i \in M_{m_i,n_i}(R)$, then $A_1 \oplus \cdots \oplus A_r \in M_{m,n}(R)$ where $m = \sum_{i=1}^{r} m_i$ and $n = \sum_{i=1}^{r} n_i$.

The following result contains some straightforward results concerning the algebra of direct sums of matrices:

(1.15) Lemma. *Let R be a ring, and let $A_1, \dots, A_r$ and $B_1, \dots, B_r$ be matrices over R of appropriate sizes. (The determination of the needed size is left to the reader.) Then*

(1) $(\oplus_{i=1}^{r} A_i) + (\oplus_{i=1}^{r} B_i) = \oplus_{i=1}^{r} (A_i + B_i),$
(2) $(\oplus_{i=1}^{r} A_i)(\oplus_{i=1}^{r} B_i) = \oplus_{i=1}^{r} (A_i B_i),$
(3) $(\oplus_{i=1}^{r} A_i)^{-1} = \oplus_{i=1}^{r} A_i^{-1}$ *if $A_i \in \text{GL}(n_i, R)$, and*
(4) $\text{Tr}\,(\oplus_{i=1}^{r} A_i) = \sum_{i=1}^{r} \text{Tr}(A_i)$ *if $A_i \in M_{m_i}(R)$.*

Proof. Exercise.   □

The concept of partitioned matrix is particularly convenient for describing and verifying various properties of the tensor product (also called the kronecker product) of two matrices.

(1.16) Definition. *Let R be a commutative ring, let $A \in M_{m_1,n_1}(R)$, and let $B \in M_{m_2,n_2}(R)$. Then the* **tensor product** *or* **kronecker product** *of A and B, denoted $A \otimes B \in M_{m_1 m_2, n_1 n_2}(R)$, is the partitioned matrix*

$$(1.3) \qquad A \otimes B = \begin{bmatrix} C_{11} & \cdots & C_{1n_1} \\ \vdots & \ddots & \vdots \\ C_{m_1 1} & \cdots & C_{m_1 n_1} \end{bmatrix}$$

where each block $C_{ij} \in M_{m_2 n_2}$ is defined by

$$C_{ij} = (\mathrm{ent}_{ij}(A))\,B = a_{ij} B.$$

There is a second possibility for the tensor product. $A \otimes B$ could be defined as the partitioned matrix

$$(1.4) \qquad \begin{bmatrix} D_{11} & \cdots & D_{1n_2} \\ \vdots & \ddots & \vdots \\ D_{m_2 1} & \cdots & D_{m_2 n_2} \end{bmatrix}$$

where each block $D_{ij} \in M_{m_1 n_1}(R)$ is defined by

$$D_{ij} = A(\mathrm{ent}_{ij}(B)) = A b_{ij}.$$

We shall see in Section 7.2 that the two versions of the tensor product arise because of different possibilities of ordering standard bases on the tensor product of free modules of finite rank. Since there is no intrinsic difference between the two possibilities, we shall use the definition in Equation (1.3) as the definition of the tensor product of matrices.

(1.17) Examples.

(1) $I_m \otimes I_n = I_{mn}.$

(2) $I_m \otimes B = \oplus_{i=1}^{m} B.$

(3) $\begin{bmatrix} a & b \\ c & d \end{bmatrix} \otimes I_2 = \begin{bmatrix} a & 0 & b & 0 \\ 0 & a & 0 & b \\ c & 0 & d & 0 \\ 0 & c & 0 & d \end{bmatrix}.$

The following result is an easy consequence of the partitioned multiplication formula (Proposition 1.14):

(1.18) Lemma. *Let R be a commutative ring and let $A_1 \in M_{m_1,n_1}(R)$, $A_2 \in M_{n_1,r_1}(R)$, $B_1 \in M_{m_2,n_2}(R)$, and $B_2 \in M_{n_2,r_2}(R)$. Then*

$$(A_1 \otimes B_1)(A_2 \otimes B_2) = (A_1 A_2) \otimes (B_1 B_2).$$

Proof. By Proposition 1.14, C_{ij}, the (i,j) block of $(A_1 \otimes B_1)(A_2 \otimes B_2)$, is given by

$$C_{ij} = \sum_{k=1}^{n_1} (\text{ent}_{ik}(A_1)B_1)(\text{ent}_{kj}(A_2)B_2)$$

$$= \left(\sum_{k=1}^{n_1} (\text{ent}_{ik}(A_1)\,\text{ent}_{kj}(A_2)) \right) B_1 B_2,$$

which one recognizes as the (i,j) block of $(A_1 A_2) \otimes (B_1 B_2)$. $\square$

(1.19) Corollary. *Let R be a commutative ring, let $A \in M_m(R)$, and let $B \in M_n(R)$. Then*

$$A \otimes B = (A \otimes I_n)(I_m \otimes B).$$

Proof. $\square$

(1.20) Lemma. *If $A \in M_m(R)$ and $B \in M_n(R)$, then*

$$\text{Tr}(A \otimes B) = \text{Tr}(A)\,\text{Tr}(B).$$

Proof. Exercise. $\square$

4.2 Determinants and Linear Equations

Throughout this section, we will assume that R is a commutative ring.

(2.1) Definition. *Let R be a commutative ring and let $D : M_n(R) \to R$ be a function. We say that D is n-**linear** (on rows) if the following two conditions are satisfied.*

(1) *If B is obtained from A by multiplying a single row of A by $a \in R$ then*

$$D(B) = aD(A).$$

(2) *If A, B, $C \in M_n(R)$ are identical in all rows except for the ith row and*

$$\text{row}_i(C) = \text{row}_i(A) + \text{row}_i(B),$$

then

$$D(C) = D(A) + D(B).$$

*Furthermore, we say that D is **alternating** if $D(A) = 0$ for any matrix A that has two rows equal. $D : M_n(A) \to R$ is said to be a **determinant function** if D is n-linear and alternating.*

Note that property (2) does *not* say that $D(A + B) = D(A) + D(B)$. This is definitely not true if $n > 1$.

One may also speak of n-linearity on columns, but Proposition 2.9 will show that there is no generality gained in considering both types of n-linearity. Therefore, we shall concentrate on rows.

(2.2) Examples.

(1) Let D_1 and D_2 be n-linear functions. Then for any choice of a and b in R, the function $D : M_n(R) \to R$ defined by $D(A) = aD_1(A) + bD_2(A)$ is also an n-linear function. That is, the set of n-linear functions on $M_n(R)$ is closed under addition and scalar multiplication of functions, i.e., it is an R-module.

(2) Let $\sigma \in S_n$ be a permutation and define $D_\sigma : M_n(R) \to R$ by the formula

$$D_\sigma(A) = a_{1\,\sigma(1)}a_{2\,\sigma(2)} \cdots a_{n\,\sigma(n)}$$

where $A = [a_{ij}]$. It is easy to check that D_σ is an n-linear function, but it is not a determinant function since it is not alternating.

(3) Let $f : S_n \to R$ be any function and define $D_f : M_n(R) \to R$ by the formula $D_f = \sum_{\sigma \in S_n} f(\sigma)D_\sigma$. Applying this to a specific $A = [a_{ij}] \in M_n(R)$ gives

$$D_f(A) = \sum_{\sigma \in S_n} f(\sigma)a_{1\,\sigma(1)}a_{2\,\sigma(2)} \cdots a_{n\,\sigma(n)}.$$

By examples (1) and (2), D_f is an n-linear function.

(4) If $n = 2$ and $c \in R$, then $D_c(A) = c(a_{11}a_{22} - a_{12}a_{21})$ defines a determinant function on $M_2(A)$.

The first order of business is to prove that there is a determinant function for every n and for every commutative ring R and that this determinant function is essentially unique. More precisely, any determinant function is completely determined by its value on the identity matrix I_n. Note that for $n = 1$ this is clear since $D([a]) = D(a[1]) = aD([1])$ for every 1×1 matrix $[a]$ by property (1) of n-linearity. The strategy for verifying existence and essential uniqueness for determinant functions is to first verify a number of basic properties that any determinant function must satisfy and then from these properties to derive a formula that must be used to define any determinant function. It will then only remain to check that this formula, in fact, defines a determinant function.

(2.3) Lemma. *Let $D : M_n(R) \to R$ be n-linear. If $\mathrm{row}_i(A) = 0$ for some i, then $D(A) = 0$.*

Proof. Since $\mathrm{row}_i(A) = 0 \cdot \mathrm{row}_i(A)$, property (1) of n-linearity applies to show that $D(A) = 0 \cdot D(A) = 0$. $\square$

(2.4) Lemma. *Let $D : M_n(R) \to R$ be a determinant function. If $i \neq j$ and P_{ij} is the elementary permutation matrix determined by i and j, then*

$$D(P_{ij}A) = -D(A)$$

for all $A \in M_n(R)$. (That is, interchanging two rows of a matrix multiplies $D(A)$ by -1.)

Proof. Let $A_k = \mathrm{row}_k(A)$ for $1 \leq k \leq n$, and let B be the matrix with $\mathrm{row}_k(B) = \mathrm{row}_k(A)$ whenever $k \neq i, j$ while $\mathrm{row}_i(B) = \mathrm{row}_j(B) = A_i + A_j$. Then since D is n-linear and alternating,

$$0 = D(B) = D\begin{bmatrix} \vdots \\ A_i + A_j \\ \vdots \\ A_i + A_j \\ \vdots \end{bmatrix} = D\begin{bmatrix} \vdots \\ A_i \\ \vdots \\ A_i + A_j \\ \vdots \end{bmatrix} + D\begin{bmatrix} \vdots \\ A_j \\ \vdots \\ A_i + A_j \\ \vdots \end{bmatrix}$$

$$= D\begin{bmatrix} \vdots \\ A_i \\ \vdots \\ A_i \\ \vdots \end{bmatrix} + D\begin{bmatrix} \vdots \\ A_i \\ \vdots \\ A_j \\ \vdots \end{bmatrix} + D\begin{bmatrix} \vdots \\ A_j \\ \vdots \\ A_i \\ \vdots \end{bmatrix} + D\begin{bmatrix} \vdots \\ A_j \\ \vdots \\ A_j \\ \vdots \end{bmatrix}$$

$$= D\begin{bmatrix} \vdots \\ A_i \\ \vdots \\ A_j \\ \vdots \end{bmatrix} + D\begin{bmatrix} \vdots \\ A_j \\ \vdots \\ A_i \\ \vdots \end{bmatrix}$$

$$= D(A) + D(P_{ij}A)$$

by Proposition 1.12. Thus, $D(P_{ij}A) = -D(A)$, and the lemma is proved.

$\square$

(2.5) *Remark.* Lemma 2.4 is the reason for giving the name "alternating" to the property that $D(A) = 0$ for a matrix A that has two equal rows. Indeed, suppose D has the property given by Lemma 2.4, and let A be a matrix with rows i and j equal. Then $P_{ij}A = A$, so from the property of

Lemma 2.4 we conclude $D(A) = -D(A)$, i.e., $2D(A) = 0$. Thus, if R is a ring in which 2 is not a zero divisor, the two properties are equivalent, but in general the property of being alternating (as given in Definition 2.1) is stronger (Exercise 16).

(2.6) Lemma. *Let $D : M_n(R) \to R$ be a determinant function. If $i \neq j$, $\alpha \in R$, and $T_{ij}(\alpha)$ is an elementary transvection, then*

$$D(T_{ij}(\alpha)A) = D(A).$$

(That is, adding a multiple of one row of A to another row does not change the value of $D(A)$.)

Proof. Let B be the matrix that agrees with A in all rows except row i, and assume that $\text{row}_i(B) = \alpha\, \text{row}_j(A)$. Let A' be the matrix that agrees with A in all rows except row i and assume that $\text{row}_i(A') = \text{row}_j(A)$. Then D is alternating so $D(A') = 0$ since $\text{row}_i(A') = \text{row}_j(A) = \text{row}_j(A')$, and thus, $D(B) = aD(A') = 0$ since D is n-linear. Since $T_{ij}(\alpha)A$ agrees with A except in row i and

$$\text{row}_i(T_{ij}(\alpha)A) = \text{row}_i(A) + \alpha\, \text{row}_j(A) = \text{row}_i(A) + \text{row}_i(B)$$

(see Proposition 1.12), it follows from property (2) of n-linearity that

$$D(T_{ij}(\alpha)A) = D(A) + D(B) = D(A).$$

$\square$

Let $E_i = \text{row}_i(I_n)$ for $1 \leq i \leq n$, and consider all $n \times n$ matrices formed by using the matrices E_i as rows. To develop a convenient notation, let Ω_n denote the set of all functions $\omega : \{1, 2, \ldots, n\} \to \{1, 2, \ldots, n\}$, and if $\omega \in \Omega_n$, let P_ω denote the $n \times n$ matrix with $\text{row}_i(P_\omega) = E_{\omega(i)}$. For example, if $n = 3$ and $\omega(1) = 2$, $\omega(2) = 1$, and $\omega(3) = 2$, then

$$P_\omega = \begin{bmatrix} 0 & 1 & 0 \\ 1 & 0 & 0 \\ 0 & 1 & 0 \end{bmatrix}.$$

If $\omega \in \Omega_n$ is bijective so that $\omega \in S_n$, then P_ω is called a **permutation matrix**. If $\omega = (i\ j)$ is a transposition, then $P_\omega = P_{ij}$ is an elementary permutation matrix as defined in Section 4.1. In general, observe that the product $P_\omega A$ is the matrix defined by

$$\text{row}_i(P_\omega A) = \text{row}_{\omega(i)}(A).$$

According to Proposition 1.12 (3), if $\omega \in S_n$ is written as a product of transpositions, say $\omega = (i_1\, j_1)(i_2\, j_2) \cdots (i_t\, j_t)$, then

$$P_\omega = P_{i_1 j_1} P_{i_2 j_2} \cdots P_{i_t j_t} I_n$$
$$= I_n P_{i_1 j_1} P_{i_2 j_2} \cdots P_{i_t j_t}.$$

Since $\omega^{-1} = (i_t\, j_t)(i_{t-1}\, j_{t-1}) \cdots (i_1\, j_1)$, the second equality, together with Proposition 1.13 (3), shows that

$$\text{col}_i(P_\omega) = E_{\omega^{-1}(i)}.$$

Therefore, again using Proposition 1.13 (3), we see that AP_ω is the matrix defined by

$$\text{col}_i(AP_\omega) = \text{col}_{\omega^{-1}(i)}(A).$$

To summarize, left multiplication by P_ω permutes the rows of A, following the permutation ω, while right multiplication by P_ω permutes the columns of A, following the permutation ω^{-1}.

Recall that the sign of a permutation σ, denoted $\text{sgn}(\sigma)$, is $+1$ if σ is a product of an even number of transpositions and -1 if σ is a product of an odd number of transpositions.

(2.7) Lemma. *Let R be a commutative ring and let $D : M_n(R) \to R$ be a determinant function. If $\omega \in \Omega_n$, then*

(1) $D(P_\omega) = 0$ *if* $\omega \notin S_n$*, and*
(2) $D(P_\omega) = \text{sgn}(\omega)D(I_n)$ *if* $\omega \in S_n$*.*

Proof. (1) If $\omega \notin S_n$ then $\omega(i) = \omega(j)$ for some $i \neq j$ so that P_ω has two rows that are equal. Thus, $D(P_\omega) = 0$.

(2) If $\omega \in S_n$, write $\omega = (i_1\, i_j) \cdots (i_t\, j_t)$ as a product of transpositions to get (by Proposition 1.12)

$$P_\omega = P_{i_1 j_1} \cdots P_{i_t j_t} I_n.$$

By Lemma 2.4, we conclude that

$$D(P_\omega) = (-1)^t D(I_n) = \text{sgn}(\omega)D(I_n),$$

and the lemma is proved.   □

Now let $A = [a_{ij}] \in M_n(R)$ and partition A by its rows, i.e.,

$$A = \begin{bmatrix} A_1 \\ \vdots \\ A_n \end{bmatrix}$$

where $A_i = \text{row}_i(A)$ for $1 \leq i \leq n$. Note that

$$
\begin{aligned}
A_i &= [a_{i1} \cdots a_{in}] \\
&= [a_{i1}\, 0 \cdots 0] + [0\, a_{i2}\, 0 \cdots 0] + \cdots + [0 \cdots 0\, a_{in}] \\
&= a_{i1}E_1 + a_{i2}E_2 + \cdots + a_{in}E_n \\
&= \sum_{j=1}^{n} a_{ij}E_j,
\end{aligned}
$$

and thus,

$$
(2.1) \qquad A = \begin{bmatrix} \sum_{j_1=1}^n a_{1j_1} E_{j_1} \\ \sum_{j_2=1}^n a_{2j_2} E_{j_2} \\ \vdots \\ \sum_{j_n=1}^n a_{nj_n} E_{j_n} \end{bmatrix}.
$$

If D is a determinant function, we may compute $D(A)$ using (2.1), the n-linearity of D, and Lemma 2.7 as follows:

$$
D(A) = D \begin{bmatrix} \sum_{j_1=1}^n a_{1j_1} E_{j_1} \\ \sum_{j_2=1}^n a_{2j_2} E_{j_2} \\ \vdots \\ \sum_{j_n=1}^n a_{nj_n} E_{j_n} \end{bmatrix}
$$

$$
= \sum_{j_1=1}^n a_{1j_1} D \begin{bmatrix} E_{j_1} \\ \sum_{j_2=1}^n a_{2j_2} E_{j_2} \\ \vdots \\ \sum_{j_n=1}^n a_{nj_n} E_{j_n} \end{bmatrix}
$$

$$
= \sum_{j_1=1}^n \sum_{j_2=1}^n a_{1j_1} a_{2j_2} D \begin{bmatrix} E_{j_1} \\ E_{j_2} \\ \vdots \\ \sum_{j_n=1}^n a_{nj_n} E_{j_n} \end{bmatrix}
$$

$$
= \sum_{j_1=1}^n \cdots \sum_{j_n=1}^n a_{1j_1} a_{2j_2} \cdots a_{nj_n} D \begin{bmatrix} E_{j_1} \\ E_{j_2} \\ \vdots \\ E_{j_n} \end{bmatrix}
$$

$$
= \sum_{\omega \in \Omega_n} a_{1\,\omega(1)} a_{2\,\omega(2)} \cdots a_{n\,\omega(n)} D(P_\omega)
$$

$$
(2.2) \qquad = \left[\sum_{\omega \in S_n} \mathrm{sgn}(\omega) a_{1\,\omega(1)} \cdots a_{n\,\omega(n)} \right] D(I_n).
$$

Thus, we have arrived at the uniqueness part of the following result since formula (2.2) is the formula which must be used to define any determinant function.

(2.8) Theorem. *Let R be a commutative ring and let $a \in R$. Then there is exactly one determinant function $D_a : M_n(R) \to R$ such that $D_a(I_n) = a$. Thus $D_a = aD_1$, and we let $\det = D_1$.*

Proof. If there is a determinant function D_a, then according to Equation (2.2), D_a must be given by

$$(2.3) \qquad D_a(A) = a \left(\sum_{\sigma \in S_n} \text{sgn}(\sigma) a_{1\,\sigma(1)} \cdots a_{n\,\sigma(n)} \right).$$

It remains to check that Equation (2.3) in fact defines a determinant function on $M_n(R)$. It is sufficient to check this with $a = 1$ since a scalar multiple of a determinant function is also a determinant function. If we let $f(\sigma) = \text{sgn}(\sigma) \cdot 1$, then $D_1 = D_f$ as defined in Example 2.2 (3), and as observed in that example, D_f is an n-linear function for each $f : S_n \to R$. In particular, D_1 is n-linear and it remains to check that it is alternating. To verify this, suppose that $A \in M_n(R)$ has $\text{row}_i(A) = \text{row}_j(A)$ with $i \neq j$. Then for $1 \leq k \leq n$ we have $a_{ik} = a_{jk}$. If $\sigma \in S_n$, let $\sigma' = \sigma \circ (i\,j)$. We claim that

$$(2.4) \qquad a_{1\,\sigma(1)} \cdots a_{n\,\sigma(n)} = a_{1\,\sigma'(1)} \cdots a_{n\,\sigma'(n)}.$$

This is because $\sigma(k) = \sigma'(k)$ if $k \neq i, j$ so that $a_{k\,\sigma(k)} = a_{k\,\sigma'(k)}$, while $\sigma'(i) = \sigma(j)$ and $\sigma'(j) = \sigma(i)$ so that $a_{i\,\sigma'(i)} = a_{i\,\sigma(j)} = a_{j\,\sigma(j)}$ and $a_{j\,\sigma'(j)} = a_{j\,\sigma(i)} = a_{i\,\sigma(i)}$. Hence Equation (2.4) is valid. But $\text{sgn}(\sigma') = -\text{sgn}(\sigma)$, so

$$\text{sgn}(\sigma) a_{1\,\sigma(1)} \cdots a_{n\,\sigma(n)} + \text{sgn}(\sigma') a_{1\,\sigma'(1)} \cdots a_{n\,\sigma'(n)} = 0.$$

But $\sigma \leftrightarrow \sigma'$ gives a pairing of the even and odd permutations in S_n, and hence, we conclude

$$D_1(A) = \sum_{\sigma \in S_n} \text{sgn}(\sigma) a_{1\,\sigma(1)} \cdots a_{n\,\sigma(n)}$$
$$= \sum_{\text{sgn}(\sigma)=1} \left(\text{sgn}(\sigma) a_{1\,\sigma(1)} \cdots a_{n\,\sigma(n)} + \text{sgn}(\sigma') a_{1\,\sigma'(1)} \cdots a_{n\,\sigma'(n)} \right)$$
$$= 0.$$

Therefore, D_1 is a determinant function on $M_n(R)$ and Theorem 2.8 is proved. $\qquad \square$

(2.9) Corollary. Let $A = [a_{ij}] \in M_n(R)$ be an upper (resp., lower) triangular matrix, i.e., $a_{ij} = 0$ for $i > j$ (resp., $i < j$). Then

$$\det(A) = a_{11} \cdots a_{nn} = \prod_{i=1}^{n} a_{ii}.$$

Proof. If $\sigma \in S_n$ is not the identity permutation, then for some i and j, $\sigma(i) > i$ and $\sigma(j) < j$, so in either case,

$$a_{1\sigma(1)} \cdots a_{n\sigma(n)} = 0,$$

and the result follows by Equation (2.3). $\qquad \square$

Our determinant functions have been biased towards functions on the rows of matrices. But, in fact, we can equally well consider functions on the columns of matrices. Perhaps the simplest way to see this is via the following result:

(2.10) Proposition. *If $A \in M_n(R)$ then $\det(A) = \det(A^t)$.*

Proof. Let $A = [a_{ij}]$. Then

$$
\begin{aligned}
\det(A) &= \sum_{\sigma \in S_n} \text{sgn}(\sigma) a_{1\,\sigma(1)} \cdots a_{n\,\sigma(n)} \\
&= \sum_{\sigma \in S_n} \text{sgn}(\sigma) a_{\sigma^{-1}(\sigma(1))\,\sigma(1)} \cdots a_{\sigma^{-1}(\sigma(n))\,\sigma(n)} \\
&= \sum_{\sigma \in S_n} \text{sgn}(\sigma) a_{\sigma^{-1}(1)\,1} \cdots a_{\sigma^{-1}(n)\,n} \\
&= \sum_{\sigma \in S_n} \text{sgn}(\sigma^{-1}) a_{\sigma^{-1}(1)\,1} \cdots a_{\sigma^{-1}(n)\,n} \\
&= \sum_{\tau \in S_n} \text{sgn}(\tau) a_{\tau(1)\,1} \cdots a_{\tau(n)\,n} \\
&= \det(A^t).
\end{aligned}
$$

Here we have used the fact that $\text{sgn}(\sigma^{-1}) = \text{sgn}(\sigma)$ and that

$$
a_{\sigma^{-1}(\sigma(1))\,\sigma(1)} \cdots a_{\sigma^{-1}(\sigma(n))\,\sigma(n)} = a_{\sigma^{-1}(1)\,1} \cdots a_{\sigma^{-1}(n)\,n}.
$$

This last equation is valid because R is commutative and $\{\sigma(1), \ldots, \sigma(n)\}$ is just a reordering of $\{1, \ldots, n\}$ for any $\sigma \in S_n$. $\qquad \square$

(2.11) Theorem. *If $A, B \in M_n(R)$ then*

$$
\det(AB) = \det(A)\det(B).
$$

Proof. Define $D_B : M_n(R) \to R$ by $D_B(A) = \det(AB)$. Since $\text{row}_i(AB) = [\text{row}_i(A)]B$, it is easy to check (do it) that D_B is n-linear and alternating. Thus D_B is a determinant function, so by Theorem 2.8, $D_B(A) = \alpha \det(A)$ where $\alpha = D_B(I_n) = \det(B)$. $\qquad \square$

This result is an example of the payoff from the abstract approach to determinants. To prove Theorem 2.11 directly from the definition of determinant as given by Equation (2.3) is a somewhat messy calculation. However, the direct calculation from Equation (2.3) is still beneficial in that a more general product formula is valid. This approach will be pursued in Theorem 2.34.

(2.12) Corollary. *If R is a commutative ring, $A \in M_n(R)$, and $S \in GL(n, R)$, then*

$$\det(S^{-1}AS) = \det(A).$$

That is, similar matrices have the same determinant.

Proof. Exercise. □

Similarly to the proof of Theorem 2.11, one can obtain the formula for the determinant of a direct sum of two matrices.

(2.13) Theorem. *If R is a commutative ring, $A_i \in M_{n_i}(R)$ for $1 \le i \le r$, and $A = \oplus_{i=1}^r A_i$, then*

$$\det(A) = \prod_{i=1}^r \det(A_i).$$

Proof. It is clearly sufficient to prove the result for $r = 2$. Thus suppose that $A = A_1 \oplus A_2$. We may write $A = (A_1 \oplus I_{n_2})(I_{n_1} \oplus A_2)$. Then Theorem 2.11 gives

$$\det(A) = (\det(A_1 \oplus I_{n_2}))(\det(I_{n_1} \oplus A_2)).$$

Therefore, if $\det(A_1 \oplus I_{n_2}) = \det(A_1)$ and $\det(I_{n_1} \oplus A_2) = \det(A_2)$, then we are finished. We shall show the first equality; the second is identical. Define $D_1 : M_{n_1}(R) \to R$ by $D_1(B) = \det(B \oplus I_{n_2})$. Since det is $(n_1 + n_2)$-linear and alternating on $M_{n_1+n_2}(R)$, it follows that D_1 is n_1-linear and alternating on $M_{n_1}(R)$. Hence, (by Theorem 2.8), $D_1(A_1) = \alpha \det(A_1)$ for all $A_1 \in M_{n_1}(R)$, where $\alpha = D_1(I_{n_1}) = \det(I_{n_1} \oplus I_{n_2}) = 1$, and the theorem is proved. □

Remark. Theorem 2.13 and Corollary 2.9 can both be generalized to a formula for block triangular matrices. (See Exercise 13.)

There is also a simple formula for the determinant of the tensor product of two square matrices.

(2.14) Proposition. *Let R be a commutative ring, let $A \in M_m(R)$, and let $B \in M_n(R)$. Then*

$$\det(A \otimes B) = (\det(A))^n (\det(B))^m.$$

Proof. By Corollary 1.19, $A \otimes B = (A \otimes I_n)(I_m \otimes B)$. By Example 1.17 (2), $I_m \otimes B = \oplus_{i=1}^m B$, so $\det(I_m \otimes B) = \det(B)^m$ by Theorem 2.13. We leave it as an exercise to verify that the rows and columns of the matrix $A \otimes I_n$ can be permuted to obtain that $A \otimes I_n$ is (permutation) similar to $\oplus_{i=1}^n A$ (Exercise 50). The proof is then completed by another application of Theorem 2.13. □

We will now consider cofactor expansions of determinants. If $A \in M_n(R)$ and $1 \leq i, j \leq n$, let A_{ij} be the $(n-1) \times (n-1)$ matrix obtained by deleting the ith row and j^{th} column of A.

(2.15) Theorem. (Laplace expansion) *Let R be a commutative ring and let $A \in M_n(R)$. Then*

$$(1) \qquad \sum_{k=1}^{n} (-1)^{k+j} a_{ki} \det(A_{kj}) = \delta_{ij} \det(A)$$

and

$$(2) \qquad \sum_{k=1}^{n} (-1)^{k+j} a_{ik} \det(A_{jk}) = \delta_{ij} \det(A).$$

Proof. If $A \in M_n(R)$ and $1 \leq i, j \leq n$, let

$$D_{ij}(A) = \sum_{k=1}^{n} (-1)^{k+j} a_{ki} \det(A_{kj}).$$

That is, $D_{ij}(A) \in R$ is defined by the left-hand side of equation (1). We claim that the function $D_{ij} : M_n(R) \to R$ is a determinant function for all i, j. Note that the function $A \mapsto a_{ki} \det(A_{kj})$ is n-linear on $M_n(R)$. Since a linear combination of n-linear functions is n-linear, it follows that D_{ij} is n-linear. It remains to check that it is alternating. To see this, suppose that $A \in M_n(R)$ is a matrix with rows p and q equal. If $k \neq p, q$, then A_{kj} has two rows equal, so $\det(A_{kj}) = 0$ in this case. Thus,

$$(2.5) \qquad D_{ij}(A) = (-1)^{p+j} a_{pi} \det(A_{pj}) + (-1)^{q+j} a_{qi} \det(A_{qj}).$$

Note that the assumption that $\text{row}_p(A) = \text{row}_q(A)$ means that $a_{pi} = a_{qi}$. To be explicit in our calculation, suppose that $p < q$. Then the matrix A_{qj} is obtained from A_{pj} by moving row $q-1$ of A_{pj} to row p and row t of A_{pj} to row $t+1$ for $t = p, \ldots, q-2$. In other words,

$$A_{qj} = P_\omega A_{pj}$$

where $\omega \in S_{n-1}$ is defined by $\omega(t) = t$ for $t < p$ and $t \geq q$, while $\omega(q-1) = p$ and $\omega(t) = t+1$ for $t = p, \ldots, q-2$. That is, ω is the $(q-p)$-cycle $(p, p+1, \ldots, q-1)$. Then by Lemma 2.7

$$\det(A_{qj}) = \det(P_\omega A) = \text{sgn}(\omega) \det(A_{pj}) = (-1)^{q-p-1} \det(A_{pj}).$$

Equation (2.5) then shows that $D_{ij}(A) = 0$, and hence D_{ij} is alternating. Therefore, D_{ij} is a determinant function, so by Theorem 2.8

$$(2.6) \qquad D_{ij}(A) = D_{ij}(I_n) \det(A)$$

for all $A \in M_n(R)$. A direct calculation shows that $D_{ij}(I_n) = \delta_{ij}$, so that Equation (2.6) yields (1) of the theorem.

Formula (2) (cofactor expansion along row j) is obtained by applying formula (1) to the matrix A^t and using the fact (Proposition 2.10) that $\det(A) = \det(A^t)$. $\qquad\square$

(2.16) Definition. *If $A \in M_n(R)$, then we define the **cofactor matrix of** A, denoted* $\mathrm{Cofac}(A)$, *by the formula*

$$\mathrm{ent}_{ij}(\mathrm{Cofac}(A)) = (-1)^{i+j} \det(A_{ij}),$$

*and we define the **adjoint** of A, denoted* $\mathrm{Adj}(A)$, *by*

$$\mathrm{Adj}(A) = (\mathrm{Cofac}(A))^t.$$

The following result should be familiar for matrices with entries in a field, and the proof is the same:

(2.17) Theorem. *Let $A \in M_n(R)$.*

(1) $A(\mathrm{Adj}(A)) = (\mathrm{Adj}(A))A = \det(A)I_n$.
(2) A *is invertible if and only if* $\det(A)$ *is a unit in R, and in this case,*

$$A^{-1} = (\det(A))^{-1} \mathrm{Adj}(A).$$

Proof. Formula (1) follows immediately from Theorem 2.15. If $A \in M_n(R)$ is invertible, then there is a matrix B such that $AB = BA = I_n$. Thus $(\det(A))(\det(B)) = 1$ so that $\det(A) \in R^*$. Conversely, if $\det(A)$ is a unit, then $B = (\det(A))^{-1} \mathrm{Adj}(A)$ satisfies $AB = BA = I_n$ by formula (1). $\qquad\square$

(2.18) Remark. The definition of inverse of a matrix requires that $AB = BA = I_n$, but as a consequence of Theorem 2.17, we can conclude that a matrix $A \in M_n(R)$ is invertible provided that there is a matrix $B \in M_n(R)$ such that $AB = I_n$. Indeed, if $AB = I_n$ then $(\det(A))(\det(B)) = 1$, so $\det(A)$ is a unit in R, and hence, A is invertible.

(2.19) Examples.

(1) A matrix $A \in M_n(\mathbf{Z})$ is invertible if and only if $\det(A) = \pm 1$.
(2) If F is a field, a matrix $A \in M_n(F[X])$ is invertible if and only if $\det(A) \in F^* = F \setminus \{0\}$.
(3) If F is a field and $A \in M_n(F[X])$, then for each $a \in F$, evaluation of each entry of A at $a \in F$ gives a matrix $A(a) \in M_n(F)$. If $\det(A) = f(X) \neq 0 \in F[X]$, then $A(a)$ is invertible whenever $f(a) \neq 0$. Thus, $A(a)$ is invertible for all but finitely many $a \in F$.

As an application of our results on determinants, we shall present a determinantal criterion for the solvability of **homogeneous** linear equations when the coefficients can be from an arbitrary commutative ring. This criterion will involve some ideals of R generated by various determinants of submatrices of matrices over R. We will start by defining these ideals.

To state the results in a reasonably compact form it is convenient to introduce some appropriate notation. If p, $m \in \mathbf{N}$ with $p \leq m$, let $Q_{p,m}$ denote the set of all sequences $\alpha = (i_1, \ldots, i_p)$ of p integers with $1 \leq i_1 < i_2 < \cdots < i_p \leq m$. Note that the cardinality of the set $Q_{p,m}$ is $|Q_{p,m}| = \binom{m}{p}$. Suppose $A \in M_{m,n}(R)$. If $\alpha \in Q_{p,m}$ and $\beta \in Q_{j,n}$, let $A[\alpha \mid \beta]$ denote the submatrix of A consisting of the elements whose row index is in α and whose column index is in β. If $\alpha \in Q_{p,m}$ then there is a complementary sequence $\hat{\alpha} \in Q_{p-m,m}$ consisting of the integers in $\{1, 2, \ldots, m\}$ not included in α and listed in increasing order. To give some examples of these notations, let

$$A = \begin{bmatrix} a_{11} & a_{12} & a_{13} & a_{14} \\ a_{21} & a_{22} & a_{23} & a_{24} \\ a_{31} & a_{32} & a_{33} & a_{34} \end{bmatrix}.$$

If $\alpha = (1, 3) \in Q_{2,3}$ and $\beta = (2, 3) \in Q_{2,4}$, then

$$A[\alpha \mid \beta] = \begin{bmatrix} a_{12} & a_{13} \\ a_{32} & a_{33} \end{bmatrix}$$

while $\hat{\alpha} = 2 \in Q_{1,3}$ and $\hat{\beta} = (1, 4) \in Q_{2,4}$ so that

$$A[\hat{\alpha} \mid \hat{\beta}] = [a_{21} \quad a_{24}].$$

(2.20) Definition. *If R is a commutative ring and $A \in M_{m,n}(R)$, then a $t \times t$ **minor** of A is the determinant of any submatrix $A[\alpha \mid \beta]$ where $\alpha \in Q_{t,m}$, $\beta \in Q_{t,n}$. The **determinantal rank** of A, denoted D-rank(A), is the largest t such that there is a nonzero $t \times t$ minor of A.*

(2.21) Definition. *If R is any commutative ring, $A \in M_{m,n}(R)$, and $1 \leq t \leq \min\{m, n\}$, let*

$$F_t(A) = \langle \{\det A[\alpha \mid \beta] : \alpha \in Q_{t,m}, \beta \in Q_{t,n}\} \rangle \subseteq R.$$

*That is, $F_t(A)$ is the ideal of R generated by all the $t \times t$ minors of A. We set $F_0(A) = R$ and $F_t(A) = 0$ if $t > \min\{m, n\}$. $F_t(A)$ is called the tth-**Fitting ideal** of A. The Laplace expansion of determinants (Theorem 2.15) shows that $F_{t+1}(A) \subseteq F_t(A)$. Thus there is a decreasing chain of ideals*

$$R = F_0(A) \supseteq F_1(A) \supseteq F_2(A) \supseteq \cdots.$$

*If R is a PID, then $F_t(A)$ is a principal ideal, say $F_t(A) = \langle d_t(A) \rangle$ where $d_t(A)$ is the greatest common divisor of all the $t \times t$ minors of A. In this case, a generator of $F_t(A)$ is called the tth-**determinantal divisor** of A.*

(2.22) Definition. *If* $A \in M_{m,n}(R)$, *then the* M-rank(A) *is defined to be the largest* t *such that* $\mathrm{Ann}(F_t(A)) = 0$.

(2.23) Remarks.

(1) M-rank$(A) = 0$ means that $\mathrm{Ann}(F_1(A)) \neq 0$. That is, there is a nonzero $a \in R$ with $a\,a_{ij} = 0$ for all a_{ij}. Note that this is stronger than saying that every element of A is a zero divisor. For example, if $A = [\,2 \quad 3\,] \in M_{1,2}(\mathbf{Z}_6)$, then every element of A is a zero divisor in $\mathbf{Z}_6$ but there is no single nonzero element of $\mathbf{Z}_6$ that annihilates both entries in the matrix.

(2) If $A \in M_n(R)$, then M-rank$(A) = n$ means that $\det(A)$ is not a zero divisor of R.

(3) To say that M-rank$(A) = t$ means that there is an $a \neq 0 \in R$ with $a \cdot D = 0$ for all $(t+1) \times (t+1)$ minors D of A, but there is no nonzero $b \in R$ which annihilates all $t \times t$ minors of A by multiplication. In particular, if $\det A[\alpha \mid \beta]$ is not a zero divisor of R for some $\alpha \in Q_{s,m}$, $\beta \in Q_{s,n}$, then M-rank$(A) \geq s$.

(2.24) Lemma. *If* $A \in M_{m,n}(R)$, *then*

$$0 \leq \text{M-rank}(A) \leq \text{D-rank}(A) \leq \min\{m, n\}.$$

Proof. Exercise. $\square$

We can now give a criterion for solvability of the homogeneous linear equation $AX = 0$, where $A \in M_{m,n}(R)$. This equation always has the trivial solution $X = 0$, so we are looking for solutions $X \neq 0 \in M_{n,1}(R)$.

(2.25) Theorem. *Let* R *be a commutative ring and let* $A \in M_{m,n}(R)$. *The matrix equation* $AX = 0$ *has a nontrivial solution* $X \neq 0 \in M_{n,1}(R)$ *if and only if*

$$\text{M-rank}(A) < n.$$

Proof. Suppose that M-rank$(A) = t < n$. Then $\mathrm{Ann}(F_{t+1}(A)) \neq 0$, so choose $b \neq 0 \in R$ with $b \cdot F_{t+1}(A) = 0$. Without loss of generality, we may assume that $t < m$ since, if necessary, we may replace the system $AX = 0$ by an equivalent one (i.e., one with the same solutions) by adding some rows of zeroes to the bottom of A. If $t = 0$ then $ba_{ij} = 0$ for all a_{ij} and we may take $X = \begin{bmatrix} b \\ \vdots \\ b \end{bmatrix}$. Then $X \neq 0 \in M_{n,1}(R)$ and $AX = 0$.

Thus, suppose $t > 0$. Then $b \notin \mathrm{Ann}(F_t(A)) = 0$, so $b \det A[\alpha \mid \beta] \neq 0$ for some $\alpha \in Q_{t,m}, \beta \in Q_{t,n}$. By permuting rows and columns, which does not affect whether $AX = 0$ has a nontrivial solution, we can assume $\alpha =$

$(1, \ldots, t)$, $\beta = (1, \ldots, t)$. For $1 \leq i \leq t + 1$ let $\beta_i = (1, 2, \ldots, \widehat{i}, \ldots, t + 1) \in$ $Q_{t,t+1}$ where $\widehat{i}$ indicates that i is deleted. Let $d_i = (-1)^{t+1+i} \det A[\alpha \mid \beta_i]$. Thus $d_1, \ldots, d_{t+1}$ are the cofactors of the matrix

$$A_1 = A[(1, \ldots, t+1) \mid (1, \ldots, t+1)]$$

obtained by deleting row $t + 1$ and column i. Hence the Laplace expansion theorem (Theorem 2.15) gives

$$(2.7) \quad \begin{cases} \sum_{j=1}^{t+1} a_{ij} d_j = 0 & 1 \leq i \leq t, \\ \sum_{j=1}^{t+1} a_{ij} d_j = \det A[(1, \ldots, t, i) \mid (1, \ldots, t, t+1)] & t < i \leq m. \end{cases}$$

Let $X = \begin{bmatrix} x_1 \\ \vdots \\ x_n \end{bmatrix}$ where

$$\begin{cases} x_i = bd_i & \text{if } 1 \leq i \leq t+1, \\ x_i = 0 & \text{if } t+2 \leq i \leq n. \end{cases}$$

Then $X \neq 0$ since $x_{t+1} = b \det A[\alpha \mid \beta] \neq 0$. But Equation (2.7) and the fact that $b \in \text{Ann}(F_{t+1}(A))$ show that

$$AX = \begin{bmatrix} b \sum_{j=1}^{t+1} a_{1j} d_j \\ \vdots \\ b \sum_{j=1}^{t+1} a_{mj} d_j \end{bmatrix}$$

$$= \begin{bmatrix} 0 \\ \vdots \\ 0 \\ b \det A[(1, \ldots, t, t+1) \mid (1, \ldots, t, t+1)] \\ \vdots \\ b \det A[(1, \ldots, t, m) \mid (1, \ldots, t, t+1)] \end{bmatrix}$$

$$= 0.$$

Thus, X is a nontrivial solution to the equation $AX = 0$.

Conversely, assume that $X \neq 0 \in M_{n,1}(R)$ is a nontrivial solution to $AX = 0$, and choose k with $x_k \neq 0$. We claim that $\text{Ann}(F_n(A)) \neq 0$. If $n > m$, then $F_n(A) = 0$, and hence, $\text{Ann}(F_n(A)) = R \neq (0)$. Thus, we may assume that $n \leq m$. Let $\alpha = (1, \ldots, n)$ and for each $\beta \in Q_{n,m}$, let $B_\beta = A[\alpha \mid \beta]$. Then, since $AX = 0$ and since each row of B_β is a full row of A, we conclude that $B_\beta X = 0$. The adjoint matrix formula (Theorem 2.17) then shows that

$$(\det B_\beta) X = (\text{Adj } B_\beta) B_\beta X = 0,$$

from which we conclude that $x_k \det B_\beta = 0$. Since $\beta \in Q_{n,m}$ is arbitrary, we conclude that $x_k \cdot F_n(A) = 0$, i.e., $x_k \in \mathrm{Ann}(F_n(A))$. But $x_k \neq 0$ so $\mathrm{Ann}(F_n(A)) \neq 0$, and we conclude that M-rank$(A) < n$ and the proof is complete. $\square$

In case R is an integral domain we may replace the M-rank by the ordinary determinantal rank to conclude the following:

(2.26) Corollary. *If R is an integral domain and $A \in M_{m,n}(R)$, then $AX = 0$ has a nontrivial solution if and only if* D-rank$(A) < n$.

Proof. If $I \subseteq R$, then $\mathrm{Ann}(I) \neq 0$ if and only if $I = 0$ since an integral domain has no nonzero zero divisors. Therefore, in an integral domain, D-rank$(A) =$ M-rank(A). $\square$

The results for n equations in n unknowns are even simpler:

(2.27) Corollary. *Let R be a commutative ring.*

(1) *If $A \in M_n(R)$, then $AX = 0$ has a nontrivial solution if and only if* $\det A$ *is a zero divisor of R.*
(2) *If R is an integral domain and $A \in M_n(R)$, then $AX = 0$ has a nontrivial solution if and only if* $\det A = 0$.

Proof. If $A \in M_n(R)$ then $F_n(A) = \langle \det A \rangle$, so M-rank$(A) < n$ if and only if $\det A$ is a zero divisor. In particular, if R is an integral domain then M-rank$(A) < n$ if and only if $\det A = 0$. $\square$

There are still two other concepts of rank which can be defined for matrices with entries in a commutative ring.

(2.28) Definition. *Let R be a commutative ring and let $A \in M_{m,n}(R)$. Then we will define the **row rank** of A, denoted* row-rank(A), *to be the maximum number of linearly independent rows in A, while the **column rank** of A, denoted* col-rank(A) *is the maximum number of linearly independent columns.*

(2.29) Corollary.

(1) *If R is a commutative ring and $A \in M_{m,n}(R)$, then*

$$\max\{\text{row-rank}(A),\ \text{col-rank}(A)\} \leq \text{M-rank}(A) \leq \text{D-rank}(A).$$

(2) *If R is an integral domain, then*

$$\text{row-rank}(A) = \text{col-rank}(A) = \text{M-rank}(A) = \text{D-rank}(A).$$

Proof. Exercise. $\square$

(2.30) *Remarks.*

(1) If R is an integral domain, then all four ranks of a matrix $A \in M_{m,n}(R)$ are equal, and we may speak unambiguously of the **rank** of A. This will be denoted by rank(A).

(2) The condition that R be an integral domain in Corollary 2.29 (2) is necessary. As an example of a matrix that has all four ranks different, consider $A \in M_4(\mathbf{Z}_{210})$ defined by

$$A = \begin{bmatrix} 0 & 2 & 3 & 5 \\ 2 & 0 & 6 & 0 \\ 3 & 0 & 3 & 0 \\ 0 & 0 & 0 & 7 \end{bmatrix}.$$

We leave it as an exercise to check that

$$\text{row-rank}(A) = 1$$
$$\text{col-rank}(A) = 2$$
$$\text{M-rank}(A) = 3$$
$$\text{D-rank}(A) = 4.$$

As a simple application of solvability of homogeneous equations, we note the following result:

(2.31) Proposition. *Let R be a commutative ring, let M be a finitely generated R-module, and let $S \subseteq M$ be a subset. If $|S| > \mu(M) = \text{rank}(M)$, then S is not R-linearly independent.*

Proof. Let $\mu(M) = m$ and let $T = \{w_1, \ldots, w_m\}$ be a generating set for M consisting of m elements. Choose n distinct elements $\{v_1, \ldots, v_n\}$ of S for some $n > m$, which is possible by our hypothesis $|S| > \mu(M) = m$. Since $M = \langle w_1, \ldots, w_m \rangle$, we may write

$$v_j = a_{1j}w_1 + \cdots + a_{mj}w_m$$

with $a_{ij} \in R$. Let $A = [a_{ij}] \in M_{m,n}(R)$. Since $n > m$, it follows that M-rank$(A) \le m < n$, so Theorem 2.25 shows that there is an $X \ne 0 \in M_{n,1}(R)$ such that $AX = 0$. Then

$$(2.8) \qquad \sum_{j=1}^{n} x_j v_j = \sum_{j=1}^{n} x_j \left(\sum_{i=1}^{m} a_{ij}w_i \right)$$

$$= \sum_{i=1}^{m} \left(\sum_{j=1}^{n} a_{ij}x_j \right) w_i$$

$$= 0$$

since $AX = 0$. Therefore, S is R-linearly dependent. $\qquad \square$

(2.32) Corollary. *Let R be a commutative ring, let M be an R-module, and let $N \subseteq M$ be a free submodule. Then $\mathrm{rank}(N) \leq \mathrm{rank}(M)$.*

Proof. If $\mathrm{rank}(M) = \infty$, there is nothing to prove, so assume that $\mathrm{rank}(M) = m < \infty$. If $\mathrm{rank}(N) > m$, then there is a linearly independent subset of M, namely, a basis of N, with more than m elements, which contradicts Proposition 2.31. $\square$

(2.33) *Remark.* Corollary 2.32 should be compared with Theorem 3.6.2, concerning submodules of free modules over a PID. Also, note that the condition that N be free is necessary (see Remark 3.6.5).

The following result, known as the Cauchy–Binet theorem, generalizes the formula for the determinant of a product (Theorem 2.11) to allow for products of possibly nonsquare matrices. One use of this formula is to investigate the behavior of the rank of matrices under products.

(2.34) Theorem. (Cauchy–Binet formula) *Let R be a commutative ring and let $A \in M_{m,n}(R)$, $B \in M_{n,p}(R)$. Assume that $1 \leq t \leq \min\{m, n, p\}$ and let $\alpha \in Q_{t,m}$, $\beta \in Q_{t,p}$. Then*

$$\det(AB[\alpha \mid \beta]) = \sum_{\gamma \in Q_{t,n}} \det(A[\alpha \mid \gamma]) \det(B[\gamma \mid \beta]).$$

Proof. Suppose that $\alpha = (\alpha_1, \ldots, \alpha_t)$, $\beta = (\beta_1, \ldots, \beta_t)$ and let $C = AB[\alpha \mid \beta]$. Thus,

$$\mathrm{ent}_{ij}(C) = \mathrm{row}_{\alpha_i}(A)\,\mathrm{col}_{\beta_j}(B)$$
$$= \sum_{k=1}^{n} a_{\alpha_i k} b_{k\beta_j},$$

so that

$$C = \begin{bmatrix} \sum_{k=1}^{n} a_{\alpha_1 k} b_{k\beta_1} & \cdots & \sum_{k=1}^{n} a_{\alpha_1 k} b_{k\beta_t} \\ \vdots & \ddots & \vdots \\ \sum_{k=1}^{n} a_{\alpha_t k} b_{k\beta_1} & \cdots & \sum_{k=1}^{n} a_{\alpha_t k} b_{k\beta_t} \end{bmatrix}.$$

Using n-linearity of the determinant function, we conclude that

$$(2.9) \quad \det C = \sum_{k_1=1}^{n} \cdots \sum_{k_t=1}^{n} a_{\alpha_1 k_1} \cdots a_{\alpha_t k_t} \det \begin{bmatrix} b_{k_1 \beta_1} & \cdots & b_{k_1 \beta_t} \\ \vdots & \ddots & \vdots \\ b_{k_t \beta_1} & \cdots & b_{k_t \beta_t} \end{bmatrix}.$$

If $k_i = k_j$ for $i \neq j$, then the ith and j^{th} rows of the matrix on the right are equal so the determinant is 0. Thus the only possible nonzero determinants on the right occur if the sequence $(k_1, \ldots, k_t)$ is a permutation of a sequence

$\gamma = (\gamma_1, \ldots \gamma_t) \in Q_{t,n}$. Let $\sigma \in S_t$ be the permutation of $\{1, \ldots, t\}$ such that $\gamma_i = k_{\sigma(i)}$ for $1 \le i \le t$. Then

$$(2.10) \qquad \det \begin{bmatrix} b_{k_1 \beta_1} & \cdots & b_{k_1 \beta_t} \\ \vdots & \ddots & \vdots \\ b_{k_t \beta_1} & \cdots & b_{k_t \beta_t} \end{bmatrix} = \operatorname{sgn}(\sigma) \det B[\gamma \mid \beta].$$

Given $\gamma \in Q_{t,n}$, all possible permutations of γ are included in the summation in Equation (2.9). Therefore, Equation (2.9) may be rewritten, using Equation (2.10), as

$$\det C = \sum_{\gamma \in Q_{t,n}} \left(\sum_{\sigma \in S_t} \operatorname{sgn}(\sigma) a_{\alpha_1 \gamma_{\sigma(1)}} \cdots a_{\alpha_t \gamma_{\sigma(t)}} \right) \det B[\gamma \mid \beta]$$

$$= \sum_{\gamma \in Q_{t,n}} \det A[\alpha \mid \gamma] \det B[\gamma \mid \beta],$$

which is the desired formula.   □

(2.35) Examples.

(1) The Cauchy–Binet formula gives another verification of the fact that $\det(AB) = \det A \det B$ for square matrices A and B. In fact, the only element of $Q_{n,n}$ is the sequence $\gamma = (1, 2, \ldots, n)$ and $A[\gamma \mid \gamma] = A$, $B[\gamma \mid \gamma] = B$, and $AB[\gamma \mid \gamma] = AB$, so the product formula for determinants follows immediately from the Cauchy–Binet formula.

(2) As a consequence of the Cauchy–Binet theorem, if $A \in M_{m,n}(R)$, and $B \in M_{n,p}(R)$ then

$$(2.11) \qquad \text{D-rank}(AB) \le \min\{\text{D-rank}(A), \text{D-rank}(B)\}.$$

To see this, let $t > \min\{\text{D-rank}(A), \text{D-rank}(B)\}$ and suppose that $\alpha \in Q_{t,m}$, $\beta \in Q_{t,p}$. Then by the Cauchy–Binet formula

$$\det(AB[\alpha \mid \beta]) = \sum_{\gamma \in Q_{t,n}} \det(A[\alpha \mid \gamma]) \det(B[\gamma \mid \beta]).$$

Since $t > \min\{\text{D-rank}(A), \text{D-rank}(B)\}$, at least one of the determinants $\det A[\alpha \mid \gamma]$ or $\det B[\gamma \mid \beta]$ must be 0 for each $\gamma \in Q_{t,n}$. Thus, $\det(AB[\alpha \mid \beta]) = 0$, and since α and β are arbitrary, it follows that $\text{D-rank}(AB) < t$, as required.

Equation (2.11) easily gives the following result, which shows that determinantal rank is not changed by multiplication by nonsingular matrices.

(2.36) Proposition. *Let $A \in M_{m,n}(R)$ and let $U \in \text{GL}(m, R)$, $V \in \text{GL}(n, R)$. Then*

$$\text{D-rank}(UAV) = \text{D-rank}(A).$$

Proof. Any matrix $B \in M_{m,n}(R)$ satisfies $\text{D-rank}(B) \leq \min\{m, n\}$ and since $\text{D-rank}(U) = n$ and $\text{D-rank}(V) = m$, it follows from Equation (2.11) that

$$\text{D-rank}(UAV) \leq \min\{\text{D-rank}(A), n, m\} = \text{D-rank}(A)$$

and

$$\text{D-rank}(A) = \text{D-rank}(U^{-1}(UAV)V^{-1}) \leq \text{D-rank}(UAV).$$

This proves the result.   □

We will conclude this section by giving a version of the Laplace expansion theorem that allows for expansion of $\det A$ along a given set of rows rather than a single row. The choice of rows along which expansion takes place is given by an element $\alpha \in Q_{t,n}$. Recall that if $\gamma \in Q_{t,n}$ then $\widehat{\gamma} \in Q_{n-t,n}$ denotes the complementary sequence. With these preliminaries out of the way, the general Laplace expansion theorem can be stated as follows:

(2.37) Theorem. (Laplace expansion) *Let $A \in M_n(R)$ and let $\alpha \in Q_{t,n}$ ($1 \leq t \leq n$) be given. Then*

$$(2.12) \qquad \det A = \sum_{\gamma \in Q_{t,n}} (-1)^{s(\alpha)+s(\gamma)} \det(A[\alpha \mid \gamma]) \det(A[\widehat{\alpha} \mid \widehat{\gamma}])$$

where

$$s(\gamma) = \sum_{j=1}^{t} \gamma_j$$

for any $\gamma = (\gamma_1, \ldots, \gamma_t) \in Q_{t,n}$.

Proof. The proof is essentially similar to the proof of Theorem 2.15 (which is a special case of the current theorem). If $A \in M_n(R)$, define

$$(2.13) \qquad D_\alpha(A) = \sum_{\gamma \in Q_{t,n}} (-1)^{s(\alpha)+s(\gamma)} \det(A[\alpha \mid \gamma]) \det(A[\widehat{\alpha} \mid \widehat{\gamma}]).$$

Then $D_\alpha : M_n(R) \to R$ is easily shown to be n-linear as a function on the *columns* of $A \in M_n(R)$. To complete the proof, it is only necessary to show that D_α is alternating and that $D_\alpha(I_n) = 1$. Thus, suppose that $\text{col}_p(A) = \text{col}_q(A)$, and to be specific, assume that $p < q$. If p and q are both in $\gamma \in Q_{t,n}$, then $A[\alpha \mid \gamma]$ will have two columns equal so that

$$\det A[\alpha \mid \gamma] = 0,$$

while, if both p and q are in $\widehat{\gamma} \in Q_{n-t,n}$, then $\det A[\widehat{\alpha} \mid \widehat{\gamma}] = 0$. Thus, in the evaluation of $D_\alpha(A)$ it is only necessary to consider those $\gamma \in Q_{t,n}$ such that $p \in \gamma$ and $q \in \widehat{\gamma}$, or vice-versa. Thus, suppose that $p \in \gamma$, $q \in \widehat{\gamma}$ and define a new sequence $\gamma' \in Q_{t,n}$ by replacing $p \in \gamma$ by q. Then $\widehat{\gamma}'$ agrees with $\widehat{\gamma}$ except that q has been replaced by p. Thus

$$(2.14) \qquad s(\gamma') - s(\gamma) = q - p.$$

Now consider the sum

$$(-1)^{s(\gamma)} \det(A[\alpha \mid \gamma]) \det(A[\widehat{\alpha} \mid \widehat{\gamma}]) + (-1)^{s(\gamma')} \det(A[\alpha \mid \gamma']) \det(A[\widehat{\alpha} \mid \widehat{\gamma}']),$$

which we denote by $S(A)$. We claim that this sum is 0. Assuming this, since γ and γ' appear in pairs in $Q_{t,n}$, it follows that $D_\alpha(A) = 0$ whenever two columns of A agree; thus D_α is alternating. It remains to check that $S(A) = 0$.

Suppose that $p = \gamma_k$ and $q = \widehat{\gamma}_\ell$. Then γ and γ' agree except in the range from p to q, as do $\widehat{\gamma}$ and $\widehat{\gamma}'$. This includes a total of $q - p + 1$ entries. If r of these entries are included in γ, then

$$\gamma_1 < \cdots < \gamma_k = p < \gamma_{k+1} < \cdots < \gamma_{k+r-1} < q < \gamma_{k+r} < \cdots < \gamma_t$$

and

$$A[\alpha \mid \gamma'] = A[\alpha \mid \gamma]P_{\omega^{-1}}$$

where ω is the r-cycle $(k + r - 1, k + r - 2, \ldots, k)$. Similarly,

$$A[\widehat{\alpha} \mid \widehat{\gamma}'] = A[\widehat{\alpha} \mid \widehat{\gamma}]P_{\omega'}$$

where ω' is a $(q - p + 1 - r)$-cycle. Thus,

$$(-1)^{s(\gamma')} \det(A[\alpha \mid \gamma']) \det(A[\widehat{\alpha} \mid \widehat{\gamma}'])$$
$$= (-1)^{s(\gamma')+(r-1)+(q-p)-r} \det(A[\alpha \mid \gamma]) \det(A[\widehat{\alpha} \mid \widehat{\gamma}]).$$

Since $s(\gamma') + (q - p) - 1 - s(\gamma) = 2(q - p) - 1$ is odd, we conclude that $S(A) = 0$. Thus D_α is alternating, and since it is straightforward to check that $D_\alpha(I_n) = 1$, the result is proved.   □

Applying formula (2.12) to A^t in place of A, gives the Laplace expansion in columns:

$$(2.15) \qquad \det A = \sum_{\gamma \in Q_{t,n}} (-1)^{s(\alpha)+s(\gamma)} \det(A[\gamma \mid \alpha]) \det(A[\widehat{\gamma} \mid \widehat{\alpha}]).$$

Note that if $t = 1$ then $Q_{1,n} = \{1, 2, \ldots, n\}$ so that $A[i \mid j] = a_{ij}$ while $A[\widehat{i} \mid \widehat{j}] = A_{ij}$, so Theorem 2.37 includes Theorem 2.15 as a special case.

4.3 Matrix Representation of Homomorphisms

Before beginning with the procedure of associating a matrix with a homomorphism between free modules, we would like to make some remarks about the ticklish situation that arises for a noncommutative ring. We will only need this once, at the very end of Section 7.1, so the reader who is only interested in the commutative case may (and is well advised to) skip the more general situation.

Difficulties already arise in the simplest case. Let R be a ring and let us consider a free R-module M of rank 1 with basis $\mathcal{B} = \{v\}$. Then $M = \{rv : r \in R\}$. We wish to give "coordinates" to the elements of M, i.e., identify the elements of M with the elements of R, and clearly, there is only one reasonable choice here, namely, that, in the basis $\mathcal{B}$, rv should have coordinate $[rv]_\mathcal{B} = [r]$. Now consider $f \in \mathrm{End}_R(M)$. We wish to represent f by a matrix with respect to the basis $\mathcal{B}$, and again there is only one reasonable choice: if $f(v) = sv$, then f should have coordinate matrix in the basis $\mathcal{B}$ given by $[f]_\mathcal{B} = [s]$. (Note that f is *not* "left-multiplication by s" unless $s \in C(R)$, the center of R. Indeed, $g : M \to M$ defined by $g(m) = sm$ is *not* an R-endomorphism unless $s \in C(R)$, as then $g(rm) = srm \neq rsm = rg(m)$ in general. Of course, there is no problem if R is commutative.) Now, the theory we are about to develop will tell us that for any $m \in M$, we may calculate $f(m)$ by

$$[f(m)]_\mathcal{B} = [f]_\mathcal{B}[m]_\mathcal{B}.$$

However, when we try to apply this to $m = rv$, we get $f(m) = f(rv) = rf(v) = r(sv) = (rs)v$, so $[f(m)]_\mathcal{B} = [rs]$ while $[f]_\mathcal{B}[m]_\mathcal{B} = [s][r] = [sr]$. If R is commutative, these are equal, but in general they are not.

On the other hand, this formulation of the problem points the way to its solution. Namely, recall that we have the ring R^{op} (Remark 3.1.2 (3)) whose elements are the elements of R, whose addition is the same as that of R, but whose multiplication is given by $r \cdot s = sr$, where on the right-hand side we have the multiplication of R. Then, indeed, the equation

$$[rs] = [s] \cdot [r]$$

is valid, and we may hope that this modification solves our problem. This hope is satisfied, and this is indeed the way to approach coordinatization of R-module homomorphisms when R is not commutative.

Now we come to a slight notational point. We could maintain the above notation for multiplication in R^{op} throughout. This has two disadvantages: the practical—that we would often be inserting the symbol ".", which is easy to overlook, and the theoretical—that it makes the ring R^{op} look special (i.e., that for any "ordinary" ring we write multiplcation simply by juxtaposing elements, whereas in R^{op} we do not), whereas R^{op} is a perfectly good ring, neither better nor worse than R itself. Thus, we adopt a second solution. Let op $: R \to R^{\mathrm{op}}$ be the function that is the identity on elements, i.e., $\mathrm{op}(t) = t$ for every $t \in R$. Then we have $\mathrm{op}(sr) = \mathrm{op}(r)\,\mathrm{op}(s)$, where the multiplication on the right-hand side, written as usual as juxtaposition, is the multiplication in R^{op}. This notation also has the advantage of reminding us that $t \in R$, but $\mathrm{op}(t) \in R^{\mathrm{op}}$.

Note that if R is commutative then $R^{\mathrm{op}} = R$ and op is the identity, which in this case is a ring homomorphism. In fact, op $: R \to R^{\mathrm{op}}$ is a ring homomorphism if and only if R is commutative. In most applications of matrices, it is the case of commutative rings that is of primary importance. If you are just interested in the commutative case (as you may well be), we advise you to simply mentally (not physically) erase "op" whenever it appears, and you will have formulas that are perfectly legitimate for a commutative ring R.

After this rather long introduction, we will now proceed to the formal mathematics of associating matrices with homomorphisms between free modules.

If R is a ring, M is a free R-module of rank n, and $\mathcal{B} = \{v_1, \dots, v_n\}$ is a basis of M, then we may write $v \in M$ as $v = a_1 v_1 + \cdots + a_n v_n$ for unique $a_1, \dots, a_n \in R$. This leads us to the definition of coordinates. Define $\psi : M \to M_{n,1}(R^{\mathrm{op}})$ by

$$\psi(v) = \begin{bmatrix} \mathrm{op}(a_1) \\ \vdots \\ \mathrm{op}(a_n) \end{bmatrix} = [v]_{\mathcal{B}}.$$

The $n \times 1$ matrix $[v]_{\mathcal{B}}$ is called the **coordinate matrix of v with respect to the basis $\mathcal{B}$**.

Suppose that $\mathcal{B}' = \{v'_1, \dots, v'_n\}$ is another basis of M and define the matrix $P^{\mathcal{B}}_{\mathcal{B}'} \in M_n(R^{\mathrm{op}})$ by the formula

(3.1) $P^{\mathcal{B}}_{\mathcal{B}'} = [\, [v_1]_{\mathcal{B}'} \quad [v_2]_{\mathcal{B}'} \quad \cdots \quad [v_n]_{\mathcal{B}'} \,].$

That is, $\mathrm{col}_j\left(P^{\mathcal{B}}_{\mathcal{B}'}\right) = [v_j]_{\mathcal{B}'}$. The matrix $P^{\mathcal{B}}_{\mathcal{B}'}$ is called the **change of basis matrix** from the basis $\mathcal{B}$ to the basis $\mathcal{B}'$. Since $v_j = \sum_{i=1}^{n} \delta_{ij} v_i$, it follows that if $\mathcal{B} = \mathcal{B}'$ then $P^{\mathcal{B}}_{\mathcal{B}} = I_n$.

(3.1) Proposition. *Let M be a free R-module of rank n, and let $\mathcal{B}$, $\mathcal{B}'$, and $\mathcal{B}''$ be bases of M. Then*

(1) *for any $v \in M$, $[v]_{\mathcal{B}'} = P^{\mathcal{B}}_{\mathcal{B}'} [v]_{\mathcal{B}}$;*

(2) *$P^{\mathcal{B}}_{\mathcal{B}''} = P^{\mathcal{B}'}_{\mathcal{B}''} P^{\mathcal{B}}_{\mathcal{B}'}$; and*

(3) *$P^{\mathcal{B}}_{\mathcal{B}'}$ is invertible and $\left(P^{\mathcal{B}}_{\mathcal{B}'}\right)^{-1} = P^{\mathcal{B}'}_{\mathcal{B}}$.*

Proof. (1) Note that $M_{n,1}(R^{\mathrm{op}})$ is an R-module where the operation of R on $M_{n,1}(R^{\mathrm{op}})$ is given by $r\,A = A\,\mathrm{op}(r)$ for $r \in R$, $A \in M_{n,1}(R^{\mathrm{op}})$. Then

$$\psi' : M \to M_{n,1}(R^{\mathrm{op}}),$$

defined by $\psi'(v) = [v]_{\mathcal{B}'}$, is an R-module homomorphism, as is

$$\psi'' : M \to M_{n,1}(R^{\mathrm{op}}),$$

defined by $\psi''(v) = P_{\mathcal{B}'}^{\mathcal{B}}[v]_{\mathcal{B}}$. To show that $\psi'' = \psi'$ we need only show that they agree on elements of the basis $\mathcal{B}$. If $\mathcal{B} = \{v_1, \ldots, v_n\}$, then $[v_i]_{\mathcal{B}} = e_i = E_{i1} \in M_{n,1}(R)$ since $v_i = \sum_{j=1}^{n} \delta_{ij} v_j$. Thus,

$$\psi''(v_i) = P_{\mathcal{B}'}^{\mathcal{B}} e_i = \mathrm{col}_i(P_{\mathcal{B}'}^{\mathcal{B}}) = [v_i]_{\mathcal{B}'} = \psi'(v_i)$$

as required.

(2) For any $v \in M$ we have

$$\left(P_{\mathcal{B}''}^{\mathcal{B}'} P_{\mathcal{B}'}^{\mathcal{B}} \right) [v]_{\mathcal{B}} = P_{\mathcal{B}''}^{\mathcal{B}'} \left(P_{\mathcal{B}'}^{\mathcal{B}} [v]_{\mathcal{B}} \right)$$
$$= P_{\mathcal{B}''}^{\mathcal{B}'} [v]_{\mathcal{B}'}$$
$$= [v]_{\mathcal{B}''}$$
$$= P_{\mathcal{B}''}^{\mathcal{B}} [v]_{\mathcal{B}}.$$

Therefore, $P_{\mathcal{B}''}^{\mathcal{B}'} P_{\mathcal{B}'}^{\mathcal{B}} = P_{\mathcal{B}''}^{\mathcal{B}}$.

(3) Take $\mathcal{B}'' = \mathcal{B}$ in part (2). Then

$$P_{\mathcal{B}}^{\mathcal{B}'} P_{\mathcal{B}'}^{\mathcal{B}} = P_{\mathcal{B}}^{\mathcal{B}} = I_n.$$

Thus, $P_{\mathcal{B}'}^{\mathcal{B}}$ is invertible and $\left(P_{\mathcal{B}'}^{\mathcal{B}} \right)^{-1} = P_{\mathcal{B}}^{\mathcal{B}'}$.   □

(3.2) Lemma. *Let R be a ring and let M be a free R-module. If $\mathcal{B}'$ is any basis of M with n elements and $P \in \mathrm{GL}(n, R^{\mathrm{op}})$ is any invertible matrix, then there is a basis $\mathcal{B}$ of M such that*

$$P = P_{\mathcal{B}'}^{\mathcal{B}}.$$

Proof. Let $\mathcal{B}' = \{v_1', \ldots, v_n'\}$ and suppose that $P = [\mathrm{op}(p_{ij})]$. Let $v_j = \sum_{i=1}^{n} p_{ij} v_i'$. Then $\mathcal{B} = \{v_1, \ldots, v_n\}$ is easily checked to be a basis of M, and by construction, $P_{\mathcal{B}'}^{\mathcal{B}} = P$.   □

Remark. Note that the choice of notation for the change of basis matrix $P_{\mathcal{B}'}^{\mathcal{B}}$ has been chosen so that the formula in Proposition 3.1 (2) is easy to remember. That is, a superscript and an adjacent (on the right) subscript that are equal cancel, as in

$$P_{\mathcal{B}''}^{\mathcal{B}'} P_{\mathcal{B}'}^{\mathcal{B}} = P_{\mathcal{B}''}^{\mathcal{B}}.$$

The same mnemonic device will be found to be useful in keeping track of superscripts and subscripts in Propositions 3.5 and 3.6.

(3.3) Definition. *Let M and N be finite rank free R-modules with bases $\mathcal{B} = \{v_1, \ldots, v_m\}$ and $\mathcal{C} = \{w_1, \ldots, w_n\}$ respectively. Fo, each $f \in \operatorname{Hom}_R(M, N)$ define the* **matrix of f with respect to $\mathcal{B}$, $\mathcal{C}$**, *denoted $[f]_{\mathcal{C}}^{\mathcal{B}}$, by*

$$(3.2) \qquad \operatorname{col}_j [f]_{\mathcal{C}}^{\mathcal{B}} = [f(v_j)]_{\mathcal{C}} \qquad for \quad 1 \le j \le m.$$

Thus, if $f(v_j) = \sum_{i=1}^n a_{ij} w_i$ for $1 \le j \le m$, then

$$[f]_{\mathcal{C}}^{\mathcal{B}} = [\operatorname{op}(a_{ij})] \in M_{n,m}(R^{\operatorname{op}}).$$

If $f \in \operatorname{End}_R(M)$ and $\mathcal{B}$ is a basis of M, then we will write $[f]_{\mathcal{B}}$ in place of $[f]_{\mathcal{B}}^{\mathcal{B}}$.

(3.4) Remarks.

(1) Note that *every* matrix $A \in M_{n,m}(R^{\operatorname{op}})$ is the matrix $[f]_{\mathcal{C}}^{\mathcal{B}}$ for a unique $f \in \operatorname{Hom}_R(M, N)$. Indeed, if $\mathcal{B} = \{v_1, \ldots, v_m\}$ and $\mathcal{C} = \{w_1, \ldots, w_n\}$ are bases of M and N respectively, and if $A = [\operatorname{op}(a_{ij})]$, then define $f \in \operatorname{Hom}_R(M, N)$ by $f(v_j) = \sum_{i=1}^n a_{ij} w_i$. Such an f exists and is unique by Proposition 3.4.9; it is clear from the construction that $A = [f]_{\mathcal{C}}^{\mathcal{B}}$. Thus the mapping $f \mapsto [f]_{\mathcal{C}}^{\mathcal{B}}$ gives a bijection between $\operatorname{Hom}_R(M, N)$ and $M_{n,m}(R^{\operatorname{op}})$.

(2) Suppose that R is a commutative ring. Then we already know (see Theorem 3.4.11) that $\operatorname{Hom}_R(M, N)$ is a free R-module of rank mn, as is the R-module $M_{n,m}(R)$; hence they are isomorphic as R-modules. A choice of basis $\mathcal{B}$ for M and $\mathcal{C}$ for N provides an explicit isomorphism

$$\Phi_{\mathcal{C}}^{\mathcal{B}} : \operatorname{Hom}_R(M, N) \to M_{n,m}(R),$$

defined by $\Phi_{\mathcal{C}}^{\mathcal{B}}(f) = [f]_{\mathcal{C}}^{\mathcal{B}}$. We leave it as an exercise for the reader to check that $\Phi_{\mathcal{C}}^{\mathcal{B}}(f_{ij}) = E_{ji}$ where $\{E_{ji}\}_{j=1\,i=1}^{n\quad m}$ is the standard basis of $M_{n,m}(R)$, while $\{f_{ij}\}_{i=1\,j=1}^{m\quad n}$ is the basis of $\operatorname{Hom}_R(M, N)$ constructed in the proof of Theorem 3.4.11.

Note that if $1_M : M \to M$ denotes the identity transformation and $\mathcal{B}$, $\mathcal{B}'$ are two bases of M, then

$$[1_M]_{\mathcal{B}'}^{\mathcal{B}} = P_{\mathcal{B}'}^{\mathcal{B}}$$

so that the change of basis matrix from the basis $\mathcal{B}$ to the basis $\mathcal{B}'$ is just the matrix of the identity homomorphism with respect to the matrix $\mathcal{B}$ on the domain and the basis $\mathcal{B}'$ on the range.

(3.5) Proposition. *With the above notation, if $v \in M$ then*

$$[f(v)]_{\mathcal{C}} = [f]_{\mathcal{C}}^{\mathcal{B}} [v]_{\mathcal{B}}.$$

Proof. If $[v]_\mathcal{B} = \begin{bmatrix} \operatorname{op}(b_1) \\ \vdots \\ \operatorname{op}(b_m) \end{bmatrix}$ then

$$f(v) = f\left(\sum_{j=1}^{m} b_j v_j\right)$$

$$= \sum_{j=1}^{m} b_j f(v_j)$$

$$= \sum_{j=1}^{m} b_j \left(\sum_{i=1}^{n} a_{ij} w_i\right)$$

$$= \sum_{i=1}^{n} \left(\sum_{j=1}^{m} b_j a_{ij}\right) w_i.$$

Therefore, $[f(v)]_\mathcal{C} = [\sum_{j=1}^{m} \operatorname{op}(b_j a_{1j}) \quad \cdots \quad \sum_{j=1}^{m} \operatorname{op}(b_j a_{nj})]^t = [f]_\mathcal{C}^\mathcal{B} [v]_\mathcal{B}.$ $\square$

(3.6) Proposition. *If M, N, and P are free R-modules with bases $\mathcal{B}$, $\mathcal{C}$, and $\mathcal{D}$, respectively, and $f : M \to N$ and $g : N \to P$ are R-module homomorphisms, then*

$$[g \circ f]_\mathcal{D}^\mathcal{B} = [g]_\mathcal{D}^\mathcal{C} [f]_\mathcal{C}^\mathcal{B}.$$

Proof. By Proposition 3.5, if $v \in M$ then

$$[g]_\mathcal{D}^\mathcal{C} \left([f]_\mathcal{C}^\mathcal{B} [v]_\mathcal{B}\right) = [g]_\mathcal{D}^\mathcal{C} [f(v)]_\mathcal{C}$$

$$= [g(f(v))]_\mathcal{D}$$

$$= [(g \circ f)(v)]_\mathcal{D}$$

$$= [g \circ f]_\mathcal{D}^\mathcal{B} [v]_\mathcal{B}.$$

Choosing $v = v_j = j^{th}$ element of the basis $\mathcal{B}$ so that

$$[v]_\mathcal{B} = E_{j1} \in M_{n,1}(R^{\operatorname{op}}),$$

we obtain

$$[g]_\mathcal{D}^\mathcal{C} \left(\operatorname{col}_j([f]_\mathcal{C}^\mathcal{B})\right) = [g]_\mathcal{D}^\mathcal{C} \left([f]_\mathcal{C}^\mathcal{B} [v_j]_\mathcal{B}\right)$$

$$= [g \circ f]_\mathcal{D}^\mathcal{B} [v_j]_\mathcal{B}$$

$$= \operatorname{col}_j \left([g \circ f]_\mathcal{D}^\mathcal{B}\right)$$

for $1 \leq j \leq n$. Applying Lemma 1.1 (6), we conclude that

$$[g \circ f]_\mathcal{D}^\mathcal{B} = [g]_\mathcal{D}^\mathcal{C} [f]_\mathcal{C}^\mathcal{B}$$

as claimed. $\square$

(3.7) *Remark.* From this proposition we can see that matrix multiplication is associative. Let M, N, P, and Q be free R-modules with bases $\mathcal{B}$, $\mathcal{C}$, $\mathcal{D}$, and $\mathcal{E}$ respectively, and let $f : M \to N$, $g : N \to P$, and $h : P \to Q$ be R-module homomorphisms. Then, by Proposition 3.6,

$$
\begin{aligned}
[h]^{\mathcal{D}}_{\mathcal{E}} \left([g]^{\mathcal{C}}_{\mathcal{D}} [f]^{\mathcal{B}}_{\mathcal{C}} \right) &= [h]^{\mathcal{D}}_{\mathcal{E}} \left([g \circ f]^{\mathcal{B}}_{\mathcal{D}} \right) \\
&= [h \circ (g \circ f)]^{\mathcal{B}}_{\mathcal{E}} \\
&= [(h \circ g) \circ f]^{\mathcal{B}}_{\mathcal{E}} \\
&= [h \circ g]^{\mathcal{C}}_{\mathcal{E}} [f]^{\mathcal{B}}_{\mathcal{C}} \\
&= \left([h]^{\mathcal{D}}_{\mathcal{E}} [g]^{\mathcal{C}}_{\mathcal{D}} \right) [f]^{\mathcal{B}}_{\mathcal{C}}.
\end{aligned}
$$

By Remark 3.4 (1), every matrix is the matrix of a homomorphism, so associativity of matrix multiplication follows from the associativity of functional composition. (Actually, this proves associativity for matrices with entries in R^{op}, but then associativity for matrices with entries in R follows from the observation that $R = (R^{\mathrm{op}})^{\mathrm{op}}$. Also observe that we used associativity in the proof of Proposition 3.1, but we did not use this proposition in the proof of Proposition 3.6, so our derivation here is legitimate.)

(3.8) Corollary. *Let M and N be free R-modules of rank n. Let $\mathcal{B}$ be a basis of M and let $\mathcal{C}$ be a basis of N. Then a homomorphism $f \in \mathrm{Hom}_R(M, N)$ is an isomorphism if and only if the matrix $[f]^{\mathcal{B}}_{\mathcal{C}} \in M_n(R^{\mathrm{op}})$ is invertible.*

Proof. Suppose $g = f^{-1} \in \mathrm{Hom}_R(N, M)$. Then, by Proposition 3.6,

$$
I_n = [1_M]^{\mathcal{B}}_{\mathcal{B}} = [g \circ f]^{\mathcal{B}}_{\mathcal{B}} = [g]^{\mathcal{C}}_{\mathcal{B}} [f]^{\mathcal{B}}_{\mathcal{C}},
$$

and similarly,

$$
I_n = [f]^{\mathcal{B}}_{\mathcal{C}} [g]^{\mathcal{C}}_{\mathcal{B}}.
$$

Thus, $[f]^{\mathcal{B}}_{\mathcal{C}}$ is an invertible matrix. The converse is left as an exercise. $\square$

(3.9) Corollary. *Let R be a ring and let M be a free R-module of rank n. Then $\mathrm{End}_R(M)$ is isomorphic (as a ring) to $M_n(R^{\mathrm{op}})$. If R is a commutative ring, then this isomorphism is an isomorphism of R-algebras.*

Proof. If $\mathcal{B}$ is a basis of M, let

$$
\Phi_{\mathcal{B}} : \mathrm{End}_R(M) \to M_n(R^{\mathrm{op}})
$$

be defined by $\Phi_{\mathcal{B}}(f) = [f]_{\mathcal{B}}$. According to Proposition 3.6, $\Phi_{\mathcal{B}}$ is a ring homomorphism, while it is a bijective map by Remark 3.4. If R is commutative, it is an R-algebra isomorphism by Lemma 1.1 (3). $\square$

(3.10) *Remark.* From Lemma 1.3 and Corollary 3.9 we immediately see that if R is a commutative ring and M is a free R-module of finte rank, then the center of its endomorphism ring is

$$C(\mathrm{End}_R(M)) = R \cdot 1_M.$$

That is, a homomorphism $f : M \to M$ commutes with every other homomorphism $g : M \to M$ if and only if $f = r \cdot 1_M$ for some $r \in R$.

For the remainder of this section we shall assume that the ring R is commutative.

(3.11) Proposition. *Let R be a commutative ring, let M and N be free R-modules, and let $f \in \mathrm{Hom}_R(M, N)$.*

(1) *If $\mathrm{rank}(M) < \mathrm{rank}(N)$, then f is not surjective.*
(2) *If $\mathrm{rank}(M) > \mathrm{rank}(N)$, then f is not injective.*
(3) *If $\mathrm{rank}(M) = \mathrm{rank}(N)$ is finite and f is injective, then $N/\mathrm{Im}(f)$ is a torsion R-module.*
(4) *If $\mathrm{rank}(M) = \mathrm{rank}(N)$ is finite and R is an integral domain, then f is injective if and only if $N/\mathrm{Im}(f)$ is a torsion R-module.*

Proof. (1) By the definition of rank (Definition 3.2.9), if f were surjective, then we would have $\mathrm{rank}(N) \le \mathrm{rank}(M)$.

(2) If f were injective, N would contain a free submodule $\mathrm{Im}(f)$ of

$$\mathrm{rank}(\mathrm{Im}(f)) = \mathrm{rank}(M) > \mathrm{rank}(N),$$

contradicting Corollary 2.32.

(3) Let f be an injection, and let $\pi : N \to N/\mathrm{Im}(f)$ be the projection. Suppose that $N/\mathrm{Im}(f)$ is not torsion, and let $\overline{n} \in N/\mathrm{Im}(f)$ be an element with $\mathrm{Ann}(\overline{n}) = \langle 0 \rangle$. Let $n \in N$ with $\pi(n) = \overline{n}$. Then $\mathrm{Im}(f) \cap Rn = \langle 0 \rangle$, and hence,

$$N \supseteq \mathrm{Im}(f) \oplus Rn,$$

which is a free module of $\mathrm{rank}(M) + 1 = \mathrm{rank}(N) + 1$, contradicting Corollary 2.32.

(4) Let R be an integral domain and assume that $N/\mathrm{Im}(f)$ is a torsion module. Pick a basis $\{w_1, \dots, w_n\}$ of N. Since $N/\mathrm{Im}(f)$ is a torsion module, there exists $v_i \in M$ and $c_i \ne 0 \in R$ with $f(v_i) = c_i w_i$ for $1 \le i \le n$. Suppose $v \in M$ and $f(v) = 0$. Then the set

$$\{v, v_1, \dots, v_n\}$$

is linearly dependent by Proposition 2.31, so let

$$av + a_1 v_1 + \cdots + a_n v_n = 0$$

be an equation of linear dependence with not all of $\{a, a_1, \dots, a_n\}$ equal to zero. Then

$$0 = af(v)$$
$$= f(av)$$
$$= f\left(-\sum_{i=1}^{n} a_i v_i\right)$$
$$= -\sum_{i=1}^{n} a_i f(v_i)$$
$$= -\sum_{i=1}^{n} c_i a_i w_i.$$

Since $\{w_1, \ldots, w_n\}$ is a basis of N and R is an integral domain, it follows that $a_i = 0$ for all i. Hence $av = 0$ and thus $v = 0$ (similarly), and we conclude that f is an injection.   □

(3.12) *Remark.* The assumption that R is an integral domain in Proposition 3.11 (4) is necessary. Let $R = \mathbf{Z}_{mn}$ and set $M = N = R$. Let $f : M \to N$ be defined by $f(v) = mv$. Then $\mathbf{Z}_{mn}/\mathrm{Im}(f) \cong \mathbf{Z}_n$ is a torsion $\mathbf{Z}_{mn}$-module, but f is not injective.

The relationship between invertibility of homomorphisms and invertibility of matrices allows one to conclude that a homomorphism between free R-modules of the same finite rank is invertible if it has either a left or a right inverse.

(3.13) *Proposition.* *Let M and N be free R-modules of finite rank n, let $\mathcal{B}$ be a basis of M, and let $\mathcal{C}$ be a basis of N. If $f \in \mathrm{Hom}_R(M, N)$, then the following are equivalent.*

(1) *f is an isomorphism.*
(2) *f has a right inverse, i.e., there is a homomorphism $g \in \mathrm{Hom}_R(N, M)$ such that $fg = 1_N$.*
(3) *f has a left inverse, i.e., there is a homomorphism $h \in \mathrm{Hom}_R(N, M)$ such that $hf = 1_M$.*
(4) *f is a surjection.*
(5) *$[f]_{\mathcal{C}}^{\mathcal{B}}$ is an invertible matrix.*
(6) *$[f]_{\mathcal{C}}^{\mathcal{B}}$ has a right inverse.*
(7) *$[f]_{\mathcal{C}}^{\mathcal{B}}$ has a left inverse.*

Proof. The equivalence of (5), (6), and (7) follows from Remark 2.18, while the equivalence of (1), (2), and (3) to (5), (6), and (7), respectively, is a consequence of Corollary 3.8.

Now clearly (1) implies (4). On the other hand, assume that f is a surjection. Then there is a short exact sequence

$$0 \longrightarrow \mathrm{Ker}(f) \longrightarrow M \xrightarrow{\ f\ } N \longrightarrow 0.$$

This sequence splits since N is free so that there is an R-module homomorphism $g : N \to M$ such that $fg = 1_N$, i.e., g is a right inverse for f. Thus (4) implies (2), and the proof is complete. $\qquad\square$

(3.14) *Remark.* In Proposition 3.13 (4) it is not possible to replace surjective by injective. It is true that if f has a left inverse, then f is injective, but the converse need not be true. For example, $f : \mathbf{Z} \to \mathbf{Z}$ by $f(x) = 2x$ is injective, but it is not left invertible. However, in case the ring R is a field, the converse is valid. This is the content of the following result.

(3.15) Proposition. *Let F be a field and let M and N be vector spaces over F of dimension n. Then the following are equivalent.*

(1) *f is an isomorphism.*
(2) *f is injective.*
(3) *f is surjective.*

Proof. This is simply a restatement of Corollary 3.8.10. $\qquad\square$

(3.16) Proposition. *Let M and N be free R-modules with bases $\mathcal{B}$, $\mathcal{B}'$ and $\mathcal{C}$, $\mathcal{C}'$ respectively. If $f : M \to N$ is an R-module homomorphism, then $[f]_{\mathcal{C}}^{\mathcal{B}}$ and $[f]_{\mathcal{C}'}^{\mathcal{B}'}$ are related by the formula*

(3.3) $$[f]_{\mathcal{C}'}^{\mathcal{B}'} = P_{\mathcal{C}'}^{\mathcal{C}} [f]_{\mathcal{C}}^{\mathcal{B}} \left(P_{\mathcal{B}'}^{\mathcal{B}} \right)^{-1} .$$

Proof. Since $f = 1_N \circ f \circ 1_M$, Proposition 3.6 shows that

(3.4) $$[f]_{\mathcal{C}'}^{\mathcal{B}'} = [1_N]_{\mathcal{C}'}^{\mathcal{C}} [f]_{\mathcal{C}}^{\mathcal{B}} [1_M]_{\mathcal{B}}^{\mathcal{B}'} .$$

But $[1_N]_{\mathcal{C}'}^{\mathcal{C}} = P_{\mathcal{C}'}^{\mathcal{C}}$ and $[1_M]_{\mathcal{B}}^{\mathcal{B}'} = P_{\mathcal{B}}^{\mathcal{B}'} = \left(P_{\mathcal{B}'}^{\mathcal{B}} \right)^{-1}$, so Equation (3.3) follows from Equation (3.4). $\qquad\square$

We now give a determinantal criterion for the various properties of a homomorphism.

(3.17) Proposition. *Let M and N be free R-modules with $\mathrm{rank}(M) = \mathrm{rank}(N)$ finite.*

(1) *f is surjective (and hence an isomorphism) if and only if in some (and, hence, in any) pair of bases $\mathcal{B}$ of M and $\mathcal{C}$ of N, $\det([f]_{\mathcal{C}}^{\mathcal{B}})$ is a unit of R.*
(2) *f is injective if and only if in some (and, hence, in any) pair of bases $\mathcal{B}$ of M and $\mathcal{C}$ of N, $\det([f]_{\mathcal{C}}^{\mathcal{B}})$ is not a zero divisor in R.*

Proof. (1) is immediate from Proposition 3.13 and Theorem 2.17 (2), while part (2) follows from Corollary 2.27. $\qquad\square$

(3.18) Definition.

(1) *Let R be a commutative ring. Matrices A, $B \in M_{n,m}(R)$ are said to be **equivalent** if and only if there are invertible matrices $P \in \mathrm{GL}(n, R)$ and $Q \in \mathrm{GL}(m, R)$ such that*

$$B = PAQ.$$

Equivalence of matrices is an equivalence relation on $M_{n,m}(R)$.

(2) *If M and N are finite rank free R-modules, then we will say that R-module homomorphisms f and g in $\mathrm{Hom}_R(M, N)$ are **equivalent** if there are invertible endomorphisms $h_1 \in \mathrm{End}_R(M)$ and $h_2 \in \mathrm{End}_R(N)$ such that $h_2 f h_1^{-1} = g$. That is, f and g are equivalent if and only if there is a commutative diagram*

$$
\begin{array}{ccc}
M & \xrightarrow{\ f\ } & N \\
\downarrow{\scriptstyle h_1} & & \downarrow{\scriptstyle h_2} \\
M & \xrightarrow{\ g\ } & N
\end{array}
$$

where the vertical maps are isomorphisms. Again, equivalence of homomorphisms is an equivalence relation on $\mathrm{Hom}_R(M, N)$.

(3.19) Proposition.

(1) *Two matrices A, $B \in M_{n,m}(R)$ are equivalent if and only if there are bases $\mathcal{B}$, $\mathcal{B}'$ of a free module M of rank m and bases $\mathcal{C}$, $\mathcal{C}'$ of a free module N of rank n such that $A = [f]_{\mathcal{C}}^{\mathcal{B}}$ and $B = [f]_{\mathcal{C}'}^{\mathcal{B}'}$. That is, two matrices are equivalent if and only if they represent the same R-module homomorphism with respect to different bases.*

(2) *If M and N are free R-modules of rank m and n respectively, then homomorphisms f and $g \in \mathrm{Hom}_R(M, N)$ are equivalent if and only if there are bases $\mathcal{B}$, $\mathcal{B}'$ of M and $\mathcal{C}$, $\mathcal{C}'$ of N such that*

$$[f]_{\mathcal{C}}^{\mathcal{B}} = [g]_{\mathcal{C}'}^{\mathcal{B}'}.$$

That is, f is equivalent to g if and only if the two homomorphisms are represented by the same matrix with respect to appropriate bases.

Proof. (1) Since every invertible matrix is a change of basis matrix (Lemma 3.2), the result is immediate from Proposition 3.16.

(2) Suppose that $[f]_{\mathcal{C}}^{\mathcal{B}} = [g]_{\mathcal{C}'}^{\mathcal{B}'}$. Then Equation (3.3) gives

$$(3.5) \qquad [f]_{\mathcal{C}}^{\mathcal{B}} = [g]_{\mathcal{C}'}^{\mathcal{B}'} = P_{\mathcal{C}'}^{\mathcal{C}} [g]_{\mathcal{C}}^{\mathcal{B}} \left(P_{\mathcal{B}'}^{\mathcal{B}} \right)^{-1}.$$

The matrices $P_{\mathcal{B}'}^{\mathcal{B}}$ and $P_{\mathcal{C}'}^{\mathcal{C}}$ are invertible so that we may write (by Corollary 3.9) $P_{\mathcal{B}'}^{\mathcal{B}} = [h_1]_{\mathcal{B}}$ and $P_{\mathcal{C}'}^{\mathcal{C}} = [h_2]_{\mathcal{C}}$ where $h_1 \in \mathrm{End}_R(M)$ and $h_2 \in \mathrm{End}_R(N)$ are invertible. Thus Equation (3.5) gives

$$[f]_{\mathcal{C}}^{\mathcal{B}} = [h_2]_{\mathcal{C}} [g]_{\mathcal{C}}^{\mathcal{B}} \left([h_1]_{\mathcal{B}} \right)^{-1} = [h_2 g h_1^{-1}]_{\mathcal{C}}^{\mathcal{B}}.$$

Hence, $f = h_2 g h_1^{-1}$ and f and g are equivalent.

The converse statement is left as an exercise.   □

Using the invariant factor theorem for submodules (Theorem 3.6.23), it is possible to explicitly describe the equivalence classes under the equivalence relations of equivalence of homomorphisms and equivalence of matrices if one restricts the ring R to be a PID. This is the content of the following result.

(3.20) Proposition. *Let R be a PID and let $f : M \to N$ be an R-module homomorphism between a free R-module M of rank m and a free R-module N of rank n. Then there is a basis $\mathcal{B} = \{v_1, \ldots, v_m\}$ of M, a basis $\mathcal{C} = \{w_1, \ldots, w_n\}$ of N, and nonzero elements $s_1, \ldots, s_r$ of R, where $r = \operatorname{rank} \operatorname{Im}(f)$, such that $s_i \mid s_{i+1}$ for $1 \le i \le r - 1$ and such that*

$$(3.6) \qquad \begin{cases} f(v_i) = s_i w_i & \text{if } 1 \le i \le r, \\ f(v_i) = 0 & \text{if } i > r. \end{cases}$$

That is, the matrix of f with respect to the bases $\mathcal{B}$ and $\mathcal{C}$ is

$$(3.7) \qquad [f]_{\mathcal{C}}^{\mathcal{B}} = \begin{bmatrix} D_r & 0 \\ 0 & 0 \end{bmatrix}$$

where $D_r = \operatorname{diag}(s_1, \ldots, s_r)$.

Proof. By the invariant factor theorem for submodules (Theorem 3.6.23), there is a basis $\mathcal{C} = \{w_1, \ldots, w_n\}$ of N and elements $s_1, \ldots, s_r \in R$ such that $s_i \mid s_{i+1}$ for $1 \le i \le r - 1$ and $\{s_1 w_1, \ldots, s_r w_r\}$ is a basis for the submodule $\operatorname{Im}(f) \subseteq N$. Now choose any subset $\{v_1, \ldots, v_r\} \subseteq M$ such that $f(v_i) = s_i w_i$ for $1 \le i \le r$. By Proposition 3.8.8, $\operatorname{Ker}(f)$ is a submodule of M of rank $m - r$. Thus, we may choose a basis $\{v_{r+1}, \ldots, v_m\}$ of $\operatorname{Ker}(f)$.

Claim. $\mathcal{B} = \{v_1, \ldots, v_r, v_{r+1}, \ldots, v_m\} \subseteq M$ *is a basis of M.*

To verify the claim, suppose that $v \in M$. Then $f(v) \in \operatorname{Im}(f)$, so we may write

$$f(v) = \sum_{i=1}^{r} a_i f(v_i)$$

$$= \sum_{i=1}^{r} f(a_i v_i).$$

Therefore, $f(v - \sum_{i=1}^{r} a_i v_i) = 0$ so that $v - \sum_{i=1}^{r} a_i v_i \in \operatorname{Ker}(f)$, and hence, we may write

$$v - \sum_{i=1}^{r} a_i v_i = \sum_{i=r+1}^{m} a_i v_i.$$

It follows that $\mathcal{B}$ generates M as an R-module.

To check linear independence, suppose that $\sum_{i=1}^{m} a_i v_i = 0$. Then

$$0 = \sum_{i=1}^{m} a_i f(v_i) = \sum_{i=1}^{r} a_i(s_i w_i).$$

Since $\{s_1 w_1, \ldots, s_r w_r\}$ is linearly independent, this implies that $a_i = 0$ for $1 \leq i \leq r$. But then

$$\sum_{i=r+1}^{m} a_i v_i = 0,$$

and since $\{v_{r+1}, \ldots, v_m\}$ is a basis of $\text{Ker}(f)$, we conclude that $a_i = 0$ for all i, and the claim is verified.

It is clear from the construction that

$$[f]_{\mathcal{C}}^{\mathcal{B}} = \begin{bmatrix} D_r & 0 \\ 0 & 0 \end{bmatrix}$$

where $D_r = \text{diag}(s_1, \ldots, s_r)$. This completes the proof. $\square$

(3.21) Remark. In the case where the ring R is a field, the invariant factors are all 1. Therefore, if $f : M \to N$ is a linear transformation between finite-dimensional vector spaces, then there is a basis $\mathcal{B}$ of M and a basis $\mathcal{C}$ of N such that the matrix of f is

$$[f]_{\mathcal{C}}^{\mathcal{B}} = \begin{bmatrix} I_r & 0 \\ 0 & 0 \end{bmatrix}.$$

The number r is the dimension (= rank) of the subspace $\text{Im}(f)$.

(3.22) Corollary. *Let R be a PID and let M and N be free R-modules of rank m and n respectively. Then homomorphisms f and $g \in \text{Hom}_R(M, N)$ are equivalent if and only if the subspaces $\text{Im}(f)$ and $\text{Im}(g)$ have the same invariant factors as submodules of N. In particular, if R is a field, then f and g are equivalent if and only if*

$$\text{rank}(\text{Im}(f)) = \text{rank}(\text{Im}(g)).$$

Proof. Exercise. $\square$

If $M = N$, then Proposition 3.16 becomes the following result:

(3.23) Proposition. *Let $f \in \text{End}_R(M)$ and let $\mathcal{B}$, $\mathcal{B}'$ be two bases for the free R-module M. Then*

$$[f]_{\mathcal{B}'} = P_{\mathcal{B}'}^{\mathcal{B}} [f]_{\mathcal{B}} \left(P_{\mathcal{B}'}^{\mathcal{B}}\right)^{-1}.$$

Proof.   □

Proposition 3.23 applies to give a result analogous to Proposition 3.19 for similarity of matrices in $M_n(R)$. Recall (Definition 1.7) that two matrices A and B in $M_n(R)$ are similar if there is a matrix $S \in \mathrm{GL}(n, R)$ such that $B = S^{-1}AS$. For homomorphisms, the definition is the following:

(3.24) Definition. *Let R be a commuative ring and let M be a free R-module of rank n. If $f, g \in \mathrm{End}_R(M)$, then we say that f and g are* **similar** *if there is an invertible homomorphism $h \in \mathrm{End}_R(M)$ such that $g = h^{-1}fh$.*

In this situation, Proposition 3.19 becomes the following result:

(3.25) Corollary.

(1) *Two matrices $A, B \in M_n(R)$ are similar if and only if there are bases $\mathcal{B}$ and $\mathcal{B}'$ of a free R-module M of rank n and $f \in \mathrm{Hom}_R(M)$ such that $A = [f]_{\mathcal{B}}$ and $B = [f]_{\mathcal{B}'}$. That is, two $n \times n$ matrices are similar if and only if they represent the same R-module homomorphism with respect to different bases.*

(2) *Let M be a free R-module of rank n and let $f, g \in \mathrm{End}_R(M)$ be endomorphisms. Then f and g are similar if and only if there are bases $\mathcal{B}$ and $\mathcal{B}'$ of M such that*
$$[f]_{\mathcal{B}} = [g]_{\mathcal{B}'}.$$

That is, f is similar to g if and only if the two homomorphisms are represented by the same matrix with respect to appropriate bases.

Proof. Exercise.   □

(3.26) *Remark.* Let R be a commutative ring and T a set. A function $\phi : M_n(R) \to T$ will be called a **class function** if $\phi(A) = \phi(B)$ whenever A and B are similar matrices. If M is a free R-module of rank n, then the class function ϕ naturally yields a function $\widetilde{\phi} : \mathrm{End}_R(M) \to T$ defined by

$$\widetilde{\phi}(f) = \phi([f]_{\mathcal{B}})$$

where $\mathcal{B}$ is a basis of M. According to Corollary 3.25, the definition of $\widetilde{\phi}$ is independent of the choice of basis of M because ϕ is a class function. The most important class functions that we have met so far are the trace and the determinant (Lemma 1.6 (2) and Corollary 2.12). Thus, the trace and the determinant can be defined for any endomorphism of a free R-module of finite rank. We formally record this observation.

(3.27) Proposition. *Let R be a commutative ring and let M be a finite rank free R-module.*

(1) *There is an R-module homomorphism* $\mathrm{Tr} : \mathrm{End}_R(M) \to R$ *defined by*

$$\mathrm{Tr}(f) = \mathrm{Tr}\left([f]_{\mathcal{B}}\right)$$

where $\mathcal{B}$ is any basis of M. $\mathrm{Tr}(f)$ will be called the trace of the homomorphism f; it is independent of the choice of basis $\mathcal{B}$.

(2) *There is a multiplicative function* $\det : \mathrm{End}_R(M) \to R$ *defined by*

$$\det(f) = \det\left([f]_{\mathcal{B}}\right)$$

where $\mathcal{B}$ is any basis of M. $\det(f)$ will be called the determinant of the homomorphism f; it is independent of the choice of basis $\mathcal{B}$.

Proof. $\square$

Note that multiplicativity of the determinant means

$$\det(fg) = \det(f)\det(g).$$

Since $\det(1) = 1$, it follows that f is invertible if and only if $\det(f)$ is a unit in R.

Since similar matrices represent the same endomorphism with respect to different bases, one goal of linear algebra is to find a matrix B similar to a given matrix A such that B is as simple as possible. This is equivalent (by Corollary 3.25) to finding a basis of a free R-module so that the matrix $[f]_{\mathcal{B}}$ of a given homomorphism f is as simple as possible. When R is a field, this is the subject of canonical form theory that will be developed in detail in the next section. For now we will only indicate the relationship between direct sum decompositions of free R-modules and decompositions of matrices.

(3.28) Proposition. *Let R be a commutative ring, and let M_1, M_2, N_1, and N_2 be finite rank free R-modules. If $\mathcal{B}_i$ is a basis of M_i and $\mathcal{C}_i$ is a basis of N_i ($i = 1, 2$), then let $\mathcal{B}_1 \cup \mathcal{B}_2$ and $\mathcal{C}_1 \cup \mathcal{C}_2$ be the natural bases of $M_1 \oplus M_2$ and $N_1 \oplus N_2$ respectively (see Example 3.4.6 (7)). If $f_i \in \mathrm{Hom}_R(M_i, N_i)$ for $i = 1, 2$, then $f_1 \oplus f_2 \in \mathrm{Hom}_R(M_1 \oplus M_2, N_1 \oplus N_2)$ and*

$$[f_1 \oplus f_2]_{\mathcal{C}_1 \cup \mathcal{C}_2}^{\mathcal{B}_1 \cup \mathcal{B}_2} = [f_1]_{\mathcal{C}_1}^{\mathcal{B}_1} \oplus [f_2]_{\mathcal{C}_2}^{\mathcal{B}_2}.$$

Proof. Exercise. $\square$

We now specialize to the case of endomorphisms.

(3.29) Definition. *Let M be an R-module and let $f \in \mathrm{End}_R(M)$. A submodule $N \subseteq M$ is said to be **invariant** under f (or an **invariant submodule of** f) if $f(x) \in N$ whenever $x \in N$.*

(3.30) Proposition. *Let R be a commutative ring, let M be a free R module of rank m, and let $f \in \mathrm{Hom}_R(M)$. If $\mathcal{B} = \{v_1, \ldots, v_m\}$ is a basis of M then the matrix $[f]_{\mathcal{B}}$ has the block form*

$$[f]_{\mathcal{B}} = \begin{bmatrix} A & B \\ 0 & D \end{bmatrix}$$

where $A \in M_r(R)$ if and only if the submodule $N = \langle v_1, \ldots, v_r \rangle$ is an invariant submodule of f.

Proof. If $[f]_{\mathcal{B}} = [t_{ij}]$ then the block form means that $t_{ij} = 0$ for $r+1 \leq i \leq n$, $1 \leq j \leq r$. Thus, if $1 \leq j \leq r$ it follows that

$$f(v_j) = \sum_{i=1}^{n} t_{ij} v_i = \sum_{i=1}^{r} t_{ij} v_i \in N.$$

Since N is generated by $v_1, \ldots, v_r$, the result follows. $\square$

(3.31) *Remark.* As a special case of this result, a matrix $[f]_{\mathcal{B}}$ is upper triangular if and only if the submodule $\langle v_1, \ldots, v_k \rangle$ (where $\mathcal{B} = \{v_1, \ldots, v_m\}$) is invariant under f for every k ($1 \leq k \leq m$).

In Proposition 3.30, if the block $B = 0$, then not only is $\langle v_1, \ldots, v_k \rangle$ an invariant submodule, but the complementary submodule $\langle v_{k+1}, \ldots, v_m \rangle$ is also invariant under f. From this observation, extended to an arbitrary number of blocks, we conclude:

(3.32) Proposition. *Let M be a free R-module of rank m, let $f \in \mathrm{End}_R(M)$, and let $A = [f]_{\mathcal{B}} \in M_m(R)$. Then A is similar to a block diagonal matrix $B = A_1 \oplus \cdots \oplus A_k$ where $A_i \in M_{r_i}(R)$ ($r_1 + \cdots + r_k = n$) if and only if there are free submodules $M_1, \ldots, M_k$ of M such that*

(1) *M_i is an invariant submodule of f which is free of rank r_i, and*
(2) *$M \cong M_1 \oplus \cdots \oplus M_k$.*

Proof. $\square$

The case of this result when $r_i = 1$ for all i is of particular importance. In this case we are asking when A is similar to a diagonal matrix. To state the result in the manner we wish, it is convenient to make the following definition.

(3.33) Definition. *If M is a free R-module of rank m and $f \in \mathrm{End}_R(M)$, then a nonzero $x \in M$ is called an **eigenvector** of f if the cyclic submodule $\langle x \rangle$ of M is invariant under f. That is, $x \neq 0 \in M$ is an eigenvector of M if and only if $f(x) = \alpha x$ for some $\alpha \in R$. The element $\alpha \in R$ is called* *an* **eigenvalue** *of* f. *The* **eigenmodule** *(or* **eigenspace***) of* f *corresponding to the eigenvalue* α *is the submodule* $\mathrm{Ker}(f - \alpha 1_M)$.

If $A \in M_n(R)$ then by an eigenvalue or eigenvector of A, we mean an eigenvalue or eigenvector of the R-module homomorphism

$$T_A : M_{n,1}(R) \to M_{n,1}(R)$$

defined by $T_A(v) = Av$. We shall usually identify $M_{n,1}(R)$ with R^n via the standard basis $\{E_{i1} : 1 \le i \le n\}$ and speak of T_A as a map from R^n to R^n.

In practice, in studying endomorphisms of a free R-module, eigenvalues and eigenvectors play a key role (for the matrix of f depends on a choice of basis, while eigenvalues and eigenvectors are intrinsically defined). We shall consider them further in Section 4.4.

(3.34) Corollary. *Let* M *be a free* R-*module of rank* m, *let* $f \in \mathrm{End}_R(M)$, *and let* $A = [f]_B \in M_m(R)$. *Then* A *is similar to a diagonal matrix* $\mathrm{diag}(a_1, \dots, a_m)$ *if and only if there is a basis of* M *consisting of eigenvectors of* A.

Proof.   □

(3.35) Definition. *A matrix which is similar to a diagonal matrix is said to be* **diagonalizable**. *An endomorphism* $f \in \mathrm{End}_R(M)$ *is* **diagonalizable** *if* M *has a basis* B *such that* $[f]_B$ *is a diagonal matrix. A set* $S = \{f_i\}_{i \in I}$ *of endomorphisms of* M *is said to be* **simultaneously diagonalizable** *if* M *has a basis* B *such that* $[f_i]_B$ *is diagonal for all* $i \in I$.

The concept of diagonalizability of matrices with entries in a field will be studied in some detail in Section 4.4. For now we will conclude this section with the following result, which we shall need later.

(3.36) Theorem. *Let* R *be a* PID, *let* M *be a free* R-*module of finite rank* n, *and let* $S = \{f_i : M \to M\}_{i \in I}$ *be a set of diagonalizable* R-*module endomorphisms. Then the set* S *is simultaneously diagonalizable if and only if* S *consists of commuting endomorphisms, i.e., there is a basis* B *of* M *with* $[f_i]_B$ *diagonal for all* $i \in I$ *if and only if* $f_i f_j = f_j f_i$ *for all* i, j.

Proof. For simplicity, we assume that we are dealing with a pair of R-module endomorphisms $\{f, g\}$; the general case is no more difficult. First suppose that $B = \{v_1, \dots, v_n\}$ is a basis of M in which both $[f]_B$ and $[g]_B$ are diagonal. Then $f(v_i) = \lambda_i v_i$ and $g(v_i) = \mu_i v_i$ for $1 \le i \le n$. Then

$$\begin{aligned}
f(g(v_i)) = f(\mu_i v_i) &= \mu_i g(v_i) = \mu_i \lambda_i v_i \\
&= \lambda_i \mu_i v_i = \lambda_i g(v_i) = g(\lambda_i v_i) \\
&= g(f(v_i)),
\end{aligned}$$

so fg and gf agree on a basis of M and hence are equal.

Conversely, suppose that $fg = gf$ and let $\lambda_1, \ldots, \lambda_k$ be the *distinct* eigenvalues of f, and let $\mu_1, \ldots, \mu_\ell$ be the *distinct* eigenvalues of g. Let

$$(3.8) \qquad M_i = \operatorname{Ker}(f - \lambda_i 1_M) \qquad (1 \le i \le k)$$
$$(3.9) \qquad N_j = \operatorname{Ker}(g - \mu_j 1_M) \qquad (1 \le j \le \ell).$$

Then the hypothesis that f and g are diagonalizable implies that

$$(3.10) \qquad M \cong M_1 \oplus \cdots \oplus M_k$$

and

$$(3.11) \qquad M \cong N_1 \oplus \cdots \oplus N_\ell.$$

First, we observe that M_i is g-invariant and N_j is f-invariant. To see this, suppose that $v \in M_i$. Then

$$f(g(v)) = g(f(v)) = g(\lambda_i v) = \lambda_i g(v),$$

i.e., $g(v) \in \operatorname{Ker}(f - \lambda_i 1_M) = M_i$. The argument that N_j is f-invariant is the same.

Claim. $M_i = \oplus_{j=1}^{\ell} (M_i \cap N_j)$ *for* $1 \le i \le k$.

To prove this, let $v \in M_i$. Then, by Equation (3.11) we may uniquely write

$$(3.12) \qquad v = w_1 + \cdots + w_\ell$$

where $w_j \in N_j$. The claim will be proved once we show that $w_j \in M_i$ for all j. Since $v \in M_i$,

$$(3.13) \qquad \lambda_i v = f(v) = f(w_1) + \cdots + f(w_\ell).$$

But N_j is f-invariant, so $f(w_j) \in N_j$ for all j. But

$$(3.14) \qquad \lambda_i v = \lambda_i w_1 + \cdots + \lambda_i w_\ell.$$

Comparing Equations (3.13) and (3.14) and using Equation (3.11), we see that $\lambda_i w_j = f(w_j)$ for $1 \le j \le \ell$, i.e., $w_j \in M_i$ for all j, and the claim is proved.

To complete the proof of the theorem, note that since R is a PID, each of the submodules $M_i \cap N_j$ is a free R-module (Theorem 3.6.2), so let $\mathcal{B}_{ij} = \{v_{ij}\}$ be a basis of $M_i \cap N_j$. According to the claim, we have

$$M \cong \bigoplus_{i=1}^{k} \bigoplus_{j=1}^{\ell} (M_i \cap N_j)$$

so that $\mathcal{B} = \cup_{i,j} \mathcal{B}_{ij}$ is a basis of M consisting of common eigenvectors of f and g, i.e., $[f]_\mathcal{B}$ and $[g]_\mathcal{B}$ are both diagonal. $\qquad \square$

(3.37) *Remark.* The only place in the above proof where we used that R is a PID is to prove that the joint eigenspaces $M_i \cap N_j$ are free submodules of M. If R is an arbitrary commutative ring and f and g are commuting diagonalizable endomorphisms of a free R-module M, then the proof of Theorem 3.36 shows that the joint eigenspaces $M_i \cap N_j$ are **projective** R-modules. There is a basis of common eigenvectors if and only if these submodules are in fact free. We will show by example that this need not be the case.

Thus, let R be a commutative ring for which there exists a finitely generated projective R-module P which is not free (see Example 3.5.6 (3) or Theorem 3.5.11). Let Q be an R-module such that $P \oplus Q = F$ is a free R-module of finite rank n, and let

$$M = F \oplus F = P_1 \oplus Q_1 \oplus P_2 \oplus Q_2$$

where $P_1 = P_2 = P$ and $Q_1 = Q_2 = Q$. Then M is a free R-module of rank $2n$. Let $\lambda_1 \neq \lambda_2$ and $\mu_1 \neq \mu_2$ be elements of R and define $f, g \in \mathrm{End}_R(M)$ by

$$f(x_1, y_1, x_2, y_2) = (\lambda_1 x_1, \lambda_1 y_1, \lambda_2 x_2, \lambda_2 y_2)$$

and

$$g(x_1, y_1, x_2, y_2) = (\mu_1 x_1, \mu_2 y_1, \mu_2 x_2, \mu_1 y_2)$$

where $x_i \in P_i$ and $y_i \in Q_i$ for $i = 1, 2$. Then f is diagonalizable with eigenspaces

$$M_1 = \mathrm{Ker}(f - \lambda_1 1_M) = P_1 \oplus Q_1 \cong F$$
$$M_2 = \mathrm{Ker}(f - \lambda_2 1_M) = P_2 \oplus Q_2 \cong F$$

and g is diagonalizable with eigenspaces

$$N_1 = \mathrm{Ker}(g - \mu_1 1_M) = P_1 \oplus Q_2 \cong F$$
$$N_2 = \mathrm{Ker}(g - \mu_2 1_M) = Q_1 \oplus P_2 \cong F.$$

Moreover, $fg = gf$. However, there is no basis of common eigenvectors of f and g since the joint eigenspace $M_1 \cap N_1 = P_1$ is not free.

4.4 Canonical Form Theory

The goal of the current section is to apply the theory of finitely generated modules over a PID to the study of a linear transformation T from a finite-dimensional vector space V to itself. The emphasis will be on showing how this theory allows one to find a basis of V with respect to which the matrix representation of T is as simple as possible. According to the properties of matrix representation of homomorphisms developed in Section 4.3, this is equivalent to the problem of finding a matrix B which is similar to a given matrix A, such that B has a form as simple as possible. In other words, we are looking for simply described representatives of each equivalence class of matrices under the equivalence relation of similarity.

We will start by carefully defining the module structures that are determined by linear transformations. This has already been mentioned in Example 3.1.5 (12), but we will repeat it here because of the fundamental importance of this construction.

Let F be a field, let V be a finite-dimensional vector space over F (i.e., V is a free F-module of finite rank), and let $T : V \to V$ be a linear transformation (i.e., T is an F-module homomorphism). Let $R = F[X]$ be the polynomial ring with coefficients from F. Recall (Theorem 2.4.12) that R is a principal ideal domain. Let

$$\phi : R \to \operatorname{Hom}_F(V)$$

be the F-algebra homomorphism determined by $\phi(X) = T$ (see Section 2.4). To be explicit, if $f(X) = a_0 + a_1 X + \cdots + a_n X^n$ then

$$(4.1) \qquad \phi(f(X)) = a_0 1_V + a_1 T + \cdots + a_n T^n.$$

Then V becomes an R-module via the scalar multiplication

$$(4.2) \qquad (f(X)) \cdot v = \phi(f(X))(v)$$

for each $f(X) \in R = F[X]$ and $v \in V$. Combining Equations (4.1) and (4.2) we see that the R-module structure on V determined by T is given by

$$(4.3) \qquad (f(X)) v = a_0 v + a_1 T(v) + \cdots + a_n T^n(v).$$

Note that each $T \in \operatorname{End}_F(V)$ will induce a different R-module structure on the same abelian group V. To distinguish these different module structures, we will write V_T for the vector space V with the R-module structure described by Equation (4.3). When there is no chance of confusion, we will sometimes write V for the R-module V_T. Again we note that the module V_T has been previously introduced in Example 3.1.5 (12).

Note that scalar multiplication of a vector $v \in V_T$ by the constant polynomial $a_0 \in F[X]$ is the same as the scalar multiplication $a_0 v$, where a_0 is considered as an element of the field F and $v \in V$. This is an immediate observation based on Equation (4.3). It is also worth pointing out explicitly that an R-submodule N of the R-module V_T is just a subspace of V that is T-invariant. Recall (Definition 3.29 and also Example 3.2.2 (4)) that this means that $T(v) \in N$ for all $v \in N$.

We will begin our study by computing the R-module homomorphisms between two R-modules V_T and W_S and by relating this computation to the similarity of linear transformations.

(4.1) Proposition. *Let V and W be vector spaces over the field F, and suppose that $T \in \operatorname{End}_F(V)$, $S \in \operatorname{End}_F(W)$. Then*

(4.4) $\mathrm{Hom}_{F[X]}(V_T, W_S) = \{U \in \mathrm{Hom}_F(V, W) : UT = SU\}$.

Proof. Suppose $U \in \mathrm{Hom}_{F[X]}(V_T, W_S)$. As we observed above, the $F[X]$-module action on V and W reduces to the F-module action (i.e., scalar multiplication). Thus $U \in \mathrm{Hom}_F(V, W)$. Let $v \in V$ and $w \in W$. Then $X \cdot v = T(v)$ and $X \cdot w = S(w)$. Then, since U is an $F[X]$- module homomorphism, we have

$$U(T(v)) = U(X \cdot v) = X \cdot U(v) = S(U(v)).$$

Since $v \in V$ is arbitrary, we conclude that $UT = SU$.

Conversely, suppose that $U : V \to W$ is a linear transformation such that $UT = SU$. We claim that

(4.5) $U(f(X) \cdot v) = f(X) \cdot U(v)$

for all $v \in V$ and $f(X) \in F[X]$. But Equation (4.5) is satisfied for polynomials of degree 0 since U is a linear transformation, and it is satisfied for $f(X) = X$ since

$$U(X \cdot v) = U(T(v)) = S(U(v)) = X \cdot U(v).$$

Since $F[X]$ is generated by the constant polynomials and X, it follows that Equation (4.5) is satisfied for all polynomials $f(X)$, and hence, $U \in \mathrm{Hom}_{F[X]}(V_T, W_S)$. $\square$

(4.2) Theorem. *Let V be a vector space over the field F, and let T_1, $T_2 \in \mathrm{End}_F(V)$. Then the R-modules V_{T_1} and V_{T_2} are isomorphic if and only if T_1 and T_2 are similar.*

Proof. By Proposition 4.1, an R-module isomorphism (recall $R = F[X]$)

$$P : V_{T_2} \to V_{T_1}$$

consists of an invertible linear transformation $P : V \to V$ such that $PT_2 = T_1 P$, i.e., $T_1 = PT_2P^{-1}$. Thus V_{T_1} and V_{T_2} are isomorphic (as R-modules) if and only if the linear transformations T_1 and T_2 are similar. Moreover, we have seen that the similarity transformation P produces the R-module isomorphism V_{T_2} to V_{T_1}. $\square$

This theorem, together with Corollary 3.25, gives the theoretical underpinning for our approach in this section. We will be studying linear transformations T by studying the R-modules V_T, so Theorem 4.2 says that, on the one hand, similar transformations are indistinguishable from this point of view, and on the other hand, any result, property, or invariant we derive in this manner for a linear transformation T holds for any transformation similar to T. Let us fix T. Then by Corollary 3.25, as we vary the basis $\mathcal{B}$ of V, we obtain similar matrices $[T]_{\mathcal{B}}$. Our objective will be to find bases in which the matrix of T is particularly simple, and hence the structure and properties of T are particularly easy to understand.

(4.3) Proposition. *Let V be a vector space over the field F and suppose that $\dim_F(V) = n < \infty$. If $T \in \text{End}_F(V)$, then the R-module ($R = F[X]$) V_T is a finitely generated torsion R-module.*

Proof. Since the action of constant polynomials on elements of V_T is just the scalar multiplication on V determined by F, it follows that any F-generating set of V is a priori an R-generating set for V_T. Thus $\mu(V_T) \leq n = \dim_F(V)$. (Recall that $\mu(M)$ (Definition 3.2.9) denotes the minimum number of generators of the R-module M.)

Let $v \in V$. We need to show that $\text{Ann}(v) \neq \langle 0 \rangle$. Consider the elements $v, T(v), \ldots, T^n(v) \in V$. These are $n+1$ elements in an n-dimensional vector space V, and hence they must be linearly dependent. Therefore, there are scalars $a_0, a_1, \ldots, a_n \in F$, not all zero, such that

$$(4.6) \qquad a_0 v + a_1 T(v) + \cdots + a_n T^n(v) = 0.$$

If we let $f(X) = a_0 + a_1 X + \cdots + a_n X^n$, then Equation (4.6) and the definition of the R-module structure on V_T (Equation (4.3)) shows that $f(X)v = 0$, i.e., $f(X) \in \text{Ann}(v)$. Since $f(X) \neq 0$, this shows that $\text{Ann}(v) \neq \langle 0 \rangle$. $\square$

This innocent looking proposition has far reaching consequences, for it means that we may apply our results on the structure of finitely generated torsion modules over a PID R to the study of V_T. Henceforth, we will fix a finite-dimensional vector space V over F and $T \in \text{End}_F(V)$. We begin with the following observation.

(4.4) Corollary. *There is a polynomial $f(X) \in F[X]$ of degree at most n with $f(T) = 0$.*

Proof. By Theorem 3.7.1, $\text{Ann}(V_T) = \text{Ann}(v)$ for some $v \in V_T$. But the proof of Proposition 4.3 shows that $\text{Ann}(v)$ contains a polynomial $f(X)$ of degree at most n. Thus $f(X) \in \text{Ann}(V_T)$ so that $f(X)w = 0$ for all $w \in V_T$, i.e., $f(T)(w) = 0$ for all $w \in V$. Hence $f(T) = 0$ as required. $\square$

(4.5) Remark. It is worth pointing out that the proofs of Proposition 4.3 and Corollary 4.4 show that a polynomial $g(X)$ is in $\text{Ann}(V_T)$ if and only if $g(T) = 0 \in \text{End}_F(V)$.

In Section 3.7 we had two decompositions of finitely generated torsion R-modules, namely, the cyclic decomposition (Theorem 3.7.1) and the cyclic primary decomposition (Theorem 3.7.13). Each of these decompositions will produce a canonical form for T, namely, the first will produce the rational canonical form and the second will give the Jordan canonical form. Moreover, Theorem 3.7.12 applied to the R-module V_T produces an important direct sum decomposition, which we will refer to as the primary decomposition theorem for the linear transformation T. We will begin by studying the cyclic decomposition of V_T.

According to Theorem 3.7.1, the torsion R-module V_T can be written as a direct sum of $k = \mu(V_T)$ cyclic R-submodules

$$(4.7) \qquad V_T \cong Rv_1 \oplus \cdots \oplus Rv_k$$

such that $\text{Ann}(v_i) = \langle f_i(X) \rangle$ for $1 \le i \le k$ and

$$(4.8) \qquad \langle f_1(X) \rangle \supseteq \langle f_2(X) \rangle \supseteq \cdots \supseteq \langle f_k(X) \rangle.$$

Equation (4.8) is equivalent to the condition

$$(4.9) \qquad f_i(X) \mid f_{i+1}(X) \qquad \text{for} \quad 1 \le i < k.$$

Theorem 3.7.3 shows that the ideals $\langle f_i(X) \rangle$ for $1 \le i \le k$ are uniquely determined by the R-module V_T (although the generators $\{v_1, \ldots, v_k\}$ are not uniquely determined), and since $R = F[X]$, we know that every ideal contains a unique monic generator. Thus we shall always suppose that $f_i(X)$ has been chosen to be monic.

(4.6) Definition.

(1) *The monic polynomials $f_1(X), \ldots, f_k(X)$ in Equation (4.8) are called the* **invariant factors** *of the linear transformation T.*
(2) *The invariant factor $f_k(X)$ of T is called the* **minimal polynomial** $m_T(X)$ *of T.*
(3) *The* **characteristic polynomial** $c_T(X)$ *of T is the product of all the invariant factors of T, i.e.,*

$$(4.10) \qquad c_T(X) = f_1(X) f_2(X) \cdots f_k(X).$$

(4.7) *Remark.* In the language of Definition 3.7.8, we have

$$(4.11) \qquad m_T(X) = \text{me}(V_T) \qquad \text{and} \qquad c_T(X) = \text{co}(V_T).$$

(4.8) Lemma. $\text{Ann}(V_T) = \langle m_T(X) \rangle$.

Proof. This is immediate from Equation (3.7.1). $\qquad\qquad\qquad\qquad\square$

(4.9) Corollary. $m_T(X)$ *is the unique monic polynomial of lowest degree with*

$$m_T(T) = 0.$$

Proof. An ideal I of $F[X]$ is generated by a polynomial of lowest degree in I. Apply this observation to $\mathrm{Ann}(V_T)$, recalling the description of $\mathrm{Ann}(V_T)$ in Remark 4.5. $\qquad\square$

(4.10) Corollary.

(1) *If $q(X) \in F[X]$ is any polynomial with $q(T) = 0$, then*

$$m_T(X) \mid q(X).$$

(2) $m_T(X)$ *divides* $c_T(X)$.
(3) *If $p(X)$ is any irreducible polynomial dividing $c_T(X)$, then $p(X)$ divides $m_T(X)$.*

Proof. This is a special case of Corollary 3.7.9. $\qquad\square$

(4.11) Lemma.

(1) *If V_T is cyclic, say $V_T = R/\langle f(X) \rangle$, then*

$$\dim_F(V) = \deg(f(X)).$$

(2) *If $V_T \cong Rv_1 \oplus \cdots \oplus Rv_k$ where $\mathrm{Ann}(v_i) = \langle f_i(X) \rangle$ as in Equation (4.7), then*

$$(4.12) \qquad \sum_{i=1}^{k} \deg(f_i(X)) = \dim(V) = \deg(c_T(X)).$$

(3) *The following are equivalent:*

   (a) V_T *is cyclic.*
   (b) $\deg(m_T(X)) = \dim(V)$.
   (c) $m_T(X) = c_T(X)$.

Proof. (1) Suppose that $V_T = Rv$. Then the map $\eta : R \to V_T$ defined by

$$\eta(q(X)) = q(T)(v)$$

is surjective and $v_T \cong R/\mathrm{Ker}(\eta)$ as F-modules. But $\mathrm{Ker}(\eta) = \langle f(X) \rangle$, and as F-modules, $R/\langle f(X) \rangle$ has a basis $\{1, X, \dots, X^{n-1}\}$ where $n = \deg(f(X))$. Thus, $\dim_F(V) = n = \deg(f(X))$.
    (2) and (3) are immediate from (1) and the definitions. $\qquad\square$

(4.12) Definition. *Let $f(X) = X^n + a_{n-1}X^{n-1} + \cdots + a_1 X + a_0 \in F[X]$ be a monic polynomial. Then the* **companion matrix** *$C(f(X)) \in M_n(F)$ of $f(X)$ is the $n \times n$ matrix*

$$(4.13) \qquad C(f(X)) = \begin{bmatrix} 0 & 0 & \cdots & 0 & 0 & -a_0 \\ 1 & 0 & \cdots & 0 & 0 & -a_1 \\ 0 & 1 & \cdots & 0 & 0 & -a_2 \\ \vdots & \vdots & \ddots & \vdots & \vdots & \vdots \\ 0 & 0 & \cdots & 1 & 0 & -a_{n-2} \\ 0 & 0 & \cdots & 0 & 1 & -a_{n-1} \end{bmatrix}.$$

(4.13) Examples.

(1) For each $\lambda \in F$, $C(X - \lambda) = [\lambda] \in M_1(F)$.

(2) $\mathrm{diag}(a_1, \ldots, a_n) = \oplus_{i=1}^n C(X - a_i)$.

(3) $C(X^2 + 1) = \begin{bmatrix} 0 & -1 \\ 1 & 0 \end{bmatrix}$.

(4) If $A = C(X - a) \oplus C(X^2 - 1)$, then

$$A = \begin{bmatrix} a & 0 & 0 \\ 0 & 0 & 1 \\ 0 & 1 & 0 \end{bmatrix}.$$

(4.14) Proposition. *Let $f(X) \in F[X]$ be a monic polynomial of degree n, and let $T : F^n \to F^n$ be defined by multiplication by $A = C(f(X))$, i.e., $T(v) = Av$ where we have identified F^n with $M_{n,1}(F)$. Then $m_T(X) = f(X)$.*

Proof. Let $e_j = \mathrm{col}_j(I_n)$. Then from the definition of $A = C(f(X))$ (Equation (4.13)), we see that $T(e_1) = e_2$, $T(e_2) = e_3$, ..., $T(e_{n-1}) = e_n$. Therefore, $T^r(e_1) = e_{r+1}$ for $0 \le r \le n - 1$ so that $\{e_1, T(e_1), \ldots, T^{n-1}(e_1)\}$ is a basis of F^n and hence $(F^n)_T$ is a cyclic R-module generated by e_1. Thus, by Lemma 4.11, $\deg(m_T(X)) = n$ and

$$\langle m_T(X) \rangle = \mathrm{Ann}((F^n)_T) = \mathrm{Ann}(e_1).$$

But

$$T^n(e_1) = -a_0 e_1 - a_1 T(e_1) - \cdots - a_{n-1} T^{n-1}(e_1),$$

i.e.,

$$T^n(e_1) + a_{n-1} T^{n-1}(e_1) + \cdots + a_1 T(e_1) + a_0 e_1 = 0.$$

Therefore, $f(T)(e_1) = 0$ so that $f(X) \in \mathrm{Ann}(e_1) = \langle m_T(X) \rangle$. But $\deg(f(X)) = \deg(m_T(X))$ and both polynomials are monic, so $f(X) = m_T(X)$. $\qquad \square$

(4.15) Corollary. *In the situation of Proposition 4.14, let*

$$\mathcal{B} = \{v, T(v), \dots, T^{n-1}(v)\}$$

where $v = e_1 = \mathrm{col}_1(I_n)$. Then $\mathcal{B}$ is a basis of F^n and $[T]_{\mathcal{B}} = C(f(X))$.

Proof. This is clear from the calculations in the proof of Proposition 4.14.
□

(4.16) Corollary. *Let V be any finite-dimensional vector space over the field F, let $T \in \mathrm{End}_F(V)$, and suppose that the R-module V_T is cyclic with $\mathrm{Ann}(V_T) = \langle f(X) \rangle$. Then there is a basis $\mathcal{B}$ of V such that*

$$[T]_{\mathcal{B}} = C(f(X)).$$

Proof. If $V_T = Rv$ then we may take

$$\mathcal{B} = \{v, T(v), \dots, T^{n-1}(v)\}$$

where $n = \dim(V)$.
□

(4.17) Theorem. (Rational canonical form) *Let V be a vector space of dimension n over a field F and let $T \in \mathrm{End}_F(V)$ be a linear transformation. If $\{f_1(X), \dots, f_k(X)\}$ is the set of invariant factors of the $F[X]$-module V_T, then V has a basis $\mathcal{B}$ such that*

$$(4.14) \qquad [T]_{\mathcal{B}} = C(f_1(X)) \oplus C(f_2(X)) \oplus \cdots \oplus C(f_k(X)).$$

Proof. Let $V_T \cong Rv_1 \oplus \cdots \oplus Rv_k$ where $R = F[X]$ and $\mathrm{Ann}(v_i) = \langle f_i(X) \rangle$ and where $f_i(X) \mid f_{i+1}(X)$ for $1 \le i < k$. Let $\deg(f_i) = n_i$. Then $\mathcal{B}_i = \{v_i, T(v_i), \dots, T^{n_i-1}(v_i)\}$ is a basis of the cyclic submodule Rv_i. Since submodules of V_T are precisely the T-invariant subspaces of V, it follows that $T|_{Rv_i} \in \mathrm{End}_F(Rv_i)$ and Corollary 4.16 applies to give

$$(4.15) \qquad\qquad [T|_{Rv_i}]_{\mathcal{B}_i} = C(f_i(X)).$$

By Equation (4.12), $n_1 + \cdots + n_k = n$, and hence, $\mathcal{B} = \cup_{i=1}^{k} \mathcal{B}_i$ is a basis of V and Proposition 3.32 and Equation (4.15) apply to give Equation (4.14).
□

(4.18) Corollary. *Two linear transformations T_1 and T_2 on V have the same rational canonical form if and only if they are similar.*

Proof. Two linear transformations T_1 and T_2 have the same rational canonical form if and only if they have the same invariant factors, which occurs if and only if the R-modules V_{T_1} and V_{T_2} are isomorphic. Now apply Theorem 4.2.
□

(4.19) Corollary. *Every matrix $A \in M_n(F)$ is similar to a unique matrix in rational canonical form.*

Proof. Regard $A \in M_n(F)$ as defining a linear transformation $T_A :$ $F^n \to F^n$ by $T_A(v) = Av$. Then A is similar to $B \in M_n(F)$ as matrices if and only if T_A is similar to T_B as elements of $\mathrm{End}_F(F^n)$. Thus we may apply Theorem 4.17 and Corollary 4.18.   ∎

We now pause in our general development in order to see how to compute $c_T(X)$ and to prove a famous result. We will need the following simple lemma:

(4.20) Lemma. *Let F be a field and let $f(X) \in F[X]$ be a monic polynomial of degree n. Then the matrix $XI_n - C(f(X)) \in M_n(F[X])$ and*

(4.16) $$\det (XI_n - C(f(X))) = f(X).$$

Proof. Let $f(X) = X^n + a_{n-1}X^{n-1} + \cdots + a_1 X + a_0 \in F[X]$. The proof is by induction on $n = \deg(f)$. If $n = 1$, the result is clear. Now suppose that $n > 1$, and compute the determinant of Equation (4.16) by cofactor expansion along the first row; applying the induction hypothesis to the first summand.

$$\det(XI_n - C(f(X))) = \det \begin{bmatrix} X & 0 & \cdots & 0 & 0 & a_0 \\ -1 & X & \cdots & 0 & 0 & a_1 \\ 0 & -1 & \cdots & 0 & 0 & a_2 \\ \vdots & \vdots & \ddots & \vdots & \vdots & \vdots \\ 0 & 0 & \cdots & -1 & X & a_{n-2} \\ 0 & 0 & \cdots & 0 & -1 & X + a_{n-1} \end{bmatrix}$$

$$= X \det \begin{bmatrix} X & \cdots & 0 & 0 & a_1 \\ -1 & \cdots & 0 & 0 & a_2 \\ \vdots & \ddots & \vdots & \vdots & \vdots \\ 0 & \cdots & -1 & X & a_{n-2} \\ 0 & \cdots & 0 & -1 & X + a_{n-1} \end{bmatrix}$$

$$+ a_0 (-1)^{n+1} \det \begin{bmatrix} -1 & X & \cdots & 0 & 0 \\ 0 & -1 & \cdots & 0 & 0 \\ \vdots & \vdots & \ddots & \vdots & \vdots \\ 0 & 0 & \cdots & -1 & X \\ 0 & 0 & \cdots & 0 & -1 \end{bmatrix}$$

$$= X \left(X^{n-1} + a_{n-1}X^{n-2} + \cdots + a_1 \right)$$
$$+ a_0 (-1)^{n+1} (-1)^{n-1}$$
$$= X^n + a_{n-1}X^{n-1} + \cdots + a_1 X + a_0$$

$$= f(X),$$

and the lemma is proved. □

(4.21) Definition. If $A \in M_n(F)$ then we will denote the polynomial $\det(XI_n - A) \in F[X]$ by $c_A(X)$ and we will call $c_A(X)$ the **characteristic polynomial** of the matrix A.

We will prove that the characteristic polynomial of a linear transformation T (as defined in Definition 4.6 (3)) is the characteristic polynomial of any matrix representation of T.

(4.22) Lemma. If A and $B \in M_n(F)$ are similar, then $c_A(X) = c_B(X)$.

Proof. Suppose that $B = P^{-1}AP$ for some $P \in \mathrm{GL}(n, F)$. Then $\det P \neq 0$ and $(\det P)^{-1} = \det P^{-1}$. Hence,

$$
\begin{aligned}
c_B(X) &= \det(XI_n - B) \\
&= \det(XI_n - P^{-1}AP) \\
&= \det(P^{-1}(XI_n - A)P) \\
&= (\det P^{-1})(\det(XI_n - A))(\det P) \\
&= \det(XI_n - A) \\
&= c_A(X).
\end{aligned}
$$

□

(4.23) Proposition. Let V be a finite-dimensional vector space over a field F and let $T \in \mathrm{End}_F(V)$. If $\mathcal{B}$ is any basis of V, then

$$(4.17) \qquad c_T(X) = c_{[T]_{\mathcal{B}}}(X) = \det(XI_n - [T]_{\mathcal{B}}).$$

Proof. By Lemma 4.22, if Equation (4.17) is true for one basis, it is true for any basis. Thus, we may choose the basis $\mathcal{B}$ so that $[T]_{\mathcal{B}}$ is in rational canonical form, i.e.,

$$(4.18) \qquad [T]_{\mathcal{B}} = \bigoplus_{i=1}^{k} C(f_i(X))$$

where $f_1(X), \ldots, f_k(X)$ are the invariant factors of T. If $\deg(f_i(X)) = n_i$, then Equation (4.18) gives

$$(4.19) \qquad XI_n - [T]_{\mathcal{B}} = \bigoplus_{i=1}^{k}(XI_{n_i} - C(f_i(X))).$$

Equation (4.19) and Theorem 2.11 imply

$$c_{[T]_\mathcal{B}}(X) = \det(XI_n - [T]_\mathcal{B})$$

$$= \prod_{i=1}^{k} \det(XI_{n_i} - C(f_i(X)))$$

$$= \prod_{i=1}^{k} f_i(X) \qquad \text{by Lemma 4.20}$$

$$= c_T(X),$$

which gives Equation (4.17). $\qquad\qquad\qquad\qquad\qquad\qquad\qquad\qquad\square$

(4.24) Corollary. (Cayley–Hamilton theorem) *Let $T \in \mathrm{End}_F(V)$ be any linear transformation on a finite-dimensional vector space V and let $\mathcal{B}$ be any basis of V. Let $A = [T]_\mathcal{B} \in F[X]$. Then*

$$c_A(T) = 0.$$

Proof. By Proposition 4.23, $c_A(X) = c_T(X)$ and $m_T(X) \mid c_T(X)$. Since $m_T(T) = 0$, it follows that $c_T(T) = 0$, i.e., $c_A(T) = 0$. $\qquad\qquad\square$

(4.25) *Remark.* The Cayley–Hamilton theorem is often phrased as *A linear transformation satisfies its characteristic polynomial.* From our perspective, the fact that $c_T(T) = 0$ is a triviality, but it is a nontrivial result that $c_T(X) = c_A(X)$ where $A = [T]_\mathcal{B}$. However, there is an alternate approach in which the characteristic polynomial of a linear transformation is defined to be $c_A(X)$ where A is some matrix representation of T. From this perspective, the Cayley–Hamilton theorem becomes a nontrivial result. It is worth pointing out that the Cayley–Hamilton theorem is valid for matrices with entries in any commutative ring; of course, the invariant factor theory is not valid for general rings so a different proof is needed (see Exercise 56). In fact, we shall sketch a second proof of the Cayley–Hamilton theorem in the exercises which is valid for any commutative ring R. From the point of view of the current section, the utility of Proposition 4.23 and Corollary 4.24 is that we have an independent method of calculating the characteristic polynomial. Further techniques for computing the invariant factors of a given linear transformation will be presented in Chapter 5.

(4.26) Example. Let V be an n-dimensional vector space over the field F, and if $\lambda \in F$, define a linear transformation $T : V \to V$ by $T(v) = \lambda v$ for all $v \in V$. Then, considering the $F[X]$-module V_T, we have $Xv = \lambda v$, i.e., $(X - \lambda)v = 0$ for every $v \in V$. Thus $X - \lambda = m_T(X)$ since $X - \lambda$ is the monic polynomial of lowest possible degree (namely, degree 1) with $m_T(T) = 0$. Then $c_T(X) = (X - \lambda)^n$ since $\deg(c_T(X)) = n$ and the only prime factor of $c_T(X)$ is $(X - \lambda)$ by Corollary 4.10 (3). Then the rational canonical form of T is

$$\bigoplus_{i=1}^{n} C(X - \lambda) = \lambda I_n.$$

Of course, this is the matrix of T in *any* basis of V. Also note that $V_T \cong Rv_1 \oplus \cdots \oplus Rv_n$ has rank n over R with each cyclic submodule $Rv_i \cong R/\langle X - \lambda \rangle$ having dimension 1 over the field F.

(4.27) Example. Let $\mathcal{B} = \{v_1, \ldots, v_n\}$ be a basis of the n-dimensional vector space V, and let $T : V \to V$ be defined by $T(v_i) = \lambda_i v_i$ where $\lambda_i \in F$. Assume that the λ_i are all distinct, i.e., $\lambda_i \neq \lambda_j$ for $i \neq j$. Then each subspace $\langle v_i \rangle$ is a T-invariant subspace and hence a submodule of the $F[X]$-module V_T. Therefore,

$$V_T \cong Rv_1 \oplus \cdots \oplus Rv_n$$

where $\mathrm{Ann}(v_i) = \langle X - \lambda_i \rangle$. Note that

$$\mathrm{me}(Rv_i) = \mathrm{Ann}(v_i) = \langle X - \lambda_i \rangle,$$

so Proposition 3.7.21 implies that

$$\begin{aligned}
m_T(X) &= \mathrm{me}(V_T) \\
&= \mathrm{lcm}\{\mathrm{me}(Rv_1), \ldots, \mathrm{me}(Rv_n)\} \\
&= \prod_{i=1}^{n}(X - \lambda_i) \\
&= f(X).
\end{aligned}$$

Also by Proposition 3.7.21, we see that $c_T(X) = f(X)$. Therefore $m_T(X) = c_T(X)$ and Lemma 4.11 (3) shows that the R-module V_T is cyclic with annihilator $\langle f(X) \rangle$. Thus the rational canonical form of T is

$$[T]_{\mathcal{B}_0} = C(f(X)) = \begin{bmatrix}
0 & 0 & \cdots & 0 & 0 & -a_0 \\
1 & 0 & \cdots & 0 & 0 & -a_1 \\
0 & 1 & \cdots & 0 & 0 & -a_2 \\
\vdots & \vdots & \ddots & \vdots & \vdots & \vdots \\
0 & 0 & \cdots & 1 & 0 & -a_{n-2} \\
0 & 0 & \cdots & 0 & 1 & -a_{n-1}
\end{bmatrix}$$

where $f(X) = X^n + a_{n-1}X^{n-1} + \cdots + a_1 X + a_0$ and the basis $\mathcal{B}_0$ is chosen appropriately.

This example actually illustrates a defect of the rational canonical form. Note that $[T]_{\mathcal{B}} = \mathrm{diag}(\lambda_1, \lambda_2, \ldots, \lambda_n)$, which is a diagonal matrix. By comparison, we see that $[T]_{\mathcal{B}}$ is much simpler than $[T]_{\mathcal{B}_0}$ and it reflects the geometry of the linear transformation much more clearly. Our next goal is to find a canonical form that is as "simple" as possible, the Jordan canonical form. When a transformation has a diagonal matrix in some basis, this will indeed be its Jordan canonical form. This special case is important enough to investigate first.

(4.28) Definition.

(1) *A linear transformation* $T : V \to V$ *is* **diagonalizable** *if* V *has a basis such that* $[T]_\mathcal{B}$ *is a diagonal matrix.*

(2) *A matrix* $A \in M_n(F)$ *is* **diagonalizable** *if it is similar to a diagonal matrix.*

(4.29) *Remark.* Recall that we have already introduced the concept of diagonalizability in Definition 3.35. Corollary 3.34 states that T is diagonalizable if and only if V posseses a basis of eigenvectors of T. Recall that $v \neq 0 \in V$ is an **eigenvector** of T if the subspace $\langle v \rangle$ is T-invariant, i.e., $T(v) = \lambda v$ for some $\lambda \in F$. The element $\lambda \in F$ is an **eigenvalue** of T. We will consider criteria for diagonalizability of a linear transformation based on properties of the invariant factors.

(4.30) Theorem. *Let* $T : V \to V$ *be a linear transformation. Then* T *is diagonalizable if and only if* $m_T(X)$ *is a product of distinct linear factors, i.e.,*

$$m_T(X) = \prod_{i=1}^{t}(X - \lambda_i)$$

where $\lambda_1, \ldots, \lambda_t$ *are distinct elements of the field* F.

Proof. Suppose that T is diagonalizable. Then there is a basis $\mathcal{B}$ of V such that $[T]_\mathcal{B} = \text{diag}(\alpha_1, \ldots, \alpha_n)$. By reordering the basis $\mathcal{B}$, if necessary, we can assume that the diagonal entries that are equal are grouped together. That is,

$$[T]_\mathcal{B} = \bigoplus_{i=1}^{t} \lambda_i I_{n_i}$$

where $n = n_1 + \cdots + n_t$ and the λ_i are distinct. If

$$\mathcal{B} = \{v_{11}, \ldots, v_{1n_1}, v_{21}, \ldots, v_{2n_2}, \ldots, v_{t1}, \ldots, v_{tn_t}\},$$

then let $\mathcal{B}_i = \{v_{i1}, \ldots, v_{in_i}\}$ and let $V_i = \langle \mathcal{B}_i \rangle$. Then $T(v) = \lambda_i v$ for all $v \in V_i$, so V_i is a T-invariant subspace of V and hence an $F[X]$-submodule of V_T. From Example 4.26, we see that $\text{me}(V_i) = X - \lambda_i$, and, as in Example 4.27,

$$m_T(X) = \text{me}(V_T) = \prod_{i=1}^{t}(X - \lambda_i)$$

as claimed.

Conversely, suppose that $m_T(X) = \prod_{i=1}^{t}(X - \lambda_i)$, where the λ_i are distinct. Since the $X - \lambda_i$ are distinct irreducible polynomials, Theorem 3.7.13 applied to the torsion $F[X]$-module provides a direct sum decomposition

$$(4.20) \qquad V_T \cong V_1 \oplus \cdots \oplus V_t$$

where $\text{Ann}(V_i) = \langle X - \lambda_i \rangle$. In other words, V_i is a T-invariant subspace of V and $T_i = T|_{V_i}$ satisfies $T_i - \lambda_i = 0$, i.e., $T(v) = \lambda_i v$ for all $v \in V_i$. Then, by Example 4.26, if V_i has a basis $\mathcal{B}_i = \{v_{i1}, \dots, v_{in_i}\}$, then $[T_i]_{\mathcal{B}_i} = \lambda_i I_{n_i}$, and if $\mathcal{B} = \cup_{i=1}^{t} \mathcal{B}_i$, then $\mathcal{B}$ is a basis of V by Equation (4.20) and $[T]_{\mathcal{B}} = \oplus_{i=1}^{t} \lambda_i I_{n_i}$, so T is diagonalizable. $\qquad \square$

(4.31) Corollary. *Let $T \in \text{End}_F(V)$ be diagonalizable. Then the exponent of $(X - \lambda_i)$ in the characteristic polynomial $c_T(X)$ is equal to the number of times that λ_i appears on the diagonal in any diagonal matrix $[T]_{\mathcal{B}}$.*

Proof. If λ_i appears n_i times on the diagonal in a diagonal matrix representation of T, then

$$n_i = \dim_F \text{Ker}(T - \lambda_i 1_V).$$

This number depends only on T and not on any particular diagonalization of T. Now suppose that $[T]_{\mathcal{B}} = \oplus_{i=1}^{t} \lambda_i I_{n_i}$, let $V_i = \text{Ker}(T - \lambda_i 1_V)$, and let $T_i = T|_{V_i}$ as in the proof of Theorem 4.30. Then by Proposition 3.7.21

$$
\begin{aligned}
c_T(X) &= \text{co}(T) \\
&= \text{co}(T_1) \cdots \text{co}(T_t) \\
&= c_{T_1}(X) \cdots c_{T_t}(X) \\
&= (X - \lambda_1)^{n_1} \cdots (X - \lambda_t)^{n_t}
\end{aligned}
$$

as claimed. $\qquad \square$

Since, by Proposition 4.23, we have an independent method for calculating the characteristic polynomial $c_T(X)$, the following result is a useful sufficient (but not necessary) criterion for diagonalizability.

(4.32) Corollary. *Let V be an n-dimensional vector space and let $T \in \text{End}_F(V)$ be a linear transformation. If the characteristic polynomial $c_T(X)$ is a product of distinct linear factors, then $m_T(X) = c_T(X)$ and hence T is diagonalizable.*

Proof. Suppose that $c_T(X) = \prod_{i=1}^{n}(X - \lambda_i)$ where the λ_i are distinct. Since every irreducible factor of $c_T(X)$ is also a factor of $m_T(X)$, it follows that $c_T(X)$ divides $m_T(X)$. But since $m_T(X)$ always divides $c_T(X)$ (Corollary 4.10), it follows that $m_T(X) = c_T(X)$. The diagonalizability of T then follows from Theorem 4.30. $\qquad \square$

(4.33) *Remark.* It is interesting to determine the cyclic decomposition of V_T when T is diagonalizable. By Theorem 4.30, we have

$$m_T(X) = (X - \lambda_1) \cdots (X - \lambda_t)$$

and

$$c_T(X) = (X - \lambda_1)^{n_1} \cdots (X - \lambda_t)^{n_k}$$

where the λ_i are distinct and $n = \dim(V) = n_1 + \cdots + n_t$. If $k = \max\{n_1, \ldots, n_t\}$ then $\operatorname{rank}(V_T) = k$, $V_T \cong Rv_1 \oplus \cdots \oplus Rv_k$, and the invariant factors of the torsion R-module V_T are the polynomials

$$f_{k+1-i}(X) = (X - \lambda)^{\epsilon(i,n_1)} \cdots (X - \lambda_t)^{\epsilon(i,n_t)} \qquad \text{for} \quad 1 \leq i \leq k,$$

where

$$\epsilon(i,j) = \begin{cases} 1 & \text{if } i \leq j, \\ 0 & \text{if } i > j. \end{cases}$$

The following special case of Theorem 4.30 will be useful to us later.

(4.34) **Corollary.** *Suppose that V is a finite-dimensional vector space over a field F and let $T : V \to V$ be a linear transformation such that $T^k = 1_V$. Suppose that F is a field in which the equation $z^k = 1$ has k distinct solutions. Then T is diagonalizable.*

Proof. Let the solutions of $z^k - 1$ be $1 = \zeta_0, \ldots, \zeta_{k-1}$. Then $X^k - 1 = \prod_{i=0}^{k-1}(X - \zeta_i)$ is a product of distinct linear factors. By hypothesis, $T^k - 1_V = 0$, so T satisfies the polynomial equation $X^k - 1 = 0$, and hence $m_T(X)$ divides $X^k - 1$. But then $m_T(X)$ is also a product of distinct linear factors, and hence T is diagonalizable. $\qquad \square$

(4.35) *Remark.* Note that the hypothesis on the field F is certainly satisfied for any k if the field F is the field $\mathbf{C}$ of complex numbers.

From Theorem 4.30, we see that there are essentially two reasons why a linear transformation may fail to be diagonalizable. The first is that $m_T(X)$ may factor into linear factors, but the factors may fail to be distinct; the second is that $m_T(X)$ may have an irreducible factor that is not linear. For example, consider the linear transformations $T_i : F^2 \to F^2$, which are given by multiplication by the matrices

$$A_1 = \begin{bmatrix} 0 & 1 \\ 0 & 0 \end{bmatrix} \qquad \text{and} \qquad A_2 = \begin{bmatrix} 0 & 1 \\ -1 & 0 \end{bmatrix}.$$

Note that $m_{T_1}(X) = X^2$, so T_1 illustrates the first problem, while $m_{T_2}(X) = X^2 + 1$. Then if $F = \mathbf{R}$, the real numbers, $X^2 + 1$ is irreducible, so T_2 provides an example of the second problem. Of course, if F is algebraically closed (and in particular if $F = \mathbf{C}$), then the second problem never arises.

We shall concentrate our attention on the first problem and deal with the second one later. The approach will be via the primary decomposition theorem for finitely generated torsion modules over a PID (Theorems 3.7.12 and 3.7.13). We will begin by concentrating our attention on a single primary cyclic R-module.

(4.36) Definition. *Let λ be in the field F and let $n \in \mathbf{N}$. An $n \times n$ Jordan block with value λ is the matrix*

$$(4.21) \qquad J_{\lambda,n} = \begin{bmatrix} \lambda & 1 & 0 & \cdots & 0 & 0 \\ 0 & \lambda & 1 & \cdots & 0 & 0 \\ \vdots & \vdots & \vdots & \ddots & \vdots & \vdots \\ 0 & 0 & 0 & \cdots & \lambda & 1 \\ 0 & 0 & 0 & \cdots & 0 & \lambda \end{bmatrix} \in M_n(F).$$

Note that $J_{\lambda,n} = \lambda I_n + H_n$ where

$$H_n = \sum_{i=1}^{n-1} E_{i,i+1} \in M_n(F).$$

That is, H_n has a 1 directly above each diagonal element and 0 elsewhere. Calculation shows that

$$(J_{\lambda,n} - \lambda I_n)^k = H_n^k = \sum_{i=1}^{n-k} E_{i,i+k} \neq 0 \qquad \text{for} \quad 1 \le k \le n-1,$$

but

$$(J_{\lambda,n} - \lambda I_n)^n = H_n^n = 0.$$

Therefore, if we let $T_{\lambda,n} : F^n \to F^n$ be the linear transformation obtained by multiplying by $J_{\lambda,n}$, we conclude that

$$m_{T_{\lambda,n}}(X) = (X - \lambda)^n = c_{T_{\lambda,n}}$$

(since $\deg(c_{T_{\lambda,n}}(X)) = n$) so that Lemma 4.11 (3) shows that the $F[X]$-module $(F^n)_{T_{\lambda,n}}$ is cyclic.

(4.37) Proposition. *Let V be a finite-dimensional vector space over the field F, and suppose that $T \in \mathrm{End}_F(V)$ is a linear transformation such that the R-module V_T is a primary cyclic R-module. Suppose that $\mathrm{Ann}(V_T) = \langle (X - \lambda)^n \rangle$, and let $v \in V$ be any element such that $V_T = Rv$. Then*

$$\mathcal{B} = \{v_k = (T - \lambda 1_V)^{n-k}(v) : 1 \le k \le n\}$$

is a basis of V over F and

$$[T]_{\mathcal{B}} = J_{\lambda,n}.$$

Proof. First we show that $\mathcal{B}$ is a basis of V. Since $\dim V = n$ by Lemma 4.11 (3), and since $\mathcal{B}$ has n elements, it is only necessary to show that $\mathcal{B}$ is linearly independent. To see this, suppose that $\sum_{k=1}^{n} a_k v_k = 0$ where $a_1, \ldots, a_n \in F$. Then $\sum_{k=1}^{n} a_k (T - \lambda I_V)^{n-k}(v) = 0$, i.e.,

$$g(X) = \sum_{k=1}^{n} a_k (X - \lambda)^{n-k} \in \text{Ann}(v) = \text{Ann}(V).$$

But $\deg(g(X)) < n$, so this can only occur if $g(X) = 0$, in which case $a_1 = \cdots = a_n = 0$. Thus $\mathcal{B}$ is linearly independent and hence a basis of V.

Now we compute the matrix $[T]_{\mathcal{B}}$. To do this note that

$$\begin{aligned}
T(v_k) &= T\left((T - \lambda)^{n-k}(v)\right) \\
&= (T - \lambda)(T - \lambda)^{n-k}(v) + \lambda(T - \lambda)^{n-k}(v) \\
&= (T - \lambda)^{n-(k-1)}(v) + \lambda(T - \lambda)^{n-k}(v) \\
&= \begin{cases} v_{k-1} + \lambda v_k & \text{if } k \geq 2, \\ \lambda v_k & \text{if } k = 1. \end{cases}
\end{aligned}$$

Therefore, $[T]_{\mathcal{B}} = J_{\lambda, n}$, as required. $\qquad\square$

(4.38) Theorem. (Jordan canonical form) *Let V be a vector space of dimension n over a field F and let $T : V \to V$ be a linear transformation. Assume that the minimal polynomial $m_T(X)$ of T factors into a product of (not necessarily distinct) linear factors. (Note that this hypothesis is automatically satisfied in case F is an algebraically closed field and, in particular if $F = \mathbf{C}$.) Then V has a basis $\mathcal{B}$ such that*

$$[T]_{\mathcal{B}} = J = \bigoplus_{i=1}^{s} J_i$$

*where each J_i is a Jordan block. Furthermore, J is unique up to the order of the blocks. (The matrix J is said to be in **Jordan canonical form**.)*

Proof. Let $V_T \cong \oplus_{i=1}^{t} V_i$ be the primary decomposition of the torsion $F[X]$-module V_T (see Theorem 3.7.12). According to the proof of Theorem 3.7.12, each V_i is the $p_i(X)$-primary component of V_T for some irreducible polynomial $p_i(X)$ dividing $\text{me}(V_T) = m_T(X)$, so by the assumption on $m_T(X)$, each $p_i(X)$ is linear, i.e., $p_i(X) = X - \lambda_i$ for some $\lambda_i \in F$. According to Theorem 3.7.13, each module V_i has a decomposition into primary cyclic submodules

$$V_i \cong W_{i1} \oplus \cdots \oplus W_{is_i} \qquad \text{for } 1 \leq i \leq t.$$

By Proposition 4.37, there is a basis $\mathcal{B}_{ij}$ of W_{ij} in which the restriction of T to W_{ij} (recall that submodules of V_T are T-invariant subspaces of V) is a Jordan block. Let $\mathcal{B} = \cup_{i,j} \mathcal{B}_{ij}$. Then $\mathcal{B}$ is a basis of V and $[T]_{\mathcal{B}}$ is in Jordan canonical form.

It remains to show uniqueness of the Jordan canonical form, but this is immediate from the fact that the blocks are in one-to-one correspondence with the elementary divisors of the module V_T—the elementary divisor corresponding to $J_{\lambda,q}$ is $(X - \lambda)^q$. $\square$

We have already briefly encountered eigenvalues of linear transformations (see Definition 3.35 and Remark 4.29). We now consider this concept in more detail and relate it to the canonical form theory just developed. Recall that $\lambda \in F$ is an eigenvalue of $T : V \to V$ if $T(v) = \lambda v$ for some nonzero $v \in V$. The nonzero element v is an eigenvector of T corresponding to λ.

(4.39) Lemma. *Let V be a finite-dimensional vector space over a field F and let $T \in \mathrm{End}_F(V)$ be a linear transformation. If $\lambda \in F$, then the following are equivalent.*

(1) *λ is an eigenvalue of T.*
(2) *$X - \lambda$ divides $m_T(X)$.*
(3) *$X - \lambda$ divides $c_T(X)$.*

Proof. (1) $\Rightarrow$ (2). Let v be an eigenvector of T corresponding to λ. Then

$$(X - \lambda) \in \mathrm{Ann}(v) \supseteq \mathrm{Ann}(V_T) = \langle m_T(X) \rangle,$$

so $X - \lambda$ divides $m_T(X)$.
    (2) $\Rightarrow$ (1). Immediate from Theorem 3.7.1 and Lemma 3.7.17.
    (2) $\Leftrightarrow$ (3). $m_T(X)$ and $c_T(X)$ have the same irreducible factors (Corollary 4.10). $\square$

For convenience, we restate Corollary 3.34 and Theorem 4.30 in the current context.

(4.40) Theorem. *Let $T : V \to V$ be a linear transformation of a finite-dimensional vector space over the field F. Then the following are equivalent:*

(1) *T is diagonalizable.*
(2) *$m_T(X)$ is a product of distinct linear factors.*
(3) *V has a basis consisting of eigenvectors of T.*

Proof. (1) $\Leftrightarrow$ (2) is Theorem 4.30, while (1) $\Leftrightarrow$ (3) is Corollary 3.34. $\square$

(4.41) Remark. Since $\mathrm{diag}(\lambda_1, \ldots, \lambda_n) = \oplus_{i=1}^n J_{\lambda_i,1}$, we see that if T is diagonalizable, then the Jordan canonical form of T is diagonal.

(4.42) Definition. *Let $T \in \mathrm{End}_F(V)$. Then a nonzero vector $v \in V$ is a **generalized eigenvector** of T corresponding to the eigenvalue $\lambda \in F$ if $p(T)(v) = 0$ for $p(X) = (X - \lambda)^k$ for some $k > 0$.*

In other words, $v \neq 0$ is a generalized eigenvector of T corresponding to the eigenvalue λ if v is in the $(X - \lambda)$-primary component of the $F[X]$-module V_T.

(4.43) Lemma. *If $\{v_i\}_{i=1}^{r}$ are generalized eigenvectors of T corresponding to distinct eigenvalues, then they are linearly independent.*

Proof. Decompose

$$(4.22) \qquad V_T \cong \bigoplus_{i=1}^{s} V_i$$

by the primary decomposition theorem (Theorem 3.7.12). After reordering, if necessary, we may assume that $v_i \in V_i$ for $1 \leq i \leq r$. If $\sum_{i=1}^{r} a_i v_i = 0$ then it follows that $a_i v_i = 0$ since Equation (4.22) is a direct sum, so $a_i = 0$ since $v_i \neq 0$. $\qquad \square$

(4.44) Theorem. *The following are equivalent for a linear transformation $T : V \to V$.*

(1) T *has a Jordan canonical form.*
(2) V *has a basis $\mathcal{B}$ consisting of generalized eigenvectors of T.*
(3) $m_T(X)$ *is a product of (not necessarily distinct) linear factors.*

Proof. (1) $\Rightarrow$ (2). If $[T]_{\mathcal{B}}$ is in Jordan canonical form then the basis $\mathcal{B}$ consists of generalized eigenvectors.

(2) $\Rightarrow$ (3). Let $\mathcal{B} = \{v_1, \ldots, v_n\}$ be a basis of V, and assume

$$(T - \lambda_i)^{k_i}(v_i) = 0,$$

i.e., each v_i is assumed to be a generalized eigenvector of T. Then

$$m_T(X) = \mathrm{me}(V_T) = \mathrm{lcm}\{(X - \lambda_1)^{k_1}, \ldots, (X - \lambda_n)^{k_n}\}$$

is a product of linear factors.

(3) $\Rightarrow$ (1). This is Theorem 4.38. $\qquad \square$

Now we define some important invariants.

(4.45) Definition. *Let $T : V \to V$ be a linear transformation and let $\lambda \in F$ be an eigenvalue of T.*

(1) *The **algebraic multiplicity** of the eigenvalue λ, denoted $\nu_{\mathrm{alg}}(\lambda)$, is the highest power of $X - \lambda$ dividing $c_T(X)$.*
(2) *The **geometric multiplicity** of the eigenvalue λ, denoted $\nu_{\mathrm{geom}}(\lambda)$ is the dimension (as a vector space over F) of $\mathrm{Ker}(T - \lambda 1_V)$.*

(4.46) *Remarks.*

(1) $\text{Ker}(T - \lambda 1_V) = \{v \in V : T(v) = \lambda v\}$ is called the **eigenspace** of λ.

(2) $\{v \in V : (T - \lambda)^k(v) = 0 \text{ for some } k \in \mathbf{N}\}$ is called the **generalized eigenspace** of the eigenvalue λ. Note that the generalized eigenspace corresponding to the eigenvalue λ is nothing more than the $(X - \lambda)$-primary component of the torsion module V_T. Moreover, it is clear from the definition of primary component of V_T that the generalized eigenspace of T corresponding to λ is $\text{Ker}(T - \lambda 1_V)^r$ where r is the exponent of $X - \lambda$ in $m_T(X) = \text{me}(V_T)$.

(3) Note that Lemma 4.43 implies that distinct (generalized) eigenspaces are linearly independent.

(4.47) Proposition. *Let λ be an eigenvalue of the linear transformation $T \in \text{End}_F(V)$ where V is a finite-dimensional vector space over F. Then the geometric multiplicity of λ is the number of elementary divisors of V_T that are powers of $(X - \lambda)$. In particular,*

$$(4.23) \qquad 1 \le \nu_{\text{geom}}(\lambda) \le \nu_{\text{alg}}(\lambda).$$

Proof. First note that $1 \le \nu_{\text{geom}}(\lambda)$ since λ is an eigenvalue. Now let $V_T \cong \oplus_{i=1}^t V_i$ be the primary decomposition of the $F[X]$-module V_T, and assume (by reordering, if necessary) that V_1 is $(X - \lambda)$-primary. Then $T - \lambda : V_i \to V_i$ is an isomorphism for $i > 1$ because $\text{Ann}(V_i)$ is relatively prime to $X - \lambda$ for $i > 1$. Now write

$$V_1 \cong W_1 \oplus \cdots \oplus W_r$$

as a sum of cyclic submodules. Since V_1 is $(X - \lambda)$-primary, it follows that $\text{Ann}(W_k) = \langle (X - \lambda)^{q_k} \rangle$ for some $q_k \ge 1$. The Jordan canonical form of $T|_{W_k}$ is J_{λ, q_k} (by Proposition 4.37). Thus we see that W_k contributes 1 to the geometric multiplicity of λ and q_k to the algebraic multiplicity of λ. $\quad\square$

(4.48) Corollary.

(1) *The geometric multiplicity of λ is the number of Jordan blocks $J_{\lambda, q}$ in the Jordan canonical form of T with value λ.*

(2) *The algebraic multiplicity of λ is the sum of the sizes of the Jordan blocks of T with value λ.*

(3) *If $m_T(X) = (X - \lambda)^q p(X)$ where $(X - \lambda)$ does not divide $p(X)$, then q is the size of the largest Jordan block of T with value λ.*

(4) *If $\lambda_1, \ldots, \lambda_k$ are the distinct eigenvalues of T, then*

$$\mu(V_T) = \max\{\nu_{\text{geom}}(\lambda_i) : 1 \le i \le k\}.$$

Proof. (1), (2), and (3) are immediate from the above proposition, while (4) follows from the algorithm of Theorem 3.7.15 for recovering the invariant factors from the elementary divisors of a torsion R-module. $\quad\square$

(4.49) Corollary. *The following are equivalent for a linear transformation* $T \in \operatorname{End}_F(V)$.

(1) T *is diagonalizable.*
(2) *If* $\lambda_1, \ldots, \lambda_k$ *are the distinct eigenvalues of* T, *then*

$$\sum_{i=1}^{k} \nu_{\text{geom}}(\lambda_i) = \dim_F(V).$$

(3) $m_T(X)$ *is a product of linear factors, and for each eigenvalue* λ,

$$\nu_{\text{geom}}(\lambda) = \nu_{\text{alg}}(\lambda).$$

Proof. $\square$

(4.50) Remark. The importance of the Jordan canonical form of a linear transformation can hardly be over-emphasized. All of the most important invariants of a linear transformation: its characteristic and minimal polynomials, its elementary divisors, its eigenvalues, and their algebraic and geometric multiplicities, can all be read off from the Jordan canonical form, and the (generalized) eigenspaces may be read off from the basis with respect to which it is in this form.

We have seen that in order for T to have a Jordan canonical form, the minimal polynomial $m_T(X)$ must be a product of linear factors. We will conclude this section by developing a generalized Jordan canonical form to handle the situation when this is not the case. In addition, the important case $F = \mathbf{R}$, the real numbers, will be developed in more detail, taking into account the special relationship between the real numbers and the complex numbers. First we will do the case of a general field F. Paralleling our previous development, we begin by considering the case of a primary cyclic R-module (compare Definition 4.36 and Proposition 4.37).

(4.51) Definition. *Let* $f(X) \in F[X]$ *be a monic polynomial of degree* d. *The* $nd \times nd$ **generalized Jordan block** *corresponding to* $f(X)$ *is the* $nd \times nd$ *matrix (given in blocks of* $d \times d$ *matrices)*

$$J_{f(X),n} = \begin{bmatrix} C(f(X)) & N & 0 & \cdots & 0 & 0 \\ 0 & C(f(X)) & N & \cdots & 0 & 0 \\ \vdots & \vdots & \vdots & \ddots & \vdots & \vdots \\ 0 & 0 & 0 & \cdots & C(f(X)) & N \\ 0 & 0 & 0 & \cdots & 0 & C(f(X)) \end{bmatrix}$$

where N *is the* $d \times d$ *matrix with* $\operatorname{ent}_{1d}(N) = 1$ *and all other entries* 0. $J_{f(X),n}$ *is called* **irreducible** *if* $f(X)$ *is an irreducible polynomial. Note, in* *particular, that the Jordan matrix* $J_{\lambda,n}$ *is the same as the irreducible Jordan block* $J_{(X-\lambda),n}$.

(4.52) Proposition. *Let* $T : V \to V$ *be a linear transformation, and assume that* $V_T = Rv$ *is a (primary) cyclic R-module such that* $\text{Ann}(v) = \text{Ann}(V_T) = \langle f(X)^n \rangle$, *where* $f(X)$ *is an (irreducible) polynomial of degree d. Define* $\mathcal{B}$ *as follows: Let* $v_k = f(T)^{n-k}(v)$ *for* $1 \leq k \leq n$ *and let* $v_{kj} = T^{j-1}(v_k)$ *for* $1 \leq j \leq d$. *Then let*

$$\mathcal{B} = \{v_{11}, \ldots, v_{1k}, v_{21}, \ldots, v_{2d}, \ldots, v_{n1}, \ldots, v_{nd}\}.$$

Then $\mathcal{B}$ *is a basis of V over F and* $[T]_{\mathcal{B}}$ *is an (irreducible) nd × nd generalized Jordan block corresponding to* $f(X)$.

Proof. This is a tedious computation, entirely paralleling the proof of Proposition 4.37; we shall leave it to the reader.   □

(4.53) Remark. Note that if $f(X) = X - \lambda$, then Proposition 4.52 reduces to Propositon 4.37. (Of course, every linear polynomial is irreducible.)

(4.54) Theorem. (Generalized Jordan canonical form) *Let V be a vector space of dimension n over a field F and let* $T : V \to V$ *be a linear transformation. Then V has a basis* $\mathcal{B}$ *such that*

$$[T]_{\mathcal{B}} = J' = \bigoplus_{i=1}^{s} J_i'$$

where each J_i' *is an irreducible generalized Jordan block. Furthermore, J' is unique up to the order of the blocks. (The matrix J' is said to be in* **generalized Jordan canonical form.***)*

Proof. Almost identical to the proof of Theorem 4.38; we leave it for the reader.   □

Now let $F = \mathbf{R}$ be the field of real numbers. We will produce a version of the Jordan canonical form theorem which is valid for all matrices with entries in $\mathbf{R}$. This canonical form will be somewhat different than the generalized Jordan canonical form of Theorem 4.54.

Recall (Theorem 4.44) that a linear transformation $T : V \to V$, where V is a finite-dimensional vector space over a field F, has a Jordan canonical form if and only if the minimal polynomial $m_T(X)$ is a product of (not necessarily distinct) linear factors. Of course, this says that all of the roots of $m_T(X)$, i.e., all eigenvalues of T, are included in the field F. This condition will be satisfied for all polynomials over F if and only if the field F is algebraically closed. The field $\mathbf{C}$ of complex numbers is algebraically closed (although we shall not present a proof of this fact in this book), but the field $\mathbf{R}$ of real numbers is not. In fact, $X^2 + 1$ is a polynomial over $\mathbf{R}$ that does not have any roots in $\mathbf{R}$. Therefore, any linear transformation $T \in \mathrm{End}_{\mathbf{R}}(\mathbf{R}^2)$ such that $m_T(X) = X^2 + 1$ cannot have a Jordan canonical form. There is, however, a very simple variant of the Jordan canonical form which is valid over $\mathbf{R}$ and which takes into account the special nature of $\mathbf{R}$ as it relates to the complex numbers. We will start by analyzing polynomials over $\mathbf{R}$.

Recall that if $z \in \mathbf{C}$ then $\bar{z}$ denotes the complex conjugate of z, i.e., if $z = a + bi$, then $\bar{z} = a - bi$. Then $\frac{1}{2}(z + \bar{z}) = a$, while $\frac{1}{2i}(z - \bar{z}) = b$ and $z \in \mathbf{R}$ if and only if $z = \bar{z}$. The conjugation map on $\mathbf{C}$ extends in a natural way to an involution $\sigma : \mathbf{C}[X] \to \mathbf{C}[X]$ defined by $\sigma(f(X)) = \bar{f}(X)$ where $\bar{f}(X)$ is the polynomial determined by conjugating all the coefficients of $f(X)$, i.e., if $f(X) = a_0 + a_1 X + \cdots + a_n X^n$ then $\bar{f}(X) = \bar{a}_0 + \bar{a}_1 X + \cdots + \bar{a}_n X^n$. Thus $\sigma : \mathbf{C}[X] \to \mathbf{C}[X]$ is a ring homomorphism such that $\sigma^2 = 1_{\mathbf{C}[X]}$, and $f(X) \in \mathbf{R}[X]$ if and only if $\bar{f}(X) = f(X)$. If $f(X) \in \mathbf{C}[X]$ and $z \in \mathbf{C}$ is a root of $f(X)$, then a simple calculation shows that $\bar{z}$ is a root of $\bar{f}(X)$. Thus if $f(X) \in \mathbf{R}[X]$ and $z \in \mathbf{C}$ is a root of $f(X)$, then $\bar{z}$ is also a root of $f(X)$. By Corollary 2.4.6 it follows that whenever $z \in \mathbf{C} \setminus \mathbf{R}$ is a root of $f(X)$, then the polynomial $(X - z)(X - \bar{z})$ divides $f(X)$ in $\mathbf{C}[X]$. But if $z = a + bi$, where $a, b \in \mathbf{R}$ and $b \neq 0$, then

$$(X - z)(X - \bar{z}) = ((X - a) - bi)((X - a) + bi)$$
$$= ((X - a)^2 + b^2).$$

Since the polynomial $(X - a)^2 + b^2$ has real coefficients and divides the real polynomial $f(X)$ in the ring $\mathbf{C}[X]$, it follows from the uniqueness part of the division algorithm (Theorem 2.4.4) that the quotient of $f(X)$ by $(X - a)^2 + b^2$ is in $\mathbf{R}[X]$. From this observation and the fact that $\mathbf{C}$ is algebraically closed, we obtain the following factorization theorem for real polynomials.

(4.55) Lemma. *Let $f(X) \in \mathbf{R}[X]$. Then $f(X)$ factors in $\mathbf{R}[X]$ as*

$$(4.24) \qquad f(X) = \alpha \left(\prod_{i=1}^{r} (X - c_i)^{m_i} \right) \left(\prod_{j=1}^{s} ((X - a_j)^2 + b_j^2)^{n_j} \right)$$

where $c_i \in \mathbf{R}$, a_j, $b_j \in \mathbf{R}$ with $b_j \neq 0$, and $\alpha \in \mathbf{R}$.

Proof. The real roots $c_1, \ldots, c_r$ of $f(X)$ give the first factor, while if $z_j = a_j + ib_j$ is a complex root, then $(X - a_j)^2 + b_j^2$ divides $f(X)$, so we may argue by induction on the degree of $f(X)$. $\qquad \square$

Note that the factorization of Equation (4.24) implies that if z_j is a nonreal root of $f(X)$ of multiplicity n_j, then $\bar{z}_j$ is a root of the same multiplicity n_j.

(4.56) Definition. *Given $z = a + bi \in \mathbf{C}$ with $b \neq 0$ and $r \in \mathbf{N}$, let $J^{\mathbf{R}}_{z,r} \in M_{2r}(\mathbf{R})$ be defined by*

$$(4.25) \qquad J^{\mathbf{R}}_{z,r} = \begin{bmatrix} A & I_2 & 0 & \cdots & 0 & 0 \\ 0 & A & I_2 & \cdots & 0 & 0 \\ \vdots & \vdots & \vdots & \ddots & \vdots & \vdots \\ 0 & 0 & 0 & \cdots & A & I_2 \\ 0 & 0 & 0 & \cdots & 0 & A \end{bmatrix}$$

where $A = \begin{bmatrix} a & -b \\ b & a \end{bmatrix}$. $J^{\mathbf{R}}_{z,r}$ is said to be a **real Jordan block** corresponding to $z \in \mathbf{C} \setminus \mathbf{R}$.

(4.57) Theorem. (real Jordan canonical form) *Let V be a finite-dimensional vector space over $\mathbf{R}$ and let $T : V \to V$ be a linear transformation. Then there is a basis $\mathcal{B}$ of V such that*

$$[T]_{\mathcal{B}} = \bigoplus_{i=1}^{s} J_i$$

where $J_i = J_{\lambda_i, k_i}$ is an ordinary Jordan block corresponding to $\lambda_i \in \mathbf{R}$, or $J_i = J^{\mathbf{R}}_{z_i, k_i}$ is a real Jordan block corresponding to some $z_i \in \mathbf{C} \setminus \mathbf{R}$.

Proof. By the primary decomposition theorem (Theorem 3.7.13), we may suppose that the $\mathbf{R}[X]$-module V_T is a primary cyclic $\mathbf{R}[X]$-module, say $V_T = \mathbf{R}[X]u$. Thus, $\mathrm{Ann}(V_T) = \langle p(X)^r \rangle$ where (by Lemma 4.55)

$$(4.26) \qquad p(X) = X - c$$

or

$$(4.27) \qquad p(X) = (X - a)^2 + b^2 \quad \text{with } b \neq 0.$$

In case (4.26), T has a Jordan canonical form $J_{c,r}$; in case (4.27) we will show that T has a real Jordan canonical form $J^{\mathbf{R}}_{z,r}$ where $z = a + bi$.

Thus, suppose that $p(X) = (X-a)^2+b^2$ and let $z = a+bi$. By hypothesis V_T is isomorphic as an $\mathbf{R}[X]$-module to the $\mathbf{R}[X]$-module $\mathbf{R}[X]/\langle p(X)^r \rangle$. Recall that the module structure on V_T is given by $Xu = T(u)$, while the module structure on $\mathbf{R}[X]/\langle p(X)^r \rangle$ is given by polynomial multiplication. Thus we can analyze how T acts on the vector space V by studying how X acts on $\mathbf{R}[X]/\langle p(X)^r \rangle$ by multiplication. This will be the approach we will follow, without explicitly carrying over the basis to V.

Consider the $\mathbf{C}[X]$-module $W = \mathbf{C}[X]/\langle p(X)^r \rangle$. The annihilator of W is $p(X)^r$, which factors in $\mathbf{C}[X]$ as $p(X)^r = q(X)^r \overline{q}(X)^r$ where $q(X) = ((X - a) + bi)$. Since $q(X)^r$ and $\overline{q}(X)^r$ are relatively prime in $\mathbf{C}[X]$,

$$(4.28) \qquad 1 = q(X)^r g_1(X) + \overline{q}(X)^r g_2(X)$$

for $g_1(X)$ and $g_2(X)$ in $\mathbf{C}[X]$. Averaging Equation (4.28) with it's complex conjugate gives

$$1 = q(X)^r \left(\frac{g_1(X) + \overline{g}_2(X)}{2} \right) + \overline{q}(X)^r \left(\frac{g_2(X) + \overline{g}_1(X)}{2} \right)$$
$$= f(X) + \overline{f}(X).$$

Thus, by Theorem 3.7.12, we may write

(4.29) $$W = \mathbf{C}[X]/\langle p(X)^r \rangle = \langle v \rangle \oplus \langle w \rangle,$$

where $v = f(X) + \langle p(X)^r \rangle$, $w = \overline{f}(X) + \langle p(X)^r \rangle$, $\text{Ann}(v) = \langle \overline{q}(X)^r \rangle$ and
$\text{Ann}(w) = \langle q(X)^r \rangle$. Equation (4.29) provides a primary cyclic decomposition of $\mathbf{C}[X]/\langle p(X)^r \rangle$ and, as in the proof of Theorem 4.37,

$$\mathcal{C} = \{ q(X)^{r-k} w : 1 \le k \le r \} \cup \{ \overline{q}(X)^{r-k} v : 1 \le k \le r \}$$

is a basis of $W = \mathbf{C}[X]/\langle p(X)^r \rangle$ over $\mathbf{C}$. For $1 \le k \le r$, let

$$v_k = \frac{1}{2} \left(q(X)^{r-k} w + \overline{q}(X)^{r-k} v \right) \in W$$

and

$$w_k = \frac{1}{2i} \left(q(X)^{r-k} w - \overline{q}(X)^{r-k} v \right) \in W.$$

Note that $\overline{v}_k = v_k$ and $\overline{w}_k = w_k$ since $\overline{v} = w$, where the conjugation map on $W = \mathbf{C}[X]/\langle p(X)^r \rangle$ is induced from that on $\mathbf{C}[X]$. Therefore v_k and w_k are in $\mathbf{R}[X]/\langle p(X)^r \rangle$. Moreover, $v_k + i w_k = q(X)^{r-k} w$ and $v_k - i w_k = \overline{q}(X)^{r-k} v$. Therefore, the set

$$\mathcal{B} = \{ v_1, w_1, v_2, w_2, \dots, v_r, w_r \} \subseteq \mathbf{R}[X]/\langle p(X)^r \rangle$$

spans $\mathbf{C}[X]/\langle p(X)^r \rangle$, and hence, it is a basis. Since $\mathcal{B}$ is linearly independent over $\mathbf{C}$, it is also linearly independent over $\mathbf{R}$, and since

$$\dim_{\mathbf{R}} \mathbf{R}[X]/\langle p(X)^r \rangle = \deg p(X)^r = 2r,$$

it follows that $\mathcal{B}$ is a basis of $\mathbf{R}[X]/\langle p(X)^r \rangle$ over $\mathbf{R}$.

Now we compute the effect of multiplication by X on the basis $\mathcal{B}$. Since $X = q(X) + (a - bi)$ and $X = \overline{q}(X) + (a + bi)$,

$$Xv_k = X \left(\frac{1}{2} (q(X)^{r-k} w + \overline{q}(X)^{r-k} v) \right)$$
$$= \frac{1}{2} \left(q(X) + (a - bi) \right) q(X)^{r-k} w + \frac{1}{2} \left(\overline{q}(X) + (a + bi) \right) \overline{q}(X)^{r-k} v$$
$$= \frac{1}{2} \left(q(X)^{r-(k-1)} w + \overline{q}(X)^{r-(k-1)} v \right)$$
$$\quad + a \left(\frac{1}{2} (q(X)^{r-k} w + \overline{q}(X)^{r-k} v) \right) + b \left(\frac{1}{2i} (q(X))^{r-k} w - \overline{q}(X)^{r-k} v) \right)$$
$$= \begin{cases} v_{k-1} + a v_k + b w_k & \text{if } k \ge 2, \\ a v_k + b w_k & \text{if } k = 1. \end{cases}$$

Similarly,

$$Xw_k = \begin{cases} w_{k-1} - bv_k + aw_k & \text{if } k \geq 2, \\ -bv_k + aw_k & \text{if } k = 1. \end{cases}$$

Thus, if $S : \mathbf{R}[X]/\langle p(X)^r \rangle \to \mathbf{R}[X]/\langle p(X)^r \rangle$ denotes multiplication by X, then $[S]_{\mathcal{B}} = J_{z,r}^{\mathbf{R}}$, and since S corresponds to T under the isomorphism $V_T \cong \mathbf{R}[X]/\langle p(X)^r \rangle$, the proof is complete. $\qquad \square$

(4.58) *Remark.* It is worthwhile to compare the real Jordan canonical form derived in Theorem 4.57 with the generalized Jordan canonical form determined by Theorem 4.54. Suppose that $f(X) = ((X - a)^2 + b^2)^2$. Then the real Jordan block determined by $f(X)$ is

$$J_{a+bi,2}^{\mathbf{R}} = \begin{bmatrix} a & -b & 1 & 0 \\ b & a & 0 & 1 \\ 0 & 0 & a & -b \\ 0 & 0 & b & a \end{bmatrix}$$

while the generalized Jordan block determined by $f(X)$ is

$$J_{f(X),2} = \begin{bmatrix} 0 & -(a^2 + b^2) & 0 & 1 \\ 1 & 2a & 0 & 0 \\ 0 & 0 & 0 & -(a^2 + b^2) \\ 0 & 0 & 1 & 2a \end{bmatrix}.$$

(4.59) Corollary. *If V is a finite-dimensional real vector space of dimension at least 2, then every linear transformation $T : V \to V$ has a 2-dimensional invariant subspace.*

Proof. T has an elementary divisor of the form

$$\begin{cases} (X - c)^r & \text{for } r \geq 2, \\ ((X - a)^2 + b^2)^r & \text{for } r \geq 1, \end{cases}$$

or T is diagonalizable. If T is diagonalizable, the result is clear, while if the real Jordan canonical form of T contains a block $J_{c,r}$ $(r \geq 2)$ the vectors corresponding to the first two columns of $J_{c,r}$ generate a two-dimensional invariant subspace. Similarly, if the real Jordan canonical form contains a block $J_{z,r}^{\mathbf{R}}$ $(r \geq 1)$, then the vectors corresponding to v_1 and w_1 constructed in the proof of Theorem 4.57 generate a T-invariant subspace. $\qquad \square$

(4.60) *Remark.* If a linear transformation possesses an eigenvalue, then it has an invariant subspace of dimension 1. Corollary 4.59 is of interest in that it states that if a linear transformation of a real vector space does not have any eigenvalues, then at least there is the next best thing, namely, a two-dimensional invariant subspace. If F is a field that has an irreducible polynomial $f(X)$ of degree n, then multiplication by $A = C(f(X))$ on F^n

is a linear transformation with no invariant subspace of dimension less than n. Thus Corollary 4.59 depends on the special nature of the real numbers.

At this point we would like to reemphasize the remark prior to Proposition 4.3.

(4.61) *Remark.* Our entire approach in this section has been to analyze a linear transformation T by analyzing the structure of the $F[X]$-module V_T. Since T_1 and T_2 are similar precisely when V_{T_1} and V_{T_2} are isomorphic, each and every invariant we have derived in this section—characteristic and minimal polynomials, rational, Jordan, and generalized Jordan canonical forms, (generalized) eigenvalues and their algebraic and geometric multiplicities, etc.—is the same for similar linear transformations T_1 and T_2.

4.5 Computational Examples

This section will be devoted to several numerical examples to illustrate the general canonical form theory developed in Section 4.4. In Chapter 5, further techniques will be presented using an analysis of equivalence of matrices over a PID. Throughout this section it will be assumed that the reader is familiar with the procedure for solving a system $AX = B$ of linear equations over a field F by the process of row reduction of the matrix $[A\, B]$. This procedure is a standard topic in elementary linear algebra courses; it will be developed in the more general context of matrices with entries in a PID in Section 5.2.

(5.1) Example. *Construct, up to similarity, all linear transformations T : $F^6 \to F^6$ with minimal polynomial*

$$m_T(X) = (X - 5)^2(X - 6)^2.$$

Solution. If $m_T(X) = (X - 5)^2(X - 6)^2$, then we must have

$$c_T(X) = (X - 5)^i(X - 6)^j$$

where $i + j = 6 = \dim(F^6)$ and $2 \le i$, $2 \le j$. This is because $m_T(X)$ divides $c_T(X)$ (Corollary 4.10 (2)), and $m_T(X)$ and $c_T(X)$ have the same irreducible factors (Corollary 4.10 (3)).

Suppose $i = 2$ and $j = 4$. Then 5 has algebraic multiplicity 2, and 2 is the size of the largest Jordan block with eigenvalue 5, so there is exactly one such block. Also 6 has algebraic multiplicity 4, and 2 is the size of its largest Jordan block, so either it has two blocks of size 2, or one block of size 2 and two of size 1. Thus the possibilities for the Jordan canonical form of T are

$$J_{5,2} \oplus J_{6,2} \oplus J_{6,2} = \begin{bmatrix} 5 & 1 & 0 & 0 & 0 & 0 \\ 0 & 5 & 0 & 0 & 0 & 0 \\ 0 & 0 & 6 & 1 & 0 & 0 \\ 0 & 0 & 0 & 6 & 0 & 0 \\ 0 & 0 & 0 & 0 & 6 & 1 \\ 0 & 0 & 0 & 0 & 0 & 6 \end{bmatrix}$$

and

$$J_{5,2} \oplus J_{6,2} \oplus J_{6,1} \oplus J_{6,1} = \begin{bmatrix} 5 & 1 & 0 & 0 & 0 & 0 \\ 0 & 5 & 0 & 0 & 0 & 0 \\ 0 & 0 & 6 & 1 & 0 & 0 \\ 0 & 0 & 0 & 6 & 0 & 0 \\ 0 & 0 & 0 & 0 & 6 & 0 \\ 0 & 0 & 0 & 0 & 0 & 6 \end{bmatrix}.$$

If $i = 4$ and $j = 2$, then there is a similar analysis, with the roles of 5 and 6 reversed. The possibilities for the Jordan canonical form of T are then

$$J_{6,2} \oplus J_{5,2} \oplus J_{5,2} \qquad \text{or} \qquad J_{6,2} \oplus J_{5,2} \oplus J_{5,1} \oplus J_{5,1}.$$

If $i = 3$ and $j = 3$, then for each eigenvalue we must have one block of size 2 and one of size 1. This gives the single possibility

$$J_{5,2} \oplus J_{5,1} \oplus J_{6,2} \oplus J_{6,1} = \begin{bmatrix} 5 & 1 & 0 & 0 & 0 & 0 \\ 0 & 5 & 0 & 0 & 0 & 0 \\ 0 & 0 & 5 & 0 & 0 & 0 \\ 0 & 0 & 0 & 6 & 1 & 0 \\ 0 & 0 & 0 & 0 & 6 & 0 \\ 0 & 0 & 0 & 0 & 0 & 6 \end{bmatrix}.$$

for the Jordan canonical form of T.   □

(5.2) **Example.** *Construct, up to similarity, all linear transformations T with characteristic polynomial $c_T(X) = (X-8)^7$, minimal polynomial $m_T(X) = (X-8)^3$, and 8 an eigenvalue of geometric multiplicity 4.*

Solution. Since $\deg(c_T(X)) = 7$, T is a linear transformation of a 7-dimensional vector space V, which we may take to be F^7, since we are working up to similarity, and 8 is the only eigenvalue of T. Thus the Jordan blocks in the Jordan canonical form of T are $J_{8,r}$. Since $m_T(X)$ has degree 3, the largest Jordan block has size 3, so the block sizes must be 3, 3, 1, or 3, 2, 2, or 3, 2, 1, 1, or 3, 1, 1, 1, 1. Since the geometric multiplicity of 8 is equal to 4, there must be 4 Jordan blocks (Corollary 4.48 (1)), and the only ways to partition 7 into 4 parts are $7 = 4+1+1+1 = 3+2+1+1 = 2+2+2+1$, these being the possible block sizes (Corollary 4.48 (2)). Comparing these last two lists, there is only only coincidence; thus there is only one possibility for T:

$$J_{8,3} \oplus J_{8,2} \oplus J_{8,1} \oplus J_{8,1} = \begin{bmatrix} 8 & 1 & 0 & 0 & 0 & 0 & 0 \\ 0 & 8 & 1 & 0 & 0 & 0 & 0 \\ 0 & 0 & 8 & 0 & 0 & 0 & 0 \\ 0 & 0 & 0 & 8 & 1 & 0 & 0 \\ 0 & 0 & 0 & 0 & 8 & 0 & 0 \\ 0 & 0 & 0 & 0 & 0 & 8 & 0 \\ 0 & 0 & 0 & 0 & 0 & 0 & 8 \end{bmatrix}.$$

□

In contrast to the first two examples, the problems which generally arise are not to construct matrices in Jordan canonical form, but rather to find the Jordan canonical form of a given linear transformation (or matrix). In this, Proposition 4.23 is our starting point, telling us how to find the characteristic polynomial. (We do not wish to address at this point the problem of practical methods for computing the characteristic polynomial, but simply assume that this is done. Chapter 5 will contain further information, which will be of computational interest in computing canonical forms.)

(5.3) Example. Let $V = F^3$ and let $T : V \to V$ be the linear transformation with matrix

$$[T]_C = A = \begin{bmatrix} 1 & -1 & 1 \\ -2 & 1 & 2 \\ -2 & -1 & 4 \end{bmatrix}$$

where C is the standard basis on F^3. Find out "everything" about T.

Solution. A calculation of the characteristic polynomial $c_T(X)$ shows that

$$\begin{aligned} c_T(X) &= \det(XI_3 - A) \\ &= X^3 - 6X^2 + 11X - 6 \\ &= (X-1)(X-2)(X-3). \end{aligned}$$

Since this is a product of distinct linear factors, it follows that T is diagonalizable (Corollary 4.32), $m_T(X) = c_T(X)$, and the Jordan canonical form of T is $J = \text{diag}(1, 2, 3)$. Then the eigenvalues of T are 1, 2, and 3, each of which has algebraic and geometric multiplicity 1.

Since $m_T(X) = c_T(X)$, it follows that V_T is a cyclic $F[X]$-module and the rational canonical form of T is

$$C(m_T(X)) = R = \begin{bmatrix} 0 & 0 & 6 \\ 1 & 0 & -11 \\ 0 & 1 & 6 \end{bmatrix}.$$

We now compute the eigenspaces of T. This is a straightforward matter of solving systems of linear equations. The details are left to the reader. If $V_\lambda = \text{Ker}(T - \lambda 1_V)$, then we find

$$V_1 = \left\langle \begin{bmatrix} 1 \\ 1 \\ 1 \end{bmatrix} \right\rangle, \qquad V_2 = \left\langle \begin{bmatrix} 1 \\ 0 \\ 1 \end{bmatrix} \right\rangle, \qquad V_3 = \left\langle \begin{bmatrix} 0 \\ 1 \\ 1 \end{bmatrix} \right\rangle.$$

Thus, if

$$\mathcal{B} = \left\{ \begin{bmatrix} 1 \\ 1 \\ 1 \end{bmatrix}, \begin{bmatrix} 1 \\ 0 \\ 1 \end{bmatrix}, \begin{bmatrix} 0 \\ 1 \\ 1 \end{bmatrix} \right\},$$

then $[T]_\mathcal{B} = J$. Also, if

$$P = \begin{bmatrix} 1 & 1 & 0 \\ 1 & 0 & 1 \\ 1 & 1 & 1 \end{bmatrix}$$

is the matrix whose columns are the vectors of $\mathcal{B}$, then $P^{-1}AP = J = \mathrm{diag}(1, 2, 3)$. Finally, if we let

$$v = \begin{bmatrix} 1 \\ 1 \\ 1 \end{bmatrix} + \begin{bmatrix} 1 \\ 0 \\ 1 \end{bmatrix} + \begin{bmatrix} 0 \\ 1 \\ 1 \end{bmatrix} = \begin{bmatrix} 2 \\ 2 \\ 3 \end{bmatrix},$$

then

$$\mathrm{Ann}(v) = \langle (X - 1)(X - 2)(X - 3) \rangle = \langle m_T(X) \rangle = \mathrm{Ann}(V_T).$$

Thus, v is a cyclic vector for the $F[X]$-module V_T. If we let

$$\mathcal{B}' = \{v, Tv, T^2 v\} = \left\{ \begin{bmatrix} 2 \\ 2 \\ 3 \end{bmatrix}, \begin{bmatrix} 3 \\ 4 \\ 6 \end{bmatrix}, \begin{bmatrix} 5 \\ 10 \\ 14 \end{bmatrix} \right\}$$

be the basis of V given by Corollary 4.16, then $[T]_{\mathcal{B}'} = R$, and if Q is the matrix whose columns are the column vectors of $\mathcal{B}'$, then $Q^{-1}AQ = R$.   □

(5.4) Example. Let $V = F^3$ and let $T : V \to V$ be the linear transformation with matrix with respect to the standard basis C

$$[T]_C = A = \begin{bmatrix} 6 & -2 & -2 \\ 6 & -1 & -3 \\ 4 & -2 & 0 \end{bmatrix}.$$

Find out "almost everything" about T.

Solution. We compute

$$\begin{aligned} c_T(X) &= \det(XI_3 - A) \\ &= X^3 - 5X^2 + 8X - 4 \\ &= (X - 1)(X - 2)^2. \end{aligned}$$

We see at this point that we have two possibilities:

(a) $\nu_{\text{alg}}(1) = \nu_{\text{geom}}(1) = 1$ and $\nu_{\text{alg}}(2) = \nu_{\text{geom}}(2) = 2$.

In this case T is diagonalizable and the Jordan canonical form is $J = \text{diag}(1, 2, 2)$. Also $m_T(X) = (X - 1)(X - 2)$, so the $F[X]$-module V_T has invariant factors $f_1(X) = X - 2$ and $f_2(X) = (X - 1)(X - 2)$, and V_T has rank 2 as an $F[X]$-module. Thus, T has the rational canonical form

$$R = C(f_1(X)) \oplus C(f_2(X)) = \begin{bmatrix} 2 & 0 & 0 \\ 0 & 0 & -2 \\ 0 & 1 & 3 \end{bmatrix}.$$

(b) $\nu_{\text{alg}}(1) = \nu_{\text{geom}}(1) = 1$ and $\nu_{\text{alg}}(2) = 2$, $\nu_{\text{geom}}(2) = 1$.

In this case T is not diagonalizable, and the eigenvalue 2 has a single Jordan block, so its Jordan canonical form is

$$J = J_{1,1} \oplus J_{2,2} = \begin{bmatrix} 1 & 0 & 0 \\ 0 & 2 & 1 \\ 0 & 0 & 2 \end{bmatrix}.$$

Also, $m_T(X) = (X-1)(X-2)^2 = c_T(X)$ so that V_T is a cyclic $F[X]$-module (Lemma 4.11 (3)) and T has the rational canonical form

$$R = C(m_T(X)) = \begin{bmatrix} 0 & 0 & 4 \\ 1 & 0 & -8 \\ 0 & 1 & 5 \end{bmatrix}.$$

To decide between these two alternatives, we may proceed in either of two ways:

(1) Compute $\nu_{\text{geom}}(2)$. If it is 2, we are in case (a), and if it is (1), we are in case (b).
(2) Compute $(T - 1)(T - 2) = (T - 1_V)(T - 2 \cdot 1_V)$. If it is 0 we are in case (a), if it is not, we are in case (b).

Since we are in any event interested in (generalized) eigenvectors, we choose the first method. We find

$$\text{Ker}(T - 2 \cdot 1_V) = \left\langle \begin{bmatrix} 1 \\ 2 \\ 0 \end{bmatrix}, \begin{bmatrix} 1 \\ 0 \\ 2 \end{bmatrix} \right\rangle = \langle v_2, v_3 \rangle,$$

so we are in case (a) here. Also,

$$\text{Ker}(T - 1_V) = \left\langle \begin{bmatrix} 2 \\ 3 \\ 2 \end{bmatrix} \right\rangle = \langle v_1 \rangle,$$

so in the basis $\mathcal{B} = \{v_1, v_2, v_3\}$ (note the order), $[T]_{\mathcal{B}} = J$. $\square$

(5.5) **Example.** *Let $V = F^3$ and let $V \to V$ be the linear transformation with the matrix*

$$[T]_C = A = \begin{bmatrix} 5 & -5 & 7 \\ -4 & 7 & -6 \\ -5 & 7 & -7 \end{bmatrix}$$

in the standard basis C. Find out "almost everything" about T.

Solution. We compute that $c_T(X) = (X - 1)(X - 2)^2$, exactly the same as in Example 5.4. Thus, we have the same two possibilities and we will proceed in the same way. Now we find that

$$\text{Ker}(T - 2 \cdot 1_V) = \left\langle \begin{bmatrix} 1 \\ 2 \\ 1 \end{bmatrix} \right\rangle = \langle v_2 \rangle,$$

so in this example we are in case (b). To find the basis for the Jordan canonical form, we find a vector $v_2 \in F^3$ with $(T - 2)(v_3) = v_2$. By solving a system of linear equations we find that we may take

$$v_3 = \begin{bmatrix} -3 \\ -2 \\ 0 \end{bmatrix}.$$

Also,

$$\text{Ker}(T - 1_V) = \left\langle \begin{bmatrix} -3 \\ -1 \\ 1 \end{bmatrix} \right\rangle = \langle v_1 \rangle.$$

Thus, if $\mathcal{B} = \{v_1, v_2, v_3\}$ (note the order), then $\mathcal{B}$ is a basis of F^3 and $[T]_{\mathcal{B}} = J$.

As a practical matter, there is a second method for finding a suitable basis of F^3. Note that $\dim(\text{Ker}(T - 2 \cdot 1_V)^2) = 2$. Pick any vector v_3', which is in $\text{Ker}(T - 2 \cdot 1_V)^2$ but not in $\text{Ker}(T - 2 \cdot 1_V)$, say $v_3' = [2 \quad 0 \quad -1]^t$. Then let $v_2' = (T - 2)(v_3') = [-1 \quad -2 \quad -1]^t$. If $\mathcal{B}' = \{v_1, v_2', v_3'\}$, then we also have $[T]_{\mathcal{B}'} = J$. $\square$

(5.6) **Example.** *Let V be a vector space over F of dimension 8, let $\mathcal{B} = \{v_i\}_{i=1}^{8}$ be a basis of V, and suppose that $T \in \text{End}_F(V)$ is a linear transformation such that*

$$[T]_{\mathcal{B}} = A$$
$$= J_{2,2} \oplus J_{2,2} \oplus J_{2,1} \oplus J_{1,2} \oplus J_{1,1}$$
$$= \begin{bmatrix} 2 & 1 & 0 & 0 & 0 & 0 & 0 & 0 \\ 0 & 2 & 0 & 0 & 0 & 0 & 0 & 0 \\ 0 & 0 & 2 & 1 & 0 & 0 & 0 & 0 \\ 0 & 0 & 0 & 2 & 0 & 0 & 0 & 0 \\ 0 & 0 & 0 & 0 & 2 & 0 & 0 & 0 \\ 0 & 0 & 0 & 0 & 0 & 1 & 1 & 0 \\ 0 & 0 & 0 & 0 & 0 & 0 & 1 & 0 \\ 0 & 0 & 0 & 0 & 0 & 0 & 0 & 1 \end{bmatrix}.$$

That is, $[T]_{\mathcal{B}}$ is already in Jordan canonical form. Compute $\mu(V_T)$, the cyclic decomposition of V_T, and the rational canonical form of T.

Solution. Since $\nu_{\text{geom}}(2) = 3$ and $\nu_{\text{geom}}(1) = 2$, it follows from Corollary 4.48 (4) that $\mu(V_T) = 3$. Moreover, the elementary divisors of T are $(X-2)^2$, $(X-2)^2$, $(X-2)$, $(X-1)^2$, and $(X-1)$, so the invariant factors are (see the proof of Theorem 3.7.15):

$$f_3(X) = (X-2)^2(X-1)^2$$
$$f_2(X) = (X-2)^2(X-1)$$
$$f_1(X) = (X-2).$$

Since

$$\text{Ann}(v_2) = \text{Ann}(v_4) = \langle (X-2)^2 \rangle$$
$$\text{Ann}(v_5) = \langle (X-2) \rangle$$
$$\text{Ann}(v_7) = \langle (X-1)^2 \rangle$$
$$\text{Ann}(v_8) = \langle (X-1) \rangle,$$

it follows (by Lemma 3.7.18) that

$$\text{Ann}(w_3) = \langle f_3(X) \rangle$$
$$\text{Ann}(w_2) = \langle f_2(X) \rangle$$
$$\text{Ann}(w_1) = \langle f_1(X) \rangle$$

where $w_3 = v_2 + v_7$, $w_2 = v_4 + v_8$, and $w_1 = v_5$. Thus

$$V_T \cong Rw_1 \oplus Rw_2 \oplus Rw_3$$

and the rational canonical form R of T is the matrix

$$R = C(f_1(X)) \oplus C(f_2(X)) \oplus C(f_3(X)).$$

Moreover, $Q^{-1}AQ = R$ if $Q \in M_8(F)$ is the invertible matrix

$$Q = P_{\mathcal{B}}^{\mathcal{C}} = \begin{bmatrix} 0 & 0 & 0 & 0 & 0 & 1 & 4 & 12 \\ 0 & 0 & 0 & 0 & 1 & 2 & 4 & 8 \\ 0 & 0 & 1 & 4 & 0 & 0 & 0 & 0 \\ 0 & 1 & 2 & 4 & 0 & 0 & 0 & 0 \\ 1 & 0 & 0 & 0 & 0 & 0 & 0 & 0 \\ 0 & 0 & 0 & 0 & 0 & 1 & 2 & 3 \\ 0 & 0 & 0 & 0 & 1 & 1 & 1 & 1 \\ 0 & 1 & 1 & 1 & 0 & 0 & 0 & 0 \end{bmatrix}$$

where

$$\mathcal{C} = \{w_1, w_2, Tw_2, T^2w_2, w_3, Tw_3, T^2w_3, T^3w_3\}.$$

□

(5.7) *Remark.* We have given information about the rational canonical form in the above examples in order to fully illustrate the situation. However, as we have remarked, it is the Jordan canonical form that is really of interest. In particular, we note that while the methods of the current section produce the Jordan canonical form via computation of generalized eigenvectors, and then the rational canonical form is computed (via Lemma 3.7.18) from the Jordan canonical form (e.g., Examples 5.3 and 5.6), it is the other direction that is of more interest. This is because the rational canonical form of T can be computed via elementary row and column operations from the matrix $XI_n - [T]_{\mathcal{B}}$. The Jordan canonical form can then be computed from this information. This approach to computations will be considered in Chapter 5.

We will conclude this survey of computations by sketching the algorithm that has been used in the previous examples to compute a basis of V in which a given linear transformation $T \in \mathrm{End}_F(V)$ is in Jordan canonical form. The Jordan canonical form can be computed solely by solving systems of homogeneous linear equations, provided the eigenvalues of T are available. It should, however, be emphasized again that the computation of eigenvalues involves solving a polynomial equation.

Suppose that $T \in \mathrm{End}_F(V)$ has $m_T(X) = (X - \lambda)^r q(X)$ where $q(\lambda) \neq 0$. Then the primary decomposition theorem allows one to write

$$V_T \cong W_1 \oplus \cdots \oplus W_t \oplus \widetilde{V}$$

where $W_i = F[X]w_i$ is cyclic with $\mathrm{Ann}(w_i) = \langle (X - \lambda)^{r_i} \rangle$ and

$$W = W_1 \oplus \cdots \oplus W_t$$

is the generalized eigenspace of the eigenvalue λ. By rearranging the order of the r_i if necessary we may assume that

$$r = r_1 \geq r_2 \geq \cdots \geq r_t$$

(and of course $r_1 + \cdots + r_t = \nu_{\mathrm{alg}}(\lambda)$.) We will let

$$r_1 = \tilde{r}_1 > \tilde{r}_2 > \cdots > \tilde{r}_s = r_t$$

denote the distinct r_i. It is the vectors w_i that are crucial in determining the Jordan blocks corresponding to the eigenvalue λ in the Jordan canonical form of T. We wish to see how these vectors can be picked out of the generalized eigenspace W corresponding to λ. First observe that

$$\mathcal{B} = \{w_1, Sw_1, \ldots, S^{r_1 - 1}w_1, w_2, \ldots, S^{r_2 - 1}w_2, \ldots, S^{r_t - 1}w_t\},$$

where $S = T - \lambda 1_V$, is a basis of W. This is just the basis (in a different order) produced in Theorem 4.38. It is suggestive to write this basis in the following table:

Table 5.1. Jordan basis of T

$$
\begin{array}{ccccccc}
w_1 & \cdots & w_{k_1} & & & & \\
Sw_1 & \cdots & Sw_{k_1} & & & & \\
\vdots & & \vdots & w_{k_1+1} & \cdots & w_{k_2} & \\
& & & Sw_{k_1+1} & \cdots & Sw_{k_2} & \\
\vdots & & \vdots & \vdots & & \vdots & \cdots \quad w_t \\
\vdots & & \vdots & \vdots & & \vdots & \\
S^{r_1-1}w_1 & \cdots & S^{r_1-1}w_{k_1} & S^{\tilde{r}_2-1}w_{k_1+1} & \cdots & S^{\tilde{r}_2-1}w_{k_2} & \cdots \quad S^{r_t-1}w_t
\end{array}
$$

(Of course, W is an $F[X]$-module of rank t, and as we observe from Table 5.1, $t = \dim_F \mathrm{Ker}(T - \lambda 1_V)$, as we expect from Corollary 4.48 (4).) Note that the linear span of each column is the cyclic submodule $W_i = Rw_i$ while the last k rows form a basis of $\mathrm{Ker}(T - \lambda 1_V)^k$. As a concrete example of this scheme, suppose that $T : F^{13} \to F^{13}$ is multiplication by the matrix

$$
A = J_{2,4} \oplus J_{2,4} \oplus J_{2,2} \oplus J_{2,2} \oplus J_{2,1}.
$$

Using the standard basis on F^{13} the above table becomes

Table 5.2. Jordan basis of $T : F^{13} \to F^{13}$

$$
\begin{array}{cccc}
e_4 & e_8 & & \\
e_3 & e_7 & & \\
e_2 & e_6 & e_{10} & e_{12} \\
e_1 & e_5 & e_9 & e_{11} \quad e_{13}
\end{array}
$$

Thus, the bottom row is a basis of $\mathrm{Ker}(T - 2 \cdot 1_V)$, the bottom two rows are a basis of $\mathrm{Ker}(T - 2 \cdot 1_V)^2$, etc. These tables suggest that the top vectors of each column are vectors that are in $\mathrm{Ker}(T - \lambda 1_V)^k$ for some k, but they are not also $(T - \lambda 1_V)w$ for some other generalized eigenvector w. This observation can then be formalized into a computational scheme to produce the generators w_i of the primary cyclic submodules W_i.

We first introduce some language to describe entries in Table 5.1. It is easiest to do this specifically in the example considered in Table 5.2, as otherwise the notation is quite cumbersome, but the generalization is clear. We will number the rows of Table 5.1 *from the bottom up*. We then say (in the specific case of Table 5.2) that $\{e_{13}\}$ is the **tail** of row 1, $\{e_{10}, e_{12}\}$ is the tail of row 2, the tail of row 3 is empty, and $\{e_4, e_8\}$ is the tail of row 4. Thus, the tail of any row consists of those vectors in the row that are at the top of some column.

Clearly, in order to determine the basis $\mathcal{B}$ of W it suffices to find the tails of each row. We do this as follows:

For $i = 0, 1, 2, \ldots$, let $V_\lambda^{(i)} = \text{Ker}(S^i)$ (where, as above, we let $S = T - \lambda 1_V$). Then

$$\{0\} = V_\lambda^{(0)} \subseteq V_\lambda^{(1)} \subseteq V_\lambda^{(2)} \subseteq \cdots \subseteq V_\lambda^{(r)} = W.$$

Let $d_i = \dim(V_\lambda^{(i)})$ so that

$$0 = d_0 < d_1 \leq d_2 \leq \cdots \leq d_r = \nu_{\text{alg}}(\lambda).$$

Note that r is the smallest value of i for which $d_i = \nu_{\text{alg}}(\lambda)$, where r is as above, i.e., where $m_T(X) = (X - \lambda)^r q(X)$ with $q(\lambda) \neq 0$. This observation is useful for determining r without first computing $m_T(X)$ explicitly. In particular, we will have $r = 1$, i.e., $d_1 = \nu_{\text{alg}}(\lambda)$, if and only if the generalized eigenspace of T corresponding to λ is the eigenspace of T corresponding to λ. Note that in any case $V_\lambda^{(r)} = W$, so $V_\lambda^{(r+1)} = V_\lambda^{(r)}$, and $d_{r+1} = d_r$.

Observe that for $1 \leq i \leq r$, row i of Table 5.1 has length $d_i - d_{i-1}$, and hence, its tail has length

$$(d_i - d_{i-1}) - (d_{i+1} - d_i).$$

We make the following observation:

Let $\overline{V}_\lambda^{(i+1)}$ be any complement of $V_\lambda^{(i)}$ in $V_\lambda^{(i+1)}$, i.e., any subspace of $V_\lambda^{(i+1)}$ such that

$$V_\lambda^{(i+1)} = V_\lambda^{(i)} \oplus \overline{V}_\lambda^{(i+1)}.$$

Then $S(\overline{V}_\lambda^{(i+1)})$ and $V_\lambda^{(i-1)}$ are both subspace of $V_\lambda^{(i)}$ and

(5.8) Claim. $S(\overline{V}_\lambda^{(i+1)}) \cap V_\lambda^{(i-1)} = \{0\}$.

To see this, let $v \in S(\overline{V}_\lambda^{(i+1)}) \cap V_\lambda^{(i-1)}$. Then $v = S(\overline{v})$ for some $\overline{v} \in \overline{V}_\lambda^{(i+1)}$, and also $S^{i-1}(v) = 0$. But

$$0 = S^{i-1}(v) = S^{i-1}(S(\overline{v})) = S^i(\overline{v})$$

implies that $\overline{v} = 0$, by the definition of $\overline{V}_\lambda^{(i+1)}$.

Now for our algorithm. We begin at the top, with $i = r$, and work down. By Claim 5.8, we may choose a complement $\overline{V}_\lambda^{(i)}$ of $V_\lambda^{(i-1)}$ in $V_\lambda^{(i)}$, which contains the subspace $S(\overline{V}_\lambda^{(i+1)})$. Let $\overline{\overline{V}}_\lambda^{(i)}$ be any complement of

$S(\overline{V}_\lambda^{(i+1)})$ in $\overline{V}_\lambda^{(i)}$. Then a basis for $\overline{\overline{V}}_\lambda^{(i)}$ gives the tail of row i of Table 5.1. (Note that $\overline{V}_\lambda^{(r+1)} = \{0\}$, so at the first stage of the process $\overline{\overline{V}}_\lambda^{(r)} = \overline{V}_\lambda^{(r)}$ is any complement of $V_\lambda^{(r-1)}$ in $V_\lambda^{(r)} = W$. Also, at the last stage of the process $V_\lambda^{(0)} = \{0\}$, so $\overline{V}_\lambda^{(1)} = V_\lambda^{(1)}$ and $\overline{\overline{V}}_\lambda^{(1)}$ is any complement of $S(\overline{V}_\lambda^{(2)})$ in $V_\lambda^{(1)}$, the eigenspace of T corresponding to λ.)

(5.9) Example. We will illustrate this algorithm with the linear transformation $T : F^{13} \to F^{13}$ whose Jordan basis is presented in Table 5.2. Of course, this transformation is already in Jordan canonical form, so the purpose is just to illustrate how the various subspace $V_\lambda^{(i)}$, $\overline{V}_\lambda^{(i)}$, and $\overline{\overline{V}}_\lambda^{(i)}$ relate to the basis $\mathcal{B} = \{e_1, \dots, e_{13}\}$. Since there is only one eigenvalue, for simplicity, let $V_2^{(i)} = V_i$, with a similar convention for the other spaces. Then

$$V_5 = V_4 = F^{13}$$
$$V_4 = \langle e_1, \dots, e_{13} \rangle$$
$$V_3 = \langle e_1, e_2, e_3, e_5, e_6, e_7, e_9, e_{10}, e_{11}, e_{12}, e_{13} \rangle$$
$$V_2 = \langle e_1, e_2, e_5, e_6, e_9, e_{10}, e_{11}, e_{12}, e_{13} \rangle$$
$$V_1 = \langle e_1, e_5, e_9, e_{11}, e_{13} \rangle$$
$$V_0 = \{0\},$$

while, for the complementary spaces we may take

$$\overline{V}_5 = \{0\}$$
$$\overline{V}_4 = \langle e_4, e_8 \rangle$$
$$\overline{V}_3 = \langle e_3, e_7 \rangle$$
$$\overline{V}_2 = \langle e_2, e_6, e_{10}, e_{12} \rangle$$
$$\overline{V}_1 = \langle e_1, e_5, e_9, e_{11}, e_{13} \rangle.$$

Since $V_4 = F^{13}$, we conclude that there are 4 rows in the Jordan table of T, and since $\overline{V}_4 = \overline{\overline{V}}_4$, we conclude that the tail of row 4 is $\{e_4, e_8\}$. Since $S(e_4) = e_3$ and $S(e_8) = e_7$, we see that $S(\overline{V}_4) = \overline{V}_3$ and hence the tail of row 3 is empty. Now $S(e_3) = e_2$ and $S(e_7) = e_6$ so that we may take $\overline{\overline{V}}_2 = \langle e_{10}, e_{12} \rangle$, giving $\{e_{10}, e_{12}\}$ as the tail of row 2. Also, $S(e_2) = e_1$, $S(e_6) = e_5$, $S(e_{10}) = e_9$, and $S(e_{12}) = e_{11}$, so we conclude that $\overline{\overline{V}}_1 = \langle e_{13} \rangle$, i.e., the tail of row 1 is $\{e_{13}\}$.

Examples 5.3 and 5.4 illustrate simple cases of the algorithm described above for producing the Jordan basis. We will present one more example, which is (slightly) more complicated.

(5.10) Example. *Let $V = F^4$ and let $T : V \to V$ be the linear transformation with the matrix*

$$[T]_C = A = \begin{bmatrix} -1 & 1 & 1 & 1 \\ -3 & 3 & 1 & 1 \\ -8 & 2 & 5 & 3 \\ 2 & 0 & -1 & 1 \end{bmatrix}$$

in the standard basis C. Find the Jordan canonical form of T and a basis B of V such that $[T]_B$ is in Jordan canonical form.

Solution. We compute that

$$c_T(X) = c_A(X) = (X - 2)^4.$$

Thus, there is only one eigenvalue, namely, 2, and $\nu_{\mathrm{alg}}(2) = 4$. Now find the eigenspace of 2, i.e.,

$$V_2^{(1)} = \mathrm{Ker}(T - 2 \cdot 1_V) = \left\langle \begin{bmatrix} 1 \\ 1 \\ 2 \\ 0 \end{bmatrix}, \begin{bmatrix} 1 \\ 1 \\ 0 \\ 2 \end{bmatrix} \right\rangle,$$

so $d_1 = 2$. We also find that

$$V_2^{(2)} = \mathrm{Ker}((T - 2 \cdot 1_V)^2) = V,$$

so $d_2 = 4$. Hence, we choose $\{w_1, w_2\} \subseteq V$ so that

$$V_2^{(1)} \oplus \langle w_1, w_2 \rangle = V$$

in order to obtain a Jordan basis of V.

Take

$$w_1 = \begin{bmatrix} 1 \\ 0 \\ 0 \\ 0 \end{bmatrix} \quad \text{and} \quad w_2 = \begin{bmatrix} 0 \\ 1 \\ 0 \\ 0 \end{bmatrix}.$$

Then compute

$$v_1 = (T - 2 \cdot 1_V)(w_1) = \begin{bmatrix} -3 \\ -3 \\ -8 \\ 2 \end{bmatrix}$$

and

$$v_3 = (T - 2 \cdot 1_V)(w_2) = \begin{bmatrix} 1 \\ 1 \\ 2 \\ 0 \end{bmatrix}.$$

Setting

$$v_1 = (T - 2 \cdot 1_V)(w_1), \qquad v_2 = w_1,$$
$$v_3 = (T - 2 \cdot 1_V)(w_2), \qquad v_4 = w_2,$$

we obtain a basis $\mathcal{B} = \{v_1, v_2, v_3, v_4\}$ such that $[T]_\mathcal{B} = J$. The table corresponding to Table 5.1 is

$$\begin{array}{cc} v_2 & v_4 \\ v_1 & v_3 \end{array}$$

We also note that $V_T \cong Rv_2 \oplus Rv_4$ so that $\mu(V_T) = 2$.   □

4.6 Inner Product Spaces and Normal Linear Transformations

In this section we will study vector spaces possessing an additional structure, that of an inner product, as well as endomorphisms of these spaces which are "normal," normality being a property defined in terms of an inner product.

(6.1) Definition. *Let V be a vector space over $F = \mathbf{R}$ or $\mathbf{C}$. An **inner product** on V is a function $(:) : V \times V \to F$ such that for all $u, v, w \in V$ and $a \in F$,*

(1) $(u + v : w) = (u : w) + (v : w)$,
(2) $(au : v) = a(u : v)$,
(3) $(u : v) = \overline{(v : u)}$, *and*
(4) $(u : u) > 0$ *if $u \neq 0$.*

*A vector space V, together with an inner product $(:)$ is called an **inner product space**.*

The bar in Definition 6.1 (3) refers to conjugation in the field F. Of course, if $F = \mathbf{R}$, conjugation in F is trivial, but it is convenient to handle the real and complex cases simultaneously.

(6.2) Examples.

(1) The standard inner product on F^n is defined by

$$(u : v) = \sum_{j=1}^{n} u_j \overline{v}_j.$$

(2) If $A \in M_{m,n}(F)$, then let $A^* = \overline{A}^t$ where $\overline{A}$ denotes the matrix obtained from A by conjugating all the entries. A^* is called the **Hermitian transpose** of A. If we define

$$(A : B) = \text{Tr}(AB^*)$$

then (:) defines an inner product on $M_{m,n}(F)$. If we identify F^n with $M_{1,n}(F)$, then this inner product agrees with the inner product in (1).

(3) Suppose that $T : V \to W$ is an injective linear transformation. If W is an inner product space, then T induces an inner product on V by the formula

$$(v_1 : v_2) = (T(v_1) : T(v_2)).$$

In particular, every subspace of an inner product space inherits an inner product.

(4) Let $V = C([0, 1], F)$ be the F-vector space of continuous F-valued functions on the unit interval $[0, 1]$. Then an inner product on V is defined by

$$(f : g) = \int_0^1 f(x)\overline{g(x)}\,dx.$$

(6.3) Definition. *If V is an inner product space, then $\|v\| = (v : v)^{1/2} \in \mathbf{R}$ is called the **norm** of v.*

The norm of a vector v is well defined by Definition 6.2 (4). There are a number of standard inequalities related to the norm.

(6.4) Proposition. *If V is an inner product space with inner product (:), if $u, v \in V$ and $a \in F$, then*

(1) $\|au\| = |a|\|u\|$,
(2) $\|u\| > 0$ *if $u \neq 0$,*
(3) **(Cauchy–Schwartz)** $|(u : v)| \le \|u\|\|v\|$, *and*
(4) **(Triangle inequality)** $\|u + v\| \le \|u\| + \|v\|$.

Proof. (1) and (2) are immediate from the definitions.

(3) Let $u, v \in V$ be arbitrary and let $x \in \mathbf{R}$. By Definition 6.1 (4),

$$(u + xv : u + xv) \ge 0$$

(and the inequality is strict unless u is a multiple of v). Thus,

$$
\begin{aligned}
0 \le (u + xv : u + xv) &= \|u\|^2 + ((u : v) + (v : u))x + \|v\|^2 x^2 \\
&= \|u\|^2 + 2(\mathrm{Re}(u : v))x + \|v\|^2 x^2 \\
&\le \|u\|^2 + 2|(u : v)|x + \|v\|^2 x^2.
\end{aligned}
$$

Since this holds for all $x \in \mathbf{R}$, we conclude that

$$f(x) = \|u\|^2 + 2|(u : v)|x + \|v\|^2 x^2$$

is a real quadratic function, which is always nonnegative. Thus, the discriminant of $f(x)$ must be nonpositive, i.e.,

$$4(u:v)^2 - 4\|u\|^2\|v\|^2 \le 0,$$

and the Cauchy–Schwartz inequality follows immediately from this.

(4) By the Cauchy–Schwartz inequality,

$$\operatorname{Re}(u:v) \le |(u:v)| \le \|v\|\|v\|.$$

Hence,

$$\begin{aligned}
\|u+v\|^2 &= (u+v:u+v) \\
&= \|u\|^2 + (u:v) + (v:u) + \|v\|^2 \\
&= \|u\|^2 + 2\operatorname{Re}(u:v) + \|v\|^2 \\
&\le \|u\|^2 + 2\|u\|\|v\| + \|v\|^2 \\
&= (\|u\| + \|v\|)^2.
\end{aligned}$$

Taking square roots gives the triangle inequality. $\qquad\square$

(6.5) *Remark.* The Cauchy–Schwartz inequality implies that, if u and v are nonzero, then

$$-1 \le \frac{(u:v)}{\|u\|\,\|v\|} \le 1.$$

Thus, we can define the angle between u and v by means of the equation

$$\cos\theta = \frac{(u:v)}{\|u\|\,\|v\|}, \qquad 0 \le \theta \le \pi.$$

It is this equation that allows us to introduce the geometric idea of angle into an arbitrary real or complex vector space by means of the algebraic notion of inner product.

(6.6) **Definition.** *If V is an inner product space, then vectors u, $v \in V$ are said to be* **orthogonal** *if $(u:v) = 0$. A subset $S \subseteq V$ is said to be* **orthogonal** *if every pair of distinct vectors in S is orthogonal. S is said to be* **orthonormal** *if S is orthogonal and if every vector in S has norm 1.*

(6.7) **Proposition.** *Let V be an inner product space and let $S \subseteq V$ be an orthogonal set of nonzero vectors. Then S is linearly independent.*

Proof. Suppose that $u_1, \ldots, u_k$ are distinct elements of S and suppose that there is an equation

$$a_1 u_1 + \cdots + a_k u_k = 0$$

where $a_1, \ldots, a_k \in F$. Then, for $1 \le i \le k$,

$$\begin{aligned}
0 &= (0:u_i) \\
&= (a_1 u_1 + \cdots + a_k u_k : u_i) \\
&= a_1(u_1:u_i) + \cdots + a_k(u_k:u_i) \\
&= a_i(u_i:u_i).
\end{aligned}$$

But $(u_i:u_i) > 0$, so we conclude that $a_i = 0$ for all i. $\qquad\square$

Proved in exactly the same manner as Proposition 6.7 is the following useful fact.

(6.8) Lemma. *Let V be an inner product space and suppose that $\mathcal{B} = \{v_1, \ldots, v_n\}$ is an orthonormal basis of V. Then for any $u \in V$, the coordinate matrix of u is*

$$[u]_{\mathcal{B}} = \begin{bmatrix} (u : v_1) \\ \vdots \\ (u : v_n) \end{bmatrix}.$$

Proof. Exercise. □

The classical Gram–Schmidt orthogonalization process allows one to produce an orthonormal basis of any finite-dimensional inner product space; this is the content of the next result.

(6.9) Theorem. *Every finite-dimensional inner product space V has an orthonormal basis.*

Proof. Let $\mathcal{C} = \{u_1, \ldots, u_n\}$ be any basis of V. We will produce an orthonormal basis inductively. Let $v_1 = u_1/\|u_1\|$. Now assume that an orthonormal set $\{v_1, \ldots, v_k\}$ has been chosen so that

$$\langle v_1, \ldots, v_t \rangle = \langle u_1 \ldots, u_t \rangle$$

for $1 \le t \le k$. Then define

$$\widehat{v}_{k+1} = u_{k+1} - \sum_{j=1}^{k} (u_{k+1} : v_j) v_j,$$

and set $v_{k+1} = \widehat{v}_{k+1}/\|\widehat{v}_{k+1}\|$. We leave to the reader the details of verifying the validity of this construction and the fact that an orthonormal basis has been produced. □

(6.10) Definition. *Suppose that V is an inner product space and $W \subseteq V$ is a subspace. Let*

$$W^{\perp} = \{u \in V : (u : w) = 0 \quad \text{for all } w \in W\}.$$

The subspace $W^{\perp}$ of V is called the **orthogonal complement** *of W.*

(6.11) Proposition. *Let V be a finite-dimensional inner product space and let $W \subseteq V$ be a subspace. Then*

(1) $W \cap W^{\perp} = \langle 0 \rangle$,

(2) $(W^\perp)^\perp = W$,
(3) $\dim W + \dim W^\perp = \dim V$, and
(4) $V \cong W \oplus W^\perp$.

Proof. Choose an orthonormal basis $\mathcal{B} = \{v_1, \ldots, v_n\}$ of V in which $\{v_1, \ldots, v_k\}$ is a basis of W. This is possible by the algorithm of Theorem 6.9. Then

$$W^\perp = \langle v_{k+1}, \ldots, v_n \rangle.$$

This equation immediately implies (2) and (3). (1) is clear and (4) follows from (1) and (3). $\square$

(6.12) *Remark.* If V is a vector space over F, its **dual space** V^* is defined to be $V^* = \mathrm{Hom}_F(V, F)$. If V is finite-dimensional, then, by Corollary 3.4.10, V and V^* have the same dimension and so are isomorphic, but in general there is no natural isomorphism between the two. However, an inner product on V gives a *canonical* isomorphism $\phi : V \to V^*$ defined as follows: For $y = V$, $\phi(y) \in V^*$ is the homomorphism $\phi(y)(x) = (x : \overline{y})$. To see that ϕ is an isomorphism, one only needs to observe that ϕ is injective since $\dim V = \dim V^*$. But if $y \neq 0$ then $\phi(y)(\overline{y}) = (\overline{y} : \overline{y}) > 0$, so $\phi(y) \neq 0$ and $\mathrm{Ker}(\phi) = \langle 0 \rangle$.

(6.13) **Theorem.** *Let V be a finite-dimensional inner product space. Then for every $T \in \mathrm{End}_F(V)$ there exists a unique $T^* \in \mathrm{End}_F(V)$ such that*

$$(Tv : w) = (v : T^*w)$$

*for all $v, w \in V$. T^* is called the **adjoint** of T.*

Proof. Let $w \in V$. Then $h_w : V \to F$ defined by $h_w(v) = (Tv : w)$ is an element of the dual space V^*. Thus (by Remark 6.12), there exists a unique $\widehat{w} \in V$ such that

$$(Tv : w) = (v : \widehat{w})$$

for all $v \in V$. Let $T^*(w) = \widehat{w}$. We leave it as an exercise to verify that $T^* \in \mathrm{End}_F(V)$. $\square$

(6.14) **Lemma.** *Let V be an inner product space and let $S, T \in \mathrm{End}_F(V)$. Then $(ST)^* = T^*S^*$.*

Proof. Exercise. $\square$

(6.15) **Lemma.** *Let V be a finite-dimensional inner product space and let $\mathcal{B}$ be an orthonormal basis of V. If $T : V \to V$ is a linear transformation and $[T]_\mathcal{B} = A = [a_{ij}]$, then*

$$[T^*]_\mathcal{B} = A^* = \overline{A}^t.$$

Proof. Let $\mathcal{B} = \{v_1, \ldots, v_n\}$. If $T^*(v_i) = \sum_{k=1}^n b_{ki}v_k$, then, according to Lemma 6.8,

$$
\begin{aligned}
b_{ji} &= \overline{(T^*(v_i) : v_j)} \\
&= \overline{(v_j : T^*(v_i))} \\
&= \overline{(T(v_j) : v_i)} \\
&= \overline{\left(\sum_{k=1}^n a_{kj}v_k : v_i\right)} \\
&= \bar{a}_{ij}.
\end{aligned}
$$

Thus, $[T^*]_{\mathcal{B}} = \overline{A}^t$, as required.   □

(6.16) Definition. *Let V be an inner product space, and let $T : V \to V$ be a linear transformation.*

(1) *T is **normal** if $TT^* = T^*T$, i.e., if T commutes with its adjoint.*
(2) *T is **self-adjoint** if $T = T^*$, i.e., if T is its own adjoint.*
(3) *T is **unitary** if $T^* = T^{-1}$.*

Let $A \in M_n(F)$ be an $n \times n$ matrix.

(1′) *A is **normal** if $AA^* = A^*A$.*
(2′) *A is **self-adjoint** if $A = A^*$.*
(3′) *A is **unitary** if $AA^* = I_n$.*

(6.17) *Remarks.*

(1) If T is self-adjoint or unitary, then it is normal.
(2) If $F = \mathbf{C}$, then a self-adjoint linear transformation (or matrix) is called **Hermitian**, while if $F = \mathbf{R}$, then a self-adjoint transformation (or matrix) is called **symmetric**. If $F = \mathbf{R}$, then a unitary transformation (or matrix) is called **orthogonal**.
(3) Lemma 6.15 shows that the concept of normal is essentially the same for transformations on finite-dimensional vector spaces and for matrices. A similar comment applies for self-adjoint and unitary.

The importance of unitary transformations arises because of the following geometric property which characterizes them.

(6.18) Proposition. *Let $T : V \to V$ be a linear transformation on the finite-dimensional inner product space V. The following are equivalent:*

(1) *T is unitary.*
(2) *$(Tu : Tv) = (u : v)$ for all $u, v \in V$.*
(3) *$\|Tv\| = \|v\|$ for all $v \in V$.*

Proof. (1) $\Rightarrow$ (2). Let $u, v \in V$. Then

$$
\begin{aligned}
(Tu : Tv) &= (u : T^*Tv) \\
&= (u : T^{-1}Tv) \\
&= (u : v).
\end{aligned}
$$

(2) $\Rightarrow$ (1). $(u : v) = (Tu : Tv) = (u : T^*Tv)$, so

$$(u : T^*Tv - v) = 0 \qquad \text{for all} \quad u, v \in V.$$

Taking $u = T^*Tv - v$ shows that $(u : u) = 0$. Thus $u = 0$, i.e.,

$$T^*Tv = v \quad \text{for all } v \in V.$$

Therefore, $T^*T = 1_V$, and since V is finite dimensional, this shows that $T^* = T^{-1}$.

(2) $\Rightarrow$ (3). Recall that $\|Tv\|^2 = (Tv : Tv)$.

(3) $\Rightarrow$ (2). Let $u, v \in V$. Then

$$
\begin{aligned}
\|T(u+v)\|^2 &= (T(u+v) : T(u+v)) \\
&= (Tu : Tu) + (Tu : Tv) + (Tv : Tu) + (Tv : Tv) \\
&= (u : u) + 2\mathrm{Re}\,((Tu : Tv)) + (v : v),
\end{aligned}
$$

while similarly

$$(u + v : v + v) = (u : u) + 2\mathrm{Re}\,((u : v)) + (v : v).$$

Thus, we conclude that

$$\mathrm{Re}((Tu : Tv)) = \mathrm{Re}(u : v).$$

Applying the same argument to $u + iv$, we obtain

$$\mathrm{Re}\,((Tu : iTv)) = \mathrm{Re}\,((Tu : T(iv))) = \mathrm{Re}(u : iv).$$

But it is easy to check that

$$\mathrm{Re}((x : iy)) = \mathrm{Im}((x : y))$$

for any $x, y \in V$, so we have

$$(Tu : Tv) = (u : v)$$

for all $u, v \in V$. $\qquad\qquad\square$

The following result collects some useful properties of normal transformations.

(6.19) Lemma. *Let V be a finite-dimensional inner product space and let $T : V \to V$ be a normal transformation.*

(1) *If $f(X) \in F[X]$ then $f(T)$ is normal.*
(2) $\|Tv\| = \|T^*v\|$ *for all $v \in V$.*
(3) $\operatorname{Ker} T = (\operatorname{Im} T)^{\perp}$.
(4) *If $T^2v = 0$ then $Tv = 0$.*
(5) *$v \in V$ is an eigenvector for T with eigenvalue λ if and only if v is an eigenvector for T^* with eigenvalue $\overline{\lambda}$.*

Proof. (1) This follows immediately from the fact that $(aT^n)^* = \overline{a}(T^*)^n$ and the definition of normality.

(2)
$$\|Tv\|^2 = (Tv : Tv) = (v : T^*Tv)$$
$$= (v : TT^*v) = (T^*v : T^*v)$$
$$= \|T^*v\|^2.$$

(3) Suppose that $(u : Tv) = 0$ for all $v \in V$. Then $(T^*u : v) = 0$ for all $v \in V$. Thus $T^*u = 0$. By (2) this is true if and only if $Tu = 0$, i.e., $u \in (\operatorname{Im} T)^{\perp}$ if and only if $u \in \operatorname{Ker} T$.

(4) If $T^2v = 0$ then $Tv \in \operatorname{Ker} T \cap \operatorname{Im} T = \langle 0 \rangle$ by (3).

(5) By (1), the linear transformation $T - \lambda I$ is normal. Then by (2), $\|(T - \lambda I)v\| = 0$ if and only if $\|(T^* - \overline{\lambda}I)v\| = 0$, i.e., v is an eigenvector of T with eigenvalue λ if and only if v is an eigenvector of T^* with eigenvalue $\overline{\lambda}$. $\square$

(6.20) Theorem. *Suppose that V is a finite-dimensional inner product space and $T : V \to V$ is a normal linear transformation.*

(1) *The minimal polynomial $m_T(X)$ is a product of distinct irreducible factors.*
(2) *Eigenspaces of distinct eigenvalues of T are orthogonal.*
(3) *If V is a complex inner product space, then T is diagonalizable.*

Proof. (1) Let $p(X)$ be an irreducible factor of $m_T(X)$. We need to show that $p^2(X)$ does not divide $m_T(X)$. Suppose it did and let $v \in V$ with $p^2(T)(v) = 0$ but $p(T)(v) \neq 0$ (such a $v \in V$ exists by Theorem 3.7.1 and Lemma 3.7.17, or see Exercise 43 in Chapter 3). By Lemma 6.19 (1), $U = p(T)$ is normal, and $U^2(v) = 0$ but $U(v) \neq 0$, contradicting Lemma 6.19 (4).

(2) Suppose that $Tv_1 = \lambda_1 v_1$ and $Tv_2 = \lambda_2 v_2$ where v_1 and v_2 are nonzero and $\lambda_1 \neq \lambda_2$. Then

$$(v_1 : v_2) = ((\lambda_1 - \lambda_2)^{-1}(T - \lambda_2 I)v_1 : v_2)$$
$$= (\lambda_1 - \lambda_2)^{-1}(v_1 : (T - \lambda_2)^* v_2)$$
$$= 0,$$

since v_2 is a eigenvector of T^* with eigenvalue $\overline{\lambda}_2$ by Lemma 6.19 (5).

(3) Every irreducible polynomial over $\mathbf{C}$ is linear, so by (1), $m_T(X)$ is a product of distinct linear factors and so T is diagonalizable. $\quad\square$

(6.21) Corollary. (Spectral theorem) *If V is a finite-dimensional complex inner product space and $T : V \to V$ is a normal transformation, then V has an orthonormal basis of eigenvectors of T.*

Proof. By Theorem 6.20 (3), T is diagonalizable and by (2) the eigenspaces are orthogonal. It is only necessary to choose an orthonormal basis of each eigenspace. $\quad\square$

(6.22) *Remark.* If V is a real vector space and T is normal, T may not be diagonalizable, but from Theorem 6.20 it follows that the real Jordan canonical form of T (cf. Theorem 4.57) will consist of 1-by-1 ordinary Jordan blocks or 2-by-2 real Jordan blocks. For example, the second case occurs

The case of self-adjoint linear transformations (which are automatically normal) is of particular importance; such transformations are diagonalizable even in the real case.

(6.23) Theorem. (Spectral theorem, self-adjoint case) *Suppose that V is an inner product space and $T : V \to V$ is a self-adjoint linear transformation.*

(1) *All of the eigenvalues of T are real.*
(2) *V has an orthonormal basis of eigenvectors of T.*

Proof. We consider first the case $F = \mathbf{C}$. In this case (1) is immediate from Lemma 6.19 (5) and then (2) follows as in Corollary 6.21.

Now consider the case $F = \mathbf{R}$. In this case, (1) is true by definition. To prove part (2), we imbed V in a complex inner product space and apply part (1). Let $V_{\mathbf{C}} = V \oplus V$ and make $V_{\mathbf{C}}$ into a complex vector space by defining the scalar multiplication

(6.1) $$(a + bi)(u, v) = (au - bv, bu + av).$$

That is, $V \oplus 0$ is the real part of $V_{\mathbf{C}}$ and $0 \oplus V$ is the imaginary part. Define an inner product on $V_{\mathbf{C}}$ by

(6.2) $$((u_1, v_1) : (u_2, v_2)) = (u_1 : u_2) + (v_1 : v_2) + i((v_1 : u_2) - (u_1 : v_2)).$$

We leave it for the reader to check that Equations (6.1) and (6.2) make $V_{\mathbf{C}}$ into a complex inner product space. Now, extend the linear transformation T to a linear transformation $T_{\mathbf{C}} : V_{\mathbf{C}} \to V_{\mathbf{C}}$ by

$$T_{\mathbf{C}}(u, v) = (T(u), T(v)).$$

(In the language to be introduced in Section 7.2, $V_{\mathbf{C}} = \mathbf{C} \otimes_{\mathbf{R}} V$ and $T_{\mathbf{C}} = 1_{\mathbf{C}} \otimes_{\mathbf{R}} T$.) It is easy to check that $T_{\mathbf{C}}$ is a complex linear transformation of

$V_{\mathbf{C}}$, and in fact, $T_{\mathbf{C}}$ is self-adjoint. By part (1) applied to $T_{\mathbf{C}}$ we see that all the eigenvalues of $T_{\mathbf{C}}$ are real and $m_{T_{\mathbf{C}}}(X)$ is a product of distinct (real) linear factors. Thus, $m_{T_{\mathbf{C}}}(X) \in \mathbf{R}[X]$. If $f(X) \in \mathbf{R}[X]$, then

$$(6.3) \qquad f(T_{\mathbf{C}})(u, v) = (f(T)(u), f(T)(v)).$$

Equation (6.3) shows that $m_T(X) = m_{T_{\mathbf{C}}}(X)$ and we conclude that T is diagonalizable. Part (2) is completed exactly as in Corollary 6.21.   □

4.7 Exercises

1. Suppose R is a finite ring with $|R| = s$. Then show that $M_{m,n}(R)$ is finite with $|M_{m,n}(R)| = s^{mn}$. In particular, $|M_n(R)| = s^{n^2}$.

2. Prove Lemma 1.1.

3. Prove Lemma 1.2.

4. (a) Suppose that $A = [a_1 \ \cdots \ a_m]$ and $B \in M_{m,n}(R)$. Then show that $AB = \sum_{i=1}^m a_i \operatorname{row}_i(B)$.

   (b) Suppose

   $$C = \begin{bmatrix} c_1 \\ \vdots \\ c_n \end{bmatrix}$$

   and $B \in M_{m,n}(R)$. Then show that $BC = \sum_{i=1}^n c_i \operatorname{col}_i(B)$.
   This exercise shows that left multiplication of B by a row matrix produces a linear combination of the rows of B and right multiplication of B by a column matrix produces a linear combination of the columns of B.

5. Let $S \subseteq M_2(\mathbf{R})$ be defined by

   $$S = \left\{ \begin{bmatrix} a & -b \\ b & a \end{bmatrix} : a, b \in \mathbf{R} \right\}.$$

   Verify that S is a subring of $M_2(\mathbf{R})$ and show that S is isomorphic to the field of complex numbers $\mathbf{C}$.

6. Let R be a commutative ring.
   (a) If $1 \le j \le n$ prove that $E_{jj} = P_{1j}^{-1} E_{11} P_{1j}$. Thus, the matrices E_{jj} are all similar.
   (b) If $A, B \in M_n(R)$ define $[A, B] = AB - BA$. The matrix $[A, B]$ is called the *commutator* of A and B and we will say that a matrix $C \in M_n(R)$ is a commutator if $C = [A, B]$ for some $A, B \in M_n(R)$. If $i \ne j$ show that E_{ij} and $E_{ii} - E_{jj}$ are commutators.
   (c) If C is a commutator, show that $\operatorname{Tr}(C) = 0$. Conclude that I_n is not a commutator in any $M_n(R)$ for which n is not a multiple of the characteristic of R. What about $I_2 \in M_2(\mathbf{Z}_2)$?

7. If S is a ring and $a \in S$ then the centralizer of a, denoted $C(a)$, is the set $C(a) = \{b \in S : ab = ba\}$. That is, it is the subset of S consisting of elements which commute with a.
   (a) Verify that $C(a)$ is a subring of S.
   (b) What is $C(1)$?

(c) Let R be a commutative ring and let $S = M_n(R)$. If $A = D_i(\beta)$ for $\beta \neq 1 \in R^*$ then compute $C(A)$.

8. A matrix $N \in M_n(R)$ is nilpotent if $N^k = 0$ for some k.
    (a) If F is field of characteristic 0 and $N \in M_n(F)$ is nilpotent, show that there is a matrix $A \in M_n(F)$ such that $A^m = I_n + N$ for any natural number m. (Hint: The binomial series may be helpful.) Are there problems if we do not assume that char$(F) = 0$?
    (b) Let $N = \begin{bmatrix} 0 & 1 \\ 0 & 0 \end{bmatrix} \in M_2(\mathbf{Q})$. Show that there is no matrix $A \in M_2(\mathbf{Q})$ such that $A^2 = N$.

9. Show that there are infinitely many $A \in M_2(\mathbf{R})$ such that $A^2 = 0$.

10. A matrix $P \in M_n(R)$ is *idempotent* if $P^2 = P$. Give an example, other than 0 or I_n, of a diagonal idempotent matrix. Give an example of a nondiagonal 2×2 idempotent matrix. Show that if P is idempotent, then $T^{-1}PT$ is also idempotent for all $T \in \mathrm{GL}(n, R)$.

11. Let $A \in M_n(R)$. We say that A has constant row sums if the sum of the entries in each row is a constant $\alpha \in R$, i.e $\sum_{j=1}^{n} a_{ij} = \alpha$ for $1 \leq i \leq n$. We define constant column sums similarly.
    (a) Show that A has constant row sums if and only if

$$A \begin{bmatrix} 1 \\ \vdots \\ 1 \end{bmatrix} = \alpha \begin{bmatrix} 1 \\ \vdots \\ 1 \end{bmatrix}$$

for $\alpha \in R$ and that A has constant column sums if and only if $\begin{bmatrix} 1 & \cdots & 1 \end{bmatrix} A = \beta \begin{bmatrix} 1 & \cdots & 1 \end{bmatrix}$ for $\beta \in R$.
    (b) Prove that if A and $B \in M_n(R)$ both have constant row sums, then so does AB.
    (c) Prove that if A and $B \in M_n(R)$ both have constant column sums, then so does AB.

12. Prove Proposition 1.13.

13. Prove Lemma 1.15.

14. Let R be a commutative ring. Verify the following formulas for the kronecker product of matrices:
    (a) $A \otimes (B + C) = A \otimes B + A \otimes C$.
    (b) $(A \otimes B)^t = A^t \otimes B^t$.

15. Prove Lemma 1.20.

16. Give an example of a function $D : M_n(\mathbf{Z}_2) \to \mathbf{Z}_2$ which is n-linear and satisfies $D(P_{ij}A) = -D(A) = D(A)$, but which is not alternating.

17. A matrix $A \in M_n(R)$ is *symmetric* if $A^t = A$ and it is *skew-symmetric* if $A^t = -A$ and all the diagonal entries of A are zero.
    (a) Let V_1 be the set of symmetric matrices and V_2 the set of skew-symmetric matrices. Show that V_1 and V_2 are both submodules of $V = M_n(R)$. If 2 is a unit in R show that $V = V_1 \oplus V_2$.
    (b) Let $A \in M_n(R)$ be skew-symmetric. If n is odd, show that $\det(A) = 0$. If n is even, show $\det(A)$ is a square in R.

18. If P_ω is a permutation matrix, show that $P_\omega^{-1} = P_\omega^t$.

19. Let R and S be commutative rings and let $f : R \to S$ be a ring homomorphism. If $A = [a_{ij}] \in M_n(R)$, define $f(A) = [f(a_{ij})] \in M_n(S)$. Show that $\det f(A) = f(\det A)$.

20. Let R be a commutative ring and let H be a subgroup of the group of units R^* of R. Let $N = \{A \in \mathrm{GL}(n, R) : \det A \in H\}$. Prove that N is a normal subgroup of $\mathrm{GL}(n, R)$ and that $\mathrm{GL}(n, R)/N \cong R^*/H$.

21. (a) Suppose that A has the block decomposition $A = \begin{bmatrix} A_1 & 0 \\ B & A_2 \end{bmatrix}$. Prove that $\det A = (\det A_1)(\det A_2)$.

(b) More generally, suppose that $A = [A_{ij}]$ is a block upper triangular (respectively, lower block triangular) matrix, i.e., A_{ii} is square and $A_{ij} = 0$ if $i > j$ (respectively, $i < j$). Show that $\det A = \prod_i (\det A_{ii})$.

22. If R is a commutative ring, then a *derivation* on R is a function $\delta : R \to R$ such that $\delta(a + b) = \delta(a) + \delta(b)$ and $\delta(ab) = a\delta(b) + \delta(a)b$.

(a) Prove that $\delta(a_1 \cdots a_n) = \sum_{i=1}^{n}(a_1 \cdots a_{i-1}\delta(a_i)a_{i+1} \cdots a_n)$.

(b) If δ is a derivation on R and $A \in M_n(R)$ let A_i be the $n \times n$ matrix obtained from A by applying δ to the elements of the ith row. Show that $\delta(\det A) = \sum_{i=1}^{n} \det A_i$.

23. If $A \in M_n(\mathbf{Q}[X])$ then $\det A \in \mathbf{Q}[X]$. Use this observation and your knowledge of polynomials to calculate $\det A$ for each of the following matrices, without doing any calculation.

(a) $A = \begin{bmatrix} 1 & 1 & 2 & 3 \\ 1 & 2 - X^2 & 2 & 3 \\ 2 & 3 & 1 & 5 \\ 2 & 3 & 1 & 9 - X^2 \end{bmatrix}$.

(b) $A = \begin{bmatrix} 1 & 1 & 1 & \cdots & 1 \\ 1 & 1 - X & 1 & \cdots & 1 \\ 1 & 1 & 2 - X & \cdots & 1 \\ \vdots & \vdots & \vdots & \ddots & \vdots \\ 1 & 1 & 1 & \cdots & m - X \end{bmatrix}$.

24. Let F be a field and consider the "generic" matrix $[X_{ij}]$ with entries in the polynomial ring $F[X_{ij}]$ in the n^2 indeterminants X_{ij} $(1 \le i, j \le n)$. Show that $\det[X_{ij}]$ is an irreducible polynomial in $F[X_{ij}]$. (Hint: Use Laplace's expansion to argue by induction on n.)

25. Let $A \in M_n(\mathbf{Z})$ be the matrix with $\mathrm{ent}_{ii}(A) = 2$ $(1 \le i \le n)$, $\mathrm{ent}_{ij}(A) = 1$ if $|i - j| = 1$, and $\mathrm{ent}_{ij}(A) = 0$ if $|i - j| > 1$. Compute $\det(A)$.

26. Let $A_n \in M_n(\mathbf{Z})$ be the matrix with $\mathrm{ent}_{ii}(A_n) = i$ for $1 \le i \le n$ and $a_{ij} = 1$ if $i \ne j$. Show that $\det(A_n) = (n - 1)!$.

27. Let $A \in M_n(\mathbf{Z})$ be a matrix such that $\mathrm{ent}_{ij}(A) = \pm 1$ for all i and j. Show that 2^{n-1} divides $\det(A)$.

28. Let R be a commutative ring and let $a, b \in R$. Define a matrix $A(a, b) \in M_n(R)$ by $\mathrm{ent}_{ii}(A(a, b)) = a$ for all i and $\mathrm{ent}_{ij}(A(a, b)) = b$ if $i \ne j$. Compute $\det(A(a, b))$. (Hint: First find the Jordan canonical form of $A(a, b)$.)

29. Let $V(x_1, \ldots, x_n)$ be the *Vandermonde determinant*:

$$V(x_1, \ldots, x_n) = \begin{bmatrix} 1 & x_1 & x_1^2 & \cdots & x_1^{n-1} \\ 1 & x_2 & x_2^2 & \cdots & x_2^{n-1} \\ \vdots & \vdots & \vdots & \ddots & \vdots \\ 1 & x_n & x_n^2 & \cdots & x_n^{n-1} \end{bmatrix}.$$

(a) Prove that

$$\det V(x_1, \ldots, x_n) = \prod_{1 \le i < j \le n} (x_i - x_j).$$

(b) Suppose that $t_1, \ldots, t_{n+1}$ are $n + 1$ distinct elements of a field F. Let $P_i(X)$ $(1 \le i \le n + 1)$ be the Lagrange interpolation polynomials determined by $t_1, \ldots, t_{n+1}$. Thus,

$$\mathcal{B} = \{P_1(X), \ldots, P_{n+1}(X)\}$$

is a basis of the vector space $\mathcal{P}_n(F)$ of polynomials in $F[X]$ of degree at most n. But $\mathcal{A} = \{1, X, \ldots, X^n\}$ is also a basis of $\mathcal{P}_n$. Show that

$$V(t_1, \ldots, t_{n+1}) = P_{\mathcal{B}}^{\mathcal{A}}.$$

30. If R is an integral domain let $A \in \mathrm{GL}(n, R)$ and let $B \in M_n(R)$.
    (a) Prove that there are at most n elements $a \in R$ such that $\det(aA + B) = 0$. If R is a field, conclude that $aA + B$ is invertible except for finitely many values of $a \in R$.
    (b) If $R = \mathbf{Z}_2 \times \mathbf{Z}_2$ verify that every $a \in R$ satisfies the equation $a^2 - a = 0$. If $A = I_2$ and $B = \begin{bmatrix} 0 & 0 \\ 1 & 1 \end{bmatrix}$ show that part (a) is false without the assumption that R is an integral domain.

31. Recall that $Q_{p,n}$ is the set of sequences $\alpha = (i_1, \ldots, i_p)$ of p integers with $1 \leq i_1 < i_2 < \cdots < i_p \leq n$. Thus there are $\binom{n}{p}$ elements of $Q_{p,n}$ and we can order these elements lexicographically, i.e., $(i_1, \ldots, i_p) < (j_1, \ldots, j_p)$ if $i_1 < j_1$ or $i_1 = j_1, \ldots, i_{r-1} = j_{r-1}$ and $i_r < j_r$ for some $1 < r \leq p$. For example, the lexicographic ordering of $Q_{2,4}$ is $(1, 2) < (1, 3) < (1, 4) < (2, 3) < (2, 4) < (3, 4)$. If $A \in M_n(R)$ then the set of $\binom{n}{p}^2$ elements $\{\det A[\alpha \mid \beta] : \alpha, \beta \in Q_{p,n}\}$ can be arranged, using the lexicographic ordering on $Q_{p,n}$ into a matrix $C_p(A)$ called the pth compound matrix of A.
    (a) If $A, B \in M_n(R)$ then verify (using the Cauchy–Binet theorem) that

    $$C_p(AB) = C_p(A)C_p(B).$$

    (b) Show that $C_p(I_n) = I_{\binom{n}{p}}$.
    (c) Prove that if A is invertible, then $C_p(A)$ is also invertible, and give a formula for $C_p(A)^{-1}$.

32. If $A \in M_{m,n}(\mathbf{R})$ then $\det(AA^t) \geq 0$. (Hint: Use the Cauchy–Binet theorem.)

33. If $A = [a_{ij}] \in M_n(\mathbf{C})$, then let $\overline{A} = [\overline{a}_{ij}]$ where $\overline{a}_{ij}$ denotes the complex conjugate of a_{ij} and let $A^* = \overline{A}^t$. Show that $\det(AA^*) \geq 0$.

34. Let F and K be fields with $F \subseteq K$. Then $M_n(F) \subseteq M_n(K)$. If $A \in M_n(F)$ and A is invertible in $M_n(K)$, prove that A is invertible in $M_n(F)$. That is,

    $$\mathrm{GL}(n, K) \cap M_n(F) = \mathrm{GL}(n, F).$$

35. (a) If $A \in M_{m,n}(\mathbf{R})$ then $\mathrm{rank}(A) = \mathrm{rank}(A^t A)$. Give an example to show that this statement may be false for matrices $A \in M_{m,n}(\mathbf{C})$.
    (b) If $A \in M_n(\mathbf{C})$ show that $\mathrm{rank}(A) = \mathrm{rank}(A^* A)$.

36. If $A \in M_n(R)$ has a submatrix $A[\alpha \mid \gamma] = 0_t$ where $t > n/2$, then $\det(A) = 0$.

37. Prove Lemma 2.24.

38. Let R be any subring of the complex numbers $\mathbf{C}$ and let $A \in M_{m,n}(R)$. Then show that the matrix equation $AX = 0$ has a nontrivial solution $X \in M_{n,1}(R)$ if and only if it has a nontrivial solution $X \in M_{n,1}(\mathbf{C})$.

39. Prove Corollary 2.29.

40. Let R be a commutative ring, let $A \in M_{m,n}(R)$, and let $B \in M_{n,p}(R)$. Prove that

    $$\text{M-rank}(AB) \leq \min\{\text{M-rank}(A), \text{M-rank}(B)\}.$$

41. Verify the claims made in Remark 2.30.

42. Let F be a field with a derivation D, i.e., $D : F \to F$ satisfies $D(a+b) = a+b$ and $D(ab) = aD(b) + D(a)b$. Let $K = \mathrm{Ker}(D)$, i.e., $K = \{a \in F : D(a) = 0\}$. K is called the field of constants of D.
    (a) Show that K is a field.

(b) If $u_1, \ldots, u_n \in F$, define the *Wronskian* of $u_1, \ldots, u_n \in F$ by

$$W(u_1, \ldots, u_n) = \det \begin{bmatrix} u_1 & \cdots & u_n \\ D(u_1) & \cdots & D(u_n) \\ \vdots & \ddots & \vdots \\ D^{n-1}(u_1) & \cdots & D^{n-1}(u_n) \end{bmatrix},$$

where $D^i = D \circ \cdots \circ D$ (i times). Show that $u_1, \ldots, u_n \in F$ are linearly dependent over K if and only if $W(u_1, \ldots, u_n) = 0$.

43. Let R be a commutative ring and let $A = [a_{ij}] \in M_n(R)$ be a matrix such that a_{11} is not a zero divisor. If $n \geq 2$ prove that $a_{11}^{n-2} \det(A) = \det(B)$ where $B \in M_{n-1}(R)$ is the matrix with $\text{ent}_{ij} = A[(1, i+1) \mid (1, j+1)]$ for $1 \leq i, j \leq n - 1$. (This formula is sometimes called *Choi's pivotal condensation formula*. It provides another inductive procedure for the computation of determinants.)

44. Prove the following facts about the adjoint of a matrix $A \in M_n(R)$ (in part (c), assume that $R = \mathbf{C}$):
    (a) If A is diagonal, then $\text{Adj}(A)$ is diagonal.
    (b) If A is symmetric, then $\text{Adj}(A)$ is symmetric.
    (c) If A is Hermitian, then $\text{Adj}(A)$ is Hermitian.
    (d) If A is skew-symmetric, then $\text{Adj}(A)$ is symmetric or skew-symmetric according as n is odd or even.

45. Let R be an arbitrary ring.
    (a) If $A \in M_{m,n}(R)$ and $B \in M_{n,p}(R)$, show that

    $$(\text{op}(AB))^t = \text{op}(B^t)\,\text{op}(A^t).$$

    (b) If $A \in M_n(R)$ is invertible, show that

    $$\text{op}(A^{-1})^t = (\text{op}(A^t))^{-1}.$$

46. Let $\mathcal{P}_3 = \{f(X) \in \mathbf{Z}[X] : \deg f(X) \leq 3\}$. Let $\mathcal{A} = \{1, X, X^2, X^3\}$ and

    $$\mathcal{B} = \{1, X, X^{(2)}, X^{(3)}\}$$

    where $X^{(i)} = X(X-1)\cdots(X-i+1)$.
    (a) Verify that $\mathcal{A}$ and $\mathcal{B}$ are bases of the $\mathbf{Z}$-module $\mathcal{P}_3$.
    (b) Compute the change of basis matrices $P_{\mathcal{B}}^{\mathcal{A}}$ and $P_{\mathcal{A}}^{\mathcal{B}}$.
    (c) Let $D : \mathcal{P}_3 \to \mathcal{P}_3$ be differentiation, i.e., $D(f(X)) = f'(X)$; e.g.,

    $$D(X^3 + 2X) = 3X^2 + 2.$$

    Compute $[D]_{\mathcal{A}}$ and $[D]_{\mathcal{B}}$.
    (d) Let $\Delta : \mathcal{P}_3 \to \mathcal{P}_3$ be defined by $\Delta(f(X)) = f(X+1) - f(X)$, e.g.,

    $$\Delta(X^3 + 2X) = ((X+1)^3 + 2(X+1)) - (X^3 + 2X) = 3X^2 + 3X + 3.$$

    Verify that Δ is a $\mathbf{Z}$-module homomorphism and compute $[\Delta]_{\mathcal{A}}$ and $[\Delta]_{\mathcal{B}}$.

47. Show that the vectors $(1, 2, 1)$, $(2, 3, 3)$, and $(3, 2, 1)$ form a basis of $\mathbf{R}^3$ and that the vectors $(3, 2, 4)$, $(5, 2, 3)$, and $(1, 1, -6)$ form a second basis. Calculate the matrix of transition from the first basis to the second.

48. If $A \in M_n(R)$, then the columns of A form a basis of R^n if and only if $\det(A)$ is a unit in R.

49. Let R be a commutative ring. We will say that A and $B \in M_n(R)$ are *permutation similar* if there is a permutation matrix P such that $P^{-1}AP = B$. Show that A and B are permutation similar if and only if there is a free R-module M of rank n, a basis $\mathcal{B} = \{v_1, \ldots, v_n\}$ of M, and a permutation $\sigma \in S_n$ such that $A = [f]_{\mathcal{B}}$ and $B = [f]_{\mathcal{C}}$ where $f \in \text{End}_R(M)$ and $\mathcal{C} = \{v_{\sigma(1)}, \ldots, v_{\sigma(n)}\}$.

50. Let R be a commutative ring, and let

$$\mathcal{B} = \{E_{11}, \ldots, E_{1n}, E_{21}, \ldots, E_{2n}, \ldots, E_{m1}, \ldots, E_{mn}\}$$

be the basis of $M_{m,n}(R)$ consisting of the matrix units in the given order. Another basis of $M_{m,n}(R)$ is given by the matrix units in the following order:

$$\mathcal{C} = \{E_{11}, \ldots, E_{m1}, E_{12}, \ldots, E_{m2}, \ldots, E_{1n} \ldots, E_{mn}\}.$$

If $A \in M_m(R)$ then $\mathcal{L}_A \in \text{End}_R(M_{m,n}(R))$ will denote left multiplication by A, while $\mathcal{R}_B$ will denote right multiplication by B, where $B \in M_n(R)$.
(a) Show that $AE_{ij} = \sum_{k=1}^m a_{ki}E_{kj}$ and $E_{ij}B = \sum_{\ell=1}^n b_{j\ell}E_{i\ell}$.
(b) Show that $[\mathcal{L}_A]_{\mathcal{B}} = A \otimes I_n$ and $[\mathcal{L}_A]_{\mathcal{C}} = I_n \otimes A$. Conclude that $A \otimes I_n$ and $I_n \otimes A$ are permutation similar.
(c) Show that $[\mathcal{R}_B]_{\mathcal{B}} = I_m \otimes B^t$ and $[\mathcal{R}_A]_{\mathcal{C}} = B^t \otimes I_m$.
(d) Show that $[\mathcal{L}_A \circ \mathcal{R}_B]_{\mathcal{B}} = A \otimes B^t$.
This exercise provides an interpretation of the tensor product of matrices as the matrix of a particular R-module endomorphism. Another interpretation, using the tensor product of R-modules, will be presented in Section 7.2 (Proposition 7.2.35).

51. Give an example of a vector space V (necessarily of infinite dimension) over a field F and endomorphisms f and g of V such that
(a) f has a left inverse but not a right inverse, and
(b) g has a right inverse but not a left inverse.

52. Let F be a field and let A and B be matrices with entries in F.
(a) Show that $\text{rank}(A \oplus B) = (\text{rank}(A)) + (\text{rank}(B))$.
(b) Show that $\text{rank}(A \otimes B) = (\text{rank}(A))(\text{rank}(B))$.
(c) Show that $\text{rank}(AB) \geq \text{rank}(A) + \text{rank}(B) - n$ if $B \in M_{n,m}(F)$.

53. Let M be a free R-module of finite rank, let $f \in \text{End}_R(M)$, and let $g \in \text{End}_R(M)$ be invertible. If $v \in M$ is an eigenvector of f with eigenvalue λ, show that $g(v)$ is an eigenvector of gfg^{-1} with eigenvalue λ.

54. Let M be a finite rank free R-module over a PID R and let $f \in \text{End}_R(M)$. Suppose that f is diagonalizable and that $M = N_1 \oplus N_2$ where N_1 and N_2 are f-invariant submodules of M. Show that $g = f|_{N_i}$ is diagonalizable for $i = 1, 2$.

55. Let $A \in M_n(R)$, and suppose

$$c_A(X) = \det(XI_n - A) = X^n + a_1 X^{n-1} + \cdots + a_{n-1}X + a_n.$$

(a) Show that $a_1 = -\text{Tr}(A)$ and $a_n = (-1)^n \det(A)$.
(b) More generally, show that

$$a_r = (-1)^r \sum_{\alpha \in Q_{r,n}} \det A[\alpha \mid \alpha].$$

56. Prove the Cayley–Hamilton theorem for matrices $A \in M_n(R)$ where R is any commutative ring. (Hint: Apply the noncommutative division algorithm (Exercise 44 of Chapter 2) and the adjoint formula (Theorem 2.17).)

57. Let $A \in M_n(F)$ where F is a field.

(a) Show that $\det(\mathrm{Adj}(A)) = (\det(A))^{n-1}$ if $n \geq 2$.
(b) Show that $\mathrm{Adj}(\mathrm{Adj}(A)) = (\det(A))^{n-2}A$ if $n > 2$ and that

$$\mathrm{Adj}(\mathrm{Adj}(A)) = A \qquad \text{if } n = 2.$$

58. Let F be a field and let $A \in M_n(F)$. Show that A and A^t have the same minimal polynomial.

59. Let K be a field and let F be a subfield. Let $A \in M_n(F)$.
    Show that the minimal polynomial of A is the same whether A is considered in $M_n(F)$ or $M_n(K)$.

60. An *algebraic integer* is a complex number which is a root of a monic polynomial with integer coefficients. Show that every algebraic integer is an eigenvalue of a matrix $A \in M_n(\mathbf{Z})$ for some n.

61. Let $A \in M_n(R)$ be an invertible matrix.
    (a) Show that $\det(X^{-1}I_n - A^{-1}) = (-X)^{-n}\det(A^{-1})c_A(X)$.
    (b) If

    $$C_A(X) = X^n + a_1 X^{n-1} + \cdots + a_{n-1}X + a_n$$

    and

    $$C_{A^{-1}}(X) = X^n + b_1 X^{n-1} + \cdots + b_{n-1}X + b_n,$$

    then show that $b_i = (-1)^n \det(A^{-1})a_{n-i}$ for $1 \leq i \leq n$ where we set $a_0 = 1$.

62. If $A \in M_n(F)$ (where F is a field) is nilpotent, i.e., $A^k = 0$ for some k, prove that $A^n = 0$. Is the same result true if F is a general commutative ring rather than a field?

63. Let F be an infinite field and let $\mathcal{F} = \{A_j\}_{j \in J} \subseteq M_n(F)$ be a commuting family of diagonalizable matrices. Show that there is a matrix $B \in M_n(F)$ and a family $\{f_j(X)\}_{j \in J} \subseteq F[X]$ of polynomials of degree $\leq n-1$ such that $A_j = f_j(B)$. (Hint: By Theorem 3.36 there is a matrix $P \in \mathrm{GL}(n, F)$ such that

    $$P^{-1}A_j P = \mathrm{diag}(\lambda_{1j}, \ldots, \lambda_{nj}).$$

    Let $t_1, \ldots, t_n$ be n distinct points in F, and let

    $$B = P\,\mathrm{diag}(t_1, \ldots, t_n)P^{-1}.$$

    Use Lagrange interpolation to get a polynomial $f_j(X)$ of degree $\leq n-1$ such that $f_j(t_i) = \lambda_{ij}$ for all i, j. Show that $\{f_j(X)\}_{j \in J}$ works.)

64. Let F be a field and V a finite-dimensional vector space over F. Let $S \in \mathrm{End}_F(V)$ and define $\mathrm{Ad}_S : \mathrm{End}_F(V) \to \mathrm{End}_F(V)$ by

    $$\mathrm{Ad}_S(T) = [S, T] = ST - TS.$$

    (a) If S is nilpotent, show that Ad_S is nilpotent.
    (b) If S is diagonalizable, show that Ad_S is diagonalizable.

65. Let $N_1, N_2 \in M_n(F)$ be nilpotent matrices. Show that N_1 and N_2 are similar if and only if

    $$\mathrm{rank}(N_1^k) = \mathrm{rank}(N_2^k) \qquad \text{for all } k \geq 1.$$

66. Let F be an algebraically closed field and V a finite-dimensional vector space over F.
    (a) Suppose that $T, S \in \mathrm{End}_F(V)$. Prove that T and S are similar if and only if

    $$\dim(\mathrm{Ker}(T - \lambda 1_V)^k) = \dim(\mathrm{Ker}(S - \lambda 1_V)^k)$$

    for all $\lambda \in F$ and $k \in \mathbf{N}$. (This result is known as Weyr's theorem.)

(b) Suppose that $T \in \text{End}_F(V)$ has Jordan canonical form $\oplus_{i=1}^{k} J_{\lambda_i, n_i}$. If T is invertible, show that T^{-1} has Jordan canonical form $\oplus_{i=1}^{k} J_{\lambda_i^{-1}, n_i}$.

67. Let F be an algebraically closed field and V a finite-dimensional vector space over F.
   (a) If $T \in \text{End}_F(V)$, show that there is a basis $\mathcal{B} = \{v_1, \ldots, v_n\}$ of V such that $V_i = \langle v_1, \ldots, v_i \rangle$ is a T-invariant subspace of V. Conclude that $[T]_\mathcal{B}$ is an upper triangular matrix. (Hint: Since F is algebraically closed T has an eigenvalue, say λ_1 with associated eigenvector v_1. Choose a complementary subspace W to $\langle v_1 \rangle$, define $T_1 \in \text{End}_F(W)$ by $T_1 = \pi \circ T$ where $\pi : V \to W$ is the projection. Now argue by induction.)
   (b) If $T, S \in \text{End}_F(V)$ commute then there is a basis $\mathcal{B}$ of V such that both $[T]_\mathcal{B}$ and $[S]_\mathcal{B}$ are upper triangular.
   (c) Show that the converse of part (b) is false by finding two upper triangular matrices which do not commute.
   (d) While the converse of part (b) is not true, show that two upper triangular matrices are "almost" commutative in the following sense. Verify that the commutator matrix $[A, B] = AB - BA$ of two upper triangular matrices is a nilpotent matrix.

68. Find matrices in $M_3(\mathbf{Q})$ with minimal polynomials X, X^2, and X^3.

69. Find the characteristic and minimal polynomial of each of the following matrices:

   (a) $\begin{bmatrix} 0 & 0 & a \\ 1 & 0 & b \\ 0 & 1 & c \end{bmatrix}$;   (b) $\begin{bmatrix} 0 & 1 & 0 \\ 1 & 0 & 0 \\ 0 & 0 & 1 \end{bmatrix}$;   (c) $\begin{bmatrix} 0 & 1 & 0 & 1 \\ 1 & 0 & 1 & 0 \\ 0 & 1 & 0 & 1 \\ 1 & 0 & 1 & 0 \end{bmatrix}$.

70. If $A \in M_n(F)$ (F a field) has characteristic polynomial

$$c_A(X) = X^2(X-1)^2(X^2-1),$$

what are the possibilities for the minimal polynomial $m_A(X)$?

71. Let V be a vector space and $T : V \to V$ a linear transformation. Assume that $m_T(X)$ is a product of linear factors. Show that T can be written as a sum $T = D + N$ where D is a diagonalizable linear transformation, N is a nilpotent linear transformation, and $DN = ND$. Note that the hypotheses are always satisfied for an algebraically closed field (e.g., $\mathbf{C}$).

72. Show that the matrices

$$\begin{bmatrix} 0 & 1 & 0 \\ 0 & 0 & 1 \\ 1 & 0 & 0 \end{bmatrix} \quad \text{and} \quad \begin{bmatrix} 1 & 1 & 0 \\ 0 & 1 & 1 \\ 0 & 0 & 1 \end{bmatrix}$$

are similar in $M_3(\mathbf{Z}_3)$.

73. For each of the following matrices with entries in $\mathbf{Q}$, find
   (1) the characteristic polynomial;
   (2) the eigenvalues, their algebraic and geometric multiplicities;
   (3) bases for the eigenspaces and generalized eigenspaces;
   (4) Jordan canonical form (if it exists) and basis for $V = \mathbf{Q}^n$ with respect to which the associated linear transformation has this form;
   (5) rational canonical form and minimal generating set for V_T as $\mathbf{Q}[X]$-module.

   (a) $\begin{bmatrix} 0 & -4 \\ 1 & -4 \end{bmatrix}$ (b) $\begin{bmatrix} 6 & -9 \\ 4 & -6 \end{bmatrix}$

   (c) $\begin{bmatrix} 3 & -2 & -4 \\ 0 & 2 & 4 \\ 0 & -1 & -2 \end{bmatrix}$ (d) $\begin{bmatrix} 1 & 2 & 2 \\ 2 & 1 & 2 \\ 2 & 2 & 1 \end{bmatrix}$

$$\text{(e)} \quad \begin{bmatrix} -2 & 0 & 0 & 1 \\ 1 & 1 & 0 & 1 \\ 2 & 0 & 1 & -2 \\ -1 & 0 & 0 & 0 \end{bmatrix} \qquad \text{(f)} \quad \begin{bmatrix} 1 & 0 & 1 & 0 \\ 4 & 3 & -2 & 0 \\ -2 & 1 & 5 & 0 \\ 2 & 0 & -1 & 3 \end{bmatrix}.$$

74. In each case below, you are given some of the following information for a linear transformation $T : V \to V$, V a vector space over the complex numbers $\mathbf{C}$: (1) characteristic polynomial for T; (2) minimal polynomial for T; (3) algebraic multiplicity of each eigenvalue; (4) geometric multiplicity of each eigenvalue; (5) rank(V_T) as an $\mathbf{C}[X]$-module; (6) the elementary divisors of the module V_T. Find all possibilities for T consistent with the given data (up to similarity) and for each possibility give the rational and Jordan canonical forms and the rest of the data.
    (a) $c_T(X) = (X - 2)^4 (X - 3)^2$.
    (b) $c_T(X) = X^2(X - 4)^7$ and $m_T(X) = X(X - 4)^3$.
    (c) $\dim V = 6$ and $m_T(X) = (X + 3)^2 (X + 1)^2$.
    (d) $c_T(X) = X(X - 1)^4 (X - 2)^5$, $\nu_{\text{geom}}(1) = 2$, and $\nu_{\text{geom}}(2) = 2$.
    (e) $c_T(X) = (X - 5)(X - 7)(X - 9)(X - 11)$.
    (f) $\dim V = 4$ and $m_T(X) = X - 1$.

75. Recall that a matrix $A \in M_n(F)$ is idempotent if $A^2 = A$.
    (a) What are the possible minimal polynomials of an idempotent A?
    (b) If A is idempotent and rank $A = r$, show that A is similar to $B = I_r \oplus 0_{n-r}$.

76. If $T : \mathbf{C}^n \to \mathbf{C}^n$ denotes a linear transformation, find all possible Jordan canonical forms of a T satisfying the given data:
    (a) $c_T(X) = (X - 4)^3 (X - 5)^2$.
    (b) $n = 6$ and $m_T(X) = (X - 9)^3$.
    (c) $n = 5$ and $m_T(X) = (X - 6)^2(X - 7)$.
    (d) T has an eigenvalue 9 with algebraic multiplicity 6 and geometric multiplicity 3 (and no other eigenvalues).
    (e) T has an eigenvalue 6 with algebraic multiplicity 3 and geometric multiplicity 3, and eigenvalue 7 with algebraic multiplicity 3 and geometric multiplicity 1 (and no other eigenvalues).

77. (a) Show that the matrix $A \in M_3(F)$ (F a field) is uniquely determined up to similarity by $c_A(X)$ and $m_A(X)$.
    (b) Give an example of two matrices $A, B \in M_4(F)$ with the same characteristic and minimal polynomials, but with A and B not similar.

78. Let $A \in M_n(\mathbf{C})$. If all the roots of the characteristic polynomial $c_A(X)$ are real numbers, show that A is similar to a matrix $B \in M_n(\mathbf{R})$.

79. Let F be an algebraically closed field, and let $A \in M_n(F)$.
    (a) Show that A is nilpotent if and only if all the eigenvalues of A are zero.
    (b) Show that $\text{Tr}(A^r) = \lambda_1^r + \cdots + \lambda_n^r$ where $\lambda_1, \ldots, \lambda_n$ are the eigenvalues of A counted with multiplicity.
    (c) If $\text{char}(F) = 0$ show that A is nilpotent if and only if $\text{Tr}(A^r) = 0$ for all $r \in \mathbf{N}$. (Hint: Use Newton's identities, Exercise 61 of Chapter 2.)

80. Prove that every normal complex matrix has a normal square root, i.e., if $A \in M_n(\mathbf{C})$ is normal, then there is a normal $B \in M_n(\mathbf{C})$ such that $B^2 = A$.

81. Prove that every Hermitian matrix with nonnegative eigenvalues has a Hermitian square root.

82. Show that the following are equivalent:
    (a) $U \in M_n(\mathbf{C})$ is unitary.
    (b) The columns of U are orthonormal.
    (c) The rows of U are orthonormal.

83. (a) Show that every matrix $A \in M_n(\mathbf{C})$ is unitarily similar to an upper triangular matrix T, i.e., $UAU^* = T$, where U is unitary.
    (b) Show that a normal complex matrix is unitarily similar to a diagonal matrix.

84. Show that a commuting family of normal matrices has a common basis of orthogonal eigenvectors, i.e., there is a unitary U such that $U A_j U^* = D_j$ for all A_j in the commuting family. (D_j denotes a diagonal matrix.)

85. A complex matrix $A \in M_n(\mathbf{C})$ is normal if and only if there is a polynomial $f(X) \in \mathbf{C}[X]$ of degree at most $n - 1$ such that $A^* = f(A)$. (Hint: Apply Lagrange interpolation.)

86. Show that $B = \oplus A_i$ is normal if and only if each A_i is normal.

87. Prove that a normal complex matrix is Hermitian if and only if all its eigenvalues are real.

88. Prove that a normal complex matrix is unitary if and only if all its eigenvalues have absolute value 1.

89. Prove that a normal complex matrix is skew-Hermitian ($A^* = -A$) if and only if all its eigenvalues are purely imaginary.

90. If $A \in M_n(\mathbf{C})$, let $H(A) = \frac{1}{2}(A + A^*)$ and let $S(A) = \frac{1}{2}(A - A^*)$. $H(A)$ is called the Hermitian part of A and $S(A)$ is called the skew-Hermitian part of A. These should be thought of as analogous to the real and imaginary parts of a complex number. Show that A is normal if and only if $H(A)$ and $S(A)$ commute.

91. Let A and B be self-adjoint linear transformations. Then AB is self-adjoint if and only if A and B commute.

92. Give an example of an inner product space V and a linear transformation $T : V \to V$ with $T^*T = 1_V$, but T not invertible. (Of course, V will necessarily be infinite dimensional.)

93. (a) If S is a skew-Hermitian matrix, show that $I - S$ is nonsingular and the matrix
$$U = (I + S)(I - S)^{-1}$$
is unitary.
    (b) Every unitary matrix U which does not have -1 as an eigenvalue can be written as
$$U = (I + S)(I - S)^{-1}$$
for some skew-Hermitian matrix S.

94. This exercise will develop the spectral theorem from the point of view of projections.
    (a) Let V be a vector space. A linear transformation $E : V \to V$ is called a *projection* if $E^2 = E$. Show that there is a one-to-one correspondence between projections and ordered pairs of subspaces (V_1, V_2) of V with $V_1 \oplus V_2 = V$ given by
$$E \leftrightarrow (\mathrm{Ker}(E), \mathrm{Im}(E)).$$

    (b) If V is an inner product space, a projection E is called *orthogonal* if $E = E^*$. Show that if E is an orthogonal projection, then $\mathrm{Im}(E)^{\perp} = \mathrm{Ker}(E)$ and $\mathrm{Ker}(E)^{\perp} = \mathrm{Im}(E)$, and conversely.
    (c) A set of (orthogonal) projections $\{E_1, \dots, E_r\}$ is called *complete* if $E_i E_j = 0$ for $i \neq j$ and
$$1_V = E_1 + \cdots + E_r.$$

    Show that any set of (orthogonal) projections $\{E_1, \dots, E_r\}$ with $E_i E_j = 0$ for $i \neq j$ is a subset of a complete set of (orthogonal) projections.
    (d) Prove the following result.

    Let $T : V \to T$ be a diagonalizable (resp., normal) linear transformation on the finite-dimensional vector space V. Then there is a unique set of *distinct scalars* $\{\lambda_1, \ldots, \lambda_r\}$ *and a unique complete set of projections (resp., orthogonal projections)* $\{E_1, \ldots, E_r\}$ *with*

$$T = \lambda_1 E_1 + \cdots + \lambda_r E_r.$$

Also, show that $\{\lambda_i\}_{i=1}^r$ are the eigenvalues of T and $\{\text{Im}(E_i)\}_{i=1}^r$ are the associated eigenspaces.

(e) Let T and $\{E_i\}$ be as in part (d). Let $U : V \to V$ be an arbitrary linear transformation. Show that $TU = UT$ if and only if $E_iU = UE_i$ for $1 \leq i \leq r$.

(f) Let F be an infinite field. Show that there are polynomials $p_i(X) \in F[X]$ for $1 \leq i \leq r$ with $p_i(T) = E_i$. (Hint: See Exercise 63.)

Chapter 5

Matrices over PIDs

5.1 Equivalence and Similarity

Recall that if $T_i : V \to V$ $(i = 1, 2)$ are linear transformations on a finite-dimensional vector space V over a field F, then T_1 and T_2 are similar if and only if the $F[X]$-modules V_{T_1} and V_{T_2} are isomorphic (Theorem 4.4.2). Since the structure theorem for finitely generated torsion $F[X]$-modules gives a criterion for isomorphism in terms of the invariant factors (or elementary divisors), one has a powerful tool for studying linear transformations, up to similarity. Unfortunately, in general it is difficult to obtain the invariant factors or elementary divisors of a given linear transformation. We will approach the problem of computation of invariant factors in this chapter by studying a specific presentation of the $F[X]$-module V_T. This presentation will be used to transform the search for invariant factors into performing elementary row and column operations on a matrix with polynomial entries. We begin with the following definition.

(1.1) Definition. *If R is a ring and M is a finitely generated R-module, then a **finite free presentation** of M is an exact sequence*

$$(1.1) \qquad R^m \xrightarrow{\phi} R^n \longrightarrow M \longrightarrow 0.$$

Note that $M \cong \mathrm{Coker}(\phi)$ and the free presentation is essentially an explicit way to write M as a quotient of a finite-rank free R-module by a finite-rank submodule.

While every finitely generated R-module has a free presentation (Definition 3.4.15), it need not be true that every finitely generated R-module has a finite free presentation as in Definition 1.1; however, if the ring R is a PID, then this is true.

(1.2) Lemma. *Let R be a PID and let M be a finitely generated R-module with $\mu(M) = n$. Then there is a finite free presentation*

$$(1.2) \qquad 0 \longrightarrow R^m \xrightarrow{\phi} R^n \longrightarrow M \longrightarrow 0$$

with $m \leq n$.

Proof. By Proposition 3.4.14, there is a free presentation

$$0 \longrightarrow K \longrightarrow R^n \longrightarrow M \longrightarrow 0.$$

Since R is a PID, Theorem 3.6.2 implies that K is a free R-module of rank $m \le n$, so Equation (1.2) is valid.   ☐

Recall (Definition 4.3.14) that two R-module homomorphisms f and g from R^m to R^n are **equivalent** if and only if there is a commutative diagram

$$\begin{array}{ccc} R^m & \xrightarrow{\ f\ } & R^n \\ \downarrow{h_1} & & \downarrow{h_2} \\ R^m & \xrightarrow{\ g\ } & R^n \end{array}$$

where h_1 and h_2 are R-module isomorphisms. That is, $g = h_2 f h_1^{-1}$. Also, matrices $A, B \in M_{n,m}(R)$ are **equivalent** if $B = PAQ$ for some $P \in \mathrm{GL}(n, R)$ and $Q \in \mathrm{GL}(m, R)$.

(1.3) Proposition. *Let R be a commutative ring and let $f, g : R^m \to R^n$ be R-module homomorphisms. If f is equivalent to g, then*

$$\mathrm{Coker}(f) \cong \mathrm{Coker}(g).$$

Proof. Let $M = \mathrm{Coker}\, f$ and $N = \mathrm{Coker}(g)$. Since f and g are equivalent there is a commutative diagram of R-modules and homomorphisms

$$\begin{array}{ccccccc} R^m & \xrightarrow{\ f\ } & R^n & \xrightarrow{\ \pi_1\ } & M & \longrightarrow & 0 \\ \downarrow{h_1} & & \downarrow{h_2} & & & & \\ R^m & \xrightarrow{\ g\ } & R^n & \xrightarrow{\ \pi_2\ } & N & \longrightarrow & 0 \end{array}$$

where the h_i are isomorphisms and π_i are the canonical projections. Define $\phi : M \to N$ by $\phi(x) = \pi_2(h_2(y))$ where $\pi_1(y) = x$. It is necessary to check that this definition is consistent; i.e., if $\pi_1(y_1) = \pi_1(y_2)$, then $\pi_2(h_2(y_1)) = \pi_2(h_2(y_2))$. But if $\pi_1(y_1) = \pi_1(y_2)$, then $y_1 - y_2 \in \mathrm{Ker}(\pi_1) = \mathrm{Im}(f)$, so $y_1 - y_2 = f(z)$ for some $z \in R^m$. Then

$$h_2(y_1 - y_2) = h_2 f(z) = g h_1(z) \in \mathrm{Im}(g) = \mathrm{Ker}(\pi_2)$$

so $\pi_2 h_2(y_1 - y_2) = 0$ and the definition of ϕ is consistent. We will leave it to the reader to check that ϕ is an R-module isomorphism.   ☐

In order to apply Proposition 1.3 to our $F[X]$-modules V_T, it is necessary to produce a finite free presentation of V_T. We now show how to use a basis of V to produce a finite free presentation of the $F[X]$-module V_T. Thus, we assume that V is a finite-dimensional vector space over a field F and $T : V \to V$ is a linear transformation. Define

$$\psi_{\mathcal{B}} : F[X]^n \to V$$

by $\psi_{\mathcal{B}}(e_j) = v_j$ for $1 \le j \le n$ where $\mathcal{A} = \{e_1, \ldots, e_n\}$ is the standard basis of the free $F[X]$-module $F[X]^n$ and $\mathcal{B} = \{v_1, \ldots, v_n\}$ is a given basis of V. Thus,

$$(1.3) \qquad \psi_{\mathcal{B}}(f_1(X), \ldots, f_n(X)) = f_1(T)(v_1) + \cdots + f_n(T)(v_n)$$

for all $(f_1(X), \ldots, f_n(X)) \in F[X]^n$. Let $K = \mathrm{Ker}(\psi_{\mathcal{B}})$. If $A = [T]_{\mathcal{B}}$ then $A = [a_{ij}] \in M_n(F)$ where

$$(1.4) \qquad T(v_j) = \sum_{i=1}^{n} a_{ij} v_i.$$

Let

$$(1.5) \qquad p_j(X) = X e_j - \sum_{i=1}^{n} a_{ij} e_i \in F[X]^n \quad \text{for } 1 \le j \le n.$$

(1.4) Lemma. $K \subseteq F[X]^n$ *is free of rank* n *and* $\mathcal{C} = \{p_1(X), \ldots, p_n(X)\}$ *is a basis of* K.

Proof. It is sufficient to show that $\mathcal{C}$ is a basis. First note that

$$\psi_{\mathcal{B}}(p_j(X)) = \psi_{\mathcal{B}}\left(X e_j - \sum_{i=1}^{n} a_{ij} e_i\right)$$

$$= X \psi_{\mathcal{B}}(e_j) - \sum_{i=1}^{n} a_{ij} \psi_{\mathcal{B}}(e_i)$$

$$= T(v_j) - \sum_{i=1}^{n} a_{ij} v_i$$

$$= 0$$

by Equation (1.4). Thus $p_j(X) \in K = \mathrm{Ker}(\psi_{\mathcal{B}})$ for all j. By Equation (1.5) we can write

$$(1.6) \qquad X e_j = p_j(X) + \sum_{i=1}^{n} a_{ij} e_i.$$

By repeated uses of this equation, any

$$H(X) = \sum_{j=1}^{n} g_j(X) e_j \in F[X]^n$$

can be written as

$$(1.7) \qquad H(X) = \sum_{j=1}^{n} g_j(X) e_j = \sum_{j=1}^{n} h_j(X) p_j(X) + \sum_{j=1}^{n} b_j e_j$$

where $b_j \in F$. If $H(X) \in K$ then it follows from Equation (1.7) that $\sum_{j=1}^{n} b_j e_j \in K$, and applying the homomorphism ψ we conclude that $\sum_{j=1}^{n} b_j v_j = 0 \in V$; but $\{v_1, \ldots, v_n\}$ is a basis of V, so we must have $b_1 = \cdots = b_n = 0$. Therefore,

$$\mathcal{C} = \{p_1(X), \ldots, p_n(X)\}$$

generates K as an $F[X]$-submodule.

To show that $\mathcal{C}$ is linearly independent, suppose that

$$\sum_{j=1}^{n} h_j(X) p_j(X) = 0.$$

Then

$$\sum_{j=1}^{n} h_j(X) X e_j = \sum_{i,j=1}^{n} h_j(X) a_{ij} e_i$$

and since $\{e_1, \ldots, e_n\}$ is a basis of $F[X]^n$, it follows that

$$(1.8) \qquad h_i(X) \cdot X = \sum_{j=1}^{n} h_j(X) a_{ij}.$$

If some $h_i(X) \neq 0$, choose i so that $h_i(X)$ has maximal degree, say,

$$\deg h_i(X) = r \geq \deg h_j(X) \qquad (1 \leq j \leq n).$$

Then the left-hand side of Equation (1.8) has degree $r + 1$ while the right-hand side has degree $\leq r$. Thus $h_i(X) = 0$ for $1 \leq i \leq n$ and $\mathcal{C}$ is a basis of K. $\square$

We can summarize the above discussion as follows:

(1.5) Proposition. *Let V be a vector space of dimension n over the field F, let $T : V \to V$ be a linear transformation, and let $\mathcal{B}$ be a basis of V. Then there is a finite free presentation of V_T*

$$(1.9) \qquad 0 \longrightarrow F[X]^n \xrightarrow{\phi_\mathcal{B}} F[X]^n \xrightarrow{\psi_\mathcal{B}} V_T \longrightarrow 0$$

in which the matrix of $\phi_\mathcal{B}$ in the standard basis of $F[X]^n$ is

$$X I_n - [T]_\mathcal{B}.$$

Proof. If $\phi_\mathcal{B}(e_j) = p_j(X)$ as in Equation (1.5), then $p_j(X) = \mathrm{col}_j(X I_n - A)$ where $A = [T]_\mathcal{B}$. Thus $[\phi_\mathcal{B}]_\mathcal{A} = X I_n - A$, and sequence (1.9) is exact by Lemma 1.4. $\square$

According to our analysis, $V_T \cong \mathrm{Coker}(\phi_\mathcal{B})$. It is worthwhile to see this isomorphism in a simple explicit example.

(1.6) Example. Let F be a field, let $V = F^2$, and let $T : V \to V$ be defined by $T(u_1, u_2) = (u_2, 0)$ (cf. Example 3.1.5 (13)). If $\mathcal{B} = \{v_1, v_2\}$ is the standard basis on F^2, then

$$\psi_{\mathcal{B}}(f(X), g(X)) = f(T)(v_1) + g(T)(v_2) = a_0 v_1 + b_0 v_2 + b_1 v_1$$

where

$$f(X) = a_0 + a_1 X + \cdots + a_n X^n$$
$$g(X) = b_0 + b_1 X + \cdots + b_m X^m.$$

Since $A = [T]_{\mathcal{B}} = \begin{bmatrix} 0 & 1 \\ 0 & 0 \end{bmatrix}$, we have $X I_2 - A = \begin{bmatrix} X & -1 \\ 0 & X \end{bmatrix}$ so that

$$\phi_{\mathcal{B}}\left(\begin{bmatrix} f(X) \\ g(X) \end{bmatrix}\right) = \begin{bmatrix} X f(X) - g(X) \\ X g(X) \end{bmatrix}.$$

Note that

$$\begin{bmatrix} f(X) \\ g(X) \end{bmatrix} = \begin{bmatrix} a_0 + b_1 \\ b_0 \end{bmatrix} + \begin{bmatrix} X \tilde{f}(X) - \tilde{g}(X) \\ X \tilde{g}(X) \end{bmatrix}$$

where

$$\tilde{g}(X) = \frac{g(X) - b_0}{X}$$
$$\tilde{f}(X) = \frac{f(X) + \tilde{g}(X) - a_0 - b_1}{X}.$$

Since $f(X)$ and $g(X)$ are arbitrary, we see that

(1.10) $$F[X]^2 / \operatorname{Im}(\phi_{\mathcal{B}}) \cong F^2$$

as F-modules, while as an $F[X]$-module,

(1.11) $$F[X]^2 / \operatorname{Im}(\phi_{\mathcal{B}}) \cong F[X] \cdot \begin{bmatrix} 0 \\ 1 \end{bmatrix}.$$

Equation (1.11) follows from Equation (1.10), the observation

$$X \cdot \begin{bmatrix} 0 \\ 1 \end{bmatrix} = \begin{bmatrix} 1 \\ 0 \end{bmatrix} + \begin{bmatrix} -1 \\ X \end{bmatrix},$$

and the fact that $\begin{bmatrix} -1 \\ X \end{bmatrix} \in \operatorname{Im}(\phi_{\mathcal{B}})$. It follows immediately that

$$F[X]^2 / \operatorname{Im}(\phi_{\mathcal{B}}) \cong V_T$$

as $F[X]$-modules.

(1.7) Theorem. *Let F be a field and let $A, B \in M_n(F)$ be matrices. Then A is similar to B if and only if the polynomial matrices $X I_n - A$ and $X I_n - B \in M_n(F[X])$ are equivalent.*

Proof. If A and B are similar, then $B = P^{-1}AP$ for some $P \in \mathrm{GL}(n, F)$. Then

$$(XI_n - B) = (XI_n - P^{-1}AP) = P^{-1}(XI_n - A)P$$

so $(XI_n - A)$ and $(XI_n - B)$ are equivalent in $M_n(F[X])$.

Conversely, suppose that $(XI_n - A)$ and $(XI_n - B)$ are equivalent in $M_n(F[X])$. Thus there are matrices $P(X)$, $Q(X) \in \mathrm{GL}(n, F[X])$ such that

$$(1.12) \qquad (XI_n - B) = P(X)(XI_n - A)Q(X).$$

Now, if V is a vector space of dimension n over F and $\mathcal{B}$ is a basis of V, then there are linear transformations T_1 and T_2 on V such that $[T_1]_{\mathcal{B}} = A$ and $[T_2]_{\mathcal{B}} = B$. By Proposition 1.5, there are injective homomorphisms

$$\phi_i : F[X]^n \to F[X]^n$$

such that $\mathrm{Coker}(\phi_i) \cong V_{T_i}$ and the matrix of ϕ_i with respect to the standard basis on $F[X]^n$ is $(XI_n - A)$ and $(XI_n - B)$ for $i = 1$, 2 respectively. By Equation (1.12), the $F[X]$-module homomorphisms ϕ_1 and ϕ_2 are equivalent. Therefore, Proposition 1.3 applies to give an isomorphism of V_{T_1} and V_{T_2} as $F[X]$-modules. By Theorem 4.4.2, the linear transformations T_1 and T_2 are similar, and taking matrices with respect to the basis $\mathcal{B}$ shows that A and B are also similar, and the theorem is proved.   □

This theorem suggests that a careful analysis of the concept of equivalence of matrices with entries in $F[X]$ is in order. Since it is no more difficult to study equivalence of matrices over any PID R, the next two sections will be devoted to such a study, after which, the results will be applied to the computation of invariant factors and canonical forms for linear transformations. We will conclude this section by carefully studying the relationship between generating sets of a module M that has two different finite free presentations.

(1.8) Example. Let R be a commutative ring and let M be an R-module with two equivalent finite free presentations. That is, there is a commutative diagram of R-modules and homomorphisms

$$(1.13) \qquad \begin{array}{ccccccc} R^m & \xrightarrow{f} & R^n & \xrightarrow{\pi_1} & M & \longrightarrow & 0 \\ \downarrow{h_1} & & \downarrow{h_2} & & \downarrow{1_M} & & \\ R^m & \xrightarrow{g} & R^n & \xrightarrow{\pi_2} & M & \longrightarrow & 0 \end{array}$$

where h_1 and h_2 are isomorphisms. If $\mathcal{A} = \{e_1, \ldots, e_n\}$ is the standard basis on R^n, then $\mathcal{V} = \{v_1, \ldots, v_n\}$ and $\mathcal{W} = \{w_1, \ldots, w_n\}$ are generating sets of the R-module M, where $v_i = \pi_1(e_i)$ and $w_j = \pi_2(e_j)$. Note that we are not assuming that these generating sets are minimal, i.e., we do not assume that $\mu(M) = n$. From the diagram (1.13) we see that the generators v_i and w_j are related by

(1.14) $w_j = \pi_2(e_j) = \pi_1(h_2^{-1}(e_j)).$

Let us analyze this situation in terms of matrix representations of the homomorphisms. All matrices are computed with respect to the standard bases on R^m and R^n. Thus, if $A = [f]$, $B = [g]$, $Q = [h_1]$, and $P = [h_2]$, then the commutativity of diagram (1.13) shows that $B = PAQ^{-1}$. Furthermore, if $P^{-1} = [p_{ij}^*]$, then Equation (1.14) becomes

$$w_j = \pi_1(h_2^{-1}(e_j))$$
$$= \pi_1(\sum_{i=1}^{n} p_{ij}^* e_i)$$
(1.15) $$= \sum_{i=1}^{n} p_{ij}^* v_i.$$

That is, w_j is a linear combination of the v_i where the scalars come from the jth column of P^{-1} in the equivalence equation $B = PAQ^{-1}$.

(1.9) Example. We will apply the general analysis of Example 1.8 to a specific numerical example. Let M be an abelian group with three generators v_1, v_2, and v_3, subject to the relations

$$6v_1 + 4v_2 + 2v_3 = 0$$
$$-2v_1 + 2v_2 + 6v_3 = 0.$$

That is, $M = \mathbf{Z}^3/K$ where K is the subgroup of $\mathbf{Z}^3$ generated by $y_1 = (6, 4, 2)$ and $y_2 = (-2, 2, 6)$, so there is a finite free presentation of M

(1.16) $0 \longrightarrow \mathbf{Z}^2 \xrightarrow{T_A} \mathbf{Z}^3 \xrightarrow{\pi} M \longrightarrow 0$

where T_A denotes multiplication by the matrix

$$A = \begin{bmatrix} 6 & -2 \\ 4 & 2 \\ 2 & 6 \end{bmatrix}.$$

We wish to find $B = PAQ$ equivalent to A where B is as in Proposition 4.3.20. If

$$P = \begin{bmatrix} 0 & 0 & 1 \\ 0 & 1 & -2 \\ 1 & -2 & 1 \end{bmatrix} \quad \text{and} \quad Q = \begin{bmatrix} 1 & 3 \\ 0 & -1 \end{bmatrix},$$

then

$$B = PAQ = \begin{bmatrix} 2 & 0 \\ 0 & 10 \\ 0 & 0 \end{bmatrix}.$$

(We will learn in Section 3 how to compute P and Q.) Then

$$P^{-1} = \begin{bmatrix} 3 & 2 & 1 \\ 2 & 1 & 0 \\ 1 & 0 & 0 \end{bmatrix}.$$

Therefore, $w_1 = 3v_1 + 2v_1 + v_3$, $w_2 = 2v_1 + v_2$, and $w_3 = v_3$ are new generators of M, and the structure of the matrix B shows that $2w_1 = 0$ (since $2w_1 = 6v_1 + 4v_2 + 2v_3 = 0$), $10w_2 = 0$, and w_3 generates an infinite cyclic subgroup of M, i.e.,

$$M \cong \mathbf{Z}_2 \oplus \mathbf{Z}_{10} \oplus \mathbf{Z}.$$

5.2 Hermite Normal Form

In this section and the next, we will be concerned with determining the simplest form to which an $m \times n$ matrix with entries in a PID can be reduced using multiplication by unimodular matrices. Recall that a unimodular matrix is just an invertible matrix. Thus, to say that $A \in M_n(R)$ is unimodular is the same as saying that $A \in \mathrm{GL}(n, R)$. Left and right multiplications by unimodular matrices gives rise to some basic equivalence relations, which we now define.

(2.1) Definition. *Let R be a commutative ring and let A, $B \in M_{m,n}(R)$.*

(1) *We say that B is **left equivalent** to A, denoted $B \overset{\mathrm{L}}{\sim} A$, if there is a unimodular matrix $U \in \mathrm{GL}(m, R)$ such that $B = UA$.*

(2) *We say that B is **right equivalent** to A, denoted $B \overset{\mathrm{R}}{\sim} A$, if there is a unimodular matrix $V \in \mathrm{GL}(n, R)$ such that $B = AV$.*

(3) *We say that B is **equivalent** to A, denoted $B \overset{\mathrm{E}}{\sim} A$, if there are unimodular matrices $U \in \mathrm{GL}(m, R)$ and $V \in \mathrm{GL}(n, R)$ such that $B = UAV$.*

Each of these relations is an equivalence relation, and we would like to compute a simple form for a representative of each equivalence class in an algorithmic manner. (Equivalence of matrices has been introduced in Definition 4.3.18, and, in fact, if R is a PID then we have described the equivalence classes of matrices under the relation of equivalence in Proposition 4.3.20. What we will concentrate on at the present time are algorithmic aspects of arriving at these representatives.) This will require the ability to construct a number of unimodular matrices. The elementary matrices introduced in Definition 4.1.8 provide some examples of unimodular matrices over any commutative ring, and we will see that for a Euclidean domain, *every* unimodular, i.e., invertible, matrix is a product of elementary matrices. The following fundamental result describes an inductive procedure for constructing unimodular matrices with a prescribed row or column.

(2.2) Theorem. *Let R be a PID, let $a_1, \ldots, a_n \in R$, and let $d = \gcd\{a_1, \ldots, a_n\}$. Then there is a matrix $A \in M_n(R)$ such that*

(1) $\mathrm{row}_1(A) = [a_1 \ \cdots \ a_n]$, *and*
(2) $\det(A) = d$.

Proof. The proof is by induction on n. If $n = 1$ the theorem is trivially true. Suppose the theorem is true for $n - 1$ and let $A_1 \in M_{n-1}(R)$ be a matrix with $\mathrm{row}_1(A_1) = [a_1 \ \cdots \ a_{n-1}]$ and $\det(A_1) = d_1 = \gcd\{a_1, \ldots, a_{n-1}\}$. Since

$$
\begin{aligned}
d &= \gcd\{a_1, \ldots, a_n\} \\
&= \gcd\{\gcd\{a_1, \ldots, a_{n-1}\}, a_n\} \\
&= \gcd\{d_1, a_n\},
\end{aligned}
$$

it follows that there are $u, v \in R$ such that $ud_1 - va_n = d$. Now define A by

$$
A = \begin{bmatrix}
& & & & a_n \\
& A_1 & & & 0 \\
& & & & \vdots \\
& & & & 0 \\
\frac{a_1 v}{d_1} & \frac{a_2 v}{d_1} & \cdots & \frac{a_{n-1} v}{d_1} & u
\end{bmatrix}.
$$

Since $d_1 \mid a_i$ for $1 \leq i \leq n - 1$, it follows that $A \in M_n(R)$ and $\mathrm{row}_1(A) = [a_1 \ \cdots \ a_n]$. Now compute the determinant of A by cofactor expansion along the last column. Thus,

$$
\det(A) = u \det(A_1) + (-1)^{n+1} a_n \det(A_{1n})
$$

where A_{1n} denotes the minor of A obtained by deleting row 1 and column n. Note that A_{1n} is obtained from A_1 by moving $\mathrm{row}_1(A_1) = [a_1 \ \cdots \ a_{n-1}]$ to the $n-1$ row, moving all other rows up by one row, and then multiplying the new row $n-1$ $(= [a_1 \ \cdots \ a_{n-1}])$ by v/d_1. That is, using the language of elementary matrices

$$
A_{1n} = D_{n-1}(v/d_1) P_{n-1, n-2} \cdots P_{3,2} P_{2,1} A_1.
$$

Thus,

$$
\begin{aligned}
\det A_{1n} &= \left(\frac{v}{d_1}\right) (-1)^{n-2} \det A_1 \\
&= (-1)^{n-2} v.
\end{aligned}
$$

Hence,

$$
\begin{aligned}
\det A &= u \det A_1 + (-1)^{n+1} a_n \det A_{1n} \\
&= ud_1 + (-1)^{n+1} a_n (-1)^{n-2} v \\
&= ud_1 - va_n \\
&= d.
\end{aligned}
$$

Therefore, the theorem is true for all n by induction. $\square$

(2.3) *Remark.* Note that the proof of Theorem 2.2 is completely algorithmic, except perhaps for finding u, v with $ud_1 - va_n = d$; however, if R is a Euclidean domain, that is algorithmic as well. It is worth noting that the existence part of Theorem 2.2 follows easily from Theorem 3.6.16. The details are left as an exercise; however, that argument is not at all algorithmic.

(2.4) **Corollary.** *Suppose that R is a PID, that a_1, ..., a_n are relatively prime elements of R, and $1 \le i \le n$. Then there is a unimodular matrix A_i with $\mathrm{row}_i(A_i) = [\,a_1 \ \cdots \ a_n\,]$ and a unimodular matrix B_i with $\mathrm{col}_i(B_i) = [\,a_1 \ \cdots \ a_n\,]^t$.*

Proof. Let $A \in M_n(R)$ be a matrix with

$$\mathrm{row}_1(A) = [\,a_1 \ \cdots \ a_n\,]$$

and $\det A = 1$, which is guaranteed by the Theorem 2.2. Then let $A_i = P_{1i}A$ and let $B_i = A^t P_{1i}$ where P_{1i} is the elementary permutation matrix that interchanges rows 1 and i (or columns 1 and i). $\qquad\square$

(2.5) **Example.** We will carry through the construction of Theorem 2.2 in a specific numerical example. Thus let $R = \mathbf{Z}$ and construct a unimodular matrix $A \in \mathrm{GL}(3, \mathbf{Z})$ with

$$\mathrm{row}_1(A) = [\,25 \ \ 15 \ \ 7\,].$$

Since $2 \cdot 25 - 3 \cdot 15 = 5$, we may take $A_1 = \left[\begin{smallmatrix} 25 & 15 \\ 3 & 2 \end{smallmatrix}\right]$. Then $3 \cdot 5 - 2 \cdot 7 = 1$, so the induction step will give

$$A = \begin{bmatrix} 25 & 15 & 7 \\ 3 & 2 & 0 \\ 10 & 6 & 3 \end{bmatrix}.$$

Then $\det A = 1$, so A is unimodular. Furthermore, if we wish to compute a unimodular matrix $B \in \mathrm{GL}(4, \mathbf{Z})$ with

$$\mathrm{row}_1(B) = [\,25 \ \ 15 \ \ 7 \ \ 9\,]$$

then we may use the matrix A in the induction step. Observe that $10 - 9 = 1$, so the algorithm gives us a matrix

$$B = \begin{bmatrix} 25 & 15 & 7 & 9 \\ 3 & 2 & 0 & 0 \\ 10 & 6 & 3 & 0 \\ 25 & 15 & 7 & 10 \end{bmatrix}$$

that is unimodular and has the required first row.

(2.6) Lemma. *Let R be a PID, let $A \neq 0 \in M_{m,1}(R)$, and let $d = \gcd(A)$ (i.e., d is the gcd of all the entries of A). Then there is a unimodular matrix $U \in \mathrm{GL}(m, R)$ such that*

$$UA = \begin{bmatrix} d \\ 0 \\ \vdots \\ 0 \end{bmatrix}.$$

Moreover, if R is a Euclidean domain, then U can be taken to be a product of elementary matrices.

Proof. We may write $b_1 a_1 + \cdots + b_m a_m = d \neq 0$ where $A = [\, a_1 \ \cdots \ a_m \,]^t$ and $b_1, \ldots, b_m \in R$. Then

$$b_1 \left(\frac{a_1}{d} \right) + \cdots + b_m \left(\frac{a_m}{d} \right) = 1$$

so $\{b_1, \ldots, b_m\}$ is relatively prime. By Theorem 2.2 there is a matrix $U_1 \in \mathrm{GL}(m, R)$ such that $\mathrm{row}_1(U_1) = [\, b_1 \ \cdots \ b_m \,]$. Then

$$U_1 A = \begin{bmatrix} d \\ c_2 \\ \vdots \\ c_m \end{bmatrix}$$

and $c_i = u_{i1} a_1 + \cdots + u_{im} a_m$ so that $d \mid c_i$ for all $i \geq 2$. Hence $c_i = \alpha_i d$ for $\alpha_i \in R$. Now, if U is defined by

$$U = T_{21}(-\alpha_2) T_{31}(-\alpha_3) \cdots T_{m1}(-\alpha_m) U_1$$

then Proposition 4.1.12 shows that

$$UA = \begin{bmatrix} d \\ 0 \\ \vdots \\ 0 \end{bmatrix}.$$

This completes the proof in the case of a general PID.

Now suppose that R is a Euclidean domain with Euclidean function $v : R \setminus \{0\} \to \mathbf{Z}^+$. We shall present an argument, which is essentially a second proof of Lemma 2.6 in this case. This argument is more constructive in that only elementary row operations, i.e., left multiplications by elementary matrices, are used. Hence the U constructed will be a product of elementary matrices. Let $v(A) = \min\{v(a_j) : 1 \leq j \leq m\}$ and suppose that $v(A) = v(a_i)$. Then $P_{1i} A = [\, \beta_1 \ \cdots \ \beta_m \,]^t$ has $v(\beta_1) \leq v(\beta_j)$ for $j \geq 2$. Each β_i can be written as $\beta_i = \gamma_i \beta_1 + r_i$ where $r_i = 0$ or $v(r_i) < v(\beta_1)$. Therefore, subtracting $\gamma_i \, \mathrm{row}_1(P_{1i} A)$ from $\mathrm{row}_i(P_{1i} A)$ gives a matrix

$$A_1 = T_{21}(-\gamma_2) \cdots T_{m1}(-\gamma_m)P_{1i}A = \begin{bmatrix} \beta_1 \\ r_1 \\ \vdots \\ r_m \end{bmatrix}$$

where $r_j = 0$ or $v(r_j) < v(\beta_1)$. If some $r_j \neq 0$ then we have found a matrix A_1 left equivalent to A via a product of elementary matrices for which $v(A_1) < v(A)$. We can repeat the above process to find a sequence of matrices

$$A_i = \begin{bmatrix} a_1^{(i)} \\ \vdots \\ a_m^{(i)} \end{bmatrix}$$

left equivalent to A via a product of elementary matrices such that

(1) $v(a_1^{(i-1)}) > v(a_1^{(i)})$ for all $i > 1$, and
(2) $a_j^{(i)} = 0$ or $v(a_j^{(i)}) < v(a_1^{(i)})$ for $j \geq 2$.

Since condition (1) can only occur for finitely many A_i, it follows that we must have $a_j^{(i)} = 0$ for $j \geq 2$ for some A_i, i.e.,

$$UA = A_i = \begin{bmatrix} a_1^{(i)} \\ 0 \\ \vdots \\ 0 \end{bmatrix}$$

for some $U \in \mathrm{GL}(m, R)$, which is a product of elementary matrices. It remains to observe that $a_1^{(i)} = b$ is a gcd of the set $\{a_1, \ldots, a_m\}$. But the equation

$$U \begin{bmatrix} a_1 \\ \vdots \\ a_m \end{bmatrix} = \begin{bmatrix} b \\ 0 \\ \vdots \\ 0 \end{bmatrix}$$

shows that $b \in (a_1, \ldots, a_m)$ and

$$\begin{bmatrix} a_1 \\ \vdots \\ a_m \end{bmatrix} = U^{-1} \begin{bmatrix} b \\ 0 \\ \vdots \\ 0 \end{bmatrix}$$

shows that $(a_1, \ldots, a_m) \subseteq (b)$, i.e., $(b) = (a_1, \ldots, a_m)$ and the lemma is proved. $\quad\square$

Given an equivalence relation $\sim$ on a set X, a **complete set of representatives** of $\sim$ is a subset $P \subseteq X$ such that P has exactly one element from each equivalence class. Thus, P is a complete set of representatives of $\sim$ if each $x \in X$ is equivalent to a unique $a \in P$. The cases we wish to consider concern some equivalence relations on a commutative ring R. A **complete set of nonassociates** of R is a subset $P \subseteq R$ such that each element of R is an associate of a unique $b \in P$, i.e., if $a \in R$ then there is a unique $b \in P$ and a unit $u \in R^*$ such that $b = au$. Similarly, if $I \subseteq R$ is an ideal, then a **complete set of residues modulo** I consists of a subset of R, which contains exactly one element from each coset $a + I$. If I is the principal ideal Ra for some $a \in R$, then we speak of a **complete set of residues modulo** a.

(2.7) Examples.

(1) If $R = \mathbf{Z}$, then a complete set of nonassociates consists of the nonnegative integers; while if $m \in \mathbf{Z}$ is a nonzero integer, then a complete set of residues modulo m consists of the m integers $0, 1, \ldots, |m| - 1$. A complete set of residues modulo 0 consists of all of $\mathbf{Z}$.

(2) If F is a field, then a complete set of nonassociates consists of $\{0, 1\}$; while if $a \in F \setminus \{0\}$, a complete set of residues modulo a is $\{0\}$.

(3) If F is a field and $R = F[X]$ then a complete set of nonassociates of R consists of the monic polynomials together with 0.

(2.8) Definition. *Let R be a commutative ring, let $P \subseteq R$ be a complete set of nonassociates of R, and for each $a \in R$ let $P(a)$ be a complete set of residues modulo a. Then a matrix $A = [a_{ij}] \in M_{m,n}(R)$ is said to be in* **Hermite normal form** *(following P, $P(a)$) if $A = 0$ or $A \neq 0$ and there is an integer r with $1 \leq r \leq m$ such that*

(1) *$\mathrm{row}_i(A) \neq 0$ for $1 \leq i \leq r$, $\mathrm{row}_i(A) = 0$ for $r + 1 \leq i \leq m$; and*

(2) *there is a sequence of integers $1 \leq n_1 < n_2 < \cdots < n_r \leq m$ such that $a_{ij} = 0$ for $j < n_i$ $(1 \leq i \leq r)$, $a_{in_i} \in P \setminus \{0\}$ $(1 \leq i \leq r)$, and $a_{jn_i} \in P(a_{in_i})$ for $1 \leq j < i$.*

Thus, if the matrix A is in Hermite normal form, then A looks like the following matrix:

Table 2.1. Hermite normal form

$$
\begin{bmatrix}
0 & \cdots & 0 & a_{1n_1} & * & \cdots & * & a_{1n_2} & \cdots & * & a_{1n_3} & * & \cdots & a_{1n_r} & \cdots & * \\
0 & \cdots & 0 & 0 & 0 & \cdots & 0 & a_{2n_2} & \cdots & * & a_{2n_3} & * & \cdots & a_{2n_r} & \cdots & * \\
0 & \cdots & 0 & 0 & 0 & \cdots & 0 & 0 & \cdots & 0 & a_{3n_3} & * & \cdots & a_{3n_r} & \cdots & * \\
\vdots & \ddots & \vdots & & \ddots & & \vdots & & \ddots & & \vdots & & \ddots & \vdots & & \vdots \\
0 & \cdots & & 0 & & \cdots & & 0 & \cdots & & 0 & & \cdots & a_{rn_r} & \cdots & * \\
0 & \cdots & & 0 & & \cdots & & 0 & \cdots & & 0 & & \cdots & 0 & \cdots & 0 \\
\vdots & & & \vdots & & \ddots & & \vdots & & \ddots & & \vdots & & \vdots & \ddots & \vdots \\
0 & \cdots & & 0 & & \cdots & & 0 & \cdots & & 0 & & \cdots & 0 & \cdots & 0
\end{bmatrix}
$$

where $*$ denotes an entry that can be any element of R. If R is a field and $P = \{0, 1\}$ while $P(a) = \{0\}$ for every $a \neq 0$, then in the Hermite normal form we will have $a_{in_i} = 1$ while $a_{jn_i} = 0$ if $j < i$. The resulting matrix is what is usually called a **reduced row echelon matrix** and it is used to solve systems of linear equations $Ax = b$ over a field.

Our main result on left equivalence of matrices is that if R is a PID then every matrix is left equivalent to one in Hermite normal form.

(2.9) Theorem. *Let R be a PID, $P \subseteq R$ a complete set of nonassociates, and $P(a)$ $(a \in R)$ a complete set of residues modulo a. Then any $A \in M_{m,n}(R)$ is left equivalent to a matrix H in Hermite normal form. If R happens to be Euclidean, then $H = UA$ where U is a product of elementary matrices.*

Proof. The proof is by induction on the number of rows m. If $m = 1$ and $A \neq 0$ let n_1 be the first index with $a_{1n_1} \neq 0$. Then let $ua_{1n_1} = b_{1n_1} \in P$. Then $B = uA$ is in Hermite normal form. Now suppose that $m > 1$ and that every matrix in $M_{m-1,n}(R)$ (for arbitrary n) is left equivalent (using a product of elementary matrices if R is Euclidean) to a matrix in Hermite normal form. Let n_1 be the smallest integer such that $\mathrm{col}_{n_1}(A) \neq 0$. Let $(\mathrm{col}_{n_1}(A))^t = [\alpha_1 \ \cdots \ \alpha_m]$ and let $d = \gcd\{\alpha_1, \ldots, \alpha_m\}$. Then $d \neq 0$ and, by Lemma 2.6, there is an invertible matrix $U_1 \in \mathrm{GL}(m, R)$ (which may be taken as a product of elementary matrices if R is Euclidean) such that

$$A_1 = U_1 A = \begin{bmatrix} d & * & \cdots & * \\ 0 & & & \\ \vdots & & B_1 & \\ 0 & & & \end{bmatrix}$$

where $B_1 \in M_{m-1,n-1}(R)$. By the induction hypothesis, there is an invertible matrix $V \in \mathrm{GL}(m-1, R)$ (which may be taken as a product of elementary matrices if R is Euclidean) such that $V B_1$ is in Hermite normal form. Let

$$U_2 = \begin{bmatrix} 1 & 0 \\ 0 & V \end{bmatrix}.$$

Then $U_2 A_1 = A_2$ is in Hermite normal form except that the entries a_{1n_i} $(i > 1)$ may not be in $P(a_{in_i})$. This can be arranged by first adding a multiple of row 2 to row 1 to arrange that $a_{1n_2} \in P(a_{2n_2})$, then adding a multiple of row 3 to row 1 to arrange that $a_{1n_3} \in P(a_{3n_3})$, etc. Since $a_{ij} = 0$ if $j < n_i$, a later row operation does not change the columns before n_i, so at the end of this sequence of operations A will have been reduced to Hermite normal form, and if R was Euclidean then only elementary row operations will have been used. $\square$

If we choose a complete set of nonassociates for R so that it contains 1 (as the unique representative for the units) and a complete set of representatives modulo 1 to be $\{0\}$, then the Hermite form of any $U \in \mathrm{GL}(n, R)$ is

I_n. This is easy to see since a square matrix in Hermite normal form must be upper triangular and the determinant of such a matrix is the product of the diagonal elements. Thus, if a matrix in Hermite normal form is invertible, then it must have units on the diagonal, and by our choice of 1 as the representative of the units, the matrix must have all 1's on the diagonal. Since the only representative modulo 1 is 0, it follows that all entries above the diagonal must also be 0, i.e., the Hermite normal form of any $U \in \mathrm{GL}(n, R)$ is I_n.

If we apply this observation to the case of a unimodular matrix U with entries in a Euclidean domain R, it follows from Theorem 2.9 that U can be reduced to Hermite normal form, i.e., I_n, by a finite sequence of elementary row operations. That is,

$$E_\ell \cdots E_1 U = I_n$$

where each E_j is an elementary matrix. Hence, $U = E_\ell^{-1} \cdots E_1^{-1}$ is itself a product of elementary matrices. Therefore, we have arrived at the following result.

(2.10) Theorem. *Let R be a Euclidean domain. Then every invertible matrix over R is a product of finitely many elementary matrices.*

Proof.   □

(2.11) *Remark.* If R is not Euclidean then the conclusion of Theorem 2.10 need not hold. Some explicit examples of matrices in $\mathrm{GL}(2, R)$ (R a PID), which cannot be written as a product of elementary matrices, have been given by P. M. Cohn in the paper *On the structure of the GL_2 of a ring*, Institut des Hautes Études Scientifiques, Publication #30 (1966), pp. 5–54. A careful study of Lemma 2.6 shows that the crucial ingredient, which Euclidean domains have that general PIDs may not have, is the Euclidean algorithm for producing the gcd of a finite subset of elements.

(2.12) Example. Let $R = \mathbf{Z}$, $P = \mathbf{Z}^+$, and for each $m \neq 0 \in \mathbf{Z}$, let

$$P(m) = \{0, 1, \ldots, |m| - 1\}.$$

Let $P(0) = \mathbf{Z}$. Thus we have chosen a complete set of nonassociates for $\mathbf{Z}$ and a complete set of residues modulo m for each $m \in \mathbf{Z}$. We will compute a Hermite normal form for the integral matrix

$$A = \begin{bmatrix} 4 & 2 & 9 & 5 \\ 6 & 3 & 4 & 3 \\ 8 & 4 & 1 & -1 \end{bmatrix}.$$

The left multiplications used in the reduction are $U_1, \ldots, U_7 \in \mathrm{GL}(3, \mathbf{Z})$, while $A_1 = U_1 A$ and $A_i = U_i A_{i-1}$ for $i > 1$. Then

$$U_1 = \begin{bmatrix} -1 & 1 & 0 \\ 0 & 1 & 0 \\ 0 & 0 & 1 \end{bmatrix} \qquad A_1 = U_1 A = \begin{bmatrix} 2 & 1 & -5 & -2 \\ 6 & 3 & 4 & 3 \\ 8 & 4 & 1 & -1 \end{bmatrix}$$

$$U_2 = \begin{bmatrix} 1 & 0 & 0 \\ -3 & 1 & 0 \\ -4 & 0 & 1 \end{bmatrix} \qquad A_2 = U_2 A_1 = \begin{bmatrix} 2 & 1 & -5 & -2 \\ 0 & 0 & 19 & 9 \\ 0 & 0 & 21 & 7 \end{bmatrix}$$

$$U_3 = \begin{bmatrix} 1 & 0 & 0 \\ 0 & 10 & -9 \\ 0 & -1 & 1 \end{bmatrix} \qquad A_3 = U_3 A_2 = \begin{bmatrix} 2 & 1 & -5 & -2 \\ 0 & 0 & 1 & 27 \\ 0 & 0 & 2 & -2 \end{bmatrix}$$

$$U_4 = \begin{bmatrix} 1 & 0 & 0 \\ 0 & 1 & 0 \\ 0 & -2 & 1 \end{bmatrix} \qquad A_4 = U_4 A_3 = \begin{bmatrix} 2 & 1 & -5 & -2 \\ 0 & 0 & 1 & 27 \\ 0 & 0 & 0 & -56 \end{bmatrix}$$

$$U_5 = \begin{bmatrix} 1 & 0 & 0 \\ 0 & 1 & 0 \\ 0 & 0 & -1 \end{bmatrix} \qquad A_5 = U_5 A_4 = \begin{bmatrix} 2 & 1 & -5 & -2 \\ 0 & 0 & 1 & 27 \\ 0 & 0 & 0 & 56 \end{bmatrix}$$

$$U_6 = \begin{bmatrix} 1 & 5 & 0 \\ 0 & 1 & 0 \\ 0 & 0 & 1 \end{bmatrix} \qquad A_6 = U_6 A_5 = \begin{bmatrix} 2 & 1 & 0 & 133 \\ 0 & 0 & 1 & 27 \\ 0 & 0 & 0 & 56 \end{bmatrix}$$

$$U_7 = \begin{bmatrix} 1 & 0 & -2 \\ 0 & 1 & 0 \\ 0 & 0 & 1 \end{bmatrix} \qquad A_7 = U_7 A_6 = \begin{bmatrix} 2 & 1 & 0 & 21 \\ 0 & 0 & 1 & 27 \\ 0 & 0 & 0 & 56 \end{bmatrix}.$$

The matrix A_7 is the Hermite normal form associated to the matrix A (using the system of representatives P and $P(m)$).

A natural question to ask is whether the Hermite normal form of A guaranteed by Theorem 2.9 is unique. Certainly one can get a different Hermite normal form by changing the complete set of nonassociates or the complete set of residues modulo $a \in R$, but if we fix these items then the Hermite normal form is uniquely determined, independent of the precise sequence of operations needed to achieve this form. This is the content of the next result.

(2.13) Theorem. *Let R be a PID, $P \subseteq R$ a complete set of nonassociates and $P(a)$ a complete set of residues modulo a for each $a \in R$. If $A \in M_{m,n}(R)$ then the Hermite normal form of A is unique.*

Proof. Without loss of generality we may assume that $A \neq 0$. First note that the number of nonzero rows in any Hermite normal matrix H, which is left equivalent to A, is just $\mathrm{rank}(A)$. To see this, suppose the Hermite normal matrix H has r nonzero rows, let $1 \leq n_1 < \cdots < n_r \leq n$ be the integers guaranteed by Definition 2.8, and let $\alpha = (1, 2, \ldots, r)$ and $\beta = (n_1, n_2, \ldots, n_r)$. Then $\det H[\alpha \mid \beta] = a_{1n_1} a_{2n_2} \cdots a_{rn_r} \neq 0$. Thus, $\mathrm{rank}(H) \geq r$ and any submatrix of H with more than r rows will have a row of zeros. Hence, $\mathrm{rank}(H) = r$, and since rank is preserved by equivalence

(Proposition 4.2.36), it follows that

$$r = \text{rank}(H) = \text{rank}(A).$$

Now suppose that $A \overset{L}{\sim} H$ and $A \overset{L}{\sim} K$ where H and K are both in Hermite normal form. Then there is a unimodular matrix U such that $H = UK$, and by the above paragraph, both H and K have $r = \text{rank}(A)$ nonzero rows. Let $1 \le n_1 < n_2 < \cdots < n_r \le n$ be the column indices for K given by the definition of Hermite normal form and let $1 \le t_1 < t_2 < \cdots < t_r \le n$ be the column indices for H. We claim that $t_i = n_i$ for $1 \le i \le r$. Indeed, n_1 is the first nonzero column of K and t_1 is the first nonzero column of H; but $\text{col}_j(H) = U \text{col}_j(K)$ and U is invertible, so $n_1 = t_1$. Then we conclude that for $j = n_1 = t_1$, $h_{1n_1} \ne 0$, $k_{1n_1} \ne 0$, and

$$\begin{bmatrix} h_{1n_1} \\ 0 \\ \vdots \\ 0 \end{bmatrix} = U \begin{bmatrix} k_{1n_1} \\ 0 \\ \vdots \\ 0 \end{bmatrix};$$

so $k_{1n_1} u_{s1} = 0$ for $s > 1$, and hence, $u_{s1} = 0$ for $s > 1$. Therefore,

$$U = \begin{bmatrix} u_{11} & * \\ 0 & U_1 \end{bmatrix}.$$

If $H = \begin{bmatrix} \text{row}_1(H) \\ H_1 \end{bmatrix}$ and $K = \begin{bmatrix} \text{row}_1(K) \\ K_1 \end{bmatrix}$ where $H_1, K_1 \in M_{m-1,n}(R)$ then $H_1 = U_1 K_1$ and H_1 and K_1 are in Hermite normal form. By induction on the number of rows we can conclude that $n_j = t_j$ for $2 \le j \le r$. Moreover, by partitioning U in the block form $U = \begin{bmatrix} U_{11} & U_{12} \\ U_{21} & U_{22} \end{bmatrix}$ where $U_{11} \in M_r(R)$ and by successively comparing $\text{col}_{n_j}(H) = U \text{col}_{n_j}(K)$ we conclude that $U_{21} = 0$ and that U_{11} is upper triangular. Thus,

$$U = \begin{bmatrix} u_{11} & u_{12} & \cdots & u_{1r} & \\ 0 & u_{22} & \cdots & u_{2r} & U_{12} \\ \vdots & & \ddots & \vdots & \\ 0 & 0 & \cdots & u_{rr} & \\ & & 0 & & U_{22} \end{bmatrix},$$

and $\det U = u_{11} \cdots u_{rr}(\det(U_{22}))$ is a unit of R. Therefore, each u_{ii} is a unit of R; but $h_{in_i} = u_{ii} k_{in_i}$ for $1 \le i \le r$ so that h_{in_i} and k_{in_i} are associates. But h_{in_i} and k_{in_i} are both in the given complete system of nonassociates of R, so we conclude that $h_{in_i} = k_{in_i}$ and hence that $u_{ii} = 1$ for $1 \le i \le r$. Therefore, each diagonal element of U_{11} must be 1. Now suppose that $1 \le s \le r - 1$. Then

$$h_{s,n_{s+1}} = \sum_{\gamma=1}^{m} u_{s,\gamma} k_{\gamma,n_{s+1}}$$

$$= u_{ss} k_{s,n_{s+1}} + u_{s,s+1} k_{s+1,n_{s+1}}$$

since $u_{s\gamma} = 0$ for $\gamma < s$ while $k_{\gamma n_{s+1}} = 0$ if $\gamma > s + 1$. Since $u_{ss} = 1$, we conclude that

$$h_{s,n_{s+1}} = k_{s,n_{s+1}} + u_{s,s+1} k_{s+1,n_{s+1}}.$$

Therefore,

$$h_{s,n_{s+1}} \equiv k_{s,n_{s+1}} \pmod{k_{s+1,n_{s+1}}},$$

and since $h_{s,n_{s+1}}$ and $k_{s,n_{s+1}}$ are both in $P(k_{s+1,n_{s+1}})$, it follows that $h_{s,n_{s+1}} = k_{s,n_{s+1}}$ and, hence, $u_{s,s+1} = 0$ for $1 \leq s \leq r - 1$.

We now proceed by induction. Suppose that $u_{s,s+j} = 0$ for $1 \leq j < r-1$ (we just verified that this is true for $j = 1$) and consider

$$
\begin{aligned}
h_{s,n_{s+j+1}} &= \sum_{\gamma=1}^{m} u_{s\gamma} k_{\gamma, n_{s+j+1}} \\
&= u_{s,s+j+1} k_{s+j+1, n_{s+j+1}} + u_{ss} k_{s,n_{s+j+1}} \\
&= u_{s,s+j+1} k_{s+j+1, n_{s+j+1}} + k_{s,n_{s+j+1}}.
\end{aligned}
$$

Therefore, $h_{s,n_{s+j+1}} = k_{s,n_{s+j+1}}$ since they both belong to the same residue class modulo $k_{s+j+1,n_{s+j+1}}$. Hence $u_{s,s+j+1} = 0$. Therefore, we have shown that U has the block form

$$U = \begin{bmatrix} I_r & U_{12} \\ 0 & U_{22} \end{bmatrix}$$

and, since the last $m - r$ rows of H and K are zero, it follows that $H = UK = K$ and the uniqueness of the Hermite normal form is proved.   □

We will conclude this section with the following simple application of the Hermite normal form. Recall that if R is any ring, then the (two-sided) ideals of the matrix ring $M_n(R)$ are precisely the sets $M_n(J)$ where J is an ideal of R (Theorem 2.2.26). In particular, if R is a division ring then the only ideals of $M_n(R)$ are $\langle 0 \rangle$ and the full ring $M_n(R)$. There are, however, many left ideals of the ring $M_n(R)$, and if R is a PID, then the Hermite normal form allows one to compute explicitly all the left ideals of $M_n(R)$, namely, they are all principal.

(2.14) Theorem. *Let R be a PID and let $J \subseteq M_n(R)$ be a left ideal. Then there is a matrix $A \in M_n(R)$ such that $J = \langle A \rangle$, i.e., J is a principal left ideal of $M_n(R)$.*

Proof. $M_n(R)$ is finitely generated as an R-module (in fact, it is free of rank n^2), and the left ideal J is an R-submodule of $M_n(R)$. By Theorem 3.6.2, J is finitely generated as an R-module. Suppose that (as an R-module)

$$J = \langle B_1, \ldots, B_k \rangle$$

where $B_i \in M_n(R)$. Consider the matrix

$$B = \begin{bmatrix} B_1 \\ \vdots \\ B_k \end{bmatrix} \in M_{nk,n}(R).$$

There is an invertible matrix $P \in M_{nk}(R)$ such that PB is in Hermite normal form. Thus,

(2.1) $$PB = \begin{bmatrix} A \\ 0 \\ \vdots \\ 0 \end{bmatrix}$$

and if we partition $P = [P_{ij}]$ into blocks where $P_{ij} \in M_n(R)$, it follows from Equation (2.1) that

$$A = P_{11}B_1 + \cdots + P_{1k}B_k.$$

Therefore, $A \in J$ and hence the left ideal $\langle A \rangle \subseteq J$. Since

$$B = Q \begin{bmatrix} A \\ 0 \\ \vdots \\ 0 \end{bmatrix}$$

where $Q = P^{-1} = [Q_{ij}]$, it follows that $B_i = Q_{i1}A$. Therefore, $J \subseteq \langle A \rangle$, and the proof is complete.   □

5.3 Smith Normal Form

In contrast to the relatively complicated nature of the Hermite normal form, if multiplication by nonsingular matrices is allowed on both the left and the right, then one can reduce a matrix A over a PID R to a particularly simple form. The existence of this simple diagonal form, known as the Smith normal form, has essentially already been proved in Proposition 4.3.20. Combining this with Theorem 2.10 will provide a reasonably efficient procedure for the computation of the invariant factors of a linear transformation. Additionally, this same computational procedure produces the change of basis map that puts a matrix (or linear transformation) in rational canonical form. We will also consider applications of the Smith normal form to the solution of linear diophantine equations.

(3.1) **Theorem.** Let R be a PID and let $A \in M_{m,n}(R)$. Then there is a $U \in \mathrm{GL}(m, R)$ and a $V \in \mathrm{GL}(n, R)$ such that

(3.1) $$UAV = \begin{bmatrix} D_r & 0 \\ 0 & 0 \end{bmatrix}$$

where $r = \text{rank}(A)$ and $D_r = \text{diag}(s_1, \dots, s_r)$ with $s_i \neq 0$ $(1 \leq i \leq r)$ and $s_i \mid s_{i+1}$ for $1 \leq i \leq r - 1$. Furthermore, if R is a Euclidean domain, then the matrices U and V can be taken to be a product of elementary matrices.

Proof. Consider the homomorphism $T_A : R^n \to R^m$ defined by multiplication by the matrix A. By Proposition 4.3.20, there is a basis $\mathcal{B}$ of R^n and a basis $\mathcal{C}$ of R^m such that

(3.2) $$[T_A]_{\mathcal{C}}^{\mathcal{B}} = \begin{bmatrix} D_r & 0 \\ 0 & 0 \end{bmatrix}$$

where $D_r = \text{diag}(s_1, \dots, s_r)$ with $s_i \neq 0$ $(1 \leq i \leq r)$ and $s_i \mid s_{i+1}$ for $1 \leq i \leq r - 1$. If $\mathcal{C}'$ and $\mathcal{B}'$ denote the standard bases on R^m and R^n respectively, then the change of basis formula (Proposition 4.3.16) gives

(3.3) $$\begin{bmatrix} D_r & 0 \\ 0 & 0 \end{bmatrix} = [T_A]_{\mathcal{C}}^{\mathcal{B}} = P_{\mathcal{C}}^{\mathcal{C}'} [T_A]_{\mathcal{C}'}^{\mathcal{B}'} P_{\mathcal{B}'}^{\mathcal{B}}.$$

Since $A = [T_A]_{\mathcal{C}'}^{\mathcal{B}'}$ and since the change of basis matrices are invertible, Equation (3.3) implies Equation (3.1). The last statement is a consequence of Theorem 2.10.   □

(3.2) Remarks.

(1) The matrix $\begin{bmatrix} D_r & 0 \\ 0 & 0 \end{bmatrix}$ is called the **Smith normal form** of A after H. J. Smith, who studied matrices over **Z**.
(2) Since the elements $s_1, \dots, s_r$ are the invariant factors of the submodule $\text{Im}(T_A)$, they are unique (up to multiplication by units of R). We shall call these elements the **invariant factors** of the matrix A. Thus, two matrices $A, B \in M_{m,n}(R)$ are equivalent if and only if they have the same invariant factors. This observation combined with Theorem 1.7 gives the following criterion for the similarity of matrices over fields.

(3.3) Theorem. *Let F be a field and let $A, B \in M_n(F)$. Then A and B are similar if and only if the matrices $X I_n - A$ and $X I_n - B \in M_n(F[X])$ have the same invariant factors.*

Proof.   □

(3.4) Remark. If R is a Euclidean domain then A can be transformed into Smith normal form by a finite sequence of elementary row and column operations since every unimodular matrix over R is a finite product of elementary matrices (Theorem 2.10). It is worthwhile to describe explicitly an algorithm by which the reduction to Smith normal form can be accomplished in the case of a Euclidean domain R with degree function $v : R \setminus 0 \to \mathbf{Z}^+$. The algorithm is an extension (to allow both row and column operations) of the second proof of Lemma 2.6. If $A \in M_{m,n}(R)$ is a nonzero matrix, let

$$v(A) = \min\{v(a_{ij}) : 1 \leq i \leq m; \; 1 \leq j \leq n\}.$$

By using a sequence of row and column exchanges we can assume that $v(a_{11}) = v(A)$. Then if $i > 1$, we may write $a_{i1} = a_{11}b_i + b_{i1}$ where $b_{i1} = 0$ or $v(b_{i1}) < v(a_{11})$. By subtracting $b_i \, \mathrm{row}_1(A)$ from $\mathrm{row}_i(A)$ we obtain a matrix $A^{(1)}$ in which every element of $\mathrm{col}_1(A^{(1)})$ is divisible by a_{11} or $v(A^{(1)}) < v(A)$. If we are not in the first case, repeat the process with A replaced by $A^{(1)}$. Since $v(A)$ is a positive integer, this process cannot go on indefinitely, so we must eventually arrive at a matrix B with $b_{11} \mid b_{i1}$ for $2 \leq i \leq m$. By applying a similar process to the elements of the first row, we may also assume that elementary row and column operations have produced a matrix B in which $b_{11} \mid b_{i1}$ and $b_{11} \mid b_{1j}$ for $2 \leq i \leq m$ and $2 \leq j \leq n$. Then subtracting multiples of the first row and column of B produces an equivalent matrix $\widetilde{B} = [b_{11}] \oplus C$ where $C \in M_{m-1,n-1}$. We may arrange that b_{11} divides every element of C. If this is not the case already, then simply add a row of C to the first row of $\widetilde{B}$, producing an equivalent matrix to which the previous process can be applied. Since each repetition reduces $v(\widetilde{B})$, only a finite number of repetitions are possible before we achieve a matrix $B = [b_{11}] \oplus C$ in which b_{11} divides every entry of C. If C is not zero, repeat the process with C. This process will end with the production, using only elementary row and column operations, of the Smith normal form.

(3.5) Example. It is worthwhile to see this algorithm in practice, so we will do a complete example of the computation of the Smith normal form of an integral matrix. A simple method for producing $U \in \mathrm{GL}(m, R)$ and $V \in \mathrm{GL}(n, R)$ so that UAV is in Smith normal form is to keep track of the elementary row and column operations used in this reduction. This can be conveniently accomplished by simultaneouly applying to I_m each row operation done to A and to I_n each column operation performed on A. This process is best illustrated by a numerical example. Thus let

$$A = \begin{bmatrix} 2 & 1 & -3 & -1 \\ 1 & -1 & -3 & 1 \\ 4 & -4 & 0 & 16 \end{bmatrix} \in M_{3,4}(\mathbf{Z}).$$

We shall reduce A to Smith normal form by a sequence of row and column operations, and we shall keep track of the net effect of these operations by simultaneously performing them on I_3 (for the row operations) and I_4 (for the column operations). We will use the arrow $\mapsto$ to indicate the passage from one operation to the next.

$$
\begin{array}{ccccc}
& I_3 & \vdots & A & \vdots & I_4 \\
\end{array}
$$

$$
=
\begin{bmatrix} 1 & 0 & 0 \\ 0 & 1 & 0 \\ 0 & 0 & 1 \end{bmatrix}
\;\vdots\;
\begin{bmatrix} 2 & 1 & -3 & -1 \\ 1 & -1 & -3 & 1 \\ 4 & -4 & 0 & 16 \end{bmatrix}
\;\vdots\;
\begin{bmatrix} 1 & 0 & 0 & 0 \\ 0 & 1 & 0 & 0 \\ 0 & 0 & 1 & 0 \\ 0 & 0 & 0 & 1 \end{bmatrix}
$$

$$
\mapsto
\begin{bmatrix} 0 & 1 & 0 \\ 1 & 0 & 0 \\ 0 & 0 & 1 \end{bmatrix}
\;\vdots\;
\begin{bmatrix} 1 & -1 & -3 & 1 \\ 2 & 1 & -3 & -1 \\ 4 & -4 & 0 & 16 \end{bmatrix}
\;\vdots\;
\begin{bmatrix} 1 & 0 & 0 & 0 \\ 0 & 1 & 0 & 0 \\ 0 & 0 & 1 & 0 \\ 0 & 0 & 0 & 1 \end{bmatrix}
$$

$$
\mapsto
\begin{bmatrix} 0 & 1 & 0 \\ 1 & -2 & 0 \\ 0 & -4 & 1 \end{bmatrix}
\;\vdots\;
\begin{bmatrix} 1 & -1 & -3 & 1 \\ 0 & 3 & 3 & -3 \\ 0 & 0 & 12 & 12 \end{bmatrix}
\;\vdots\;
\begin{bmatrix} 1 & 0 & 0 & 0 \\ 0 & 1 & 0 & 0 \\ 0 & 0 & 1 & 0 \\ 0 & 0 & 0 & 1 \end{bmatrix}
$$

$$
\mapsto
\begin{bmatrix} 0 & 1 & 0 \\ 1 & -2 & 0 \\ 0 & -4 & 1 \end{bmatrix}
\;\vdots\;
\begin{bmatrix} 1 & 0 & 0 & 0 \\ 0 & 3 & 3 & -3 \\ 0 & 0 & 12 & 12 \end{bmatrix}
\;\vdots\;
\begin{bmatrix} 1 & 1 & 3 & -1 \\ 0 & 1 & 0 & 0 \\ 0 & 0 & 1 & 0 \\ 0 & 0 & 0 & 1 \end{bmatrix}
$$

$$
\mapsto
\begin{bmatrix} 0 & 1 & 0 \\ 1 & -2 & 0 \\ 0 & -4 & 1 \end{bmatrix}
\;\vdots\;
\begin{bmatrix} 1 & 0 & 0 & 0 \\ 0 & 3 & 0 & 0 \\ 0 & 0 & 12 & 12 \end{bmatrix}
\;\vdots\;
\begin{bmatrix} 1 & 1 & 2 & 0 \\ 0 & 1 & -1 & 1 \\ 0 & 0 & 1 & 0 \\ 0 & 0 & 0 & 1 \end{bmatrix}
$$

$$
\mapsto
\begin{bmatrix} 0 & 1 & 0 \\ 1 & -2 & 0 \\ 0 & -4 & 1 \end{bmatrix}
\;\vdots\;
\begin{bmatrix} 1 & 0 & 0 & 0 \\ 0 & 3 & 0 & 0 \\ 0 & 0 & 12 & 0 \end{bmatrix}
\;\vdots\;
\begin{bmatrix} 1 & 1 & 2 & -2 \\ 0 & 1 & -1 & 2 \\ 0 & 0 & 1 & -1 \\ 0 & 0 & 0 & 1 \end{bmatrix}
$$

$$
\begin{array}{ccccc}
= & U & \vdots & S & \vdots & V
\end{array}
$$

Then $UAV = S$ and S is the Smith normal form of A.

(3.6) Remark. Theorem 3.3 and the algorithm of Remark 3.4 explain the origin of the adjective rational in rational canonical form. Specifically, the invariant factors of a linear transformation can be computed by "rational" operations, i.e., addition, subtraction, multiplication, and division of polynomials. Contrast this with the determination of the Jordan canonical form, which requires the complete factorization of polynomials. This gives an indication of why the rational canonical form is of some interest, even though the Jordan canonical form gives greater insight into the geometry of linear transformations.

The ability to compute the invariant factors of a linear transformation by reduction of the characteristic matrix $XI_n - [T]_\mathcal{B}$ to Smith normal form has the following interesting consequence.

(3.7) Proposition. Let F be a field and let $A \in M_n(F)$. Then A is similar to the transposed matrix A^t.

Proof. Consider the matrix $XI_n - A \in M_n(F[X])$. Then there are invertible matrices $P(X)$, $Q(X) \in \mathrm{GL}(n, F[X])$ such that

$$(3.4) \qquad P(X)(XI_n - A)Q(X) = \mathrm{diag}(s_1(X), \ldots, s_n(X))$$

where $s_i(X) \mid s_{i+1}(X)$ for $1 \leq i \leq n - 1$. Taking the transpose of Equation (3.4) shows that

$$Q(X)^t(XI_n - A^t)P(X)^t = \mathrm{diag}(s_1(X), \ldots, s_n(X)).$$

Thus, A and A^t have the same invariant factors and hence are similar. $\square$

This result cannot be extended to matrices over arbitrary rings, even PIDs, as the following example shows:

(3.8) Example. *Let R be a PID, which is not a field, and let $p \in R$ be a prime. Consider the matrix*

$$A = \begin{bmatrix} 0 & p & 0 \\ 0 & 0 & 1 \\ 0 & 0 & 0 \end{bmatrix} \in M_3(R).$$

Claim. A *is* **not** *similar to* A^t.

Proof. Suppose that $T = [t_{ij}] \in M_3(R)$ satisfies the matrix equation $AT = TA^t$. Then

$$(3.5) \qquad \begin{bmatrix} 0 & p & 0 \\ 0 & 0 & 1 \\ 0 & 0 & 0 \end{bmatrix} \begin{bmatrix} t_{11} & t_{12} & t_{13} \\ t_{21} & t_{22} & t_{23} \\ t_{31} & t_{32} & t_{33} \end{bmatrix} = \begin{bmatrix} t_{11} & t_{12} & t_{13} \\ t_{21} & t_{22} & t_{23} \\ t_{31} & t_{32} & t_{33} \end{bmatrix} \begin{bmatrix} 0 & 0 & 0 \\ p & 0 & 0 \\ 0 & 1 & 0 \end{bmatrix},$$

which implies that

$$(3.6) \qquad \begin{bmatrix} pt_{21} & pt_{22} & pt_{23} \\ t_{31} & t_{32} & t_{33} \\ 0 & 0 & 0 \end{bmatrix} = \begin{bmatrix} pt_{12} & t_{13} & 0 \\ pt_{22} & t_{23} & 0 \\ pt_{32} & t_{33} & 0 \end{bmatrix}.$$

From Equation (3.6), we conclude that $t_{32} = t_{23} = t_{33} = 0$, $t_{12} = t_{21}$, $t_{13} = t_{31} = pt_{22} = ps$ for some $s \in R$. Therefore, T must have the form

$$T = \begin{bmatrix} t_{11} & t_{12} & ps \\ t_{12} & s & 0 \\ ps & 0 & 0 \end{bmatrix},$$

and hence, $\det(T) = -p^2 s^3$. Since this can never be a unit of the ring R, it follows that the matrix equation $AT = TA^t$ has no invertible solution T. Thus, A is not similar to A^t. $\square$

(3.9) *Remark.* Theorem 1.7 is valid for any commutative ring R. That is, if A, $B \in M_n(R)$, then A and B are similar if and only if the polynomial matrices $XI_n - A$ and $XI_n - B$ are equivalent in $M_n(R[X])$. The proof we have given for Theorem 1.7 goes through with no essential modifications. With this in mind, a consequence of Example 3.8 is that the polynomial matrix

$$XI_3 - A = \begin{bmatrix} X & -p & 0 \\ 0 & X & -1 \\ 0 & 0 & X \end{bmatrix} \in M_3(R[X])$$

is not equivalent to a diagonal matrix. This is clear since the proof of Proposition 3.7 would show that A and A^t were similar if $XI_3 - A$ was equivalent to a diagonal matrix.

What this suggests is that the theory of equivalence for matrices with entries in a ring that is not a PID (e.g., $R[X]$ when R is a PID that is not a field) is not so simple as the theory of invariant factors. Thus, while Theorem 1.7 (extended to $A \in M_n(R)$) translates the problem of similarity of matrices in $M_n(R)$ into the problem of equivalence of matrices in $M_n(R[X])$, this merely replaces one difficult problem with another that is equally difficult, except in the fortuitous case of $R = F$ a field, in which case the problem of equivalence in $M_n(F[X])$ is relatively easy to handle.

The invariant factors of a matrix $A \in M_{m,n}(R)$ (R a PID) can be computed from the determinantal divisors of A. Recall (see the discussion prior to Definition 4.2.20) that $Q_{p,m}$ denotes the set of all sequences $\alpha = (i_1, \ldots, i_p)$ of p integers with $1 \le i_1 < i_2 < \cdots < i_p \le m$. If $\alpha \in Q_{p,m}$, $\beta \in Q_{j,n}$, and $A \in M_{m,n}(R)$, then $A[\alpha \mid \beta]$ denotes the submatrix of A whose row indices are in α and whose column indices are in β. Also recall (Definition 4.2.20) that the determinantal rank of A, denoted D-rank(A), is the largest t such that there is a submatrix $A[\alpha \mid \beta]$ (where $\alpha \in Q_{t,m}$ and $\beta \in Q_{t,n}$) with $\det A[\alpha \mid \beta] \ne 0$. Since R is an integral domain, all the ranks of a matrix are the same, so we will write rank(A) for this common number. For convenience, we will repeat the following definition (see Definition 4.2.21):

(3.10) **Definition.** *Let R be a PID, let $A \in M_{m,n}(R)$, and let k be an integer such that $1 \le k \le \min\{m, n\}$. If $\det A[\alpha \mid \beta] = 0$ for all $\alpha \in Q_{k,m}$, $\beta \in Q_{k,n}$, then we set $d_k(A) = 0$. Otherwise, we set*

$$d_k(A) = \gcd\{\det A[\alpha \mid \beta] : \alpha \in Q_{k,m}, \ \beta \in Q_{k,n}\}.$$

$d_k(A)$ *is called the kth* **determinantal divisor** *of A. For convenience in some formulas, we set $d_0(A) = 1$.*

(3.11) **Lemma.** *Let R be a PID and let A, $B \in M_{m,n}(R)$. Suppose that A is equivalent to B and that $1 \le k \le \min\{m, n\}$. Then $d_k(A)$ is an associate of $d_k(B)$.*

Proof. Suppose $UAV = B$ where $U \in GL(m, R)$ and $V \in GL(n, R)$. If $\alpha \in Q_{k,m}$ and $\beta \in Q_{k,n}$, then the Cauchy–Binet theorem (Theorem 4.2.34) shows that

$$(3.7) \qquad \det B[\alpha \mid \beta] = \sum_{\substack{\omega \in Q_{k,m} \\ \tau \in Q_{k,n}}} \det U[\alpha \mid \omega] \det A[\omega \mid \tau] \det V[\tau \mid \beta].$$

Thus, if $d_k(A) = 0$ then $\det B[\alpha \mid \beta] = 0$ for all $\alpha \in Q_{k,n}$, $\beta \in Q_{k,n}$ and hence $d_k(B) = 0$. If $d_k(A) \neq 0$ then $d_k(A) \mid \det A[\omega \mid \tau]$ for all $\omega \in Q_{k,m}$, $\tau \in Q_{k,n}$, so Equation (3.7) shows that $d_k(A) \mid \det B[\alpha \mid \beta]$ for all $\alpha \in Q_{k,m}$, $\beta \in Q_{k,n}$. Therefore, $d_k(A) \mid d_k(B)$.

Since it is also true that $A = U^{-1}BV^{-1}$, we conclude that $d_k(A) = 0$ if and only if $d_k(B) = 0$ and if $d_k(A) \neq 0$ then

$$d_k(A) \mid d_k(B) \quad \text{and} \quad d_k(B) \mid d_k(A)$$

so $d_k(A)$ and $d_k(B)$ are associates. $\qquad\qquad\qquad\qquad\qquad\qquad\square$

Now suppose that R is a PID and that $A \in M_{m,n}(R)$ is in Smith normal form. That is, we suppose that $\text{rank}(A) = r$ and

$$A = \begin{bmatrix} D_r & 0 \\ 0 & 0 \end{bmatrix}$$

where $D_r = \text{diag}(s_1, \dots, s_r)$ with $s_i \neq 0$ $(1 \leq i \leq r)$ and $s_i \mid s_{i+1}$ for $1 \leq i \leq r - 1$. If $\alpha = (i_1, i_2 \dots, i_k) \in Q_{k,r}$ then $\det A[\alpha \mid \alpha] = s_{i_1} \cdots s_{i_k}$, while $\det A[\beta \mid \gamma] = 0$ for all other $\beta \in Q_{k,m}$, $\gamma \in Q_{k,n}$. Then, since $s_i \mid s_{i+1}$ for $1 \leq i \leq r - 1$, it follows that

$$(3.8) \qquad d_k(A) = \begin{cases} s_1 \cdots s_k & \text{if } 1 \leq k \leq r \\ 0 & \text{if } r + 1 \leq k \leq \min\{m, n\}. \end{cases}$$

From Equation (3.8) we see that the diagonal entries of A, i.e., $s_1, \dots, s_r$, can be computed from the determinantal divisors of A. Specifically,

$$s_1 = d_1(A)$$
$$s_2 = \frac{d_2(A)}{d_1(A)}$$

$$(3.9) \qquad\qquad\qquad \vdots$$

$$s_r = \frac{d_r(A)}{d_{r-1}(A)}.$$

By Lemma 3.11, Equations (3.9) are valid for computing the invariant factors of any matrix $A \in M_{m,n}(R)$.

(3.12) Examples.

(1) Let

$$A = \begin{bmatrix} -2 & 0 & 10 \\ 0 & -3 & -4 \\ 1 & 2 & -1 \end{bmatrix} \in M_3(\mathbf{Z}).$$

Then the Smith normal form of A is $\text{diag}(1, 1, 8)$. To see this note that $\text{ent}_{31}(A) = 1$, so $d_1(A) = 1$;

$$\det A[(1, 2) \mid (1, 2)] = 6 \quad \text{and} \quad \det A[(2, 3) \mid (2, 3)] = 11,$$

so $d_2(A) = 1$, while $\det A = 8$, so $d_3(A) = 8$. Thus, $s_1(A) = 1$, $s_2(A) = 1$, and $s_3(A) = 8$.

(2) Let

$$B = \begin{bmatrix} X(X-1)^3 & 0 & 0 \\ 0 & X-1 & 0 \\ 0 & 0 & X \end{bmatrix} \in M_3(\mathbf{Q}[X]).$$

Then $d_1(B) = 1$, $d_2(B) = X(X-1)$, and $d_3(X) = X^2(X-1)^4$. Therefore, the Smith normal form of B is $\text{diag}(1, X(X-1), X(X-1)^3)$.

Let M be a finitely generated R-module (where R is a PID) and choose a finite free presentation of M

$$0 \longrightarrow R^n \xrightarrow{T_A} R^m \longrightarrow M \longrightarrow 0$$

where T_A denotes multiplication by the matrix $A \in M_{m,n}(R)$. If A is put in Smith normal form

$$UAV = \begin{bmatrix} D_r & 0 \\ 0 & 0 \end{bmatrix} = B$$

where $D_r = \text{diag}(s_1, \ldots, s_r)$ with $s_i \neq 0$ for all i and $s_i \mid s_{i+1}$ for $1 \leq i \leq r-1$, then by Proposition 1.3

(3.10) $M \cong \text{Coker}(T_B) \cong (R/\langle s_1 \rangle) \oplus \cdots \oplus (R/\langle s_r \rangle) \oplus R^{m-r}.$

Therefore, we see that the $s_i \neq 1$ are precisely the invariant factors of the torsion submodule M_τ of M. This observation combined with Equation (3.9) provides a determinantal formula for the invariant factors of M. We record the results in the following theorem.

(3.13) Theorem. *Let R be a PID and let $A \in M_{m,n}(R)$. Suppose that the Smith normal form of A is $\begin{bmatrix} D_r & 0 \\ 0 & 0 \end{bmatrix}$. Suppose that $s_i = 1$ for $1 \leq i \leq k$ (take $k = 0$ if $s_1 \neq 1$) and $s_i \neq 1$ for $k < i \leq r$. If $M = \text{Coker}(T_A)$ where $T_A : R^n \to R^m$ is multiplication by A, then*

(1) $\mu(M) = m - k;$

(2) $\mathrm{rank}(M/M_\tau) = m - r$;

(3) the invariant factors of M_τ are $t_i = s_{k+i}$ for $1 \leq i \leq r - k$; and

(4) $t_i = \dfrac{d_{k+1}(A)}{d_{k+i-1}(A)}$ for $1 \leq i \leq r - k$.

Proof. All parts follow from the observations prior to the theorem; details are left as an exercise. □

If we apply this theorem to the presentation of V_T from Proposition 1.5, we arrive at a classical description of the minimal polynomial $m_T(X)$ due to Frobenius.

(3.14) Theorem. (Frobenius) *Let V be a finite dimensional vector space over a field F and let $T : V \rightarrow V$ be a linear transformation. Let $\mathcal{B}$ be any basis of V, let $[T]_{\mathcal{B}} = A \in M_n(F)$, and let $d(X) = d_{n-1}(XI_n - A)$. Then*

(3.11) $$m_T(X) = \frac{c_T(X)}{d(X)}.$$

Proof. $c_T(X) = \det(XI_n - A) = d_n(X)$. Since $m_T(X)$ is the highest degree invariant factor of V_T (Definition 4.4.6 (2)), formula (3.11) follows immediately from Theorem 3.13 (4). □

The determinantal criterion for invariant factors also allows one to prove the following fact.

(3.15) Theorem. *Let F be a field and let A, $B \in M_n(F)$. If K is field that contains F as a subfield, then A and B are also in $M_n(K)$. If A and B are similar in $M_n(K)$, then they are similar in $M_n(F)$.*

Proof. This follows immediately from Theorem 3.3 and the following observations.

(1) If $f(X)$ and $g(X) \neq 0$ are in $F[X]$, then the quotient and remainder upon division of $f(X)$ by $g(X)$ in $K[X]$ are, in fact, in $F[X]$. To see this, divide $f(X)$ by $g(X)$ in $F[X]$ to get

$$f(X) = g(X)q(X) + r(X)$$

where $q(X)$, $r(X) \in F[X]$ and $\deg r(X) < \deg g(X)$. The uniqueness of division in $K[X]$ shows that this is also the division of $f(X)$ by $g(X)$ in $K[X]$.

(2) Let $f_1(X)$, ..., $f_k(X) \in F[X]$. Then the greatest common divisor of these polynomials is the same whether they are considered in $F[X]$ or in $K[X]$. This follows from (1) because the greatest common divisor can be computed by the Euclidean algorithm, which only uses the division algorithm.

If A and $B \in M_n(F)$ are similar in $M_n(K)$, then the polynomial matrices $XI_n - A$ and $XI_n - B$ have the same invariant factors in $K[X]$. But since the invariant factors are computed as quotients of determinantal divisors, and since $XI_n - A$ and $XI_n - B$ are in $M_n(F[X])$, we conclude from items (1) and (2) above that the invariant factors of both polynomial matrices are in $F[X]$. Hence A and B are similar in $M_n(F)$. □

(3.16) Remark. The content of Theorem 3.15 is that in order to determine if two matrices are similar, we may, without loss of generality, assume that they are matrices over a large (e.g., algebraically closed) field. This observation is useful, for example, in Theorem 5.13 (see Remark 5.14).

Let $A \in M_{m,n}(R)$ (R a PID) be a matrix of rank r and let $s_1, \ldots, s_r$ be the invariant factors of A. Then, by definition, $s_i \neq 0$, $s_i \mid s_{i+1}$ for $1 \leq i \leq r - 1$ and A is equivalent to $\begin{bmatrix} D_r & 0 \\ 0 & 0 \end{bmatrix}$ where $D_r = \mathrm{diag}(s_1, \ldots, s_r)$. Let $p_1, \ldots, p_k$ be a complete set of nonassociate primes that occur as prime divisors of some invariant factor. Then for appropriate nonnegative integers e_{ij} and units u_i of R, we have

$$\begin{aligned} s_1 &= u_1 p_1^{e_{11}} p_2^{e_{12}} \cdots p_k^{e_{1k}} \\ (3.12) \qquad s_2 &= u_2 p_1^{e_{21}} p_2^{e_{22}} \cdots p_k^{e_{2k}} \\ &\vdots \\ s_r &= u_r p_1^{e_{r1}} p_2^{e_{r2}} \cdots p_r^{e_{rk}}. \end{aligned}$$

Since $s_i \mid s_{i+1}$, it follows that

$$(3.13) \qquad 0 \leq e_{1j} \leq e_{2j} \leq \cdots \leq e_{rj} \qquad (1 \leq j \leq k).$$

The prime power factors $\{p_j^{e_{ij}} : e_{ij} > 0\}$, counted according to the number of times each occurs in the Equation (3.12), are called the **elementary divisors** of the matrix A. Of course, the elementary divisors of A are nothing more than the elementary divisors of the torsion submodule of $\mathrm{Coker}(T_A)$, where, as usual, $T_A : R^n \to R^m$ is multiplication by the matrix A (Theorem 3.13 (3)). For example, let

$$A = \mathrm{diag}(12, 36, 360, 0, 0).$$

A is already in Smith normal form, so the invariant factors of A are $12 = 2^2 \cdot 3$, $36 = 2^2 \cdot 3^2$, and $360 = 2^3 \cdot 3^2 \cdot 5$. Hence the elementary divisors of A are

$$\{2^2, 3, 2^2, 3^2, 2^3, 3^2, 5\}.$$

(3.17) Theorem. *Let R be a PID and let $A, B \in M_{m,n}(R)$. Then A and B are equivalent if and only if they have the same rank and the same set of elementary divisors (up to multiplication by a unit).*

Proof. According to Remark 3.2 (2) (uniqueness of the invariant factors), A and B are equivalent if and only if they have the same invariant factors. Since the invariant factors determine the elementary divisors, it follows that equivalent matrices have the same set of elementary divisors. Conversely, we will show that the set of elementary divisors and the rank determine the invariant factors. Indeed, if

$$e_j = \max_{1 \le i \le r} e_{ij} \qquad 1 \le j \le k$$

then s_r is an associate of $p_1^{e_1} \cdots p_k^{e_k}$. Delete $\{p_1^{e_1}, \ldots, p_k^{e_k}\}$ from the set of elementary divisors and repeat the process with the set of remaining elementary divisors to obtain s_{r-1}. Continue this process until all the elementary divisors have been used. At this point the remaining s_i are 1 and we have recovered the invariant factors from the set of elementary divisors.

$\square$

(3.18) Remark. The argument of Theorem 3.17 is essentially a reproduction of the proof of Theorem 3.7.15.

(3.19) Example. It is worthwhile to present a complete example illustrating the process of recovering the invariant factors from the elementary divisors. Thus, suppose that $A \in M_{7,6}(\mathbf{Z})$ is a rank 5 matrix with elementary divisors

$$\{2, 2^2, 2^2, 2^3, 3^2, 3^2, 3^2, 7, 11, 11^2\}.$$

Then $s_5 = 2^3 \cdot 3^2 \cdot 7 \cdot 11^2 = 60984$. Deleting 2^3, 3^2, 7, 11^2 from the set of elementary divisors leaves the set

$$\{2, 2^2, 2^2, 3^2, 3^2, 11\}.$$

Thus, $s_4 = 2^2 \cdot 3^2 \cdot 11 = 396$. Deleting 2^2, 3^2, 11 leaves the set $\{2, 2^2, 3^2\}$ so that $s_3 = 2^2 \cdot 3^2 = 36$. Deleting 2^2 and 3^2 gives a set $\{2\}$ so that $s_2 = 2$. Since the set obtained by deleting 2 from $\{2\}$ is empty, we must have that $s_1 = 1$. Therefore, A is equivalent to the matrix

$$\begin{bmatrix} 1 & 0 & 0 & 0 & 0 & 0 \\ 0 & 2 & 0 & 0 & 0 & 0 \\ 0 & 0 & 36 & 0 & 0 & 0 \\ 0 & 0 & 0 & 396 & 0 & 0 \\ 0 & 0 & 0 & 0 & 60984 & 0 \\ 0 & 0 & 0 & 0 & 0 & 0 \\ 0 & 0 & 0 & 0 & 0 & 0 \end{bmatrix}.$$

The next result is useful if an equivalent diagonal matrix (not necessarily in Smith normal form) is known.

(3.20) Proposition. *Suppose that R is a PID and $A \in M_{m,n}(R)$ is a matrix of rank r, which is equivalent to the matrix*

$$\begin{bmatrix} D_r & 0 \\ 0 & 0 \end{bmatrix}$$

where $D_r = \mathrm{diag}(t_1, \ldots, t_r)$. Then the prime power factors of the t_i ($1 \le i \le r$) are the elementary divisors of A.

Proof. Let p be any prime that divides some t_i and arrange the t_i according to ascending powers of p, i.e.,

$$t_{i_1} = p^{e_1} q_1$$
$$\vdots$$
$$t_{i_r} = p^{e_r} q_r$$

where $(p, q_i) = 1$ for $1 \le i \le r$ and $0 \le e_1 \le e_2 \le \cdots \le e_r$. Then the exact power of p that divides the determinantal divisor $d_k(A)$ is $p^{e_1 + \cdots + e_k}$ for $1 \le k \le r$ and hence the exact power of p that divides the kth invariant factor $s_k(A) = d_k(A)/d_{k-1}(A)$ is p^{e_k} for $1 \le k \le r$. (Recall that $d_0(A)$ is defined to be 1.) Thus p^{e_k} is an elementary divisor for $1 \le k \le r$. Applying this process to all the primes that divide some t_i completes the proof. $\quad\square$

(3.21) Remark. Proposition 3.20 is a matrix theoretic version of Proposition 3.7.19. The proof presented above is simpler than the proof of Proposition 3.7.19, because we now have the determinantal divisor description of the invariant factors. The following result is a consequence of Proposition 3.20 in exactly the same way that Corollary 3.7.20 is a consequence of Proposition 3.7.19.

(3.22) Corollary. *Let $B \in M_{m,n}(R)$, $C \in M_{p,q}(R)$, and let*

$$A = B \oplus C = \begin{bmatrix} B & 0 \\ 0 & C \end{bmatrix}.$$

Then the elementary divisors of A are the union of the elementary divisors of B and C.

Proof. If $U_1 B V_1$ and $U_2 C V_2$ are in Smith normal form, then setting $U = U_1 \oplus U_2$ and $V = V_1 \oplus V_2$ we see that

$$UAV = \begin{bmatrix} D_r & 0 & 0 & 0 \\ 0 & 0 & 0 & 0 \\ 0 & 0 & E_s & 0 \\ 0 & 0 & 0 & 0 \end{bmatrix}$$

where $D_r = \mathrm{diag}(d_1, \ldots, d_r)$ and $E_s = \mathrm{diag}(t_1, \ldots, t_s)$. Therefore, A is equivalent to the block matrix

$$\begin{bmatrix} D_r & 0 & 0 \\ 0 & E_s & 0 \\ 0 & 0 & 0 \end{bmatrix},$$

and Proposition 3.20 applies to complete the proof. $\qquad\qquad\square$

(3.23) Example. Let F be a field, and let

$$A = \text{diag}(X^2(X-1)^2, \ X(X-1)^3, \ X-1, \ X) \in M_4(F[X]).$$

Then by Proposition 3.20, the elementary divisors of A are

$$X^2, (X-1)^2, \qquad X, \qquad (X-1)^3, \qquad (X-1), \qquad X$$

so the invariant factors are given by

$$s_4(A) = X^2(X-1)^3$$
$$s_3(A) = X(X-1)^2$$
$$s_2(A) = X(X-1)$$
$$s_1(A) = 1.$$

Therefore, A is equivalent to the matrix

$$\text{diag}(1, \ X(X-1), \ X(X-1)^2, \ X^2(X-1)^3).$$

5.4 Computational Examples

This section will be devoted to some computational examples related to the Smith normal form. Specific computations to be considered include the reduction of matrices to rational and Jordan canonical form, generators and relations for finitely generated abelian groups, and linear diophantine equations. We will start with some examples of reduction of matrices to canonical form. These calculations are supplemental to those of Section 4.5, and it is recommended that the reader review the discussion there. In particular, note that the use of elementary row and column operations to produce the Smith normal form is also a particularly efficient technique if one is only interested in producing the characteristic polynomial of a linear transformation, which was the starting point for the calculations in Section 4.5.

(4.1) Example. *Let V be a four-dimensional vector space over $\mathbf{Q}$ with basis $\mathcal{B} = \{v_1, v_2, v_3, v_4\}$ and let $T : V \to V$ be a linear transformation with matrix $[T]_{\mathcal{B}}$ given by*

$$[T]_\mathcal{B} = A = \begin{bmatrix} 2 & -4 & 1 & 3 \\ 2 & -3 & 0 & 2 \\ 0 & -1 & 1 & 2 \\ 1 & -1 & -1 & 0 \end{bmatrix} \in M_4(\mathbf{Q}).$$

Compute the rational canonical form of T, and if it exists, compute the Jordan canonical form.

Solution. To compute the rational canonical form of T we compute the Smith normal form of the matrix $XI_4 - A \in M_4(\mathbf{Q}[X])$. Since $\mathbf{Q}[X]$ is a Euclidean domain, we may compute the Smith normal form by means of a finite sequence of elementary row and column operations on $XI_4 - A$. We will use the symbol $\mapsto$ to indicate the passage from one matrix to another by means of finitely many row or column operations. We will write the row operations to the left of the matrix, and the column operations to the right to keep a record of the row and column operations performed. They will be recorded by means of the elementary matrices, which were left (or right) multiplied to obtain the given matrix. Thus, the symbol

$$\mapsto A_n \qquad \begin{matrix} T_{ij}(\alpha) \\ D_2(\alpha') \end{matrix}$$

$$\begin{matrix} D_1(\beta) \\ P_{23}(\gamma) \end{matrix} \qquad \mapsto A_{n+1}$$

indicates that A_n is obtained from A_{n-1} by multiplying on the right by $T_{ij}(\alpha)D_2(\alpha')$, while A_{n+1} is obtained from A_n by multiplying on the left by $D_1(\beta)P_{23}(\gamma)$. (See Propositions 4.1.12 and 4.1.13.)

With these preliminaries out of the way, our calculations are as follows:

$$XI_4 - A = \begin{bmatrix} X-2 & 4 & -1 & -3 \\ -2 & X+3 & 0 & -2 \\ 0 & 1 & X-1 & -2 \\ -1 & 1 & 1 & X \end{bmatrix}$$

$$\begin{matrix} D_1(-1) \\ P_{14} \end{matrix} \quad \mapsto \begin{bmatrix} 1 & -1 & -1 & -X \\ -2 & X+3 & 0 & -2 \\ 0 & 1 & X-1 & -2 \\ X-2 & 4 & -1 & -3 \end{bmatrix}$$

$$\begin{matrix} T_{21}(2) \\ T_{41}(-(X-2)) \end{matrix} \quad \mapsto \begin{bmatrix} 1 & -1 & -1 & -X \\ 0 & X+1 & -2 & -2X-2 \\ 0 & 1 & X-1 & -2 \\ 0 & X+2 & X-3 & X^2-2X-3 \end{bmatrix}$$

$$\mapsto \begin{bmatrix} 1 & 0 & 0 & 0 \\ 0 & X+1 & -2 & -2X-2 \\ 0 & 1 & X-1 & -2 \\ 0 & X+2 & X-3 & X^2-2X-3 \end{bmatrix} \quad \begin{matrix} T_{12}(1) \\ T_{13}(1) \\ T_{14}(X) \end{matrix}$$

$$P_{23} \quad \mapsto \begin{bmatrix} 1 & 0 & 0 & 0 \\ 0 & 1 & X-1 & -2 \\ 0 & X+1 & -2 & -2X-2 \\ 0 & X+2 & X-3 & X^2-2X-3 \end{bmatrix}$$

$$
\begin{matrix} T_{42}(-(X+2)) \\ T_{32}(-(X+1)) \end{matrix} \mapsto
\begin{bmatrix}
1 & 0 & 0 & 0 \\
0 & 1 & X-1 & -2 \\
0 & 0 & -X^2-1 & 0 \\
0 & 0 & -X^2-1 & X^2+1
\end{bmatrix}
$$

$$
\mapsto
\begin{bmatrix}
1 & 0 & 0 & 0 \\
0 & 1 & 0 & 0 \\
0 & 0 & X^2+1 & 0 \\
0 & 0 & X^2+1 & X^2+1
\end{bmatrix}
\begin{matrix} T_{23}(-(X-1)) \\ T_{24}(2) \\ D_3(-1) \end{matrix}
$$

$$
\mapsto
\begin{bmatrix}
1 & 0 & 0 & 0 \\
0 & 1 & 0 & 0 \\
0 & 0 & X^2+1 & 0 \\
0 & 0 & 0 & X^2+1
\end{bmatrix}
\quad T_{43}(-1).
$$

Therefore, $XI_4 - A$ is equivalent to $\mathrm{diag}(1, 1, X^2 + 1, X^2 + 1)$ so that the nonunit invariant factors of A are $X^2 + 1$ and $X^2 + 1$. The minimal polynomial of T is $s_4(X) = X^2 + 1$ while $c_T(X) = (X^2 + 1)^2$. Since the companion matrix of $X^2 + 1$ is

$$
C(X^2 + 1) = \begin{bmatrix} 0 & -1 \\ 1 & 0 \end{bmatrix},
$$

it follows that the rational canonical form of T is

$$
(4.1) \qquad R = C(X^2 + 1) \oplus C(X^2 + 1) =
\begin{bmatrix}
0 & -1 & 0 & 0 \\
1 & 0 & 0 & 0 \\
0 & 0 & 0 & -1 \\
0 & 0 & 1 & 0
\end{bmatrix}.
$$

Since $m_T(X)$ does not split into linear factors over the field $\mathbf{Q}$, it follows that T does not possess a Jordan canonical form.

Our next goal is to produce a basis $\mathcal{B}'$ of V such that $[T]_{\mathcal{B}'} = R$. Our calculation will be based on Example 1.8, particularly Equation (1.15). That is, supposing that

$$
P(X)(XI_4 - A)Q(X) = \mathrm{diag}(1, 1, X^2 + 1, X^2 + 1)
$$

then

$$
V_T \cong \mathbf{Q}[X]w_1 \oplus \mathbf{Q}[x]w_2
$$

where $w_j = \sum_{i=1}^{4} p^*_{i,j+2}v_i$ if $P(X)^{-1} = [p^*_{ij}]$. But from our caclulations above (and Lemma 4.1.11), we conclude that

$$
P(X)^{-1} = P_{14}D_1(-1)T_{21}(-2)T_{41}(X-2)P_{23}T_{32}(X+1)T_{42}(X+2).
$$

Therefore,

$$
(4.2) \qquad P(X)^{-1} =
\begin{bmatrix}
X-2 & X+2 & 0 & 1 \\
-2 & X+1 & 1 & 0 \\
0 & 1 & 0 & 0 \\
-1 & 0 & 0 & 0
\end{bmatrix}.
$$

Thus, $w_1 = v_2$ and $w_2 = v_1$ each have annihilator $\langle X^2 + 1 \rangle$, and hence,

$$\mathcal{B}' = \{w_1, T(w_1) = -4v_1 - 3v_2 - v_3 - v_4, w_2, T(w_2) = 2v_1 + 2v_2 + v_4\}$$

is a basis of V such that $[T]_{\mathcal{B}'} = R$. Moreover, $S^{-1}AS = R$ where

$$S = \begin{bmatrix} 0 & -4 & 1 & 2 \\ 1 & -3 & 0 & 2 \\ 0 & -1 & 0 & 0 \\ 0 & -1 & 0 & 1 \end{bmatrix} = P_{\mathcal{B}}^{\mathcal{B}'}.$$

□

(4.2) Remark. Continuing with the above example, if the field in Example 4.1 is $\mathbf{Q}[i]$, rather than $\mathbf{Q}$, then $m_T(X) = X^2 + 1 = (X + i)(X - i)$, so T is diagonalizable. A basis of each eigenspace can be read off from the caclulations done above. In fact, it follows immediately from Lemma 3.7.17 that $\mathrm{Ann}((X - i)w_j) = \langle X + i \rangle$ for $j = 1, 2$. Thus, the eigenspace corresponding to the eigenvalue $-i$, that is, $\mathrm{Ker}(T + i)$, has a basis

$$\{(X - i)w_1, (X - i)w_2\}$$

and similarly for the eigenvalue i. Therefore, $\mathrm{diag}(i, i, -i, -i) = S_1^{-1}AS_1$, where

$$S_1 = \begin{bmatrix} -4 & 2+i & -4 & 2-i \\ -3+i & 2 & -3-i & 2 \\ -1 & 0 & -1 & 0 \\ -1 & 1 & -1 & 1 \end{bmatrix}.$$

(4.3) Example. *Let V be a vector space over $\mathbf{Q}$ of dimension 4, let $\mathcal{B} = \{v_1, v_2, v_3, v_4\}$ be a basis of V, and let $T : V \to V$ be a linear transformation such that*

$$[T]_{\mathcal{B}} = \begin{bmatrix} 0 & 2 & 1 & -1 \\ 1 & -1 & -1 & 1 \\ 2 & 4 & -1 & -2 \\ 1 & -2 & -1 & 2 \end{bmatrix}.$$

Compute the Jordan canonical form of T.

Solution. As in Example 4.1, the procedure is to compute the Smith normal form of $XI_4 - A$ by means of elementary row and column operations. We will leave it as an exercise for the reader to perform the actual calculations, and we will be content to record what is needed for the remainder of our computations. Thus,

(4.3) $P(X)(XI_4 - A)Q(X) = \mathrm{diag}(1, 1, X - 1, (X - 1)(X + 1)^2)$

where

$$P(X) = T_{43}(X - 1)D_3(1/4)T_{34}(1)T_{42}(-(X - 1))T_{32}(X - 1)D_2(1/4)$$
$$\cdot P_{23}T_{41}(1)T_{31}(2)T_{21}(-X)D_1(-1)P_{12}$$

and

$$Q(X) = T_{12}(X+1)T_{13}(1)T_{14}(-1)P_{24}D_3(4)D_4(2)$$
$$\cdot\, T_{23}(-(X-1))T_{24}(X+3)D_4(2)T_{34}(-(X+1)).$$

Therefore, we can immediately read off the rational and Jordan canonical forms of T:

$$R = \begin{bmatrix} 1 & 0 & 0 & 0 \\ 0 & 0 & 0 & 1 \\ 0 & 1 & 0 & 1 \\ 0 & 0 & 1 & -1 \end{bmatrix} \quad \text{and} \quad J = \begin{bmatrix} 1 & 0 & 0 & 0 \\ 0 & 1 & 0 & 0 \\ 0 & 0 & -1 & 1 \\ 0 & 0 & 0 & -1 \end{bmatrix}.$$

It remains to compute the change of basis matrices which transform A into R and J, respectively. As in Example 4.1, the computation of these matrices is based upon Equation (1.15) and Lemma 3.7.17. We start by computing $P(X)^{-1}$:

$$P(X)^{-1} = P_{12}D_1(-1)T_{21}(X)T_{31}(-2)T_{41}(-1)P_{23}D_2(4)$$
$$\cdot\, T_{32}(-(X-1))T_{42}(X-1)T_{34}(-1)D_3(4)T_{43}(-(X-1))I_4$$

$$= \begin{bmatrix} X & -(X-1) & X+3 & -1 \\ -1 & 0 & 0 & 0 \\ -2 & 4 & 0 & 0 \\ -1 & X-1 & -(X-1) & 1 \end{bmatrix}.$$

Therefore, we see from Equation (1.15) that the vector

$$v = (X+3)v_1 - (X-1)v_4$$

is a cyclic vector with annihilator $\langle X-1 \rangle$, i.e., v is an eigenvector of T with eigenvalue 1. We calculate

$$v = (T+3)(v_1) - (T-1)(v_4)$$
$$= (3v_1 + v_2 + 2v_3 + v_4) - (-v_1 + v_2 - 2v_3 + v_4)$$
$$= 4(v_1 + v_3).$$

Let $w_1 = (1/4)v = v_1 + v_4$.

Also, the vector $w_2 = -v_1 + v_4$ is a cyclic vector with annihilator $\langle (X-1)(X+1)^2 \rangle$ (again by Equation (1.15)). If $w_3 = T(w_2) = -v_1 - 4v_3 + v_4$ and $w_4 = T^2(w_2) = -5v_1 + 4v_2 + 5v_4$, then $\{w_2, w_3, w_4\}$ is a basis for $\mathbf{Q}[X]w_2$, and hence $\mathcal{B}' = \{w_1, w_2, w_3, w_4\}$ is a basis of V in which T is in rational canonical form, i.e.,

$$[T]_{\mathcal{B}'} = R = \begin{bmatrix} 1 & 0 & 0 & 0 \\ 0 & 0 & 0 & 1 \\ 0 & 1 & 0 & 1 \\ 0 & 0 & 1 & -1 \end{bmatrix}.$$

Moreover, $S^{-1}AS = R$ where

$$S = \begin{bmatrix} 1 & -1 & -1 & -5 \\ 0 & 0 & 0 & 4 \\ 1 & 0 & -4 & 0 \\ 0 & 1 & 1 & 5 \end{bmatrix} = P_{\mathcal{B}}^{\mathcal{B}'}.$$

Now for the Jordan canonical form. From Lemma 3.7.17, we see that

$$(4.4) \qquad \mathrm{Ann}((X-1)w_2) = \langle (X+1)^2 \rangle$$

and

$$(4.5) \qquad \mathrm{Ann}((X+1)^2 w_2) = \langle (X-1) \rangle.$$

Equation (4.5) implies that

$$w_2' = (X+1)^2 w_2 = -8v_1 + 4v_2 - 8v_3 + 8v_4$$

is an eigenvector of T with eigenvalue 1. Let $w_4' = (X-1)w_2 = -4v_3$ and let $w_3' = (X+1)w_4' = -4v_1 + 4v_2 + 4v_4$ (see Proposition 4.4.37). Then

$$\mathcal{B}'' = \{w_1, \, w_2', \, w_3,' \, w_4'\}$$

is a basis of V in which T is in Jordan canonical form. In particular, $U^{-1}AU = J = [T]_{\mathcal{B}''}$ where

$$U = P_{\mathcal{B}}^{\mathcal{B}''} = \begin{bmatrix} 1 & -8 & 4 & 0 \\ 0 & 4 & -4 & 0 \\ 1 & -8 & 0 & -4 \\ 0 & 8 & -4 & 0 \end{bmatrix}.$$

(4.4) *Remark.* The theory developed in this chapter, as illustrated by the above examples, allows one to compute the rational canonical form of any matrix in $M_n(F)$ (and hence any linear transformation $T : V \to V$) using only the operations of the field F. Nothing more involved than the division algorithm for polynomials is needed to be able to reduce the characteristic matrix $XI_n - A \in M_n(F[X])$ to Smith normal form. Once one has the rational canonical form (and the transforming matrix), there are two steps to computing the Jordan canonical form. First, one must be able to factor each invariant factor into irreducible factors (which must be linear if A has a Jordan normal form). This step is the difficult one; factorization of polynomials is known to be difficult. To get an appreciation for this, see any book on Galois theory. Assuming the first step has been completed, the second step in computing the Jordan canonical form is the application of Lemma 3.7.17 and Proposition 4.4.37, as in the above example.

The main applications of the Smith normal form and the description of equivalence of matrices via invariant factors and elementary divisors are to the similarity theory of matrices over fields, as illustrated by the above examples. There are, however, some other applications of the Smith normal form. Two of these applications are to the computation of the structure of a finitely generated abelian group given by generators and relations and the use of the Smith normal form in the problem of solving systems of linear equations over PIDs. We will consider examples of both of these problems, starting with abelian groups defined by generators and relations.

(4.5) Example. One explicit way to describe an abelian group is by giving generators and relations. This is expressed by saying that $M = \langle x_1, \ldots, x_n \rangle$ where the generators x_i are subject to the relations

$$a_{11}x_1 + \cdots + a_{1n}x_n = 0$$

(4.6)
$$\vdots$$

$$a_{m1}x_1 + \cdots + a_{mn}x_n = 0$$

where $A = [a_{ij}] \in M_{m,n}(\mathbf{Z})$. We can express this more formally by saying that $M = \mathbf{Z}^n/K$ where K is the subgroup of $\mathbf{Z}^n$ generated by

$$y_1 = a_{11}e_1 + \cdots + a_{1n}e_n$$

(4.7)
$$\vdots$$

$$y_m = a_{m1}e_1 + \cdots + a_{mn}e_n.$$

Here, $\{e_1, \ldots, e_n\}$ is the standard basis of $\mathbf{Z}^n$. If $\pi : \mathbf{Z}^n \to M$ is the natural projection map, then $\pi(e_i) = x_i$. We can compute the structure of an abelian group given by generators $x_1, \ldots, x_n$ subject to relations (4.6) by using the invariant factor theorem for submodules of a free module (Theorem 3.6.23). That is, we find a basis $\{v_1, \ldots, v_n\}$ of $\mathbf{Z}^n$ and natural numbers $s_1, \ldots, s_r$ such that $\{s_1v_1, \ldots, s_rv_r\}$ is a basis of the relation subgroup $K \subseteq \mathbf{Z}^n$. Then

$$M = \mathbf{Z}^n/K \cong \mathbf{Z}_{s_1} \oplus \cdots \oplus \mathbf{Z}_{s_r} \oplus \mathbf{Z}^{n-r}.$$

Note that some of the factors $\mathbf{Z}_{s_i}$ may be 0. This will occur precisely when $s_i = 1$ and it is a reflection of the fact that it may be possible to use fewer generators than were originally presented. The elements $s_1, \ldots, s_r$ are precisely the invariant factors of the integral matrix $A \in M_{m,n}(\mathbf{Z})$. To see this, note that there is an exact sequence

$$\mathbf{Z}^m \xrightarrow{\phi} \mathbf{Z}^n \xrightarrow{\pi} M \longrightarrow 0$$

where $\phi(c) = A^t c$. This follows from Equation (4.7). Then the conclusion follows from the analysis of Example 1.8, and Example 1.9 provides a numerical example of computing the new generators of M.

(4.6) Example. By a system of m linear equations over $\mathbf{Z}$ in n unknowns we mean a system

(4.8)

$$a_{11}x_1 + a_{12}x_2 + \cdots + a_{1n}x_n = b_1$$
$$a_{21}x_1 + a_{22}x_2 + \cdots + a_{2n}x_n = b_2$$
$$\vdots$$
$$a_{m1}x_1 + a_{m2}x_2 + \cdots + a_{mn}x_n = b_m$$

where $a_{ij} \in \mathbf{Z}$ and $b_i \in \mathbf{Z}$ for all i, j. System (4.8) is also called a linear diophantine system. We let $A = [a_{ij}] \in M_{m,n}(\mathbf{Z})$, $X = [\,x_1 \quad \cdots \quad x_n\,]^t$, and $B = [\,b_1 \quad \cdots \quad b_m\,]^t$. Then the system of Equations (4.8) can be written in matrix form as

$$(4.9) \qquad\qquad\qquad AX = B.$$

Now transform A to Smith normal form

$$(4.10) \qquad\qquad UAV = \begin{bmatrix} D_r & 0 \\ 0 & 0 \end{bmatrix}$$

where $U \in \mathrm{GL}(m, \mathbf{Z})$, $V \in \mathrm{GL}(n, \mathbf{Z})$, and $D_r = \mathrm{diag}(s_1, \ldots, s_r)$ with $s_i \neq 0$ and $s_i \mid s_{i+1}$ for $1 \leq i \leq r - 1$. Thus, Equation (4.9) becomes

$$(4.11) \qquad\qquad UAV(V^{-1}X) = UB = C = \begin{bmatrix} c_1 \\ \vdots \\ c_m \end{bmatrix}.$$

Setting $Y = V^{-1}X = [\,y_1 \quad \cdots \quad y_m\,]^t$ gives the equivalent system of equations

$$s_1 y_1 = c_1$$
$$s_2 y_2 = c_2$$
$$\vdots$$
$$(4.12) \qquad\qquad s_r y_r = c_r$$
$$0 = c_{r+1}$$
$$\vdots$$
$$0 = c_m.$$

The solution of the system (4.12) can be easily read off; there is a solution if and only if $s_i \mid c_i$ for $1 \leq i \leq r$ and $c_i = 0$ for $r + 1 \leq i \leq m$. If there is a solution, all other solutions are obtained by arbitrarily specifying the $n - r$ parameters $y_{r+1}, \ldots, y_n$. Observing that $X = VY$ we can then express the solutions in terms of the original variables $x_1, \ldots, x_n$.

We will illustrate the method just described with a numerical example.

(4.7) Example. Consider the linear diophantine system $AX = B$ where

$$A = \begin{bmatrix} 2 & 1 & -3 & -1 \\ 1 & -1 & -3 & 1 \\ 4 & -4 & 0 & 16 \end{bmatrix} \quad \text{and} \quad B = \begin{bmatrix} 8 \\ 1 \\ 16 \end{bmatrix}.$$

We leave it as an exercise for the reader to verify (via elementary row and column operations) that if

$$U = \begin{bmatrix} 0 & 1 & 0 \\ 1 & -2 & 0 \\ 0 & -4 & 1 \end{bmatrix} \quad \text{and} \quad V = \begin{bmatrix} 1 & 1 & 2 & -2 \\ 0 & 1 & -1 & 2 \\ 0 & 0 & 1 & -1 \\ 0 & 0 & 0 & 1 \end{bmatrix}$$

then

$$U A V = \begin{bmatrix} 1 & 0 & 0 & 0 \\ 0 & 3 & 0 & 0 \\ 0 & 0 & 12 & 0 \end{bmatrix} = S.$$

Let

$$C = UB = \begin{bmatrix} 1 \\ 6 \\ 12 \end{bmatrix}.$$

Then the system $AX = B$ is transformed into the system $SY = C$, i.e.,

$$y_1 = 1$$
$$3y_2 = 6$$
$$12y_3 = 12.$$

This system has the solutions

$$Y = \begin{bmatrix} 1 \\ 2 \\ 1 \\ t \end{bmatrix}$$

where $t = y_4$ is an arbitrary integer. We conclude that the solutions of the original equation $AX = B$ are given by $X = VY$, i.e.,

$$X = VY = \begin{bmatrix} 1 & 1 & 2 & -2 \\ 0 & 1 & -1 & 2 \\ 0 & 0 & 1 & -1 \\ 0 & 0 & 0 & 1 \end{bmatrix} \begin{bmatrix} 1 \\ 2 \\ 1 \\ t \end{bmatrix} = \begin{bmatrix} 5 - 2t \\ 1 + 2t \\ 1 - t \\ t \end{bmatrix}$$

where t is an arbitrary integer.

(4.8) Remark. The method just described will work equally well to solve systems of linear equations with coefficients from any PID.

5.5 A Rank Criterion for Similarity

Given two matrices A and $B \in M_n(F)$ (F a field) one way to determine if A and B are similar is to determine the invariant factors of both A and B, e.g., by reducing both $XI_n - A$ and $XI_n - B$ to Smith normal form. If the invariant factors are the same, then the matrices are similar. This approach, however, is not particularly amenable to providing an explicit description of the set

$$\mathcal{O}_A = \{B \in M_n(F) : B \text{ is similar to } A\}$$

by means of polynomial equations and inequations in the entries of the matrix B. Note that $\mathcal{O}_A$ is just the orbit of A under the group action

$$(P, A) \mapsto PAP^{-1}$$

of $GL(n, F)$ on $M_n(F)$.

Another approach is via Weyr's theorem (Chapter 4, Exercise 66), which states that if F is algebraically closed, then A is similar to B if and only if

$$(5.1) \qquad \operatorname{rank}((A - \lambda I_n)^k) = \operatorname{rank}((B - \lambda I_n)^k)$$

for all $\lambda \in F$ and $k \in \mathbf{N}$. This can be reduced to a finite number of rank conditions if the eigenvalues of A are known. But knowledge of the eigenvalues involves solving polynomial equations, which is intrinsically difficult.

(5.1) Example. As a simple example of the type of equations and inequations that can be derived from Equation (5.1) to describe an orbit under similarity, one can show that $A = \begin{bmatrix} 1 & 1 \\ 0 & 1 \end{bmatrix}$ is similar to $B = \begin{bmatrix} a & b \\ c & d \end{bmatrix}$ if and only if B is in one of the following sets of matrices:

$$S_1 = \left\{ \begin{bmatrix} 1 & b \\ 0 & 1 \end{bmatrix} : b \neq 0 \right\}$$

$$S_2 = \left\{ \begin{bmatrix} 1 & 0 \\ c & 1 \end{bmatrix} : c \neq 0 \right\}$$

$$S_3 = \left\{ \begin{bmatrix} a & b \\ c & d \end{bmatrix} : bc \neq 0,\ a + d = 2,\ ad - bc = 1 \right\}.$$

We leave the verification of this description of the orbit of A as an exercise.

In this section we will present a very simple criterion for the similarity of two matrices A and B (linear transformations), which depends only on the computation of three matrices formed from A and B. This has the effect of providing explicit (albeit complicated) equations and inequations for the orbit $\mathcal{O}_A$ of A under similarity. Unlike the invariant factor and elementary divisor theory for linear transformations, which was developed in the nineteenth century, the result we present now is of quite recent vintage. The original condition (somewhat more complicated than the one we present) was proved by C. I. Byrnes and M. A. Gauger in a paper published in 1977 (Decidability criteria for the similarity problem, with applications to the moduli of linear dynamical systems, *Advances in Mathematics*, Vol. 25, pp 59–90). The approach we will follow is due to J. D. Dixon (An isomorphism criterion for modules over a principal ideal domain, *Linear and Multilinear Algebra*, Vol. 8, pp. 69–72 (1979)) and is based on a numerical criterion for two finitely generated torsion modules over a PID R to be isomorphic. This result is then applied to the $F[X]$-modules V_T and V_S, where $S, T \in \text{End}_F(V)$, to get the similarity criterion.

(5.2) Lemma. *Let R be a PID and let a and b be nonzero elements of R. If $d = (a, b) = \gcd\{a, b\}$, then*

$$\text{Hom}_R(R/\langle a\rangle, R/\langle b\rangle) \cong R/\langle d\rangle.$$

Proof. This is essentially the same calculation as Example 3.3.11. We leave it to the reader. □

(5.3) Lemma. *Let R be a PID, and let*

$$M \cong R/\langle s_1\rangle \oplus \cdots \oplus R/\langle s_n\rangle$$

and

$$N \cong R/\langle t_1\rangle \oplus \cdots \oplus R/\langle t_m\rangle$$

be two finitely generated torsion R-modules. Then

$$\text{Hom}_R(M, N) \cong \bigoplus_{i=1}^{n}\bigoplus_{j=1}^{m} R/\langle s_i, t_j\rangle.$$

Proof. This follows immediately from Lemma 5.2 and Proposition 3.3.15. □

(5.4) Definition. *If R is a PID and M is a finitely generated torsion R-module, then let*

$$\ell(M) = \sum_{i=1}^{n} k_i$$

where $\{p_1^{k_1}, \ldots, p_n^{k_n}\}$ is the set of elementary divisors of M.

In the language of Section 7.1, $\ell(M)$ is the **length** of the R-module M.

(5.5) Definition. *Let M and N be finitely generated torsion R-modules $(R$ a PID). Then let*
$$\langle M : N \rangle = \ell(\mathrm{Hom}_R(M, N)).$$

The notation $\langle M : N \rangle$ is suggestive of an inner product, and it is precisely for this reason that the notation was chosen. The following result gives some facts concerning $\langle M : N \rangle$, which are reminiscent of basic properties of inner products, and the main theorem (Theorem 5.7) is analogous to the Cauchy–Schwartz inequality.

(5.6) Proposition. *Let R be a PID and let M, N, and P be finitely generated torsion R-modules. Then*

(1) $\langle M \oplus N : P \rangle = \langle M : P \rangle + \langle N : P \rangle$;
(2) $\langle M : M \rangle \geq 0$ *with equality if and only if $M = \{0\}$;*
(3) $\langle M : N \rangle = \langle N : M \rangle$.

Proof. All three parts are immediate consequences of Lemma 5.3. $\qquad\square$

(5.7) Theorem. *Let M and N be finitely generated torsion modules over a PID R. Then*
$$\langle M : N \rangle^2 \leq \langle M : M \rangle \langle N : N \rangle.$$

Equality holds if and only if $M^s = N^t$ for some relatively prime integers s and t.

Proof. We may write

(5.2) $$M \cong R/\langle s_1 \rangle \oplus \cdots \oplus R/\langle s_n \rangle$$

and

(5.3) $$N \cong R/\langle t_1 \rangle \oplus \cdots \oplus R/\langle t_m \rangle.$$

Let $p_1, \ldots, p_r$ be the distinct primes of R that divide some elementary divisor of either M or N; by Proposition 3.7.19, these are the distinct prime divisors of the s_i and t_j. Let
$$c = \mathrm{lcm}\{\mathrm{Ann}(M), \mathrm{Ann}(N)\} = p_1^{k_1} \cdots p_r^{k_r}$$

and let $k = k_1 + \cdots + k_r$. Let $\mathbf{R}^k$ be identified with the vector space of $1 \times k$ matrices over $\mathbf{R}$, and consider each $A \in \mathbf{R}^k$ as a block matrix
$$A = [\, A_1 \quad \cdots \quad A_r \,]$$

where $A_i \in \mathbf{R}^{k_i}$. Given any divisor a of c, write
$$a = p_1^{e_1} \cdots p_r^{e_r},$$

and define $v(a) \in \mathbf{R}^k$ by
$$v(a) = [\, v(a_1) \quad \cdots \quad v(a_r) \,]$$

where

$$v(a_i) = [1 \quad 1 \quad \cdots \quad 1 \quad 0 \quad \cdots \quad 0]$$

with e_i ones. Then define

$$v(M) = \sum_{i=1}^{n} v(s_i)$$

and

$$v(N) = \sum_{j=1}^{m} v(t_j).$$

Notice that the matrix $v(M)$ determines M up to isomorphism since one can recover the elementary divisors of M from $v(M)$. To see this, choose the largest $t_i \geq 0$ such that $v(M) - v(p_i^{t_i})$ has nonnegative entries. Then $p_i^{t_i}$ is an elementary divisor of M. Subtract $v(p_i^{t_i})$ from $v(M)$ and repeat the process until the zero vector is obtained. (See the proof of Proposition 3.7.19.)

Let $s_i = p_1^{e_{i1}} \cdots p_r^{e_{ir}}$, $t_j = p_1^{f_{j1}} \cdots p_r^{f_{jr}}$, and define $d_{ijl} = \min\{e_{il}, f_{jl}\}$ for $1 \leq l \leq r$. Then

$$\langle s_i, t_j \rangle = \langle p_1^{d_{ij1}} \cdots p_r^{d_{ijr}} \rangle.$$

If $(:)$ denotes the standard inner product on $\mathbf{R}^k$, then

$$(5.4) \qquad (v(s_i) : v(t_j)) = \sum_{l=1}^{r} d_{ijl}$$

$$= \ell(R/\langle s_i, t_j \rangle).$$

Therefore,

$$(v(M) : v(N)) = \sum_{i=1}^{n} \sum_{j=1}^{m} (v(s_i) : v(t_j))$$

$$= \sum_{i=1}^{n} \sum_{j=1}^{m} \ell(R/\langle s_i, t_j \rangle)$$

$$= \langle M : N \rangle.$$

Similarly, $\langle M : M \rangle = (v(M) : v(M))$ and $\langle N : N \rangle = (v(N) : v(N))$. By the Cauchy–Schwartz inequality in $\mathbf{R}^k$ we conclude that

$$\langle M : N \rangle^2 = (v(M) : v(N))^2$$
$$\leq (v(M) : v(M))(v(N) : v(N))$$
$$= \langle M : M \rangle \langle N : N \rangle,$$

as required. Moreover, equality holds if and only if $v(M)$ and $v(N)$ are linearly dependent over $\mathbf{R}$, and since the vectors have integral coordinates, it follows that we must have $v(M)$ and $v(N)$ linearly dependent over $\mathbf{Q}$, i.e.,

$$sv(M) = tv(N)$$

where s and t are relatively prime natural numbers. But

$$v(M^s) = sv(M) = tv(N) = v(N^t)$$

so that $M^s \cong N^t$ since we observed above that $v(W)$ determines the elementary divisors, and hence, the isomorphism class of a finitely generated torsion R-module W.   □

(5.8) Corollary. *Let M and N be finitely generated torsion R-modules. Then $M \cong N$ if and only if*

(1) $\ell(M) = \ell(N)$, *and*
(2) $\langle M : N \rangle^2 = \langle M : M \rangle \langle N : N \rangle$.

Proof. $M \cong N$ certainly implies (1) and (2). Conversely, suppose that (1) and (2) are satisfied. Then by Theorem 5.7, $M^s \cong N^t$ for relatively prime integers s and t. But

$$\begin{aligned}
s\ell(M) &= \ell(M^s) \\
&= \ell(N^t) \\
&= t\ell(N) \\
&= t\ell(M)
\end{aligned}$$

so $s = t$. Since s and t are relatively prime, it follows that $s = t = 1$, and hence, $M \cong N$.   □

(5.9) Remark. If $M^s \cong N^t$ for relatively prime integers s and t, then it is an easy consequence of the uniqueness of the primary cyclic decomposition of a finitely generated torsion R-module (Theorem 3.7.15) that $M \cong P^t$ and $N \cong P^s$ where P is a finitely generated torsion R-module.

We now wish to apply Corollary 5.8 to derive a simply stated rank criterion for the similarity of two linear transformations. We will start by computing matrix representations for some basic linear transformations. Let V be a finite-dimensional vector space over a field F and let $\mathcal{B} = \{v_1, \ldots, v_n\}$ be a basis of V. Then a basis of $\mathrm{End}_F(V)$ is given by $\mathcal{C} = \{f_{ij}\}_{i=1}^{n}{}_{j=1}^{n}$ where

$$f_{ij}(v_k) = \delta_{ik} v_j.$$

Under the F-algebra isomorphism $\Phi : \mathrm{End}_F(V) \to M_n(F)$ given by $\Phi(f) = [f]_{\mathcal{B}}$, we have $\Phi(f_{ij})$ is the matrix unit E_{ji}.

Given a linear transformation $T \in \mathrm{End}_F(V)$, define two linear transformations $\mathcal{L}_T \in \mathrm{End}_F(\mathrm{End}_F(V))$ and $\mathcal{R}_T \in \mathrm{End}_F(\mathrm{End}_F(V))$ by

(5.5) $$\mathcal{L}_T(U) = TU$$

and

(5.6) $$\mathcal{R}_T(U) = UT.$$

That is, $\mathcal{L}_T$ is left multiplication by T and $\mathcal{R}_T$ is right multiplication by T in the F-algebra $\mathrm{End}_F(V)$. Let us now order the basis $\mathcal{C}$ of $\mathrm{End}_F(V)$ as follows:

$$\mathcal{C} = \{f_{11}, f_{21}, \ldots, f_{n1}, f_{12}, \ldots, f_{22}, \ldots, f_{n2}, \ldots, f_{nn}\}.$$

With these notations there is the following result. (Recall that the tensor product (or kronecker product) of matrices A and B was defined in Definition 4.1.16 as a block matrix $[C_{ij}]$ where $C_{ij} = a_{ij}B$.)

(5.10) Lemma. *Let* $T \in \mathrm{End}_F(V)$ *with* $[T]_\mathcal{B} = A = [a_{ij}]$. *Then*

(5.7) $$[\mathcal{L}_T]_\mathcal{C} = A \otimes I_n$$

and

(5.8) $$[\mathcal{R}_T]_\mathcal{C} = I_n \otimes A^t.$$

Proof. Note that

$$\begin{aligned} Tf_{ij}(v_k) &= T(\delta_{ik}v_j) \\ &= \delta_{ik}T(v_j) \\ &= \delta_{ik}\sum_{l=1}^{n} a_{lj}v_l \\ &= \sum_{l=1}^{n} a_{lj}\delta_{ik}v_l \\ &= \left(\sum_{l=1}^{n} a_{lj}f_{il}\right)(v_k). \end{aligned}$$

This equation immediately gives $[\mathcal{L}_T]_\mathcal{C} = A \otimes I_n$.

A similar calculation gives $f_{ij}T = \sum_{l=1}^{n} a_{il}f_{lj}$, so Equation (5.8) is also satisfied. $\square$

(5.11) Corollary. *Let* $S, T \in \mathrm{End}_F(V)$, *and define*

$$\mathcal{T}_{S,T} \in \mathrm{End}_F(\mathrm{End}_F(V))$$

by

$$\mathcal{T}_{S,T}(U) = SU - UT.$$

If $[S]_\mathcal{B} = A$ *and* $[T]_\mathcal{B} = B$, *then*

(5.9) $$[\mathcal{T}_{S,T}]_\mathcal{C} = A \otimes I_n - I_n \otimes B^t.$$

Proof. Since $\mathcal{T}_{S,T} = \mathcal{L}_S - \mathcal{R}_T$, the result is immediate from Lemma 5.10. $\square$

If $S, T \in \mathrm{End}_F(V)$, we will let (as usual) V_S and V_T denote the $F[X]$-module structures on V determined by S and T respectively. Then by Proposition 4.4.1, we know that

$$\mathrm{Hom}_{F[X]}(V_T, V_S) = \{U \in \mathrm{End}_F(V) : UT = SU\}.$$

Thus, we have an identification

(5.10) $\mathrm{Hom}_{F[X]}(V_T, V_S) = \mathrm{Ker}(T_{S,T}),$

and hence:

(5.12) Lemma. $\dim_F \mathrm{Hom}_{F[X]}(V_T, V_S) = n^2 - \mathrm{rank}(A \otimes I_n - I_n \otimes B^t).$

Proof. By Proposition 3.8.8, we have

$$\dim_F(\mathrm{Ker}(T_{S,T})) + \dim_F(\mathrm{Im}(T_{S,T})) = \dim_F \mathrm{End}_F(V) = n^2.$$

The result then follows from Equation (5.10).   □

We can now give the proof of Dixon's theorem.

(5.13) Theorem. (Dixon) *Let V be a vector space of finite dimension n over an algebraically closed field F, and let S, $T \in \mathrm{End}_F(V)$ be linear transformations. Let $\mathcal{B}$ be a basis of V and let $A = [S]_{\mathcal{B}}$, $B = [T]_{\mathcal{B}}$. Then S and T are similar if and only if*

$$(\mathrm{rank}(A \otimes I_n - I_n \otimes B^t))^2 = (\mathrm{rank}(A \otimes I_n - I_n \otimes A^t))(\mathrm{rank}(B \otimes I_n - I_n \otimes B^t)).$$

Proof. To simplify the notation, we will let

$$r_{AA} = \mathrm{rank}(A \otimes I_n - I_n \otimes A^t),$$

with a similar definition for r_{AB} and r_{BB}. Since $V_T \cong V_S$ as $F[X]$-modules if and only if S and T are similar (Proposition 4.4.2), it follows that if S and T are similar then

$$\mathrm{Hom}_{F[X]}(V_T, V_S) \cong \mathrm{Hom}_{F[X]}(V_T, V_T) \cong \mathrm{Hom}_{F[X]}(V_S, V_S)$$

as $F[X]$-modules. Hence, they have the same rank as F-modules, and thus $r_{AA} = r_{AB} = r_{BB}$ follows immediately from Lemma 5.12.
    Conversely, assume that

(5.11) $r_{AB}^2 = r_{AA} r_{BB}.$

Since F is assumed to be algebraically closed, the elementary divisors of any finitely generated torsion $F[X]$-module W are of the form $(X - \lambda_i)^{k_i}$. Since

$$W \cong \bigoplus_{i=1}^{t} F[X]/\langle(X - \lambda_i)^{k_i}\rangle,$$

it follows that

(5.12) $$\ell(W) = \sum_{i=1}^{t} k_i = \dim_F W.$$

In particular, Lemma 5.12 shows that

$$\langle V_T : V_S \rangle = n^2 - r_{AB}$$
$$\langle V_T : V_T \rangle = n^2 - r_{BB}$$

and

$$\langle V_S : V_S \rangle = n^2 - r_{AA}.$$

Equation (5.11) then gives

$$\langle V_T : V_S \rangle^2 - \langle V_T : V_T \rangle \langle V_S : V_S \rangle = n^2(r_{AA} + r_{BB} - 2r_{AB})$$
$$+ (r_{AB}^2 - r_{AA}r_{BB})$$
$$= n^2(\sqrt{r_{AA}} - \sqrt{r_{BB}})^2$$
$$\geq 0.$$

By Theorem 5.7

$$\langle V_T : V_S \rangle^2 \leq \langle V_T : V_T \rangle \langle V_S : V_S \rangle.$$

Thus,

$$\langle V_T : V_S \rangle^2 = \langle V_T : V_T \rangle \langle V_S : V_S \rangle.$$

Since $\ell(V_T) = \ell(V_S) = n$, Corollary 5.8 then shows that $V_T \cong V_S$ as $F[X]$-modules. Hence T and S are similar. $\qquad\square$

(5.14) *Remark.* The restriction that F be algebraically closed in Theorem 5.13 is not necessary. Indeed, let K be an algebraically closed field containing F (see Remark 2.4.18 (3)) and consider $A = [S]_{\mathcal{B}}$ and $B = [T]_{\mathcal{B}} \in M_n(F)$. Then A is similar to B in $M_n(F)$ if and only if A is similar to B in $M_n(K)$ (Theorem 3.15) and the rank condition in Theorem 5.13 does not depend upon which field we are using.

The computation of $\mathrm{Hom}_R(M, N)$ where M and N are finitely generated torsion R-modules over a PID R (Lemma 5.3) is also useful for some applications other than Theorem 5.13. We will give one such example. Suppose that V is a finite-dimensional vector space over a field F and $T \in \mathrm{End}_F(V)$. The centralizer of T in the ring $\mathrm{End}_F(V)$ is

$$C(T) = \{U \in \mathrm{End}_F(V) : TU = UT\}.$$

Note that, according to Proposition 4.4.1,

$$C(T) = \mathrm{End}_{F[X]}(V_T).$$

The F-algebra generated by T, namely, $F[T]$, is certainly contained in the centralizer $C(T)$. There is a theorem of Frobenius, which computes the dimension of $C(T)$ over F. This result is an easy corollary of Lemma 5.3.

(5.15) Theorem. (Frobenius) *Let F be a field, V a finite-dimensional vector space over F, and $T \in \mathrm{End}_F(V)$. If $f_1(X), \ldots, f_k(X)$ (where $f_i(X)$ divides $f_{i+1}(X)$ for $1 \le i \le k-1$) are the invariant factors of T, then*

$$(5.13) \qquad \dim_F C(T) = \sum_{i=1}^{k}(2k - 2i + 1)\deg(f_i(X)).$$

Proof. By Lemma 5.3,

$$C(T) = \mathrm{End}_{F[X]}(V_T) \cong \bigoplus_{i=1}^{k}\bigoplus_{j=1}^{k} F[X]/\langle f_i(X),\, f_j(X)\rangle.$$

But $\langle f_i(X),\, f_j(X)\rangle = \langle f_{\min\{i,j\}}(X)\rangle$, so

$$\dim_F C(T) = \sum_{i,j=1}^{k} \deg f_{\min\{i,j\}}(X).$$

But

$$|\{(i, j) : 1 \le i,\, j \le k \quad \text{and} \quad \min\{i,\, j\} = t\}| = 2k - 2t + 1,$$

so

$$\dim_F C(T) = \sum_{t=1}^{k}(2k - 2t + 1)\deg f_t(X)$$

as required. $\qquad \square$

We have observed above that $F[T] \subseteq C(T)$. As a corollary of Frobenius's theorem, there is a simple criterion for when they are equal, i.e., a criterion for when every linear transformation that commutes with T is a polynomial in T.

(5.16) Corollary. *Let $T \in \mathrm{End}_F(V)$. Then $F[T] = C(T)$ if and only if $m_T(X) = c_T(X)$, i.e., if and only if V_T is a cyclic $F[X]$-module.*

Proof. First note that $\dim_F F[T] = \deg m_T(X)$ and if $\{f_i(X)\}_{i=1}^{k}$ are the invariant factors of T, then $m_T(X) = f_k(X)$. By Equation (5.13)

$$\dim_F C(T) = \sum_{i=1}^{k}(2k - 2i + 1)\deg f_i(X)$$

$$= \dim_F F[T] + \sum_{i=1}^{k-1}(2k - 2i + 1)\deg f_i(X).$$

From this we see that $C(T) = F[T]$ if and only if $k = 1$, i.e., if and only if V_T is a cyclic $F[X]$-module. $\qquad \square$

(5.17) Corollary. *If* $T \in \text{End}_F(V)$ *then* $\dim_F C(T) \geq n = \dim_F(V)$.

Proof. Since $\prod_{i=1}^k f_i(X) = c_T(X)$, it follows that

$$\dim_F C(T) = \sum_{i=1}^k (2k - 2i + 1) \deg f_i(X) \geq \sum_{i=1}^k \deg f_i(X) = n.$$

$\square$

(5.18) Example. The Jordan matrix $J_{\lambda, n}$ is cyclic, so

$$C(J_{\lambda, n}) = F[J_{\lambda, n}].$$

It is easily checked that a basis of $F[J_{\lambda, n}]$ consists of the n matrices $A_i = J_{0, n}^i$ for $0 \leq i \leq n - 1$. That is, a matrix $A = [a_{ij}]$ commutes with $J_{\lambda, n}$ if and only if it is upper triangular and constant on the lines parallel to the main diagonal.

5.6 Exercises

1. Compute a finite free presentation of the **Z**-module

    $$M = \mathbf{Z}_2 \oplus \mathbf{Z}_{30} \oplus \mathbf{Z}.$$

2. Compute two distinct finite free presentations of the $\mathbf{R}[X]$-module $(\mathcal{P}_3)_D$, where $\mathcal{P}_3$ denotes the real vector space of polynomials of degree at most 3 and $D \in \text{End}_{\mathbf{R}}(\mathcal{P}_3)$ is the differentiation map.

3. Let M be an abelian group with three generators v_1, v_2, and v_3, subject to the relations

    $$\begin{aligned} 2v_1 - 4v_2 - 2v_3 &= 0 \\ 10v_1 - 6v_2 + 4v_3 &= 0 \\ 6v_1 - 12v_2 - 6v_3 &= 0. \end{aligned}$$

    Assuming the matrix identity

    $$\begin{bmatrix} 1 & 0 & 0 \\ 2 & 1 & 0 \\ -1 & -1 & 1 \end{bmatrix} \begin{bmatrix} 2 & 10 & 6 \\ -4 & -6 & -12 \\ -2 & 4 & -6 \end{bmatrix} \begin{bmatrix} 1 & -5 & -3 \\ 0 & 1 & 0 \\ 0 & 0 & 1 \end{bmatrix} = \begin{bmatrix} 2 & 0 & 0 \\ 0 & 14 & 0 \\ 0 & 0 & 0 \end{bmatrix},$$

    show that $M \cong \mathbf{Z}_2 \oplus \mathbf{Z}_{14} \oplus \mathbf{Z}$, and find new generators w_1, w_2, and w_3 such that $2w_1 = 0$, $14w_2 = 0$, and w_3 has infinite order.

4. Use Theorem 3.6.16 to give an alternative proof of Theorem 2.2.

5. Construct a matrix $A \in GL(4, \mathbf{Z})$ with

    $$\text{row}_1(A) = \begin{bmatrix} 12 & -10 & 9 & 8 \end{bmatrix}.$$

6. Construct a matrix $A \in GL(3, \mathbf{Q}[X])$ with

$$\text{col}_1(A) = \begin{bmatrix} X(X-1) & X^2 & X+1 \end{bmatrix}^t.$$

7. Construct a matrix $A \in M_3(\mathbf{Z}[i])$ with

$$\text{row}_2(A) = \begin{bmatrix} 1-2i & 1+3i & 3-i \end{bmatrix}$$

and with $\det A = 2+i$.

8. Reduce each of the following matrices to Hermite normal form:

(a) $A = \begin{bmatrix} 2 & 6 & 9 \\ -2 & 0 & 4 \\ 2 & 1 & -1 \end{bmatrix} \in M_3(\mathbf{Z})$.

(b) $B = \begin{bmatrix} 0 & 0 & 2-X \\ 0 & 1+X & 2X \\ 2-X & 0 & 0 \end{bmatrix} \in M_3(\mathbf{Q}[X])$.

(c) $C = \begin{bmatrix} 2-i & 2 \\ 7-i & 3+i \end{bmatrix} \in M_2(\mathbf{Z}[i])$.

9. Write the unimodular matrix

$$A = \begin{bmatrix} 5 & 3 & 4 \\ 3 & 1 & 3 \\ 6 & 3 & 5 \end{bmatrix} \in M_3(\mathbf{Z})$$

as a product of elementary matrices.

10. Let $R = \mathbf{Z}[\sqrt{-3}]$. Show that no matrix $\begin{bmatrix} a & b \\ c & d \end{bmatrix}$ with $a = 2$ and $c = 1 - \sqrt{-3}$ is left equivalent to a matrix in Hermite normal form.

11. Same as Exercise 10 with $R = \mathbf{Q}[X^2, X^3]$ and $a = X^2$, $c = X^3$.

12. Let R be a PID. Show that there is a one-to-one correspondence between the left ideals of $M_n(R)$ and the R-submodules of $M_{1,n}(R)$.

13. Find the Smith normal form for each of the following matrices:

(a) $\begin{bmatrix} -2 & 0 & 10 \\ 0 & -3 & -4 \\ 1 & 2 & -1 \end{bmatrix} \in M_3(\mathbf{Z})$.

(b) $\begin{bmatrix} 2 & 6 & -8 \\ 12 & 14 & 6 \\ 4 & -4 & 8 \end{bmatrix} \in M_3(\mathbf{Z})$.

(c) $\begin{bmatrix} X(X-1)^3 & 0 & 0 \\ 0 & (X-1) & 0 \\ 0 & 0 & X \end{bmatrix} \in M_3(\mathbf{Q}[X])$.

(d) $\begin{bmatrix} -1+8i & -23+2i \\ -5+i & 13i \end{bmatrix} \in M_2(\mathbf{Z}[i])$.

14. Find the invariant factors and elementary divisors of each of the following matrices:

(a)

$$\begin{bmatrix} 0 & 0 & 7-6X \\ 0 & -4+X & 2X \\ 2+4X & 5 & 0 \end{bmatrix} \in M_3(\mathbf{Z}_5[X]).$$

(b) $\text{diag}(20, 18, 75, 42) \in M_4(\mathbf{Z})$.

(c) $\text{diag}\left(X(X-1)^2, X(X-1)^3, (X-1), X\right)$.

15. Let R be a PID and let $A \in M_{m,n}(R)$ with $m < n$. Extend Theorem 2.2 by proving that there is a matrix $B = \begin{bmatrix} A \\ A_1 \end{bmatrix} \in M_n(R)$ (so that $A_1 \in$

$M_{n-m,n}(R)$) such that det B is an associate of $d_m(A)$, the mth determinantal divisor of A. (Hint: First put A in Smith normal form.)

16. Let $S = \{v_1, \ldots, v_k\} \subseteq M_{n,1}(R)$ where R is a PID. Show that S can be extended to a basis of $M_{n,1}(R)$ if and only if $d_k(A) = 1$, where $A = [v_1 \; \cdots \; v_k]$.

17. Suppose that $A \in M_3(\mathbf{Z})$ and det $A = 210$. Compute the Smith normal form of A. More generally, suppose that R is a PID and $A \in M_n(R)$ is a matrix such that det A is square-free. Then compute the Smith normal form of A.

18. Let $A \in M_n(\mathbf{Z})$ and assume that det $A \neq 0$. Then the inverse of A exists in $M_n(\mathbf{Q})$, and by multiplying by a common denominator t of all the nonzero entries of A^{-1}, we find that $tA^{-1} \in M_n(\mathbf{Z})$. Show that the least positive integer t such that $tA^{-1} \in M_n(\mathbf{Z})$ is $t = |s_n(A)|$ where $s_n(A)$ is the nth invariant factor of A.

19. Let $A, B \in M_n(\mathbf{Z})$ such that $AB = kI_n$ for some $k \neq 0$. Show that the invariant factors of A are divisors of k.

20. Let R be a PID and let $A \in M_n(R)$, $B \in M_m(R)$. Show that the elementary divisors of $A \otimes B$ are the product of elementary divisors of A and of B. More precisely, if p^r is an elementary divisor of $A \otimes B$ where $p \in R$ is a prime, then $p^r = p^k p^l$ where p^k is an elementary divisor of A and p^l is an elementary divisor of B, and conversely, if p^k is an elementary divisor of A and p^l is an elementary divisor of B, then p^{k+l} is an elementary divisor of $A \otimes B$.

21. Let $A \in M_4(F)$ where F is a field. If A has an invariant factor $s(X)$ of degree 2 show that the Smith normal form of $XI_4 - A$ is diag$(1, 1, s(X), s(X))$. Conclude that $c_A(X)$ is a perfect square in $F[X]$.

22. Find all integral solutions to the following systems $AX = B$ of equations:

(a) $A = \begin{bmatrix} 1 & -1 & 1 \\ 1 & 0 & 2 \end{bmatrix}$, $\quad B = \begin{bmatrix} 4 \\ 5 \end{bmatrix}$.

(b) $A = \begin{bmatrix} 0 & 2 & -1 \\ 1 & -1 & 0 \\ 2 & 0 & -1 \end{bmatrix}$, $\quad B = \begin{bmatrix} 5 \\ 1 \\ 7 \end{bmatrix}$.

(c) $A = \begin{bmatrix} 8 & 19 & 30 \\ 6 & 14 & 22 \end{bmatrix}$, $\quad B = \begin{bmatrix} 5 \\ 7 \end{bmatrix}$.

23. Show that the matrices $A = \begin{bmatrix} 0 & 1 \\ 8 & 1 \end{bmatrix}$ and $B = \begin{bmatrix} 16 & -1 \\ 232 & -15 \end{bmatrix}$ in $M_2(\mathbf{Q})$ are similar.

24. Show that the matrices

$$\begin{bmatrix} 0 & 1 & 0 \\ 0 & 0 & 2 \\ 3 & 4 & 0 \end{bmatrix} \quad \text{and} \quad \begin{bmatrix} 3 & 4 & 0 \\ 2 & 4 & 5 \\ 0 & 1 & 0 \end{bmatrix}$$

are similar in $M_3(\mathbf{Z}_7)$.

25. Show that the matrices

$$\begin{bmatrix} 0 & 1 & 0 \\ 0 & 0 & 1 \\ 1 & 0 & 0 \end{bmatrix} \quad \text{and} \quad \begin{bmatrix} 1 & 1 & 0 \\ 0 & 1 & 1 \\ 0 & 0 & 1 \end{bmatrix}$$

are similar in $M_3(\mathbf{Z}_3)$.

26. Find the characteristic polynomial, invariant factors, elementary divisors, rational canonical form, and Jordan canonical form (when possible) of each of the matrices from Exercise 73 of Chapter 4. Additionally, find bases of $\mathbf{Q}^n$ with respect to which the matrix (or linear transformation) is in rational or Jordan canonical form. Do this exercise by reducing $XI_n - A$ to Smith

canonical form and compare your results with the same calculations done in Chapter 4.

27. Find an example of a unimodular matrix $A \in M_3(\mathbf{Z})$ such that A is not similar to A^t. (Compare with Example 3.8.)

28. Show that the matrix $A = \begin{bmatrix} 2X & X \\ 0 & 2 \end{bmatrix}$ is not equivalent in $M_2(\mathbf{Z}[X])$ to a diagonal matrix. (Hint: Use Fitting ideals.)

29. Let $\mathbf{Z}^n$ have the standard basis $\{e_1, \ldots, e_n\}$ and let $K \subseteq \mathbf{Z}^n$ be the submodule generated by $f_i = \sum_{j=1}^{n} a_{ij} e_j$ where $a_{ij} \in \mathbf{Z}$ and $1 \le i \le n$. Let $A = [a_{ij}] \in M_n(\mathbf{Z})$ and let $d = \det A$. Show that $\mathbf{Z}/K$ is torsion if and only if $\det A = d \ne 0$ and if $d \ne 0$ show that $|\mathbf{Z}/K| = |d|$.

30. Suppose that an abelian group G has generators x_1, x_2, and x_3 subject to the relations $x_1 - 3x_3 = 0$ and $x_1 + 2x_2 + 5x_3 = 0$. Determine the invariant factors of G and $|G|$ if G is finite.

31. Suppose that an abelian group G has generators x_1, x_2, and x_3 subject to the relations $2x_1 - x_2 = 0$, $x_1 - 3x_2 = 0$, and $x_1 + x_2 + x_3 = 0$. Determine the invariant factors of G and $|G|$ if G is finite.

32. Verify the claim of Example 5.1.

33. Let F be a field and let $A \in M_n(F)$, $B \in M_m(F)$. Show that the matrix equation
$$AX - XB = 0$$
for $X \in M_{n,m}(F)$ has only the trivial solution $X = 0$ if and only if the characteristic polynomials $c_A(X)$ and $c_B(X)$ are relatively prime in $F[X]$. In particular, if F is algebraically closed, this equation has only the trivial solution if and only if A and B have no eigenvalues in common.

34. Let F be a field. Suppose that $A = A_1 \oplus A_2 \in M_n(F)$ where $A_1 \in M_k(F)$ and $A_2 \in M_m(F)$ and assume that $c_{A_1}(X)$ and $c_{A_2}(X)$ are relatively prime. Prove that if $B \in M_n(F)$ commutes with A, then B is also a direct sum $B = B_1 \oplus B_2$ where $B_1 \in M_k(F)$ and $B_2 \in M_m(F)$.

35. Let F be a field. Recall that $C(f(X))$ denotes the companion matrix of the monic polynomial $f(X) \in F[X]$. If $\deg(f(X)) = n$ and $\deg(g(X)) = n$, show that
$$\mathrm{rank}(C(f(X)) \otimes I_m - I_n \otimes C(g(X))) = \deg(\mathrm{lcm}\{f(X), g(X)\}).$$

36. Let V be a finite-dimensional vector space over a field F and let $T \in \mathrm{End}_F(V)$. Prove that the center of $C(T)$ is $F[T]$.

Chapter 6

Bilinear and Quadratic Forms

6.1 Duality

Recall that if R is a commutative ring, then $\mathrm{Hom}_R(M, N)$ denotes the set of all R-module homomorphisms from M to N. It has the structure of an R-module by means of the operations $(f + g)(x) = f(x) + g(x)$ and $(af)(x) = a(f(x))$ for all $x \in M$, $a \in R$. Moreover, if $M = N$ then $\mathrm{Hom}_R(M, M) = \mathrm{End}_R(M)$ is a ring under the multiplication $(fg)(x) = f(g(x))$. An R-module A, which is also a ring, is called an R-algebra if it satisfies the extra axiom $a(xy) = (ax)y = x(ay)$ for all $x, y \in A$ and $a \in R$. Thus $\mathrm{End}_R(M)$ is an R-algebra. Recall (Theorem 3.4.11) that if M and N are finitely generated free R-modules (R a commutative ring) of rank m and n respectively, then $\mathrm{Hom}_R(M, N)$ is a free R-module of rank mn.

In this section R will always denote a commutative ring so that $\mathrm{Hom}_R(M, N)$ will always have the structure of an R-module.

(1.1) Definition. *If M is an R-module, then $\mathrm{Hom}_R(M, R)$ is called the **dual module** of M and is denoted M^*.*

(1.2) Remark. If M if free of rank n then $M^* = \mathrm{Hom}_R(M, R) \cong R^n \cong M$ by Corollary 3.4.10. Note, however, that this isomorphism is obtained by choosing a basis of M^* and M. One particular choice of basis for M^* is the following, which is that described in the proof of Theorem 3.4.11 if the basis $\{1\}$ is chosen for R.

(1.3) Definition. *If M is a free R-module and $\mathcal{B} = \{v_1, \ldots, v_n\}$ is a basis of M, then the **dual basis** of M^* is defined by $\mathcal{B}^* = \{v_1^*, \ldots, v_n^*\}$ where $v_i^* \in M^*$ is defined by*

$$v_i^*(v_j) = \delta_{ij} = \begin{cases} 1 & \text{if } i = j \\ 0 & \text{if } i \neq j. \end{cases}$$

(1.4) Example. Let $R = \mathbf{Z}$ and $M = \mathbf{Z}^2$. Consider the basis $\mathcal{B} = \{v_1 = (1,0), v_2 = (0,1)\}$. Then $v_1^*(a, b) = a$ and $v_2^*(a, b) = b$. Now consider the basis $\mathcal{C} = \{w_1 = (1, 1),\ w_2 = (1, 2)\}$. Then $(a, b) = (2a - b)w_1 + (b - a)w_2$ so that $w_1^*(a, b) = 2a - b$ and $w_2^*(a, b) = b - a$. Therefore, $v_1^* \neq w_1^*$ and $v_2^* \neq w_2^*$. Moreover, if $\mathcal{D} = \{u_1 = (1, 0),\ u_2 = (1, 1)\}$ then $u_1^*(a, b) = a - b$ and $u_2^*(a, b) = b$ so that $u_1^* \neq v_1^*$ even though $u_1 = v_1$. The point is that an element v_i^* in a dual basis depends on the entire basis and not just the single element v_i.

(1.5) Proposition. *Let M be a free R-module of finite rank n and let $\mathcal{B} = \{v_1, \ldots, v_n\}$ be a basis of M. Then $\mathcal{B}^* = \{v_1^*, \ldots, v_n^*\}$ is a basis of M^*.*

Proof. $\mathcal{B}^*$ is the basis produced in the proof of Theorem 3.4.11. $\square$

(1.6) Corollary. *Let M be a free R-module of finite rank n and let $\mathcal{B} = \{v_1, \ldots, v_n\}$ be a basis of M. Then the map $\omega : M \to M^*$ defined by*

$$\omega\left(\sum_{i=1}^{n} a_i v_i\right) = \sum_{i=1}^{n} a_i v_i^*$$

is an R-module isomorphism.

Proof. $\square$

The isomorphism given in Corollary 1.6 depends upon the choice of a basis of M. However, if we consider the double dual of M, the situation is much more intrinsic, i.e., it does not depend upon a choice of basis. Let M be any R-module. Then define the **double dual** of M, denoted M^{**}, by $M^{**} = (M^*)^* = \mathrm{Hom}_R(M^*, R)$. There is a natural homomorphism $\eta : M \to M^{**} = \mathrm{Hom}_R(M^*, R)$ defined by

$$\eta(v)(\omega) = \omega(v)$$

for all $v \in M$ and $\omega \in M^* = \mathrm{Hom}_R(M, R)$.

(1.7) Theorem. *If M is a free R-module, then the map $\eta : M \to M^{**}$ is injective. If $\mathrm{rank}(M) < \infty$ then η is an isomorphism.*

Proof. Suppose that $v \neq 0 \in M$. Let $\mathcal{B}$ be a basis of M. If $v = a_1 v_1 + \cdots + a_n v_n$ where $a_1 \neq 0$ and $\{v_1, \ldots, v_n\} \subseteq \mathcal{B}$, then we can define an element $\omega \in M^*$ by $\omega(v_1) = 1$ and $\omega(w) = 0$ for all $w \neq v_1 \in \mathcal{B}$. Then $\eta(v)(\omega) = \omega(a_1 v_1 + \cdots + a_n v_n) = a_1 \neq 0$. Hence, η is injective.

Now suppose that $\mathrm{rank}(M) < \infty$ and let $\mathcal{B} = \{v_1, \ldots, v_n\}$ be a basis of M. Let $\mathcal{B}^* = \{v_1^*, \ldots, v_n^*\}$ be the dual basis of M^* and let $\mathcal{B}^{**} = \{v_1^{**}, \ldots, v_n^{**}\}$ be the basis of M^{**} dual to the basis $\mathcal{B}^*$ of M^*. We claim that $\eta(v_i) = v_i^{**}$ for $1 \leq i \leq n$. To see this, note that

$$\eta(v_i)(v_j^*) = v_j^*(v_i)$$
$$= \delta_{ij}$$
$$= v_i^{**}(v_j^*).$$

Since $\eta(v_i)$ and v_i^{**} agree on a basis of M^*, they are equal. Hence, $\eta(M) \supseteq \langle v_1^{**}, \ldots, v_n^{**} \rangle = M^{**}$ so that η is surjective, and hence, is an isomorphism.

$\square$

(1.8) *Remark.* For general R-modules M, the map $\eta : M \to M^{**}$ need not be either injective or surjective. (See Example 1.9 below.) When η happens to be an isomorphism, the R-module M is said to be **reflexive**. According to Theorem 1.7, free R-modules of finite rank are reflexive. We shall prove below that finitely generated projective modules are also reflexive, but first some examples of nonreflexive modules are presented.

(1.9) Examples.

(1) Let R be a PID that is not a field and let M be any finitely generated nonzero torsion module over R. Then according to Exercise 9 of Chapter 3, $M^* = \mathrm{Hom}_R(M, R) = \langle 0 \rangle$. Thus, $M^{**} = \langle 0 \rangle$ and the natural map $\eta : M \to M^{**}$ is clearly not injective.

(2) Let $R = \mathbf{Q}$, and let $M = \oplus_{n \in \mathbf{N}} \mathbf{Q}$ be a vector space over $\mathbf{Q}$ of countably infinite dimension. Then $M^* \cong \prod_{n \in \mathbf{N}} \mathbf{Q}$. Since

$$M = \bigoplus_{n \in \mathbf{N}} \mathbf{Q} \subseteq \prod_{n \in \mathbf{N}} \mathbf{Q},$$

we see that $M^* \cong M \oplus M'$ where M' is a vector space complement of $M = \oplus_{n \in \mathbf{N}} \mathbf{Q}$ in M^*. Then

$$M^{**} \cong M^* \oplus (M')^*$$

so that M^{**} contains a subspace isomorphic to M^*. But $\oplus_{n \in \mathbf{N}} \mathbf{Q}$ is countably infinite, while the infinite product $\prod_{n \in \mathbf{N}} \mathbf{Q} \cong M^*$ is uncountable (the decimal representation identifies every real number with an element of $\prod_{n \in \mathbf{N}} \mathbf{Q}$). Therefore, $\eta : M \to M^{**}$ cannot be surjective by cardinality consideration and we conclude that M is not a reflexive $\mathbf{Q}$-module.

Let M_1 and M_2 be R-modules and let $M = M_1 \oplus M_2$. Then according to Corollary 3.3.13, $M^{**} \cong M_1^{**} \oplus M_2^{**}$. In order to study reflexivity for direct sums and summands, it is necessary to identify carefully this isomorphism. To this end, define

$$\Psi : (M_1 \oplus M_2)^{**} \to M_1^{**} \oplus M_2^{**}$$

by $\Psi(\omega) = (\omega_1, \omega_2)$, where $\omega_i \in M_i^{**}$ is defined by $\omega_i(\theta_i) = \omega(\theta_i \circ \pi_i)$ for each $\theta_i \in M_i^* = \mathrm{Hom}_R(M_i, R)$. $\pi_i : M_1 \oplus M_2 \to M_i$ is the canonical projection map. Similarly, define

$$\Phi : M_1^{**} \oplus M_2^{**} \to (M_1 \oplus M_2)^{**}$$

by
$$\Phi(\omega_1, \omega_2)(\theta) = \omega_1(\theta \circ \iota_1) + \omega_2(\theta \circ \iota_2)$$

where $\theta \in (M_1 \oplus M_2)^*$ and $\iota_i : M_i \to M_1 \oplus M_2$ is the canonical injection.

(1.10) Lemma. Ψ *and* Φ *are inverse R-module homomorphisms.*

Proof. Let $\omega \in (M_1 \oplus M_2)^{**}$ and let $\theta \in (M_1 \oplus M_2)^*$. Then

$$\begin{aligned}
\Phi \circ \Psi(\omega)(\theta) &= \Phi(\omega_1, \omega_2)(\theta) \\
&= \omega_1(\theta \circ \iota_1) + \omega_2(\theta \circ \iota_2) \\
&= \omega(\theta \circ \iota_1 \circ \pi_1) + \omega(\theta \circ \iota_2 \circ \pi_2) \\
&= \omega(\theta \circ \iota_1 \circ \pi_1 + \theta \circ \iota_2 \circ \pi_2) \\
&= \omega(\theta \circ (\iota_1 \circ \pi_1 + \iota_2 \circ \pi_2)) \\
&= \omega(\theta \circ 1_{M_1 \oplus M_2}) \\
&= \omega(\theta),
\end{aligned}$$

and

$$\begin{aligned}
(\Psi \circ \Phi(\omega_1, \omega_2))(\theta_1, \theta_2) &= \Psi(\Phi(\omega_1, \omega_2))(\theta_1, \theta_2) \\
&= (\Phi(\omega_1, \omega_2)(\theta_1 \circ \pi_1), \Phi(\omega_1, \omega_2)(\theta_2 \circ \pi_2)) \\
&= (\omega_1(\theta_1 \circ \pi_1 \circ \iota_1), \omega_2(\theta_2 \circ \pi_2 \circ \iota_2)) \\
&= (\omega_1(\theta_1), \omega_2(\theta_2)) \\
&= (\omega_1, \omega_2)(\theta_1, \theta_2).
\end{aligned}$$

Therefore,
$$\Phi \circ \Psi = 1_{(M_1 \oplus M_2)^{**}}$$

and
$$\Psi \circ \Phi = 1_{M_1^{**} \oplus M_2^{**}},$$

and the lemma is proved. □

Now let $\eta_i : M_i \to M_i^{**}$ $(i = 1, 2)$ and $\eta : M_1 \oplus M_2 \to (M_1 \oplus M_2)^{**}$ be the natural maps into the double duals.

(1.11) Lemma. *Using the notation introduced above, there is a commutative diagram*

$$
\begin{array}{ccc}
M_1 \oplus M_2 & & \\
\Big\downarrow{\scriptstyle \eta} & \overset{(\eta_1, \eta_2)}{\searrow} & \\
(M_1 \oplus M_2)^{**} & \overset{\Psi}{\longrightarrow} & M_1^{**} \oplus M_2^{**}
\end{array}
$$

That is,
$$\Psi \circ \eta = (\eta_1, \eta_2).$$

Proof.

$$
\begin{aligned}
((\psi \circ \eta)\,(v_1,\, v_2))\,(\omega_1,\, \omega_2) &= \Psi\,(\eta(v_1,\, v_2))\,(\omega_1,\, \omega_2) \\
&= (\eta(v_1,\, v_2)(\omega_1 \circ \pi_1),\ \eta(v_1,\, v_2)(\omega_2 \circ \pi_2)) \\
&= ((\omega_1 \circ \pi_1)(v_1,\, v_2)\,,\ (\omega_2 \circ \pi_2)(v_1,\, v_2)) \\
&= (\omega_1(v_1)\,,\ \omega_2(v_2)) \\
&= ((\eta_1,\, \eta_2)(v_1,\, v_2))\,(\omega_1,\, \omega_2).
\end{aligned}
$$

That is, $\Psi \circ \eta = (\eta_1,\, \eta_2)$. $\qquad\square$

(1.12) Lemma. (1) $\eta : M_1 \oplus M_2 \to (M_1 \oplus M_2)^{**}$ *is injective if and only if* $\eta_i : M_i \to M_i^{**}$ *is injective for each* $i = 1,\, 2$.

   (2) $M_1 \oplus M_2$ *is reflexive if and only if* M_1 *and* M_2 *are reflexive.*

Proof. Both results are immediate from Lemma 1.11 and the fact that Ψ is an isomorphism. $\qquad\square$

(1.13) Proposition. *If P is a projective R-module, then $\eta : P \to P^{**}$ is injective. If P is also finitely generated, then P is reflexive.*

Proof. Since P is projective, there is an R-module P' such that $P \oplus P' \cong F$ where F is a free R-module (Theorem 3.5.1), and furthermore, if P is finitely generated, then F may be taken to have finite rank (Corollary 3.5.5). The result now follows from Lemma 1.12 and Theorem 1.7. $\qquad\square$

The remainder of this section will be concerned with the relationship between submodules of an R-module M and submodules of the dual module M^*. The best results are obtained when the ring R is a PID, and the module M is a finite rank free R-module. Thus, we will make the following convention for the rest of the current section.

Convention. *For the remainder of this section R will denote a PID and M will denote a free R-module of finite rank unless explicitly stated otherwise.*

(1.14) Definition.

(1) *If N is a submodule of M, then we define the **hull** of N, denoted*

$$
\mathrm{Hull}(N) = \{x' \in M : rx' \in N \ \text{ for some } r \neq 0 \in R\ \}.
$$

   If A is a subset of M, then we define $\mathrm{Hull}(A) = \mathrm{Hull}(\langle A \rangle)$.

(2) *If A is a subset of M then define the **annihilator** of A to be the following subset of the dual module M^*:*

$$
\begin{aligned}
K(A) = \mathrm{Ann}(A) \\
= \{\omega \in M^* : \mathrm{Ker}(\omega) \supseteq A\} \\
= \{\omega \in M^* : \omega(x) = 0 \ \text{ for all } x \in A\} \\
\subseteq M^*.
\end{aligned}
$$

(3) *If B is a subset of M^* then define the* **annihilator** *of B to be the following subset of M:*

$$K^*(B) = \text{Ann}(B)$$
$$= \{x \in M : \omega(x) = 0 \quad for\ all\ \omega \in B\}$$
$$\subseteq M.$$

(1.15) Remarks.

(1) If N is a submodule of M, then $M/\text{Hull}(N)$ is torsion-free, so $\text{Hull}(N)$ is always a complemented submodule (see Proposition 3.8.2); furthermore, $\text{Hull}(N) = N$ if and only M/N is torsion-free, i.e., N itself is complemented. In particular, if R is a field then $\text{Hull}(N) = N$ for all subspaces of the vector space M.

(2) If A is a subset of M, then the annihilator of A in the current context of duality, should not be confused with the annihilator of A as an ideal of R (see Definition 3.2.13). In fact, since M is a free R-module and R is a PID, the ideal theoretic annihilator of any subset of M is automatically $\langle 0 \rangle$.

(3) Note that $\text{Ann}(A) = \text{Ann}(\text{Hull}(A))$. To see this note that $\omega(ax') = 0 \Leftrightarrow a\omega(x') = 0$. But R has no zero divisors, so $a\omega(x') = 0$ if and only if $\omega(x') = 0$. Also note that $\text{Ann}(A)$ is a complemented submodule of M^* for the same reason. Namely, $a\omega(x) = 0$ for all $x \in A$ and $a \neq 0 \in R \Leftrightarrow \omega(x) = 0$ for all $x \in A$.

(4) Similarly, $\text{Ann}(B) = \text{Ann}(\text{Hull}(B))$ and $\text{Ann}(B)$ is a complemented submodule of M for any subset $B \subseteq M^*$.

The concepts of annihilators of subsets of M and M^* will be used to get a duality between submodules of M and M^*. But since annihilators of subsets are complemented submodules, we see immediately that it is necessary to restrict any correspondence between submodules of M and M^* to the set of complemented submodules. Thus, if M if a free R-module, then we will denote the set of all complemented submodules by $\mathcal{C}(M)$. The following result is a collection of straightforward properties of annihilators. The verifications are left as an exercise.

(1.16) Proposition. *Let M be a free R-module of finite rank, let A, A_1, and A_2 be subsets of M, and let B, B_1, and B_2 be subsets of M^*. Then the following properties of annihilators are valid:*

(1) *If $A_1 \subseteq A_2$, then $K(A_1) \supseteq K(A_2)$.*
(2) $K(A) = K(\text{Hull}(A))$.
(3) $K(A) \in \mathcal{C}(M^*)$.
(4) $K(\{0\}) = M^*$ and $K(M) = \{0\}$.

(5) $K^*(K(A)) \supseteq A$.

(1^*) *If* $B_1 \subseteq B_2$, *then* $K^*(B_1) \supseteq K^*(B_2)$.
(2^*) $K^*(B) = K^*(\text{Hull}(B))$.
(3^*) $K^*(B) \in \mathcal{C}(M)$.
(4^*) $K^*(\{0\}) = M$ *and* $K^*(M^*) = \{0\}$.
(5^*) $K(K^*(B)) \supseteq B$.

Proof. Exercise.   □

The following result is true for any reflexive R-module (and not just finite rank free modules). Since the work is the same, we will state it in that context:

(1.17) Lemma. *Let M be a reflexive R-module and let $\eta : M \to M^{**}$ be the natural isomorphism. Then for every submodule T of M^*, we have $\eta(K^*(T)) = K(T) \subseteq M^{**}$.*

Proof. Let $\omega \in K(T)$. Then $\omega = \eta(x)$ for some $x \in M$. For any $t \in T$,

$$t(x) = \eta(x)(t) = \omega(t) = 0$$

because $\omega \in K(T)$. Therefore, $x \in K^*(T)$ by definition and hence $K(T) \subseteq \eta(K^*(T))$.

Conversely, if $x \in K^*(T)$ then

$$0 = t(x) = \eta(x)(t)$$

for any $t \in T$ so $\eta(x) \in K(T)$ by definition. Thus, $\eta(K^*(T)) \subseteq K(T)$, and the lemma is proved.   □

(1.18) Theorem. *Let M be a free R-module of finite rank, let S be a complemented submodule of M, and let T be a complemented submodule of M^*. Then*

$$\text{rank}(M) = \text{rank}(S) + \text{rank}(K(S)),$$

and

$$\text{rank}(M^*) = \text{rank}(T) + \text{rank}(K^*(T)).$$

Proof. Let $\mathcal{B}_1 = \{v_1, \dots, v_k\}$ be a basis of S. Since S is complemented, it follows (Corollary 3.8.4) that $\mathcal{B}_1$ extends to a basis

$$\mathcal{B} = \{v_1, \dots, v_k, v_{k+1}, \dots, v_m\}$$

of M. Let $\mathcal{B}^* = \{v_1^*, \dots, v_m^*\}$ be the basis of M^* dual to $\mathcal{B}$. If $i \leq k$ and $j > k$ then $v_j^*(v_i) = 0$. Therefore,

$$\langle v_{k+1}^*, \ldots, v_m^* \rangle \subseteq K(S).$$

If $\omega \in K(S)$, then we may write $\omega = \sum_{j=1}^{m} a_j v_j^*$, and if $1 \le i \le k$, then

$$0 = \omega(v_i) = \sum_{j=1}^{m} a_j v_j^*(v_i) = a_i.$$

Therefore, $\omega = \sum_{j=k+1}^{m} a_j v_j^*$, and hence,

$$K(S) = \langle v_{k+1}^*, \ldots, v_m^* \rangle$$

so that $\operatorname{rank}(K(S)) = m - k = \operatorname{rank}(M) - \operatorname{rank}(S)$.
    Similarly,

$$\begin{aligned}
\operatorname{rank}(M^*) &= \operatorname{rank}(T) + \operatorname{rank}(K(T)) \\
&= \operatorname{rank}(T) + \operatorname{rank}(\eta(K^*(T))) \\
&= \operatorname{rank}(T) + \operatorname{rank}(K^*(T))
\end{aligned}$$

where the last equality is valid because $\eta : M \to M^{**}$ is an isomorphism, and hence, it preserves ranks of submodules. $\square$

(1.19) Theorem. *Let M be a free R-module of finite rank. Then the function*

$$K : \mathcal{C}(M) \to \mathcal{C}(M^*)$$

is a one-to-one correspondence with inverse K^*.

Proof. We claim that for every complemented submodule $S \subseteq M$ and $T \subseteq M^*$, we have

$$K^*(K(S)) = S$$

and

$$K(K^*(T)) = T.$$

We will prove the first of these equalities; the second is similar.
    First note the $K^*(K(S)) \supseteq S$ for every complemented submodule $S \subseteq M$ by Proposition 1.16 (5), so Corollary 3.8.5 implies that it suffices to show that $\operatorname{rank}(K^*(K(S))) = \operatorname{rank}(S)$. But

$$\operatorname{rank}(S) = \operatorname{rank}(M) - \operatorname{rank}(K(S))$$

and

$$\operatorname{rank}(K(S)) = \operatorname{rank}(M^*) - \operatorname{rank}(K^*(K(S)))$$

by Theorem 1.18. Since $\operatorname{rank}(M) = \operatorname{rank}(M^*)$, the result follows. $\square$

(1.20) Definition. *If M and N are R-modules and $f \in \mathrm{Hom}_R(M, N)$ then the **adjoint** of f is the function $f^* : N^* \to M^*$ defined by $f^*(\omega) = \omega \circ f$, that is,*

$$(f^*(\omega))(x) = \omega(f(x))$$

for all $x \in M$.

(1.21) Remarks.

(1) $f^* : N^* \to M^*$ is an R-module homomorphism.
(2) $\mathrm{Ad} : \mathrm{Hom}_R(M, N) \to \mathrm{Hom}_R(N^*, M^*)$, defined by $\mathrm{Ad}(f) = f^*$, is an R-module homomorphism.
(3) If M and N are free, $\mathrm{Ker}(f)$ is always a complemented submodule of M, but $\mathrm{Im}(f)$ need not be complemented. (See Proposition 3.8.7.)

(1.22) Theorem. *Let M and N be free R-modules of finite rank and let $f \in \mathrm{Hom}_R(M, N)$. Then*

(1) $\mathrm{Ann}(\mathrm{Im}(f)) = \mathrm{Ker}(f^*) \subseteq N^*$,
(2) $\mathrm{rank}(\mathrm{Im}(f^*)) = \mathrm{rank}(\mathrm{Im}(f)))$, *and*
(3) $\mathrm{Im}(f^*) = \mathrm{Ann}(\mathrm{Ker}(f)) \subseteq M^*$ *if $\mathrm{Im}(f^*)$ is a complemented submodule of M^*.*

Proof. (1) Let $\omega \in N^*$. Then

$$\begin{aligned}
\omega \in \mathrm{Ker}(f^*) &\Leftrightarrow f^*(\omega) = 0 \\
&\Leftrightarrow \omega \circ f = 0 \\
&\Leftrightarrow \omega(f(x)) = 0 \qquad \forall x \in M \\
&\Leftrightarrow \omega(y) = 0 \qquad \forall y \in \mathrm{Im}(f) \\
&\Leftrightarrow \omega \in \mathrm{Ann}(\mathrm{Im}(f)).
\end{aligned}$$

(2) Since $f^* : N^* \to M^*$, Proposition 3.8.8 gives

$$\mathrm{rank}(N^*) = \mathrm{rank}(\mathrm{Im}(f^*)) + \mathrm{rank}(\mathrm{Ker}(f^*))$$

while Theorem 1.18 shows

$$\mathrm{rank}(N) = \mathrm{rank}(\mathrm{Im}(f)) + \mathrm{rank}(\mathrm{Ann}(\mathrm{Im}(f))).$$

Since $\mathrm{rank}(N) = \mathrm{rank}(N^*)$, (2) follows from (1).

(3) Now let $\tau \in M^*$. Then

$$\begin{aligned}
\tau \in \mathrm{Im}(f^*) &\Leftrightarrow \tau = f^*(\omega) \quad \text{for some } \omega \in N^* \\
&\Leftrightarrow \tau(x) = \omega(f(x)) \quad \forall x \in M.
\end{aligned}$$

If $x \in \mathrm{Ker}(f)$ then $f(x) = 0$, so $\omega(f(x)) = 0$. Therefore, $\tau(x) = 0$, and we conclude that $\tau \in \mathrm{Ann}(\mathrm{Ker}(f))$. Hence, $\mathrm{Im}(f^*) \subseteq \mathrm{Ann}(\mathrm{Ker}(f))$.

By Theorem 1.18 and part (2),

$$\begin{aligned}
\mathrm{rank}(\mathrm{Ann}(\mathrm{Ker}(f))) &= \mathrm{rank}(M) - \mathrm{rank}(\mathrm{Ker}(f)) \\
&= \mathrm{rank}(\mathrm{Im}(f)) = \mathrm{rank}(\mathrm{Im}(f^*)).
\end{aligned}$$

Since $\mathrm{Im}(f^*)$ is assumed to be complemented, we conclude that $\mathrm{Im}(f^*) = \mathrm{Ann}(\mathrm{Ker}(f))$. $\square$

(1.23) Corollary. *Let F be a field, let V and W be finite-dimensional vector spaces over F, and let $f \in \mathrm{Hom}_F(V, W)$. Then*

(1) *f is injective if and only if f^* is surjective;*
(2) *f is surjective if and only if f^* is injective; and*
(3) *f is an isomorphism if and only if f^* is an isomorphism.*

Proof.   □

(1.24) Proposition. *Let M and N be free R-modules of finite rank with bases $\mathcal{B}$ and $\mathcal{C}$, respectively, and let $f \in \mathrm{Hom}_R(M, N)$. Then*

$$[f^*]_{\mathcal{B}^*}^{\mathcal{C}^*} = \left([f]_{\mathcal{C}}^{\mathcal{B}}\right)^t.$$

Proof. Let $\mathcal{B} = \{v_i\}_{i=1}^n$ and $\mathcal{C} = \{w_j\}_{j=1}^m$. If $A = [a_{ij}] = [f]_{\mathcal{C}}^{\mathcal{B}}$ and $B = [b_{ij}] = [f^*]_{\mathcal{B}^*}^{\mathcal{C}^*}$, then by definition

$$f(v_j) = \sum_{k=1}^m a_{kj} w_k$$

and

$$f^*(w_i^*) = \sum_{k=1}^n b_{ki} v_k^*.$$

But then

$$
\begin{aligned}
a_{ij} &= w_i^*(f(v_j)) \\
&= (w_i^* \circ f)(v_j) \\
&= (f^*(w_i^*))(v_j) \\
&= b_{ji}.
\end{aligned}
$$

□

6.2 Bilinear and Sesquilinear Forms

In this section we present an introduction to an important branch of mathematics that is the subject of much study. Throughout this section R will be a commutative ring with 1 and all R-modules will be free.

(2.1) Definition. *A **conjugation** on R is a function $c : R \to R$ satisfying*

(1) *$c(c(r)) = r$ for all $r \in R$;*
(2) *$c(r_1 + r_2) = c(r_1) + c(r_2)$ for all $r_1, r_2 \in R$; and*
(3) *$c(r_1 r_2) = c(r_1)c(r_2)$ for all $r_1, r_2 \in R$.*

That is, a nontrivial conjugation of R is a ring automorphism, which has order 2 as an element of the group Aut(R).

(2.2) Examples.

(1) Every ring has the trivial conjugation $c(r) = r$. Since $\text{Aut}(\mathbf{Q}) = \{1_{\mathbf{Q}}\}$, it follows that the trivial conjugation is the only one on $\mathbf{Q}$. The same is true for the ring $\mathbf{Z}$.
(2) The field $\mathbf{C}$ has the conjugation $c(z) = \overline{z}$, where the right-hand side is complex conjugation. (This is where the name "conjugation" for a function c as above comes from.)
(3) The field $\mathbf{Q}[\sqrt{d}]$ and the ring $\mathbf{Z}[\sqrt{d}]$ (where d is not a square) both have the conjugation $c(a + b\sqrt{d}) = a - b\sqrt{d}$.

Because of Example 2.2 (2), we will write $\overline{r}$, instead of $c(r)$, to denote conjugation.

(2.3) Definition. *Let M be a free R-module. A **bilinear form** on M is a function $\phi : M \times M \to R$ satisfying*

(1) $\phi(r_1 x_1 + r_2 x_2, y) = r_1 \phi(x_1, y) + r_2 \phi(x_2, y)$, *and*
(2) $\phi(x, r_1 y_1 + r_2 y_2) = r_1 \phi(x, y_1) + r_2 \phi(x, y_2)$

for all x_1, x_2, y_1, $y_2 \in M$, and r_1, $r_2 \in R$.

  *A **sesquilinear form** on M is a function $\phi : M \times M \to R$ satisfying* (1) *and*

$\overline{(2)}$ $\phi(x, r_1 y_1 + r_2 y_2) = \overline{r}_1 \phi(x, y_1) + \overline{r}_2 \phi(x, y_2)$

*for a **nontrivial** conjugation $r \mapsto \overline{r}$ on R.*

Observe that this notion is a generalization of the notion of inner product space that we considered in Section 4.6. Some (but not all) authors use the term "inner product space" to refer to this more general situation. (Strictly speaking, in the second part of the definition we should say that ϕ is sesquilinear with respect to the given conjugation, but we shall assume that we have chosen a particular conjugation and use it throughout.)

(2.4) Definition. *Let R be a ring with conjugation and let M and N be R-modules. A map $f : M \to N$ is called an **antihomomorphism** if*

$$f(r_1 m_1 + r_2 m_2) = \overline{r}_1 f(m_1) + \overline{r}_2 f(m_2)$$

for all r_1, $r_2 \in R$, m_1, $m_2 \in M$.

We observed in Section 6.1 that there is no canonical isomorphism from M to its dual module M^*; however, a bilinear form produces for us a canonical map, and, conversely, a map produces a canonical form. (Here we do not necessarily have isomorphisms, but we shall investigate this point shortly.)

(2.5) Proposition.

(1) *Let ϕ be a bilinear (resp., sesquilinear) form on M. Then α_ϕ : $M \to M^*$, defined by*

$$\alpha_\phi(y)(x) = \phi(x, y)$$

is an R-homomorphism (resp., R-antihomomorphism).

(2) *Let $\alpha : M \to M^*$ be an R-homomorphism (resp., R-antihomo-morphism). Then $\phi_\alpha : M \times M \to R$, defined by*

$$\phi_\alpha(x, y) = \alpha(y)(x)$$

is a bilinear (resp., sesquilinear) form on M.

Proof. Exercise.   □

(2.6) Examples.

(1) Fix $s \in R$. Then $\phi(r_1, r_2) = r_1 s r_2$ (resp., $= r_1 s \bar{r}_2$) is a bi- (resp., sesqui-) linear form on R.
(2) $\phi(x, y) = x^t y$ is a bilinear form on $M_{n,1}(R)$, and $\phi(x, y) = x^t \bar{y}$ is a sesquilinear form on $M_{n,1}(R)$. Note that $\bar{y}$ is obtained from y by entry by entry conjugation.
(3) More generally, for any $A \in M_n(R)$, $\phi(x, y) = x^t A y$ is a bilinear form, and $\phi(x, y) = x^t A \bar{y}$ is a sesquilinear form on $M_{n,1}(R)$.
(4) Let $M = M_{n,m}(R)$. Then $\phi(A, B) = \text{Tr}(A^t B)$ (resp., $\phi(A, B) = \text{Tr}(A^t \bar{B})$) is a bi- (resp., sesqui-) linear form on M.
(5) Let M be the space of continuous real- (resp., complex-) valued functions on $[0, 1]$. Then

$$\phi(f, g) = \int_0^1 f(x) g(x) \, dx$$

is a bilinear form on the **R**- (resp., **C**-) module M. If M is the space of continuous complex-valued functions on $[0, 1]$, then

$$\phi(f, g) = \int_0^1 f(x) \overline{g(x)} \, dx$$

is a sesquilinear form on the **C**-module M.

We will often have occasion to state theorems that apply to both bilinear and sesquilinear forms. We thus, for convenience, adopt the language that ϕ is a b/s-linear form means ϕ is a bilinear or sesquilinear form. Also, the theorems will often have a common proof for both cases. We will then write the proof for the sesquilinear case, from which the proof for the bilinear case follows by taking the conjugation to be trivial (i.e., $\bar{r} = r$ for all $r \in R$).

We will start our analysis by introducing the appropriate equivalence relation on b/s-linear forms.

(2.7) Definition. *Let ϕ_1 and ϕ_2 be b/s-linear forms on free R-modules M_1 and M_2 respectively. Then ϕ_1 and ϕ_2 are* **isometric** *if there is an R-module isomorphism $f : M_1 \to M_2$ with*

$$\phi_2(f(x), f(y)) = \phi_1(x, y) \qquad \text{for all} \quad x, y \in M_1.$$

(If there is no danger of confusion, we will call M_1 and M_2 isometric.) The map f is called an **isometry***.*

Our object in this section will be to derive some general facts about b/s-linear forms, to derive canonical forms for them, and to classify them up to isometry in favorable cases. Later on we will introduce the related notion of a quadratic form and investigate it. We begin by considering the matrix representation of a b/s-linear form with respect to a given basis.

(2.8) Definition. *Let M be a free R-module of rank n with basis $\mathcal{B} = \{v_1, \ldots, v_n\}$ and ϕ a b/s-linear form on M. Define the matrix of ϕ with respect to the basis $\mathcal{B}$, denoted $[\phi]_{\mathcal{B}}$, by*

$$\mathrm{ent}_{ij}\left([\phi]_{\mathcal{B}}\right) = \phi(v_i, v_j) \qquad 1 \le i, j \le n.$$

(2.9) Proposition. (1) *Let M be a free R-module of rank n with basis $\mathcal{B}$ and let ϕ be a bilinear form on M. Then for any $x, y \in M$,*

$$(2.1) \qquad\qquad \phi(x, y) = [x]_{\mathcal{B}}^t [\phi]_{\mathcal{B}} [y]_{\mathcal{B}}.$$

(2) *If ϕ is a sesquilinear form on M and $x, y \in M$, then*

$$(2.2) \qquad\qquad \phi(x, y) = [x]_{\mathcal{B}}^t [\phi]_{\mathcal{B}} \overline{[y]}_{\mathcal{B}}.$$

Proof. Just as a linear transformation is determined by its values on a basis, a b/s-linear form is determined by its values on pairs of basis elements. According to Example 2.6 (3), the right-hand sides of equations (2.1) and (2.2) define such forms, and the two sides clearly agree on each pair (v_i, v_j). $\square$

(2.10) Definition. *If M is a free R-module, then we will denote the set of all bilinear forms on M by $\mathrm{Bilin}(M)$, and if R has a conjugation, then we will denote the set of all sesquilinear forms on M by $\mathrm{Seslin}(M)$. Each of* *these sets is an R-module in a natural way, i.e., via addition and scalar multiplication of R-valued functions.*

(2.11) Corollary. *Let M be a free R-module of rank n. Then there are R-module isomorphisms*

(2.3) $$\text{Bilin}(M) \xrightarrow{\sim} M_n(R)$$

and

(2.4) $$\text{Seslin}(M) \xrightarrow{\sim} M_n(R)$$

given by

$$\phi \mapsto [\phi]_{\mathcal{B}},$$

where $\mathcal{B}$ is any basis of M.

Proof. Proposition 2.9 gives a bijection, and it is easy to check that it is a homomorphism (in both cases). □

(2.12) Remarks.

(1) Note that this corollary says that, in the case of a free module of finite rank, *all* forms arise as in Example 2.6 (3).
(2) We have now seen matrices arise in several ways: as the matrices of linear transformations, as the matrices of bilinear forms, and as the matrices of sesquilinear forms. It is important to keep these different roles distinct, though as we shall see below, they are closely related.

One obvious question is how the matrices of a given form with respect to different bases are related. This is easy to answer.

(2.13) Theorem. *Let ϕ be a b/s-linear form on the free R-module M of rank n. Let $\mathcal{B}$ and $\mathcal{C}$ be two bases for M. If $P = P_{\mathcal{B}}^{\mathcal{C}}$ is the change of basis matrix from $\mathcal{C}$ to $\mathcal{B}$, then*

(2.5) $$[\phi]_{\mathcal{C}} = P^t [\phi]_{\mathcal{B}} \overline{P}.$$

Proof. By definition, $[\phi]_{\mathcal{C}}$ is the unique matrix with

$$\phi(x,\, y) = [x]_{\mathcal{C}}^t [\phi]_{\mathcal{C}} [\overline{y}]_{\mathcal{C}}.$$

But, also,

$$\phi(x,\, y) = [x]_{\mathcal{B}}^t [\phi]_{\mathcal{B}} [\overline{y}]_{\mathcal{B}},$$

and if $P = P_{\mathcal{B}}^{\mathcal{C}}$, then Proposition 4.3.1 gives

$$[x]_{\mathcal{B}} = P[x]_{\mathcal{C}} \quad \text{and} \quad [y]_{\mathcal{B}} = P[y]_{\mathcal{C}}.$$

Thus,

$$\phi(x, y) = (P[x]_C)^t [\phi]_B \left(\overline{P[y]_C} \right)$$
$$= [x]_C^t \left(P^t [\phi]_B \overline{P} \right) [\bar{y}]_C$$

yielding the theorem. □

(2.14) *Remark.* Note that the matrix P is nonsingular (Proposition 4.3.1) and that every nonsingular matrix arises in this way (Proposition 4.3.2). The relation $A \sim B$ if A and B are matrices of the same b/s-linear form with respect to different bases is clearly an equivalence relation, and the above theorem states that this equivalence relation is given by $A \sim B$ if and only if $A = P^t B \overline{P}$ for some invertible matrix P. In the case of a bilinear form, this relation becomes $A = P^t B P$ and is known as **congruence**. There is no generally accepted name in the sesquilinear case; we shall refer to it as **conjugate congruence**. Note that this relation is *completely different* from the relation of similarity.

(2.15) *Remark.* We shall often have occasion to speak of $\det(\phi)$ in this section. By this, we mean $\det(A)$, where A is the matrix of ϕ in some basis. Note that this is well defined up to multiplication by a unit of R of the form $r\bar{r}$, for if B is the matrix of ϕ in a different basis, then $B = P^t A \overline{P}$, for some invertible matrix P. Then $\det(B) = r\bar{r} \det(A)$ where $\det(P) = r$ is a unit, and for any r we may find such a matrix P, e.g., $P = \operatorname{diag}(r, 1, 1, \dots)$.

On the other hand, this observation gives an invariant of a b/s-linear form: If ϕ_1 and ϕ_2 are two forms with the equation $\det(\phi_1) = r\bar{r} \det(\phi_2)$ having no solution for r a unit in R, then ϕ_1 and ϕ_2 are *not* isometric.

(2.16) **Lemma.** *Let M be a free R-module of rank n and ϕ a b/s-linear form on M. Let $\mathcal{B}$ be a basis of M and $\mathcal{B}^*$ the dual basis of M^*. Then $[\phi]_\mathcal{B} = [\alpha_\phi]_{\mathcal{B}^*}^\mathcal{B}$.*

Proof. Let $\mathcal{B} = \{v_1, \dots, v_n\}$. We claim that

$$\alpha_\phi(v_j) = \sum_{i=1}^n \phi(v_i, v_j) v_i^*.$$

In order to see that this is true we need only check that $\alpha_\phi(v_j)(v_k) = \phi(v_k, v_j)$, which is immediate from the definition of α_ϕ (see Proposition 2.5). Then from the definition of $A = [\alpha_\phi]_{\mathcal{B}^*}^\mathcal{B}$ (Definition 4.3.3), we see that A is the matrix with $\operatorname{ent}_{ij}(A) = \phi(v_i, v_j)$, and this is precisely the definition of $[\phi]_\mathcal{B}$ (Definition 2.8). □

We shall restrict the forms we wish to consider.

(2.17) **Definition.** *Let M be a free R-module.*

(1) *A bilinear form ϕ on M is said to be* **symmetric** *if $\phi(x, y) = \phi(y, x)$ for all x, $y \in M$.*

(2) *A bilinear form ϕ on M is said to be* **skew-symmetric** *if $\phi(x, y) = -\phi(y, x)$ for every x, $y \in M$, and $\phi(x, x) = 0$ for every $x \in M$.*

(3) *A sesquilinear form ϕ on M is said to be* **Hermitian** *if $\phi(x, y) = \overline{\phi(y, x)}$ for every x, $y \in M$.*

(4) *If 2 is not a zero divisor in the ring R, then a sesquilinear form ϕ on M is said to be* **skew-Hermitian** *if $\phi(x, y) = -\overline{\phi(y, x)}$ for every x, $y \in M$.*

(2.18) *Remarks.*

(1) We do not define skew-Hermitian if 2 divides 0 in R.

(2) Let ϕ be a b/s-linear form on M and let A be the matrix of ϕ (with respect to any basis). Then the conditions on ϕ in Definition 2.17 correspond to the following conditions on A:

  (a) ϕ is symmetric if and only if $A^t = A$;

  (b) ϕ is skew-symmetric if and only if $A^t = -A$ and all the diagonal entries of A are zero;

  (c) ϕ is Hermitian if and only if $A \neq \overline{A}$ and $A^t = \overline{A}$; and

  (d) ϕ is skew-Hermitian if and only if $A \neq \overline{A}$ and $A^t = -\overline{A}$ (and hence every diagonal entry of A satisfies $a = -\overline{a}$).

(3) In practice, most forms that arise are one of these four types.

   We introduce a bit of terminological shorthand. A symmetric bilinear form will be called $(+1)$-**symmetric** and a skew-symmetric bilinear form will be called (-1)-**symmetric**; when we wish to consider both possibilities simultaneously we will refer to the form as ε-**symmetric**. Similar language applies with ε-Hermitian. When we wish to consider a form that is either symmetric (bilinear) or Hermitian (sesquilinear) we will refer to it as $(+1)$-symmetric b/s-linear, with a similar definition for (-1)-symmetric b/s-linear. When we wish to consider all four cases at once we will refer to an ε-symmetric b/s-linear form.

(2.19) Definition. *The ε-symmetric b/s-linear form ϕ is called* **non-singular** *if the map $\alpha_\phi : M \to M^*$ is bijective. It is called* **non-degenerate** *if $\alpha_\phi : M \to M^*$ is injective. Note that if R is a field and M has finite rank, then these notions are equivalent.*

(2.20) Proposition. *Let V be a finite-dimensional vector space over a field F, and let ϕ be an ε-symmetric b/s-linear form on V. Then ϕ is non-singular if and only if it is non-degenerate, which is the case if and only if, for every $y \neq 0 \in V$, there is an $x \in V$ with*

$$\alpha_\phi(y)(x) = \phi(x, y) \neq 0.$$

Proof. Since α_ϕ is a homomorphism (or antihomomorphism) between vector spaces of the same dimension, in order to show that it is an isomorphism (or antiisomorphism) we need only show that it is injective, i.e., that $\mathrm{Ker}(\alpha_\phi) \neq \langle 0 \rangle$. But

$$\mathrm{Ker}(\alpha_\phi) = \{y \in V : \alpha_\phi(y)(x) = \phi(x,\, y) = 0 \quad \text{for every } x \in V\}.$$

$\square$

To give a criterion for non-singularity over a ring, we need to use the matrix representation of a form.

(2.21) Theorem. *Let M be a free R-module of finite rank, and let ϕ be an ε-symmetric b/s-linear form on M.*

(1) *The form ϕ is non-singular if and only if in some (and hence in every) basis $\mathcal{B}$ of M, $\det([\phi]_\mathcal{B})$ is a unit in R.*

(2) *The form ϕ is non-degenerate if and only if in some (and hence in every) basis $\mathcal{B}$ of M, $\det([\phi]_\mathcal{B})$ is not a zero divisor in R.*

Proof. This follows immediately from Lemma 2.16 and Proposition 4.3.17.

$\square$

Note that if N is any submodule of M, the restriction $\phi_N = \phi|_N$ of any ε-symmetric b/s-linear form on M to N is an ε-symmetric b/s-linear form on N. However, the restriction of a non-singular b/s-linear form is not necessarily non-singular. For example, let ϕ be the b/s-linear form on $M_{2,1}(R)$ with matrix $\begin{bmatrix} 1 & 1 \\ 1 & 0 \end{bmatrix}$. If $N_1 = \langle \begin{bmatrix} 1 \\ 0 \end{bmatrix} \rangle$ and $N_2 = \langle \begin{bmatrix} 0 \\ 1 \end{bmatrix} \rangle$, then $\phi|_{N_1}$ is non-singular, but $\phi|_{N_2}$ is singular and, indeed, degenerate.

The following is standard terminology:

(2.22) Definition. *Let ϕ be an ε-symmetric b/s-linear form on M. A submodule $N \subseteq M$ is **totally isotropic** if $\phi|_N$ is identically zero.*

Thus, in the above example, N_2 is a totally isotropic subspace of M.

Recall that in studying free modules, we found it useful to decompose them into direct sums, and in studying a vector space with a linear transformation, we found it useful to decompose it into a direct sum of invariant subspaces. There is an analogous, and similarly useful notion in our present context, which we now introduce.

(2.23) Definition. *Let ϕ be an ε-symmetric b/s-linear form on M.*

(1) *Two submodules N_1 and N_2 of M are **orthogonal** if $\phi(n_1,\, n_2) = 0$ for every $n_1 \in N_1$, $n_2 \in N_2$.*

(2) *M is the **orthogonal direct sum** of two submodules N_1 and N_2, written $M = N_1 \perp N_2$, if $M = N_1 \oplus N_2$ and N_1 and N_2 are orthogonal.*

(2.24) *Remark.* Let N_1 have a basis $\mathcal{B}_1$, N_2 a basis $\mathcal{B}_2$, and let $M = N_1 \oplus N_2$, in which case $\mathcal{B} = \mathcal{B}_1 \cup \mathcal{B}_2$ is a basis of M. Then $M = N_1 \perp N_2$ if and only if

$$[\phi]_{\mathcal{B}} = \begin{bmatrix} A & 0 \\ 0 & B \end{bmatrix}.$$

Conversely, if $[\phi]_{\mathcal{B}}$ is of this form, and if N_1 (resp., N_2) denotes the span of $\mathcal{B}_1$ (resp., $\mathcal{B}_2$), then $M = N_1 \perp N_2$. In this case we will also say $\phi = \phi_1 \perp \phi_2$ where $\phi_i = \phi|_{M_i}$.

(2.25) Definition. *Let ϕ be an ε-symmetric b/s-linear form on M. The* **kernel** *of ϕ, denoted $M^{\circ} = M^{\circ}(\phi) \subseteq M$, is defined to be*

$$M^{\circ} = \mathrm{Ker}(\alpha_{\phi}) = \{y \in M : \phi(x, y) = 0 \quad \text{for all } x \in M\}.$$

As a first step in studying forms, we have the following decomposition:

(2.26) Proposition. *Let M be a finite rank free module over a PID R, and let ϕ be an ε-symmetric b/s-linear form on M. Then ϕ is isometric to $\phi_0 \perp \phi_1$ defined on $M^{\circ} \perp M_1$, where ϕ_0 is identically zero on M° and ϕ_1 is non-degenerate on M_1. Furthermore, ϕ_0 and ϕ_1 are uniquely determined up to isometry.*

Proof. Note that M° is a pure submodule of M (since it is the kernel of a homomorphism (Proposition 3.8.7)) and so it is complemented. Choose a complement M_1. Then M_1 is free and $M \cong M^{\circ} \oplus M_1$. We let $\phi_0 = \phi|_{M^{\circ}}$ and $\phi_1 = \phi|_{M_1}$. Of course M° and M_1 are orthogonal since M° is orthogonal to all of M, so we have $M = M^{\circ} \perp M_1$ with $\phi = \phi_0 \perp \phi_1$. Also, if $m_1 \in M_1$ with $\phi_1(m_1', m_1) = 0$ for all $m_1' \in M_1$, then $\phi(m, m_1) = 0$ for all $m \in M$, i.e., $m_1 \in M^{\circ}$. Since $M = M^{\circ} \oplus M_1$, $M^{\circ} \cap M_1 = \langle 0 \rangle$ and so ϕ_1 is non-degenerate.

The construction in the above paragraph is well defined except for the choice of M_1. We now show that different choices of M_1 produce isometric forms. Let $\pi : M \to M/M^{\circ} = M'$. Then M' has a form ϕ' defined as follows: If x', $y' \in M'$, choose x, $y \in M$ with $\pi(x) = x'$ and $\pi(y) = y'$. Set $\phi'(x', y') = \phi(x, y)$, and note that this is independent of the choice of x and y. But now note that regardless of the choice of M_1, not only is $\pi|_{M_1} : M_1 \to M'$ an isomorphism, but is in fact an isometry between ϕ_1 and ϕ'. $\qquad\square$

The effect of this proposition is to reduce the problem of classifying ε-symmetric b/s-linear forms to that of classifying non-degenerate ones. It also says that the following definition does indeed give an invariant of such a form.

(2.27) Definition. *Let ϕ be an ε-symmetric b/s-linear form on a finite rank free module M over a PID R. If ϕ is isometric to $\phi^{\circ} \perp \phi_1$ with ϕ° identically* zero and ϕ_1 non-degenerate on M_1, then ϕ_1 is called the **non-degenerate part** of ϕ and we set $\mathrm{rank}(\phi) = \mathrm{rank}(M_1)$.

(2.28) Example. Let ϕ be the symmetric bilinear form on $M = M_{3,1}(\mathbf{Z})$ with matrix (with respect to the standard basis)

$$A = \begin{bmatrix} -1 & 1 & -1 \\ 1 & -3 & -1 \\ -1 & -1 & -3 \end{bmatrix}.$$

Then $\det(A) = 0$, so ϕ is degenerate. Routine computation shows that $\mathrm{Ker}(\alpha_\phi) = \{X \in M : AX = 0\}$ is the rank 1 subspace spanned by $v_1 = (2, 1, -1)^t$. Note that

$$\det \begin{bmatrix} 2 & 1 & 0 \\ 1 & 0 & 1 \\ -1 & 0 & 0 \end{bmatrix} = 1$$

so that v_1, $v_2 = \begin{bmatrix} 1 & 0 & 0 \end{bmatrix}^t$, and $v_3 = \begin{bmatrix} 0 & 1 & 0 \end{bmatrix}^t$ form a basis $\mathcal{B}$ for M. Then we let M_1 be the submodule spanned by v_2 and v_3. The matrix of ϕ_1 in this basis is

$$\begin{bmatrix} \phi(v_2, v_2) & \phi(v_2, v_3) \\ \phi(v_3, v_2) & \phi(v_3, v_3) \end{bmatrix} = \begin{bmatrix} -1 & 1 \\ 1 & -3 \end{bmatrix}$$

and

$$[\phi]_{\mathcal{B}} = \begin{bmatrix} 0 & 0 & 0 \\ 0 & -1 & 1 \\ 0 & 1 & -3 \end{bmatrix}.$$

(2.29) Definition. Let ϕ be an ε-symmetric b/s-linear form on M and let N be a submodule of M. Then $N^\perp$, the **orthogonal complement** of N, is defined to be

$$N^\perp = \{x \in M : \phi(x, y) = 0 \quad \text{for all } y \in N\}.$$

(2.30) Examples.

(1) $M^\perp = M^\circ$.
(2) If $N \subseteq M^\circ$, then $N^\perp = M$.
(3) Let ϕ be the b/s-linear form on $M_{2,1}(R)$ whose matrix with respect to the standard basis is

(a) $\begin{bmatrix} 1 & 0 \\ 0 & 1 \end{bmatrix}$;

(b) $\begin{bmatrix} 2 & 1 \\ 1 & 0 \end{bmatrix}$;

(c) $\begin{bmatrix} 0 & 1 \\ 1 & 0 \end{bmatrix}$; and

(d) $\begin{bmatrix} 1 & 0 \\ 0 & 0 \end{bmatrix}$.

Let $N_1 = \langle \begin{bmatrix} 1 \\ 0 \end{bmatrix} \rangle$ and $N_2 = \langle \begin{bmatrix} 0 \\ 1 \end{bmatrix} \rangle$. Then in the above cases we have

(a) $N_1^\perp = N_2$, $N_2^\perp = N_1$, $M = N_1 \perp N_1^\perp$, $M = N_2 \perp N_2^\perp$;

(b) $N_1^\perp = \langle \begin{bmatrix} 1 \\ -2 \end{bmatrix} \rangle$, $N_2^\perp = M$, $M = \mathrm{Hull}(N_1 \perp N_1^\perp)$;

(c) $N_1^\perp = N_1$, $N_2^\perp = N_2$;

(d) $N_1^\perp = N_2$, $N_2^\perp = M$, $M = N_1 \perp N_1^\perp$.

(2.31) Lemma. *Let N be a submodule of M and suppose that $\phi|_N$ is non-degenerate. Then $N \cap N^\perp = \langle 0 \rangle$. Conversely, if $N \cap N^\perp = \langle 0 \rangle$, then $\phi|_N$ is non-degenerate.*

Proof. If $\psi = \phi|_N$, then

$$(2.6) \qquad\qquad N \cap N^\perp = \mathrm{Ker}(\alpha_\psi).$$

The result follows immediately from Equation (2.6). $\square$

(2.32) Proposition. *Let R be a PID, ϕ an ε-symmetric b/s-linear form on a free R-module M, and $N \subseteq M$ a submodule of finite rank. If $\phi|_N$ is non-degenerate, then $M = \mathrm{Hull}(N \perp N^\perp)$. If $\phi|_N$ is non-singular, then $M = N \perp N^\perp$.*

Proof. Let $\psi = \phi|_N$. If $m \in M$, then $f_m : N \to R$, defined by $f_m(n) = \phi(n, m)$, is an R-module homomorphism, i.e., $f \in N^*$. Since α_ψ is assumed to be injective, it follows that $\mathrm{Im}(\alpha_\psi)$ is a submodule of N^* of rank n where $n = \mathrm{rank}(N) = \mathrm{rank}(N^*)$. Thus, $N^*/\mathrm{Im}(\alpha_\psi)$ is a torsion R-module by Proposition 4.3.11, and hence there are $r \in R$ and $n_0 \in N$ such that $rf_m = \alpha_\psi(n_0)$; in case α_ψ is an isomorphism, we may take $r = 1$. If $m_1 = rm - n_0$, then for every $n \in N$, we have

$$\begin{aligned}
\phi(n, m_1) &= \phi(n, rm - n_0) \\
&= r\phi(n, m) - \phi(n, n_0) \\
&= rf_m(n) - \alpha_\psi(n_0)(n) \\
&= 0,
\end{aligned}$$

i.e., $m_1 \in N^\perp$. Thus, $rm = n_0 + m_1$, where $n_0 \in N$, and $m_1 \in N^\perp$. Thus, $M = \mathrm{Hull}(N + N^\perp)$ (or $M = N + N^\perp$ if $r = 1$). But $N \cap N^\perp = \langle 0 \rangle$ by Lemma 2.31, yielding the proposition. $\square$

(2.33) Remark. Note that in Lemma 2.31 and Proposition 2.32 there is no restriction on ϕ, just on $\phi|_N$. The reader should reexamine Examples 2.30 in light of the above lemma and proposition.

(2.34) Corollary. *Let R be a PID and M a free R-module of finite rank. Let N be a pure submodule of M with $\phi|_N$ and $\phi|_{N^\perp}$ both non-singular. Then*

$$\left(N^\perp\right)^\perp = N.$$

Proof. We have $M = N \perp N^\perp = \left(N^\perp\right)^\perp \perp N^\perp$. But it is easy to check that $\left(N^\perp\right)^\perp \supseteq N$, so they are equal. $\square$

Now, a bit of obvious notation: $n\phi$ denotes $\phi \perp \cdots \perp \phi$ (where there are n summands). We now come to the first classification result. Note that it suffices to classify non-degenerate forms.

(2.35) Theorem. *Let R be a PID, and let ϕ be a non-degenerate skew-symmetric bilinear form on a free R-module M of finite rank. Then ϕ is classified up to isometry by $M^*/\operatorname{Im}(\alpha_\phi)$, a torsion R-module. Both M and $M^*/\operatorname{Im}(\alpha_\phi)$ have even rank, say $2n$ and $2k$ respectively. The invariant factors of $M^*/\operatorname{Im}(\alpha_\phi)$ are of the form*

$$e_1, e_1, e_2, e_2, \ldots, e_k, e_k.$$

Furthermore, ϕ is isometric to

$$(n-k)\begin{bmatrix} 0 & 1 \\ -1 & 0 \end{bmatrix} \perp \begin{bmatrix} 0 & e_1 \\ -e_1 & 0 \end{bmatrix} \perp \begin{bmatrix} 0 & e_2 \\ -e_2 & 0 \end{bmatrix} \perp \cdots \perp \begin{bmatrix} 0 & e_k \\ -e_k & 0 \end{bmatrix}.$$

Proof. Let $\operatorname{rank}(M) = \operatorname{rank}(M^*) = m$. Then $Q = M^*/\operatorname{Im}(\alpha_\phi)$ is a torsion R-module of rank q, and hence it is determined by its invariant factors. We will write the invariant factors of the submodule $\operatorname{Im}(\alpha_\phi)$ as $f_1, f_2, \ldots, f_m$ where $f_1 = \cdots = f_{m-q} = 1$, $f_{m-q+1} \neq 1$ and $f_i \mid f_{i+1}$ for $m-q+1 \le i < m$. (See Proposition 3.6.23.) It is evident then that $Q \cong \oplus_{i=1}^m R/f_i R$.

Clearly, $M^*/\operatorname{Im}(\alpha_\phi)$ is an invariant of the isometry class of ϕ. We need to show, conversely, that this determines ϕ, or in other words, that the sequence $f_1, f_2, \ldots, f_m$ determines ϕ. In fact, we will show this by showing that ϕ is isometric to the form given in the statement of the theorem, with $m = 2n$, $q = 2k$, and

$$(f_{m-q+1}, f_{m-q+2}, \ldots, f_m) = (e_1, e_1, e_2, e_2, \ldots, e_k, e_k).$$

The proof is by induction on the rank of M. If $\operatorname{rank}(M) = 1$, then the only possible skew-symmetric form on M is $[0]$, which is degenerate, and hence this case is excluded. If $\operatorname{rank}(M) = 2$, then ϕ must have a matrix in some (and, hence, in any) basis of the form $\begin{bmatrix} 0 & e \\ -e & 0 \end{bmatrix}$, and so the theorem is true in this case also.

Now for the inductive step. Assume the theorem is true for all free R-modules of rank less than m. By Proposition 3.6.23, we may choose a basis $\{w_1, \ldots, w_m\}$ of M^* such that $\{f_1 w_1, \ldots, f_m w_m\}$ is a basis of the submodule $\text{Im}(\alpha_\phi)$. Since $f_1 \mid f_2 \mid \cdots \mid f_m$, we see that $\text{Im}(\alpha_\phi) \subseteq f_1 M^*$, i.e., $\phi(v_1, v_2)$ is divisible by f_1 for every $v_1, v_2 \in M$. Let $x_1, \ldots, x_m$ be the dual basis of M, that is, $w_i(x_j) = \delta_{ij}$. Let $y_1 \in M$ with $\alpha_\phi(y_1) = f_1 w_1$. If $a x_1 + b y_1 = 0$, then

$$\begin{aligned} 0 &= \phi(x_1, ax_1 + by_1) \\ &= a\phi(x_1, x_1) + b\phi(x_1, y_1) \\ &= b f_1. \end{aligned}$$

Thus, $b = 0$ and hence $a = 0$ also. Thus, $\{x_1, y_1\}$ is linearly independent. Let N be the submodule of M with basis $\mathcal{B} = \{x_1, y_1\}$. If $\psi = \phi|_N$, then ψ has the matrix

$$[\psi]_{\mathcal{B}} = \begin{bmatrix} 0 & f_1 \\ -f_1 & 0 \end{bmatrix}.$$

Note that N is a pure submodule. To see this, suppose $az = bx_1 + cy_1$ where $a, b, c \in R$; by cancelling common factors we can assume that $\gcd\{a, b, c\} = 1$. Then $\phi(x_1, az) = \phi(x_1, bx_1 + cy_1) = cf_1$, while $\phi(x_1, az) = a\phi(x, z) = adf_1$ since $\phi(v_1, v_2)$ is divisible by f_1 for all $v_1, v_2 \in M$. Thus, $ad = c$, i.e., $a \mid c$.

A similar computation with $\phi(y_1, az)$ shows that $a \mid b$, so a is a common divisor of a, b, and c, i.e., a is a unit and $z \in N$. By Proposition 2.32, $M = \text{Hull}(N \perp N^\perp)$. But, in fact, $M = N \perp N^\perp$. To see this, let us consider the form ϕ' defined by

$$\phi'(v_1, v_2) = f_1^{-1}\phi(v_1, v_2).$$

Then $N^\perp$ is also the orthogonal complement of N with respect to ϕ' and $\phi'|_N$ is non-singular, so $M = N \perp N^\perp$, i.e., $\phi = \psi \perp \phi_1$ with $\phi_1 = \phi|_{N^\perp}$.

Note that $N^*/\text{Im}(\alpha_\psi)$ has "invariant factors" (in the above sense) f_1 and f_1, so we see that $f_2 = f_1$. Then the "invariant factors" of $(N^\perp)^*/\text{Im}(\alpha_{\phi_1})$ are $f_3, \ldots, f_m$, and the theorem follows by induction. $\square$

(2.36) Corollary. *Let R be a PID, and let ϕ be a non-singular skew-symmetric bilinear form on a free R-module M of finite rank. Then $\text{rank}(M) = 2n$ is even, and ϕ is isometric to*

$$n \begin{bmatrix} 0 & 1 \\ -1 & 0 \end{bmatrix}.$$

Proof. $\square$

Remark. Recall that over a field, ϕ is non-degenerate if and only if it is non-singular, so Corollary 2.36 classifies skew symmetric forms over a field.

(2.37) Examples.

(1) Consider the skew-symmetric bilinear form ϕ over $\mathbf{Z}$ with matrix

$$A = \begin{bmatrix} 0 & 2 & 0 & -2 \\ -2 & 0 & -2 & -8 \\ 0 & 2 & 0 & 4 \\ 2 & 8 & -4 & 0 \end{bmatrix}.$$

According to the theory, to classify ϕ we need to find the "invariant factors" of $\mathbf{Z}^4/A\mathbf{Z}^4$. To do this, we apply elementary row operations:

$$\begin{bmatrix} 0 & 2 & 0 & -2 \\ -2 & 0 & -2 & -8 \\ 0 & 2 & 0 & 4 \\ 2 & 8 & -4 & 0 \end{bmatrix}$$

$$\longrightarrow \begin{bmatrix} 0 & 2 & 0 & -2 \\ -2 & 0 & -2 & -8 \\ 0 & 2 & 0 & 4 \\ 0 & 8 & -6 & -8 \end{bmatrix}$$

$$\longrightarrow \begin{bmatrix} 0 & 2 & 0 & -2 \\ -2 & 0 & -2 & -8 \\ 0 & 0 & 0 & 6 \\ 0 & 0 & -6 & 0 \end{bmatrix}$$

from which we see that the invariant factors are $(2, 2, 6, 6)$ and, hence, that ϕ is isometric to

$$\begin{bmatrix} 0 & 2 \\ -2 & 0 \end{bmatrix} \perp \begin{bmatrix} 0 & 6 \\ -6 & 0 \end{bmatrix}.$$

(2) Let A be any invertible $n \times n$ skew-symmetric matrix over a field F, i.e., $A^t = -A$ and the diagonal entries are 0. Then A is the matrix of a non-singular skew-symmetric form over F, and hence, $P^t A P = mJ$ where $m = n/2$ and $J = \begin{bmatrix} 0 & 1 \\ -1 & 0 \end{bmatrix}$. Then

$$\det(A)(\det(P))^2 = \det(nJ) = 1.$$

In particular, $\det(A)$ is a square in F. Now let $R = \mathbf{Z}[Y]$ where $Y = \{X_{ij} : 1 \le i < j \le n\}$, that is, R is the polynomial ring over $\mathbf{Z}$ in the $\binom{n}{2}$ indeterminates X_{ij} for $1 \le i < j \le n$. Let $F = \mathbf{Q}(Y)$ be the quotient field of R and let $A \in M_n(F)$ be the skew-symmetric matrix with $\text{ent}_{ij} = X_{ij}$ for $1 \le i < j \le n$, $\text{ent}_{ij} = -X_{ji}$ for $1 \le j < i \le n$, and $\text{ent}_{ii} = 0$. Then $\det(A) = P(X_{ij})^2$ for some element $P(X_{ij}) \in F$. But R is a UFD, and hence, the equation $Z^2 = \det(A)$ has a solution in F if and only if it has a solution in R. Thus, $P(X_{ij}) \in R$, and since $P(X_{ij})$ is a root of a quadratic equation, there are two possibilities for the solution. We choose the solution as follows. Choose integers x_{ij} so that the evaluation of the matrix A at the integers x_{ij} gives the matrix mJ. Then choose $P(X_{ij})$ so that the polynomial evaluation $P(x_{ij}) = \det(mJ) = 1$. Then we will call the polynomial $P(X_{ij}) \in \mathbf{Z}[X_{ij}]$ the **generic Pfaffian** and we will denote it $\mathrm{Pf}(A)$.

If S is any commutative ring with identity, then the evaluation $X_{ij} \mapsto b_{ij}$ induces a ring homomorphism

$$\eta : \mathbf{Z}[X_{ij}] \to S.$$

Under this homomorphism, the generic skew-symmetric matrix A is sent to the skew-symmetric matrix $B = [b_{ij}]$. Since determinants commute with ring homomorphisms, we find

$$\det(B) = \det(\eta(A)) = \eta(\det(A)) = \eta\left((\mathrm{Pf}(X_{ij})^2\right) = \mathrm{Pf}(b_{ij})^2.$$

We conclude that the determinant of every skew-symmetric matrix over any commutative ring is a square in the ring, and moreover, the square root can be chosen in a canonical manner via the Pfaffian.

(2.38) Remark. The case of skew-Hermitian sesquilinear forms is quite different. For example, the form $[i]$ (i refers to the complex number i) is a non-singular form over $\mathbf{C}$ (or over the PID $\mathbf{Z}[i]$), a module of odd rank. Also, consider the following two non-singular forms:

$$\phi_1 = \begin{bmatrix} i & 0 \\ 0 & i \end{bmatrix} \qquad \phi_2 = \begin{bmatrix} i & 0 \\ 0 & -i \end{bmatrix}.$$

We leave it to the reader to check the following facts:

(1) ϕ_1 and ϕ_2 are not isometric.

(2) ϕ_1 is not isometric to any form with matrix $\begin{bmatrix} 0 & z \\ -\bar{z} & 0 \end{bmatrix}$.

(3) ϕ_2 is not isometric to $\begin{bmatrix} 0 & 1 \\ -1 & 0 \end{bmatrix}$ over $\mathbf{Z}[i]$, but is isometric to it over $\mathbf{C}$.

We shall not discuss these forms any further.

Now we come to the case of symmetric bilinear or Hermitian sesquilinear forms.

(2.39) Definition. A $(+1)$-*symmetric b/s-linear form* ϕ on an R-*module* M *is called* **diagonalizable** *if* ϕ *is isometric to the form*

$$[a_1] \perp [a_2] \perp \cdots \perp [a_n]$$

for some elements $a_1, a_2, \ldots, a_n \in R$. *Here* $[a_i]$ *denotes the form on* R *whose matrix is* $[a_i]$, *i.e., the form* $\phi(r_1, r_2) = r_1 a_i \bar{r}_2$. *(The terminology "diagonalizable" is used because in the obvious basis* ϕ *has the matrix* $\mathrm{diag}(a_1, a_2, \ldots, a_n).)$

Note that the notion of diagonalizability for a form is quite different than the notion of diagonalizability for a linear transformation; in terms of matrices, a form ϕ is diagonalizable if its matrix is (conjugate)-congruent to a diagonal matrix, while a linear transformation is diagonalizable if its matrix is similar to a diagonal matrix.

(2.40) Definition. *A symmetric bilinear form ϕ on a module M over a ring R is called* **even** *if $\phi(x, x) \in 2R$ for every $x \in M$. If ϕ is not even, it is called* **odd**.

Here, $2R$ denotes the principal ideal of R generated by 2. Note that if 2 is a unit in R, then *every* form over R is even.

(2.41) Lemma. *A symmetric bilinear form ϕ on a module M is even if and only if for some (and hence for every) basis $\mathcal{B} = \{v_i\}_{i \in I}$ of M, $\phi(v_i, v_i) \in 2R$ for every $i \in I$.*

Proof. Since ϕ is symmetric, it follows that

$$\phi(x + y, x + y) = \phi(x, x) + \phi(x, y) + \phi(y, x) + \phi(y, y)$$
$$= \phi(x, x) + 2\phi(x, y) + \phi(y, y)$$

for any $x, y \in M$. Given this observation, the rest of the proof is left to the reader. $\square$

(2.42) Theorem. *Let ϕ be a non-singular $(+1)$-symmetric b/s-linear form on a module M of finite rank over a field R. If $\operatorname{char}(R) = 2$ and ϕ is symmetric bilinear, assume also that ϕ is odd. Then ϕ is diagonalizable.*

Proof. We prove this by induction on $\operatorname{rank}(M)$. If $\operatorname{rank}(M) = 1$, then $\phi = [a]$, and there is nothing to prove. Thus, suppose that the theorem is true for $\operatorname{rank}(M) < n$ and let $\operatorname{rank}(M) = n$.

First, we claim that there is an $x \in M$ with $\phi(x, x) \neq 0$. If $\operatorname{char}(R) = 2$ and ϕ is symmetric bilinear, this is true by hypothesis. Otherwise, pick $y \in M$. If $\phi(y, y) \neq 0$, set $x = y$. If $\phi(y, y) = 0$, pick $z \in M$ with $\phi(y, z) \neq 0$. Such a z exists because ϕ is assumed to be non-singular. If $\phi(z, z) \neq 0$, set $x = z$. Otherwise, note that for any $r \in R$,

$$\phi(ry + z, ry + z) = \phi(ry, ry) + \phi(ry, z) + \phi(z, ry) + \phi(z, z)$$
$$= r\phi(y, z) + \overline{r\phi(y, z)}.$$

If $\operatorname{char}(R) \neq 2$ set $x = ry + z$ with $r = \phi(y, z)^{-1}$. If $\operatorname{char}(R) = 2$ and ϕ is Hermitian, let $a \in R$ with $\overline{a} \neq a$, and set $x = ry + z$ with $r = a\phi(y, z)^{-1}$. Then, in any case, we have $\phi(x, x) \neq 0$.

Let N be the subspace of M spanned by x. By construction, $\phi|_N$ is non-singular, so $M = N \perp N^{\perp}$ by Proposition 2.32. But then $\operatorname{rank}(N^{\perp}) =$

rank$(M) - 1$ and $\psi = \phi|_{N^\perp}$ is non-singular (since $\phi(x, x)\det(\psi) = \det(\phi) \neq 0$). Thus, unless char$(R) = 2$ and ϕ is symmetric bilinear, we are done.

If char$(R) = 2$ and ϕ is symmetric bilinear, we cannot apply induction yet as we do not know that the form ψ is odd. Indeed, it is possible that ψ is even and so there is more work to be done.

First, consider the case rank$(M) = 2$. Let $a = \phi(x, x)$ and choose a basis $\{x, x_2\}$ of M. Then, in this basis ϕ has the matrix

$$\begin{bmatrix} a & b \\ b & c \end{bmatrix}$$

with $a \neq 0$. Let $e = b/a$. Then

$$\begin{bmatrix} 1 & 0 \\ e & 1 \end{bmatrix} \begin{bmatrix} a & b \\ b & c \end{bmatrix} \begin{bmatrix} 1 & e \\ 0 & 1 \end{bmatrix} = \begin{bmatrix} a & 0 \\ 0 & d \end{bmatrix}$$

(with $d = ae^2 + c$) and ϕ is diagonalized.

Now suppose that rank$(M) \geq 3$. Find x as above with $\phi(x, x) = a \neq 0$ and write $M = N \perp N^\perp$ as above. Pick $y \in N^\perp$. If $\phi(y, y) \neq 0$, then $\phi|_{N^\perp}$ is odd and we are done (because we can apply induction). Thus, suppose $\phi(y, y) = 0$. Since $\psi = \phi|_{N^\perp}$ is non-singular, there is $z \in N^\perp$ with $\phi(y, z) = b \neq 0$. If $\phi(z, z) \neq 0$, then ψ is odd and we are again done by induction. Thus suppose $\phi(z, z) = 0$. Let M_1 be the subspace of M with basis $\{x, y, z\}$ and let $\phi_1 = \phi|_{M_1}$. Then, in this basis ϕ_1 has the matrix

$$A = \begin{bmatrix} a & 0 & 0 \\ 0 & 0 & b \\ 0 & b & 0 \end{bmatrix}.$$

Let $e = b/a$ and

$$P = \begin{bmatrix} 1 & e & e \\ 1 & 1 & 0 \\ 1 & 0 & 1 \end{bmatrix},$$

and note that $\det(P) = 1$. Then

$$P^t A P = \begin{bmatrix} a & 0 & 0 \\ 0 & be & (b+a)e \\ 0 & (b+a)e & be \end{bmatrix}$$

in a basis, which we will simply denote by $\{x', y', z'\}$. Now let N be the subspace of M spanned by x', and so, as above, $M = N \perp N^\perp$. But now $\psi = \phi|_{N^\perp}$ is odd. This is because $y' \in N^\perp$ and $\psi(y', y') = be \neq 0$. Thus, we may apply induction and we are done.   □

(2.43) Example. We will diagonalize the symmetric bilinear form with matrix

$$A = \begin{bmatrix} 0 & 1 & 2 \\ 1 & 0 & 3 \\ 2 & 3 & 0 \end{bmatrix}$$

over $\mathbf{Q}$. The procedure is to apply a sequence of elementary row/column operations. If A is symmetric and E is any elementary matrix, then $E^t A E$ is also symmetric. We indicate the matrices and record the results.

$$A_1 = T_{12}(1) A T_{21}(1) = \begin{bmatrix} 2 & 1 & 5 \\ 1 & 0 & 3 \\ 5 & 3 & 0 \end{bmatrix}$$

$$A_2 = T_{21}(-1/2) A_1 T_{12}(-1/2) = \begin{bmatrix} 2 & 0 & 5 \\ 0 & -1/2 & -1/2 \\ 5 & 1/2 & 0 \end{bmatrix}$$

$$A_3 = T_{31}(-5/2) A_2 T_{13}(-5/2) = \begin{bmatrix} 2 & 0 & 0 \\ 0 & -1/2 & -1/2 \\ 0 & 1/2 & -25/2 \end{bmatrix}$$

$$A_4 = T_{32}(1) A_3 T_{23}(1) = \begin{bmatrix} 2 & 0 & 0 \\ 0 & -1/2 & 0 \\ 0 & 0 & -12 \end{bmatrix}$$

The reader should not be under the impression that, just because we have been able to diagonalize symmetric or Hermitian forms, we have been able to classify them. However, there are a number of important cases where we can achieve a classification.

(2.44) Corollary. *Let ϕ be a non-degenerate symmetric bilinear form on a module M of finite rank over a field R of characteristic not equal to 2 in which every element is a square. (Note that $R = \mathbf{C}$ satisfies this hypothesis.) Then ϕ is determined up to isometry by $\mathrm{rank}(M)$. If $\mathrm{rank}(M) = n$ then ϕ is isometric to $n[1]$.*

Proof. This follows immediately from Theorem 2.42 and Remark 2.15. $\square$

(2.45) Corollary. *Let ϕ be a non-degenerate symmetric bilinear form on a module M of finite rank over a finite field R of odd characteristic. Then ϕ is determined up to isometry by $\mathrm{rank}(M)$ and $\det(\phi)$, the latter being well defined up to multiplication by a square in R. Let $x \in R$ be any element that is not a square, and let $\mathrm{rank}(M) = n$. If $\det(\phi)$ is a square, then ϕ is isometric to $n[1]$. If $\det(\phi)$ is not a square, then ϕ is isometric to $(n - 1)[1] \perp [x]$.*

Proof. By Theorem 2.42 we know that ϕ is isometric to the form

$$[r_1] \perp [r_2] \perp \cdots \perp [r_n]$$

for some $r_i \in R$. Note that the multiplicative group R^* has even order, so the squares form a subgroup of index 2. Then $\det(\phi)$ is a square or a nonsquare accordingly as there are an even or an odd number of nonsquares among the $\{r_i\}$. Thus, the theorem will be proved once we show that the form $[r_i] \perp [r_j]$ with r_i and r_j both nonsquares is equivalent to the form $[1] \perp [s]$ for some s (necessarily a square).

Thus, let $[\phi]_\mathcal{B} = \begin{bmatrix} r_1 & 0 \\ 0 & r_2 \end{bmatrix}$ in a basis $\mathcal{B} = \{v_1, v_2\}$ of M. R has an odd number of elements, say $2k+1$, of which $k+1$ are squares. Let $A = \{a^2 r_1 : a \in R\}$ and $B = \{1 - b^2 r_2 : b \in R\}$. Then A and B both have $k+1$ elements, so $A \cap B \neq \emptyset$. Thus, for some $a_0, b_0 \in R$,

$$a_0^2 r_1 = 1 - b_0^2 r_2,$$

i.e.,

$$a_0^2 r_1 + b_0^2 r_2 = 1.$$

Let N be the subspace of M spanned by $a_0 v_1 + b_0 v_2$. Then $\phi|_N = [1]$ and $M = N \perp N^\perp$, so, in an appropriate basis, ϕ has the matrix $\begin{bmatrix} 1 & 0 \\ 0 & s \end{bmatrix}$, as claimed. $\square$

(2.46) Corollary. *Let ϕ be a non-degenerate symmetric bilinear form on a module M of finite rank over a finite field R of characteristic 2. Then ϕ is determined up to isometry by $n = \mathrm{rank}(M)$ and whether ϕ is even or odd. If n is odd, then ϕ is odd and is isometric to $n[1]$. If n is even, then either ϕ is odd and isometric to $n[1]$, or ϕ is even and ϕ is isometric to* $(n/2)\begin{bmatrix} 0 & 1 \\ 1 & 0 \end{bmatrix}$.

Proof. Since R^* has odd order, every element of this multiplicative group is a square. If ϕ is odd then, by Theorem 2.42, ϕ is diagonalizable and then ϕ is isometric to $n[1]$ as in Corollary 2.44.

Suppose that ϕ is even. Note that an even symmetric form over a field of characteristic 2 may be regarded as a skew-symmetric form. Then, by Corollary 2.36, n is even and ϕ is isometric to $(n/2)\begin{bmatrix} 0 & 1 \\ 1 & 0 \end{bmatrix}$. $\square$

(2.47) Theorem. (Witt) *Let ϕ be a $(+1)$-symmetric b/s-linear form on a module M of finite rank over a field R. If $\mathrm{char}(R) = 2$, assume also that ϕ is Hermitian. Let N_1 and N_2 be submodules of M with $\phi|_{N_1}$ and $\phi|_{N_2}$ non-singular and isometric to each other. Then $\phi|_{N_1^\perp}$ and $\phi|_{N_2^\perp}$ are isometric.*

Proof. If $N_1 = N_2$, then $N_1^\perp = N_2^\perp$, and there is nothing to prove; so we assume $N_1 \neq N_2$. Let $\phi_i = \phi|_{N_i}$, and let $f : N_1 \to N_2$ be an isometry between ϕ_1 and ϕ_2.

We prove the theorem by induction on $n = \mathrm{rank}(N_1) = \mathrm{rank}(N_2)$. Let $n = 1$. Let $m = \mathrm{rank}(M)$. If $m = 1$ the theorem is trivial.

Let $m = 2$. Let v_1 generate N_1 and $v_2 = f(v_1)$ generate N_2, so

$$\phi(v_1, v_1) = \phi(v_2, v_2) = a \neq 0.$$

Then M has bases $\mathcal{B}_i = \{v_i, v_i^\perp\}$ with $v_i^\perp \in N_i^\perp$, $i = 1, 2$, and hence $[\phi]_{\mathcal{B}_i} = \text{diag}(a, b_i)$ for $i = 1, 2$, with b_1 and b_2 both nonzero. By Theorem 2.13, there is an invertible matrix P with

$$P^t[\phi]_{\mathcal{B}_1}\overline{P} = [\phi]_{\mathcal{B}_2},$$

and taking determinants shows that $ab_2 = c\bar{c}ab_1$ (where $c = \det(P)$); so $g : N_1^\perp \to N_2^\perp$ defined by $g(v_1^\perp) = c^{-1}v_2^\perp$ gives an isometry between $\phi|_{N_1^\perp}$ and $\phi|_{N_2^\perp}$.

Next let $m \geq 3$ and consider the submodule N_{12} of M with basis $\{v_1, v_2\}$, where v_1 generates N_1, and $v_2 = f(v_1)$ generates N_2. Then

$$\phi(v_1, v_1) = \phi(v_2, v_2) \neq 0.$$

(Since we are assuming $N_1 \neq N_2$, v_1 and v_2 are linearly independent in M and so N_{12} has rank 2.) Consider $\phi|_{N_{12}}$. Either it is non-singular or it is not.

First suppose that $\phi|_{N_{12}}$ is non-singular. Then, by the case $m = 2$ we have that $\phi|_{N_1^\perp \cap N_{12}}$ and $\phi|_{N_2^\perp \cap N_{12}}$ are isometric, and hence,

$$\phi|_{(N_1^\perp \cap N_{12}) \perp N_{12}^\perp} \quad \text{and} \quad \phi|_{(N_2^\perp \cap N_{12}) \perp N_{12}^\perp}$$

are also isometric. But in this case $M = N_{12} \perp N_{12}^\perp$, from which it readily follows that

$$(N_i^\perp \cap N_{12}) \perp N_{12}^\perp = N_i^\perp,$$

yielding the theorem.

Now suppose $\phi|_{N_{12}}$ is singular. Then there is a $0 \neq w \in N_{12}$ with $\phi(v, w) = 0$ for all $v \in N_{12}$. Suppose there is a $v_3 \in M$ with $\phi(v_3, w) \neq 0$. (Such an element v_3 certainly exists if ϕ is non-singular on M.) Of course, $v_3 \notin N_{12}$, so $\{v_1, v_2, v_3\}$ form a basis for a submodule N_{123} of M of rank 3 and $\phi|_{N_{123}}$ is non-singular. (To see this, consider the matrix of ϕ in the basis $\{v_1, w, v_3\}$ of N_{123}.) Now for $i = 1, 2$,

$$N_{123} = N_i \perp (N_i^\perp \cap N_{123})$$

and $w \in N_i^\perp \cap N_{123}$ with $\phi(w, w) = 0$, so there is a basis $\{w, w_i\}$ of $N_i^\perp \cap N_{123}$ with ϕ having matrix $\begin{bmatrix} 0 & 1 \\ 1 & a_i \end{bmatrix}$ in that basis (with $a_i = \bar{a}_i$). We claim that any two such forms are isometric, and this will follow if, given any $a \in R$ with $a = \bar{a}$, there is a $b \in R$ with

$$\begin{bmatrix} 1 & 0 \\ b & 1 \end{bmatrix}\begin{bmatrix} 0 & 1 \\ 1 & a \end{bmatrix}\begin{bmatrix} 1 & \bar{b} \\ 0 & 1 \end{bmatrix} = \begin{bmatrix} 0 & 1 \\ 1 & 0 \end{bmatrix}.$$

If $\text{char}(R) \neq 2$, take $b = -a/2$ (and note that $b = \bar{b}$). If $\text{char}(R) = 2$, let $c \in R$ with $c \neq \bar{c}$ (which exists as we are assuming that ϕ is Hermitian) and let $b = ac/(c+\bar{c})$. Hence, $\phi|_{N_1^\perp \cap N_{123}}$ and $\phi|_{N_2^\perp \cap N_{123}}$ are isometric, and $M = N_{123} \perp N_{123}^\perp$ (as $\phi|_{N_{123}}$ is non-singular); so, as in the case $\phi|_{N_{12}}$ is non-singular, it follows that $\phi|_{N_1^\perp}$ and $\phi|_{N_2^\perp}$ are isometric.

It remains to deal with the case that $\phi(v, w) = 0$ for every $v \in M$. We claim that in this case $N_1^\perp = N_2^\perp$, so the theorem is true. To see this, let $\mathcal{B}_{12} = \{v_1, v_2\}$ and extend $\mathcal{B}_{12}$ to a basis $\mathcal{B} = \{v_1, \ldots, v_m\}$ of M. Let $A = [\phi]_\mathcal{B}$. Then

$$A[(1, 2) \mid (1, 2)] = [\phi|_{N_{12}}]_{\mathcal{B}_{12}} = \begin{bmatrix} a & b \\ \bar{b} & a \end{bmatrix}$$

with $a \neq 0$ and $a^2 - b\bar{b} = 0$, and we may assume $w \in N_{12}$ is given by

$$[w]_\mathcal{B} = [\,a \quad -b \quad 0 \quad \cdots \quad 0\,]^t$$

(as w is well defined up to a scalar factor). Then $\phi(v_i, w) = 0$ for $i = 3,$ $\ldots, m$ implies that there exist scalars c_i so that

$$\begin{bmatrix} \text{ent}_{1i}(A) \\ \text{ent}_{2i}(A) \end{bmatrix} = \begin{bmatrix} \phi(v_1, v_i) \\ \phi(v_2, v_i) \end{bmatrix} = c_i \begin{bmatrix} b \\ a \end{bmatrix}, \qquad i = 3, \ldots, m.$$

If we let P be the matrix defined by

$$\begin{aligned} \text{ent}_{ii}(P) &= 1 \\ \text{ent}_{2i}(P) &= -\bar{c}_i \qquad i = 3, \ldots, m \\ \text{ent}_{ij}(P) &= 0 \qquad \text{otherwise,} \end{aligned}$$

then

$$P^t A \overline{P} = \begin{bmatrix} A & 0 \\ 0 & * \end{bmatrix}$$

(the right-hand side being a block matrix). This is $[\phi]_{\mathcal{B}'}$ in the basis

$$\mathcal{B}' = \{v_1, v_2, v_3', \ldots, v_m'\}$$

and then $N_1^\perp = N_2^\perp$ is the subspace with basis

$$\{w, v_3', \ldots, v_m'\}.$$

This concludes the proof of the theorem in case $n = 1$.

Now we apply induction on n. Assume the theorem is true for

$$\text{rank}(N_1) = \text{rank}(N_2) < n,$$

and let

$$\text{rank}(N_1) = \text{rank}(N_2) = n.$$

Let $v_1 \in N_1$ with $\phi(v_1, v_1) \neq 0$, and let $v_2 = f(v_1) \in N_2$; so

$$\phi(v_2, v_2) = \phi(v_1, v_1) \neq 0.$$

Note that such an element exists by the proof of Theorem 2.42. Let N_{11} be the subspace of M generated by v_1 and let N_{21} be the subspace of M generated by v_2. Then

$$M = N_{11} \perp (N_{11}^{\perp} \cap N_1) \perp N_1^{\perp} = N_{21} \perp (N_{21}^{\perp} \cap N_2) \perp N_2^{\perp}.$$

Then the case $n = 1$ of the theorem implies that

$$(N_{11}^{\perp} \cap N_1) \perp N_1^{\perp} \quad \text{and} \quad (N_{21}^{\perp} \cap N_2) \perp N_2^{\perp}$$

are isometric, and then the inductive hypothesis implies that $N_1^{\perp}$ and $N_2^{\perp}$ are isometric, proving the theorem.   □

This theorem is often known as Witt's cancellation theorem because of the following reformulation:

(2.48) Corollary. *Let ϕ_1, ϕ_2, and ϕ_3 be forms on modules of finite rank over a field R, all three of which are either symmetric or Hermitian. If $\text{char}(R) = 2$, assume all three are Hermitian. If ϕ_1 is non-singular and $\phi_1 \perp \phi_2$ and $\phi_1 \perp \phi_3$ are isometric, then ϕ_2 and ϕ_3 are isometric.*

Proof.   □

(2.49) Remark. Witt's theorem is *false* in the case we have excluded. Note that

$$\begin{bmatrix} 1 & 0 & 0 \\ 0 & 0 & 1 \\ 0 & 1 & 0 \end{bmatrix} \quad \text{and} \quad \begin{bmatrix} 1 & 0 & 0 \\ 0 & 1 & 0 \\ 0 & 0 & 1 \end{bmatrix}$$

are isometric forms on $(\mathbf{F}_2)^3$, as they are both odd and non-singular, but

$$\begin{bmatrix} 0 & 1 \\ 1 & 0 \end{bmatrix} \quad \text{and} \quad \begin{bmatrix} 1 & 0 \\ 0 & 1 \end{bmatrix}$$

are not isometric on $(\mathbf{F}_2)^2$, as the first is even and the second is odd.

Now we come to the very important case of symmetric bilinear forms over $\mathbf{R}$ and Hermitian forms over $\mathbf{C}$.

(2.50) Definition. *If ϕ is a symmetric bilinear form over $\mathbf{R}$, or a Hermitian form over $\mathbf{C}$, on a module M, then ϕ is said to be **positive definite** if $\phi(v, v) > 0$ for every $v \neq 0 \in M$. If $\phi(v, v) < 0$ for every $v \neq 0 \in M$, then ϕ is said to be **negative definite**.*

(2.51) Theorem. (Sylvester's law of inertia) *Let ϕ be a non-degenerate symmetric bilinear form over $R = \mathbf{R}$, or Hermitian form over $R = \mathbf{C}$, on a module M of finite rank over R. Then ϕ is isometric to*

$$r[1] \perp s[-1]$$

with $r+s = n = \text{rank}(M)$. Furthermore, the integers r and s are well defined and ϕ is determined up to isometry by $\text{rank}(\phi) = n$, and $\text{signature}(\phi) = r - s$.

Proof. Except for the fact that r and s are well defined, this is all a direct corollary of Theorem 2.42. (Any two of n, r, and s determine the third, and we could use any two of these to classify ϕ. However, these determine and are determined by the rank and signature, which are the usual invariants that are used.) Thus, we need to show that r and s are well defined by ϕ.

To this end, let M_+ be a subspace of M of largest dimension with $\phi|_{M_+}$ positive definite. We claim that $r = \text{rank}(M_+)$. Let $\mathcal{B} = \{v_1, \ldots, v_n\}$ be a basis of M with

$$[\phi]_{\mathcal{B}} = I_r \oplus -I_s.$$

If $M_1 = \langle v_1, \ldots, v_r \rangle$, then $\phi|_{M_1}$ is positive definite. Thus, $\text{rank}(M_+) \geq r$. This argument also shows that if M_- is a subspace of M of largest possible dimension with $\phi|_{M_-}$ negative definite, then $\text{rank}(M_-) \geq s$.

We claim that $r = \text{rank}(M_+)$ and $s = \text{rank}(M_-)$. If not, then the above two inequalities imply

$$\text{rank}(M_+) + \text{rank}(M_-) > r + s = n,$$

so $M_+ \cap M_- \neq \{0\}$. Let $x \neq 0 \in M_+ \cap M_-$. Then $\phi(x, x) > 0$ since $x \in M_+$, while $\phi(x, x) < 0$ since $x \in M_-$, and this contradiction completes the proof.

We present an alternative proof as an application of Witt's theorem. Suppose

$$r_1[1] \perp s_1[-1] \quad \text{and} \quad r_2[1] \perp s_2[-1]$$

are isometric. We may assume $r_1 \leq r_2$. Then by Witt's theorem, $s_1[-1]$ and $(r_2 - r_1)[1] \perp s_2[-1]$ are isometric. As the first of these is negative-definite, so is the second, and so $r_1 = r_2$ (and hence $s_1 = s_2$). $\quad\square$

(2.52) Remark. If ϕ is not assumed to be non-degenerate then applying Theorem 2.51 to the non-degenerate part of ϕ (Definition 2.27) shows that a symmetric bilinear form over $R = \mathbf{R}$ or a Hermitian form over $R = \mathbf{C}$ is isometric to

$$r[1] \perp s[-1] \perp k[0]$$

with $r + s = \text{rank}(\phi) = \text{rank}(M) - k$, and with r and s well defined by ϕ. Again, we let $\text{signature}(\phi) = r - s$.

Of course, we have a procedure for determining the signature of ϕ, namely, diagonalize ϕ and inspect the result. In view of the importance of this case, we give an easier method.

(2.53) Proposition. *Let ϕ be a symmetric bilinear form over $R = \mathbf{R}$, or a Hermitian form over $R = \mathbf{C}$, on a module M of finite rank over R, and let $A = [\phi]_{\mathcal{B}}$ for some basis $\mathcal{B}$ of M. Then*

(1) *rank(ϕ) = rank(A);*
(2) *all of the eigenvalues of A are real; and*
(3) *signature$(\phi) = r - s$, where r (resp., s) is the number of positive (resp., negative) eigenvalues of A.*

Proof. It is tempting, but wrong, to try to prove this as follows: The form ϕ is diagonalizable, so just diagonalize it and inspect the diagonal entries. The mistake here is that to diagonalize the form ϕ we take $P^t A \overline{P}$, whereas to diagonalize A we take PAP^{-1}, and these will usually be quite different. For an arbitrary matrix P there is no reason to suppose that the diagonal entries of $P^t A \overline{P}$ are the eigenvalues of A, which are the diagonal entries of PAP^{-1}.

On the other hand, this false argument points the way to a correct argument: First note that we may write similarity as $(\overline{P})^{-1} A \overline{P}$. Thus if P is a matrix with $P^t = (\overline{P})^{-1}$, then the matrix $B = P^t A \overline{P}$ will have the same eigenvalues as A.

Let us regard A as the matrix of a linear transformation α on R^n where $R = \mathbf{R}$ or $R = \mathbf{C}$. Then A is either real symmetric or complex Hermitian. In other words, α is self-adjoint in the language of Definition 4.6.16. But then by Theorem 4.6.23, there is an orthonormal basis $\mathcal{B} = \{v_1, \ldots, v_n\}$ of R^n with $B = [\alpha]_{\mathcal{B}}$ diagonal, i.e., $P^{-1}AP = B$ where P is the matrix whose columns are $v_1, \ldots, v_n$, and furthermore $B \in M_n(\mathbf{R})$, where $n = \text{rank}(A)$. But then the condition that $\mathcal{B}$ is orthonormal is exactly the condition that $P^t = (\overline{P})^{-1}$, and we are done. $\square$

There is an even handier method for computing the signature. Note that it does not apply to all cases, but when it does apply it is easy to use.

(2.54) Proposition. *Let ϕ be a symmetric bilinear form over $R = \mathbf{R}$ or a Hermitian form over $R = \mathbf{C}$, on an R-module M of finite rank n. Let $\mathcal{B}$ be any basis for M and let $A = [\phi]_{\mathcal{B}}$. Set $\delta_0(A) = 1$, and for $1 \leq i \leq n$, let $\delta_i(A) = \det(A_i)$, where $A_i = A[(1, 2, \ldots, i) \mid (1, 2, \ldots, i)]$ is the ith principal submatrix of A, i.e., the $i \times i$ submatrix in the upper left-hand corner.*

(1) *If $\delta_i(A) \neq 0$ for all i, then*

(2.7)   *signature$(\phi) = |\{i : \delta_i(A) \text{ and } \delta_{i-1}(A) \text{ have the same sign}\}|$*
         $-|\{i : \delta_i(A) \text{ and } \delta_{i-1}(A) \text{ have opposite signs}\}|.$

(2) *ϕ is positive definite if and only if $\delta_i(A) > 0$ for all i.*
(3) *ϕ is negative definite if and only if $(-1)^i \delta_i(A) > 0$ for all i.*

Proof. We first prove part (1) by induction on n. The proposition is trivially true if $n = 1$. Assume it is true if $\text{rank}(M) < n$, and consider M with $\text{rank}(M) = n$. Write Equation (2.7) as

$$(2.8) \qquad\qquad \text{signature}(\phi) = r'(A) - s'(A).$$

Of course, $r'(A) + s'(A) = n$. Let $\mathcal{B} = \{v_1, \ldots, v_n\}$, let $N \subseteq M$ be the subspace with basis $\mathcal{B}' = \{v_1, \ldots, v_{n-1}\}$, and let $\psi = \phi|_N$. Set $B = [\psi]_{\mathcal{B}'}$. Then $B = A_{n-1}$, so $\det(B) \neq 0$ by hypothesis, and thus ψ is non-singular. Let $\mathcal{C}' = \{w_1, \ldots, w_{n-1}\}$ be a basis of N with $[\psi]_{\mathcal{C}'}$ diagonal, say,

$$[\psi]_{\mathcal{C}'} = \text{diag}(c_1, \ldots, c_{n-1}).$$

Then, by definition,

$$\text{signature}(\psi) = |\{i : c_i > 0\}| - |\{i : c_i < 0\}|$$
$$= r_1 - s_1$$

with $r_1 + s_1 = n - 1$. By induction we have

$$\text{signature}(\psi) = r'(B) - s'(B),$$

with $r'(B) + s'(B) = n - 1$, so $r'(B) = r_1$ and $s'(B) = s_1$. Now, since $\phi|_N = \psi$ is non-singular, we have $M = N \perp N^\perp$. Hence $\dim(N^\perp) = 1$; say $N^\perp = \langle w_n \rangle$. Then $\mathcal{C} = \{w_1, \ldots, w_n\}$ is a basis of M, and $[\phi]_{\mathcal{C}} = \text{diag}(c_1, \ldots, c_n)$ with $c_n = \phi(w_n, w_n)$. Also, by definition,

$$\text{signature}(\phi) = |\{i : c_i > 0\}| - |\{i : c_i < 0\}|$$
$$= r - s$$

with $r + s = n$. Note also that $\det(B)$ and $c_1 \cdots c_{n-1}$ have the same sign, as do $\det(A)$ and $c_1 \cdots c_n$, as they are determinants of matrices of the same form with respect to different bases. Now there are two possibilities:

(a) $c_n > 0$. In this case, $\text{signature}(\phi) = \text{signature}(\psi) + 1$, so $r = r_1 + 1$ and $s = s_1$. But also $\det(A_{n-1})$ and $\det(A)$ have the same sign, so $r'(A) = r'(B) + 1$ and $s'(A) = s'(B)$; hence, $r'(A) = r$ and $s'(A) = s$, so

$$\text{signature}(\phi) = r'(A) - s'(A).$$

(b) $c_n < 0$. Here the situation is reversed, with $r = r_1$, $s = s_1 + 1$, and also $r'(A) = r'(B)$, $s'(A) = s'(B) + 1$, again yielding

$$\text{signature}(\phi) = r'(A) - s'(A).$$

Thus, by induction, part (1) is proved.

(2) In light of (1), we need only prove that ϕ positive definite implies that all $\delta_i(A)$ are nonzero. But this is immediate as $\delta_i(A) = \det(\phi|_{N_i})$, $N_i = \langle v_1, \ldots, v_i \rangle$, and $\phi|_{N_i}$ is non-singular since $\phi(v, v) \neq 0$ for every $v \in M$ (so $\text{Ker}(\alpha_\psi) = \langle 0 \rangle$, where $\psi = \phi|_{N_i}$).

(3) Observe that ϕ is negative definite if and only if $-\phi$ is positive definite; thus (3) follows from (2).   □

(2.55) Example. Diagonalize the symmetric bilinear form over $\mathbf{R}$ with matrix

$$A = \begin{bmatrix} -2 & 3 & 1 & 0 \\ 3 & -6 & 0 & 1 \\ 1 & 0 & -3 & 1 \\ 0 & 1 & 1 & -1 \end{bmatrix}.$$

To do this, calculate that the determinants of the principal minors are

$$\delta_0(A) = 1, \quad \delta_1(A) = -2, \quad \delta_2(A) = 3, \quad \delta_3(A) = -3, \quad \delta_4(A) = -59$$

giving 3 sign changes, so this form diagonalizes to

$$\begin{bmatrix} -1 & 0 & 0 & 0 \\ 0 & -1 & 0 & 0 \\ 0 & 0 & -1 & 0 \\ 0 & 0 & 0 & 1 \end{bmatrix}.$$

Although we do not use it here, we wish to remark on a standard notion. In the following definition ϕ may be arbitrary (i.e., ϕ need not be ε-symmetric):

(2.56) Definition. *Let ϕ be a bilinear form on M such that $\alpha_\phi : M \to M^*$ is an isomorphism. Let $f \in \mathrm{End}_R(M)$. Then $f^T : M \to M$, the **adjoint of** f **with respect to** ϕ, is defined by*

$$(2.9) \qquad \phi(f(x), y) = \phi(x, f^T(y)) \quad \text{for all } x, y \in M.$$

To see that this definition makes sense, note that it is equivalent to the equation

$$\left(\alpha_\phi(f^T(y))\right)(x) = (\alpha_\phi(y))(f(x)) = (f^*(\alpha_\phi(y)))(x),$$

or in other words,

$$\alpha_\phi \circ f^T = f^* \circ \alpha_\phi,$$

i.e., $f^T = \alpha_\phi^{-1} \circ f^* \circ \alpha_\phi$ where f^* is the adjoint of f as given in Definition 1.20.

Note that f^T and f^* are quite distinct (although they both have the name "adjoint"). First, $f^* : M^* \to M^*$, while $f^T : M \to M$. Second, f^* is always defined, while f^T is only defined once we have a form ϕ as above, and it depends on ϕ. On the other hand, while distinct, they are certainly closely related.

Now when it comes to finding a matrix representative for f^T, there is a subtlety we wish to caution the reader about. There are two natural choices. Choose a basis $\mathcal{B}$ for M. Then we have the dual basis $\mathcal{B}^*$ of M^*. Recall that $[f^*]_{\mathcal{B}^*} = ([f]_{\mathcal{B}})^t$ (Proposition 1.24), so

$$[f^T]_{\mathcal{B}} = [\alpha_\phi^{-1}]_{\mathcal{B}}^{\mathcal{B}^*} ([f]_{\mathcal{B}})^t [\alpha_\phi]_{\mathcal{B}^*}^{\mathcal{B}}.$$

On the other hand, if $\mathcal{B} = \{v_i\}$, then we have a basis $\mathcal{C}^*$ of M^* given by $\mathcal{C}^* = \{\alpha_\phi(v_i)\}$. Then there is also a basis $\mathcal{C}$ dual to $\mathcal{C}^*$ (using the canonical isomorphism between M and M^{**}). By definition, $[\alpha]_{\mathcal{C}^*}^{\mathcal{B}} = I_n$, the identity matrix, where $n = \operatorname{rank}(M)$, so we have more simply

$$[f^T]_{\mathcal{B}} = [f^*]_{\mathcal{C}^*} = ([f]_{\mathcal{C}})^t.$$

The point we wish to make is that in general the bases $\mathcal{B}^*$ and $\mathcal{C}^*$ (or equivalently, $\mathcal{B}$ and $\mathcal{C}$) are distinct, so care must be taken. There is, however, one happy (and important) case where $\mathcal{B}^* = \mathcal{C}^*$ and $\mathcal{B} = \mathcal{C}$. As the reader may check, this is true if and only if the basis $\mathcal{B}$ is orthonormal with respect to ϕ, i.e., if $\phi(v_i, v_j) = \delta_{ij}$. In this case $[\phi]_{\mathcal{B}} = I_n$, so we see that ϕ has an orthonormal basis if and only if ϕ is isometric to $n[1]$, and we have seen situations when this is and is not possible.

6.3 Quadratic Forms

This section will be devoted to some aspects of the theory of quadratic forms. As in the previous section, R will denote a commutative ring with 1 and all modules are assumed to be free.

(3.1) Definition. *Let M be a free R-module. A **quadratic form** Φ on M is a function $\Phi : M \to R$ satisfying*

(1) $\Phi(rx) = r^2\Phi(x)$ *for any $r \in R$, $x \in M$; and*
(2) *the function $\phi : M \times M \to R$ defined by*

$$\phi(x, y) = \Phi(x + y) - \Phi(x) - \Phi(y)$$

is a (necessarily symmetric) bilinear form on M.

In this situation we will say that Φ and ϕ are *associated*. The quadratic form Φ is called **non-singular** or **non-degenerate** if the symmetric bilinear form ϕ is.

A basic method of obtaining quadratic forms is as follows: Let ψ be any (not necessarily symmetric) bilinear form on M and define the symmetric bilinear form ϕ on M by

$$\phi(x, y) = \psi(x, y) + \psi(y, x).$$

(Of course, if ψ is symmetric then $\phi(x, y) = 2\psi(x, y)$.) The function

$$\Phi : M \to R$$

defined by

$$\Phi(x) = \psi(x, x)$$

is a quadratic form on M with associated bilinear form ϕ.

(3.2) Theorem. *Let M be a free module over the ring R.*

(1) *Let Φ be a quadratic form on M. Then the associated bilinear form ϕ is uniquely determined by Φ.*

(2) *Let ϕ be a symmetric bilinear form on M. Then ϕ is associated to a quadratic form Φ if and only if ϕ is even (see Definition 2.40).*

(3) *If 2 is not a zero divisor in R and ϕ is an even symmetric bilinear form, then the associated quadratic form Φ is uniquely determined by ϕ.*

Proof. Part (1) is obvious from Definition 3.1 (2) and is merely stated for emphasis. To prove (2), suppose that ϕ is associated to Φ. Then for every $x \in M$,

$$4\Phi(x) = \Phi(2x) = \Phi(x + x) = 2\Phi(x) + \phi(x, x).$$

Thus,

(3.1) $$\phi(x, x) = 2\Phi(x)$$

and ϕ is even.

Conversely, suppose that ϕ is even. Let $\mathcal{B} = \{v_i\}_{i \in I}$ be a basis for M and choose an ordering of the index set I. (Of course, if $\operatorname{rank}(M) = n$ is finite, then $I = \{1, \dots, n\}$ certainly has an order, but it is a consequence of Zorn's lemma that every set has an order.) To define $\psi(x, y)$ it suffices to define $\psi(v_i, v_j)$ for $i, j \in I$. Define $\psi(v_i, v_j)$ as follows: $\psi(v_i, v_j) = \phi(v_i, v_j)$ if $i < j$, $\psi(v_i, v_j) = 0$ if $j < i$, and $\psi(v_i, v_i)$ is any solution of the equation $2\psi(v_i, v_i) = \phi(v_i, v_i)$. Since ϕ is assumed to be even, this equation has a solution. Then $\Phi(x) = \psi(x, x)$ is a quadratic form with associated bilinear form ϕ.

(3) This is a direct consequence of Equation (3.1). $\qquad\square$

(3.3) Lemma. (1) *Let Φ_i be a quadratic form on a module M_i over R with associated bilinear form ϕ_i, for $i = 1, 2$. Then $\Phi : M_1 \oplus M_2 \to R$, defined by*

$$\Phi(x_1, x_2) = \Phi_1(x_1) + \Phi_2(x_2),$$

is a quadratic form on $M_1 \oplus M_2$ with associated bilinear form $\phi_1 \perp \phi_2$. (In this situation we write $\Phi = \Phi_1 \perp \Phi_2$.)

(2) *Let ϕ_i be an even symmetric bilinear form on M_i, $i = 1, 2$, and let Φ be a quadratic form associated to $\phi_1 \perp \phi_2$. Then $\Phi = \Phi_1 \perp \Phi_2$ where Φ_i is a quadratic form associated to ϕ_i, $i = 1, 2$. Also, Φ_1 and Φ_2 are unique.*

Proof. (1) First we check that

$$\begin{aligned}
\Phi(r(x_1, x_2)) &= \Phi(rx_1, rx_2) \\
&= \Phi_1(rx_1) + \Phi_2(rx_2) \\
&= r^2\Phi_1(x_1) + r^2\Phi_2(x_2) \\
&= r^2(\Phi_1(x_1) + \Phi_2(x_2)) \\
&= r^2\Phi(x_1, x_2).
\end{aligned}$$

Next we need to check that

(3.2) $\quad \Phi(x_1 + y_1, x_2 + y_2) = \Phi(x_1, x_2) + \Phi(y_1, y_2) + \phi((x_1, x_2), (y_1, y_2))$

where $\phi = \phi_1 \perp \phi_2$. Since $\phi = \phi_1 \perp \phi_2$, we have

$$\begin{aligned}
\phi((x_1, x_2), (y_1, y_2)) &= \phi((x_1, 0), (y_1, 0)) + \phi((x_1, 0), (0, y_2)) \\
&\quad + \phi((0, x_2), (y_1, 0)) + \phi((0, x_2), (0, y_2)) \\
\text{(3.3)} \qquad &= \phi_1(x_1, y_1) + \phi_2(x_2, y_2).
\end{aligned}$$

On the other hand,

$$\begin{aligned}
\Phi(x_1 + y_1, x_2 + y_2) &= \Phi_1(x_1 + y_1) + \Phi_2(x_2 + y_2) \\
&= \Phi_1(x_1) + \Phi_1(y_1) + \phi_1(x_1, y_1) \\
&\quad + \Phi_2(x_2) + \Phi_2(y_2) + \phi_2(x_2, y_2) \\
&= \Phi(x_1, x_2) + \Phi(y_1, y_2) + \phi_1(x_1, y_1) + \phi_2(x_2, y_2),
\end{aligned}$$

which together with Equation (3.3) gives Equation (3.2), as desired.

(2) Set $\Phi_1 = \Phi|_{M_1}$ and $\Phi_2 = \Phi|_{M_2}$. Uniqueness is trivial, as Φ certainly determines Φ_1 and Φ_2. $\square$

We call Φ the **orthogonal direct sum** of Φ_1 and Φ_2. Thus Lemma 3.3 tells us that the procedures of forming orthogonal direct sums of associated bilinear and quadratic forms are compatible.

It follows from Theorem 3.2 that if 2 is not a zero divisor in R, the classification problem for quadratic forms over R reduces to that for even symmetric bilinear forms. (Recall that if 2 is a unit of R, then *every* symmetric bilinear form is even.) Thus we have already dealt with a number of important cases—$R = \mathbf{R}$, $R = \mathbf{C}$, or R a finite field of odd characteristic. We will now study the case of quadratic forms over the field $\mathbf{F}_2$ of 2 elements. In this situation it is common to call a quadratic form Φ associated to the even symmetric bilinear form ϕ a **quadratic refinement** of ϕ, and we shall use this terminology. Note that in this case condition (1) of Definition 3.1 simply reduces to the condition that $\Phi(0) = 0$, and this is implied by condition (2)—set $y = 0$. Thus, we may neglect condition (1).

(3.4) Proposition. *Let ϕ be an even symmetric bilinear form on a module M over the field $R = \mathbf{F}_2$.*

(1) *If Φ_1 and Φ_2 are two quadratic refinements of ϕ, then*

$$f(x) = \Phi_1(x) + \Phi_2(x)$$

is an R-linear function $f : M \to R$, i.e., $f \in M^*$.

(2) *If Φ_1 is any quadratic refinement of ϕ and $f \in M^*$ is arbitrary, then $\Phi_2 = \Phi_1 + f$ is also a quadratic refinement of ϕ.*

Proof. These are routine computations, which are left to the reader.   □

Suppose now that M is a module of finite rank n over $R = \mathbf{F}_2$. Then any even symmetric bilinear form ϕ on M has $|M^*| = |M| = 2^n$ quadratic refinements. We have already classified these forms in Corollary 2.46. If $\operatorname{rank}(\phi) = 2m$ (necessarily even), then ϕ is isometric to

$$(n - 2m)[0] \perp m \begin{bmatrix} 0 & 1 \\ 1 & 0 \end{bmatrix}.$$

We will now see how to classify quadratic forms on M. By Lemma 3.3, we may handle the cases ϕ identically zero and ϕ non-singular separately. (Of course, we are interested in classifying ϕ up to isometry. An isometry has the analogous definition for quadratic forms as for bilinear forms: Two quadratic forms Φ_i on R-modules M_i, $i = 1, 2$, are **isometric** if there is an R-isomorphism $f : M_1 \to M_2$ with $\Phi_2(f(x)) = \Phi_1(x)$ for every $x \in M_1$.)

First we will deal with the case where ϕ is identically zero.

(3.5) Proposition. *Let M be a free $\mathbf{F}_2$-module of rank n and let ϕ be the identically zero bilinear form on M. Then a quadratic refinement Φ of ϕ is simply an element $\Phi \in M^*$. There are two isometry classes of these. One, containing one element, consists of $0 \in M^*$. The other, containing $2^n - 1$ elements, consists of all $\Phi \in M^* \setminus \{0\}$.*

Proof. In this case, Definition 3.1 (2) says

$$\Phi(x + y) = \Phi(x) + \Phi(y),$$

i.e., $\Phi : M \to \mathbf{F}_2$ is a homomorphism, so $\Phi \in M^*$. Clearly, the zero homomorphism is not isometric to any nonzero homomorphism. On the other hand, a nonzero homomorphism Φ is uniquely determined by $\Phi^{-1}(0)$, a subspace of M of rank $n - 1$, and there is an automorphism of M taking any one of these to any other.   □

Of course, the case we are really interested in is the classification of quadratic refinements of a non-singular form. We prepare for this classification with the following elementary lemma.

(3.6) Lemma. *Define two function $e : \mathbf{N} \to \mathbf{N}$ and $o : \mathbf{N} \to \mathbf{N}$ by means of the following recursions:*

$$e(1) = 3, \qquad\qquad\qquad o(1) = 1$$

$$e(m) = 3e(m-1) + o(m-1), \qquad o(m) = 3o(m-1) + e(m-1)$$

for $m > 1$. Then for every $m \in \mathbf{N}$,

$$e(m) = 2^{m-1}(2^m + 1) \quad and \quad o(m) = 2^{m-1}(2^m - 1).$$

Proof. The proof is a routine induction. $\qquad\qquad\qquad\qquad\qquad\qquad\square$

(3.7) **Proposition.** Let Φ be a quadratic refinement of a non-singular even symmetric bilinear form on a module M of (necessarily) even rank $n = 2m$ over $\mathbf{F}_2$. Then either

(a) $|\Phi^{-1}(0)| = e(m)$ and $|\Phi^{-1}(1)| = o(m)$, or
(b) $|\Phi^{-1}(0)| = o(m)$ and $|\Phi^{-1}(1)| = e(m)$.

Furthermore, among the 2^{2m} such quadratic forms, there are $e(m)$ forms Φ with $|\Phi^{-1}(0)| = e(m)$ and $o(m)$ forms Φ with $|\Phi^{-1}(0)| = o(m)$.

Proof. The proof is by induction on m. Let $m = 1$. Then M has a basis $\mathcal{B} = \{x, y\}$ in which $[\phi]_{\mathcal{B}} = \left[\begin{smallmatrix} 0 & 1 \\ 1 & 0 \end{smallmatrix}\right]$, where ϕ is the bilinear form associated to Φ. Then $M = \{0, x, y, x + y\}$. Of course, $\Phi(0) = 0$, and then from Definition 3.1 (2),

$$1 = \phi(x, y) = \Phi(x + y) + \Phi(x) + \Phi(y)$$

so we see that $\Phi(z) = 1$ for z either exactly one of x, y, and $x + y$ (3 possibilities) or for all three of them (1 possibility). Since, in this case, the above equation is the only condition on Φ, all possibilities indeed occur. This is precisely the statement of the proposition in case $m = 1$.

Now suppose the proposition is true for $\mathrm{rank}(M) < 2m$, and let $\mathrm{rank}(M) = 2m$. Choose an element $x \neq 0$ of M, let $y \in M$ with $\phi(x, y) = 1$ (which exists since ϕ is non-singular), let M_1 be the subspace of M with basis $\mathcal{B} = \{x, y\}$, and let $M_2 = M_1^{\perp}$. Then $M = M_1 \perp M_2$, $\phi = \phi_1 \perp \phi_2$, and $\Phi = \Phi_1 \perp \Phi_2$, where $\phi_i = \phi|_{M_i}$ and $\Phi_i = \Phi|_{M_i}$. Now

$$0 = \Phi(x_1, x_2) = \Phi(x_1) + \Phi(x_2)$$

implies that either $\Phi(x_1) = \Phi(x_2) = 0$ or $\Phi(x_1) = \Phi(x_2) = 1$. If Φ_1 and Φ_2 are both as in part (a) (resp., both as in part (b)), the first (resp., second) case arises for $e(1)e(m-1) = 3e(m-1)$ values of (x_1, x_2), and the other for $o(1)o(m-1) = o(m-1)$ cases. Thus,

$$|\Phi^{-1}(0)| = 3e(m-1) + o(m-1) = e(m)$$

by Lemma 3.6. If Φ_1 is as in (a) and Φ_2 as in (b) (resp., vice-versa), then the first (resp., second) case arises for $e(1)o(m-1) = 3o(m-1)$ values of (x_1, x_2), and the other for $o(1)e(m-1)$ cases. Thus, Lemma 3.6 again gives

$$|\Phi^{-1}(0)| = 3o(m-1) + e(m-1) = o(m).$$

This proves the first claim of the proposition, but also the second, as we see that Φ is in case (a) if either Φ_1 and Φ_2 are both in (a) or both in (b), and in case (b) if one of them is in (a) and the other in (b), and this gives exactly the same recursion. (We need not check $o(m)$ separately since $e(m) + o(m) = 2^{2m} = |M|$.) $\square$

Given this proposition we may define a "democratic" invariant of Φ: We let each $x \in M$ "vote" for $\Phi(x)$ and go along with the majority. This is formalized in the following definition.

(3.8) Definition. *Let Φ be a quadratic refinement of a non-singular even symmetric bilinear form on a module M of rank $2m$ over $\mathbf{F}_2$. The Arf invariant $\mathrm{Arf}(\Phi) \in \mathbf{F}_2$ is defined by*

$$\mathrm{Arf}(\Phi) = 0 \quad if \; |\Phi^{-1}(0)| = e(m) \; and \; |\Phi^{-1}(1)| = o(m)$$

and

$$\mathrm{Arf}(\Phi) = 1 \quad if \; |\Phi^{-1}(0)| = o(m) \; and \; |\Phi^{-1}(1)| = e(m).$$

*The form Φ is called **even** or **odd** according as $\mathrm{Arf}(\Phi) = 0$ or 1.*

Note that the proof of Proposition 3.7 yields the following:

(3.9) Corollary. $\mathrm{Arf}(\Phi_1 \perp \Phi_2) = \mathrm{Arf}(\Phi_1) + \mathrm{Arf}(\Phi_2)$.

Proof. $\square$

(3.10) Theorem. *Let Φ be a quadratic refinement of a non-singular even symmetric bilinear form on a module M of rank $2m$ over $\mathbf{F}_2$. Then $\{\Phi\}$ falls into two isometry classes. One, containing $e(m)$ elements, consists of all Φ with $\mathrm{Arf}(\Phi) = 0$, and the other, containing $o(m)$ elements, consists of all Φ with $\mathrm{Arf}(\Phi) = 1$.*

Proof. $|\Phi^{-1}(0)|$ is certainly an invariant of the isometry class of Φ, so two forms with unequal Arf invariants cannot be isometric. It remains to prove that all forms Φ with $\mathrm{Arf}(\Phi) = 0$ are isometric, as are all forms Φ with $\mathrm{Arf}(\Phi) = 1$.

First let $m = 1$, and let ϕ be the bilinear form associated to Φ. Since ϕ is non-singular, there are $x, y \in M$ with $\phi(x, y) = 1$.
Then

$$\phi(x, x+y) = \phi(x, x) + \phi(x, y) = 1,$$

and similarly, $\phi(y, x+y) = 1$. Thus, ϕ is completely symmetric in x, y, and $z = x + y$. In the proof of Proposition 3.7, we saw that there are four possibilities for Φ:

$$\begin{array}{lll}
\Phi(x) = 1 & \Phi(y) = 0 & \Phi(z) = 0 \\
\Phi(x) = 0 & \Phi(y) = 1 & \Phi(z) = 0 \\
\Phi(x) = 0 & \Phi(y) = 0 & \Phi(z) = 1 \\
\Phi(x) = 1 & \Phi(y) = 1 & \Phi(z) = 1.
\end{array}$$

The first three have $\mathrm{Arf}(\Phi) = 0$; the last one has $\mathrm{Arf}(\Phi) = 1$. But then it is easy to check that $\mathrm{Aut}_{\mathbf{F}_2}(M) \cong S_3$, the permutation group on 3 elements, acting by permuting x, y, and z. We observed that ϕ is completely symmetric in x, y, and z, so S_3 leaves ϕ invariant and permutes the first three possibilites for Φ transitively; hence, they are all equivalent. (As there is only one Φ with $\mathrm{Arf}(\Phi) = 1$, it certainly forms an equivalence class by itself.)

Now let $m > 1$. We have that Φ is isometric to $\Phi_1 \perp \cdots \perp \Phi_m$ (by Lemma 3.3) and

$$\mathrm{Arf}(\Phi) = \mathrm{Arf}(\Phi_1) + \cdots + \mathrm{Arf}(\Phi_m)$$

(by Corollary 3.9), so $\mathrm{Arf}(\Phi) = 0$ or 1 accordingly as there are an even number or an odd number of the forms Φ_i with $\mathrm{Arf}(\Phi_i) = 1$. Each Φ_i has rank 2, and we have just seen that all rank 2 forms Φ_i with $\mathrm{Arf}(\Phi_i) = 0$ are isometric. Thus to complete the proof we need only show that if Ψ is the unique rank 2 form with $\mathrm{Arf}(\Psi) = 1$, then $\Psi \perp \Psi$ is isometric to $\Phi'_1 \perp \Phi'_2$ with $\mathrm{Arf}(\Phi'_i) = 0$ for $i = 1, 2$.

Let ϕ be the bilinear form associated to $\Phi = \Psi \perp \Psi$ and let M have a basis $\mathcal{B} = \{x_1, y_1, x_2, y_2\}$ in which

$$[\phi]_{\mathcal{B}} = \begin{bmatrix} 0 & 1 & 0 & 0 \\ 1 & 0 & 0 & 0 \\ 0 & 0 & 0 & 1 \\ 0 & 0 & 1 & 0 \end{bmatrix}.$$

(Such a basis is called **symplectic**.) Let $f : M \to M$ be defined by

$$f(x_i) = x'_i, \qquad f(y_i) = y'_i,$$

where

$$x'_1 = x_1 + x_2, \quad y'_1 = y_1, \quad x'_2 = x_2, \quad y'_2 = y_1 + y_2.$$

It is easy to check that f is an isometry of ϕ, i.e., that $\phi(f(u), f(v)) = \phi(u, v)$ for all $u, v \in M$. (It suffices to check this for basis elements.) This implies that f is invertible, but in any case, direct computation shows that $f^2 = 1_M$. Also, it is easy to check that

$$\mathcal{B}' = \{x'_1, y'_1, x'_2, y'_2\}$$

is a symplectic basis as well.

Then if M'_i is the subspace of M spanned by $\{x'_i, y'_i\}$, $i = 1, 2$, then $\phi = \phi'_1 \perp \phi'_2$ with ϕ'_i non-singular for $i = 1, 2$, and hence, $\Phi = \Phi'_1 \perp \Phi'_2$, where $\Phi'_i = \Phi|_{M'_i}$. But now

$$\Phi_1'(x_1') = \Phi(x_1') = \Phi(x_1 + x_2)$$
$$= \Phi(x_1) + \Phi(x_2) + \phi(x_1, x_2)$$
$$= 1 + 1 + 0 = 0$$
$$\Phi_1'(y_1') = \Phi(y_1') = \Phi(y_1) = 1$$
$$\Phi_1'(x_1' + y_1') = \Phi(x_1' + y_1')$$
$$= \Phi(x_1') + \Phi(y_1') + \phi(x_1', y_1')$$
$$= 0 + 1 + 1 = 0,$$

so Φ_1' is indeed a form with $\text{Arf}(\Phi_1') = 0$. Similarly,

$$\Phi_2'(x_2') = 1, \quad \Phi_2'(y_2') = 0, \quad \Phi_2'(x_2' + y_2') = 0$$

so that $\text{Arf}(\Phi_2') = 0$, and we are done.   □

We have been careful to give an intrinsic definition of $\text{Arf}(\Phi)$—one that does not depend on any choices. However, there is an alternate extrinsic definition, which is useful for calculations.

(3.11) Proposition. *Let ϕ be a non-singular even symmetric bilinear form on M, a module of rank $2m$ over $\mathbf{F}_2$, and let*

$$\mathcal{B} = \{x_1, y_1, \ldots, x_m, y_m\}$$

be a symplectic basis for M, i.e., a basis in which

$$\phi(x_i, x_j) = \phi(y_i, y_j) = 0, \quad \phi(x_i, y_j) = \delta_{ij}.$$

(1) *If Φ is a quadratic refinement of ϕ, then*

$$\text{Arf}(\Phi) = \sum_{i=1}^{n} \Phi(x_i)\Phi(y_i).$$

(2) *Let $c = (a_1, b_1, \ldots, a_m, b_m) \in (\mathbf{F}_2)^{2m}$ be arbitrary. If $v \in M$, then $v = \sum_{i=1}^{m}(r_i x_i + s_i y_i)$, for r_i, $s_i \in \mathbf{F}_2$, and we define*

$$\Phi_c(v) = \sum_{i=1}^{m} r_i s_i + \sum_{i=1}^{m} a_i r_i + \sum_{i=1}^{m} b_i s_i.$$

The function $\Phi_c : M \to \mathbf{F}_2$ is a quadratic refinement of ϕ, and

$$\text{Arf}(\Phi) = \sum_{i=1}^{m} a_i b_i.$$

Proof. Exercise.   □

(3.12) *Remark.* The reader should not get the impression that there is a canonical Φ_c obtained by taking $c = (0, \ldots, 0)$ of which the others are modifications; the formula for Φ_c depends on the choice of symplectic basis, and this is certainly not canonical.

We will now give a brief discussion of the concept of isometry groups. The definitions will be given in complete generality, and then we will specialize to the case of fields of characteristic $\neq 2$ and prove a theorem of Cartan and Dieudonné concerning the generation of isometry groups by reflections. The theory of reflections will also allow us to give a second (much easier) proof of Witt's theorem for quadratic forms over fields of characteristic $\neq 2$.

(3.13) Definition.

(1) *Let ϕ be an arbitrary b/s-linear form on a free module M over a ring R. Then the* **isometry group** $\mathrm{Isom}(\phi)$ *is defined by* $\mathrm{Isom}(\phi) =$

$$\{f \in \mathrm{Aut}_R(M) : \phi(f(x), f(y)) = \phi(x, y) \quad \text{for all } x, y \in M\}.$$

(2) *Let Φ be an arbitrary quadratic form on a free module M over a ring R. Then the* **isometry group** $\mathrm{Isom}(\Phi)$ *is defined by*

$$\mathrm{Isom}(\Phi) = \{f \in \mathrm{Aut}_R(M) : \Phi(f(x)) = \Phi(x) \quad \text{for all } x \in M\}.$$

Consider the situation where Φ and ϕ are associated. In many cases Φ and ϕ determine each other, so $\mathrm{Isom}(\Phi) = \mathrm{Isom}(\phi)$. (In particular, this happens if R is a field of characteristic not equal to 2.) In the other cases, Φ determines ϕ, so $\mathrm{Isom}(\Phi) \subseteq \mathrm{Isom}(\phi)$.

There is a simple matrix criterion for an R-module isomorphism of M to be an isometry:

(3.14) Proposition. *Let ϕ be an arbitrary b/s-linear form on a free module M of finite rank over a ring R, and let $\mathcal{B}$ be a basis of M. If $f \in \mathrm{Aut}_R(M)$, then $f \in \mathrm{Isom}(\phi)$ if and only if*

$$(3.4) \qquad [f]_{\mathcal{B}}^t [\phi]_{\mathcal{B}} [\overline{f}]_{\mathcal{B}} = [\phi]_{\mathcal{B}}.$$

Proof. If $x, y \in M$, then we have

$$\phi(f(x), f(y)) = [f(x)]_{\mathcal{B}}^t [\phi]_{\mathcal{B}} [\overline{f(y)}]_{\mathcal{B}}$$
$$= ([f]_{\mathcal{B}} [x]_{\mathcal{B}})^t [\phi]_{\mathcal{B}} ([\overline{f}]_{\mathcal{B}} [\overline{y}]_{\mathcal{B}})$$
$$(3.5) \qquad\qquad = [x]_{\mathcal{B}}^t ([f]_{\mathcal{B}}^t [\phi]_{\mathcal{B}} [\overline{f}]_{\mathcal{B}}) [\overline{y}]_{\mathcal{B}}.$$

But

(3.6) $$\phi(x, y) = [x]_B^t [\phi]_B [\bar{y}]_B$$

for all $x, y \in M$. Comparing Equation (3.5) and Equation (3.6) gives the result.   □

(3.15) Corollary. *Let ϕ be a non-degenerate b/s-linear form on a free module M over a ring R, and let $f \in \mathrm{End}_R(M)$ be such that*

$$\phi(f(x), f(y)) = \phi(x, y) \qquad \text{for all} \quad x, y \in M.$$

Then f is an injection. Furthermore, if M has finite rank, then f is an isomorphism (and hence $f \in \mathrm{Isom}(\phi)$).

Proof. Let $0 \neq y \in M$. Since ϕ is non-degenerate, there is an $x \in M$ with $\phi(x, y) \neq 0$. Then $\phi(f(x), f(y)) \neq 0$, so $f(y) \neq 0$.

In case M has finite rank, the proof of Proposition 3.14 applies to show that Equation (3.4) holds. Since, by Proposition 2.21, $\det(\phi)$ is not a zero divisor, $\det(f)$ is a unit, and so f is an isomorphism.   □

(3.16) Remarks.

(1) If M has infinite rank then f need not be an isomorphism in the situation of Corollary 3.15. For an example, let $M = \mathbf{Q}^\infty = \oplus_{i=1}^\infty \mathbf{Q}$ with ϕ the bilinear form on M given by

$$\phi((x_1, x_2, \ldots), (y_1, y_2, \ldots)) = \sum_{i=1}^\infty x_i y_i.$$

Then $f : M \to M$, defined by

$$f((x_1, x_2, \ldots)) = (0, x_1, x_2, \ldots),$$

satisfies the hypothesis of Corollary 3.15, but it is not an isomorphism.

(2) Some authors use the term "the orthogonal group of ϕ" for what we are calling the isometry group of ϕ, while other call the isometry group of ϕ "the orthogonal/unitary/symplectic group of ϕ" when ϕ is symmetric bilinear/Hermitian/skew-symmetric, so beware!

(3.17) Example. Let $M = \mathbf{R}^2$, and let $\phi(x, y) = (x : y)$ be the standard inner product on M. Thus, if we use the standard basis on $\mathbf{R}^2$, then $[\phi] = I_2$, so $f \in \mathrm{End}_\mathbf{R}(\mathbf{R}^2)$ is an isometry if and only if

$$[f]^t [f] = I_2.$$

Geometrically, this means that an isometry of $\mathbf{R}^2$ is determined by a pair of orthonormal vectors of $\mathbf{R}^2$, namely, the first and second columns of $[f]$. Hence, the isometries of $\mathbf{R}^2$ (with respect to the standard inner product) are one of the two types:

$$(3.7) \qquad \rho_\theta = \begin{bmatrix} \cos\theta & -\sin\theta \\ \sin\theta & \cos\theta \end{bmatrix}$$

$$(3.8) \qquad r_{\theta/2} = \begin{bmatrix} \cos\theta & \sin\theta \\ \sin\theta & -\cos\theta \end{bmatrix}.$$

The isometry ρ_θ is the counterclockwise rotation through an angle of θ, while

$$r_{\theta/2} = \rho_\theta \circ r_0$$

is the orthogonal reflection of $\mathbf{R}^2$ through a line through the origin making an angle of $\theta/2$ with the x-axis. (Check that this geometric description of $r_{\theta/2}$ is valid.) In particular, Equations (3.7) and (3.8) show that an isometry of $\mathbf{R}^2$ is a reflection if and only if its determinant is -1, while any isometry is a product of at most 2 reflections. The theorem of Cartan–Dieudonné is a far reaching generalization of this simple example.

(3.18) Definition. *Let R be a field with $\mathrm{char}(R) \neq 2$ and let Φ be a quadratic form on an R-module M. Let M_1 be a finite rank submodule of M with $\Phi|_{M_1}$ non-singular. Set $M_2 = M_1^\perp$, so $M = M_1 \perp M_2$. The **reflection** of M determined by M_1 is the element $f_{M_1} \in \mathrm{Aut}_R(M)$ defined as follows:*

Let $x \in M$ and write x uniquely as $x = x_1 + x_2$, where $x_1 \in M_1$, $x_2 \in M_2$. Then

$$f_{M_1}(x) = -x_1 + x_2.$$

*If $M_1 = \langle y \rangle$ with $\Phi(y) \neq 0$, then we write $f_{M_1} = f_y$ and call f_y the **hyperplane reflection** determined by y.*

(3.19) Lemma. *Let M, Φ, M_1, and y be as in Definition 3.18, and let ϕ be the symmetric bilinear form associated to Φ.*

(1) *$f_{M_1} \in \mathrm{Isom}(\Phi)$, and $(f_{M_1})^2 = 1_M$.*
(2) *If $g \in \mathrm{Isom}(\Phi)$, then $g f_{M_1} g^{-1} = f_{g(M_1)}$.*
(3) *f_y is given by the formula*

$$f_y(x) = x - \frac{\phi(x,y)}{\Phi(y)} y \qquad \text{for all} \quad x \in M.$$

Proof. Exercise.   □

Since a hyperplane reflection f_w is an isometry, it is certainly true that $\Phi(x) = \Phi(f_w(x))$. The following lemma gives a simple criterion for the existence of a hyperplane reflection that interchanges two given points.

(3.20) Lemma. *Let Φ be a quadratic form on M over a field R with $\mathrm{char}(R) \neq 2$. Let x, y, and $w = x - y \in M$ satisfy*

$$\Phi(x) = \Phi(y) \neq 0 \quad \text{and} \quad \Phi(w) \neq 0.$$

Then $f_w(x) = y$.

Proof. Exercise.   □

As an easy application of reflections, we will present another proof of Witt's theorem (Theorem 2.47). The difficult part of the proof of Theorem 2.47 was the $n = 1$ step in the induction. When reflections are available, this step is very easy.

(3.21) Theorem. (Witt) *Let Φ be a quadratic form on a module M of finite rank over a field R of characteristic $\neq 2$. Let N_1 and N_2 be submodules of M with $\Phi|_{N_1}$ and $\Phi|_{N_2}$ non-singular and isometric to each other. Then $\Phi|_{N_1^\perp}$ and $\Phi|_{N_2^\perp}$ are isometric.*

Proof. As in the case of Theorem 2.47, the proof is by induction on $n = \mathrm{rank}(N_1) = \mathrm{rank}(N_2)$. Thus, suppose that $n = 1$ and let v_1 generate N_1 and v_2 generate N_2. Since N_1 and N_2 are non-singular and isometric, we may assume that $\Phi(v_1) = \Phi(v_2) \neq 0$. Let ϕ be the symmetric bilinear form associated to Φ. Then $\Phi(v_1 + v_2) = 2\Phi(v_1) + \phi(v_1, v_2)$ and $\Phi(v_1 - v_2) = 2\Phi(v_1) - \phi(v_1, v_2)$. If both $\Phi(v_1 + v_2)$ and $\Phi(v_1 - v_2)$ are zero, then it follows that $\Phi(v_1) = 0$. Thus, either $\Phi(v_1 + v_2) \neq 0$ or $\Phi(v_1 - v_2) \neq 0$. Since N_2 is also generated by $-v_2$, we may thus assume that $\Phi(v_1 - v_2) \neq 0$. If $w = v_1 - v_2$ then Lemma 3.20 shows that the reflection f_w takes v_1 to v_2, and hence it takes $N_1^\perp$ to $N_2^\perp$, and the theorem is proved in case $n = 1$.

The inductive step is identical with that presented in the proof of Theorem 2.47 and, hence, will not be repeated.   □

The next two lemmas are technical results needed in the proof of the Cartan–Dieudonné theorem.

(3.22) Lemma. *Let ϕ be a non-singular symmetric bilinear form on the finite rank R-module M, where R is a field of characteristic different from 2. If $f \in \mathrm{Isom}(\phi)$, then*

$$\mathrm{Ker}(f - 1_M) = (\mathrm{Im}(f - 1_M))^\perp.$$

Proof. Suppose $x \in \mathrm{Ker}(f - 1_M)$ and $y \in M$. Then $f(x) = x$ and

$$\begin{aligned}
\phi(x, (f - 1_M)(y)) &= \phi(x, f(y) - y) \\
&= \phi(x, f(y)) - \phi(x, y) \\
&= \phi(f(x), f(y)) - \phi(x, y) \\
&= 0.
\end{aligned}$$

Thus, $x \in (\mathrm{Im}(f - 1_M))^\perp$. Conversely, if $x \in (\mathrm{Im}(f - 1_M))^\perp$, then

$$\phi(f(x) - x, f(y)) = \phi(f(x), f(y)) - \phi(x, f(y))$$
$$= \phi(x, y) - \phi(x, f(y))$$
$$= \phi(x, y - f(y))$$
$$= 0.$$

Since f is invertible, this implies that $f(x) - x \in M^\perp = \{0\}$, so $x \in \mathrm{Ker}(f - 1_M)$. $\qquad\square$

(3.23) Corollary. *With the notation of Lemma 3.22:*

(1) $(\mathrm{Ker}(f - 1_M))^\perp = \mathrm{Im}(f - 1_M)$.
(2) $(f - 1_M)^2 = 0$ *if and only if* $\mathrm{Im}(f - 1_M)$ *is a totally isotropic subspace.*

Proof. (1) is immediate from Lemma 3.22.

(2) Note that a subspace $N \subseteq M$ is totally isotropic (Definition 2.22) if and only if $N \subseteq N^\perp$. Then $\mathrm{Im}(f - 1_M)$ is totally isotropic if and only if

$$(3.9) \qquad \mathrm{Im}(f - 1_M) \subseteq (\mathrm{Im}(f - 1_M))^\perp = \mathrm{Ker}(f - 1_M).$$

But Equation (3.9) is valid if and only if $(f - 1_M)^2 = 0$. $\qquad\square$

(3.24) Theorem. (Cartan–Dieudonné) *Let Φ be a nonsingular quadratic form on a vector space M of dimension n over a field R with $\mathrm{char}(R) \neq 2$. Then any $g \in \mathrm{Isom}(\Phi)$ is a product of $\leq n$ hyperplane reflections.*

Proof. Let ϕ be the associated bilinear form. The proof is by induction on n. If $n = 1$, then $\mathrm{Isom}(\Phi) = \{\pm 1\}$, where -1 is the unique reflection. Since 1 is a product of 0 reflections, the result is clear for $n = 1$. So assume the theorem holds for nonsingular quadratic forms on vector spaces of dimension $< n$, and let $g \in \mathrm{Isom}(\Phi) = \mathrm{Isom}(\phi)$. We will consider several cases.

Case 1. *There exists $x \in \mathrm{Ker}(g - 1)$ such that $\Phi(x) \neq 0$.*

Let $N = \langle x \rangle$. If $y \in N^\perp$, then we have

$$\phi(g(y), x) = \phi(g(y), g(x)) = \phi(y, x) = 0.$$

Thus, $g|_{N^\perp} \in \mathrm{Isom}(\phi|_{N^\perp})$ and by induction $g|_{N^\perp}$ is a product of $\leq n - 1$ reflections of $N^\perp$. Each reflection of $N^\perp$ can be extended (by $x \mapsto x$) to a reflection of M and hence g is a product of $\leq n - 1$ reflections of M.

Case 2. *There is $x \in M$ with $\Phi(x) \neq 0$ and $\Phi(x - g(x)) \neq 0$.*

By Lemma 3.20, if $w = x - g(x)$, then there is a hyperplane reflection $f_w \in \mathrm{Isom}(\Phi)$ such that $f_w(g(x)) = x$. Thus $x \in \mathrm{Ker}(f_w \circ g - 1)$, and hence, by Case 1, $f_w \circ g$ is a product of $\leq n - 1$ reflections. Therefore, g is a product of $\leq n$ reflections.

Case 3. $\dim(M) = 2$.

If $\Phi(x) \neq 0$ for all $x \neq 0 \in M$, then this follows from the first two cases. Thus, suppose there exists $x \neq 0 \in M$ with $\Phi(x) = 0$. Choose $y \in M$ such that $\phi(x, y) \neq 0$, which is possible since Φ is non-singular. Since $\phi(x, rx) = r\phi(x, x)$, it is clear that $\mathcal{B} = \{x, y\}$ is a linearly independent subset of M and, hence, a basis. Replacing y by a multiple of y, we may assume that $\phi(x, y) = 1$. Furthermore, if $r \in R$, then

$$\phi(y + rx, y + rx) = \phi(y, y) + 2r\phi(x, y)$$

so by replacing x with a multiple of x we can also assume that $\phi(y, y) = 0$. That is, we have produced a basis $\mathcal{B}$ of M such that

$$[\phi]_\mathcal{B} = \begin{bmatrix} 0 & 1 \\ 1 & 0 \end{bmatrix}.$$

If $g \in \text{Isom}(\phi)$, then $[g]_\mathcal{B} = \begin{bmatrix} a & b \\ c & d \end{bmatrix}$ satisfies (Proposition 3.14)

$$\begin{bmatrix} a & c \\ b & d \end{bmatrix} \begin{bmatrix} 0 & 1 \\ 1 & 0 \end{bmatrix} \begin{bmatrix} a & b \\ c & d \end{bmatrix} = \begin{bmatrix} 0 & 1 \\ 1 & 0 \end{bmatrix}.$$

This equation implies that

$$[g]_\mathcal{B} = \begin{bmatrix} a & 0 \\ 0 & a^{-1} \end{bmatrix} \qquad \text{or} \qquad [g]_\mathcal{B} = \begin{bmatrix} 0 & b \\ b^{-1} & 0 \end{bmatrix}.$$

Since

$$\begin{bmatrix} a & 0 \\ 0 & a^{-1} \end{bmatrix} = \begin{bmatrix} 0 & a \\ a^{-1} & 0 \end{bmatrix} \begin{bmatrix} 0 & 1 \\ 1 & 0 \end{bmatrix},$$

we are finished with Case 3 when we observe that $\begin{bmatrix} 0 & b \\ b^{-1} & 0 \end{bmatrix}$ is the matrix of a reflection, for all $b \neq 0 \in R$. But if $g(x) = by$ and $g(y) = b^{-1}x$, then $x + by \in \text{Ker}(g - 1_M)$ and Case 1 implies that g is a reflection.

Case 4. $\dim(M) \geq 3$, $\Phi(x) = 0$ *for all* $x \in \text{Ker}(g - 1_M)$, *and whenever* $\Phi(x) \neq 0$, $\Phi(x - g(x)) = 0$.

Note that Case 4 simply incorporates all the situations not covered by Cases 1, 2, and 3. Our first goal is to show that in this situation, $\text{Im}(g - 1_M)$ is a totally isotropic subspace. We already know (by hypothesis) that

$$\Phi(x - g(x)) = 0 \qquad \text{whenever} \qquad \Phi(x) \neq 0.$$

Thus, suppose that $y \neq 0$ and $\Phi(y) = 0$. We want to show that

$$\Phi(y - g(y)) = 0.$$

Choose $z \in M$ with $\phi(y, z) \neq 0$ and consider the two-dimensional subspace $N = \langle y, z \rangle$ with basis $\mathcal{B} = \{y, z\}$. If $\psi = \phi|_N$, then

$$[\psi]_\mathcal{B} = \begin{bmatrix} 0 & \phi(y, z) \\ \phi(y, z) & \phi(z, z) \end{bmatrix}$$

so ψ is non-singular. Hence, we may write $\phi = \psi \perp \psi'$ where ψ' is non-singular on $N^\perp$ and $\dim N^\perp = \dim M - 2 > 0$. Thus, there exists $x \neq 0 \in N^\perp$ with $\Phi(x) = \Psi'(x) \neq 0$, but $\phi(x, y) = 0$. Then $\phi(y \pm x, y) = 0$ so that

$$\Phi(y \pm x) = \Phi(x) \neq 0.$$

The hypotheses of Case 4 then imply that

$$\Phi(x - g(x)) = \Phi((y + x) - g(y + x)) = \Phi((y - x) - g(y - x)) = 0,$$

and from this we conclude that $\Phi(y - g(y)) = 0$ and we have verified that $\mathrm{Im}(g - 1_M)$ is a totally isotropic subspace of M.

By Lemma 3.23, it follows that $(g - 1_M)^2 = 0$. Hence, the minimum polynomial of g is $(X - 1)^2$ (since $g = 1_M$ is a trivial case), and therefore, $\det g = 1$ (Chapter 4, Exercise 55). Now $\mathrm{Ker}(g - 1_M)$ is assumed to be totally isotropic (one of our hypotheses for Case 4). Thus,

$$\mathrm{Ker}(g - 1_M) \subseteq (\mathrm{Ker}(g - 1_M))^\perp = \mathrm{Im}(g - 1_M).$$

Since $(g - 1)^2 = 0$, we have $\mathrm{Im}(g - 1_M) \subseteq \mathrm{Ker}(g - 1_M)$ and conclude that $\mathrm{Im}(g - 1_M) = \mathrm{Ker}(g - 1_M)$. By the dimension formula (Propostion 3.8.8),

$$n = \dim(\mathrm{Im}(g - 1)) + \dim(\mathrm{Ker}(g - 1)) = 2 \dim(\mathrm{Ker}(g - 1)),$$

and hence, n is even.

To complete the proof, suppose that $g \in \mathrm{Isom}(\Phi)$ satisfies the hypotheses of Case 4, and let f_z be any hyperplane reflection. Then $g' = f_z \circ g$ is an isometry with $\det(g') = -1$, and hence g' does not fall under Case 4, so it must be covered by Case 1 or Case 2. Thus, it follows that g' is a product of $m \leq n$ reflections. Since m must be odd (in order to get a determinant of -1) and n is even, it follows that $m \leq n - 1$. Therefore, g is a product of $\leq n$ reflections and the proof is complete. $\square$

We conclude with some corollaries of this theorem.

(3.25) Corollary. *Suppose* $\dim M = n$.

(1) *If* $g \in \mathrm{Isom}(\Phi)$ *is the product of* $r \leq n$ *reflections, then*

$$\dim(\mathrm{Ker}(g - 1_M)) \geq n - r.$$

(2) *If* $\mathrm{Ker}(g - 1) = \langle 0 \rangle$, *then* g *cannot be written as a product of fewer than* n *reflections.*

Proof. (1) Let $g = f_{y_1} \cdots f_{y_r}$ and let $N_j = \mathrm{Ker}(f_{y_j} - 1)$. Then

$$N \cap \cdots \cap N_r \subseteq \mathrm{Ker}(g - 1_M).$$

Since $\dim N_j = n - 1$, it follows that (Proposition 3.8.10)

$$\dim(N_1 \cap \cdots \cap N_r) \geq n - r.$$

(2) This follows immediately from (1).   □

(3.26) Corollary.

(1) If $\dim M = 2$ then every isometry of determinant -1 is a reflection.
(2) If $\dim M = 3$ and g is an isometry of determinant 1, then g is the product of 2 reflections.

Proof. Exercise.   □

6.4 Exercises

1. Give an example of a field **F** with no/exactly one/more than one nontrivial conjugation.
2. Let ϕ be a b/s-linear form (or Φ a quadratic form) on a free R-module N and let $f : M \to N$ be an R-module homomorphism. Show that $f^*(\phi)$ defined by
    $$(f^*(\phi))(x, y) = \phi(f(x), f(y)),$$
    or $f^*(\Phi)$, defined by
    $$(f^*(\Phi))(x) = \Phi(f(x)),$$
    is a b/s-linear (or quadratic) form on M.
3. Let ϕ be an even symmetric bilinear form over a PID R in which 2 is prime (e.g., $R = \mathbf{Z}$) and suppose that $\det(\phi) \notin 2R$. Show that $\text{rank}(\phi)$ is even.
4. Note that $n \begin{bmatrix} 0 & 1 \\ 1 & 0 \end{bmatrix}$ is a form satisfying the conditions of Exercise 3, for any n. It is far from obvious that there are any other even symmetric non-singular forms over **Z**, but there are. Here is a famous example. Consider the form ϕ over **Z** with matrix
    $$E_8 = \begin{bmatrix} 2 & 1 & 0 & 0 & 0 & 0 & 0 & 0 \\ 1 & 2 & 1 & 0 & 0 & 0 & 0 & 0 \\ 0 & 1 & 2 & 1 & 0 & 0 & 0 & 0 \\ 0 & 0 & 1 & 2 & 1 & 0 & 0 & 0 \\ 0 & 0 & 0 & 1 & 2 & 1 & 0 & 1 \\ 0 & 0 & 0 & 0 & 1 & 2 & 1 & 0 \\ 0 & 0 & 0 & 0 & 0 & 1 & 2 & 0 \\ 0 & 0 & 0 & 0 & 1 & 0 & 0 & 2 \end{bmatrix}.$$

    (a) Show that $\det(\phi) = 1$.
    (b) Show that $\text{signature}(\phi) = 8$, where ϕ is regarded as a form over **R**.
5. (a) Let ϕ be a b/s-linear form on a module M over a ring R. Suppose R is an integral domain. If there is a subspace $H \subseteq M$ with $\text{rank}(H) > (1/2)\,\text{rank}(M)$, which is totally isotropic for ϕ, show that ϕ is degenerate.
    (b) Suppose that ϕ is either an even symmetric bilinear form on a module M over a ring R or a Hermitian form on a module M over a field R. If ϕ is non-singular and there is a totally isotropic subspace $H \subseteq M$ with $\text{rank}(H) = (1/2)\,\text{rank}(M)$, show that ϕ is isometric to
    $$(\text{rank}(M)/2) \begin{bmatrix} 0 & 1 \\ 1 & 0 \end{bmatrix}.$$

6. Let Φ be a quadratic form associated to a bilinear form ϕ. Derive the following identities:
    (a) $2\Phi(x+y) = 2\phi(x,y) + \phi(x,x) + \phi(y,y)$.
    (b) $\Phi(x+y) + \Phi(x-y) = 2(\Phi(x) + \Phi(y))$.
    The latter equation is known as the *polarization identity*.

7. Prove Proposition 2.5.

8. The following illustrates some of the behavior that occurs for forms that are not ε-symmetric. For simplicity, we take modules over a field F. Thus ϕ will denote a form on a module M over F.
    (a) Define the right/left/two-sided kernel of ϕ by
    $$M_r^\circ = \{y \in M : \phi(x,y) = 0 \quad \text{for all } x \in M\}$$
    $$M_l^\circ = \{y \in M : \phi(y,x) = 0 \quad \text{for all } x \in M\}$$
    $$M^\circ = \{y \in M : \phi(x,y) = \phi(y,x) = 0 \quad \text{for all } x \in M\}.$$
    Show that if $\text{rank}(M) < \infty$, then $M_r^\circ = \{0\}$ if and only if $M_l^\circ = \{0\}$. Give a counterexample if $\text{rank}(M) = \infty$.
    (b) Of course, $M^\circ = M_r^\circ \cap M_l^\circ$. Give an example of a form ϕ on M (with $\text{rank}(M) < \infty$) where $M_r^\circ \neq \{0\}$, $M_l^\circ \neq \{0\}$, but $M^\circ = \{0\}$. We say that ϕ is right/left non-singular if $M_r^\circ = \{0\}/M_l^\circ = \{0\}$ and non-singular if it is both right and left non-singular.
    (c) If $N \subseteq M$, define
    $$N_r^\perp = \{y \in M : \phi(x,y) = 0 \quad \text{for all } x \in N\}$$
    with an analogous definition for $N_l^\perp$. Let $N^\perp = N_r^\perp \cap N_l^\perp$. Give an example where these three subspaces of M are all distinct.
    (d) Suppose that $N \subseteq M$ is a subspace and $\phi|_N$ is non-singular. Show that $M = N \oplus N_r^\perp$ and $M = N \oplus N_l^\perp$. Give an example where $M \neq N \oplus N^\perp$. Indeed, give an example where N is a proper subspace of M but $N^\perp = \{0\}$.

9. Let R be a PID, M a free R-module of rank 4, and let ϕ be a non-degenerate skew-symmetric form on M. Show that ϕ is classified up to isometry by $\det(\phi)$ and by
    $$\gcd\{\phi(v,w) : v,w \in M\}.$$

10. Let R be a PID, and let ϕ be a skew-symmetric form on a free R-module M. If $\det(\phi)$ is square-free, show that ϕ is classified up to isometry by its rank.

11. Classify the following forms over $\mathbf{Z}$ (as in Theorem 2.35):

    (a)
    $$\begin{bmatrix} 0 & 1 & 1 & -1 & 2 & 1 \\ -1 & 0 & 5 & -3 & 2 & -3 \\ -1 & -5 & 0 & 3 & -7 & 4 \\ 1 & 3 & -3 & 0 & 6 & 5 \\ -2 & -2 & 7 & -6 & 0 & 1 \\ -1 & 3 & -4 & -5 & -1 & 0 \end{bmatrix}$$

    (b)
    $$\begin{bmatrix} 0 & 0 & -6 & -6 & -6 & -8 \\ 0 & 0 & -6 & -7 & -7 & -9 \\ 6 & 6 & 0 & -1 & -5 & -7 \\ 6 & 7 & 1 & 0 & -6 & -8 \\ 6 & 7 & 5 & 6 & 0 & 0 \\ 8 & 9 & 7 & 8 & 0 & 0 \end{bmatrix}$$

12. Find the signature of each of the following forms over $\mathbf{R}$. Note that this also gives their diagonalization over $\mathbf{R}$.

(a) $\begin{bmatrix} 5 & 40 & 9 \\ 40 & 50 & 12 \\ 9 & 12 & 3 \end{bmatrix}$

(b) $\begin{bmatrix} 5 & -11 & 3 \\ -11 & 3 & -12 \\ 3 & -12 & 0 \end{bmatrix}$

(c) $\begin{bmatrix} 2 & -4 & 6 \\ -4 & 13 & -12 \\ 6 & -12 & 18 \end{bmatrix}$

(d) $\begin{bmatrix} 1 & 3 & 4 \\ 3 & 9 & 5 \\ 4 & 5 & 0 \end{bmatrix}$

(e) $\begin{bmatrix} 2 & 1 & 0 & 0 & 0 & 0 \\ 1 & 2 & 1 & 0 & 0 & 0 \\ 0 & 1 & 2 & 1 & 0 & 0 \\ 0 & 0 & 1 & 0 & 1 & 0 \\ 0 & 0 & 0 & 1 & -2 & 1 \\ 0 & 0 & 0 & 0 & 1 & -2 \end{bmatrix}$

Also, diagonalize each of these forms over **Q**.

13. Carry out the details of the proof of Lemma 2.41.

14. Analogous to the definition of even, we could make the following definition: Let R be a PID and p a prime not dividing 2 (e.g., $R = \mathbf{Z}$ and p an odd prime). A form ϕ is p-ary if $\phi(x, x) \in pR$ for every $x \in M$. Show that if ϕ is p-ary, then $\phi(x, y) \in pR$ for every $x, y \in M$.

15. Prove Proposition 3.4.

16. Prove Lemma 3.6.

17. Prove Proposition 3.11.

18. Prove Lemma 3.19.

19. Prove Lemma 3.20.

20. Let $f \in \text{Aut}_R(M)$ where $R = \mathbf{R}$ or $\mathbf{C}$ and where rank$(M) < \infty$. If $f^k = 1_M$ for some k, prove that there is a non-singular form ϕ on M with $f \in \text{Isom}(\phi)$.

21. Diagonalize the following forms over the indicated fields:

(a) $\begin{bmatrix} 13 & 4 & 6 \\ 4 & 7 & 8 \\ 6 & 8 & 2 \end{bmatrix}$ over $\mathbf{F}_2$, $\mathbf{F}_3$, $\mathbf{F}_5$, and $\mathbf{Q}$

(b) $\begin{bmatrix} 1 & 4 & -1 \\ 4 & -2 & 10 \\ -1 & 10 & 4 \end{bmatrix}$ over $\mathbf{F}_2$, $\mathbf{F}_3$, $\mathbf{F}_5$, and $\mathbf{Q}$

22. A symmetric matrix $A = [a_{ij}] \in M_n(\mathbf{R})$ is called *diagonally dominant* if

$$a_{ii} \geq \sum_{j \neq i} |a_{ij}|$$

for $1 \leq i \leq n$. If the inequality is strict, then A is called *strictly diagonally dominant*. Let ϕ be the bilinear form on $\mathbf{R}^n$ whose matrix (in the standard basis) is A.
(a) If A is diagonally dominant, show that ϕ is positive semidefinite, i.e., $\phi(x, x) \geq 0$ for all $x \in \mathbf{R}^n$.
(b) If A is strictly diagonally dominant, show that ϕ is positive definite.

23. Let ϕ be an arbitrary positive (or negative) semidefinite form. Show that ϕ is non-degenerate if and only if it is positive (negative) definite.

24. Let R be a ring with a (possibly trivial) conjugation. Show that

$$\{P \in \text{GL}(n, R) : P^t = \left(\overline{P}\right)^{-1}\}$$

is a subgroup of GL(n, R). If $R = \mathbf{R}$, with trivial conjugation, this group is called the *orthogonal group* and denoted O(n), and if $R = \mathbf{C}$, with complex conjugation, it is called the *unitary group* and denoted U(n).

25. Let $A \in M_2(\mathbf{C})$ be the matrix of a Hermitian form, so

$$A = \begin{bmatrix} a & b \\ \overline{b} & c \end{bmatrix}$$

with $a, c \in \mathbf{R}$. Find an explicit matrix P with $P^t = (\overline{P})^{-1}$ such that $P^t A \overline{P}$ is diagonal.

26. Let $\mathbf{F}$ be an arbitrary subfield of $\mathbf{C}$ and ϕ a Hermitian form on an $\mathbf{F}$-module M of finite rank. Show how to modify the Gram–Schmidt procedure to produce an orthogonal (but not in general orthonormal) basis of M, i.e., a basis $\mathcal{B} = \{v_1, \ldots, v_n\}$ in which $\phi(v_i, v_j) = 0$ if $i \neq j$.

27. Let R be a commutative ring and let $A \in M_n(R)$ be a skew-symmetric matrix. If $P \in M_n(R)$ is any matrix, then show that $P^t A P$ is also skew-symmetric and
$$\text{Pf}(P^t A P) = \det(P)\text{Pf}(A).$$

28. Let $A \in M_n(\mathbf{C})$ be a Hermitian matrix. Show that A is positive definite if and only if A^{-1} is positive definite. More generally, show that the signature of A^{-1} is the signature of A. (We say that A is positive definite if the associated Hermitian form $\phi(x, x) = x^t A x$ is positive definite.)

29. If V is a real vector space, then a nonempty subset $C \subseteq V$ is a *cone* if a, $b \in C \Rightarrow a + b \in C$ and $a \in C$, $\alpha > 0 \in \mathbf{R} \Rightarrow \alpha a \in C$. Prove that the set of positive definite Hermitian matrices in $M_n(\mathbf{C})$ is a cone.

30. If A is a real symmetric matrix prove that there is $\alpha \in \mathbf{R}$ such that $A + \alpha I_n$ is positive definite.

31. If $A \in M_{m,n}(\mathbf{C})$, show that $A^* A$ and $A A^*$ are positive semidefinite Hermitian matrices.

32. Let $A, B \in M_n(\mathbf{C})$ be Hermitian and assume that B is positive definite. Prove that the two Hermitian forms determined by A and B can be simultaneously diagonalized. That is, prove that there is a nonsingular matrix P such that $P^t B \overline{P} = I_n$ and $P^t A \overline{P} = \text{diag}(\lambda_1, \ldots, \lambda_n)$. (Hint: B determines an inner product on $\mathbf{C}^n$.)

Chapter 7

Topics in Module Theory

This chapter will be concerned with collecting a number of results and constructions concerning modules over (primarily) noncommutative rings that will be needed to study group representation theory in Chapter 8.

7.1 Simple and Semisimple Rings and Modules

In this section we investigate the question of decomposing modules into "simpler" modules.

(1.1) Definition. *If R is a ring (not necessarily commutative) and $M \neq \langle 0 \rangle$ is a nonzero R-module, then we say that M is a **simple** or **irreducible** R-module if $\langle 0 \rangle$ and M are the only submodules of M.*

(1.2) Proposition. *If an R-module M is simple, then it is cyclic.*

Proof. Let x be a nonzero element of M and let $N = \langle x \rangle$ be the cyclic submodule generated by x. Since M is simple and $N \neq \langle 0 \rangle$, it follows that $M = N$. $\qquad \square$

(1.3) Proposition. *If R is a ring, then a cyclic R-module $M = \langle m \rangle$ is simple if and only if $\mathrm{Ann}(m)$ is a maximal left ideal.*

Proof. By Proposition 3.2.15, $M \cong R/\mathrm{Ann}(m)$, so the correspondence theorem (Theorem 3.2.7) shows that M has no submodules other than M and $\langle 0 \rangle$ if and only if R has no submodules (i.e., left ideals) containing $\mathrm{Ann}(m)$ other than R and $\mathrm{Ann}(m)$. But this is precisely the condition for $\mathrm{Ann}(m)$ to be a maximal left ideal. $\qquad \square$

(1.4) Examples.

(1) An abelian group A is a simple $\mathbf{Z}$-module if and only if A is a cyclic group of prime order.

(2) The hypothesis in Proposition 1.3 that M be cyclic is necessary. The $\mathbf{Z}$-module $A = \mathbf{Z}_2^2$ has annihilator $2\mathbf{Z}$ but the module A is not simple.

(3) Consider the vector space F^2 (where F is any field) as an $F[x]$-module via the linear transformation $T(u_1, u_2) = (u_2, 0)$. Then F^2 is a cyclic $F[X]$-module, but it is not a simple $F[X]$-module. Indeed,

$$F^2 = F[X] \cdot (0, 1)$$

but $N = \{(u, 0) : u \in F\}$ is an $F[X]$-submodule of F^2. Thus the converse of Proposition 1.2 is not true.

(4) Let $V = \mathbf{R}^2$ and consider the linear transformation $T : V \to V$ defined by $T(u, v) = (-v, u)$. Then the $\mathbf{R}[X]$-module V_T is simple. To see this let $w = (u_1, v_1) \neq 0 \in V$ and let N be the $\mathbf{R}[X]$-submodule of V_T generated by w. Then $w \in N$ and $Xw = T(w) = (-v_1, u_1) \in N$. Since any $(x, y) \in V$ can be written as $(x, y) = \alpha w + \beta X w$ where $\alpha = (xu_1 + yv_1)/(u_1^2 + v_1^2)$ and $\beta = (yu_1 - xv_1)/(u_1^2 + v_1^2)$, it follows that $N = V_T$ and hence V_T is simple.

(5) Now let $W = \mathbf{C}^2$ and consider the linear transformation $T : W \to W$ defined by $T(u, v) = (-v, u)$. Note that T is defined by the same formula used in Example 1.4 (4). However, in this case the $\mathbf{C}[X]$-module W_T is not simple. Indeed, the $\mathbf{C}$-subspace $\mathbf{C} \cdot (i, 1)$ is a T-invariant subspace of W, and hence, it is a $\mathbf{C}[X]$-submodule of W_T different from W and from $\langle 0 \rangle$.

The following lemma is very easy, but it turns out to be extremely useful:

(1.5) Proposition. (Schur's lemma)

(1) *Let M be a simple R-module. Then the ring $\mathrm{End}_R(M)$ is a division ring.*

(2) *If M and N are simple R-modules, then $\mathrm{Hom}_R(M, N) \neq \langle 0 \rangle$ if and only if M and N are isomorphic.*

Proof. (1) Let $f \neq 0 \in \mathrm{End}_R(M)$. Then $\mathrm{Im}(f)$ is a nonzero submodule of M and $\mathrm{Ker}(f)$ is a submodule of M different from M. Since M is simple, it follows that $\mathrm{Im}(f) = M$ and $\mathrm{Ker}(f) = \langle 0 \rangle$, so f is an R-module isomorphism and hence is invertible as an element of the ring $\mathrm{End}_R(M)$.

(2) The same argument as in (1) shows that any nonzero homomorphism $f : M \to N$ is an isomorphism.   □

We have a second concept of decomposition of modules into simpler pieces, with simple modules again being the building blocks.

(1.6) Definition. *If R is a ring (not necessarily commutative), then an R-module M is said to be* **indecomposable** *if it has no proper nontrivial com-*

plemented submodule M_1, i.e., if $M = M_1 \oplus M_2$ implies that $M_1 = \langle 0 \rangle$ or $M_1 = M$.

If M is a simple R-module, then M is also indecomposable, but the converse is false. For example, $\mathbf{Z}$ is an indecomposable $\mathbf{Z}$-module, but $\mathbf{Z}$ is not a simple $\mathbf{Z}$-module; note that $\mathbf{Z}$ contains the proper submodule $2\mathbf{Z}$.

One of the major classes of modules we wish to study is the following:

(1.7) Definition. *An R-module M is said to be **semisimple** if it is a direct sum of simple R-modules.*

The idea of semisimple modules is to study modules by decomposing them into a direct sum of simple submodules. In our study of groups there was also another way to construct groups from simpler groups, namely, the extension of one group by another, of which a special case was the semidirect product. Recall from Definition 1.6.6 that a group G is an extension of a group N by a group H if there is an exact sequence of groups

$$1 \longrightarrow N \longrightarrow G \longrightarrow H \longrightarrow 1.$$

If this exact sequence is a split exact sequence, then G is a semidirect product of N and H. In the case of abelian groups, semidirect and direct products coincide, but extension of N by H is still a distinct concept.

If G is an abelian group and N is a subgroup, then the exact sequence

$$\langle 0 \rangle \longrightarrow N \longrightarrow G \longrightarrow H \longrightarrow \langle 0 \rangle$$

is completely determined by the chain of subgroups $\langle 0 \rangle \subseteq N \subseteq G$. By allowing longer chains of subgroups, we can consider a group as obtained by multiple extensions. We will consider this concept within the class of R-modules.

(1.8) Definition.

(1) *If R is a ring (not necessarily commutative) and M is an R-module, then a **chain of submodules** of M is a sequence $\{M_i\}_{i=0}^n$ of submodules of M such that*

$$(1.1) \qquad \langle 0 \rangle = M_0 \subsetneqq M_1 \subsetneqq M_2 \subsetneqq \cdots \subsetneqq M_n = M.$$

*The **length** of the chain is n.*

(2) *We say that a chain $\{N_j\}_{j=0}^m$ is a **refinement** of the chain $\{M_i\}_{i=0}^n$ if each M_i is equal to N_j for some j. Refinement of chains defines a partial order on the set $\mathcal{C}$ of all chains of submodules of M.*

(3) *A maximal element of $\mathcal{C}$ (if it exists) is called a **composition series** of M.*

(1.9) *Remarks.*

(1) Note that the chain (1.1) is a composition series if and only if each of the modules M_i/M_{i-1} $(1 \le i \le n)$ is a simple module.

(2) Our primary interest will be in decomposing a module as a direct sum of simple modules. Note that if $M = \oplus_{i=1}^{n} M_i$ where M_i is a simple R-module, then M has a composition series

$$\langle 0 \rangle \subsetneqq M_1 \subsetneqq M_1 \oplus M_2 \subsetneqq \cdots \subsetneqq \bigoplus_{i=1}^{n} M_i = M.$$

On the other hand, if $M = \oplus_{i=1}^{\infty} M_i$, then M does not have a composition series. In a moment (Example 1.10 (2)) we shall see an example of a module that is not semisimple but does have a composition series. Thus, while these two properties—semisimplicity and having a composition series—are related, neither implies the other. However, our main interest in composition series is as a tool in deriving results about semisimple modules.

(1.10) Examples.

(1) Let D be a division ring and let M be a D-module with a basis $\{x_1, \ldots, x_m\}$. Let $M_0 = \langle 0 \rangle$ and for $1 \le i \le n$ let $M_i = \langle x_1, \ldots, x_i \rangle$. Then $\{M_i\}_{i=0}^{n}$ is a chain of submodules of length n, and since

$$M_i/M_{i-1} = \langle x_1, \ldots, x_i \rangle / \langle x_1, \ldots, x_{i-1} \rangle$$
$$\cong Dx_i$$
$$\cong D,$$

we conclude that this chain is a composition series because D is a simple D-module.

(2) If p is a prime, the chain

$$\langle 0 \rangle \subsetneqq p\mathbf{Z}_{p^2} \subsetneqq \mathbf{Z}_{p^2}$$

is a composition series for the $\mathbf{Z}$-module $\mathbf{Z}_{p^2}$. Note that $\mathbf{Z}_{p^2}$ is not semisimple as a $\mathbf{Z}$-module since it has no proper complemented submodules.

(3) The $\mathbf{Z}$-module $\mathbf{Z}$ does not have a composition series. Indeed, if $\{I_i\}_{i=0}^{n}$ is any chain of submodules of length n, then writing $I_1 = \langle a_1 \rangle$, we can properly refine the chain by putting the ideal $\langle 2a_1 \rangle$ between I_1 and $I_0 = \langle 0 \rangle$.

(4) If R is a PID which is not a field, then essentially the same argument as Example 1.10 (3) shows that R does not have a composition series as an R-module.

(1.11) Definition. *Let M be an R-module. If M has a composition series let $\ell(M)$ denote the minimum length of a composition series for M. If M does not have a composition series, let $\ell(M) = \infty$. $\ell(M)$ is called the **length** of the R-module M. If $\ell(M) < \infty$, we say that M has **finite length**.*

Note that isomorphic R-modules have the same length, since if $f : M \to N$ is an R-module isomorphism, the image under f of a composition series for M is a composition series for N.

(1.12) Lemma. *Let M be an R-module of finite length and let N be a proper submodule (i.e., $N \neq M$). Then $\ell(N) < \ell(M)$.*

Proof. Let

$$(1.2) \qquad \langle 0 \rangle = M_0 \subsetneq M_1 \subsetneq \cdots \subsetneq M_n = M$$

be a composition series of M of length $n = \ell(M)$ and let $N_i = N \cap M_i \subseteq N$. Let $\phi : N_i \to M_i/M_{i-1}$ be the inclusion map $N_i \to M_i$ followed by the projection map $M_i \to M_i/M_{i-1}$. Since $\mathrm{Ker}(\phi) = N_{i-1}$, it follows from the first isomorphism theorem that N_i/N_{i-1} is isomorphic to a submodule of M_i/M_{i-1}. But (1.2) is a composition series, so M_i/M_{i-1} is a simple R-module. Hence $N_i = N_{i-1}$ or $N_i/N_{i-1} = M_i/M_{i-1}$ for $i = 1, 2, \ldots, n$. By deleting the repeated terms of the sequence $\{N_i\}_{i=0}^n$ we obtain a composition series for the module N of length $\leq n = \ell(M)$. Suppose that this composition series for N has length n. Then we must have $N_i/N_{i-1} = M_i/M_{i-1}$ for all $i = 1, 2, \ldots, n$. Thus $N_1 = M_1$, $N_2 = M_2$, $\ldots$, $N_n = M_n$, i.e., $N = M$. Since we have assumed that N is a proper submodule, we conclude that the chain $\{N_i\}_{i=0}^n$ has repeated terms, and hence, after deleting repeated terms we find that N has a composition series of length $< \ell(M)$, that is, $\ell(N) < \ell(M)$. $\qquad\qquad\qquad\qquad\qquad\qquad\square$

(1.13) Proposition. *Let M be an R-module of finite length. Then every composition series of M has length $n = \ell(M)$. Moreover, every chain of submodules can be refined to a composition series.*

Proof. We first show that any chain of submodules of M has length $\leq \ell(M)$. Let

$$\langle 0 \rangle = M_0 \subsetneq M_1 \subsetneq \cdots \subsetneq M_k = M$$

be a chain of submodules of M of length k. By Lemma 1.12,

$$0 = \ell(M_0) < \ell(M_1) < \cdots < \ell(M_k) = \ell(M).$$

Thus, $k \leq \ell(M)$.

Now consider a composition series of M of length k. By the definition of composition series, $k \geq \ell(M)$ and we just proved that $k \leq \ell(M)$. Thus, $k = \ell(M)$. If a chain has length $\ell(M)$, then it must be maximal and, hence, is a composition series. If the chain has length $< \ell(M)$, then it is not a composition series and hence it may be refined until its length is $\ell(M)$, at which time it will be a composition series. $\qquad \square$

According to Example 1.10 (1), if D is a division ring and M is a D-module, then a basis $S = \{x_1, \ldots, x_n\}$ with n elements determines a composition series of M of length n. Since all composition series of M must have the same length, we conclude that any two finite bases of M must have the same length n. Moreover, if M had also an infinite basis T, then M would have a linearly independent set consisting of more than n elements. Call this set $\{y_1, \ldots, y_k\}$ with $k > n$. Then

$$\langle 0 \rangle \subsetneq \langle y_1 \rangle \subsetneq \langle y_1, y_2 \rangle \subsetneq \cdots \subsetneq \langle y_1, \ldots, y_k \rangle \subsetneq M$$

is a chain of length $> n$, which contradicts Proposition 1.13. Thus, every basis of M is finite and has n elements. We have arrived at the following result.

(1.14) Proposition. *Let D be a division ring and let M be a D-module with a finite basis. Then every basis of M is finite and all bases have the same number of elements.*

Proof. $\qquad \square$

An (almost) equivalent way to state the same result is the following. It can be made equivalent by the convention that D^∞ refers to any infinite direct sum of copies of D, without regard to the cardinality of the index set.

(1.15) Corollary. *If D is a division ring and $D^m \cong D^n$ then $m = n$.*

Proof. $\qquad \square$

We conclude our treatment of composition series with the following result, which is frequently useful in constructing induction arguments.

(1.16) Proposition. *Let $0 \longrightarrow K \overset{\phi}{\longrightarrow} M \overset{\psi}{\longrightarrow} L \longrightarrow 0$ be a short exact sequence of R-modules. If K and L are of finite length then so is M, and*

$$\ell(M) = \ell(K) + \ell(L).$$

Proof. Let

$$\langle 0 \rangle = K_0 \subsetneq K_1 \subsetneq \cdots \subsetneq K_n = K$$

be a composition series of K, and let

$$\langle 0 \rangle = L_0 \subsetneq L_1 \subsetneq \cdots \subsetneq L_m = L$$

be a composition series for L. For $0 \leq i \leq n$, let $M_i = \phi(K_i)$, and for $n + 1 \leq i \leq n + m$, let $M_i = \psi^{-1}(L_{i-n})$. Then $\{M_i\}_{i=0}^{n+m}$ is a chain of submodules of M and

$$M_i/M_{i-1} \cong \begin{cases} K_i/K_{i-1} & \text{for } 1 \leq i \leq n \\ L_{i-n}/L_{i-n-1} & \text{for } n+1 \leq i \leq n+m \end{cases}$$

so that $\{M_i\}_{i=0}^{n+m}$ is a composition series of M. Thus, $\ell(M) = n + m$. $\square$

(1.17) Example. Let R be a PID and let M be a finitely generated torsion R-module. We may write M as a finite direct sum of primary cyclic torsion modules:

$$M \cong \bigoplus_{i=1}^{k} R/\langle p_i^{e_i} \rangle.$$

Then it is an easy exercise to check that M is of finite length and

$$\ell(M) = \sum_{i=1}^{k} e_i.$$

We now return to our consideration of semisimple modules. For this purpose we introduce the following convenient notation.

If M is an R-module and s is a positive integer, then sM will denote the direct sum $M \oplus \cdots \oplus M$ (s summands). More generally, if Γ is any index set then ΓM will denote the R-module $\Gamma M = \oplus_{\gamma \in \Gamma} M_\gamma$ where $M_\gamma = M$ for all $\gamma \in \Gamma$. Of course, if $|\Gamma| = s < \infty$ then $\Gamma M = sM$, and we will prefer the latter notation.

This notation is convenient for describing semisimple modules as direct sums of simple R-modules. If M is a semisimple R-module, then

$$(1.3) \qquad\qquad M \cong \bigoplus_{i \in I} M_i$$

where M_i is simple for each $i \in I$. If we collect all the simple modules in Equation (1.3) that are isomorphic, then we obtain

$$(1.4) \qquad\qquad M \cong \bigoplus_{\alpha \in A} (\Gamma_\alpha M_\alpha)$$

where $\{M_\alpha\}_{\alpha \in A}$ is a set of pairwise distinct (i.e., $M_\alpha \not\cong M_\beta$ if $\alpha \neq \beta$) simple modules. Equation (1.4) is said to be a **simple factorization** of the semisimple module M. Notice that this is analogous to the prime factorization of elements in a PID. This analogy is made even more compelling by the following uniqueness result for the simple factorization.

(1.18) Theorem. *Suppose that M and N are semisimple R-modules with simple factorizations*

(1.5)
$$M \cong \bigoplus_{\alpha \in A} (\Gamma_\alpha M_\alpha)$$

and

(1.6)
$$N \cong \bigoplus_{\beta \in B} (\Lambda_\beta N_\beta)$$

where $\{M_\alpha\}_{\alpha \in A}$ and $\{N_\beta\}_{\beta \in B}$ are the distinct simple factors of M and N, respectively. If M is isomorphic to N, then there is a bijection $\psi : A \to B$ such that $M_\alpha \cong N_{\psi(\alpha)}$ for all $\alpha \in A$. Moreover, $|\Gamma_\alpha| < \infty$ if and only if $|\Lambda_{\psi(\alpha)}| < \infty$ and in this case $|\Gamma_\alpha| = |\Lambda_{\psi(\alpha)}|$.

Proof. Let $\phi : M \to N$ be an isomorphism and let $\alpha \in A$ be given. We may write $M \cong M_\alpha \oplus M'$ with $M' = \oplus_{\gamma \in A \setminus \{\alpha\}} (\Gamma_\gamma M_\gamma) \oplus \Gamma'_\alpha M_\alpha$ where Γ'_α is Γ_α with one element deleted. Then by Proposition 3.3.15,

$$\operatorname{Hom}_R(M, N) \cong \operatorname{Hom}_R(M_\alpha, N) \oplus \operatorname{Hom}_R(M', N)$$

(1.7)
$$\cong \left(\bigoplus_{\beta \in B} \Lambda_\beta \operatorname{Hom}_R(M_\alpha, N_\beta) \right) \oplus \operatorname{Hom}_R(M', N).$$

By Schur's lemma, $\operatorname{Hom}_R(M_\alpha, N_\beta) = \langle 0 \rangle$ unless $M_\alpha \cong N_\beta$. Therefore, in Equation (1.7) we will have $\operatorname{Hom}_R(M_\alpha, N) = 0$ or $\operatorname{Hom}_R(M_\alpha, N) \cong \Lambda_\beta \operatorname{Hom}_R(M_\alpha, N_\beta)$ for a unique $\beta \in B$. The first alternative cannot occur since the isomorphism $\phi : M \to N$ is identified with $(\phi \circ \iota_1, \phi \circ \iota_2)$ where $\iota_1 : M_\alpha \to M$ is the canonical injection (and $\iota_2 : M' \to M$ is the injection). If $\operatorname{Hom}_R(M_\alpha, N) = 0$ then $\phi \circ \iota_1 = 0$, which means that $\phi|_{M_\alpha} = 0$. This is impossible since ϕ is injective. Thus the second case occurs and we define $\psi(\alpha) = \beta$ where $\operatorname{Hom}_R(M_\alpha, N_\beta) \neq \langle 0 \rangle$. Thus we have defined a function $\psi : A \to B$, which is one-to-one by Schur's lemma. It remains to check that ψ is surjective. But given $\beta \in B$, we may write $N \cong N_\beta \oplus N'$. Then

$$\operatorname{Hom}_R(M, N) \cong \operatorname{Hom}_R(M, N_\beta) \oplus \operatorname{Hom}_R(M, N')$$

and

$$\operatorname{Hom}_R(M, N_\beta) \cong \prod_{\alpha \in A} \left(\prod_{\Gamma_\alpha} \operatorname{Hom}_R(M_\alpha, N_\beta) \right).$$

Since ϕ is surjective, we must have $\operatorname{Hom}_R(M, N_\beta) \neq \langle 0 \rangle$, and thus, Schur's lemma implies that

$$\operatorname{Hom}_R(M, N_\beta) \cong \prod_{\Gamma_\alpha} \operatorname{Hom}(M_\alpha, N_\beta)$$

for a unique $\alpha \in A$. Then $\psi(\alpha) = \beta$, so ψ is surjective.

According to Proposition 3.3.15 and Schur's lemma,

$$\operatorname{Hom}_R(M, N) \cong \prod_{\alpha \in A} \left(\bigoplus_{\beta \in B} \operatorname{Hom}_R(\Gamma_\alpha M_\alpha, \Lambda_\beta N_\beta) \right)$$

$$\cong \prod_{\alpha \in A} \operatorname{Hom}_R(\Gamma_\alpha M_\alpha, \Lambda_{\psi(\alpha)} N_{\psi(\alpha)}).$$

Therefore, $\phi \in \operatorname{Hom}_R(M, N)$ is an isomorphism if and only if

$$\phi_\alpha = \phi|_{\Gamma_\alpha M_\alpha} : \Gamma_\alpha M_\alpha \to \Lambda_{\psi(\alpha)} N_{\psi(\alpha)}$$

is an isomorphism for all $\alpha \in A$. But by the definition of ψ and Schur's lemma, M_α is isomorphic to $N_{\psi(\alpha)}$. Also, $\Gamma_\alpha M_\alpha$ has length $|\Gamma_\alpha|$, and $\Lambda_{\psi(\alpha)} N_{\psi(\alpha)}$ has length $|\Lambda_{\psi(\alpha)}|$, and since isomorphic modules have the same length, $|\Gamma_\alpha| = |\Lambda_{\psi(\alpha)}|$, completing the proof. $\square$

(1.19) Corollary. *Let M be a semisimple R-module and suppose that M has two simple factorizations*

$$M \cong \bigoplus_{\alpha \in A} (\Gamma_\alpha M_\alpha) \cong \bigoplus_{\beta \in B} (\Lambda_\beta N_\beta)$$

with distinct simple factors $\{M_\alpha\}_{\alpha \in A}$ and $\{N_\beta\}_{\beta \in B}$. Then there is a bijection $\psi : A \to B$ such that $M_\alpha \cong N_{\psi(\alpha)}$ for all $\alpha \in A$. Moreover, $|\Gamma_\alpha| < \infty$ if and only if $|\Lambda_{\psi(\alpha)}| < \infty$ and in this case $|\Gamma_\alpha| = |\Lambda_{\psi(\alpha)}|$.

Proof. Take $\phi = 1_M$ in Theorem 1.18. $\square$

(1.20) Remarks.

(1) While it is true in Corollary 1.19 that $M_\alpha \cong N_{\psi(\alpha)}$ (isomorphism as R-modules), it is not necessarily true that $M_\alpha = N_{\psi(\alpha)}$. For example, let $R = F$ be a field and let M be a vector space over F of dimension s. Then for *any* choice of basis $\{m_1, \ldots, m_s\}$ of M, we obtain a direct sum decomposition

$$M \cong Rm_1 \oplus \cdots \oplus Rm_s.$$

(2) In Theorem 1.18 we have been content to distinguish between finite and infinite index sets Γ_α, but we are not distinguishing between infinite sets of different cardinality. Using the theory of cardinal arithmetic, one can refine Theorem 1.18 to conclude that $|\Gamma_\alpha| = |\Lambda_{\psi(\alpha)}|$ for *all* $\alpha \in A$, where $|S|$ denotes the cardinality of the set S.

We will now present some alternative characterizations of semisimple modules. The following notation, which will be used only in this section, will be convenient for this purpose. Let $\{M_i\}_{i \in I}$ be a set of submodules of a module M. Then let

$$M_I = \sum_{i \in I} M_i$$

be the sum of the submodules $\{M_i\}_{i \in I}$.

(1.21) Lemma. *Let M be an R-module that is a sum of simple submodules $\{M_i\}_{i \in I}$, and let N be an arbitrary submodule of M. Then there is a subset $J \subseteq I$ such that*

$$M \cong N \oplus \left(\bigoplus_{i \in J} M_i \right).$$

Proof. The proof is an application of Zorn's lemma. Let

$$\mathcal{S} = \left\{ P \subseteq I : M_P \cong \bigoplus_{i \in P} M_i \text{ and } M_P \cap N = \langle 0 \rangle \right\}.$$

Partially order $\mathcal{S}$ by inclusion and let $\mathcal{C} = \{P_\alpha\}_{\alpha \in A}$ be an arbitrary chain in $\mathcal{S}$. If $P = \cup_{\alpha \in A} P_\alpha$, we claim that $P \in \mathcal{S}$. Suppose that $P \notin \mathcal{S}$. Since it is clear that $M_P \cap N = \langle 0 \rangle$, we must have that $M_P \ncong \oplus_{i \in P} M_i$. Then Theorem 3.3.2 shows that there is some $p_0 \in P$, such that $M_{p_0} \cap M_{P'} \neq \langle 0 \rangle$, where $P' = P \setminus \{p_0\}$. Suppose that $0 \neq x \in M_{p_0} \cap M_{P'}$. Then we may write

(1.8) $$x = x_{p_1} + \cdots + x_{p_k}$$

where $x_{p_i} \neq 0 \in M_{p_i}$ for $\{p_1, \ldots, p_k\} \subseteq P'$. Since $\mathcal{C}$ is a chain, there is an index $\alpha \in A$ such that $\{p_0, p_1, \ldots, p_k\} \subseteq P_\alpha$. Equation (1.8) shows that $M_{P_\alpha} \ncong \oplus_{i \in P_\alpha} M_i$, which contradicts the fact that $P_\alpha \in \mathcal{S}$. Therefore, we must have $P \in \mathcal{S}$, and Zorn's lemma applies to conclude that $\mathcal{S}$ has a maximal element J.

Claim. $M = N + M_J \cong N \oplus \left(\oplus_{i \in J} M_i \right)$.

If this were not true, then there would be an index $i_0 \in I$ such that $M_{i_0} \not\subseteq N + M_J$. This implies that $M_{i_0} \not\subseteq N$ and $M_{i_0} \not\subseteq M_J$. Since $M_{i_0} \cap N$ and $M_{i_0} \cap M_J$ are proper submodules of M_{i_0}, it follows that $M_{i_0} \cap N = \langle 0 \rangle$ and $M_{i_0} \cap M_J = \langle 0 \rangle$ because M_{i_0} is a simple R-module. Therefore, $\{i_0\} \cup J \in \mathcal{S}$, contradicting the maximality of J. Hence, the claim is proved. $\square$

(1.22) Corollary. *If an R-module M is a sum of simple submodules, then M is semisimple.*

Proof. Take $N = \langle 0 \rangle$ in Theorem 1.21. $\square$

(1.23) Theorem. *If M is an R-module, then the following are equivalent:*

(1)   *M is a semisimple module.*
(2)   *Every submodule of M is complemented.*
(3)   *Every submodule of M is a sum of simple R-modules.*

Proof. (1) $\Rightarrow$ (2) follows from Lemma 1.21, and (3) $\Rightarrow$ (1) is immediate from Corollary 1.22. It remains to prove (2) $\Rightarrow$ (3).

Let M_1 be a submodule of M. First we observe that every submodule of M_1 is complemented in M_1. To see this, suppose that N is any submodule of M_1. Then N is complemented in M, so there is a submodule N' of M such that $N \oplus N' \cong M$. But then $N + (N' \cap M_1) = M_1$ so that $N \oplus (N' \cap M_1) \cong M_1$, and hence N is complemented in M_1.

Next we claim that every nonzero submodule M_2 of M contains a nonzero simple submodule. Let $m \in M_2$, $m \neq 0$. Then $Rm \subseteq M_2$ and, furthermore, $R/\operatorname{Ann}(m) \cong Rm$ where $\operatorname{Ann}(m) = \{a \in R : am = 0\}$ is a left ideal of R. A simple Zorn's lemma argument (see the proof of Theorem 2.2.16) shows that there is a maximal left ideal I of R containing $\operatorname{Ann}(m)$. Then Im is a maximal submodule of Rm by the correspondence theorem. By the previous paragraph, Im is a complemented submodule of Rm, so there is a submodule N of Rm with $N \oplus Im \cong Rm$, and since Im is a maximal submodule of Rm, it follows that the submodule N is simple. Therefore, we have produced a simple submodule of M_2.

Now consider an arbitrary submodule N of M, and let $N_1 \subseteq N$ be the sum of all the simple submodules of N. We claim that $N_1 = N$. N_1 is complemented in N, so we may write $N \cong N_1 \oplus N_2$. If $N_2 \neq \langle 0 \rangle$ then N_2 has a nonzero simple submodule N', and since $N' \subseteq N$, it follows that $N' \subseteq N_1$. But $N_1 \cap N_2 = \langle 0 \rangle$. This contradiction shows that $N_2 = \langle 0 \rangle$, i.e., $N = N_1$, and the proof is complete. $\square$

(1.24) Corollary. *Sums, submodules, and quotient modules of semisimple modules are semisimple.*

Proof. Sums: This follows immediately from Corollary 1.22.

Submodules: Any submodule of a semisimple module satisfies condition (3) of Theorem 1.23.

Quotient modules: If M is a semisimple module, $N \subseteq M$ is a submodule, and $Q = M/N$, then N has a complement N' in M, i.e., $M \cong N \oplus N'$. But then $Q \cong N'$, so Q is isomorphic to a submodule of M, and hence, is semisimple. $\square$

(1.25) Corollary. *Let M be a semisimple R-module and let $N \subseteq M$ be a submodule. Then N is irreducible (simple) if and only if N is indecomposable.*

Proof. Since every irreducible module is indecomposable, we need to show that if N is not irreducible, then N is not indecomposable. Let N_1 be a nontrivial proper submodule of N. Then N is semisimple by Corollary 1.24, so N_1 has a complement by Theorem 1.23, and N is not indecomposable. $\square$

(1.26) Remark. The fact that every submodule of a semisimple R-module M is complemented is equivalent (by Theorem 3.3.9) to the statement that whenever M is a semisimple R-module, every short exact sequence

$$0 \longrightarrow N \longrightarrow M \longrightarrow K \longrightarrow 0$$

of R-modules splits.

(1.27) Definition. *A ring R is called* **semisimple** *if R is semisimple as a left R-module.*

Remark. The proper terminology should be "left semisimple," with an analogous definition of "right semisimple," but we shall see below that the two notions coincide.

(1.28) Theorem. *The following are equivalent for a ring R:*

(1) *R is a semisimple ring.*
(2) *Every R-module is semisimple.*
(3) *Every R-module is projective.*

Proof. (1) $\Rightarrow$ (2). Let M be an R-module. By Proposition 3.4.14, M has a free presentation

$$0 \longrightarrow K \longrightarrow F \longrightarrow M \longrightarrow 0$$

so that M is a quotient of the free R-module F. Since F is a direct sum of copies of R and R is assumed to be semisimple, it follows that F is semisimple, and hence M is also (Corollary 1.24).

(2) $\Rightarrow$ (3). Assume that every R-module is semisimple, and let P be an arbitrary R-module. Suppose that

(1.9) $$0 \longrightarrow K \longrightarrow M \longrightarrow P \longrightarrow 0$$

is a short exact sequence. Since M is an R-module, our assumption is that it is semisimple and then Remark 1.26 implies that sequence (1.9) is split exact. Since (1.9) is an arbitrary short exact sequence with P on the right, it follows from Theorem 3.5.1 that P is projective.

(3) $\Rightarrow$ (1). Let M be an arbitrary submodule of R (i.e., an arbitrary left ideal). Then we have a short exact sequence

$$0 \longrightarrow M \longrightarrow R \longrightarrow R/M \longrightarrow 0.$$

Since all R-modules are assumed projective, we have that R/M is projective, and hence (by Theorem 3.5.1) this sequence splits. Therefore, $R \cong M \oplus N$ for some submodule $N \subseteq R$, which is isomorphic (as an R-module) to R/M. Then by Theorem 1.23, R is semisimple.   □

(1.29) Corollary. *Let R be a semisimple ring and let M be an R-module. Then M is irreducible (simple) if and only if M is indecomposable.*

Proof.   □

(1.30) Theorem. *Let R be a semisimple ring. Then every simple R-module is isomorphic to a submodule of R.*

Proof. Let N be a simple R-module, and let $R = \oplus_{i \in I} M_i$ be a simple factorization of the semisimple R-module R. We must show that at least one of the simple R-modules M_i is isomorphic to N. If this is not the case, then

$$\operatorname{Hom}_R(R, N) \cong \operatorname{Hom}_R\Big(\bigoplus_{i\in I} M_i, N\Big) \cong \prod_{i\in I} \operatorname{Hom}_R(M_i, N) = \langle 0 \rangle$$

where the last equality is because $\operatorname{Hom}_R(M_i, N) = \langle 0 \rangle$ if M_i is not isomorphic to N (Schur's lemma). But $\operatorname{Hom}_R(R, N) \cong N \neq \langle 0 \rangle$, and this contradiction shows that we must have N isomorphic to one of the simple submodules M_i of R. $\square$

(1.31) Corollary. *Let R be a semisimple ring.*

(1) *There are only finitely many isomorphism classes of simple R-modules.*
(2) *If $\{M_\alpha\}_{\alpha \in A}$ is the set of isomorphism classes of simple R-modules and*

$$R \cong \bigoplus_{\alpha \in A} (\Gamma_\alpha M_\alpha),$$

then each Γ_α is finite.

Proof. Since R is semisimple, we may write

$$R = \bigoplus_{\beta \in B} N_\beta$$

where each N_β is simple. We will show that B is finite, and then both finiteness statements in the corollary are immediate from Theorem 1.30.

Consider the identity element $1 \in R$. By the definition of direct sum, we have

$$1 = \sum_{\beta \in B} r_\beta n_\beta$$

for some elements $r_\beta \in R$, $n_\beta \in N_\beta$, with all but finitely many r_β equal to zero. Of course, each N_β is a left R-submodule of R, i.e., a left ideal.

Now suppose that B is infinite. Then there is a $\beta_0 \in B$ for which $r_{\beta_0} = 0$. Let n be any nonzero element of N_{β_0}. Then

$$n = n \cdot 1 = n\Big(\sum_{\beta \in B} r_\beta n_\beta\Big) = \sum_{\beta \in B \setminus \{\beta_0\}} (nr_\beta) n_\beta,$$

so

$$n \in \bigoplus_{\beta \in B \setminus \{\beta_0\}} N_\beta.$$

Thus,

$$n \in N_{\beta_0} \bigcap \Big(\bigoplus_{\beta \in B \setminus \{\beta_0\}} N_\beta \Big) = \{0\},$$

by the definition of direct sum again, which is a contradiction. Hence, B is finite. $\square$

We now come to the basic structure theorem for semisimple rings.

(1.32) Theorem. (Wedderburn) *Let R be a semisimple ring. Then there is a finite collection of integers $n_1, \ldots, n_k$, and division rings $D_1, \ldots, D_k$ such that*

$$R \cong \bigoplus_{i=1}^{k} \operatorname{End}_{D_i}(D_i^{n_i}).$$

Proof. By Corollary 1.31, we may write

$$R \cong \bigoplus_{i=1}^{k} n_i M_i$$

where $\{M_i\}_{i=1}^{k}$ are the distinct simple R-modules and $n_1, \ldots, n_k$ are positive integers. Then R is anti-isomorphic to R^{op}, and

$$
\begin{aligned}
R^{\mathrm{op}} &\cong \operatorname{End}_R(R) \\
&\cong \operatorname{Hom}_R\left(\bigoplus_{i=1}^{k} n_i M_i, \bigoplus_{i=1}^{k} n_i M_i\right) \\
&\cong \bigoplus_{i=1}^{k} \operatorname{Hom}_R(n_i M_i, n_i M_i) \\
&\cong \bigoplus_{i=1}^{k} \operatorname{End}_R(n_i M_i),
\end{aligned}
$$

by Schur's lemma. Also, by Schur's lemma, $\operatorname{End}_R(M_i)$ is a division ring, which we denote by E_i, for each $i = 1, \ldots, k$. Then it is easy to check (compare the proof of Theorem 1.18) that

$$\operatorname{End}_R(n_i M_i) \cong \operatorname{End}_{E_i}(E_i^{n_i}).$$

Setting $D_i = E_i^{\mathrm{op}}$, the proof is completed by observing that $\operatorname{End}_{E_i}(E_i^{n_i})$ is anti-isomorphic to $\operatorname{End}_{D_i}(D_i^{n_i})$. $\square$

Remark. Note that by Corollary 4.3.9, $\operatorname{End}_D(D^n)$ is isomorphic to $M_n(D^{\mathrm{op}})$. Thus, Wedderburn's theorem is often stated as, *Every semisimple ring is isomorphic to a finite direct sum of matrix rings over division rings.*

(1.33) Lemma. *Let D be a division ring and n a positive integer. Then $R = \operatorname{End}_D(D^n)$ is semisimple as a left R-module and also as a right R-module. Furthermore, R is semisimple as a left D-module and as a right D-module.*

Proof. Write $D^n = D_1 \oplus D_2 \oplus \cdots \oplus D_n$ where $D_i = D$. Let

$$M_i = \left\{ f \in \mathrm{End}_D(D^n) : \mathrm{Ker}(f) \supseteq \bigoplus_{k \neq i} D_k \right\},$$

$$N_j = \{ f \in \mathrm{End}_D(D^n) : \mathrm{Im}(f) \subseteq D_j \},$$

and let

$$P_{ij} = M_i \cap N_j.$$

Note that $P_{ij} \cong D^{\mathrm{op}}$. Then

$$\mathrm{End}_D(D^n) \cong M_1 \oplus \cdots \oplus M_n$$

as a left R-module, and

$$\mathrm{End}_D(D^n) \cong N_1 \oplus \cdots \oplus N_n$$

as a right R-module. We leave it to the reader to check that each M_i (resp., N_j) is a simple left (resp., right) R-module. Also,

$$\mathrm{End}_D(D^n) \cong \bigoplus P_{ij}$$

as a right (resp., left) D-module, and each P_{ij} is certainly simple (on either side). □

(1.34) Corollary. *A ring R is semisimple as a left R-module if and only if it is semisimple as a right R-module.*

Proof. This follows immediately from Theorem 1.32 and Lemma 1.33. □

Observe that R is a simple left R-module (resp., right R-module) if and only if R has no nontrivial proper left (resp., right) ideals, which is the case if and only if R is a division algebra. Thus, to define simplicity of R in this way would bring nothing new. Instead we make the following definition:

(1.35) Definition. *A ring R with identity is **simple** if it has no nontrivial proper (two-sided) ideals.*

Remark. In the language of the next section, this definition becomes "A ring R with identity is simple if it is simple as an (R, R)-bimodule."

(1.36) Corollary. *Let D be a division ring and n a positive integer. Then $\mathrm{End}_D(D^n)$ is a simple ring that is semisimple as a left $\mathrm{End}_D(D^n)$-module.*

Conversely, if R is a simple ring that is semisimple as a left R-module, or, equivalently, as a right R-module, then

$$R \cong \mathrm{End}_D(D^n)$$

for some division ring D and positive integer n.

Proof. We leave it to the reader to check that $\mathrm{End}_D(D^n)$ is simple (compare Theorem 2.2.26 and Corollary 2.2.27), and then the first part of the corollary follows from Lemma 1.33. Conversely, if R is semisimple we have the decomposition given by Wedderburn's theorem (Theorem 1.32), and then the condition of simplicity forces $k = 1$.   □

Our main interest in semisimple rings and modules is in connection with our investigation of group representation theory, but it is also of interest to reconsider modules over a PID from this point of view. Thus let R be a PID. We wish to give a criterion for R-modules to be semisimple. The following easy lemma is left as an exercise.

(1.37) Lemma. *Let R be an integral domain. Then R is a semisimple ring if and only if R is a field. If R is a field, R is simple.*

Proof. Exercise.   □

From this lemma and Theorem 1.28, we see that if R is a field, then every R-module (i.e., vector space) is semisimple and there is nothing more to say. For the remainder of this section, we will assume that R is a PID that is not a field.

Let M be a finitely generated R-module. Then by Corollary 3.6.9, we have that $M \cong F \oplus M_\tau$, where F is free (of finite rank) and M_τ is the torsion submodule of M. If $F \neq \langle 0 \rangle$ then Lemma 1.37 shows that M is not semisimple. It remains to consider the case where $M = M_\tau$, i.e., where M is a finitely generated torsion module. Recall from Theorem 3.7.13 that each such M is a direct sum of primary cyclic R-modules.

(1.38) Proposition. *Let M be a primary cyclic R-module (where R is a PID is not a field) and assume that $\mathrm{Ann}(M) = \langle p^e \rangle$ where $p \in R$ is a prime. If $e = 1$ then M is simple. If $e > 1$, then M is not semisimple.*

Proof. First suppose that $e = 1$, so that $M \cong R/\langle p \rangle$. Then M is a simple R-module because $\langle p \rangle$ is a prime ideal in the PID R, and hence, it is a maximal ideal.

Next suppose that $e > 1$. Then

$$\langle 0 \rangle \neq p^{e-1}M \subsetneqq M,$$

and $p^{e-1}M$ is a proper submodule of M, which is not complemented; hence, M is not semisimple by Theorem 1.23 (2).   □

(1.39) Theorem. *Let M be a finitely generated torsion R-module (where R is a PID that is not a field). Then M is semisimple if and only if $\mathrm{me}(M)$*

(see Definition 3.7.8) is a product of distinct prime factors. M is a simple R-module if and only if

$$me(M) = co(M) = \langle p \rangle$$

where $p \in R$ is a prime.

Proof. First suppose that M is cyclic, and $me(M) = \langle p_1^{e_1} \ldots p_k^{e_k} \rangle$. Then the primary decomposition of M is given by

$$M \cong (R/\langle p_1^{e_1} \rangle) \oplus \cdots \oplus (R/\langle p_k^{e_k} \rangle),$$

and M is semisimple if and only if each of the summands is semisimple, which by Proposition 1.38, is true if and only if

$$e_1 = e_2 = \cdots = e_k = 1.$$

Now let M be general. Then by Theorem 3.7.1, there is a cyclic decomposition

$$M \cong Rw_1 \oplus \cdots \oplus Rw_n$$

such that $\mathrm{Ann}(w_i) = \langle s_i \rangle$ and $s_i \mid s_{i+1}$ for $1 \leq i \leq n - 1$. Then M is semisimple if and only if each of the cyclic submodules Rw_i is semisimple, which occurs (by the previous paragraph) if and only if s_i is a product of distinct prime factors. Since $s_i \mid s_{i+1}$, this occurs if and only if $s_n = me(M)$ is a product of distinct prime factors. The second assertion is then easy to verify. $\square$

(1.40) Remark. In the two special cases of finite abelian groups and linear transformations that we considered in some detail in Chapters 3 and 4, Theorem 1.39 takes the following form:

(1) A finite abelian group is semisimple if and only if it is the direct product of cyclic groups of prime order, and it is simple if and only if it is cyclic of prime order.

(2) Let V be a finite-dimensional vector space over a field F and let $T : V \to V$ be a linear transformation. Then V_T is a semisimple $F[X]$-module if and only if the minimal polynomial $m_T(X)$ of T is a product of distinct irreducible factors and is simple if and only if its characteristic polynomial $c_T(X)$ is equal to its minimal polynomial $m_T(X)$, this polynomial being irreducible (see Lemma 4.4.11.) If F is algebraically closed (so that the only irreducible polynomials are linear ones) then V_T is semisimple if and only if T is diagonalizable and simple if and only if V is one-dimensional (see Corollary 4.4.32).

7.2 Multilinear Algebra

We have three goals in this section: to introduce the notion of a bimodule, to further our investigation of "Hom," and to introduce and investigate tensor products. The level of generality of the material presented in this section is dictated by the applications to the theory of group representations. For this reason, most of the results will be concerned with modules over rings that are not commutative; frequently there will be more than one module structure on the same abelian group, and many of the results are concerned with the interaction of these various module structures. We start with the concept of bimodule.

(2.1) Definition. *Let R and S be rings. An abelian group M is an (R, S)-bimodule if M is both a left R-module and a right S-module, and the compatibility condition*

$$(2.1) \qquad\qquad r(ms) = (rm)s$$

is satisfied for every $r \in R$, $m \in M$, and $s \in S$.

(2.2) Examples.

(1) Every left R-module is an $(R, \mathbf{Z})$-bimodule, and every right S-module is a $(\mathbf{Z}, S)$-bimodule.

(2) If R is a commutative ring, then every left or right R-module is an (R, R)-bimodule in a natural way. Indeed, if M is a left R-module, then according to Remark 3.1.2 (1), M is also a right R-module by means of the operation $mr = rm$. Then Equation (2.1) is

$$r(ms) = r(sm) = (rs)m = (sr)m = s(rm) = (rm)s.$$

(3) If T is a ring and R and S are subrings of T (possibly with $R = S = T$), then T is an (R, S)-bimodule. Note that Equation (2.1) is simply the associative law in T.

(4) If M and N are left R-modules, then the abelian group $\mathrm{Hom}_R(M, N)$ has the structure of an $(\mathrm{End}_R(N), \mathrm{End}_R(M))$-bimodule, as follows. If $f \in \mathrm{Hom}_R(M, N)$, $\phi \in \mathrm{End}_R(M)$, and $\psi \in \mathrm{End}_R(N)$, then define $f\phi = f \circ \phi$ and $\psi f = \psi \circ f$. These definitions provide a left $\mathrm{End}_R(N)$-module and a right $\mathrm{End}_R(M)$-module structure on $\mathrm{Hom}_R(M, N)$, and Equation (2.1) follows from the associativity of composition of functions.

(5) Recall that a ring T is an R-algebra, if T is an R-module and the R-module structure on T and the ring structure of T are compatible, i.e., $r(t_1 t_2) = (rt_1)t_2 = t_1(rt_2)$ for all $r \in R$ and $t_1, t_2 \in T$. If T happens to be an (R, S)-bimodule, such that $r(t_1 t_2) = (rt_1)t_2 = t_1(rt_2)$ and $(t_1 t_2)s = t_1(t_2 s) = (t_1 s)t_2$ for all $r \in R$, $s \in S$, and $t_1, t_2 \in T$, then we

say that T is an (R, S)-**bialgebra**. For example, if R and S are subrings of a commutative ring T, then T is an (R, S)-bialgebra.

Suppose that M is an (R, S)-bimodule and $N \subseteq M$ is a subgroup of the additive abelian group of M. Then N is said to be an (R, S)-**bisubmodule** of M if N is both a left R-submodule and a right S-submodule of M. If M_1 and M_2 are (R, S)-bimodules, then a function $f : M_1 \to M_2$ is an (R, S)-**bimodule homomorphism** if it is both a left R-module homomorphism and a right S-module homomorphism. The set of (R, S)-bimodule homomorphisms will be denoted $\mathrm{Hom}_{(R,S)}(M_1, M_2)$. Since bimodule homomorphisms can be added, this has the structure of an abelian group, but, a priori, nothing more. If $f : M_1 \to M_2$ is an (R, S)-bimodule homomorphism, then it is a simple exercise to check that $\mathrm{Ker}(f) \subseteq M_1$ and $\mathrm{Im}(f) \subseteq M_2$ are (R, S)-bisubmodules.

Furthermore, if $N \subseteq M$ is an (R, S)-bisubmodule, then the quotient abelian group is easily seen to have the structure of an (R, S)-bimodule. We leave it as an exercise for the reader to formulate and verify the noether isomorphism theorems (see Theorems 3.2.3 to 3.2.6) in the context of (R, S)-bimodules. It is worth pointing out that if M is an (R, S)-bimodule, then there are three distinct concepts of submodule of M, namely, R-submodule, S-submodule, and (R, S)-bisubmodule. Thus, if $X \subseteq M$, then one has three concepts of submodule of M generated by the set X. To appreciate the difference, suppose that $X = \{x\}$ consists of a single element $x \in M$. Then the R-submodule generated by X is the set

$$(2.2) \qquad Rx = \{rx : r \in R\},$$

the S-submodule generated by X is the set

$$(2.3) \qquad xS = \{xs : s \in S\},$$

while the (R, S)-bisubmodule generated by X is the set

$$(2.4) \qquad RxS = \{\sum_{i=1}^{n} r_i x s_i : n \in \mathbf{N} \text{ and } r_i \in R, \ s_i \in S \text{ for } 1 \le i \le n\}.$$

(2.3) Examples.

(1) If R is a ring, then a left R-submodule of R is a left ideal, a right R-submodule is a right ideal, and an (R, R)-bisubmodule of R is a (two-sided) ideal.

(2) As a specific example, let $R = M_2(\mathbf{Q})$ and let $x = \begin{bmatrix} 1 & 0 \\ 0 & 0 \end{bmatrix}$. Then the left R-submodule of R generated by $\{x\}$ is

$$Rx = \left\{ \begin{bmatrix} a & 0 \\ b & 0 \end{bmatrix} : a,\, b \in \mathbf{Q} \right\},$$

the right R-submodule of R generated by $\{x\}$ is

$$xR = \left\{ \begin{bmatrix} a & b \\ 0 & 0 \end{bmatrix} : a,\, b \in \mathbf{Q} \right\},$$

while the (R, R)-bisubmodule of R generated by $\{x\}$ is R itself (see Theorem 2.2.26).

When considering bimodules, there are (at least) three distinct types of homomorphisms that can be considered. In order to keep them straight, we will adopt the following notational conventions. If M and N are left R-modules (in particular, both could be (R, S)-bimodules, or one could be an (R, S)-bimodule and the other a (R, T)-bimodule), then $\operatorname{Hom}_R(M, N)$ will denote the set of (left) R-module homomorphisms from M to N. If M and N are right S-modules, then $\operatorname{Hom}_{-S}(M, N)$ will denote the set of all (right) S-module homomorphisms. If M and N are (R, S)-bimodules, then $\operatorname{Hom}_{(R,S)}(M, N)$ will denote the set of all (R, S)-bimodule homomorphisms from M to N. With no additional hypotheses, the only algebraic structure that can be placed upon these sets of homomorphisms is that of abelian groups, i.e., addition of homomorphisms is a homomorphism in each situation described. The first thing to be considered is what additional structure is available.

(2.4) Proposition. *Suppose that M is an (R, S)-bimodule and N is an (R, T)-bimodule. Then $\operatorname{Hom}_R(M, N)$ can be given the structure of an (S, T)-bimodule.*

Proof. We must define compatible left S-module and right T-module structures on $\operatorname{Hom}_R(M, N)$. Thus, let $f \in \operatorname{Hom}_R(M, N)$, $s \in S$, and $t \in T$. Define sf and ft as follows:

$$(2.5) \qquad\qquad sf(m) = f(ms) \quad \text{for all } m \in M$$

and

$$(2.6) \qquad\qquad ft(m) = f(m)t \quad \text{for all } m \in M.$$

We must show that Equation (2.5) defines a left S-module structure on $\operatorname{Hom}_R(M, N)$ and that Equation (2.6) defines a right T-module structure on $\operatorname{Hom}_R(M, N)$, and we must verify the compatibility condition $s(ft) = (sf)t$.

We first verify that sft is an R-module homomorphism. To see this, suppose that $r_1, r_2 \in R$, $m_1, m_2 \in M$ and note that

$$\begin{aligned}
sft(r_1 m_1 + r_2 m_2) &= f((r_1 m_1 + r_2 m_2)s)t \\
&= f((r_1 m_1)s + (r_2 m_2)s)t \\
&= f(r_1(m_1 s) + r_2(m_2 s))t \\
&= (r_1 f(m_1 s) + r_2 f(m_2 s))\, t
\end{aligned}$$

$$= (r_1 f(m_1 s))t + (r_2 f(m_2 s))t$$
$$= r_1 (f(m_1 s)t) + r_2 (f(m_2 s)t)$$
$$= r_1 (sft)(m_1) + r_2 (sft)(m_2),$$

where the third equality follows from the (R, S)-bimodule structure on M, while the next to last equality is a consequence of the (R, T)-bimodule structure on N. Thus, sft is an R-module homomorphism for all $s \in S$, $t \in T$, and $f \in \mathrm{Hom}_R(M, N)$.

Now observe that, if s_1, $s_2 \in S$ and $m \in M$, then

$$(s_1(s_2 f))(m) = (s_2 f)(ms_1)$$
$$= f((ms_1)s_2)$$
$$= f(m(s_1 s_2))$$
$$= ((s_1 s_2)f)(m)$$

so that $\mathrm{Hom}_R(M, N)$ satisfies axiom (c_l) of Definition 3.1.1. The other axioms are automatic, so $\mathrm{Hom}_R(M, N)$ is a left S-module. Similarly, if t_1, $t_2 \in T$ and $m \in M$, then

$$((ft_1)t_2)(m) = ((ft_1)(m))t_2$$
$$= (f(m)t_1)t_2$$
$$= f(m)(t_1 t_2)$$
$$= (f(t_1 t_2))(m).$$

Thus, $\mathrm{Hom}_R(M, N)$ is a right T-module by Definition 3.1.1 (2). We have only checked axiom (c_r), the others being automatic.

It remains to check the compatibility of the left S-module and right T-module structures. But, if $s \in S$, $t \in T$, $f \in \mathrm{Hom}_R(M, N)$, and $m \in M$, then

$$((sf)t)(m) = (sf)(m)t = f(ms)t = (ft)(ms) = s(ft)(m).$$

Thus, $(sf)t = s(ft)$ and $\mathrm{Hom}_R(M, N)$ is an (S, T)-bimodule, which completes the proof of the proposition.   □

Proved in exactly the same way is the following result concerning the bimodule structure on the set of right R-module homomorphisms.

(2.5) Proposition. *Suppose that M is an (S, R)-bimodule and N is a (T, R)-bimodule. Then $\mathrm{Hom}_{-R}(M, N)$ has the structure of a (T, S)-bimodule, via the module operations*

$$(tf)(m) = t(f(m)) \quad and \quad (fs)(m) = f(sm)$$

where $s \in S$, $t \in T$, $f \in \mathrm{Hom}_{-R}(M, N)$, and $m \in M$.

Proof. Exercise.   □

Some familiar results are corollaries of these propositions. (Also see Example 3.1.5 (10).)

(2.6) Corollary.

(1) *If M is a left R-module, then $M^* = \mathrm{Hom}_R(M, R)$ is a right R-module.*
(2) *If M and N are (R, R)-bimodules, then $\mathrm{Hom}_R(M, N)$ is an (R, R)-bimodule, and $\mathrm{End}_R(M)$ is an (R, R)-bialgebra. In particular, this is the case when the ring R is commutative.*

Proof. Exercise.   □

Remark. If M and N are both (R, S)-bimodules, then the set of bimodule homomorphisms $\mathrm{Hom}_{(R,S)}(M, N)$ has only the structure of an abelian group.

Theorem 3.3.10 generalizes to the following result in the context of bimodules. The proof is identical, and hence it will be omitted.

(2.7) Theorem. *Let*

$$(2.7) \qquad\qquad 0 \longrightarrow M_1 \overset{\phi}{\longrightarrow} M \overset{\psi}{\longrightarrow} M_2$$

be a sequence of (R, S)-bimodules and (R, S)-bimodule homomorphisms. Then the sequence (2.7) is exact if and only if the sequence

$$(2.8) \qquad 0 \longrightarrow \mathrm{Hom}_R(N, M_1) \overset{\phi_*}{\longrightarrow} \mathrm{Hom}_R(N, M) \overset{\psi_*}{\longrightarrow} \mathrm{Hom}_R(N, M_2)$$

is an exact sequence of (T, S)-bimodules for all (R, T)-bimodules N.
    If

$$(2.9) \qquad\qquad M_1 \overset{\phi}{\longrightarrow} M \overset{\psi}{\longrightarrow} M_2 \longrightarrow 0$$

is a sequence of (R, S)-bimodules and (R, S)-bimodule homomorphisms, then the sequence (2.9) is exact if and only if the sequence

$$(2.10) \qquad 0 \longrightarrow \mathrm{Hom}_R(M_2, N) \overset{\psi^*}{\longrightarrow} \mathrm{Hom}_R(M, N) \overset{\phi^*}{\longrightarrow} \mathrm{Hom}_R(M_1, N)$$

is an exact sequence of (S, T)-bimodules for all (R, T)-bimodules N.

Proof.   □

Similarly, the proof of the following result is identical to the proof of Theorem 3.3.12.

(2.8) Theorem. *Let N be a fixed (R, T)-bimodule. If*

$$(2.11) \qquad\qquad 0 \longrightarrow M_1 \overset{\phi}{\longrightarrow} M \overset{\psi}{\longrightarrow} M_2 \longrightarrow 0$$

is a split short exact sequence of (R, S)-bimodules, then

$$(2.12) \quad 0 \longrightarrow \mathrm{Hom}_R(N, M_1) \xrightarrow{\phi_*} \mathrm{Hom}_R(N, M) \xrightarrow{\psi_*} \mathrm{Hom}_R(N, M_2) \longrightarrow 0$$

is a split short exact sequence of (T, S)-bimodules, and

$$(2.13) \quad 0 \longrightarrow \mathrm{Hom}_R(M_2, N) \xrightarrow{\psi^*} \mathrm{Hom}_R(M, N) \xrightarrow{\phi^*} \mathrm{Hom}_R(M_1, N) \longrightarrow 0$$

is a split short exact sequence of (S, T)-bimodules.

Proof.   □

This concludes our brief introduction to bimodules and module structures on spaces of homomorphisms; we turn our attention now to the concept of tensor product of modules. As we shall see, Hom and tensor products are closely related, but unfortunately, there is no particularly easy definition of tensor products. On the positive side, the use of the tensor product in practice does not usually require an application of the definition, but rather fundamental properties (easier than the definition) are used.

Let M be an (R, S)-bimodule and let N be an (S, T)-bimodule. Let F be the free abelian group on the index set $M \times N$ (Remark 3.4.5). Recall that this means that $F = \bigoplus_{(m,n) \in M \times N} \mathbf{Z}_{(m,n)}$ where $\mathbf{Z}_{(m,n)} = \mathbf{Z}$ for all $(m, n) \in M \times N$, and that a basis of F is given by $S = \{e_{(m,n)}\}_{(m,n) \in M \times N}$ where $e_{(m,n)} = (\delta_{mk}\delta_{n\ell})_{(k,\ell) \in M \times N}$, that is, $e_{(m,n)} = 1$ in the component of F corresponding to the element $(m, n) \in M \times N$ and $e_{(m,n)} = 0$ in all other components. As is conventional, we will identify the basis element $e_{(m,n)}$ with the element $(m, n) \in M \times N$. Thus a typical element of F is a linear combination

$$\sum_{(m,n) \in M \times N} c_{(m,n)}(m, n)$$

where $c_{(m,n)} \in \mathbf{Z}$ and all but finitely many of the integers $c_{(m,n)}$ are 0. Note that F can be given the structure of an (R, T)-bimodule via the multiplication

$$(2.14) \qquad r\left(\sum_{i=1}^k c_i(m_i, n_i)\right) t = \sum_{i=1}^k c_i(rm_i, n_i t)$$

where $r \in R$, $t \in T$, and $c_1, \ldots, c_k \in \mathbf{Z}$.

Let $K \subseteq F$ be the subgroup of F generated by the subset $H_1 \cup H_2 \cup H_3$ where the three subsets H_1, H_2, and H_3 are defined by

$$H_1 = \{(m_1 + m_2, n) - (m_1, n) - (m_2, n) : m_1, m_2 \in M, n \in N\}$$
$$H_2 = \{(m, n_1 + n_2) - (m, n_1) - (m, n_2) : m \in M, n_1, n_2 \in N\}$$
$$H_3 = \{(ms, n) - (m, sn) : m \in M, n \in N, s \in S\}.$$

Note that K is an (R,T)-submodule of F using the bimodule structure given by Equation (2.14).

With these preliminaries out of the way, we can define the tensor product of M and N.

(2.9) Definition. *With the notation introduced above, the* **tensor product** *of the (R,S)-bimodule M and the (S,T)-bimodule N, denoted $M \otimes_S N$, is the quotient (R,T)-bimodule*

$$M \otimes_S N = F/K.$$

If $\pi : F \to F/K$ is the canonical projection map, then we let $m \otimes_S n = \pi((m,n))$ for each $(m,n) \in M \times N \subseteq F$. When S is clear from the context we will frequently write $m \otimes n$ in place of $m \otimes_S n$.

Note that the set

$$(2.15) \qquad \{m \otimes_S n : (m,n) \in M \times N\}$$

generates $M \otimes_S N$ as an (R,T)-bimodule, but it is important to recognize that $M \otimes_S N$ is not (in general) equal to the set in (2.15). Also important to recognize is the fact that $m \otimes_S n = (m,n) + K$ is an equivalence class, so that $m \otimes n = m' \otimes n'$ does not necessarily imply that $m = m'$ and $n = n'$. As motivation for this rather complicated definition, we have the following proposition. The proof is left as an exercise.

(2.10) Proposition. *Let M be an (R,S)-bimodule, N an (S,T)-bimodule, and let m, $m_i \in M$, n, $n_i \in N$, and $s \in S$. Then the following identities hold in $M \otimes_S N$.*

$$(2.16) \qquad (m_1 + m_2) \otimes n = m_1 \otimes n + m_2 \otimes n$$

$$(2.17) \qquad m \otimes (n_1 + n_2) = m \otimes n_1 + m \otimes n_2$$

$$(2.18) \qquad ms \otimes n = m \otimes sn.$$

Proof. Exercise. $\qquad\qquad\qquad\qquad\qquad\qquad\qquad\qquad\qquad\qquad\qquad\quad$ $\square$

Indeed, the tensor product $M \otimes_S N$ is obtained from the cartesian product $M \times N$ by "forcing" the relations (2.16)–(2.18), but no others, to hold. This idea is formalized in Theorem 2.12, the statement of which requires the following definition.

(2.11) Definition. *Let M be an (R,S)-bimodule, N an (S,T)-bimodule, and let $M \times N$ be the cartesian product of M and N as sets. Let Q be any (R,T)-bimodule. A map $g : M \times N \to Q$ is said to be S-**middle linear** if it satisfies the following properties (where $r \in R$, $s \in S$, $t \in T$, m, $m_i \in M$ and n, $n_i \in N$):*

(1) $g(rm, nt) = rg(m, n)t$,
(2) $g(m_1 + m_2, n) = g(m_1, n) + g(m_2, n)$,
(3) $g(m, n_1 + n_2) = g(m, n_1) + g(m, n_2)$, *and*
(4) $g(ms, n) = g(m, sn)$.

Note that conditions (1), (2), and (3) simply state that for each $m \in M$ the function $g_m : N \to Q$ defined by $g_m(n) = g(m, n)$ is in $\mathrm{Hom}_{-T}(N, Q)$ and for each $n \in N$ the function $g^n : M \to Q$ defined by $g^n(m) = g(m, n)$ is in $\mathrm{Hom}_R(M, Q)$. Condition (4) is compatibility with the S-module structures on M and N.

If $\pi : F \to M \otimes_S N = F/K$ is the canonical projection map and $\iota : M \times N \to F$ is the inclusion map that sends (m, n) to the basis element $(m, n) \in F$, then we obtain a map $\theta : M \times N \to M \otimes N$. According to Proposition 2.10, the function θ is S-middle linear. The content of the following theorem is that *every* S-middle linear map "factors" through θ. This can, in fact, be taken as the fundamental defining property of the tensor product.

(2.12) Theorem. *Let M be an (R, S)-bimodule, N an (S, T)-bimodule, Q an (R, T)-bimodule, and $g : M \times N \to Q$ an S-middle linear map. Then there exists a unique (R, T)-bimodule homomorphism $\widetilde{g} : M \otimes_S N \to Q$ with $g = \widetilde{g} \circ \theta$. Furthermore, this property characterizes $M \otimes_S N$ up to isomorphism.*

Proof. If F denotes the free $\mathbf{Z}$-module on the index set $M \times N$, which is used to define the tensor product $M \otimes_S N$, then Equation (2.14) gives an (R, T)-bimodule structure on F. Since F is a free $\mathbf{Z}$-module with basis $M \times N$ and $g : M \times N \to Q$ is a function, Proposition 3.4.9 shows that there is a unique $\mathbf{Z}$-module homomorphism $g' : F \to Q$ such that $g' \circ \iota = g$ where $\iota : M \times N \to F$ is the inclusion map. The definition of the (R, T)-bimodule structure on F and the fact that g is S-middle linear implies that g' is in fact an (R, T)-bimodule homomorphism. Let $K' = \mathrm{Ker}(g')$, so the first isomorphism theorem provides an injective (R, T)-bimodule homomorphism $g'' : F/K' \to Q$ such that $g' = g'' \circ \pi'$ where $\pi' : F \to F/K'$ is the canonical projection map. Recall that $K \subset F$ is the subgroup of F generated by the sets H_1, H_2, and H_3 defined prior to Definition 2.9. Since g is an S-middle linear map, it follows that $K \subseteq \mathrm{Ker}(g') = K'$, so there is a map $\pi_2 : F/K \to F/K'$ such that $\pi_2 \circ \pi = \pi'$.

Thus, $g : M \times N \to Q$ can be factored as follows:

$$(2.19) \qquad M \times N \xrightarrow{\;\iota\;} F \xrightarrow{\;\pi\;} F/K \xrightarrow{\;\pi_2\;} F/K' \xrightarrow{\;g''\;} Q.$$

Recalling that $F/K = M \otimes_S N$, we define $\widetilde{g} = g'' \circ \pi_2$. Since $\theta = \pi \circ \iota$, Equation (2.19) shows that $g = \widetilde{g} \circ \theta$.

It remains to consider uniqueness of $\widetilde{g}$. But $M \otimes_S N$ is generated by the set $\{m \otimes_S n = \theta(m, n) : m \in M, \ n \in N\}$, and any function $\widetilde{g}$ such that $\widetilde{g} \circ \theta = g$ satisfies $\widetilde{g}(m \otimes n) = \widetilde{g}(\theta(m, n)) = g(m, n)$, so $\widetilde{g}$ is uniquely specified on a generating set and, hence, is uniquely determined.

Now suppose that we have (R, T)-bimodules P_i and S-middle linear maps $\theta_i : M \times N \to P_i$ such that, for any (R, T)-bimodule Q and any S-middle linear map $g : M \times N \to Q$, there exist unique (R, T)-bimodule homomorphisms $\widetilde{g}_i : P_i \to Q$ with $g = \widetilde{g}_i \circ \theta_i$ for $i = 1, 2$. We will show that P_1 and P_2 are isomorphic, and indeed that there is a unique (R, T)-bimodule isomorphism $\phi : P_1 \to P_2$ with the property that $\theta_2 = \phi \circ \theta_1$.

Let $Q = P_2$ and $g = \theta_2$. Then by the above property of P_1 there is a unique (R, T)-bimodule homomorphism $\phi : P_1 \to P_2$ with $\theta_2 = \phi \circ \theta_1$. We need only show that ϕ is an isomorphism. To this end, let $Q = P_1$ and $g = \theta_1$ to obtain $\psi : P_2 \to P_1$ with $\theta_1 = \psi \circ \theta_2$. Then

$$\theta_1 = \psi \circ \theta_2 = \psi \circ (\phi \circ \theta_1) = (\psi \circ \phi) \circ \theta_1.$$

Now apply the above property of P_1 again with $Q = P_1$ and $g = \theta_1$. Then there is a unique $\widetilde{g}$ with $g = \widetilde{g} \circ \theta_1$, i.e., a unique $\widetilde{g}$ with $\theta_1 = \widetilde{g} \circ \theta_1$. Obviously, $\widetilde{g} = 1_{P_1}$ satisfies this condition but so does $\widetilde{g} = \psi \circ \phi$, so we conclude that $\psi \circ \phi = 1_{P_1}$.

Similarly, $\phi \circ \psi = 1_{P_2}$, so $\psi = \phi^{-1}$, and we are done. $\square$

(2.13) Remarks.

(1) If M is a right R-module and N is a left R-module, then $M \otimes_R N$ is an abelian group.

(2) If M and N are both (R, R)-bimodules, then $M \otimes_R N$ is an (R, R)-bimodule. A particular (important) case of this occurs when R is a commutative ring. In this case every left R-module is automatically a right R-module, and vice-versa. Thus, over a commutative ring R, it is meaningful to speak of the tensor product of R-modules, without explicit attention to the subtleties of bimodule structures.

(3) Suppose that M is a left R-module and S is a ring that contains R as a subring. Then we can form the tensor product $S \otimes_R M$ which has the structure of an $(S, \mathbf{Z})$-bimodule, i.e, $S \otimes_R M$ is a left S-module. This construction is called **change of rings** and it is useful when one would like to be able to multiply elements of M by scalars from a bigger ring. For example, if V is any vector space over $\mathbf{R}$, then $\mathbf{C} \otimes_\mathbf{R} V$ is a vector space over the complex numbers. This construction has been implicitly used in the proof of Theorem 4.6.23.

(4) If R is a commutative ring, M a free R-module, and ϕ a bilinear form on M, then $\phi : M \times M \to R$ is certainly middle linear, and so ϕ induces an R-module homomorphism

$$\widetilde{\phi} : M \otimes_R M \to R.$$

(2.14) Corollary.

(1) *Let M and M' be (R, S)-bimodules, let N and N' be (S, T)-bimodules, and suppose that $f : M \to M'$ and $g : N \to N'$ are bimodule homomorphisms. Then there is a unique (R, T)-bimodule homomorphism*

$$(2.20) \qquad f \otimes g = f \otimes_S g : M \otimes_S N \longrightarrow M' \otimes_S N'$$

satisfying $(f \otimes g)(m \otimes n) = f(m) \otimes g(n)$ for all $m \in M$, $n \in N$.

(2) *If M'' is another (R, S)-bimodule, N'' is an (S, T)-bimodule, and $f'' : M' \to M''$, $g'' : N' \to N''$ are bimodule homomorphisms, then letting $f \otimes g : M \otimes N \to M' \otimes N'$ and $f' \otimes g' : M' \otimes N' \to M'' \otimes N''$ be defined as in part (1), we have*

$$(f' \otimes g')(f \otimes g) = (f'f) \otimes (g'g) : M \otimes N \longrightarrow M'' \otimes N''.$$

Proof. (1) Let F be the free abelian group on $M \times N$ used in the definition of $M \otimes_S N$, and let $h : F \to M' \otimes_S N'$ be the unique **Z**-module homomorphism such that $h(m, n) = f(m) \otimes_S g(n)$. Since f and g are bimodule homomorphisms, it is easy to check that h is an S-middle linear map, so by Theorem 2.12, there is a unique bimodule homomorphism $\tilde{h} : M \otimes N \to M' \otimes N'$ such that $h = \tilde{h} \circ \theta$ where $\theta : M \times N \to M \otimes N$ is the canonical map. Let $f \otimes g = \tilde{h}$. Then

$$(f \otimes g)(m \otimes n) = \tilde{h}(m \otimes n) = \tilde{h} \circ \theta(m, n) = h(m, n) = f(m) \otimes g(n)$$

as claimed.

(2) is a routine calculation, which is left as an exercise.   □

We will now consider some of the standard canonical isomorphisms relating various tensor product modules. The verifications are, for the most part, straightforward applications of Theorem 2.12. A few representative calculations will be presented; the others are left as exercises.

(2.15) Proposition. *Let M be an (R, S)-bimodule. Then there are (R, S)-bimodule isomorphisms*

$$R \otimes_R M \cong M \quad and \quad M \otimes_S S \cong M.$$

Proof. We check the first isomorphism; the second is similar. Let $f : R \times M \to M$ be defined by $f(r, m) = rm$. It is easy to check that f is an R-middle linear map, and thus Theorem 2.12 gives an (R, S)-bimodule homomorphism $\tilde{f} : R \otimes_R M \to M$ such that $\tilde{f}(r \otimes m) = rm$. Define $g : M \to R \otimes_R M$ by $g(m) = 1 \otimes m$. Then g is an (R, S)-bimodule homomorphism, and it is immediate that $\tilde{f}$ and g are inverses of each other.   □

(2.16) Proposition. *Let R be a commutative ring and let M and N be R-modules. Then*

$$M \otimes_R N \cong N \otimes_R M.$$

Proof. The isomorphism is given (via an application of Theorem 2.12) by $m \otimes n \mapsto n \otimes m$.   □

(2.17) Proposition. *Let M be an (R, S)-bimodule, N an (S, T)-bimodule, and P a (T, U)-bimodule. Then there is an isomorphism of (R, U)-bimodules*

$$(M \otimes_S N) \otimes_T P \cong M \otimes_S (N \otimes_T P).$$

Proof. Fix an element $p \in P$ and define a function

$$f_p : M \times N \to M \otimes_S (N \otimes_T P)$$

by

$$f_p(m, \, n) = m \otimes_S (n \otimes_T p).$$

f_p is easily checked to be S-middle linear, so Theorem 2.12 applies to give an (R, T)-bimodule homomorphism $\widetilde{f}_p : M \otimes_S N \to M \otimes_S (N \otimes_T P)$. Then we have a map $f : (M \otimes_S N) \times P \to M \otimes_S (N \otimes_T P)$ defined by

$$f((m \otimes n), \, p) = \widetilde{f}_p(m \otimes n) = m \otimes (n \otimes p).$$

But f is T-middle linear, and hence there is a map of (R, U)-bimodules

$$\widetilde{f} : (M \otimes_S N) \otimes_T P \longrightarrow M \otimes_S (N \otimes_T P)$$

satisfying $\widetilde{f}((m \otimes n) \otimes p) = m \otimes (n \otimes p)$. Similarly, there is an (R, U)-bimodule homomorphism

$$\widetilde{g} : M \otimes_S (N \otimes_T P) \longrightarrow (M \otimes_S N) \otimes_T P$$

satisfying $\widetilde{g}(m \otimes (n \otimes p)) = (m \otimes n) \otimes p$. Clearly, $\widetilde{g}\widetilde{f}$ (respectively $\widetilde{f}\widetilde{g}$) is the identity on elements of the form $(m \otimes n) \otimes p$ (respectively, $m \otimes (n \otimes p)$), and since these elements generate the respective tensor products, we conclude that $\widetilde{f}$ and $\widetilde{g}$ are isomorphisms.   □

(2.18) Proposition. *Let $M = \oplus_{i \in I} M_i$ be a direct sum of (R, S)-bimodules, and let $N = \oplus_{j \in J} N_j$ be a direct sum of (S, T)-bimodules. Then there is an isomorphism*

$$M \otimes_S N \cong \bigoplus_{i \in I} \bigoplus_{j \in J} (M_i \otimes_S N_j)$$

of (R, T)-bimodules.

Proof. Exercise.   □

(2.19) *Remark.* When one is taking Hom and tensor product of various bimodules, it can be somewhat difficult to keep track of precisely what type of module structure is present on the given Hom or tensor product. The following is a useful mnemonic device for keeping track of the various module structures when forming Hom and tensor products. We shall write $_RM_S$ to indicate that M is an (R, S)-bimodule. When we form the tensor product of an (R, S)-bimodule and an (S, T)-bimodule, then the resulting module has the structure of an (R, T)-bimodule (Definition 2.9). This can be indicated mnemonically by

$$(2.21) \qquad\qquad _RM_S \otimes_S {}_SN_T = {}_RP_T.$$

Note that the two subscripts "S" on the bimodules appear adjacent to the subscript "S" on the tensor product sign, and after forming the tensor product they all disappear leaving the outside subscripts to denote the bimodule type of the answer ($=$ tensor product).

A similar situation holds for Hom, but with one important difference. Recall from Proposition 2.4 that if M is an (R, S)-bimodule and N is an (R, T)-bimodule, then $\text{Hom}_R(M, N)$ has the structure of an (S, T)-bimodule. (Recall that $\text{Hom}_R(M, N)$ denotes the *left* R-module homomorphisms.) In order to create a simple mnemonic device similar to that of Equation (2.21), we make the following definition. If M and N are left R-modules, then we will write $M \pitchfork_R N$ for $\text{Hom}_R(M, N)$. Using $\pitchfork_R$ in place of $\otimes_R$, we obtain the same convention about matching subscripts disappearing, leaving the outer subscripts to give the bimodule type, *provided* that the order of the subscripts of the module on the left of the $\pitchfork_R$ sign are reversed. Thus, Proposition 2.4 is encoded in this context as the statement

$$_RM_S \quad \text{and} \quad _RN_T \implies {}_SM_R \pitchfork_R {}_RN_T = {}_SP_T.$$

A similar convention holds for homomorphisms of *right* T-modules. This is illustrated by

$$\text{Hom}_{-T}(_RM_T, {}_SN_T) = {}_SN_T \pitchfork_{-T} {}_TM_R = {}_SP_R,$$

the result being an (S, R)-bimodule (see Proposition 2.5). Note that we must reverse the subscripts on M and interchange the position of M and N.

We shall now investigate the connection between Hom and tensor product. This relationship will allow us to deduce the effect of tensor products on exact sequences, using the known results for Hom (Theorems 2.7 and 2.8 in the current section, which are generalizations of Theorems 3.3.10 and 3.3.12).

(2.20) Theorem. (Adjoint associativity of Hom and tensor product) *Let M_1 and M_2 be (S, R)-bimodules, N a (T, S)-bimodule, and P a (T, U)-bimodule. If $\psi : M_2 \to M_1$ is an (S, R)-bimodule homomorphism, then* *there are (R, U)-bimodule isomorphisms*

$$\Phi_i : \mathrm{Hom}_S(M_i, \mathrm{Hom}_T(N, P)) \longrightarrow \mathrm{Hom}_T(N \otimes_S M_i, P)$$

such that the following diagram commutes:

$$(2.22) \quad \begin{array}{ccc} \mathrm{Hom}_S(M_1, \mathrm{Hom}_T(N, P)) & \overset{\psi^*}{\longrightarrow} & \mathrm{Hom}_S(M_2, \mathrm{Hom}_T(N, P)) \\ \Big\downarrow{\Phi_1} & & \Big\downarrow{\Phi_2} \\ \mathrm{Hom}_T(N \otimes_S M_1, P) & \overset{(1_N \otimes_S \phi)^*}{\longrightarrow} & \mathrm{Hom}_T(N \otimes_S M_2, P) \end{array}$$

Proof. Define $\Phi_i : \mathrm{Hom}_S(M_i, \mathrm{Hom}_T(N, P)) \to \mathrm{Hom}_T(N \otimes_S M_i, P)$ by

$$\Phi_i(f)(n \otimes m) = (f(m))(n)$$

where $f \in \mathrm{Hom}_S(M_i, \mathrm{Hom}_T(N, P))$, $m \in M_i$, and $n \in N$. It is easy to check that $\Phi_i(f) \in \mathrm{Hom}_T(N \otimes_S M, P)$ and that Φ is a homomorphism of (R, U)-bimodules. The inverse map is given by

$$(\Psi_i(g)(m))(n) = g(m \otimes n)$$

where $g \in \mathrm{Hom}_T(N \otimes_S M, P)$, $m \in M$, and $n \in N$. To check the commutativity of the diagram, suppose that $f \in \mathrm{Hom}_S(M, \mathrm{Hom}_T(N, P))$, $n \in N$, and $m_2 \in M_2$. Then

$$\begin{aligned} ((\Phi_2 \circ \psi^*)(f))(n \otimes m_2) &= (\Phi_2(f \circ \psi))(n \otimes m_2) \\ &= ((f \circ \psi)(m_2))(n) \\ &= f(\psi(m_2))(n) \\ &= (\Phi_1(f))(n \otimes \psi(m_2)) \\ &= (\Phi_1(f))((1_n \otimes \psi)(n \otimes m_2)) \\ &= (1_n \otimes \psi)^*(\Phi_1(f))(n \otimes m_2) \\ &= ((1_N \otimes \psi)^* \circ \Phi_1(f))(n \otimes m_2). \end{aligned}$$

Thus, $\Phi_2 \circ \psi^* = (1_N \otimes \psi)^* \circ \Phi_1$ and diagram (2.22) is commutative. $\quad\square$

There is an analogous result concerning homomorphisms of right modules. In general we shall not state results explicitly for right modules; they can usually be obtained by obvious modifications of the left module results. However, the present result is somewhat complicated, so it will be stated precisely.

(2.21) Theorem. *Let M_1 and M_2 be (R, S)-bimodules, N an (S, T)-bimodule, and P a (U, T)-bimodule. If $\psi : M_2 \to M_1$ is an (R, S)-bimodule homomorphism, then there are (U, R)-bimodule isomorphisms*

$$\Phi_i : \mathrm{Hom}_{-S}(M_i, \mathrm{Hom}_{-T}(N, P)) \longrightarrow \mathrm{Hom}_{-T}(M_i \otimes_S N, P)$$

such that the following diagram commutes:

$$
(2.23) \quad
\begin{array}{ccc}
\mathrm{Hom}_{-S}(M_1, \mathrm{Hom}_{-T}(N, P)) & \xrightarrow{\psi^*} & \mathrm{Hom}_{-S}(M_2, \mathrm{Hom}_{-T}(N, P)) \\
\Big\downarrow{\scriptstyle \Phi_1} & & \Big\downarrow{\scriptstyle \Phi_2} \\
\mathrm{Hom}_{-T}(M_1 \otimes_S N, P) & \xrightarrow{(1_N \otimes_S \phi)^*} & \mathrm{Hom}_{-T}(M_2 \otimes_S N, P)
\end{array}
$$

Proof. The proof is the same as that of Theorem 2.20. $\square$

Remark. Note that Theorems 2.20 and 2.21 are already important results in case $M_1 = M_2 = M$ and $\psi = 1_M$.

As a simple application of adjoint associativity, there is the following result.

(2.22) Corollary. *Let M be an (R, S)-bimodule, N an (S, T)-bimodule, and let $P = M \otimes_S N$ (which is an (R, T)-bimodule). If M is projective as a left R-module (resp., as a right S-module) and N is projective as a left S-module (resp., as a right T-module), then P is projective as a left R-module (resp., as a right T-module).*

Proof. To show that P is projective as a left R-module, we must show that, given any surjection $f : A \to B$ of R-modules, the induced map

$$f_* : \mathrm{Hom}_R(P, A) \longrightarrow \mathrm{Hom}_R(P, B)$$

is also surjective. By hypothesis, M is projective as a left R-module so that

$$f_* : \mathrm{Hom}_R(M, A) \longrightarrow \mathrm{Hom}_R(M, B)$$

is surjective. Also, N is assumed to be projective as a left S-module, so the map

$$(f_*)_* : \mathrm{Hom}_S(N, \mathrm{Hom}_R(M, A)) \longrightarrow \mathrm{Hom}_S(N, \mathrm{Hom}_R(M, A))$$

is also surjective. But, by Theorem 2.20, if $C = A$ or B, then

$$\mathrm{Hom}_S(N, \mathrm{Hom}_R(M, C)) \cong \mathrm{Hom}_R(P, C).$$

It is simple to check that in fact there is a commutative diagram

$$
\begin{array}{ccc}
\mathrm{Hom}_S(N, \mathrm{Hom}_R(M, A)) & \xrightarrow{(f_*)_*} & \mathrm{Hom}_S(N, \mathrm{Hom}_R(M, B)) \\
\Big\downarrow{\scriptstyle \Phi_1} & & \Big\downarrow{\scriptstyle \Phi_2} \\
\mathrm{Hom}_R(P, A) & \xrightarrow{f_*} & \mathrm{Hom}_R(P, B)
\end{array}
$$

and this completes the proof. $\square$

One of the most important consequences of the adjoint associativity property relating Hom and tensor product is the ability to prove theorems concerning the exactness of sequences of tensor product modules by appealing to the theorems on exactness of Hom sequences, namely, Theorems 2.7 and 2.8.

(2.23) Theorem. *Let N be a fixed (R, T)-bimodule. If*

$$(2.24) \qquad\qquad M_1 \xrightarrow{\phi} M \xrightarrow{\psi} M_2 \longrightarrow 0$$

is an exact sequence of (S, R)-bimodules, then

$$(2.25) \qquad M_1 \otimes_R N \xrightarrow{\phi \otimes 1_N} M \otimes_R N \xrightarrow{\psi \otimes 1_N} M_2 \otimes_R N \longrightarrow 0$$

is an exact sequence of (S, T)-bimodules, while if (2.24) is an exact sequence of (T, S)-bimodules, then

$$(2.26) \qquad N \otimes_T M_1 \xrightarrow{1_N \otimes \phi} N \otimes_T M \xrightarrow{1_N \otimes \psi} N \otimes_T M_2 \longrightarrow 0$$

is an exact sequence of (R, S)-bimodules.

Proof. We will prove the exactness of sequence (2.26); exactness of sequence (2.25) is similar and it is left as an exercise. According to Theorem 2.7, in order to check the exactness of sequence (2.26), it is sufficient to check that the induced sequence

$$(2.27) \quad 0 \longrightarrow \text{Hom}_R(N \otimes_T M_2, \, P) \longrightarrow \text{Hom}_R(N \otimes_T M, \, P)$$
$$\longrightarrow \text{Hom}_R(N \otimes_T M_1, \, P)$$

is exact for every (R, U)-bimodule P. But Theorem 2.20 identifies sequence (2.27) with the following sequence, which is induced from sequence (2.24) by the (T, U)-bimodule $\text{Hom}_R(N, P)$:

$$(2.28) \quad 0 \longrightarrow \text{Hom}_T(M_2, \text{Hom}_R(N, \, P)) \longrightarrow \text{Hom}_T(M, \text{Hom}_R(N, \, P))$$
$$\longrightarrow \text{Hom}_T(M_1, \text{Hom}_R(N, \, P)).$$

Since (2.24) is assumed to be exact, Theorem 2.7 shows that sequence (2.28) is exact for any (R, U)-bimodule P. Thus sequence (2.27) is exact for all P, and the proof is complete. $\qquad\qquad\square$

(2.24) Examples.

(1) Consider the following short exact sequence of **Z**-modules:

$$(2.29) \qquad\qquad 0 \longrightarrow \mathbf{Z} \xrightarrow{\phi} \mathbf{Z} \xrightarrow{\psi} \mathbf{Z}_m \longrightarrow 0$$

where $\phi(i) = mi$ and ψ is the canonical projection map. If we take $N = \mathbf{Z}_n$, then exact sequence (2.25) becomes

(2.30) $\mathbf{Z} \otimes \mathbf{Z}_n \xrightarrow{\phi \otimes 1} \mathbf{Z} \otimes \mathbf{Z}_n \xrightarrow{\psi \otimes 1} \mathbf{Z}_m \otimes \mathbf{Z}_n \longrightarrow 0.$

By Proposition 2.15, exact sequence (2.30) becomes the exact sequence

(2.31) $\mathbf{Z}_n \xrightarrow{\widetilde{\phi}} \mathbf{Z}_n \xrightarrow{\widetilde{\psi}} \mathbf{Z}_m \otimes \mathbf{Z}_n \longrightarrow 0$

where $(\widetilde{\phi})(i) = mi$. Thus $\mathbf{Z}_m \otimes \mathbf{Z}_n \simeq \mathrm{Coker}(\widetilde{\phi})$. Now let $d = \gcd(m, n)$ and write $m = m'd$, $n = n'd$. Then the map $\widetilde{\phi}$ is the composite

$$\mathbf{Z}_n \xrightarrow{\phi_1} \mathbf{Z}_n \xrightarrow{\phi_2} \mathbf{Z}_n$$

where $\phi_1(i) = m'i$ and $\phi_2(i) = di$. Since $\gcd(m', n) = 1$, it follows that ϕ_1 is an isomorphism (Proposition 1.4.11), while $\mathrm{Im}(\phi_2) = d\mathbf{Z}_n$. Hence, $\mathrm{Coker}(\widetilde{\phi}) \cong \mathbf{Z}_n / d\mathbf{Z}_n \cong \mathbf{Z}_d$, i.e.,

$$\mathbf{Z}_m \otimes \mathbf{Z}_n \cong \mathbf{Z}_d.$$

(2) Suppose that M is any finite abelian group. Then

$$M \otimes_{\mathbf{Z}} \mathbf{Q} = \langle 0 \rangle.$$

To see this, consider a typical generator $x \otimes r$ of $M \otimes_{\mathbf{Z}} \mathbf{Q}$, where $x \in M$ and $r \in \mathbf{Q}$. Let $n = |M|$. Then $nx = 0$ and, according to Equation (2.18),

$$x \otimes r = x \otimes n(r/n) = xn \otimes (r/n) = 0 \otimes (r/n) = 0.$$

Since $x \in M$ and $r \in \mathbf{Q}$ are arbitrary, it follows that every generator of $M \otimes \mathbf{Q}$ is 0.

(3) Let R be a commutative ring, $I \subseteq R$ an ideal, and M any R-module. Then

(2.32) $(R/I) \otimes_R M \cong M/IM.$

To see this consider the exact sequence of R-modules

$$0 \longrightarrow I \xrightarrow{\iota} R \longrightarrow R/I \longrightarrow 0.$$

Tensor this sequence of R-modules with M to obtain an exact sequence

$$I \otimes_R M \xrightarrow{\iota \otimes 1} R \otimes_R M \longrightarrow (R/I) \otimes_R M \longrightarrow 0.$$

But according to Proposition 2.15, $R \otimes_R M \cong M$ (via the isomorphism $\Phi(r \otimes m) = rm$), and under this identification it is easy to see that $\mathrm{Im}(\iota \otimes 1) = IM$. Thus, $(R/I) \otimes_R M \cong M/IM$, as we wished to verify.

Example 2.32 (1) shows that even if a sequence

$$0 \longrightarrow M_1 \longrightarrow M \longrightarrow M_2 \longrightarrow 0$$

is short exact, the tensored sequence (2.25) need not be part of a short exact sequence, i.e., the initial map need not be injective. For a simple situation where this occurs, take $m = n$ in Example 2.32 (1). Then exact sequence (2.30) becomes

$$\mathbf{Z}_n \xrightarrow{\phi \otimes 1} \mathbf{Z}_n \longrightarrow \mathbf{Z}_n \longrightarrow 0.$$

The map $\phi \otimes 1$ is the zero map, so it is certainly not an injection.

This example, plus our experience with Hom, suggests that we consider criteria to ensure that tensoring a short exact sequence with a fixed module produces a short exact sequence. We start with the following result, which is exactly analogous to Theorem 2.8 for Hom.

(2.25) Theorem. *Let N be a fixed (R, T)-bimodule. If*

$$(2.33) \qquad\qquad 0 \longrightarrow M_1 \xrightarrow{\phi} M \xrightarrow{\psi} M_2 \longrightarrow 0$$

is a split short exact sequence of (S, R)-bimodules, then

$$(2.34) \qquad 0 \longrightarrow M_1 \otimes_R N \xrightarrow{\phi \otimes 1_N} M \otimes_R N \xrightarrow{\psi \otimes 1_N} M_2 \otimes_R N \longrightarrow 0$$

is a split short exact sequence of (S, T)-bimodules, while if (2.33) is a split short exact sequence of (T, S)-bimodules, then

$$(2.35) \qquad 0 \longrightarrow N \otimes_T M_1 \xrightarrow{1_N \otimes \phi} N \otimes_T M \xrightarrow{1_N \otimes \psi} N \otimes_T M_2 \longrightarrow 0$$

is a split short exact sequence of (R, S)-bimodules.

Proof. We will do sequence (2.34); (2.35) is similar and is left as an exercise. Let $\alpha : M \to M_1$ split ϕ, and consider the map

$$\alpha \otimes 1 : M \otimes_R N \to M_1 \otimes_R N.$$

Then

$$((\alpha \otimes 1)(\phi \otimes 1))(m \otimes n) = (\alpha\phi \otimes 1)(m \otimes n) = (1 \otimes 1)(m \otimes n) = m \otimes n.$$

so that $\phi \otimes 1$ is an injection, which is split by $\alpha \otimes 1$. The rest of the exactness is covered by Theorem 2.23. $\qquad\square$

(2.26) Remark. Theorems 2.7 and 2.23 show that given a short exact sequence, applying Hom or tensor product will give a sequence that is exact on one end or the other, but in general not on both. Thus Hom and tensor product are both called **half exact**, and more precisely, Hom is called **left exact** and tensor product is called **right exact**. We will now investigate some conditions under which the tensor product of a module with a short exact sequence always produces a short exact sequence. It was precisely this type of consideration for Hom that led us to the concept of projective module. In fact, Theorem 3.5.1 (4) shows that if P is a projective R-module and

$$0 \longrightarrow M_1 \xrightarrow{\phi} M \xrightarrow{\psi} M_2 \longrightarrow 0$$

is a short exact sequence of R-modules, then the sequence

$$0 \longrightarrow \mathrm{Hom}_R(P, M_1) \xrightarrow{\phi_*} \mathrm{Hom}_R(P, M) \xrightarrow{\psi_*} \mathrm{Hom}_R(P, M_2) \longrightarrow 0$$

is short exact. According to Theorem 3.3.10, the crucial ingredient needed is the surjectivity of ψ_* and this is what projectivity of P provides. For the case of tensor products, the crucial fact needed to obtain a short exact sequence will be the injectivity of the initial map of the sequence.

(2.27) Proposition. *Let N be an (R, T)-bimodule that is projective as a left R-module. Then for any injection $\iota : M_1 \to M$ of (S, R)-bimodules,*

$$\iota \otimes 1 : M_1 \otimes_R N \longrightarrow M \otimes_R N$$

is an injection of (S, T)-bimodules. If N is projective as a right T-module and $\iota : M_1 \to M$ is an injection of (T, S)-bimodules, then

$$1 \otimes \iota : N \otimes_T M_1 \longrightarrow N \otimes_T M$$

is an injection of (R, S)-bimodules.

Proof. First suppose that as a left R-module N is free with a basis $\{n_j\}_{j \in J}$. Then $N \cong \oplus_{j \in J} R_j$ where each summand $R_j = Rn_j$ is isomorphic to R as a left R-module. Then by Proposition 2.18

$$M_1 \otimes_R N \cong \bigoplus_{j \in J} (M_1 \otimes_R R_j) = \bigoplus_{j \in J} M_{1j}$$

where each M_{1j} is isomorphic to M_1 as a left S-module, and similarly $M \otimes_R N \cong \oplus_{j \in J} M_j$, where each M_j is isomoprhic to M as a left S-module. Furthermore, the map $\iota \otimes 1 : M_1 \otimes_R N \to M \otimes_R N$ is given as a direct sum

$$\bigoplus_{j \in J} (\iota_j : M_{1j} \to M_j)$$

where each ι_j agrees with ι under the above identifications. But then, since ι is an injection, so is each ι_j, and hence so is $\iota \otimes 1$.

Now suppose that N is projective as a left R-module. Then there is a left R-module N' such that $N \oplus N' = F$ where F is a free left R-module. We have already shown that

$$\iota \otimes 1 : M_1 \otimes_R F \longrightarrow M \otimes_R F$$

is an injection. But using Proposition 2.18 again,

$$M_1 \otimes_R F \cong (M_1 \otimes_R N) \oplus (M_1 \otimes_R N')$$

so we may write $\iota \otimes 1 = \iota_1 \oplus \iota_2$ where (in particular) $\iota_1 = \iota \otimes 1 : M_1 \otimes_R N \rightarrow M \otimes_R F$. Since $\iota \otimes 1$ is an injection, so is ι_1, as claimed. Thus the proof is complete in the case that N is projective as a left R-module. The proof in case N is projective as a right T-module is identical. $\square$

Note that we have not used the right T-module structures in the above proof. This is legitimate, since if a homomorphism is injective as a map of left S-modules, and it is an (S, T)-bimodule map, then it is injective as an (S, T)-bimodule map.

(2.28) Corollary. *Let N be a fixed (R, T)-bimodule that is projective as a left R-module. If*

$$(2.36) \qquad 0 \longrightarrow M_1 \overset{\phi}{\longrightarrow} M \overset{\psi}{\longrightarrow} M_2 \longrightarrow 0$$

is a short exact sequence of (S, R)-bimodules, then

$$(2.37) \qquad 0 \longrightarrow M_1 \otimes_R N \overset{\phi \otimes 1_N}{\longrightarrow} M \otimes_R N \overset{\psi \otimes 1_N}{\longrightarrow} M_2 \otimes_R N \longrightarrow 0$$

is a short exact sequence of (S, T)-bimodules; while if (2.36) is an exact sequence of (T, S)-bimodules and N is projective as a right T-module, then

$$(2.38) \qquad 0 \longrightarrow N \otimes_T M_1 \overset{1_N \otimes \phi}{\longrightarrow} N \otimes_T M \overset{1_N \otimes \psi}{\longrightarrow} N \otimes_T M_2 \longrightarrow 0$$

is a short exact sequence of (R, S)-bimodules.

Proof. This follows immediately from Theorem 2.23 and Proposition 2.27. $\square$

(2.29) *Remark.* A module satisfying the conclusion of Proposition 2.27 is said to be **flat**. That is, a left R-module N is flat if tensoring with all short exact sequences of right R-modules produces a short exact sequence, with a similar definition for right R-modules. Given Theorem 2.23, in order to prove that a left R-module N is flat, it is sufficient to prove that for all right R-modules M and submodules K, the inclusion map $\iota : K \rightarrow M$ induces an injective map

$$\iota \otimes 1 : K \otimes_R N \longrightarrow M \otimes_R N.$$

Thus, what we have proven is that projective modules are flat.

In Section 6.1 we discussed duality for free modules over commutative rings. Using the theory developed in the current section, we will extend portions of our discussion of duality to the context of projective (bi-)modules.

(2.30) Definition. *Let M be an (R, S)-bimodule. The* **dual module** *of M is the (S, R)-bimodule M^* defined by*

$$\text{Hom}_R(M, R).$$

In particular, if M is a left R-module, i.e., take $S = \mathbf{Z}$, then the dual module M^* is a right R-module. The **double dual** *of M is defined to be*

$$M^{**} = \mathrm{Hom}_{-R}(M^*, R).$$

As in Section 6.1, there is a homomorphism $\eta : M \to M^{**}$ of (R, S)-bimodules defined by

$$(\eta(v))(\omega) = \omega(v) \quad for\ all\ v \in M,\ \omega \in M^*$$

and if η is an isomorphism, then we will say that M is **reflexive**.

If M is an (R, S)-bimodule, which is finitely generated and free as a left R-module, then given any basis $\mathcal{B}$ of M, one may construct a basis of M^* (as a right R-module) exactly as in Definition 6.1.3 and the proof of Theorem 6.1.7 goes through verbatim to show that finitely generated free R-modules are reflexive, even when R need not be commutative. Furthermore, the proofs of Theorems 3.5.8 and 6.1.13 go through without difficulty if one keeps track of the types of modules under consideration. We will simply state the following result and leave the details of tracing through the module types as an exercise.

(2.31) Proposition. *Let M be an (R, S)-bimodule, which is finitely generated and projective as a left R-module. Then the dual module M^* is finitely generated and projective as a right R-module. Furthermore, M is reflexive as an (R, S)-bimodule.*

Proof. Exercise. See the comments above.   ☐

If M is an (R, S)-bimodule and P is an (R, T)-bimodule, then define

$$\zeta : M^* \times P \to \mathrm{Hom}_R(M, P)$$

by

$$(\zeta(\omega, p))(m) = \omega(m)p \quad for\ \ \omega \in M^*,\ p \in P,\ and\ m \in M.$$

Then ζ is S-middle linear and hence it induces an (S, T)-bimodule homomorphism

$$\widetilde{\zeta} : M^* \otimes_R P \longrightarrow \mathrm{Hom}_R(M, P)$$

given by

(2.39) $$(\widetilde{\zeta}(\omega \otimes p))(m) = \omega(m)p$$

for all $\omega \in M^*$, $p \in P$, and $m \in M$.

(2.32) Proposition. *Let M be an (R, S)-bimodule, which is finitely generated and projective as a left R-module, and let P be an arbitrary (R, T)-bimodule. Then the map*

$$\widetilde{\zeta} : M^* \otimes_R P \longrightarrow \mathrm{Hom}_R(M, P)$$

defined by Equation (2.39) is an (S, T)-bimodule isomorphism.

Proof. Since $\widetilde{\zeta}$ is an (S, T)-bimodule homomorphism, it is only necessary to prove that it is bijective. To achieve this first suppose that M is free of finite rank k as a left R-module. Let $\mathcal{B} = \{v_1, \ldots, v_k\}$ be a basis of M and let $\{v_1^*, \ldots, v_k^*\}$ be the basis of M^* dual to $\mathcal{B}$. Note that every element of $M^* \otimes_R P$ can be written as $x = \sum_{i=1}^k v_i^* \otimes p_i$ for $p_1, \ldots, p_k \in P$. Suppose that $\widetilde{\zeta}(x) = 0$, i.e., $(\widetilde{\zeta}(x))(m) = 0$ for every $m \in M$. But $\widetilde{\zeta}(x)(v_i) = p_i$ so that $p_i = 0$ for $1 \le i \le k$. That is, $x = 0$ and we conclude that $\widetilde{\zeta}$ is injective.

Given any $f \in \mathrm{Hom}_R(M, P)$, let

$$x_f = \sum_{i=1}^k v_i^* \otimes f(v_i).$$

Then $(\widetilde{\zeta}(x_f))(v_i) = f(v_i)$ for $1 \le i \le k$, i.e., $\widetilde{\zeta}(x_f)$ and f agree on a basis of M; hence, $\widetilde{\zeta}(x_f) = f$ and $\widetilde{\zeta}$ is a surjection, and the proof is complete in case M is free of rank k.

Now suppose that M is finitely generated and projective, and let N be a left R-module such that $F = M \oplus N$ is finitely generated and free. Then $\widetilde{\zeta} : F^* \otimes_R P \to \mathrm{Hom}_R(F, P)$ is a $\mathbf{Z}$-module isomorphism, and

$$F^* \otimes_R P = (M \oplus N)^* \otimes_R P \cong (M^* \oplus N^*) \otimes_R P \cong (M^* \otimes_R P) \oplus (N^* \otimes_R P)$$

while

$$\mathrm{Hom}_R(F, P) = \mathrm{Hom}_R(M \oplus N, P) \cong \mathrm{Hom}_R(M, P) \oplus \mathrm{Hom}_R(N, P)$$

where all isomorphisms are $\mathbf{Z}$-module isomorphisms. Under these isomorphisms,

$$\widetilde{\zeta}_F = \widetilde{\zeta}_M \oplus \widetilde{\zeta}_N$$
$$\widetilde{\zeta}_M : M^* \otimes_R P \longrightarrow \mathrm{Hom}_R(M, P)$$
$$\widetilde{\zeta}_N : N^* \otimes_R P \longrightarrow \mathrm{Hom}_R(N, P).$$

Since $\widetilde{\zeta}_F$ is an isomorphism, it follows that $\widetilde{\zeta}_M$ and $\widetilde{\zeta}_N$ are isomorphisms as well. In particular, $\widetilde{\zeta}_M$ is bijective and the proof is complete. $\square$

(2.33) Corollary. *Let M be an (R, S)-bimodule, which is finitely generated and projective as a left R-module, and let P be an arbitrary (T, R)-bimodule. Then*

$$M^* \otimes_R P^* \cong (P \otimes_R M)^*$$

as (S, T)-bimodules.

Proof. From Proposition 2.32, there is an isomorphism

$$M^* \otimes_R P^* \cong \operatorname{Hom}_R(M, P^*)$$
$$= \operatorname{Hom}_R(M, \operatorname{Hom}_R(P, R))$$
$$\cong \operatorname{Hom}_R(P \otimes_R M, R) \qquad \text{(by adjoint associativity)}$$
$$= (P \otimes_R M)^*.$$

$\square$

(2.34) Remark. The isomorphism of Corollary 2.33 is given explicitly by

$$\phi(f \otimes g)(p \otimes m) = f(m)g(p) \in R$$

where $f \in M^*$, $g \in P^*$, $p \in P$, and $m \in M$.

We will conclude this section by studying the matrix representation of the tensor product of R-module homomorphisms. Thus, let R be a commutative ring, let M_1, M_2, N_1, and N_2 be finite rank free R-modules, and let $f_i : M_i \to N_i$ be R-module homomorphisms for $i = 1, 2$. Let m_i be the rank of M_i and n_i the rank of N_i for $i = 1, 2$. If $M = M_1 \otimes M_2$ and $N = N_1 \otimes N_2$, then it follows from Proposition 2.18 that M and N are free R-modules of rank $m_1 n_1$ and $m_2 n_2$, respectively. Let $f = f_1 \otimes f_2 \in \operatorname{Hom}_R(M, N)$. We will compute a matrix representation for f from that for f_1 and f_2. To do this, suppose that

$$\mathcal{A} = \{a_1, \ldots, a_{m_1}\}$$
$$\mathcal{B} = \{b_1, \ldots, b_{n_1}\}$$
$$\mathcal{C} = \{c_1, \ldots, c_{m_2}\}$$
$$\mathcal{D} = \{d_1, \ldots, c_{n_2}\}$$

are bases of M_1, N_1, M_2, and N_2, respectively. Let

$$\mathcal{E} = \{a_1 \otimes c_1, \ a_1 \otimes c_2, \ \ldots, a_1 \otimes c_{m_2},$$
$$a_2 \otimes c_1, \ a_2 \otimes c_2, \ \ldots, a_2 \otimes c_{m_2},$$
$$\vdots$$
$$a_{m_1} \otimes c_1, \ a_{m_1} \otimes c_2, \ \ldots, a_{m_1} \otimes c_{m_2}\}$$

and

$$\mathcal{F} = \{b_1 \otimes d_1, \ b_1 \otimes d_2, \ \ldots, b_1 \otimes d_{n_2},$$
$$b_2 \otimes d_1, \ b_2 \otimes d_2, \ \ldots, b_2 \otimes d_{n_2},$$
$$\vdots$$
$$b_{n_1} \otimes d_1, \ b_{n_1} \otimes d_2, \ \ldots, b_{n_1} \otimes d_{n_2}\}.$$

Then $\mathcal{E}$ is a basis for M and $\mathcal{F}$ is a basis for N. With respect to these bases, there is the following result:

(2.35) Proposition. *With the notation introduced above,*

$$[f_1 \otimes f_2]_{\mathcal{F}}^{\mathcal{E}} = [f_1]_B^A \otimes [f_2]_D^C.$$

Proof. Exercise.   □

Recall that the notion of tensor product of matrices was introduced in Definition 4.1.16 and has been used subsequently in Section 5.5. If $[f_1]_B^A = A = [\alpha_{ij}]$ and $[f_2]_D^C = B = [\beta_{ij}]$, then Proposition 2.35 states that (in block matrix notation)

$$[f_1 \otimes f_2]_{\mathcal{F}}^{\mathcal{E}} = \begin{bmatrix} \alpha_{11}B & \alpha_{12}B & \cdots & \alpha_{1m_1}B \\ \vdots & \vdots & \ddots & \vdots \\ \alpha_{n_1 1}B & \alpha_{n_1 2}B & \cdots & \alpha_{n_1 m_1}B \end{bmatrix}.$$

There is another possible ordering for the bases $\mathcal{E}$ and $\mathcal{F}$. If we set

$$\mathcal{E}' = \{a_i \otimes c_j : 1 \le i \le m_1, \ 1 \le j \le m_2\}$$

and

$$\mathcal{F}' = \{b_i \otimes d_j : 1 \le i \le n_1, \ 1 \le j \le n_2\}$$

where the elements are ordered by first fixing j and letting i increase (lexicographic ordering with j the dominant letter), then we leave it to the reader to verify that the matrix of $f_1 \otimes f_2$ is given by

$$[f_1 \otimes f_2]_{\mathcal{F}'}^{\mathcal{E}'} = \begin{bmatrix} \beta_{11}A & \beta_{12}A & \cdots & \beta_{1m_2}A \\ \vdots & \vdots & \ddots & \vdots \\ \beta_{n_2 1}A & \beta_{n_2 2}A & \cdots & \beta_{n_2 m_2}A \end{bmatrix}.$$

7.3 Exercises

1. Let M be a simple R-module, and let N be any R-module.
    (a) Show that every nonzero homomorphism $f : M \to N$ is injective.
    (b) Show that every nonzero homomorphism $f : N \to M$ is surjective.

2. Let F be a field and let $R = \{\begin{bmatrix} a & b \\ 0 & c \end{bmatrix} : a, b, c \in F\}$ be the ring of upper triangular matrices over F. Let $M = F^2$ and make M into a (left) R-module by matrix multiplication. Show that $\text{End}_R(M) \cong F$. Conclude that the converse of Schur's lemma is false, i.e., $\text{End}_R(M)$ can be a division ring without M being a simple R-module.

3. Suppose that R is a D-algebra, where D is a division ring, and let M be an R-module which is of finite rank as a D-module. Show that as an R-module, $\ell(M) \le \text{rank}_D(M)$.

4. An R-module M is said to satisfy the decending chain condition (DCC) on submodules if any strictly decreasing chain of submodules of M is of finite length.

(a) Show that if M satisfies the DCC, then any nonempty set of submodules of M contains a minimal element.

(b) Show that $\ell(M) < \infty$ if and only if M satisfies both the ACC (ascending chain condition) and DCC.

5. Let $R = \{\left[\begin{smallmatrix} a & b \\ 0 & c \end{smallmatrix}\right] : a, b \in \mathbf{R}; c \in \mathbf{Q}\}$. R is a ring under matrix addition and multiplication. Show that R satisfies the ACC and DCC on left ideals, but neither chain condition is valid for right ideals. Thus R is of finite length as a left R-module, but $\ell(R) = \infty$ as a right R-module.

6. Let R be a ring without zero divisors. If R is not a division ring, prove that R does not have a composition series.

7. Let $f : M_1 \to M_2$ be an R-module homomorphism.
   (a) If f is injective, prove that $\ell(M_1) \leq \ell(M_2)$.
   (b) If f is surjective, prove that $\ell(M_2) \leq \ell(M_1)$.

8. Let M be an R-module of finite length and let K and N be submodules of M. Prove the following length formula:

$$\ell(K + N) + \ell(K \cap N) = \ell(K) + \ell(N).$$

9. (a) Compute $\ell(\mathbf{Z}_{p^n})$.
   (b) Compute $\ell(\mathbf{Z}_{p^n} \oplus \mathbf{Z}_{q^m})$.
   (c) Compute $\ell(G)$ where G is any finite abelian group.
   (d) More generally, compute $\ell(M)$ for any finitely generated torsion module over a PID R.

10. Compute the length of $M = F[X]/\langle f(X)\rangle$ as an $F[X]$-module if $f(X)$ is a polynomial of degree n with two distinct irreducible factors. What is the length of M as an F-module?

11. Let F be a field, let V be a finite-dimensional vector space over F, and let $T \in \text{End}_F(V)$. We shall say that T is semisimple if the $F[X]$-module V_T is semisimple. If $A \in M_n(F)$, we shall say that A is semisimple if the linear transformation $T_A : F^n \to F^n$ (multiplication by A) is semisimple. Let $\mathbf{F}_2$ be the field with 2 elements and let $F = \mathbf{F}_2(Y)$ be the rational function field in the indeterminate Y, and let $K = F[X]/\langle X^2 + Y\rangle$. Since $X^2 + Y \in F[X]$ is irreducible, K is a field containing F as a subfield. Now let

$$A = C(X^2 + Y) = \begin{bmatrix} 0 & Y \\ 1 & 0 \end{bmatrix} \in M_2(F).$$

Show that A is semisimple when considered in $M_2(F)$ but A is not semisimple when considered in $M_2(K)$. Thus, semisimplicity of a matrix is not necessarily preserved when one passes to a larger field. However, prove that if L is a subfield of the complex numbers $\mathbf{C}$, then $A \in M_n(L)$ is semisimple if and only if it is also semisimple as a complex matrix.

12. Let V be a vector space over $\mathbf{R}$ and let $T \in \text{End}_{\mathbf{R}}(V)$ be a linear transformation. Show that $T = S + N$ where S is a semisimple linear transformation, N is nilpotent, and $SN = NS$.

13. Prove that the modules M_i and N_j in the proof of Lemma 1.33 are simple, as claimed.

14. Prove Lemma 1.37.

15. If D is a division ring and n is a positive integer, prove that $\text{End}_D(D^n)$ is a simple ring.

16. Give an example of a semisimple commutative ring that is not a field.

17. (a) Prove that if R is a semisimple ring and I is an ideal, then R/I is semisimple.
    (b) Show (by example) that a subring of a semisimple ring need not be semisimple.

18. Let R be a ring that is semisimple as a left R-module. Show that R is simple if and only if all simple left R-modules are isomorphic.

19. Let M be a finitely generated abelian group. Compute each of the following groups:
    (a) $\mathrm{Hom}_{\mathbf{Z}}(M, \mathbf{Q}/\mathbf{Z})$.
    (b) $\mathrm{Hom}_{\mathbf{Z}}(\mathbf{Q}/\mathbf{Z}, M)$.
    (c) $M \otimes_{\mathbf{Z}} \mathbf{Q}/\mathbf{Z}$.

20. Let M be an (R, S)-bimodule and N an (S, T)-bimodule. Suppose that $\sum x_i \otimes y_i = 0$ in $M \otimes_S N$. Prove that there exists a finitely generated (R, S)-bisubmodule M_0 of M and a finitely generated (S, T)-bisubmodule N_0 of N such that $\sum x_i \otimes y_i = 0$ in $M_0 \otimes_S N_0$.

21. Let R be an integral domain and let M be an R-module. Let Q be the quotient field of R and define $\phi : M \to Q \otimes_R M$ by $\phi(x) = 1 \otimes x$. Show that $\mathrm{Ker}(\phi) = M_\tau$ = torsion submodule of M. (Hint: If $1 \otimes x = 0 \in Q \otimes_R M$ then $1 \otimes x = 0$ in $(Rc^{-1}) \otimes_R M \cong M$ for some $c \neq 0 \in R$. Then show that $cx = 0$.)

22. Let R be a PID and let M be a free R-module with N a submodule. Let Q be the quotient field and let $\phi : M \to Q \otimes_R M$ be the map $\phi(x) = 1 \otimes x$. Show that N is a pure submodule of M if and only if $Q \cdot (\phi(N)) \cap \mathrm{Im}(\phi) = \phi(N)$.

23. Let R be a PID and let M be a finitely generated R-module. If Q is the quotient field of R, show that $M \otimes_R Q$ is a vector space over Q of dimension equal to $\mathrm{rank}_R(M/M_\tau)$.

24. Let R be a commutative ring and S a multiplicatively closed subset of R containing no zero divisors. Let R_S be the localization of R at S. If M is an R-module, then the R_S-module M_S was defined in Exercise 6 of Chapter 3. Show that $M_S \cong R_S \otimes_R M$ where the isomorphism is an isomorphism of R_S-modules.

25. If S is an R-algebra, show that $M_n(S) \cong S \otimes_R M_n(R)$.

26. Let M and N be finitely generated R-modules over a PID R. Compute $M \otimes_R N$. As a special case, if M is a finite abelian group with invariant factors $s_1, \ldots, s_t$ (where as usual we assume that s_i divides s_{i+1}), show that $M \otimes_{\mathbf{Z}} M$ is a finite group of order $\prod_{j=1}^{t} s_j^{2t-2j+1}$.

27. Let F be a field and K a field containing F. Suppose that V is a finite-dimensional vector space over F and let $T \in \mathrm{End}_F(V)$. If $\mathcal{B} = \{v_i\}$ is a basis of V, then $\mathcal{C} = \{1\} \otimes \mathcal{B} = \{1 \otimes v_i\}$ is a basis of $K \otimes_F V$. Show that $[1 \otimes T]_{\mathcal{C}} = [T]_{\mathcal{B}}$. If $S \in \mathrm{End}_F(V)$, show that $1 \otimes T$ is similar to $1 \otimes S$ if and only if S is similar to T.

28. Let V be a complex inner product space and $T : V \to V$ a normal linear transformation. Prove that T is self-adjoint if and only if there is a real inner product space W, a self-adjoint linear transformation $S : W \to W$, and an isomorphism $\phi : \mathbf{C} \otimes_{\mathbf{R}} W \to V$ making the following diagram commute.

$$
\begin{array}{ccc}
\mathbf{C} \otimes_{\mathbf{R}} W & \xrightarrow{1 \otimes S} & \mathbf{C} \otimes_{\mathbf{R}} W \\
\downarrow{\phi} & & \downarrow{\phi} \\
V & \xrightarrow{T} & V
\end{array}
$$

29. Let R be a commutative ring.
    (a) If I and J are ideals of R, prove that
    $$R/I \otimes_R R/J \cong R/(I + J).$$

    (b) If S and T are R-algebras, I is an ideal of S, and J is an ideal of T, prove that
    $$S/I \otimes_R T/J \cong (S \otimes_R T)/\langle I, J \rangle,$$

where $\langle I, J \rangle$ denotes the ideal of $S \otimes_R T$ generated by $I \otimes_R T$ and $S \otimes_R J$.

30. (a) Let F be a field and K a field containing F. If $f(X) \in F[X]$, show that there is an isomorphism of K-algebras:

$$K \otimes_F (F[X]/\langle f(X) \rangle) \cong K[X]/\langle f(X) \rangle.$$

(b) By choosing F, $f(X)$, and K appropriately, find an example of two fields K and L containing F such that the F-algebra $K \otimes_F L$ has nilpotent elements.

31. Let F be a field. Show that $F[X, Y] \cong F[X] \otimes_F F[Y]$ where the isomorphism is an isomorphism of F-algebras.

32. Let G_1 and G_2 be groups, and let $\mathbf{F}$ be a field. Show that

$$\mathbf{F}(G_1 \times G_2) \cong \mathbf{F}(G_1) \otimes_{\mathbf{F}} \mathbf{F}(G_2).$$

33. Let R and S be rings and let $f : R \to S$ be a ring homomorphism. If N is an S-module, then we may make N into an R-module by restriction of scalars, i.e., $a \cdot x = f(a) \cdot x$ for all $a \in R$ and $x \in N$. Now form the S-module $N_S = S \otimes_R N$ and define $g : N \to N_S$ by

$$g(y) = 1 \otimes y.$$

Show that g is injective and $g(N)$ is a direct summand of N_S.

34. Let F be a field, V and W finite-dimensional vector spaces over F, and let $T \in \text{End}_F(V)$, $S \in \text{End}_F(W)$.
    (a) If α is an eigenvalue of S and β is an eigenvalue of T, show that the product $\alpha\beta$ is an eigenvalue of $S \otimes T$.
    (b) If S and T are diagonalizable, show that $S \otimes T$ is diagonalizable.

35. Let R be a semisimple ring, M an (R, S)-bimodule that is simple as a left R-module, and let P be an (R, T)-bimodule that is simple as a left R-module. Prove that

$$M^* \otimes_R P = \begin{cases} \text{End}_R(M) & \text{if } P \cong M \text{ as left } R\text{-modules} \\ 0 & \text{otherwise.} \end{cases}$$

36. Let R be a commutative ring and M an R-module. Let

$$M^{\otimes k} = M \otimes \cdots \otimes M,$$

where there are k copies of M, and let S be the submodule of $M^{\otimes k}$ generated by all elements of the form $m_1 \otimes \cdots \otimes m_k$ where $m_i = m_j$ for some $i \neq j$. Then $\Lambda^k(M) = M^{\otimes k}/S$ is called an *exterior algebra*.
    (a) Show that if M is free of rank n, then $\Lambda^k(M)$ is free and

$$\text{rank}(\Lambda^k(M)) = \begin{cases} \binom{n}{k} & \text{if } k \leq n \\ 0 & \text{if } k > n. \end{cases}$$

(b) As a special case of part (a),

$$\text{rank}(\Lambda^n(M)) = 1.$$

Show that $\text{Hom}_R(\Lambda^n(M), R)$ may be regarded as the space of determinant functions on M.

Chapter 8

Group Representations

8.1 Examples and General Results

We begin by defining the objects that we are interested in studying. Recall that if R is a ring and G is a group, then $R(G)$ denotes the group ring of G with coefficients from R. The multiplication on $R(G)$ is the convolution product (see Example 2.1.10 (15)).

(1.1) Definition. *Let G be a group and $\mathbf{F}$ a field. A* **(left) F-representation** *of G is a (left) $\mathbf{F}(G)$-module M.*

In other words, M is an $\mathbf{F}$ vector space, and for each $g \in G$ we have a linear transformation $\sigma(g) : M \to M$ given by the action of g regarded as $1g \in \mathbf{F}(G)$ on M. These linear transformations satisfy $\sigma(e) = 1_M$ and $\sigma(g_2 g_1)(m) = \sigma(g_2)(\sigma(g_1)(m))$ for all $m \in M$ and $g \in G$. Note then that

$$\sigma(g^{-1})\sigma(g) = \sigma(e) = 1_M$$

so that $\sigma(g^{-1}) = \sigma(g)^{-1}$. In particular, each $\sigma(g)$ is invertible, i.e., $\sigma(g) \in \text{Aut}(M)$.

Conversely, we may view an $\mathbf{F}$-representation of G on M, where M is an $\mathbf{F}$ vector space, as being given by a homomorphism $\sigma : G \to \text{Aut}(M)$. Then we define an $\mathbf{F}(G)$-module structure on M by

$$\left(\sum_{g \in G} a_g g \right)(m) = \sum_{g \in G} a_g \sigma(g)(m).$$

In this situation, we say that the representation is defined by σ.

We will denote a representation as above by M, when we wish to emphasize the underlying vector space, or (more often) by σ, when we wish to emphasize the homomorphism, and we will use the term representation instead of $\mathbf{F}$-representation, when $\mathbf{F}$ is understood. Occasionally (as is often done) we shall omit σ when it is understood and write $g(m)$ for $\sigma(g)(m)$.

(1.2) Definition. *The* **degree** $\deg(M)$ *of a representation* M *is*

$$\dim_{\mathbf{F}}(M) \in \{0, 1, 2, \dots\} \cup \{\infty\}.$$

Two $\mathbf{F}$-representations M_1 and M_2 of G, defined by $\sigma_i : G \to \text{Aut}(M_i)$ for $i = 1, 2$, are said to be **isomorphic (or equivalent)** if they are isomorphic as $\mathbf{F}(G)$-modules. Concretely, this is the case if and only if there is an invertible $\mathbf{F}$-linear transformation $f : M_1 \to M_2$ with

$$f(\sigma_1(g)(m)) = \sigma_2(g)(f(m)) \qquad \text{for every} \quad m \in M_1, \, g \in G.$$

We will be considering general groups G and fields $\mathbf{F}$, though our strongest results will be in the case G is finite (and $\mathbf{F}$ satisfies certain restrictions). *Accordingly, we will adopt the following notational convention throughout this chapter: n will always denote the order of G.* (The reader should note that many variables in this chapter will range over the set $\{0, 1, 2, \dots\} \cup \{\infty\}$, and the comment of Remark 7.1.20 (2) is relevant here. Namely, distinction is made between finite and infinite, but no distinction is made between sets of different infinite cardinality.)

Since we are considering $\mathbf{F}(G)$-modules, we make some elementary remarks about $\mathbf{F}(G)$ itself.

(1) $\mathbf{F}(G)$ is an $\mathbf{F}$-algebra.
(2) $\mathbf{F}(G)$ is an $(\mathbf{F}(G), \mathbf{F}(G))$-bimodule, as well as an $(\mathbf{F}(K), \mathbf{F}(H))$-bimodule for any pair of subgroups K and H of G.
(3) $\mathbf{F}(G)$ is commutative if and only if G is abelian.
(4)   If G has torsion (i.e., elements of finite order), then $\mathbf{F}(G)$ has zero divisors. For let $g \in G$ with $g^m = 1$. Then

$$(1 - g)(1 + g + \cdots + g^{m-1}) = 0 \in \mathbf{F}(G).$$

(5)   Any $\mathbf{F}(G)$-module is an $(\mathbf{F}(G), \mathbf{F})$-bimodule.

It will be useful to us to single out the following class of fields $\mathbf{F}$:

(1.3) Definition. *Let G be a finite group. A field $\mathbf{F}$ is called* **good for** G *(or simply* **good***) if the following conditions are satisfied.*

(1)   *The characteristic of $\mathbf{F}$ is relatively prime to the order of G.*
(2)   *If m denotes the exponent of G, then the equation $X^m - 1 = 0$ has m* **distinct** *roots in $\mathbf{F}$.*

Remark. Actually, (2) implies (1), and also, (2) implies that the equation $X^k - 1 = 0$ has k distinct roots in $\mathbf{F}$, for every k dividing m (Exercise 34, Chapter 2). Furthermore, since the roots of $X^k - 1 = 0$ form a subgroup of the multiplicative group $\mathbf{F}^*$, it follows from Theorem 3.7.24 that there is (at least) one root $\zeta = \zeta_k$ such that these roots are

$$1 = \zeta^0, \zeta, \zeta^2, \ldots, \zeta^{k-1}.$$

We shall reserve the use of the symbol ζ (or ζ_k) to denote this. We further assume that these roots have been consistently chosen, in the following sense:

$$\text{If } k_1 \text{ divides } k_2, \text{ then } (\zeta_{k_2})^{k_2/k_1} = \zeta_{k_1}.$$

Note that the field $\mathbf{C}$ (or, in fact, any algebraically closed field of characteristic zero) is good for every finite group, and in $\mathbf{C}$ we may simply choose $\zeta_k = \exp(2\pi i/k)$ for every positive integer k.

In this chapter we will have occasion to consider Hom_R, End_R, or $\otimes_R$ for various rings of the form $R = \mathbf{F}(H)$, where H is a subgroup of G (or $R = \mathbf{F}$, in which case we may identify R with $\mathbf{F}(\langle 1 \rangle)$). We adopt the notational convention that Hom, End, and $\otimes$ mean $\text{Hom}_{\mathbf{F}}$, $\text{End}_{\mathbf{F}}$, and $\otimes_{\mathbf{F}}$, respectively, and Hom_H, End_H, and $\otimes_H$ mean Hom_R, End_R, and $\otimes_R$, respectively, for $R = \mathbf{F}(H)$.

(1.4) Examples.

(1) Let M be an $\mathbf{F}$ vector space of dimension 1 and define $\sigma : G \to \text{Aut}(M)$ by $\sigma(g) = 1_M$ for all $g \in G$. We call M the **trivial representation of degree** 1, and we denote it by τ.

(2) $M = \mathbf{F}(G)$ as an $\mathbf{F}(G)$-module. This is a representation of degree n. As an $\mathbf{F}$ vector space, M has a basis $\{g : g \in G\}$, and an element $g_0 \in G$ acts on M by

$$\sigma(g_0) \left(\sum_{g \in G} a_g g \right) = \sum_{g \in G} a_g g_0 g.$$

M is called the **(left) regular representation of** G and it plays a crucial role in the theory. We denote it by $\mathcal{R} = \mathcal{R}(G)$.

(3) **Permutation representations.** Let P be a set, $P = \{p_i\}_{i \in I}$, and suppose we have a homomorphism $\sigma : G \to S_P = \text{Aut}(P)$ (see Remark 1.4.2). Let $M = \mathbf{F}(P)$ be the free $\mathbf{F}$-module with basis P. Then G acts on $\mathbf{F}(P)$ by the formula

$$\sigma(g) \left(\sum_{i \in I} a_i p_i \right) = \sum_{i \in I} a_i (\sigma(g)(p_i)),$$

giving a representation of degree $|P|$. Note that the regular representation is a special case of this construction, obtained by taking $P = G$. As an important variant, we could take $P = G/H = \{gH\}$, the set of left cosets of some subgroup H of G. (This concept was used in a preliminary way in Section 1.4.)

(4) If M_1 and M_2 are two representations of G, defined by σ_1 and σ_2, then $M_1 \oplus M_2$ is a representation of G defined by $\sigma_1 \oplus \sigma_2$. Note that $\deg(M_1 \oplus M_2) = \deg(M_1) + \deg(M_2)$.

(5) If M_1 and M_2 are two representations of G, defined by σ_1 and σ_2, then $M_1 \otimes M_2$ is a representation of G defined by $\sigma_1 \otimes \sigma_2$. Note that $\deg(M_1 \otimes M_2) = (\deg(M_1))(\deg(M_2))$.

(6) Let $G = \mathbf{Z}_n = \langle y : y^n = 1 \rangle$ be cyclic of order n and let $\mathbf{F}$ be a field that is good for G. We have the one-dimensional representations $\theta_k : G \to \mathrm{Aut}(\mathbf{F}) \cong \mathbf{F}^*$ defined by

$$\theta_k(g) = \zeta^k \quad \text{for } k = 0, \ldots, n-1.$$

These are all distinct (i.e., pairwise nonisomorphic) and $\theta_0 = \tau$.

(7) Let

$$G = D_{2m} = \langle x, y : x^m = 1, \ y^2 = 1, \ xy = yx^{-1} \rangle$$

be the dihedral group of order $2m$, and let $\mathbf{F}$ be a field that is good for G. The representations of D_{2m} are described in Tables 1.1 and 1.2, using the following matrices:

$$(1.1) \qquad A_k = \begin{bmatrix} \zeta^k & 0 \\ 0 & \zeta^{-k} \end{bmatrix} \qquad \text{and} \qquad B = \begin{bmatrix} 0 & 1 \\ 1 & 0 \end{bmatrix}.$$

(Note that it is only necessary to give the value of the representation on the two generators x and y.)

Table 1.1. Representations of D_{2m} (m odd)

Representation	x	y	degree	
$\psi_+ = \tau$	1	1	1	
ψ_-	1	-1	1	
ϕ_k	A_k	B	2	$1 \le k \le (m-1)/2$

Table 1.2. Representations of D_{2m} (m even)

Representation	x	y	degree	
$\psi_{++} = \tau$	1	1	1	
ψ_{+-}	1	-1	1	
ψ_{-+}	-1	1	1	
ψ_{--}	-1	-1	1	
ϕ_k	A_k	B	2	$1 \le k \le (m/2) - 1$

(8) Suppose that M has a basis $\mathcal{B}$ such that for every $g \in G$, $[\sigma(g)]_{\mathcal{B}}$ is a matrix with exactly one nonzero entry in every row and column. Then M is called a **monomial representation of** G. For example, the representations of D_{2m} given above are monomial. If all nonzero entries of $[\sigma(g)]_{\mathcal{B}}$ are 1, then the monomial representation is called a **permutation representation**. (The reader should check that this definition agrees with the definition of permutation representation given in Example 1.4 (3).)

(9) Let $X = \{1, x, x^2, \dots\}$ with the multiplication $x^i x^j = x^{i+j}$. Then X is a **monoid**, i.e., it satisfies all of the group axioms except for the existence of inverses. Then one can define the monoid ring exactly as in the case of the group ring. If this is done, then $\mathbf{F}(X)$ is just the polynomial ring $\mathbf{F}[x]$. Let M be an $\mathbf{F}$ vector space and let $T : M \to M$ be a linear transformation. We have already seen that M becomes an $\mathbf{F}(X)$-module via $x^i(m) = T^i(m)$, for $m \in M$ and $i \geq 0$. Thus, we have an example of a monoid representation. Now we may identify $\mathbf{Z}$ with

$$\{\dots, x^{-2}, x^{-1}, 1, x, x^2, \dots\}.$$

If the linear transformation T is invertible, then M becomes an $\mathbf{F}(\mathbf{Z})$-module via the action $x^i(m) = T^i(m)$ for $m \in M$, $i \in \mathbf{Z}$.

(10) As an example of (9), let $T : \mathbf{F}^2 \to \mathbf{F}^2$ have matrix $\left[\begin{smallmatrix} 1 & 1 \\ 0 & 1 \end{smallmatrix}\right]$ (in the standard basis). Then we obtain a representation of $\mathbf{Z}$ of degree 2. If $\mathrm{char}(\mathbf{F}) = p > 0$, then $T^p = 1_{\mathbf{F}^2}$, so in this case we obtain a representation of $\mathbf{Z}_p$ of degree 2.

(11) Let $\varepsilon : \mathcal{R}(G) \to \mathbf{F}$ be defined by

$$\varepsilon\left(\sum_{g \in G} a_g g\right) = \sum_{g \in G} a_g.$$

If we let G act trivially on $\mathbf{F}$, we may regard ε as a map of $\mathbf{F}$-representations

$$\varepsilon : \mathcal{R}(G) \to \tau.$$

We let $\mathcal{R}_0(G) = \mathrm{Ker}(\varepsilon)$. The homomorphism ε is known as the **augmentation map** and $\mathcal{R}_0(G)$ is known as the **augmentation ideal** of $\mathcal{R}(G)$. It is then, of course, an $\mathbf{F}$-representation of G.

(12) If $\mathbf{F}$ is a subfield of $\mathbf{F}'$ and M is an $\mathbf{F}$-representation of G, then $M' = \mathbf{F}' \otimes_{\mathbf{F}} M$ is an $\mathbf{F}'$-representation of G. An $\mathbf{F}'$-representation arising in this way is said to be **defined over** $\mathbf{F}$.

(13) If M_i is an $\mathbf{F}$-representation of G_i, defined by σ_i for $i = 1, 2$, then $M_1 \oplus M_2$ is a representation of $G_1 \times G_2$, defined by

$$(\sigma_1 \oplus \sigma_2)(g_1, g_2) = \sigma_1(g_1) \oplus \sigma_2(g_2) \qquad \text{for} \qquad g_i \in G_i.$$

(Compare with Example 1.4 (4).)

(14) If M_i is an **F**-representation of G_i, defined by σ_i for $i = 1, 2$, then $M_1 \otimes M_2$ is a representation of $G_1 \times G_2$, defined by

$$(\sigma_1 \otimes \sigma_2)(g_1, g_2) = \sigma_1(g_1) \otimes \sigma_2(g_2) \quad \text{for} \quad g_i \in G_i.$$

(Compare with Example 1.4 (5).)

(15) If $\sigma : G \to \text{Aut}(M)$ is injective, then σ is called **faithful**. If not, let $K = \text{Ker}(\sigma)$. Then σ determines a homomorphism $\sigma' : G/K \to \text{Aut}(M)$, which is a faithful representation of the group G/K.

(16) Let $f : G_1 \to G_2$ be a group homomorphism and let $\sigma : G_2 \to \text{Aut}(M)$ be a representation of G_2. The **pullback of σ by f**, denoted $f^*(\sigma)$, is the representation of G_1, defined by $f^*(\sigma) = \sigma \circ f$, i.e., $f^*(\sigma) : G_1 \to \text{Aut}(M)$ by

$$f^*(\sigma)(g) = \sigma(f(g)) \quad \text{for} \quad g \in G_1.$$

(17) Let M be an **F**-representation of G, so M is an $\mathbf{F}(G)$-module. Then for any subgroup H of G, M is also an $\mathbf{F}(H)$-module, i.e., an **F**-representation of H.

(18) Let H be a subgroup of G and let M be an **F**-representation of H, i.e., an $\mathbf{F}(H)$-module. Then $\mathbf{F}(G) \otimes_{\mathbf{F}(H)} M$ is an $\mathbf{F}(G)$-module, i.e., an **F**-representation of G.

(1.5) Definition.

(1) M is an **irreducible representation** of G if M is an irreducible $\mathbf{F}(G)$-module. *(See Definition 7.1.1.)*

(2) M is an **indecomposable representation of** G if M is an indecomposable $\mathbf{F}(G)$-module. *(See Definition 7.1.6.)*

One of our principal objectives will be to find irreducible representations of a group G and to show how to express a representation as a sum of irreducible representations, when possible. The following examples illustrate this theme.

(1.6) Example. *Let* **F** *be good for* $\mathbf{Z}_n$. *Then the regular representation* $\mathcal{R}(\mathbf{Z}_n)$ *is isomorphic to*

$$\theta_0 \oplus \theta_1 \oplus \cdots \oplus \theta_{n-1}.$$

Proof. Consider the **F**-basis $\{1, g, \ldots, g^{n-1}\}$ of $\mathcal{R}(\mathbf{Z}_n)$. In this basis, $\sigma(g)$ has the matrix

$$\begin{bmatrix} 0 & 0 & \cdots & 0 & 1 \\ 1 & 0 & \cdots & 0 & 0 \\ 0 & 1 & \cdots & 0 & 0 \\ \vdots & \vdots & \ddots & \vdots & \vdots \\ 0 & 0 & \cdots & 0 & 0 \\ 0 & 0 & \cdots & 1 & 0 \end{bmatrix},$$

which we recognize as $C(X^n - 1)$, the companion matrix of the polynomial $X^n - 1$. Hence,

$$m_{\sigma(g)}(X) = c_{\sigma(g)}(X) = X^n - 1.$$

By our assumption on $\mathbf{F}$, this polynomial factors into distinct linear factors

$$X^n - 1 = \prod_{k=0}^{n-1} (X - \zeta^k).$$

Hence, $\sigma(g)$ is diagonalizable, and indeed, in an appropriate basis $\mathcal{B}$, it has matrix

$$\mathrm{diag}(1, \zeta, \zeta^2, \ldots, \zeta^{n-1}).$$

Each of the eigenspaces is an $\mathbf{F}(\mathbf{Z}_n)$-submodule, so we see immediately that

$$\mathcal{R}(\mathbf{Z}_n) \cong \theta_0 \oplus \theta_1 \oplus \cdots \oplus \theta_{n-1}.$$

$\square$

(1.7) **Example.** Let $\mathbf{F} = \mathbf{Q}$, the rational numbers, and let p be a prime. Consider the augmentation ideal $\mathcal{R}_0 \subseteq \mathcal{R} = \mathbf{Q}(\mathbf{Z}_p)$. If g is a generator of $\mathbf{Z}_p$ and $T = \sigma(g) : \mathcal{R}_0 \to \mathcal{R}_0$, then

$$m_T(X) = c_T(X) = \frac{(X^p - 1)}{(X - 1)},$$

which is irreducible by Corollary 2.6.12. Hence, by Remark 7.1.40 $\mathcal{R}_0$ is an irreducible $\mathbf{Q}$-representation of $\mathbf{Z}_p$. (Strictly speaking, to apply Remark 7.1.40 we must consider not σ, but $\pi^*(\sigma)$ where $\pi : \mathbf{Z} \to \mathbf{Z}_p$ is the canonical projection, and apply Exercise 2.) Note that $\mathcal{R}_0 \otimes_{\mathbf{Q}} \mathbf{C}$ is the augmentation ideal in $\mathbf{C}(G)$, and $\mathcal{R}_0 \otimes_{\mathbf{Q}} \mathbf{C}$ is not irreducible and, in fact, is isomorphic to

$$\theta_1 \oplus \cdots \oplus \theta_{p-1}.$$

(In fact, this statement is true without the requirement that p be prime.)

(1.8) **Example.** Let $\mathbf{F}$ be good for D_{2m}.

(1) Each of the $\mathbf{F}(D_{2m})$-modules ϕ_k of Example 1.4 (7) is irreducible, and they are distinct.
(2) If, for m even, we define $\phi_{m/2}$ by $\phi_{m/2}(x) = A_{m/2}$ and $\phi_{m/2}(y) = B$, then

$$\phi_{m/2} \cong \psi_{-+} \oplus \psi_{--}.$$

(3) Let P be the set of vertices of a regular m-gon, and let D_{2m} act on P in the usual manner (Section 1.5). Then, as $\mathbf{F}(D_{2m})$-modules,

$$\mathbf{F}(P) \cong \begin{cases} \psi_+ \oplus \phi_1 \oplus \phi_2 \oplus \cdots \oplus \phi_{(m-1)/2} & \text{if } m \text{ is odd}, \\ \psi_{++} \oplus \psi_{-+} \oplus \phi_1 \oplus \phi_2 \oplus \cdots \oplus \phi_{\frac{m}{2}-1} & \text{if } m \text{ is even}. \end{cases}$$

(4)   *For the regular representation, we have*

$$\mathcal{R}(D_{2m}) \cong$$

$$\begin{cases} \psi_+ \oplus \psi_- \oplus 2\phi_1 \oplus \cdots \oplus 2\phi_{(m-1)/2} & \text{if } m \text{ is odd,} \\ \psi_{++} \oplus \psi_{+-} \oplus \psi_{-+} \oplus \psi_{--} \oplus 2\phi_1 \oplus \cdots \oplus 2\phi_{\frac{m}{2}-1} & \text{if } m \text{ is even.} \end{cases}$$

Proof. We leave this as an exercise for the reader.   □

(1.9) Definition. *Let M and N be $\mathbf{F}$-representations of G. The **multiplicity** m of M in N is the largest nonnegative integer with the property that mM is isomorphic to a submodule of N. If no such m exists, i.e., if mM is isomorphic to a submodule of N for every nonnegative integer m, we say that the multiplicity of M in N is infinite.*

If the multiplicity of M in N is m, we shall often say that N contains M m times (or that N contains m copies of M). If $m = 0$, we shall often say that N does not contain M.

(1.10) Lemma. *Let G be a group. Then the regular representation $\mathcal{R}$ contains the trivial representation τ once if G is finite, but it does not contain τ if G is infinite.*

Proof. Let M be the submodule of $\mathcal{R}$ consisting of all elements on which G acts trivially. Let $m \in M$. Then we may write

$$m = \sum_{g \in G} a_g g \quad \text{where} \quad a_g \in \mathbf{F}.$$

By assumption, $g_0 m = m$ for every $g_0 \in G$. But

$$g_0 m = \sum_{g \in G} a_g (g_0 g)$$

and

$$m = \sum_{g \in G} a_{g_0 g}(g_0 g).$$

Therefore, $a_{g_0 g} = a_g$ for every $g \in G$ and every $g_0 \in G$. In particular,

(1.2) $$a_{g_0} = a_e$$

for every $g_0 \in G$.

Now suppose that G is finite. Then by Equation (1.2)

$$m = \sum_{g \in G} a_e g = a_e \left(\sum_{g \in G} g \right) \quad \text{where} \quad a_e \in \mathbf{F},$$

i.e., $M = \langle \sum_{g \in G} g \rangle$ as an **F**-module. Therefore, M is one-dimensional over **F**, so that $M = \tau$.

On the other hand, if G is infinite, Equation (1.2) implies that $a_g \neq 0$ for all $g \in G$ whenever $a_e \neq 0$. But the group ring $\mathbf{F}(G)$ consists of *finite* sums $\sum_{g \in G} a_g g$, so this is impossible. Hence, we must have $a_e = 0$ and $M = \langle 0 \rangle$. $\square$

(1.11) Corollary. *If G is an infinite group, then $\mathcal{R}$ is not semisimple.*

Proof. Suppose that $\mathcal{R}$ were semisimple. Then by Theorem 7.1.30, $\mathcal{R}$ would contain the simple $\mathcal{R}$-module τ, but by Lemma 1.10, it does not. $\square$

(1.12) Example. Consider Example 1.4 (10). This is an **F**-representation of **Z**, which is indecomposable but not irreducible: it has τ as a subrepresentation (consisting of vectors in $\mathbf{F} \times \{0\}$). Indeed, if T is a linear transformation on M, Remark 7.1.40 gives a criterion for M to be a semisimple $\mathbf{F}(X)$-module.

The following lemma, incorporating the technique of **averaging over the group**, is crucial:

(1.13) Lemma. *Let G be a finite group and $\mathbf{F}$ a field with $\mathrm{char}(\mathbf{F}) = 0$ or prime to the order of G. Let V_1 and V_2 be $\mathbf{F}$-representations of G defined by $\sigma_i : G \to \mathrm{Aut}(V_i)$ for $i = 1, 2$, and let $n = |G|$. Let $f \in \mathrm{Hom}(V_1, V_2)$ and set*

$$(1.3) \qquad \mathrm{Av}(f) = \frac{1}{n} \sum_{g \in G} \sigma_2(g^{-1}) f(\sigma_1(g)).$$

Then $\mathrm{Av}(f) \in \mathrm{Hom}_G(V_1, V_2)$ Furthermore, if $f \in \mathrm{Hom}_G(V_1, V_2)$, then

$$(1.4) \qquad\qquad \mathrm{Av}(f) = f.$$

Proof. We need to show that for every $g_0 \in G$ and every $v_1 \in V_1$,

$$\sigma_2(g_0) \, \mathrm{Av}(f)(v_1) = \mathrm{Av}(f)(\sigma_1(g_0)(v_1)).$$

But

$$\mathrm{Av}(f)(\sigma_1(g_0)(v_1)) = \left(\frac{1}{n} \sum_{g \in G} \sigma_2(g^{-1}) f(\sigma_1(g)) \right) (\sigma_1(g_0)(v_1))$$

$$= \frac{1}{n} \sum_{g \in G} \sigma_2(g^{-1}) f(\sigma_1(g)) \, (\sigma_1(g_0)(v_1))$$

$$= \frac{1}{n} \sum_{g \in G} \sigma_2(g^{-1}) f(\sigma_1(g g_0))(v_1)$$

$$= \frac{1}{n} \sum_{g \in G} \sigma_2(g_0) \sigma_2(g_0^{-1} g^{-1}) f(\sigma_1(g g_0))(v_1)$$

$$= \frac{1}{n} \sum_{g \in G} \sigma_2(g_0) \sigma_2((gg_0)^{-1}) f(\sigma_1(gg_0))(v_1).$$

Let $g' = gg_0$. As g runs through the elements of G, so does g'. Thus,

$$\mathrm{Av}(f)(\sigma_1(g_0)(v_1)) = \frac{1}{n} \sum_{g' \in G} \sigma_2(g_0) \sigma_2(g'^{-1}) f(\sigma_1(g'))(v_1)$$

$$= \sigma_2(g_0) \left(\frac{1}{n} \sum_{g' \in G} \sigma_2(g'^{-1}) f(\sigma_1(g')) \right)(v_1)$$

$$= \sigma_2(g_0) \mathrm{Av}(f)(v_1)$$

as required, so Equation (1.3) is satisfied.

Also, if $f \in \mathrm{Hom}_G(V_1, V_2)$, then for every $g \in G$,

$$\sigma_2(g)f = f\sigma_1(g).$$

Hence, in this case

$$\mathrm{Av}(f) = \frac{1}{n} \sum_{g \in G} \sigma_2(g^{-1}) f \sigma_1(g)$$

$$= \frac{1}{n} \sum_{g \in G} f \sigma_1(g^{-1}) \sigma_1(g)$$

$$= \frac{1}{n} \sum_{g \in G} f \sigma_1(e)$$

$$= \frac{1}{n} \sum_{g \in G} f$$

$$= \frac{1}{n}(nf)$$

$$= f.$$

Hence, Equation (1.4) is satisfied.   □

Now we come to one of the cornerstones of the representation theory of finite groups.

(1.14) Theorem. (Maschke) *Let G be a finite group. Then $\mathbf{F}(G)$ is a semisimple ring if and only if $\mathrm{char}(\mathbf{F}) = 0$ or $\mathrm{char}(\mathbf{F})$ is relatively prime to the order of G.*

Proof. First, consider the case where $\mathrm{char}(\mathbf{F}) = 0$ or is relatively prime to the order of G. By Theorems 7.1.28 and 7.1.23, it suffices to show that every submodule M_1 of an arbitrary $\mathbf{F}(G)$-module M is complemented.

Let $\iota : M_1 \to M$ be the inclusion. We will construct $\pi : M \to M_1$ with $\pi\iota = 1_{M_1}$. Assuming that, we have a split exact sequence

$$0 \longrightarrow M_1 \overset{\iota}{\longrightarrow} M \longrightarrow M/M_1 \longrightarrow 0,$$

so Theorem 3.3.9 applies to show that $M \cong M_1 \oplus \text{Ker}(\pi)$, and hence, M_1 is complemented.

Now, M_1 is a subvector space of M, so as an **F**-module M_1 is complemented and there is certainly a linear map $\rho : M \to M_1$ with $\rho\iota = 1_{M_1}$. This is an **F**-module homomorphism, but there is no reason to expect it to be an $\mathbf{F}(G)$-module homomorphism. We obtain one by averaging it. (Since we are dealing with a single representation here, for simplicity we will write $g(v)$ instead of $\sigma(g)(v)$.)

Let $\pi = \text{Av}(\rho)$. By Lemma 1.13, $\pi \in \text{Hom}_G(M, M_1)$. We have $\rho\iota = 1_{M_1}$; we need $\pi\iota = 1_{M_1}$. Let $v \in M_1$. Then, since M_1 is an $\mathbf{F}(G)$-submodule, $g(v) \in M_1$ for all $g \in G$, so $\rho(\iota(g(v))) = g(v)$ for all $g \in G$. Then

$$
\begin{aligned}
\pi(\iota(v)) &= \frac{1}{n} \sum_{g \in G} g^{-1}\rho(g(\iota(v))) \\
&= \frac{1}{n} \sum_{g \in G} g^{-1}\rho(\iota(g(v))) \\
&= \frac{1}{n} \sum_{g \in G} g^{-1}(g(v)) \\
&= \frac{1}{n} \sum_{g \in G} v \\
&= \frac{nv}{n} \\
&= v
\end{aligned}
$$

as required.

Now suppose that $\text{char}(\mathbf{F})$ divides the order of G. Recall that we have the augmentation map ε as defined in Example 1.4 (11), giving a short exact sequence of $\mathbf{F}(G)$-modules

(1.5) $$0 \longrightarrow \mathcal{R}_0 \longrightarrow \mathcal{R} \overset{\varepsilon}{\longrightarrow} \tau \longrightarrow 0.$$

Suppose that $\mathcal{R}$ were semisimple. Then ε would have a splitting α, so by Theorem 3.3.9, $\mathcal{R} \cong \mathcal{R}_0 \oplus \alpha(\tau)$. Since this is a direct sum, $\mathcal{R}_0 \cap \alpha(\tau) = \langle 0 \rangle$. On the other hand, $\alpha(\tau)$ is a trivial subrepresentation in $\mathcal{R}$, so by the proof of Lemma 1.10,

$$\alpha(\tau) = \left\{ a \sum_{g \in G} g : a \in \mathbf{F} \right\}.$$

However,

$$\varepsilon\left(a \sum_{g \in G} g \right) = a\varepsilon\left(\sum_{g \in G} g \right) = an = 0 \in \mathbf{F}$$

since $\text{char}(\mathbf{F}) \mid n$. Thus, $\alpha(\tau) \subseteq \mathcal{R}_0$, contradicting $\mathcal{R}_0 \cap \alpha(\tau) = \langle 0 \rangle$. $\square$

(1.15) Example. Consider Example 1.4 (10) again, but this time with $\mathbf{F} = \mathbf{F}_p$, the field of p elements. Then $T^p = \begin{bmatrix} 1 & p \\ 0 & 1 \end{bmatrix} = \begin{bmatrix} 1 & 0 \\ 0 & 1 \end{bmatrix}$, so we may regard T as giving an $\mathbf{F}$-representation of $\mathbf{Z}_p$. As in Example 1.12, this is indecomposable but not irreducible.

Having proven that $\mathbf{F}(G)$ is semisimple in favorable cases, we collect the relevant results of Section 7.1 into an omnibus theorem.

(1.16) Theorem. *Let G be a finite group and $\mathbf{F}$ a field with $\mathrm{char}(\mathbf{F}) = 0$ or prime to the order of G. Then the following are valid.*

(1) *An $\mathbf{F}$-representation of G is indecomposable if and only if it is irreducible.*

(2) *Every irreducible $\mathbf{F}$-representation of G is isomorphic to a subrepresentation of $\mathbf{F}(G)$.*

(3) *Every $\mathbf{F}$-representation of G is a projective $\mathbf{F}(G)$-module.*

(4) *Every $\mathbf{F}$-representation of G is semisimple, and if M is written as*

$$M \cong \bigoplus_{i \in I} s_i M_i \qquad \text{for distinct irreducibles} \quad \{M_i\}_{i \in I},$$

then $s_i \in \{0, 1, 2, \ldots\} \cup \{\infty\}$ is well determined.

(5) *There are only finitely many distinct irreducible $\mathbf{F}$-representations of G (up to isomorphism), and each has finite multiplicity in $\mathbf{F}(G)$.*

Proof.   □

(1.17) Corollary. *Let G be a finite group and $\mathbf{F}$ a field with $\mathrm{char}(\mathbf{F}) = 0$ or prime to the order of G. Then every irreducible $\mathbf{F}$-representation of G has degree at most n and there are at most n distinct irreducible $\mathbf{F}$-representations of G, up to isomorphism.*

Proof. If M_i is irreducible, then Theorem 1.16 (2) implies that M_i is a subrepresentation of $\mathbf{F}(G)$, so

$$\deg(M_i) \leq \deg(\mathbf{F}(G)) = n.$$

If $\{M_i\}_{i \in I}$ are all of the distinct irreducible representations, then by Theorem 1.16 (2), $M = \oplus_{i \in I} M_i$ is a subrepresentation of $\mathbf{F}(G)$.

Hence, $\deg(M) \leq \deg(\mathbf{F}(G))$, so that

$$|I| = \sum_{i \in I} 1 \leq \sum_{i \in I} \deg(M_i) \leq \deg(\mathbf{F}(G)) = n,$$

as claimed.   □

The second basic result we have is Schur's lemma, which we amplify a bit in our situation.

(1.18) Lemma. (Schur) *Let R be an $\mathbf{F}$-algebra.*

(1) *Let M be a simple R-module. Then $\operatorname{End}_R(M)$ is a division ring containing $\mathbf{F}$ in its center. If $\mathbf{F}$ is algebraically closed, then every $\phi \in \operatorname{End}_R(M)$ is a homothety (i.e., is multiplication by an element of $\mathbf{F}$) and $\operatorname{End}_R(M) \cong \mathbf{F}$.*

(2) *Let M_1 and M_2 be two distinct (i.e., nonisomorphic) simple R-modules. Then $\operatorname{Hom}_R(M_1, M_2) = \langle 0 \rangle$.*

Proof. Much of this is a restatement of Schur's lemma (Proposition 7.1.5). If $D = \operatorname{End}_R(M)$, then clearly $D \supseteq \mathbf{F}$, where we regard $a \in \mathbf{F}$ as the endomorphism given by (left) multiplication by a. If $\phi \in D$, then $\phi(am) = a\phi(m)$ for all $m \in M$, so $\mathbf{F}$ is contained in the center of D.

Now suppose that $\mathbf{F}$ is algebraically closed and $\phi \in \operatorname{End}_R(M)$. Then the characteristic polynomial of ϕ splits into linear factors in $\mathbf{F}$, so, in particular, ϕ has an eigenvalue $a \in \mathbf{F}$. Then $\operatorname{Ker}(\phi - a)$ is a nontrivial submodule of M. Since M is simple, this implies $\operatorname{Ker}(\phi - a) = M$, i.e., $\phi = a$. $\qquad\square$

(1.19) Example. Here is an example to see that if M is a simple R-algebra and $\mathbf{F}$ is not algebraically closed, then $\operatorname{End}_R(M)$ need not consist solely of homotheties. Let $R = \mathbf{R}(\mathbf{Z}_4)$ and $M = \mathbf{R}^2$ with the action of $\mathbf{Z}_4 = \{1, g, g^2, g^3\}$ given by the matrix

$$\sigma(g) = \begin{bmatrix} 0 & -1 \\ 1 & 0 \end{bmatrix}.$$

It is easy to check that

$$\operatorname{End}_R(M) = \left\{ \begin{bmatrix} a & -b \\ b & a \end{bmatrix} : a, b \in \mathbf{R} \right\} \cong \mathbf{C}$$

under the isomorphism $\begin{bmatrix} a & -b \\ b & a \end{bmatrix} \mapsto a + bi$.

We close this section with the following lemma, which will be important later.

(1.20) Lemma. *Let G be a finite group and $\mathbf{F}$ a good field. Let $\sigma : G \to \operatorname{Aut}(M)$ be an $\mathbf{F}$-representation of G of finite degree. Then for each $g \in G$, the linear transformation $\sigma(g) : M \to M$ is diagonalizable.*

Proof. Since $g \in G$, we have that $g^k = 1$ for some k dividing $m = \operatorname{exponent}(G)$. Since $\mathbf{F}$ is good, the lemma follows immediately from Theorem 4.4.34. $\qquad\square$

8.2 Representations of Abelian Groups

Before developing the general theory, we will first directly develop the representation theory of finite abelian groups over good fields. We will then see the similarities and differences between the situation for abelian groups and that for groups in general.

(2.1) Theorem. *Let G be a finite group and $\mathbf{F}$ a good field for G. Then G is abelian if and only if every irreducible $\mathbf{F}$-representation of G is one-dimensional.*

Proof. First assume that G is abelian and let M be an $\mathbf{F}$-representation of G. By Corollary 1.17, if $\deg(M)$ is infinite, M cannot be irreducible. Thus we may assume that $\deg(M) < \infty$. Now the representation is given by a homomorphism $\sigma : G \to \operatorname{Aut}(M)$, so

$$\sigma(g)\sigma(h) = \sigma(gh) = \sigma(hg) = \sigma(h)\sigma(g) \qquad \text{for all} \quad g, h \in G.$$

By Lemma 1.20, each $\sigma(g)$ is diagonalizable, so

$$S = \{\sigma(g) : g \in G\}$$

is a set of mutually commuting diagonalizable transformations. By Theorem 4.3.36, S is simultaneously diagonalizable. If $\mathcal{B} = \{v_1, \ldots, v_k\}$ is a basis of M in which they are all diagonal, then

$$M \cong \mathbf{F}v_1 \oplus \mathbf{F}v_2 \oplus \cdots \oplus \mathbf{F}v_k$$

(thus showing directly that M is semisimple). Hence M, is simple if and only if $k = 1$, i.e., if and only if $\deg(M) = 1$.

Conversely, let M be a one-dimensional representation of G. Then the representation is defined by a homomorphism $\sigma : G \to \operatorname{Aut}(M) \cong \mathbf{F}^*$, and $\mathbf{F}^*$ is, of course, abelian, so that

$$(2.1) \qquad \sigma(g)\sigma(h) = \sigma(h)\sigma(g) \qquad \text{for all} \quad g, h \in G.$$

By assumption, every irreducible representation of G is one-dimensional, so Equation (2.1) is valid for every irreducible representation of G. Since every representation of G is a direct sum of irreducible representations, Equation (2.1) is valid for every representation of G. In particular, it is valid for $\mathcal{R}(G)$, the regular representation of G. If σ_0 is the homomorphism defining the regular representation, and we consider $1 \in \mathcal{R}(G)$, then for any $g, h \in G$,

$$(\sigma_0(g)\sigma_0(h))(1) = (\sigma_0(h)\sigma_0(g))(1)$$
$$\sigma_0(g)(h) = \sigma_0(h)(g)$$
$$gh = hg$$

and G is abelian. $\qquad\qquad\qquad\qquad\qquad\qquad\qquad\qquad\qquad\qquad\square$

(2.2) Theorem. *Let G be a finite group and $\mathbf{F}$ a good field for G. Then G is abelian if and only if G has n distinct irreducible $\mathbf{F}$-representations.*

Proof. Let G be abelian. We shall construct n distinct $\mathbf{F}$-representations of G. By Theorem 3.7.22, we know that we may write G as a direct sum of cyclic groups

$$(2.2) \qquad G \cong \mathbf{Z}_{n_1} \oplus \mathbf{Z}_{n_2} \oplus \cdots \oplus \mathbf{Z}_{n_s}$$

with $n = n_1 \cdots n_s$. Since $\mathbf{F}$ is good for G, it is also good for each of the cyclic groups $\mathbf{Z}_{n_i}$, and by Example 1.4 (6), the cyclic group $\mathbf{Z}_{n_i}$ has the n_i distinct $\mathbf{F}$-representations θ_k for $0 \le k \le n_i - 1$; to distinguish these representations for different i, we shall denote them $\theta_k^{n_i}$. Thus, $\theta_k^{n_i} : \mathbf{Z}_{n_i} \to \mathbf{F}^*$. If $\pi_i : G \to \mathbf{Z}_{n_i}$ denotes the projection, then

$$(2.3) \qquad \theta_k^{n_i} \pi_i : G \to \mathbf{F}^* = \mathrm{Aut}(\mathbf{F})$$

defines a one-dimensional representation (and, hence, an irreducible representation) of G. Thus,

$$\{\theta_k^{n_i} \pi_i : \ 1 \le i \le s, \ 0 \le k \le n_i - 1\}$$

is a collection of n irreducible $\mathbf{F}$-representations of G; by Corollary 1.17, this is all of them.

On the other hand, suppose that G is not abelian, and let $\{M_i\}_{i \in I}$ be the set of irreducible representations of G. Since G is not abelian, Theorem 2.1 implies that $\deg(M_i) > 1$ for some i. Then, as in the proof of Corollary 1.17,

$$|I| = \sum_{i \in I} 1 < \sum_{i \in I} \deg(M_i) \le \deg(\mathbf{F}(G)) = n,$$

so $|I| < n$, as claimed. $\square$

(2.3) Corollary. *Let G be a finite abelian group and $\mathbf{F}$ a good field for G. If $M_1, \dots, M_n$ denote the distinct irreducible $\mathbf{F}$-representations of G, then*

$$\mathbf{F}(G) = \bigoplus_{i=1}^{n} M_i.$$

(In other words, every irreducible representation of G appears in the regular representation with multiplicity one.)

Proof. If $M = \oplus_{i=1}^{n} M_i$, then by Theorem 1.16 (2), M is a subrepresentation of $\mathcal{R}$. But $\deg(M) = n = \deg(\mathcal{R})$, so $M = \mathcal{R}$. $\square$

(2.4) Corollary. *Let G be a finite abelian group and $\mathbf{F}$ a good field for G. If M is an irreducible representation of G, then*

$$\text{End}_G(M) = \mathbf{F}.$$

Proof. Clearly, $\mathbf{F} \subseteq \text{End}_G(M) \subseteq \text{End}(M) = \mathbf{F}$. (Every $\mathbf{F}(G)$-homomorphism is an $\mathbf{F}$-homomorphism and M is one-dimensional.) $\square$

(2.5) Proposition. *Let G be a group and suppose that $Q = G/[G,G]$ is finite. Let $\mathbf{F}$ be a good field for Q. Then G has $|Q|$ distinct one-dimensional $\mathbf{F}$-representations.*

Proof. If π is the canonical projection $\pi : G \to Q$, then for any one-dimensional $\mathbf{F}$-representation $\sigma : Q \to \text{Aut}(\mathbf{F})$, its pullback $\pi^*(\sigma) = \sigma\pi$ is a one-dimensional $\mathbf{F}$-representation of G, and by Theorem 2.2, we obtain $|Q|$ distinct representations in this way. On the other hand, if $\sigma' : G \to \text{Aut}(\mathbf{F}) = \mathbf{F}^*$ is any one-dimensional representation of G, then $\sigma'|_{[G,G]}$ is trivial (as $\mathbf{F}^*$ is abelian). Thus, σ' factors as $\sigma\pi$, so it is one of the representations constructed above. $\square$

8.3 Decomposition of the Regular Representation

(3.1) Definition. *Let G be a finite group. A field $\mathbf{F}$ is called **excellent** for G or simply **excellent** if it is algebraically closed of characteristic zero or prime to the order of G.*

Observe that an excellent field is good and that the field $\mathbf{C}$ is excellent for every G. Our objective in this section is to count the number of irreducible representations of a finite group G over an excellent field $\mathbf{F}$, and to determine their multiplicities in the regular representation $\mathbf{F}(G)$. The answers turn out to be both simple and extremely useful.

(3.2) Definition. *A representation P of G is called **isotypic (of type M)** if for some positive integer m and some irreducible representation M of G, P is isomorphic to mM.*

(3.3) Lemma. *Let $\mathbf{F}$ be an excellent field for G, and let P and Q be isotypic representations of G of the same type. If $P \cong m_1 M$ and $Q \cong m_2 M$, then as $\mathbf{F}$-algebras*

$$(3.1) \qquad\qquad \text{Hom}_G(P, Q) \cong M_{m_2, m_1}(\mathbf{F}).$$

Proof. Let $\phi : P \to m_1 M$ and $\psi : Q \to m_2 M$ be the isomorphisms (of $\mathbf{F}(G)$-modules). Let $\alpha_i : M \to m_1 M$ be the inclusion of M as the ith summand, and let $\beta_j : m_2 M \to M$ be the projection onto the jth summand.

If $f \in \operatorname{Hom}_G(P, Q)$, consider the following composition of $\mathbf{F}(G)$-module homomorphisms (which we will call f_{ji}):

$$M \xrightarrow{\alpha_i} m_1 M \xrightarrow{\phi^{-1}} P \xrightarrow{f} Q \xrightarrow{\psi} m_2 M \xrightarrow{\beta_j} M.$$

Then $f_{ji} \in \operatorname{End}_G(M)$, and since $\mathbf{F}$ is excellent for G, Lemma 1.18 (Schur's lemma) implies that f_{ji} is given by multiplication by some element $a_{ji} \in \mathbf{F}$. Then the isomorphism of the lemma is given by

$$f \mapsto A = [a_{ji}].$$

$\square$

Remark. Note that this isomorphism depends on the choice of isomorphisms ϕ and ψ. This dependence is nothing more than the familiar fact that the matrix of a linear transformation depends on a choice of bases.

The following theorem, proved by Frobenius in 1896, is fundamental:

(3.4) Theorem. (Frobenius) *Let $\mathbf{F}$ be an excellent field for the finite group G.*

(1) *The number of distinct irreducible $\mathbf{F}$-representations of G is equal to the number t of distinct conjugacy classes of elements of G.*

(2) *If $\{M_i\}_{i=1}^t$ are the distinct irreducible $\mathbf{F}$-representations of G, then the multiplicity of M_i in the regular representation $\mathcal{R}$ of G is equal to its degree $d_i = \deg(M_i)$ for $1 \le i \le t$.*

(3) $\sum_{i=1}^t d_i^2 = n = |G|.$

Convention. *We adopt the notational convention henceforth that t will always denote the number of conjugacy classes of the group G.*

Proof. If $\{M_i\}_{i=1}^q$ are the distinct irreducible representations of G (the number of these being finite by Theorem 1.16 (5)), then by Theorem 1.16 (2) we have

$$(3.2) \qquad\qquad \mathcal{R} \cong \bigoplus_{i=1}^q m_i M_i$$

for some positive integers $m_1, \ldots, m_q$.

(1) We shall prove this by calculating $\dim_{\mathbf{F}} \mathcal{C}$ in two ways, where $\mathcal{C}$ is the center of $\mathcal{R}$, i.e.,

$$\mathcal{C} = \{r \in \mathcal{R} : rr' = r'r \quad \text{for all } r' \in \mathcal{R}\}.$$

$\mathcal{C}$ is clearly an $\mathbf{F}$-algebra.

First we compute $\dim_{\mathbf{F}}(\mathcal{C})$ directly. Let $\{C_i\}_{i=1}^t$ be the sets of mutually conjugate elements of G, i.e., for each i, and each g_1 and $g_2 \in C_i$, there is a $g \in G$ with $g_1 = gg_2g^{-1}$. Since conjugacy is an equivalence relation, $\{C_i\}_{i=1}^t$ is a partition of G. For $1 \le i \le t$, let $c_i \in \mathcal{R}$ be defined by

$$(3.3) \qquad c_i = \sum_{g \in C_i} g.$$

For each element g of G, we have the following equality in $\mathcal{R}$:

$$\begin{aligned}
c_i g &= \left(\sum_{g_i \in C_i} g_i \right) g \\
&= \sum_{g_i \in C_i} g_i g \\
&= \sum_{g_i \in C_i} g(g^{-1} g_i g) \\
&= \sum_{g_i' \in C_i} g g_i' \\
&= g \left(\sum_{g_i' \in C_i} g_i' \right) \\
&= g c_i
\end{aligned}$$

where the fourth equality holds because C_i is a conjugacy class. This immediately implies that

$$(3.4) \qquad \mathcal{C} \supseteq \langle c_1, \ldots, c_t \rangle,$$

the **F**-vector space spanned by these elements.

On the other hand, suppose that we have an element

$$x = \sum_{g \in G} a_g g \in \mathcal{C}.$$

Then for any $g_0 \in G$, we have $g_0 x g_0^{-1} = x g_0 g_0^{-1} = x$. But

$$g_0 x g_0^{-1} = \sum_{g \in G} a_g g_0 g g_0^{-1} = \sum_{g' \in G} a_{g_0^{-1} g' g_0} g',$$

which implies that

$$a_g = a_{g_0^{-1} g g_0} \qquad \text{for all} \quad g, g_0 \in G.$$

That is, any two mutually conjugate elements have the same coefficient, and hence, $\mathcal{C} \subseteq \langle c_1, \ldots, c_t \rangle$. Together with Equation (3.4), this implies that $\mathcal{C} = \langle c_1, \ldots, c_t \rangle$, and since $c_1, \ldots, c_t$ are obviously **F**-linearly independent elements of $\mathcal{R}$, it follows that

(3.5) $$\dim_{\mathbf{F}}(\mathcal{C}) = t.$$

Now for our second calculation of $\dim_{\mathbf{F}}(\mathcal{C})$. We know in general that

(3.6) $$\mathcal{R} \cong \operatorname{Hom}_{\mathcal{R}}(\mathcal{R}, \mathcal{R}) = \operatorname{End}_{\mathcal{R}}(\mathcal{R}).$$

We will calculate $\dim_{\mathbf{F}}(\mathcal{C}')$ where $\mathcal{C}'$ is the center of $\operatorname{End}_{\mathcal{R}}(\mathcal{R})$. Of course, $\dim_{\mathbf{F}}(\mathcal{C}') = \dim_{\mathbf{F}}(\mathcal{C})$ by Equation (3.6). But

$$\operatorname{End}_{\mathcal{R}}(\mathcal{R}) = \operatorname{Hom}_{\mathcal{R}}(\mathcal{R}, \mathcal{R})$$
$$\cong \operatorname{Hom}_{\mathcal{R}}\left(\bigoplus_{i=1}^{q} m_i M_i, \bigoplus_{j=1}^{q} m_j M_j\right)$$
$$\cong \bigoplus_{i=1}^{q}\bigoplus_{j=1}^{q} \operatorname{Hom}_{\mathcal{R}}(m_i M_i, m_j M_j).$$

By Schur's lemma, $\operatorname{Hom}_{\mathcal{R}}(M_i, M_j) = \langle 0 \rangle$ for $i \neq j$, and by Lemma 3.3,

$$\operatorname{Hom}_{\mathcal{R}}(m_i M_i, m_i M_i) \cong M_{m_i}(\mathbf{F})$$

so that

(3.7) $$\mathcal{R} \cong \operatorname{End}_{\mathcal{R}}(\mathcal{R}) \cong \bigoplus_{i=1}^{q} M_{m_i}(\mathbf{F})$$

as $\mathbf{F}$-algebras. It is easy to see that

(3.8) $$\mathcal{C}' \cong \mathcal{C}_1 \oplus \cdots \oplus \mathcal{C}_q$$

where $\mathcal{C}_i$ is the center of the matrix algebra $M_{m_i}(\mathbf{F})$; but by Lemma 4.1.3, $\mathcal{C}_i = \mathbf{F} I_{m_i} \subseteq M_{m_i}(\mathbf{F})$ so that $\dim_{\mathbf{F}}(\mathcal{C}_i) = 1$. By Equation (3.8), it follows that

$$\dim_{\mathbf{F}}(\mathcal{C}') = q.$$

Hence, $q = t$, as claimed.

(2) We shall prove this by calculating $\dim_{\mathbf{F}}(M_i)$ in two ways. First, by definition,

(3.9) $$\dim_{\mathbf{F}}(M_i) = d_i.$$

Second, we have $\operatorname{Hom}_{\mathcal{R}}(\mathcal{R}, M_i) \cong M_i$ as $\mathcal{R}$-modules, and we calculate the dimension of this. Now

$$\operatorname{Hom}_{\mathcal{R}}(\mathcal{R}, M_i) \cong \operatorname{Hom}_{\mathcal{R}}\left(\bigoplus_{j=1}^{q} m_j M_j, M_i\right)$$
$$\cong \operatorname{Hom}_{\mathcal{R}}(m_i M_i, M_i)$$
$$\cong M_{1, m_i}(\mathbf{F})$$

by Schur's lemma and Lemma 3.3 again. This matrix space has dimension m_i over $\mathbf{F}$, so $m_i = d_i$, as claimed.

(3) By part (2) and Equation (3.7),

$$
\begin{aligned}
n &= \dim_{\mathbf{F}}(\mathcal{R}) \\
&= \dim_{\mathbf{F}}\left(\bigoplus_{i=1}^{q} M_{m_i}(\mathbf{F})\right) \\
&= \sum_{i=1}^{q} \dim_{\mathbf{F}}(M_{m_i}(\mathbf{F})) \\
&= \sum_{i=1}^{q} m_i^2 \\
&= \sum_{i=1}^{t} d_i^2,
\end{aligned}
$$

as claimed. $\square$

(3.5) Remark. Note that this theorem generalizes the results of Section 8.2. For a group G is abelian if and only if every conjugacy class of G consists of a single element, in which case there are $n = |G|$ conjugacy classes. Then G has n distinct irreducible $\mathbf{F}$-representations, each of degree 1 and appearing with multiplicity 1 in the regular representation (and $n = \sum_{i=1}^{n} 1^2$).

(3.6) Warning. Although the number of conjugacy classes of elements of G is equal to the number of distinct irreducible $\mathbf{F}$-representations of G (if $\mathbf{F}$ is an excellent field for G), there is **no** natural one-to-one correspondence between the set of conjugacy classes and the set of irreducible $\mathbf{F}$-representations.

(3.7) Example. Consider the dihedral group

$$
G = D_{2m} = \langle x, y : x^m = 1, y^2 = 1, xy = yx^{-1}\rangle.
$$

For m odd, G has the following conjugacy classes:

$$
\{1\}, \{x, x^{m-1}\}, \{x^2, x^{m-2}\}, \dots,
$$
$$
\{x^{(m-1)/2}, x^{(m+1)/2}\}, \{y, xy, x^2y, \dots, x^{m-1}y\}.
$$

There are $(m+3)/2$ conjugacy classes and in Example 1.4 (7), we constructed $(m+3)/2$ distinct irreducible representations over a good field $\mathbf{F}$, so by Theorem 3.4 we have found all of them.

For m even G has the following conjugacy classes:

$$
\{1\}, \{x, x^{m-1}\}, \{x^2, x^{m-2}\}, \dots, \{x^{\frac{m}{2}-1}, x^{\frac{m}{2}+1}\}, \{x^{\frac{m}{2}}\},
$$
$$
\{x^i y : i \text{ is even}\}, \{x^i y : i \text{ is odd}\}.
$$

There are $\frac{m}{2} + 3$ of these and in Example 1.4 (7) we also constructed $\frac{m}{2} + 3$ irreducible representations over a good field $\mathbf{F}$, so again we have found all of them.

Note also that the decomposition in Example 1.8 (4) is as predicted by Theorem 3.4.

(3.8) Example. Let us construct all irreducible $\mathbf{C}$-representations of the quaternion group $G = Q_8$. Recall that

$$Q_8 = \{\pm 1, \pm \mathbf{i}, \pm \mathbf{j}, \pm \mathbf{k}\},$$

and it is straightforward to compute that it has conjugacy classes

$$\{1\}, \quad \{-1\}, \quad \{\pm \mathbf{i}\}, \quad \{\pm \mathbf{j}\}, \quad \{\pm \mathbf{k}\}.$$

(See Example 2.1.10 (10), where we denote Q_8 by Q.)

Thus, we have five irreducible representations whose degrees satisfy $\sum_{i=1}^{5} d_i^2 = 8$, which forces $d_1 = d_2 = d_3 = d_4 = 1$ and $d_5 = 2$. (Actually, it is unnecessary here to find the number of conjugacy classes, for the Equation $\sum_{i=1}^{t} d_i^2 = 8$ only has the solutions

$$(2, 2), \quad (1, 1, 1, 1, 1, 1, 1, 1), \quad \text{and} \quad (1, 1, 1, 1, 2).$$

The first of these is impossible since we must have some $d_i = 1 = \deg(\tau)$, and the second cannot be right because G is nonabelian, so the third must apply.)

Note that $C = \{\pm 1\}$ is the center of G, and so $C \lhd G$, and we have an exact sequence

$$1 \longrightarrow C \longrightarrow G \overset{\pi}{\longrightarrow} V \longrightarrow 1$$

where $V \cong \mathbf{Z}_2 \oplus \mathbf{Z}_2$. Since V is abelian of order 4, it has four 1-dimensional representations and their pullbacks give four 1-dimensional (and hence certainly irreducible) representations of G.

To be precise, let

$$V = \{1, I, J, K : I^2 = J^2 = K^2 = 1, \ IJ = K\}.$$

Then $\pi : G \to V$ is defined by $\pi(\pm 1) = 1$, $\pi(\pm \mathbf{i}) = I$, $\pi(\pm \mathbf{j}) = J$, and $\pi(\pm \mathbf{k}) = K$. The representations of V are the trivial representation $\sigma_0 = \tau$ (and $\pi^*(\tau) = \tau$) and the representations σ_i for $i = 1, 2, 3$ given by

$$
\begin{array}{llll}
\sigma_1 : & \sigma_1(I) = 1, & \sigma_1(J) = -1, & \sigma_1(K) = -1 \\
\sigma_2 : & \sigma_2(I) = -1, & \sigma_2(J) = 1, & \sigma_2(K) = -1 \\
\sigma_3 : & \sigma_3(I) = -1, & \sigma_3(J) = -1, & \sigma_3(K) = 1.
\end{array}
$$

We also need to find a two-dimensional representation ρ of Q_8. Here it is:

$$\rho(\pm 1) = \begin{bmatrix} \pm 1 & 0 \\ 0 & \pm 1 \end{bmatrix}$$

$$\rho(\pm i) = \begin{bmatrix} \pm i & 0 \\ 0 & \mp i \end{bmatrix}$$

$$\rho(\pm j) = \begin{bmatrix} 0 & \pm i \\ \pm i & 0 \end{bmatrix}$$

$$\rho(\pm k) = \begin{bmatrix} 0 & \mp 1 \\ \pm 1 & 0 \end{bmatrix}.$$

Note that in the matrices, i is the complex number i. We must check that ρ is irreducible. This can be done directly, but it is easier to make the following observation. If ρ were not irreducible it would have to be isomorphic to $\pi^*(\sigma_i) \oplus \pi^*(\sigma_j)$ for some $i, j \in \{0, 1, 2, 3\}$, but it cannot be, for $\rho(-1)$ is nontrivial, but $(\pi^*(\sigma_i) \oplus \pi^*(\sigma_j))(-1)$ is trivial for any choice of i and j.

(3.9) Example. Let us construct all irreducible **C**-representations of the alternating group A_4 of order 12. Recall that A_4 is the subgroup of the symmetric group S_4 consisting of the even permutations, and it is a semidirect product

$$1 \longrightarrow V \longrightarrow A_4 \overset{\pi}{\longrightarrow} S \longrightarrow 1$$

where

$$V \cong \mathbf{Z}_2 \oplus \mathbf{Z}_2$$
$$= \{1, (1\,2)(3\,4), (1\,3)(2\,4), (1\,4)(2\,3)\}$$
$$= \{1, I, J, K\}$$

and

$$S \cong \mathbf{Z}_3 = \{1, (1\,2\,3), (1\,3\,2)\} = \{1, T, T^2\}.$$

We compute that A_4 has 4 conjugacy classes

$$\{1\}, \quad \{I, J, K\}, \quad \{T, TI, TJ, TK\}, \quad \{T^2, T^2I, ,T^2J, T^2K\},$$

so we expect 4 irreducible representations whose degrees satisfy $\sum_{i=1}^{4} d_i^2 = 12$, giving $d_1 = d_2 = d_3 = 1$ and $d_4 = 3$. (Alternatively, we find that V is the commutator subgroup of G, and so we have exactly 3 one-dimensional representations of G by Proposition 2.5. Then the equation $\sum_{i=1}^{t} d_i^2 = 12$ and $d_1 = d_2 = d_3 = 1$ with $d_i > 1$ for $i > 3$ forces $t = 4$ and $d_4 = 3$.)

The three one-dimensional representations of G are $\pi^*(\theta_i)$ for $0 \le i \le 2$, where θ_i are the representations of the cyclic group S constructed in Example 1.4 (6).

Now we need to find a three-dimensional representation. Let

$$M = \mathbf{C}^4 = \{(z_1, z_2, z_3, z_4) : z_i \in \mathbf{C}, \ 1 \le i \le 4\}.$$

Then S_4, and hence, A_4, acts on $\mathbf{C}^4$ by permuting the coordinates, i.e.,

$$g(z_1, z_2, z_3, z_4) = (z_{g(1)}, z_{g(2)}, z_{g(3)}, z_{g(4)}) \qquad \text{for} \quad g \in S_4.$$

Consider

$$M_0 = \{(z_1, z_2, z_3, z_4) : z_1 + z_2 + z_3 + z_4 = 0\}.$$

This subspace of M is invariant under S_4, so it gives a three-dimensional representation α of S_4, and we consider its restriction to A_4, which we still denote by α. We claim that this representation is irreducible, and the argument is the same as the final observation in Example 3.8: The subgroup V acts trivially in each of the representations $\pi^*(\theta_i)$ for $i = 0, 1, 2$, but nontrivially in the representation α.

(3.10) Example. Let us construct all irreducible **C**-representations of the symmetric group S_4. We have a semidirect product

$$1 \longrightarrow V \longrightarrow S_4 \overset{\pi}{\longrightarrow} W \longrightarrow 1$$

where V is the same as in Example 3.9 and

$$W = \langle T, U : T^3 = U^2 = 1, \, TU = UT^2 \rangle \cong D_6 \cong S_3$$

where $T = (1\,2\,3)$ as before and $U = (1\,2)$. By Corollary 1.5.10, S_4 has 5 conjugacy classes, so we look for a solution of $\sum_{i=1}^{5} d_i^2 = 24 = |S_4|$. This has the unique solution

$$d_1 = d_2 = 1, \qquad d_3 = 2, \qquad d_4 = d_5 = 3.$$

As before, we obtain an irreducible representation of S_4 from every irreducible representation of W, and by Example 1.4 (7), W has the irreducible representations $\psi_+ = \tau$, ψ_-, and ϕ_1, so we have irreducible representations $\pi^*(\tau) = \tau$, $\pi^*(\psi_+)$, and $\pi^*(\phi_1)$ of degrees 1, 1, and 2, respectively.

We need to find two three-dimensional irreducible representations. For the first we simply take the representation α constructed in Example 3.9. Since the restriction of α to A_4 is irreducible, α itself is certainly irreducible.

For the second, we take $\alpha' = \pi^*(\psi_-) \otimes \alpha$. This is also a three-dimensional representation of S_4 (which we may now view as acting on M_0 by

$$\alpha'(g)(v) = \psi_- \pi(g) \alpha(v) \qquad \text{for} \quad v \in M_0$$

since $\psi_- \pi(g) = \pm 1$). Since α' restricted to A_4 agrees with α restricted to A_4 and the latter is irreducible, so is the former. To complete our construction, we need only show that α and α' are inequivalent. We see this from the fact that $\alpha(U)$ has characteristic polynomial $(X - 1)^2(X + 1)$, while $\alpha'(U)$ has characteristic polynomial $(X - 1)(X + 1)^2$. (The reader may wonder about the representation $\pi^*(\psi_-) \otimes \pi^*(\phi_1)$, but this is isomorphic to $\pi^*(\phi_1)$, so it gives nothing new.)

We conclude this section with the following result:

(3.11) Theorem. (Burnside) *Let* $\mathbf{F}$ *be an excellent field for the finite group* G *and let* $\rho : G \rightarrow \mathrm{Aut}(V)$ *be an irreducible* $\mathbf{F}$*- representation of* G. *Then*

$$\{\rho(g) : g \in G\}$$

spans $\mathrm{End}_{\mathbf{F}}(V)$.

Proof. We first claim that for any field $\mathbf{F}$, if $\rho_0 : G \rightarrow \mathrm{Aut}(\mathbf{F}(G))$ is the regular representation of G, then $\{\rho_0(g)\}$ is a linearly independent set in $\mathrm{Aut}(\mathbf{F}(G))$. For suppose

$$a = \sum_{g \in G} a_g \rho_0(g) = 0.$$

Then

$$0 = a(1) = \sum_{g \in G} a_g g$$

so $a_g = 0$ for each $g \in G$.

Now let $\mathbf{F}$ be an excellent field for G. Then by Theorem 3.4 we have an isomorphism $\phi : V_0 \rightarrow \mathbf{F}(G)$ with $V_0 = \oplus_{i=1}^t d_i V_i$, $\rho_0 : G \rightarrow \mathrm{Aut}(V_0)$, where $\rho_i : G \rightarrow \mathrm{Aut}(V_i)$ are the distinct irreducible $\mathbf{F}$-representations of G. Choose a basis $\mathcal{B}_i$ for each V_i and let $\mathcal{B}$ be the basis of V_0 that is the union of these bases. If $M_i(g) = [\rho_i(g)]_{\mathcal{B}_i}$, then for each $g \in G$, $[\rho_0(g)]_{\mathcal{B}}$ is the block diagonal matrix

$$\mathrm{diag}(M_1(g), M_2(g), \ldots, M_2(g), \ldots, M_t(g), \ldots, M_t(g))$$

where $M_i(g)$ is repeated $d_i = \dim(V_i)$ times. (Of course, $M_1(g) = [1]$ appears once.) By the first paragraph of the proof, we have that the dimension of $\{\rho_0(g) : g \in G\}$ is equal to n, so we see that

$$n \le \sum_{i=1}^t q_i$$

where q_i is the dimension of the span of $\{\rho_i(g) : g \in G\}$. But this span is a subspace of $\mathrm{End}(V_i)$, a space of dimension d_i^2. Thus, we have

$$n \le \sum_{i=1}^t q_i \le \sum_{i=1}^t d_i^2 = n$$

where the latter equality is Theorem 3.4 (3), so we have $q_i = d_i^2$ for $1 \le i \le t$, proving the theorem. □

8.4 Characters

In this section we develop the theory of characters. In practice, characters are a tool whose usefulness, especially in characteristic zero, can hardly be overemphasized. We will begin without restricting the characteristic.

(4.1) Definition. *Let* $\sigma : G \to \mathrm{Aut}(M)$ *be an* **F**-*representation of* G *of finite degree, and let* $\mathcal{B}$ *be a basis of* M. *The* **character** *of the representation* σ *is the function* $\chi_\sigma : G \to \mathbf{F}$ *defined by*

$$(4.1) \qquad\qquad \chi_\sigma(g) = \mathrm{Tr}([\sigma(g)]_\mathcal{B}).$$

Recall that Tr denotes the trace of a matrix or a linear transformation. This is independent of the choice of the basis $\mathcal{B}$ by Proposition 4.3.27. By the same logic, it is the case that if two representations are equivalent, then their characters are equal. It is one of the great uses of characters that, under the proper circumstances, the converse of this is true as well.

(4.2) Examples.

(1) If $\sigma = d\tau$, then $\chi_\sigma(g) = d$ for every $g \in G$.
(2) If σ is any representation of degree d, then $\chi_\sigma(1) = d$.
(3) If σ is the regular representation of G, then $\chi_\sigma(1) = n = |G|$ and $\chi_\sigma(g) = 0$ for all $g \neq 1$. (To see this, consider $[\sigma(g)]$ in the basis $\{g : g \in G\}$ of $\mathbf{F}(G)$.)

(4.3) Lemma. *If* g_1 *and* g_2 *are conjugate elements of* G, *then for any representation* σ,

$$(4.2) \qquad\qquad \chi_\sigma(g_1) = \chi_\sigma(g_2).$$

Proof. If $g_2 = gg_1g^{-1}$, then

$$
\begin{aligned}
\chi_\sigma(g_2) &= \chi_\sigma(gg_1g^{-1}) \\
&= \mathrm{Tr}\left([\sigma(gg_1g^{-1})]_\mathcal{B}\right) \\
&= \mathrm{Tr}\left([\sigma(g)]_\mathcal{B}[\sigma(g_1)]_\mathcal{B}[\sigma(g)]_\mathcal{B}^{-1}\right) \\
&= \mathrm{Tr}\left([\sigma(g_1)]_\mathcal{B}\right) \\
&= \chi_\sigma(g_1).
\end{aligned}
$$

$\square$

(4.4) Proposition. *Let* σ_1 *and* σ_2 *be two representations of the group* G. *Then*

(1) $\chi_{\sigma_1 \oplus \sigma_2} = \chi_{\sigma_1} + \chi_{\sigma_2}$, *and*

(2) $\chi_{\sigma_1 \otimes \sigma_2} = \chi_{\sigma_1} \chi_{\sigma_2}$.

Proof. (1) is obvious, and (2) follows immediately from Proposition 7.2.35 and Lemma 4.1.20.   □

Our next goal is to derive the basic orthogonality results for characters. Along the way, we will derive a bit more: orthogonality for matrix coefficients.

(4.5) Proposition. *Let* **F** *be an excellent field for the group* G, *let* V_i *be irreducible representations of* G, *given by* $\sigma_i : G \to \operatorname{Aut}(V_i)$, *and let* $\mathcal{B}_i$ *be a basis of* V_i, *for* $i = 1, 2$.
For $g \in G$, *let*

$$P(g) = [p_{i_1 j_1}(g)] = [\sigma_1(g)]_{\mathcal{B}_1}$$

and

$$Q(g) = [q_{i_2 j_2}(g)] = [\sigma_2(g)]_{\mathcal{B}_2}.$$

(1) *Suppose that* V_1 *and* V_2 *are distinct. Then for any* i_1, j_1, i_2, j_2,

(4.3)
$$\frac{1}{n} \sum_{g \in G} p_{i_1 j_1}(g) q_{i_2 j_2}(g^{-1}) = 0.$$

(2) *Suppose that* $V_1 = V_2$ *(so that* $\sigma_1 = \sigma_2$*) and* $\mathcal{B}_1 = \mathcal{B}_2$. *Let* $d = \deg(V_1)$. *Then*

(4.4)
$$\frac{1}{n} \sum_{g \in G} p_{i_1 j_1}(g) q_{i_2 j_2}(g^{-1}) = \begin{cases} 1/d & \text{if } i_1 = j_2 \text{ and } j_1 = i_2, \\ 0 & \text{otherwise.} \end{cases}$$

(Note that in this case $p_{ij}(g) = q_{ij}(g)$, *of course.)*

Proof. Let β_i be the projection of V_1 onto its ith summand **F** (as determined by the basis $\mathcal{B}_1$) and let α_j be the inclusion of **F** onto the jth summand of V_2 (as determined by the basis $\mathcal{B}_2$). Then $f = \alpha_j \beta_i \in \operatorname{Hom}(V_1, V_2)$. Note that $[\alpha_j \beta_i]_{\mathcal{B}_2}^{\mathcal{B}_1} = E_{ji}$ where E_{ji} is the matrix with 1 in the ji^{th} position and 0 elsewhere (see Section 5.1). Let us compute $\operatorname{Av}(f)$. By definition

$$\operatorname{Av}(f) = \frac{1}{n} \sum_{g \in G} \sigma_2(g^{-1})(\alpha_j \beta_i) \sigma_1(g).$$

Direct matrix calculation shows

(4.5) $$[\sigma_2(g^{-1})(\alpha_j \beta_i)\sigma_1(g)]_{\mathcal{B}_2}^{\mathcal{B}_1} = \begin{bmatrix} p_{1j}(g)q_{i1}(g^{-1}) & p_{1j}(g)q_{i2}(g^{-1}) & \cdots \\ p_{2j}(g)q_{i1}(g^{-1}) & p_{2j}(g)q_{i2}(g^{-1}) & \cdots \\ \vdots & \vdots & \ddots \end{bmatrix}$$

so the sums in question are just the entries of $[\mathrm{Av}(f)]^{\mathcal{B}_1}_{\mathcal{B}_2}$ (as we vary i, j and the entry of the matrix.)

Consider case (1). Then $\mathrm{Av}(f) \in \mathrm{Hom}_G(V_1, V_2) = \langle 0 \rangle$ by Schur's lemma, so f is the zero map and all matrix entries are 0, as claimed.

Now for case (2). Then $\mathrm{Av}(f) \in \mathrm{Hom}_G(V, V) = \mathbf{F}$, by Schur's lemma, with every element a homothety, represented by a scalar matrix. Thus all the off-diagonal entries of $[\mathrm{Av}(f)]_{\mathcal{B}_1}$ are 0, showing that the sum in Equation (4.4) is zero if $i_1 \neq j_2$. Since $\sigma_1 = \sigma_2$, we may rewrite the sum (replacing g by g^{-1}) as

$$\frac{1}{n} \sum_{g \in G} p_{i_2 j_2}(g) q_{i_1 j_1}(g^{-1})$$

showing that it is zero if $j_1 \neq i_2$.

Consider the remaining case, where $i_1 = j_2 = i$ and $j_1 = i_2 = j$. As $\mathrm{Av}(f)$ is a homothety, all of the diagonal entries of its matrix are equal, so we obtain a common value, say x, for

$$\frac{1}{n} \sum_{g \in G} p_{ij}(g) q_{ji}(g^{-1})$$

for any i, j (by varying the choice of f and the diagonal element in question).

Now consider

$$f_0 = \alpha_1 \beta_1 + \alpha_2 \beta_2 + \cdots + \alpha_d \beta_d.$$

Since there are d summands, we see that the diagonal entries of $\mathrm{Av}(f_0)$ are all equal to dx. But f_0 is the identity! Hence, $\mathrm{Av}(f_0) = f_0$ has its diagonal entries equal to one, so $dx = 1$ and $x = 1/d$, as claimed.   □

(4.6) Corollary. *Let* $\mathbf{F}$ *be an excellent field for* G *with* $\mathrm{char}(\mathbf{F}) = p \neq 0$, *and let* V *be an irreducible* $\mathbf{F}$-*representation of* G. *Then* $\deg(V)$ *is not divisible by* p.

Proof. If $d = \deg(V)$, then the above proof shows that $dx = 1 \in \mathbf{F}$, so that $d \neq 0 \in \mathbf{F}$.   □

(4.7) Corollary. (Orthogonality of characters) *Let* $\mathbf{F}$ *be an excellent field for* G, *and let* V_1 *and* V_2 *be irreducible representations of* G *defined by* $\sigma_i : G \to \mathrm{Aut}(V_i)$ *for* $i = 1, 2$. *Then*

$$\frac{1}{n} \sum_{g \in G} \chi_{\sigma_1}(g) \chi_{\sigma_2}(g^{-1}) = \begin{cases} 0 & \text{if } V_1 \text{ and } V_2 \text{ are distinct,} \\ 1 & \text{if } V_1 \text{ and } V_2 \text{ are isomorphic.} \end{cases}$$

Proof. If V_i has degree d_i, this sum is equal to

$$\frac{1}{n}\sum_{g\in G}\left(\sum_{i=1}^{d_1}\sum_{j=1}^{d_2}p_{ii}(g)q_{jj}(g^{-1})\right)=\sum_{i=1}^{d_1}\sum_{j=1}^{d_2}\left(\frac{1}{n}\sum_{g\in G}p_{ii}(g)q_{jj}(g^{-1})\right)$$

which is 0 if V_1 and V_2 are distinct. If V_1 and V_2 are isomorphic of degree d, then, since isomorphic representations have the same character, we may assume that $V_1 = V_2$. The terms with $i \neq j$ are all zero, so the sum is

$$\sum_{j=1}^{d}p_{jj}(g)q_{jj}(g^{-1})=d(1/d)=1.$$

$\square$

Proposition 4.5 and Corollary 4.7 have a generalization, as follows:

(4.8) Proposition. *Given the same hypothesis and notation as Proposition 4.5, let h be a fixed element of G.*

(1) *Suppose that V_1 and V_2 are distinct. Then for any i_1, j_1, i_2, and j_2, we have*

$$\frac{1}{n}\sum_{g\in G}p_{i_1 j_1}(hg)q_{i_2 j_2}(g^{-1})=0.$$

(2) *Suppose that $V_1 = V_2$. Then if $i_1 \neq j_2$,*

$$\frac{1}{n}\sum_{g\in G}p_{i_1 j_1}(hg)q_{i_2 j_2}(g^{-1})=0.$$

(3) *For any two representations V_1 and V_2,*

$$\frac{1}{n}\sum_{g\in G}\chi_{\sigma_1}(hg)\chi_{\sigma_2}(g^{-1})=\begin{cases}0 & \text{if } V_1 \text{ and } V_2 \text{ are distinct,}\\(1/d)\chi_{\sigma_1}(h) & \text{if } V_1 = V_2 \text{ is of degree } d.\end{cases}$$

Proof. The proof follows that of Proposition 4.5, with $f = \alpha_j\beta_i\sigma_1(h)$. Then the matrix corresponding to that in Equation (4.5) is

(4.6)

$$[\sigma_2(g^{-1})(\alpha_j\beta_i)\sigma_1(hg)]_{B_2}^{B_1}=\begin{bmatrix}p_{1j}(hg)q_{i1}(g^{-1}) & p_{1j}(hg)q_{i2}(g^{-1}) & \cdots\\p_{2j}(hg)q_{i1}(g^{-1}) & p_{2j}(hg)q_{i2}(g^{-1}) & \cdots\\\vdots & \vdots & \ddots\end{bmatrix}.$$

First suppose that V_1 and V_2 are distinct. Then, again, $\mathrm{Av}(f) = 0$, so all the matrix entries are zero, yielding (1), and then the first assertion of (3) follows as in the proof of Corollary 4.7.

Now suppose that $V_1 = V_2$. Then, again, $\text{Av}(f)$ is a homothety, so the off-diagonal matrix entries are zero, yielding (2), while all the diagonal entries are equal. Set $i = j$ and call the common value in this case x_j, i.e.,

$$x_j = \frac{1}{n} \sum_{g \in G} p_{ij}(hg)q_{ji}(g^{-1}),$$

this sum being independent of i. As in the proof of Corollary 4.7, the sum we are interested in is

$$\sum_{i=1}^{d} \sum_{j=1}^{d} \left(\frac{1}{n} \sum_{g \in G} p_{ii}(hg)q_{jj}(g^{-1}) \right) = \sum_{j=1}^{d} \left(\frac{1}{n} \sum_{g \in G} p_{jj}(hg)q_{jj}(g^{-1}) \right)$$
$$= x_1 + \cdots + x_d$$

where the first equality follows from part (2).

Now consider

$$f_0 = (\alpha_1 \beta_1 + \cdots + \alpha_d \beta_d)\sigma_1(h).$$

Then $\text{Av}(f_0)$ is a homothety, and all of its diagonal entries are equal to $x_1 + \cdots + x_d$, so

$$\text{Tr}(\text{Av}(f_0)) = d(x_1 + \cdots + x_d).$$

But $f_0 = \sigma_1(h)$, so

$$\text{Av}(f_0) = \frac{1}{n} \sum_{g \in G} \sigma_1(g^{-1}hg).$$

Then

$$d(x_1 + \cdots + x_d) = \text{Tr}(\text{Av}(f_0))$$
$$= \frac{1}{n} \sum_{g \in G} \text{Tr}(\sigma_1(g^{-1}hg))$$
$$= \frac{1}{n} \sum_{g \in G} \text{Tr}(\sigma_1(h))$$
$$= \text{Tr}(\sigma_1(h))$$
$$= \chi_{\sigma_1}(h),$$

yielding the desired equality. $\square$

(4.9) *Remarks.*

(1) We should caution the reader that there are in fact three alternatives in Proposition 4.5: that V_1 and V_2 are distinct, that they are equal, or that they are isomorphic but unequal. In the latter case the sum in Equation (4.4) and the corresponding sum in the statement of Proposition 4.8 may vary (see Exercise 14). Note that we legitimately reduced the third

case to the second in the proof of Corollary 4.7; the point there is that while the individual matrix entries of isomorphic representations will differ, their traces will be the same.

(2) We should caution the reader that the sum in Proposition 4.8 (2) with $i_1 = j_2$ but $j_1 \neq i_2$ may well be nonzero and that the quantities $x_1, \ldots, x_d$ in the proof of the proposition may well be unequal (see Exercise 15).

Given a finite group G and an excellent field $\mathbf{F}$, let G have conjugacy classes

$$C_1, \ldots, C_t$$

(in some order) and irreducible representations

$$\sigma_1, \ldots, \sigma_t$$

(in some order).

(4.10) Definition. *The* **character table** *of G is the matrix $A \in M_t(\mathbf{F})$ defined by $A = [a_{ij}] = [\chi_i(c_j)]$ where $\chi_i = \chi_{\sigma_i}$ and $c_j \in C_j$.*

Let $|C_i|$ be the number of elements of C_i. Then Corollary 4.7 gives a sort of orthogonality relation on the rows of A; namely,

$$(4.7) \qquad \frac{1}{n} \sum_{g \in G} \chi_i(g)\chi_j(g^{-1}) = \frac{1}{n} \sum_{k=1}^{t} |C_k|\chi_i(c_k)\chi_j(c_k)^{-1} = \delta_{ij}.$$

(Recall that $\delta_{ij} = 1$ if $i = j$ and 0 if not.)

From this, we can immediately write down $B = A^{-1}$; namely,

(4.11) Lemma. $B = [b_{ij}] = \left[\dfrac{|C_i|}{n} \chi_j(c_i^{-1}) \right].$

Proof. The equation $AB = I$ is the equation $\sum_{k=1}^{t} a_{ik}b_{kj} = \delta_{ij}$, which is immediate from Equation (4.7). $\qquad\square$

Now let us interpret B.

(4.12) Definition. *Let $f : G \to \mathbf{F}$ be a function with the property that whenever g_1 and $g_2 \in G$ are conjugate, then $f(g_1) = f(g_2)$. Then f is called a* **class function** *on G.*

Clearly, $f : G \to \mathbf{F}$ is a class function if and only if it is constant on each conjugacy class C_i $(1 \leq i \leq t)$. Thus the space of class functions is clearly an $\mathbf{F}$-vector space of dimension t, with basis $\mathcal{B} = \{f_1, \ldots, f_t\}$ where

$$f_i(g) = \begin{cases} 1 & \text{if } g \in C_i, \\ 0 & \text{otherwise.} \end{cases}$$

(4.13) Lemma. *Let* **F** *be an excellent field for* G. *Then*

$$\mathcal{A} = \{\chi_1, \cdots, \chi_t\}$$

is a basis for the space of class functions on G.

Proof. By Lemma 4.3, the characters are class functions. There are t of them, so to show that they are a basis it suffices to show that they are linearly independent. Suppose

$$a_1\chi_1 + a_2\chi_2 + \cdots + a_t\chi_t = 0 \qquad \text{where} \qquad a_i \in \mathbf{F}$$

where, as in Definition 4.10, we have written χ_i for χ_{σ_i}. Note that χ_i^*, defined by $\chi_i^*(g) = \chi_i(g^{-1})$, is also a class function. Then for each i,

$$(a_1\chi_1 + a_2\chi_2 + \cdots + a_t\chi_t)\chi_i^* = 0$$

and averaging over the group

$$\frac{1}{n} \sum_{g \in G} a_1\chi_1(g)\chi_i^*(g) + a_2\chi_2(g)\chi_i^*(g) + \cdots + a_t\chi_t(g)\chi_i^*(g) = 0.$$

But, by the orthogonality relations, this sum is just a_i/n, so that $a_i/n = 0$ and $a_i = 0$ for each i, as required. $\qquad\square$

(4.14) Remark. If χ is the character of the representation V defined by $\sigma : G \to \text{Aut}(V)$, then $\chi^*(g) = \chi(g^{-1})$ is indeed the character of a representation; namely, with the given action of G, V is a *left* $\mathbf{F}(G)$-module. The action of G given by $g \mapsto \sigma(g^{-1})$ gives V the structure of a *right* $\mathbf{F}(G)$-module. Then $V^* = \text{Hom}_G(V, \mathbf{F}(G))$ is a *left* $\mathbf{F}(G)$-module with character χ^*. (In terms of matrices, if $\mathcal{B}$ is a basis for V, $\mathcal{B}^*$ is the dual basis of V^*, and χ^* is defined by $\sigma^* : G \to \text{Aut}(V^*)$, then

$$[\sigma^*(g)]_{\mathcal{B}^*} = [\sigma(g^{-1})]_{\mathcal{B}}^t.)$$

(4.15) Proposition. *Let* A *and* B *be the above matrices, and let* χ_i *and* f_i *be defined as above. Then for every* $1 \le i \le t$,

(1) $\chi_i = \sum_{j=1}^{t} a_{ij}f_j$, *and*

(2) $f_i = \sum_{j=1}^{t} b_{ij}\chi_j$.

Proof. (1) is easy. We need only show that both sides agree on c_k for $1 \le k \le t$. Since $f_j(c_k) = \delta_{jk}$,

$$\sum_{j=1}^{t} a_{ij}f_j(c_k) = a_{ik} = \chi_i(c_k)$$

as claimed.

Now (1) says that the change of basis matrix $P_{\mathcal{B}}^{\mathcal{A}}$ is just A. Then

$$P_{\mathcal{A}}^{\mathcal{B}} = \left(P_{\mathcal{B}}^{\mathcal{A}}\right)^{-1} = A^{-1} = B,$$

giving (2). □

As a corollary, we may derive another set of orthogonality relations.

(4.16) Corollary. *Let g_1, $g_2 \in G$. Then*

$$\sum_{j=1}^{t} \chi_j(g_1)\chi_j(g_2^{-1}) = \begin{cases} 0 & \text{if } g_1 \text{ and } g_2 \text{ are not conjugate,} \\ \dfrac{n}{|C_i|} & \text{if } g_1, g_2 \in C_i. \end{cases}$$

Proof. We may assume $g_1 = c_i$ and $g_2 = c_k$ for some i, k. Then by the definition of f_k and Proposition 4.15 (2),

$$\delta_{ki} = f_k(c_i)$$

$$= \left(\sum_{j=1}^{t} b_{kj}\chi_j\right)(c_i)$$

$$= \sum_{j=1}^{t} \left(\frac{|C_k|}{n}\chi_j(c_k^{-1})\right)\chi_j(c_i)$$

$$= \frac{|C_k|}{n}\sum_{j=1}^{t}\chi_j(c_i)\chi_j(c_k^{-1})$$

yielding the corollary. □

We now define an important quantity.

(4.17) Definition. *Let G and $\mathbf{F}$ be arbitrary and let V and W be two $\mathbf{F}$-representations of G. The* **intertwining number of V and W** *is*

$$i(V, W) = \langle V, W \rangle = \dim_{\mathbf{F}} \operatorname{Hom}_G(V, W).$$

(4.18) Lemma. *Let G be finite, $\mathbf{F}$ a field of characteristic 0 or prime to the order of G, and let V and W be two f-representations of G. Then*

(1) $i(V, W) = i(W, V)$;

(2) *if $V \cong \oplus_{i \in I} p_i M_i$ and $W \cong \oplus_{i \in I} q_i M_i$ are the decompositions of V and W into irreducibles, then*

$$i(V, W) = \sum_{i \in I} p_i q_i \dim_{\mathbf{F}}(\operatorname{End}(M_i));$$

(3) *if* **F** *is excellent for* G *and* V *is irreducible, then* $i(V, W)$ *is the multiplicity of* V *in* W;

(4) *if* **F** *is excellent for* G, *then* V *is irreducible if and only if* $i(V, V) = 1$;

(5) *if* **F** *is excellent for* G *and* char(**F**) $= 0$, *then*

$$i(V, W) = \langle V, W \rangle = \langle \chi_V, \chi_W \rangle = \frac{1}{n} \sum_{g \in G} \chi_V(g)\chi_W(g^{-1}).$$

Proof. Note that the hypotheses imply that $\mathcal{R}$ is semisimple, and recall Schur's lemma. The proof then becomes straightforward, and we leave it for the reader. □

Remark. If G is finite, **F** is of characteristic zero or prime to the order of G, and V is of finite degree, then we have

$$i(V, W) = \dim_{\mathbf{F}} V^* \otimes_{\mathbf{F}(G)} W.$$

This follows directly from Theorem 7.1.28 and Proposition 7.2.32.

Here we see the great utility of characters—if **F** is excellent for G and of characteristic zero, we may use them to compute intertwining numbers. In particular, we have:

(4.19) Corollary. *If* **F** *is excellent for* G *and is of characteristic zero and* W_1 *and* W_2 *are two* **F**-*representations of* G *of finite degree, then* W_1 *and* W_2 *are isomorphic if and only if* $\chi_{W_1} = \chi_{W_2}$.

Proof. Clearly, W_1 and W_2 are isomorphic if and only if $i(V, W_1) = i(V, W_2)$ for every irreducible V. But, by Lemma 4.18 (3), this is true if $\chi_{W_1} = \chi_{W_2}$. The converse is trivial. □

We could continue to work over a suitable excellent field of characteristic 0, but instead, for the sake of simplicity, we will take **F** = **C** (the field of complex numbers). We note that **C** has the field automorphism $z \mapsto \bar{z}$ of complex conjugation, with $\overline{z_1 + z_2} = \bar{z}_1 + \bar{z}_2$, $\overline{z_1 z_2} = \overline{z_1}\,\overline{z_2}$, and $\bar{\bar{z}} = z$.

If $\sigma : G \to \text{Aut}(V)$ is any complex representation, its conjugate $\bar{\sigma} : G \to \text{Aut}(V)$ is another representation, and $\chi_{\bar{\sigma}} = \overline{\chi}_\sigma$. By Lemma 4.18 (3), if G is finite, then σ is isomorphic to $\bar{\sigma}$ if and only if

$$\chi_\sigma = \chi_{\bar{\sigma}} = \overline{\chi}_\sigma,$$

i.e., if and only if χ_σ is real valued.

We have the following important result:

(4.20) Lemma. *Let* $\sigma : G \to \text{Aut}(V)$ *be a complex representation of the finite group* G. *Then for every* $g \in G$,

$$\chi_\sigma(g^{-1}) = \overline{\chi}_\sigma(g).$$

Proof. Let $g \in G$. Then g has finite order k, say. By Lemma 1.20, V has a basis $\mathcal{B}$ with $[\sigma(g)]_\mathcal{B}$ diagonal, and in fact,

$$[\sigma(g)]_\mathcal{B} = \mathrm{diag}(\zeta^{a_1}, \zeta^{a_2}, \dots, \zeta^{a_d})$$

where $\zeta = \exp(2\pi i/k)$, $d = \deg(V)$, and $a_1, \dots, a_d \in \mathbf{Z}$. Then

$$[\sigma(g^{-1})]_\mathcal{B} = ([\sigma(g)]_\mathcal{B})^{-1} = \mathrm{diag}(\zeta^{-a_1}, \zeta^{-a_2}, \dots, \zeta^{-a_d}).$$

But $\zeta^{-a} = \overline{\zeta^a}$ for any a, so

$$\chi_\sigma(g^{-1}) = \mathrm{Tr}\left([\sigma(g^{-1})]_\mathcal{B}\right) = \sum_{i=1}^{d} \zeta^{-a_i} = \sum_{i=1}^{d} \overline{\zeta^{a_i}} = \overline{\mathrm{Tr}\left([\sigma(g)]_\mathcal{B}\right)} = \overline{\chi_\sigma(g)}.$$

$\square$

(4.21) Consequence. *If* $\mathbf{F} = \mathbf{C}$*, then* $\chi_\sigma(g^{-1})$ *may be replaced by* $\overline{\chi}_\sigma(g)$ *in results* 4.7, 4.8, 4.11, 4.16, *and* 4.18.

A further consequence is that complex characters suffice to distinguish conjugacy classes. To be precise:

(4.22) Proposition. *Let* g_1 *and* $g_2 \in G$*. Then* g_1 *and* g_2 *are conjugate if and only if for every complex character* χ *of* G*,*

$$\chi(g_1) = \chi(g_2).$$

Proof. The only if is trivial. As for the if part, suppose that $\chi(g_1) = \chi(g_2)$ for every complex character. Then by Lemma 4.20,

$$\chi(g_2^{-1}) = \overline{\chi(g_1)},$$

so if $\chi_1, \dots, \chi_t$ are the irreducible characters,

$$\sum_{j=1}^{t} \chi_j(g_1)\chi_j(g_2^{-1}) = \sum_{j=1}^{t} \chi_j(g_1)\overline{\chi_j(g_1)} = \sum_{j=1}^{t} |\chi_j(g_1)|^2 > 0$$

since each term is nonnegative and $\chi_1(g_1) = 1$ (χ_1 being the character of the trivial representation τ). Then by Corollary 4.16, g_1 and g_2 are conjugate.

$\square$

(4.23) Proposition. *Let* G *be a finite group. Then every complex character of* G *is real valued if and only if every element in* G *is conjugate to its own inverse.*

Proof. Suppose that every g is conjugate to g^{-1}. Then for every complex character χ, $\chi(g) = \chi(g^{-1}) = \overline{\chi(g)}$, so $\chi(g)$ is real.

Conversely, suppose that $\chi_i(g)$ is real for every irreducible complex representation σ_i and every $g \in G$. Since $\sigma_1 = \tau$, $\chi_1(g) = 1$, so

$$0 < \sum_{j=1}^{t} \chi_j(g)^2 = \sum_{j=1}^{t} \chi_j(g)\chi_j((g^{-1})^{-1})$$

so g and g^{-1} are conjugate by Corollary 4.16 again. $\square$

Lemma 4.20 also gives a handy way of encoding the orthogonality relations (Corollaries 4.7 and 4.16) for complex characters. Recall the character table $A = [a_{ij}] = [\chi_i(c_j)]$ of Definition 4.10. Let

$$C = [c_{ij}] = \left[\sqrt{\frac{|C_j|}{n}} a_{ij} \right] \in M_t(\mathbf{C}).$$

(4.24) Proposition. *The above matrix C is unitary, i.e., $C^{-1} = \overline{C}^t$. (See Definition 4.6.16.)*

Proof. We leave this for the reader. $\square$

Using this result, we may generalize Proposition 4.23.

Let us call a conjugacy class C_i **self-inversive** if for some (and hence for every) $c_i \in C_i$, we also have $c_i^{-1} \in C_i$.

(4.25) Proposition. *The number of self-inversive conjugacy classes of G is equal to the number of irreducible complex characters of G that are real valued.*

Proof. We outline the proof and leave the details to the reader. We make use of the above matrix C. Consider the diagonal elements of CC^t. From the fact that $\overline{\chi}_i = \chi_i$ if χ_i is real valued and is the character of a distinct irreducible if not, we see that the ith diagonal entry of CC^t is 1 if χ_i is real valued and zero if not. Thus $\mathrm{Tr}(CC^t)$ is equal to the number of real-valued characters.

Now consider the diagonal elements of $C^t C$. From the fact that $c_i^{-1} \in C_i$ if C_i is self-inversive, and is in a different conjugacy class if not, we see that the ith diagonal entry of $C^t C$ is 1 if C_i is self-inversive and zero if not. Thus $\mathrm{Tr}(C^t C)$ is equal to the number of self-inversive conjugacy classes, and since $\mathrm{Tr}(C^t C) = \mathrm{Tr}(CC^t)$, the proposition follows. $\square$

(4.26) Corollary. *If G has odd order, no nontrivial irreducible character of G is real valued.*

Proof. Suppose that $ghg^{-1} = h^{-1}$. Then $g^i hg^{-i} = h$ if i is even or h^{-1} if i is odd. In particular, $g^n hg^{-n} = h^{-1}$ (recall that $n = |G|$). But $g^n = 1$, so this gives $h = h^{-1}$, i.e., $h^2 = 1$. Since n is odd, this give $h = 1$, and the corollary follows from Proposition 4.25. $\qquad\square$

We now determine the idempotents in $\mathbf{C}(G)$. In our approach the irreducible representations and their characters have been central, and the idempotents are of peripheral importance, but there is an alternative approach in which they play a leading role.

(4.27) Proposition. *For $i = 1, \ldots, t$ let $e_i \in \mathbf{C}(G)$ be defined by*

$$(4.8) \qquad e_i = \frac{d_i}{n} \sum_{g \in G} \chi_i(g^{-1}) g$$

where χ_i are the characters of the irreducible representations σ_i of G and $d_i = \deg(\sigma_i) = \chi_i(1)$. Then

(1) $e_i^2 = e_i$ and $e_j e_i = 0$ for $i \neq j$, and
(2) $e_1 + \cdots + e_t = 1$.

Proof. (1) We compute

$$(4.9) \qquad e_j e_i = \left(\frac{d_j}{n}\right)\left(\frac{d_i}{n}\right) \sum_{g_1, g_2 \in G} \chi_j(g_2^{-1})\chi_i(g_1^{-1}) g_2 g_1.$$

Setting $g_1 = g$, $g_2 = hg^{-1}$, and recalling that $\chi_j(gh^{-1}) = \chi_j(h^{-1}g)$, Equation (4.9) becomes

$$e_j e_i = \left(\frac{d_j}{n}\right)\left(\frac{d_i}{n}\right) \sum_{h \in G} \left(\sum_{g \in G} \chi_i(h^{-1}g)\chi_j(g^{-1})\right) h.$$

The interior sum, and hence the double sum, is zero if $i \neq j$ by Proposition 4.8. If $i = j$ the interior sum is $n\chi_i(h^{-1})/d_i$ (again by Proposition 4.8), and so in this case the double sum is

$$\frac{d_i^2}{n^2} \sum_{h \in G} \left(n\chi_i(h^{-1})/d_i\right) h = e_i.$$

(2) Let χ be the character of the regular representation. If $e_1 + \cdots + e_t = \sum_{h \in G} a_h h$, then by Example 4.2 (3)

$$a_h = \frac{1}{n}\chi(h^{-1}(e_1 + \cdots + e_t)).$$

Now

$$\chi(h^{-1}(e_1 + \cdots + e_t)) = \sum_{j=1}^{t} \sum_{g \in G} \frac{d_j}{n} \chi_j(g^{-1}) \chi(h^{-1}g)$$

$$= \sum_{j=1}^{t} \sum_{g \in G} \frac{d_j}{n} \chi_j(g^{-1}) \sum_{i=1}^{t} d_i \chi_i(h^{-1}g)$$

$$= \sum_{j=1}^{t} \sum_{i=1}^{t} \frac{d_i d_j}{n} \sum_{g \in G} \chi_i(h^{-1}g) \chi_j(g^{-1})$$

$$= \sum_{i=1}^{n} \left(\frac{d_i^2}{n} \right) \left(n \chi_i(h^{-1})/d_i \right)$$

$$= \sum_{i=1}^{n} d_i \chi_i(h^{-1})$$

$$= \sum_{i=1}^{n} \chi_i(1) \chi_i(h^{-1})$$

$$= \begin{cases} 0 & \text{if } h \neq 1 \\ n & \text{if } h = 1 \end{cases}$$

where the third equality is by Proposition 4.8 and the last is by Corollary 4.16. Thus, $a_1 = 1$ and $a_h = 0$ for $h \neq 1$, giving $e_1 + \cdots + e_t = 1$. $\square$

We will use idempotents to prove the following result:

(4.28) Theorem. *Let $d_i = \deg(\sigma_i)$ for $i = 1, \ldots, t$ be the degrees of the irreducible complex representations of G. Then for each i, d_i divides n, the order of G.*

Proof. We begin with the following observation: Let M be a finitely generated torsion-free (and, hence, free) $\mathbf{Z}$-module, and set $N = \mathbf{Q} \otimes M$. If $r \in \mathbf{Q}$ has the property that $rM \subseteq M$, then $r \in \mathbf{Z}$. To see this, let b be a primitive element of M, which exists by Lemma 3.6.14. If $r = p/q$ with p and q relatively prime, then $rb \in M$ implies that $b = qb'$ for some $b' \in M$, which by primitivity implies that $q = \pm 1$.

Let $\zeta = \exp(2\pi i/m)$, with m the exponent of G. For fixed i, let M_i be the $\mathbf{Z}$-submodule of $\mathbf{C}(G)$ spanned by

$$\{\zeta^k g e_i : k = 0, \ldots, m-1, \quad g \in G\}.$$

(Here we consider $\mathbf{C}(G)$ as an additive abelian group.) Since $e_i^2 = e_i$, we have

$$\frac{n}{d_i} e_i = \frac{n}{d_i} e_i^2 = \left(\chi_i(g^{-1})g \right) e_i \in M_i.$$

This immediately implies $(n/d_i)(M_i) \subseteq M_i$, so $n/d_i \in \mathbf{Z}$ and d_i divides n, as claimed. $\square$

We now present some typical examples of the use of characters and their properties. In the first two examples we show how to use the basic Theorem 3.4 (and the orthogonality relations (Corollary 4.7)) to find all complex characters of a group (without first finding all the irreducible representations), and in the third we show how to use Lemma 4.18 to find the decomposition of a complex representation into irreducibles.

(4.29) Example. *Let us determine all the irreducible complex characters of the alternating group A_5, of order 60.*

First we determine that A_5 has 5 conjugacy classes: $\{1\}$, { all products of two disjoint 2-cycles }, { all 3-cycles } each constitute a conjugacy class, while { all 5-cycles } splits into two conjugacy classes (with the property that if g is in one of them, g^2 is in the other). Thus, as representatives for the conjugacy classes we may take $c_1 = 1$, $c_2 = (1\,2)(3\,4)$, $c_3 = (1\,2\,3)$, $c_4 = (1\,2\,3\,4\,5)$, and $c_5 = (1\,3\,5\,2\,4)$. It may be checked that the conjugacy classes have sizes 1, 15, 20, 12, and 12 respectively. Thus there are five irreducible representations whose degrees d_i satisfy $\sum_{i=1}^{5} d_i^2 = 60$, which has the unique solution $d_1 = 1$, $d_2 = 3$, $d_3 = 3$, $d_4 = 4$, and $d_5 = 5$. Thus, we have determined the degrees of the irreducible representations.

Denote the irreducible representations by α_1, α_3, α_3', α_4, and α_5 with characters χ_1, χ_3, χ_3', χ_4, and χ_5. We of course have $\alpha_1 = \tau$, the trivial representation, with $\chi_1(g) = 1$ for all $g \in G$.

Consider the representation β on $\mathbf{C}^5$ given by letting A_5 act by permuting the coordinates, i.e.,

$$g(z_1, \ldots, z_5) = (z_{g(1)}, \ldots, z_{g(5)}).$$

This is a permutation representation and the trace of any element is easy to compute. For $g \in A_5$, $\chi_\beta(g) = |\{i : g(i) = i\}|$. Of course $\chi_\beta(1) = 5$, and we observe that $\chi_\beta(c_2) = 1$, $\chi_\beta(c_3) = 2$, and $\chi_\beta(c_4) = \chi_\beta(c_5) = 0$. We compute $\langle \chi_\beta, \chi_\beta \rangle$ (noting that $\chi_\beta^* = \chi_\beta$) and obtain

$$\langle \chi_\beta, \chi_\beta \rangle = \frac{1}{60}(1 \cdot 5^2 + 15 \cdot 1^2 + 20 \cdot 2^2$$
$$+ 12 \cdot 0^2 + 12 \cdot 0^2)$$
$$= 2$$

so β has two irreducible components. We compute

$$\langle \chi_1, \chi_\beta \rangle = \frac{1}{60}(1 \cdot 1 \cdot 5 + 15 \cdot 1 \cdot 1 + 20 \cdot 1 \cdot 2$$
$$+ 12 \cdot 1 \cdot 0 + 12 \cdot 1 \cdot 0)$$
$$= 1$$

so $\tau = \alpha_1$ of degree one is one of them. The complement of τ is then an irreducible representation of degree 4, so is α_4. Furthermore, since $\beta =$

$\tau \oplus \alpha_4$, we have $\chi_\beta = \chi_1 + \chi_4$, i.e., $\chi_4 = \chi_\beta - \chi_1$. We thus compute $\chi_4(1) = 4$, $\chi_4(c_2) = 0$, $\chi_4(c_3) = 1$, and $\chi_4(c_4) = \chi_4(c_5) = -1$.

Next we consider the representation γ on $\mathbf{C}^{10}$ given by the action of A_5 permuting coordinates, where now $\mathbf{C}^{10}$ is coordinatized by the 10 sets $\{i, j\}$ of unordered pairs of distinct elements of $\{1, \ldots, 5\}$. Again,

$$\chi_\gamma(g) = |\{\{i, j\} : g(\{i, j\}) = \{i, j\}\}|.$$

Then $\chi_\gamma(1) = 10$, $\chi_\gamma(c_2) = 2$, $\chi_\gamma(c_3) = 1$, and $\chi_\gamma(c_4) = \chi_\gamma(c_5) = 0$. Again, $\chi_\gamma^* = \chi_\gamma$, and

$$\langle \chi_\gamma, \chi_\gamma \rangle = \frac{1}{60}(1 \cdot 10^2 + 15 \cdot 2^2 + 20 \cdot 1^2$$
$$+ 12 \cdot 0^2 + 12 \cdot 0^2)$$
$$= 3,$$

so γ has three irreducible components. We compute

$$\langle \tau, \chi_\gamma \rangle = \frac{1}{60}(1 \cdot 1 \cdot 10 + 15 \cdot 1 \cdot 2 + 20 \cdot 1 \cdot 1$$
$$+ 12 \cdot 1 \cdot 0 + 12 \cdot 1 \cdot 0)$$
$$= 1,$$

and since we have already computed χ_4, we may compute

$$\langle \chi_4, \chi_\gamma \rangle = \frac{1}{60}(1 \cdot 4 \cdot 10 + 15 \cdot 0 \cdot 2 + 20 \cdot 1 \cdot 1$$
$$+ 12 \cdot (-1) \cdot 0 + 12 \cdot (-1) \cdot 0)$$
$$= 1,$$

so the complement of $\tau \oplus \alpha_4$ in γ is an irreducible representation of degree 5, so it is α_5. Also, $\chi_5 = \chi_\gamma - \chi_4 - \chi_1$, and we compute

$$\chi_5(1) = 5, \qquad \chi_5(c_2) = 1, \qquad \chi_5(c_3) = -1, \qquad \chi_5(c_4) = \chi_5(c_5) = 0.$$

We are left with the task of finding χ_3 and χ_3'. Let us at this point write down what we know of the character table of A_5:

	C_1	C_2	C_3	C_4	C_5
α_1	1	1	1	1	1
α_3	3	x_2	x_3	x_4	x_4
α_3'	3	y_2	y_3	y_4	y_5
α_4	4	0	1	-1	-1
α_5	5	1	-1	0	0

Our task is to determine the unknown entries. Set $z_i = x_i + y_i$ for $2 \le i \le 5$. First we note that if $\mathcal{R}$ is the regular representation

$$\mathcal{R} = \alpha_1 \oplus 3\alpha_3 \oplus 3\alpha_3' \oplus 4\alpha_4 \oplus 5\alpha_5,$$

so

$$\chi_{\mathcal{R}} = \chi_1 + 3\chi_3 + 3\chi_3' + 4\chi_4 + 5\chi_5.$$

Evaluating this on c_2 gives (using Example 4.2 (3))

$$0 = 1 + 3z_2 + 4 \cdot 0 + 5 \cdot 1$$

so $z_2 = -2$, and by evaluation on c_3, c_4, and c_5 we obtain $z_3 = 0$, $z_4 = z_5 = 1$. Now let us use orthogonality.

$$0 = \langle \chi_3, \chi_1 \rangle = 1 \cdot 3 \cdot 1 + 15 \cdot x_2 \cdot 1 + 20 \cdot x_3 \cdot 1$$
$$+ 12 \cdot x_4 \cdot 1 + 12 \cdot x_5 \cdot 1$$
$$0 = \langle \chi_3, \chi_4 \rangle = 1 \cdot 3 \cdot 4 + 15 \cdot x_2 \cdot 0 + 20 \cdot x_3 \cdot 1$$
$$+ 12 \cdot x_4 \cdot (-1) + 12 \cdot x_5 \cdot (-1)$$
$$0 = \langle \chi_3, \chi_5 \rangle = 1 \cdot 3 \cdot 5 + 15 \cdot x_2 \cdot 1 + 20 \cdot x_3 \cdot (-1)$$
$$+ 12 \cdot x_4 \cdot 0 + 12 \cdot x_5 \cdot 0$$

so we obtain a linear system

$$
\begin{array}{rcrcrcl}
15x_2 & + & 20x_3 & + & 12(x_4 + x_5) & = & -3 \\
      &   & 20x_3 & - & 12(x_4 + x_5) & = & -12 \\
15x_2 & - & 20x_3 &   &   & = & -15
\end{array}
$$

with solution $x_2 = -1$, $x_3 = 0$, $x_4 + x_5 = 1$. Then also $y_2 = -1$, $y_3 = 0$, $y_4 + y_5 = 1$, and so, since $z_4 = 1$, $x_4 = x_5 = x$, say, $y_4 = x_5 = y$, say, with $x + y = 1$. Thus, it remains to determine x and y. We observe that A_5 is a group in which each element is conjugate to its inverse, so by Proposition 4.23 all of its characters are real valued, and so

$$\chi_3 = \chi_3^*, \qquad \chi_3' = (\chi_3')^*.$$

We use orthogonality once more to get

$$0 = \langle \chi_3, \chi_3' \rangle = 1 \cdot 3 \cdot 3 + 15 \cdot (-1) \cdot (-1) + 20 \cdot 0 \cdot 0$$
$$+ 12 \cdot x \cdot y + 12 \cdot y \cdot x,$$

so $xy = -1$, which together with $x + y = 1$ gives

$$x = (1 + \sqrt{5})/2 \quad \text{and} \quad y = (1 - \sqrt{5})/2$$

(or vice-versa, but there is no order on χ_3 and χ_3', so we make this choice). Hence, we find the complete "expanded" character table of A_5:

	C_1	C_2	C_3	C_4	C_5
	1	15	20	12	12
α_1	1	1	1	1	1
α_3	3	-1	0	$(1+\sqrt{5})/2$	$(1-\sqrt{5})/2$
α_3'	3	-1	0	$(1-\sqrt{5})/2$	$(1+\sqrt{5})/2$
α_4	4	0	1	-1	-1
α_5	5	1	-1	0	0

(We call this the "expanded" character table of A_5 because we have included on the second line, as is often but not always done, the number of elements in each conjugacy class. Note that there is **no** canonical order for either conjugacy classes or representations, so different character tables for the same group may "look" different.)

(4.30) Example. *Let us show how to determine all the irreducible complex characters of the symmetric group S_5 of order 120.*

First we see from Corollary 1.5.10 that S_5 has seven conjugacy classes, with representatives 1, (12), (123), (12)(34), (1234), (123)(45), and (12345), so we expect seven irreducible representations. One of them is τ, of course, and another, also of degree 1, is ε given by $\varepsilon(g) = \mathrm{sgn}(g) = \pm 1 \in \mathrm{Aut}(\mathbf{C})$, the sign of the permutation g.

Observe that the representations β and γ of A_5 constructed in Example 4.29 are actually restrictions of representations of S_5, so the representations α_4 and α_5 of A_5 constructed there are restrictions of representations $\widetilde{\alpha}_4$ and $\widetilde{\alpha}_5$ of S_5. Since α_4 and α_5 are irreducible, so are $\widetilde{\alpha}_4$ and $\widetilde{\alpha}_5$. Furthermore, since $\widetilde{\alpha}_4$ and $\widetilde{\alpha}_5$ are irreducible, so are $\varepsilon \otimes \widetilde{\alpha}_4$ and $\varepsilon \otimes \widetilde{\alpha}_5$ (by Exercise 13), and we may compute that the characters of $\widetilde{\alpha}_4$ and $\varepsilon \otimes \widetilde{\alpha}_4$ (respectively, $\widetilde{\alpha}_5$ and $\varepsilon \otimes \widetilde{\alpha}_5$) are unequal, so these representations are distinct.

Hence, we have found six irreducible representations, of degrees 1, 1, 4, 4, 5, 5, so we expect one more of degree d, with

$$1^2 + 1^2 + 4^2 + 4^2 + 5^2 + 5^2 + d^2 = 120$$

so $d = 6$. If we call this $\widetilde{\alpha}_6$, then we have (using $\chi(\sigma)$ for χ_σ, for convenience),

$$\chi(\mathcal{R}) = \chi(\tau) + \chi(\varepsilon) + 4\chi(\widetilde{\alpha}_4) + 4\chi(\varepsilon \otimes \widetilde{\alpha}_4)$$
$$+ 5\chi(\widetilde{\alpha}_5) + 5\chi(\varepsilon \otimes \widetilde{\alpha}_5) + 6\chi(\widetilde{\alpha}_6)$$

enabling us to determine $\chi(\widetilde{\alpha}_6)$. We leave the details for the reader.

(4.31) Example. It is easy to check from Example 3.9 that A_4 has the following expanded character table (where we have listed the conjugacy classes in the same order as there):

	C_1	C_2	C_3	C_4
	1	3	4	4
$\tau = \pi^*(\theta_0)$	1	1	1	1
$\pi^*(\theta_1)$	1	1	ζ	ζ^2
$\pi^*(\theta_2)$	1	1	ζ^2	ζ
α	3	-1	0	0

with $\zeta = \exp(2\pi i/3)$. We wish to find the decomposition of the tensor product of any two irreducible representations. The only nontrivial case is $\alpha \otimes \alpha$. Recall that $\chi_{\alpha \otimes \alpha} = (\chi_\alpha)(\chi_\alpha)$, and thus has values 9, 1, 0, 0 on the four conjugacy classes. We compute

$$\langle \alpha \otimes \alpha, \alpha \rangle = \frac{1}{12}(1 \cdot 3 \cdot 9 + 3 \cdot (-1) \cdot 1) = 2,$$

and for $i = 0, 1,$ or 2

$$\langle \alpha \otimes \alpha, \pi^*(\theta_i) \rangle = \frac{1}{12}(1 \cdot 1 \cdot 9 + 3 \cdot 1 \cdot 1) = 1,$$

so

$$\alpha \otimes \alpha = 2\alpha \oplus \pi^*(\theta_0) \oplus \pi^*(\theta_1) \oplus \pi^*(\theta_2).$$

(4.32) Remark. It is perhaps natural to conjecture that two groups with identical character tables must be isomorphic. This is, in fact, false! The groups D_8 and Q_8 (the dihedral and quaternion groups of order 8) are two distinct groups with the same character table. We leave the verification to the reader.

(4.33) Remark. It is clearly a necessary condition for a representation to be defined over a field $\mathbf{F}$ (cf. Example 1.4 (12)) that its character takes its values in $\mathbf{F}$. However, this condition is not sufficient. For example, if σ is the representation ρ of Q_8 defined in Example 3.8, then χ_σ is real valued but σ cannot be defined over $\mathbf{R}$. (It is known in this situation that some finite multiple of the given representation can be defined over $\mathbf{F}$. For example, here 2σ can be defined over $\mathbf{R}$.) Again we leave the verification to the reader.

8.5 Induced Representations

In this section we develop an important and powerful method of constructing representations, the technique of induction. First, however, we consider restriction.

(5.1) Definition. *Let G be a group, H a subgroup of G, and V an $\mathbf{F}(G)$-module. The* **restriction of V to H**, $\mathrm{Res}_H^G(V)$, *is V regarded as an $\mathbf{F}(H)$-module. (In other words, we have the same underlying vector space, but we restrict our attention to the action of the subring $\mathbf{F}(H)$ of $\mathbf{F}(G)$.)*

If G is understood, we will often write $\mathrm{Res}_H(V)$; if both G and H are understood, we will often write $\mathrm{Res}(V)$. As well, we may often write $V_H = \mathrm{Res}_H^G(V_G)$, using subscripts to denote the group that is operating. Finally, we may sometimes simply write V for $\mathrm{Res}_H^G(V)$ when it is clear from the context what is meant.

Let V_G be a representation of G and $H \subseteq G$ a subgroup. Clearly, if V_H is irreducible, then so is V_G, but the converse need not hold. For example, if $H = \langle 1 \rangle$, $\mathrm{Res}_H^G(V) = \deg(V)\tau$. As a nontrivial example of restriction, let $G = D_{2m}$ and $H = \mathbf{Z}_m$. Then, in the notation of Examples 1.4 (6) and (7),

$$\mathrm{Res}(\phi_i) = \theta_i \oplus \theta_{m-i}.$$

(5.2) Example. Let H be a subgroup of G. Then

$$\mathrm{Res}_H^G(\mathbf{F}(G)) = [G : H]\mathbf{F}(H),$$

since $\mathbf{F}(G)$ is a free left $\mathbf{F}(H)$-module of rank $[G : H]$ (with a basis given by a set of right coset representatives).

Now we come to induction.

(5.3) Definition. *Let G be a group, H a subgroup of G, and W an $\mathbf{F}(H)$-module. Then the* **induction of W to G**, *denoted* $\mathrm{Ind}_H^G(W)$, *is given by*

$$\mathrm{Ind}_H^G(W) = \mathbf{F}(G) \otimes_{\mathbf{F}(H)} W.$$

Note that this makes sense as we may regard $\mathbf{F}(G)$ as an $(\mathbf{F}(G), \mathbf{F}(H))$-bimodule, and then the result of induction is an $\mathbf{F}(G)$-module. If $V = \mathrm{Ind}_H^G(W)$, we say that V is induced from W and call V an induced representation.

(5.4) Lemma. $\deg(\mathrm{Ind}_H^G(W)) = [G : H]\deg(W)$.

Proof. As a right $\mathbf{F}(H)$-module, $\mathbf{F}(G)$ is free of rank $[G : H]$ (with basis given by a set of left coset representatives). $\qquad\square$

(5.5) Example. Let G be an arbitrary group and H a subgroup of G. Then for the regular representation $\mathbf{F}(H)$ of H, we have

$$\mathrm{Ind}_H^G(\mathbf{F}(H)) = \mathbf{F}(G),$$

the regular representation of G. This is immediate, for it is just the equality

$$\mathbf{F}(G) \otimes_{\mathbf{F}(H)} \mathbf{F}(H) = \mathbf{F}(G).$$

In particular, setting $H = \langle 1 \rangle$, we have $\mathrm{Ind}_H^G(\tau) = \mathbf{F}(G)$.

Let us now give a criterion that will enable us to recognize induced representations.

(5.6) Theorem. *Let* $V = \mathrm{Ind}_H^G(W)$ *and identify* W *with*

$$W_1 = \mathbf{F}(H) \otimes_{\mathbf{F}(H)} W \subseteq \mathbf{F}(G) \otimes_{\mathbf{F}(H)} W = V.$$

Let $\{g_i\}_{i \in I}$ *be a complete set of left coset representatives of* H *in* G, *with* $g_1 = 1$, *and let* $W_i = g_i(W_1)$. *Then*

(1) $V = \oplus_{i \in I} W_i$;
(2) *the action of* G *on* V *permutes the* W_i, *i.e., for every* $g \in G$ *and for every* $i \in I$, $g(W_i) = W_j$ *for some* $j \in I$;
(3) *this permutation is transitive, i.e., for every* i, j, *there is a* $g \in G$ *with* $g(W_i) = W_j$; *and*
(4) $H = \{g \in G : g(W_1) = W_1\}$.

Conversely, let V *be an* $\mathbf{F}(G)$-*module and suppose there are subspaces* $\{W_i\}_{i \in I}$ *of* V *such that* (1), (2), *and* (3) *hold. Define* H *by* (4), *and set* $W = W_1$. *Then*

$$V = \mathrm{Ind}_H^G(W).$$

Proof. The first statement of the proposition is clear from the isomorphism

$$\mathbf{F}(G) = \bigoplus_{i \in I} g_i \mathbf{F}(H)$$

of right $\mathbf{F}(H)$-modules. As for the converse, note that there is a one-to-one correspondence

$$\{W_i\} \longleftrightarrow \{\text{ left cosets of } H \}$$

given by

$$W_i \longleftrightarrow \{g \in G : g(W_1) = W_i\}.$$

Pick coset representatives $\{g_i\}_{i \in I}$ with $g_1 = 1$. Define a function $\alpha : G \times W \to V$ by $\alpha(g, w) = g(w)$. This clearly extends to an $\mathbf{F}$-linear transformation $\alpha : \mathbf{F}(G) \times W \to V$, and since

$$\alpha(gh, w) = gh(w) = g(hw) = \alpha(g, hw)$$

it readily follows that α is $\mathbf{F}(H)$-middle linear and so defines

$$\tilde{\alpha} : \mathbf{F}(G) \otimes_{\mathbf{F}(H)} W \to V.$$

Now we define $\beta : V \to \mathbf{F}(G) \otimes_{\mathbf{F}(H)} W$. Since $V = \oplus W_i$, it suffices to define $\beta : W_i \to \mathbf{F}(G) \otimes_{\mathbf{F}(H)} W$ for each $i \in I$. We let $\beta(w_i) = g_i \otimes g_i^{-1}(w_i)$, for $w_i \in W_i$. Let us check that $\tilde{\alpha}$ and β are inverses of each other, establishing the claimed isomorphism.

First,

$$\tilde{\alpha}\beta(w_i) = \tilde{\alpha}(g_i \otimes g_i^{-1}(w_i)) = g_i(g_i^{-1}(w_i)) = w_i.$$

Now each $g \in G$ can be written as $g = g_i h$ for a unique $i \in I$ and $h \in H$. Thus, for $w \in W$,

$$\begin{aligned}
\beta\tilde{\alpha}(g \otimes w) &= \beta(g(w)) \\
&= \beta(g_i h(w)) \\
&= \beta(g_i(hw)) \\
&= g_i \otimes hw \\
&= g_i h \otimes w \\
&= g \otimes w,
\end{aligned}$$

as required.   □

We make two observations. First, if V itself is irreducible, the condition of transitivity is automatic (for $\sum_{g \in G} g(W_1)$ is a subrepresentation of V, which must then be V itself). Second, our choice of W_1 was arbitrary; we could equally well have chosen some other W_i.

This second observation yields the following useful result:

(5.7) Corollary. *Let H be a subgroup of G. Let $g_0 \in G$ be fixed and set $K = g_0^{-1} H g_0$. Let $\sigma_H : H \to \mathrm{Aut}(W)$ be a representation of H, and let $\sigma_K : K \to \mathrm{Aut}(W)$ be the representation of K defined by $\sigma_K(g) = \sigma_H(g_0 g g_0^{-1})$ for $g \in K$. Then*

$$\mathrm{Ind}_H^G(\sigma_H) = \mathrm{Ind}_K^G(\sigma_K).$$

Proof. Let $V = \mathrm{Ind}_H^G(W)$. Then we may identify W with W_1 in the statement of Theorem 5.6. Then $K = \{g \in G : g(W_i) = W_i\}$ acting on W by the above formula, so $V = \mathrm{Ind}_K^G(W)$ as well.   □

As a corollary of Theorem 5.6, we may identify two types of representations as induced representations.

(5.8) Corollary.

(1) *Let V be a transitive permutation representation on the set $P = \{p_i\}_{i \in I}$ and let $H = \{g \in G : g(p_1) = p_1\}$. Then $V = \mathrm{Ind}_H^G(\tau)$.*
(2) *Let V be a transitive monomial representation with respect to the basis $\mathcal{B} = \{b_i\}$, and let $H = \{g \in G : g(\mathbf{F}b_1) = \mathbf{F}b_1\}$. Let $\sigma = \mathbf{F}b_1$ as a representation of H. Then $V = \mathrm{Ind}_H^G(\sigma)$.*

Proof.   □

Observe from Theorem 5.6 that W is certainly a subrepresentation of $\operatorname{Res}_H^G \operatorname{Ind}_H^G(W)$. We will consider this point in more detail later. However, this observation is enough to enable us to identify some induced representations already.

For example, if $G = D_{2m}$ and ϕ_i is one of the 2-dimensional representations of G defined in Example 1.4 (7), then each ϕ_i, which is a monomial representation, is induced from $H = \mathbf{Z}_m$, a subgroup of G of index two. If θ_i is the representation of that name of H in Example 1.4 (6), $i \leq m/2$, then $\operatorname{Ind}_H^G(\theta_i) = \phi_i$, and if $i > m/2$, $\operatorname{Ind}_H^G(\theta_i) = \phi_{m-i}$. This also illustrates Corollary 5.7, for if $g = y$ (in the notation of Example 1.4 (7)) and $\sigma_H = \theta_i$, then $K = H$ and $\sigma_K = \theta_{m-i}$ and $\operatorname{Ind}_H^G(\theta_i) = \operatorname{Ind}_H^G(\theta_{m-i})$. (Of course, $\operatorname{Res}_H^G(\phi_i) = \theta_i \oplus \theta_{m-i}$.)

Let us now concentrate on the case of a normal subgroup H of G and a representation $\sigma : H \to \operatorname{Aut}(W)$. If σ' is defined by $\sigma'(h) = \sigma(ghg^{-1})$ for some fixed $g \in G$, we call σ' a conjugate of σ, or more precisely, the conjugate of σ by g.

Let $\{\sigma^j\}$ be a complete set of conjugates of $\sigma = \sigma^1$. Note that if we let

$$N(\sigma) = \{g \in G : \quad \sigma' : H \to \operatorname{Aut}(W) \quad \text{defined by } \sigma'(h) = \sigma(ghg^{-1})$$

$$\text{is a representation of } H \text{ isomorphic to } \sigma\}$$

then $N(\sigma)$ is a subgroup of G containing H and $[G : N(\sigma)]$ is the number of conjugates of σ. (Note also that all the subgroups $N(\sigma^j)$ are mutually conjugate.) The subgroup $N(\sigma)$ is known as the **inertia group of** σ.

(5.9) Corollary. Let H be a normal subgroup of G and $\sigma : H \to \operatorname{Aut}(W)$ a representation of H. Then if σ' is any conjugate of σ,

$$\operatorname{Ind}_H^G(\sigma') = \operatorname{Ind}_H^G(\sigma).$$

Proof. This is a special case of Corollary 5.7.   □

(5.10) Corollary. Let H be a normal subgroup of G and $\sigma : H \to \operatorname{Aut}(W)$ a representation of H. Let $\{\sigma^j\}$ be a complete set of conjugates of $\sigma = \sigma^1$. Then

$$\operatorname{Res}_H^G \operatorname{Ind}_H^G(\sigma) = [N(\sigma) : H] \oplus_j \sigma^j.$$

Proof. Let $\{g_i\}$ be a set of right coset representatives of H, with $g_1 = 1$, and let $\sigma_i : H \to \operatorname{Aut}(W)$ be defined by $\sigma_i(h) = \sigma(g_i^{-1} h g_i)$. Then

$$\operatorname{Res}_H^G \operatorname{Ind}_H^G(\sigma) = \bigoplus_i \sigma_i$$

by Theorem 5.6. But in the statement of the corollary we have just grouped the σ_i into isomorphism classes, there being $[N(\sigma) : H]$ of these in each class.   □

(5.11) Theorem. (Clifford) *Let H be a normal subgroup of G. Let $\rho :$ $G \to \mathrm{Aut}(V)$ be an irreducible representation of G, and let $\sigma : H \to$ $\mathrm{Aut}(W)$ be any irreducible component of $\mathrm{Res}_H^G(\rho)$. Let $\{\sigma^j\}$ be a complete set of conjugates of $\sigma = \sigma^1$, and set $V_j = \sum_{g \in G} g(W_1)$, where the sum is taken over the left coset representatives of H such that the conjugate of σ by g is isomorphic to σ^j. Set $K = \{g \in G : g(V_1) = V_1\}$. Then V_1 is an irreducible representation of K and*

$$V = \mathrm{Ind}_K^G(V_1).$$

Furthermore, $K = N(\sigma)$.

Proof. First consider $V' = \sum_{g \in G} g(W_1)$. This is an $\mathbf{F}(G)$-submodule of V, but V was assumed irreducible, so $V' = V$. Next observe that instead of summing over $g \in G$, we may instead sum over left coset representatives of H. Further note that the representation of H on $g(W_1)$ is the conjugate of σ by g, so we may certainly group the terms together to get $V = \sum_j V_j$.

Now each V_j is a sum of subspaces W_j, which are isomorphic to σ^j as an $\mathbf{F}(H)$-module, so by Lemma 7.1.20, each V_j is in fact isomorphic to a direct sum of these.

We claim that $V = \oplus_j V_j$. Consider $U = V_1 \cap \sum_{j>1} V_j$. Then U is an $\mathbf{F}(H)$-submodule of V_1, and so is isomorphic to $m_1 M_1$ for some m_1. Also, U is an $\mathbf{F}(H)$-submodule of $\sum_{j>1} V_j$, and so is isomorphic to $\oplus_{j>1} m_j W_j$ for some $\{m_j\}$. By Corollary 7.1.19, $m_1 = m_2 = \cdots = 0$, so $U = \langle 0 \rangle$, as required.

If σ^j is the conjugate of σ by $g_j \in G$, then $g_j(V_1) = V_j$, so G permutes the subspaces V_j transitively. Then by Theorem 5.6, we obtain

$$V = \mathrm{Ind}_K^G(V_1).$$

Also, V_1 is irreducible, for if it contained a nonzero proper subrepresentation V_1', then $\mathrm{Ind}_K^G(V_1')$ would be a nonzero proper subrepresentation of V, contradicting the irreducibility of V.

Finally, $g \in K$ if and only if the conjugate by g of W_1 is isomorphic to W_1. But this is exactly the condition for $g \in N(\sigma)$. $\square$

(5.12) Corollary. *Let H be a normal subgroup of G. Let $\rho : G \to \mathrm{Aut}(V)$ be an irreducible representation of G, and let $\sigma = \sigma^1 : H \to \mathrm{Aut}(W_1)$ be any irreducible component of $\mathrm{Res}_H^G(\rho)$. Then*

(1) *$\mathrm{Res}_H^G(\rho)$ is a semisimple representation of H.*

(2) *In the decomposition of $\mathrm{Res}_H^G(\rho)$ into a direct sum of irreducible representations $\oplus m_j U_j$, all of the U_j are conjugate to σ^1 (and so are mutually conjugate), all conjugates of σ^1 appear among the U_j, and all the multiplicities m_j are equal.*

(3) *$\mathrm{Res}_H^G(\rho) = m \oplus_j \sigma^j$ for some $m \in \{1, 2, 3, \ldots\} \cup \{\infty\}$, where $\{\sigma^j\}$ is a complete set of conjugates of σ^1.*

(4) Let $H^1 = \{g \in G : g(W_1) = W_1\}$. Then $H \subseteq H^1 \subseteq N(\sigma)$, and $m \le [N(\sigma) : H^1]$.

Proof. (1) In the notation of the proof of Theorem 5.11, we have that, as an $\mathbf{F}(H)$-module, $V = \sum g(W_1)$, so V is a sum of simple $\mathbf{F}(H)$-modules and hence is semisimple by Lemma 4.3.20.

(2) Clearly, all the U_j are conjugate to σ^1 and all conjugates appear. Let V_1 be as in Theorem 5.11, with $V_1 = \oplus_{i=1}^{m_1} k_i(W_1)$ for some group elements $k_i \in G$. Then if g_j is as in the proof of Theorem 5.11,

$$\begin{aligned}
m_j W_j &\cong V_j \\
&= g_j(V_1) \\
&= \bigoplus_{i=1}^{m_1} g_j(k_i(W_1)) \\
&\cong m_1 W_j
\end{aligned}$$

so $m_j = m_1$ by Corollary 7.1.18.

(3) This is merely a restatement of (2).

(4) Since $g(W_1) = W_1$ for $g \in H^1$, $V_1 = \sum_j g(W_1)$ where the summation is over left coset representatives of H^1 in $N(\sigma)$.   □

Remark. It is not true that $m = [N(\sigma) : H^1]$, or even that m divides $[N(\sigma) : H^1]$, in general. Let $G = \mathbf{Z}_3$ have the representation $\rho = \tau \oplus \theta_1$ on $V = \mathbf{C}^2$ and let $H = \langle 1 \rangle$. Let $W = \{(z, z) \in \mathbf{C}^2\}$. Then $N(\sigma) = G$, $H^1 = H$, so $[N(\sigma) : H^1] = 3$, but $m = 2$.

As a consequence of Clifford's theorem we have the following result:

(5.13) Corollary. *Let H be a normal subgroup of G, and let $\rho : G \to \mathrm{Aut}(V)$ be an irreducible representation of G. Then if $\mathrm{Res}_H^G(V)$ is not isotypic, there is a proper subgroup K of G containing H and an irreducible representation V_1 of K such that $V = \mathrm{Ind}_K^G(V_1)$.*

Proof. Suppose $\mathrm{Res}_H^G(V)$ is not isotypic. Let W_1 be an irreducible $\mathbf{F}(H)$-submodule of $\mathrm{Res}_H^G(V)$, and set $V_1 = \sum_g g(W_1)$ where the sum is over all $g \in G$ with $g(W_1)$ isomorphic to W_1 as an $\mathbf{F}(H)$-module, and let $K = \{g \in G : g(V_1) = V_1\}$. Then, as above, V_1 is an irreducible $\mathbf{F}(K)$-module. Note $K \subseteq H$ and $K \ne G$ as $\mathrm{Res}_H^G(V)$ is not isotypic. Then $\mathrm{Ind}_K^G(V_1)$ is an $\mathbf{F}(G)$-submodule of V, but as V is irreducible, it is equal to V.   □

We may use Corollary 5.13 to sharpen Theorem 4.28.

(5.14) Theorem. *Let G be a finite group and A an abelian normal subgroup of G. If V is an irreducible complex representation of G, then $d = \deg(V)$ divides $[G : A]$.*

Proof. We prove this by induction on $n = |G|$. Let V be defined by $\sigma : G \to \operatorname{Aut}(V)$. Let $W = \operatorname{Res}_A^G(V)$. By Corollary 5.13, if W is not isotypic, $V = \operatorname{Ind}_K^G(V_1)$ for some proper subgroup K of G containing H and some irreducible representation V_1 of K. Then $\deg(V_1)$ divides $[K : H]$, so $\deg(V) = [G : K] \deg(V_1)$ divides $[G : K][K : H] = [G : H]$.

If W is isotypic, then $W = dW_1$ for some one-dimensional representation W_1 of A, or, in other words, $\sigma : A \to \operatorname{Aut}(W) = \operatorname{Aut}(V)$ is given by $\sigma(g) = $ multiplication by some complex number $\lambda(g)$ for each $g \in A$. (Also, for any $h \in G$, $\lambda(hgh^{-1}) = \lambda(g)$, $g \in A$.)

Now for each m consider $\sigma^{\otimes m}$ on $V \otimes \cdots \otimes V$. By Exercise 20, this is an irreducible representation of $G^m = G \times \cdots \times G$. Let

$$H = \{(g_1, \ldots, g_m) \in A^m : \sigma^{\otimes m}(g_1, \ldots, g_m) = \lambda(g_1) \cdots \lambda(g_m) = 1\}.$$

H is a subgroup of A^m and hence a normal subgroup of G^m, and it acts trivially on $V^{\otimes m}$, so we obtain an irreducible representation of G^m/H, which is irreducible as $\sigma^{\otimes m}$ is. Hence, by Theorem 4.28, the degree d^m of this representation divides the order of G^m/H.

Note, however, that H has a subgroup

$$\{(g_1, \ldots, g_{m-1}, g_1^{-1} \cdots g_{m-1}^{-1}) : g_i \in A\}$$

isomorphic to A^{m-1}, so if $a = |A|$, then d^m divides $n^m/a^{m-1} = a(n/a)^m$ for every m, which implies that d divides $n/a = [G : A]$, as claimed. $\square$

We now return to the general study of induction.

(5.15) Lemma. *Let $K \subseteq H \subseteq G$ be subgroups.*

(1) (Transitivity of restriction) *For any representation V of G,*

$$\operatorname{Res}_K^H \operatorname{Res}_H^G(V) = \operatorname{Res}_K^G(V).$$

(2) (Transitivity of induction) *For any representation W of K,*

$$\operatorname{Ind}_H^G \operatorname{Ind}_K^H(W) = \operatorname{Ind}_K^G(W).$$

Proof. (1) is trivial. As for (2),

$$\begin{aligned}
\operatorname{Ind}_H^G \operatorname{Ind}_K^H(W) &= \mathbf{F}(G) \otimes_{\mathbf{F}(H)} (\mathbf{F}(H) \otimes_{\mathbf{F}(K)} W) \\
&= (\mathbf{F}(G) \otimes_{\mathbf{F}(H)} \mathbf{F}(H)) \otimes_{\mathbf{F}(K)} W \\
&= \mathbf{F}(G) \otimes_{\mathbf{F}(K)} W \\
&= \operatorname{Ind}_K^G(W)
\end{aligned}$$

where the second equality is just the associativity of the tensor product (Theorem 7.2.17). $\square$

The next formula turns out to be tremendously useful, and we will see many examples of its use. Recall that we defined the intertwining number

$$i(V, W) = \langle V, W \rangle$$

of two representations in Definition 4.17.

(5.16) Theorem. (Frobenius reciprocity) *Let* **F** *be an arbitrary field,* G *an arbitrary group, and* H *a subgroup of* G. *Let* W *be an* **F**-*representation of* H *and* V *an* **F**-*representation of* G. *Then*

$$\langle \text{Ind}_H^G(W), V \rangle = \langle W, \text{Res}_H^G(V) \rangle.$$

Proof. By definition,

$$\langle \text{Ind}_H^G(W), V \rangle = \dim_{\mathbf{F}} \text{Hom}_{\mathbf{F}(G)}(\text{Ind}_H^G(W), V).$$

But

$$\begin{aligned}
\text{Hom}_{\mathbf{F}(G)}(\text{Ind}_H^G(W), V) &= \text{Hom}_{\mathbf{F}(G)}(\mathbf{F}(G) \otimes_{\mathbf{F}(H)} W, V) \\
&= \text{Hom}_{\mathbf{F}(H)}(W, \text{Hom}_{\mathbf{F}(G)}(\mathbf{F}(G), V)) \\
&= \text{Hom}_{\mathbf{F}(H)}(W, V)
\end{aligned}$$

where the second equality is the adjoint associativity of Hom and tensor product (Theorem 7.2.20). Again, by definition,

$$\dim_{\mathbf{F}} \text{Hom}_{\mathbf{F}(H)}(W, V) = \langle W, \text{Res}_H^G(V) \rangle.$$

$\square$

One direct consequence of Frobenius reciprocity occurs so often that it is worth stating explicitly.

(5.17) Corollary. *Let* **F** *be an excellent field for* G. *Let* W *be an irreducible* **F**-*representation of* H *and* V *an irreducible* **F**-*representation of* G. *Then the multiplicity of* V *in* $\text{Ind}_H^G(W)$ *is equal to the multiplicity of* W *in* $\text{Res}_H^G(V)$.

Proof. Immediate from Theorem 5.16 and Lemma 4.18 (3). $\square$

As an example of the use of Frobenius reciprocity let us use it to provide an alternate proof of part (2) of the fundamental Theorem 3.4. Let **F** be an excellent field for G and M an irreducible **F**-representation of G of degree d. Let m be the multiplicity of M in $\mathbf{F}(G)$. We wish to show $m = d$.

Let $H = \langle 1 \rangle$ and recall from Example 5.5 that $\mathbf{F}(G) = \text{Ind}_H^G(\tau)$. Then

$$m = \langle M, \mathbf{F}(G) \rangle$$
$$= \langle \mathbf{F}(G), M \rangle$$
$$= \langle \mathrm{Ind}_H^G(\tau), M \rangle$$
$$= \langle \tau, \mathrm{Res}_H^G(M) \rangle$$
$$= \langle \tau, d\tau \rangle$$
$$= d$$

as claimed.

(5.18) Example. Consider A_4 from Example 3.9, and let $\mathbf{F} = \mathbf{C}$. We have a split extension

$$1 \longrightarrow V \longrightarrow A_4 \stackrel{\pi}{\longrightarrow} S \longrightarrow 1$$

with $V \cong \mathbf{Z}_2 \oplus \mathbf{Z}_2$. V has 4 irreducible representations, all of degree 1: τ, and three others, which we shall simply denote by λ_1, λ_2, and λ_3. A_4 has 4 irreducible representations: $\pi^*(\theta_i)$ for $i = 0, 1, 2$ of degree 1 and α of degree 3. Then

$$\langle \mathrm{Ind}_V^{A_4}(\tau), \pi^*(\theta_i) \rangle = \langle \tau, \mathrm{Res}_V^{A_4}(\pi^*(\theta_i)) \rangle$$
$$= \langle \tau, \tau \rangle$$
$$= 1,$$

so

$$\mathrm{Ind}_V^{A_4}(\tau) = \bigoplus_{i=0}^{2} \pi^*(\theta_i)$$

(since

$$\deg(\mathrm{Ind}_V^{A_4}(\tau)) = [A_4 : V]\deg(\tau) = 3 \cdot 1 = 3$$

and the right-hand side is a representation of degree 3). Also, for $i = 0, 1, 2$ and $j = 1, 2, 3$

$$0 = \langle \lambda_j, \tau \rangle = \langle \lambda_j, \mathrm{Res}_H^G(\pi^*(\theta_i)) \rangle = \langle \mathrm{Ind}_H^G(\lambda_j), \pi^*(\theta_i) \rangle,$$

so we must have $\mathrm{Ind}_H^G(\lambda_j) = \alpha$ (as both are of degree 3).

Continuing with this example, since we have a split extension, we have a subgroup S of A_4 isomorphic to $\mathbf{Z}_3$, and we identify S with $\mathbf{Z}_3$ via this isomorphism. (We have given S and this isomorphism explicitly in Example 3.9). Now S has three irreducible representations $\theta_0 = \tau$, θ_1, and θ_2. Because we have a splitting, $\mathrm{Res}_S^{A_4}(\pi^*(\theta_i)) = \theta_i$, or, more generally,

$$\delta_{ij} = \langle \mathrm{Res}_S^{A_4}(\pi^*(\theta_j)), \theta_i \rangle$$
$$= \langle \pi^*(\theta_j), \mathrm{Ind}_S^{A_4}(\theta_i) \rangle;$$

since $\mathrm{Ind}_S^{A_4}(\theta_i)$ has degree 4, this gives

$$\mathrm{Ind}_S^{A_4}(\theta_i) = \pi^*(\theta_i) \oplus \alpha \qquad \text{for} \quad i = 0, 1, 2.$$

Here is another useful consequence of Frobenius reciprocity:

(5.19) Proposition. *Let G be a group and H an abelian subgroup of G. Let $\mathbf{F}$ be an excellent field for G. Then every irreducible $\mathbf{F}$-representation of G has degree at most $[G : H]$.*

Proof. Let V be an irreducible representation of G. Then $\operatorname{Res}_H^G(V)$ is a representation of H and so contains an irreducible representation W, which is one-dimensional, since H is abelian. Then

$$1 \le \langle \operatorname{Res}_H^G(V), W \rangle = \langle V, \operatorname{Ind}_H^G(W) \rangle$$

the equality being Frobenius reciprocity, so V is a subrepresentation of $\operatorname{Ind}_H^G(W)$. But $\deg\left(\operatorname{Ind}_H^G(W)\right) = [G : H]$, so $\deg(V) \le [G : H]$. $\square$

Example. Note that D_{2m} has an abelian subgroup of index 2 and its irreducible complex representations all have dimension at most 2. The same is true for Q_8. Also, A_4 has an abelian subgroup of index 3 and its irreducible complex representations all have dimension at most 3.

Now we determine the character of an induced representation. In order to state this most simply, we adopt in this theorem the following nonstandard notation: If H is a subgroup of G and W an $\mathbf{F}$-representation of H with character χ_W, we let $\widetilde{\chi}_W$ be the function on G defined by

$$\widetilde{\chi}_W(g) = \begin{cases} \chi_W(g) & \text{if } g \in H, \\ 0 & \text{if } g \notin H. \end{cases}$$

(5.20) Theorem. *Let H be a subgroup of G of finite index k, and let W be an $\mathbf{F}$-representation of H of finite degree. Set $V = \operatorname{Ind}_H^G(W)$. If $\{g_i\}_{i=1}^k$ is a complete set of left coset representatives of H, then for any $g \in G$,*

$$\chi_V(g) = \sum_{i=1}^k \widetilde{\chi}_W(g_i^{-1} g g_i).$$

Proof. We know that we may write

$$V = \bigoplus_{i=1}^k g_i(W) = \bigoplus_{i=1}^k W_i$$

and that every element of G acts by permuting $\{W_i\}$. To be precise, if we let $H_i = g_i H g_i^{-1}$, then

$$H_i = \{g \in G : g(W_i) = W_i\},$$

so if $g \notin H_i$, then $g(W_i) = W_j$ for some $j \ne i$. On the other hand, if the representation of H_i on W_i is given by $\sigma_i : H_i \to \operatorname{Aut}(W_i)$, with $\sigma_1 = \sigma$, then for $g \in H_i$, $\sigma_i(g) = \sigma(g_i^{-1}gg_i) = \sigma(g')$. Now $f_i : W_1 \to W_i$ by $f_i(w) = g_i(w)$ is an isomorphism with

$$(f_i\sigma(g')f_i^{-1})(w_i) = \sigma_i(g)(w_i)$$

for all $g \in H_i$ and $w_i \in W_i$, so

$$\mathrm{Tr}(\sigma_i(g)) = \mathrm{Tr}(\sigma(g')).$$

Now $\chi_V(g)$ is the trace of a matrix representing the operation of g. Choose a basis for V that is a union of bases for the W_i. Then if $g \in G$ with $g \notin H_i$, the action of g on W_i contributes nothing to $\chi_V(g)$; while if $g \in H_i$, it contributes $\mathrm{Tr}(\sigma(g')) = \chi_W(g')$ to $\chi_V(g)$, yielding the theorem.   □

(5.21) Example. In Example 4.29, we found the character table of A_5. Let us here adopt an alternate approach, finding the irreducible characters via induced representations. We still know, of course, that they must have degrees 1, 3, 3, 4, 5 and we still denote them as in Example 4.29.

Of course, $\alpha_1 = \tau$. The construction of α_4 was so straightforward that an alternative is hardly necessary, but we shall give one anyway (as we shall need most of the work in any case). Let $G = A_5$ and let $H = A_4$ included in the obvious way (as permutations of $\{1, 2, 3, 4\} \subseteq \{1, 2, 3, 4, 5\}$). Then a system of left coset representatives for H is

$$\{1, (1\,2)(4\,5), (1\,2)(3\,5), (1\,3)(2\,5), (2\,3)(1\,5)\} = \{g_1, \dots, g_5\}.$$

We choose as representatives for the conjugacy classes of G

$$\{1, (1\,4)(2\,3), (1\,2\,3), (1\,2\,3\,4\,5), (1\,3\,5\,2\,4)\} = \{c_1, \dots, c_5\}.$$

Of course, $g_i^{-1}(c_1)g_i = g_i$ for every i. Otherwise, one can check that $g_i^{-1}(c_j)g_i \notin H$ except in the following cases:

$$g_1^{-1}(c_2)g_1 = c_2, \qquad g_1^{-1}(c_3)g_1 = c_3, \qquad \text{and} . \qquad g_1^{-1}((1\,2\,3))g_1 = (1\,3\,2).$$

(Note that $(1\,2\,3) = c_3$ and that while $(1\,3\,2)$ is conjugate to it in A_5, it is **not** conjugate to it in A_4, so we have written the permutation explicitly.)

Let $W = \tau$ and consider $\mathrm{Ind}_H^G(W) = V$. Then by Theorem 5.20, it is easy to compute χ_V:

$$\chi_V(c_1) = 5, \qquad \chi_V(c_2) = 1, \qquad \chi_V(c_3) = 2, \qquad \chi_V(c_4) = \chi_V(c_5) = 0.$$

Then

$$\langle \chi_V, \chi_V \rangle = \frac{1}{60}(5^2 + 15 \cdot 1 + 20 \cdot 2^2) = 2$$

and τ appears in V with multiplicity 1 by Frobenius reciprocity (or by calculating $\langle \chi_V, \chi_1 \rangle = 1$), so V contains one other irreducible representation, $V = \tau \oplus \alpha_4$, and $\chi_4 = \chi_V - \chi_1$, giving the character of α_4.

Now, following Example 3.9, let $W = \pi^*(\theta_1)$ (or $\pi^*(\theta_2)$) and consider $\text{Ind}_H^G(W) = V$. Again, it is easy to compute χ_V:

$$\chi_V(c_1) = 5, \qquad \chi_V(c_2) = 1,$$
$$\chi_V(c_3) = \exp(2\pi i/3) + \exp(4\pi i/3) = -1,$$
$$\chi_V(c_4) = \chi_V(c_5) = 0.$$

Now τ does not appear in W by Frobenius reciprocity, so that implies here (by considering degrees) that V is irreducible (or alternatively one may calculate that $\langle \chi_V, \chi_V \rangle = 1$) so $V = \alpha_5$ and its character is given above.

Now we are left with determining the characters of the two irreducible representations of degree 3. To find these, let $H = \mathbf{Z}_5$ be the subgroup generated by the 5-cycle $(1\,2\,3\,4\,5)$. Then H has a system of left coset representatives

$$\{1, (1\,4)(2\,3), (2\,4\,3), (1\,4\,2), (2\,3\,4), (1\,4\,3),$$
$$(1\,2)(3\,4), (1\,3)(2\,4), (1\,2\,3), (1\,3\,4), (1\,2\,4), (1\,3\,2)\}$$
$$= \{g_1, \ldots, g_{12}\}.$$

Again, $g_i^{-1}(c_1)g_i = c_1$ for every i. Otherwise, one can check that $g_i^{-1}(c_j)g_i \notin H$ except in the following cases: $g_1^{-1}(c_4)g_1 = c_4$, $g_1^{-1}(c_5)g_1 = c_5$, and $g_2^{-1}((1\,2\,3\,4\,5))g_2 = (1\,5\,4\,3\,2) = (1\,2\,3\,4\,5)^{-1}$ and $g_2^{-1}((1\,3\,5\,2\,4))g_2 = (1\,4\,2\,5\,3) = (1\,3\,5\,2\,4)^{-1}$.

Now let $W = \theta_j$ and let $V = \text{Ind}_H^G(W)$. Again by Theorem 5.20 we compute $\chi_V(c_1) = 12$, $\chi_V(c_2) = \chi_V(c_3) = 0$, and

$$\left.\begin{array}{l} \chi_V(c_4) = \exp(2\pi i/5) + \exp(8\pi i/5) \\ \chi_V(c_5) = \exp(4\pi i/5) + \exp(6\pi i/5) \end{array}\right\} \quad \text{if } j = 1 \text{ or } 4,$$

$$\left.\begin{array}{l} \chi_V(c_4) = \exp(4\pi i/5) + \exp(6\pi i/5) \\ \chi_V(c_5) = \exp(2\pi i/5) + \exp(8\pi i/5) \end{array}\right\} \quad \text{if } j = 2 \text{ or } 3.$$

In any case, one has that τ does not appear in V (by either Frobenius reciprocity or calculating $\langle \chi_V, \chi_1 \rangle = 0$) and α_4 and α_5 each appear in V with multiplicity 1 (by calculating $\langle \chi_V, \chi_4 \rangle = \langle \chi_V, \chi_5 \rangle = 1$), so their complement is an irreducible representation of degree 3 (which checks with $\langle \chi_V, \chi_V \rangle = 3$), whose character is $\chi_V - \chi_4 - \chi_5$.

Choose $j = 1$ (or 4) and denote this representation by α_3, and choose $j = 2$ (or 3) and denote this representation by α_3' (and note that they are distinct as their characters are unequal). Then we may calculate that

$$\chi_3(c_1) = \chi_3'(c_1) = 3$$
$$\chi_3(c_2) = \chi_3'(c_2) = -1$$
$$\chi_3(c_3) = \chi_3'(c_3) = 0$$
$$\chi_3(c_4) = 1 + \exp(2\pi i/5) + \exp(8\pi i/5) = (1 + \sqrt{5})/2$$
$$\chi_3(c_5) = 1 + \exp(4\pi i/5) + \exp(6\pi i/5) = (1 - \sqrt{5})/2$$

and vice-versa for χ_3', agreeing with Example 4.29.

Our last main result in this section is Mackey's theorem, which will generalize Corollary 5.10, but, more importantly, give a criterion for an induced representation to be irreducible. We begin with a pair of subgroups K, H of G. A K-H double coset is

$$KgH = \{kgh : k \in K, \quad h \in H\}.$$

It is easy to check that the K-H double cosets partition G (though, unlike for ordinary cosets, they need not have the same cardinality).

We shall also refine our previous notation slightly. Let $\sigma : H \to \mathrm{Aut}(W)$ be a representation of H. For $g \in G$, we shall set $H^g = g^{-1}Hg$ and we will let σ^g be the representation $\sigma^g : H^g \to \mathrm{Aut}(W)$ by $\sigma^g(h) = \sigma(ghg^{-1})$, for $h \in H^g$. Finally, let us set $H_g = H^g \cap K$. We regard any representation of H, given by $\sigma : H \to \mathrm{Aut}(W)$, as a representation of H^g by σ^g and, hence, as a representation of H_g by the restriction of σ^g to H_g. (In particular, this applies to the regular representation $\mathbf{F}(H)$ of H.)

(5.22) Theorem. (Mackey) *As* $(\mathbf{F}(K), \mathbf{F}(H))$-*bimodules,*

$$\mathbf{F}(G) \cong \bigoplus_g \mathbf{F}(K) \otimes_{\mathbf{F}(H_g)} \mathbf{F}(H)$$

where the sum is taken over a complete set of K-H double coset representatives.

Proof. For simplicity, let us write the right-hand side as $\oplus_g (\mathbf{F}(K) \otimes \mathbf{F}(H))_g$. Define maps α and $\widetilde{\beta}$ as follows:

For $g' \in G$, write g' as $g' = kgh$ for $k \in K$, $h \in H$, and g one of the given double coset representatives, and let $\alpha(g') = (k \otimes h)_g$. We must check that α is well defined. Suppose that $g' = \overline{k}g\overline{h}$ with $\overline{k} \in K$ and $\overline{h} \in H$. We need to show $(\overline{k} \otimes \overline{h})_g = (k \otimes h)_g$. Now $kgh = g' = \overline{k}g\overline{h}$ gives $g^{-1}k^{-1}\overline{k}g = h\overline{h}^{-1}$, and then

$$\begin{aligned}
(\overline{k} \otimes \overline{h})_g &= (k(k^{-1}\overline{k}) \otimes (\overline{h}h^{-1})h)_g \\
&= (k \otimes g^{-1}(k^{-1}\overline{k})g(\overline{h}h^{-1})h)_g \\
&= (k \otimes h)_g,
\end{aligned}$$

as required. Then α extends to a map on $\mathbf{F}(G)$ by linearity. Conversely, define β_g on $K \times H$ by $\beta_g(k, h) = kgh \in G$ and extend β_g to a map

$$\beta_g : \mathbf{F}(K) \times \mathbf{F}(H) \to \mathbf{F}(G)$$

by linearity. Then for any $x \in H_g$, we have

$$\beta_g(kx, h) = kxgh = kg(g^{-1}xg)h = \beta_g(k, g^{-1}xgh),$$

so β_g is $\mathbf{F}(H_g)$-middle linear and so defines

$$\widetilde{\beta}_g : (\mathbf{F}(K) \otimes \mathbf{F}(H))_g \to \mathbf{F}(G).$$

Set $\widetilde{\beta} = \prod \widetilde{\beta}_g$. Then it is easy to check that α and $\widetilde{\beta}$ are inverses of each other, yielding the theorem. $\square$

Note that the subgroup H_g depends not only on the double coset KgH, but on the choice of representative g. However, the modules involved in the statement of Mackey's theorem are independent of this choice. We continue to use the notation of the preceding proof.

(5.23) Proposition. *Let g and $\overline{g}$ be in the same K-H double coset. Then $(\mathbf{F}(K) \otimes \mathbf{F}(H))_g$ is isomorphic to $(\mathbf{F}(K) \otimes \mathbf{F}(H))_{\overline{g}}$ as $(\mathbf{F}(K), \mathbf{F}(H))$-bimodules.*

Proof. Let $g = \overline{k}\overline{g}\overline{h}$ with $\overline{k} \in K$, $\overline{h} \in H$, and define $\alpha : K \times H \to K \times H$ by $\alpha(k, h) = (k\overline{k}, \overline{h}h)$. Extend to

$$\alpha : \mathbf{F}(K) \times \mathbf{F}(H) \to \mathbf{F}(K) \times \mathbf{F}(H)$$

by linearity, thus giving

$$\alpha : \mathbf{F}(K) \times \mathbf{F}(H) \to (\mathbf{F}(K) \otimes \mathbf{F}(H))_{\overline{g}}.$$

We show that α is middle linear:

Let $x \in H_g$ be arbitrary. Then

$$\begin{aligned}
\alpha(kx, h) &= kx\overline{k} \otimes \overline{h}h \\
&= k(x\overline{k}) \otimes \overline{h}h \\
&= k \otimes \overline{g}^{-1}(x\overline{k})\overline{g}\overline{h}h
\end{aligned}$$

and

$$\begin{aligned}
\alpha(k, g^{-1}xgh) &= k\overline{k} \otimes \overline{h}g^{-1}xgh \\
&= k \otimes (\overline{g}^{-1}\overline{k}\overline{g})\overline{h}g^{-1}xgh.
\end{aligned}$$

But $g = \overline{k}\overline{g}\overline{h}$, so

$$\begin{aligned}
(\overline{g}^{-1}\overline{k}\overline{g})\overline{h}g^{-1}xgh &= \overline{g}^{-1}\overline{k}\overline{g}\overline{h}(\overline{h}^{-1}\overline{g}^{-1}\overline{k}^{-1})x(\overline{k}\overline{g}\overline{h})h \\
&= \overline{g}^{-1}x\overline{k}\overline{g}\overline{h}h,
\end{aligned}$$

and these are equal. Hence, we obtain a map

$$\widetilde{\alpha} : (\mathbf{F}(K) \otimes \mathbf{F}(H))_g \to (\mathbf{F}(K) \otimes \mathbf{F}(H))_{\overline{g}}.$$

Its inverse is constructed similarly, so $\widetilde{\alpha}$ is an isomorphism. $\square$

Remark. Note that $\widetilde{\alpha}$ depends on the choice of $\overline{k}$ and $\overline{h}$, which may not be unique, but we have not claimed that there is a unique isomorphism, only that the two bimodules are isomorphic.

(5.24) Corollary. *For any* **F**-*representation* W *of* H,

$$\text{Res}_K^G \text{Ind}_H^G(W) = \bigoplus_g (\mathbf{F}(K) \otimes_{F(H_g)} W) = \bigoplus_g \text{Ind}_{H_g}^K(W_g).$$

Proof. By Theorem 5.22 and associativity of the tensor product (Proposition 7.2.17), we have the following equalities among $\mathbf{F}(K)$-modules.

$$\text{Ind}_H^G(W) = \mathbf{F}(G) \otimes_{\mathbf{F}(H)} W$$

$$= \left(\bigoplus_g \mathbf{F}(K) \otimes_{\mathbf{F}(H_g)} \mathbf{F}(H) \right) \otimes_{\mathbf{F}(H)} W$$

$$= \bigoplus_g \mathbf{F}(K) \otimes_{\mathbf{F}(H_g)} \left(\mathbf{F}(H) \otimes_{\mathbf{F}(H)} W \right)$$

$$= \bigoplus_g \mathbf{F}(K) \otimes_{\mathbf{F}(H_g)} W.$$

(Note that although H_g is a subgroup of H, its action on $\mathbf{F}(H)$ is not the usual action, so one needs to check that $\mathbf{F}(H)$ is indeed an $(\mathbf{F}(H_g), \mathbf{F}(H))$-bimodule, but this is immediate.) $\square$

In the final term in the statement of the corollary, we have denoted the representation by W_g to remind the reader of the action of H_g.

(5.25) Lemma. *Let* G *be a finite group,* H *a subgroup of* G, *and* **F** *a field of characteristic zero or relatively prime to the order of* H. *Let* W *be an* **F**-*representation of* H *defined by* $\sigma : H \rightarrow \text{Aut}(W)$ *and set* $V = \text{Ind}_H^G(W)$. *Then*

$$\text{End}_G(V) \cong \bigoplus_g \text{Hom}_{H_g}(W_g, \text{Res}_{H_g}^H(W))$$

where the direct sum is over a complete set of H-H *double cosets,* $H_g = H^g \cap H$, *and* W_g *is the representation of* H_g *on* W *defined by* σ^g.

Proof. The required isomorphism is a consequence of the following chain of equalities and isomorphisms.

$$\text{End}_G(V) = \text{Hom}_G(V, V)$$

$$= \text{Hom}_H(W, V)$$

as in the proof of Frobenius reciprocity

$$\cong \text{Hom}_H(V, W)$$

as by our assumption on **F**, $\mathbf{F}(H)$ is semisimple

$$= \operatorname{Hom}_{\mathbf{F}(H)} \left(\bigoplus_g \mathbf{F}(H) \otimes_{\mathbf{F}(H_g)} W, W \right)$$

$$= \bigoplus_g \operatorname{Hom}_{\mathbf{F}(H)} \left(\mathbf{F}(H) \otimes_{\mathbf{F}(H_g)} W, W \right).$$

Now note that in the first W above, H_g is operating by σ^g; while in the second, it is operating by σ

$$= \bigoplus_g \operatorname{Hom}_{\mathbf{F}(H_g)} \left(W_g, \operatorname{Res}_{H_g}^H(W) \right)$$

by adjoint associativity of Hom and tensor product (Theorem 7.2.20).   □

(5.26) Corollary.

(1) Let G be a finite group, H a subgroup of G, and $\mathbf{F}$ a field of character-istic zero or relatively prime to the order of H. Let W be a represen-tation of H and set $V = \operatorname{Ind}_H^G(W)$. Then $\operatorname{End}_G(V) = \mathbf{F}$ if and only if $\operatorname{End}_H(W) = \mathbf{F}$ and $\operatorname{Hom}_{H_g}(W_g, \operatorname{Res}_{H_g}^H(W)) = 0$ for every $g \in G$, $g \notin H$.

(2) Let $\mathbf{F}$ be an algebraically closed field of characteristic zero or relatively prime to the order of G. Then V is irreducible if and only if W is irreducible and, for each $g \in G$, $g \notin H$, the H_g-representations W_g and $\operatorname{Res}_{H_g}^H(W)$ are disjoint (i.e., have no mutually isomorphic irre-ducible components). In particular, if H is a normal subgroup of G, V is irreducible if and only if W is irreducible and distinct from all its conjugates.

Proof. Note $\operatorname{End}_G(V)$ contains a subspace isomorphic to $\operatorname{End}_H(W)$ (given by the double coset representative $g = 1$). Also, under our assumptions V is irreducible if and only if $\operatorname{End}_G(V) = \mathbf{F}$, and similarly for W.   □

(5.27) Example. Let $H = \mathbf{Z}_m$ and $G = D_{2m}$. Then the representations θ_i and θ_{m-i} of H are conjugate, so for $i \neq m/2$, $\operatorname{Ind}_H^G(\theta_i)$ is irreducible. Of course, this representation is just ϕ_i.

(5.28) Example. Let $H = A_5$ and $G = S_5$. As a system of coset representa-tives, we choose $\{1, (1\,2\,5\,4)\} = \{g_1, g_2\}$. Note that

$$g_2(1\,2\,3\,4\,5)g_2^{-1} = (1\,3\,5\,2\,4) = (1\,2\,3\,4\,5)^2.$$

Thus, if we let σ^1 be the representation α_3 of A_5 (in the notation of Ex-ample 4.29) we see that its conjugate $\sigma^2 = \alpha_3'$, a distinct irreducible repre-sentation. Hence by Corollary 5.26, $\operatorname{Ind}_H^G(\alpha_3) = \operatorname{Ind}_H^G(\alpha_3')$ is an irreducible representation of degree 6, and by Theorem 5.20 we may compute its char-acter, giving an alternative to the method of Example 4.30.

(5.29) Example. Let $H = V$ and $G = A_4$ in the notation of Example 5.18. Then the representations λ_1, λ_2, and λ_3 of H are mutually conjugate, so $\text{Ind}_H^G(\lambda_i)$ is irreducible and is equal to α for $i = 1, 2, 3$, verifying the result of Example 5.18.

(5.30) Example. Let m be such that $p = 2^m - 1$ is prime. Then $S = \mathbf{Z}_p$ acts on $V = (\mathbf{Z}/2)^m$ by cyclically permuting the elements of V other than the identity. (This may most easily be seen by observing that $\text{GL}(m, \mathbf{F}_2)$ has an element of order p.) Thus, we may form the semidirect product

$$1 \longrightarrow V \longrightarrow G \overset{\pi}{\longrightarrow} S \longrightarrow 1$$

with G a group of order $n = 2^m(2^m - 1)$. G has the one-dimensional complex representations $\pi^*(\theta_i)$ for $i = 0, \ldots, p - 1$. Also, if σ is *any* nontrivial complex representation of V, G has the representation $\text{Ind}_V^G(\sigma)$ of degree $[G : V] = 2^m - 1$. Now σ is disjoint from all its conjugates (as $\text{Ker}(\sigma)$ may be considered to be an $\mathbf{F}_2$-vector space of dimension $m - 1$, and $\text{GL}(m - 1, \mathbf{F}_2)$ does not have an element of order p), so by Lemma 5.25, σ is irreducible. As $(2^m - 1)^2 + 2^{m-1}(1)^2 = n$, these 2^m complex representations are all of the irreducible complex representations of G. (Note that if $m = 2$, then $G = A_4$, so this is a generalization of Example 5.29.)

8.6 Permutation Representations

We have already encountered permutation representations, but because of their particular importance, we wish to discuss them here further. We shall restrict our attention to complex representations, so that we may use the full power of character theory. We begin with a bit of recapitulation.

(6.1) Definition. *Let $P = \{p_i\}_{i \in I}$ be a set and $\sigma : G \to \text{Aut}(P)$ a homomorphism. Then σ defines a representation on $\mathbf{C}P$ (=complex vector space with basis P) by*

$$\left(\sum_{g \in G} a_g g \right) \left(\sum_{i \in I} a_i p_i \right) = \sum_{g, i} a_g a_i \sigma(g)(p_i).$$

We will often call this representation σ as well.

For simplicity, we shall assume throughout that G and P are both finite, though some of our results hold more generally.

(6.2) Definition. *Let* $p \in P$. *Then*

$$\mathcal{O}_p = \text{Orbit}(p) = \{p' \in P : \sigma(g)(p) = p' \quad \text{for some } g \in G\}$$

$$G_p = \text{Stab}(p) = \{h \in G : \sigma(h)(p) = p\}.$$

Note that $\text{Orbit}(p)$ is a subset of P and $\text{Stab}(p)$ is a subgroup of G. ($\text{Orbit}(p)$ and $\text{Stab}(p)$ have been previously defined in Definition 1.4.9.)

(6.3) Definition. *A nonempty subset Q of P is a* **domain of transitivity** *for σ if $\sigma(g)(q) \in Q$ for every $g \in G$ and $q \in Q$. If P is the only domain of transitivity for σ (which holds if and only if $\mathcal{O}_p = P$ for every $p \in P$) then σ is called* **transitive**, *otherwise* **intransitive**.

The following is obvious:

(6.4) Lemma. *Let $Q_1, \ldots, Q_k$ partition P into domains of transitivity. Then*

$$CP = CQ_1 \oplus \cdots \oplus CQ_k.$$

Proof. □

Recall the following basic result of Corollary 5.8:

(6.5) Theorem. *Let σ be a transitive permutation representation of G on P. Let $p \in P$ and set $H = G_p$. Then*

$$\sigma = \text{Ind}_H^G(\tau).$$

Proof. □

(Recall that in this situation all of the subgroups G_p, for $p \in P$, are conjugate and we may choose H to be any one of them.)

Because of Lemma 6.4, we shall almost always restrict our attention to transitive representations, though we state the next two results more generally.

(6.6) Proposition. *Let P be partitioned into k domains of transitivity under the representation σ of G. Then the multiplicity of τ in CP is equal to k.*

Proof. By Lemma 6.4, we may assume $k = 1$. But then, by Theorem 6.5 and Frobenius reciprocity,

$$\langle \tau, \sigma \rangle = \langle \tau, \text{Ind}_H^G(\tau) \rangle = \langle \text{Res}_H^G(\tau), \tau \rangle = 1.$$

□

We record the following simple but useful result:

(6.7) Lemma. *Let σ_i be a permutation representation of σ on a set P_i for $i = 1, 2$. Then σ_1 and σ_2 are equivalent if and only if for each $g \in G$,*

$$|\{p_1 \in P_1 : \sigma_1(g)(p_1)\}| = |\{p_2 \in P_2 : \sigma_2(g)(p_2) = p_2\}|.$$

Proof. We know that σ_1 and σ_2 are equivalent if and only if their characters χ_1 and χ_2 are equal. But if χ is the character of a permutation representation σ of G on a set P, then its character is given by

$$\chi(g) = |\{p \in P : \sigma(g)(p) = p\}|.$$

$\square$

(6.8) Definition. *A partition of P into subsets $\{Q_i\}_{i \in I}$ is called a partition into **domains of imprimitivity** for σ if for every $g \in G$ and every $i \in I$ there exists a $j \in I$ with $\sigma(g)(Q_i) = Q_j$. If the only partitions into domains of imprimitivity are either the partition consisting of the single set $Q = P$, or the partition into subsets of P consisting of single elements, then σ is called **primitive**, otherwise **imprimitive**.*

(Note that an intransitive representation of G is certainly imprimitive.)
Let G be transitive, Q a subset of P with the property that for every $g \in G$, either $\sigma(g)(Q) = Q$ or $\sigma(g)(Q) \cap Q = \emptyset$. Set

$$H = \{g \in G : \sigma(g)(Q) = Q\}.$$

Then H is a subgroup of G, and if $\{g_i\}$ are a set of left coset representatives of H, then $Q_i = g_i(Q)$ partitions P into domains of imprimitivity. Furthermore, all partitions into domains of imprimitivity arise in this way. Note that H acts as a group of permutations on Q. We have the following result, generalizing Theorem 6.5, which also comes from Corollary 5.8.

(6.9) Theorem. *Let σ be a transitive permutation representation of G on P, Q a domain of imprimitivity for σ, and $H = \{g \in G : \sigma(g)(Q) = Q\}$. If ρ denotes the permutation representation of H on Q given by $\rho(h)(q) = \sigma(h)(q)$ for $h \in H$, $q \in Q$, then ρ is a transitive permutation representation of H on Q and*

$$\sigma = \text{Ind}_H^G(\rho).$$

Proof. $\square$

We have the following useful proposition:

(6.10) Proposition. *Let σ be a transitive permutation representation of G on a set P, and let $p \in P$.*

(1) Let P be partitioned into a set of domains of imprimitivity $\{Q_i\}$ with $p \in Q_1$, and let

$$H = \{g \in G : \sigma(g)(Q_1) = Q_1\}.$$

Then there are $[G : H]$ sets in the partition, each of which has $[H : G_p]$ elements.

(2) Let H be a subgroup of G containing G_p, and let

$$Q = \{\sigma(g)(p) : g \in H\}.$$

Then Q is one element in a partition of G into $[G : H]$ domains of imprimitivity, each of which contains $[H : G_p]$ elements.

Proof. Clear from the remark preceding Theorem 6.9, identifying domains of imprimitivity with left cosets. $\square$

(6.11) Corollary. A transitive permutation representation σ of G on a set P is primitive if and only if some (and hence every) G_p is a maximal subgroup of G.

Proof. $\square$

(6.12) Corollary. Let σ be a transitive permutation representation on a set P with $|P|$ prime. Then σ is primitive.

Proof. By Proposition 6.10, the cardinality of a domain of imprimitivity for G must divide $[G : G_p] = |P|$, and so consists of either a single element or all of P. $\square$

We now introduce another sort of property of a representation.

(6.13) Definition. A permutation representation σ of a group G on a set P is k-**fold transitive** if P has at least k elements and for any pair $(p_1, \ldots, p_k)$ and $(q_1, \ldots, q_k)$ of k-tuples of distinct elements of P, there is a $g \in G$ with $\sigma(g)(p_i) = q_i$ for $i = 1, \ldots, k$. 2-fold transitive is called doubly transitive.

(6.14) Examples.

(1) Note that 1-fold transitive is just transitive.
(2) The permutation representation of D_{2n} on the vertices of an n-gon is doubly transitive if $n = 3$, but only singly ($= 1$-fold) transitive if $n > 3$.
(3) The natural permutation representation of S_n on $\{1, \ldots, n\}$ is n-fold transitive, and of A_n on $\{1, \ldots, n\}$ is $(n-2)$-fold (but not $(n-1)$-fold) transitive.
(4) The natural permutation representation of S_n on (ordered or unordered) pairs of elements of $\{1, \ldots, n\}$ is transitive but **not** doubly

transitive for $n > 3$, for there is no $g \in S_n$ taking $\{(1,2), (2,3)\}$ to $\{(1,2), (3,4)\}$.

Doubly transitive permutation representations have two useful properties.

(6.15) Proposition. *Let σ be a doubly transitive permutation representation of G on a set P. Then σ is primitive.*

Proof. Suppose σ is not primitive, and let Q be a domain of imprimitivity, with $p_1, p_2 \in Q$ and $p_3 \notin Q$. Then there is no $g \in G$ with $\sigma(g)(p_1) = p_1$, and $\sigma(g)(p_2) = p_3$, so σ is not doubly transitive. The proposition follows by contraposition. $\qquad\square$

(6.16) Theorem. *Let σ be a transitive permutation representation of G on a set P. Then σ is doubly transitive if and only if $\sigma = \tau \oplus \sigma'$ for some irreducible representation σ' of G.*

Proof. Since σ is a permutation representation, its character χ is real valued.

Note that χ^2 is the character of the permutation representation $\sigma \otimes \sigma$ on $P \times P$. Let k be the number of orbits of $\sigma \otimes \sigma$ on $P \times P$. Note that $k = 2$ if σ is doubly transitive (the orbits being $\{(p,p) : p \in P\}$ and $\{(p,q) : p, q \in P, p \neq q\}$) and $k > 2$ otherwise. Then, by Lemma 6.4,

$$k = \langle \sigma \otimes \sigma, \tau \rangle$$
$$= \frac{1}{n} \sum_{g \in G} \chi^2(g)$$
$$= \frac{1}{n} \sum_{g \in G} \chi(g)\chi(g)$$
$$= \frac{1}{n} \sum_{g \in G} \chi(g)\overline{\chi}(g)$$
$$= \langle \chi, \chi \rangle.$$

Note that τ is a subrepresentation of σ by Lemma 6.4; also note that $\langle \chi, \chi \rangle = 2$ if and only if in the decomposition of σ into irreducibles there are exactly two distinct summands, yielding the theorem. $\qquad\square$

(6.17) Example. The representation β of Example 4.29 was doubly transitive and decomposed as $\tau \oplus \alpha_4$, α_4 irreducible. The representation γ of Example 4.29 was not doubly transitive; $\gamma \otimes \gamma$ had three orbits on $\{1, \ldots, 5\} \times \{1, \ldots, 5\}$, and γ decomposed into a sum $\tau \oplus \alpha_4 \oplus \alpha_5$ of three distinct irreducibles.

(6.18) Remark. The converse of Proposition 6.15 is false. For example, let p be a prime and consider the permutation representation of D_{2p} on the vertices of a p-gon. By Corollary 6.12, this representation is primitive, but for $p > 3$ it is not doubly transitive.

(6.19) Example. The action of a group on itself by left multiplication is a permutation representation. Of course, this is nothing other than the regular representation, which we have already extensively studied. Instead, we consider the action of a group on itself by **conjugation**, i.e., $\gamma(g)(h) - ghg^{-1}$ for all $g, h \in G$. Let us determine some examples of this representation. Of course, if G is abelian, this representation is just $n\tau$. In any case, the orbits of this representation are just the conjugacy classes. In general, if $C_1, \ldots, C_t$ are the conjugacy classes of G (in some order), we will let γ_i be γ on C_i, so

$$\gamma = \gamma_1 \oplus \cdots \oplus \gamma_t.$$

(1) Consider D_{2m} for m odd. Then (cf. Example 3.7) we have

$$C_1 = \{1\}, \quad C_2 = \{x, x^{m-1}\}, \quad \ldots, \quad C_{(m+1)/2} = \{x^{(m-1)/2}, x^{(m+1)/2}\},$$
$$C_{(m+3)/2} = \{y, xy, \ldots, x^{m-1}y\}.$$

Then $\gamma_1 = \tau$, and for $i = 2, \ldots, (m+1)/2$, γ_i is a nontrivial representation containing τ as a subrepresentation. Since D_{2m} only has one nontrivial one-dimensional representation, namely, ψ_-, we see that

$$\gamma_i = \tau \oplus \psi_- \qquad \text{for} \quad 2 \leq i \leq (m+1)/2.$$

We are left with the action on $C_{(m+3)/2}$, and we claim this is

$$\mathbf{F}P = \tau \oplus \phi_1 \oplus \cdots \oplus \phi_{(m-1)/2}$$

(cf Example 1.4 (7)), where $P = \{$ vertices of a regular m-gon $\}$. This may be seen as follows: $C_{(m+3)/2}$ consists of the elements of order exactly two in D_{2m}, and each such fixes exactly one vertex of P. Thus we have a one-to-one correspondence

$$C_{(m+3)/2} \longleftrightarrow P$$

by

$$x^i y \longrightarrow \text{vertex } p_i \text{ fixed by } x^i y$$

and further, for any $g \in D_{2m}$, $g(x^i y)g^{-1}$ fixes the vertex $g(p_i)$, so the two actions of G are isomorphic.

(2) Consider D_{2m} for m even. Then (cf. Example 3.7) we have

$$C_1 = \{1\}, \quad C_2 = \{x, x^{m-1}\}, \quad \ldots, \quad C_{m/2} = \{x^{\frac{m}{2}-1}, x^{\frac{m}{2}+1}\},$$
$$C_{\frac{m}{2}+1} = \{x^{\frac{m}{2}}\}, \quad C_{\frac{m}{2}+2} = \{x^i y : i \text{ is even}\}, \quad C_{\frac{m}{2}+3} = \{x^i y : i \text{ is odd}\}.$$

Again, $\gamma_1 = \gamma_{\frac{m}{2}+1} = \tau$ and $\gamma_i = \tau \oplus \psi_{+-}$ for $i = 2, \ldots, m/2$ (as conjugation by x is trivial on C_i but conjugation by y is not). We will determine the last two representations by computing their characters. Of course, 1 acts trivially, as does $x^{m/2}$, being central, so

$$\chi_{\frac{m}{2}+2}(1) = \chi_{\frac{m}{2}+3}(1) = \chi_{\frac{m}{2}+2}(x^{m/2}) = \chi_{\frac{m}{2}+3}(x^{m/2}) = \frac{m}{2}.$$

Also, $x^j(x^iy)x^{-j} = x^{i+2j}y$, so for $i \neq m/2$, x^j fixes no element and

$$\chi_{\frac{m}{2}+2}(x^j) = \chi_{\frac{m}{2}+3}(x^j) = 0 \qquad \text{for} \quad j \neq \frac{m}{2}.$$

Now $y(x^iy)y^{-1} = yx^i = x^{-1}y$, so y fixes x^iy when $2i = 0$, i.e., $i = 0$ or $m/2$. Hence, if $m/2$ is odd,

$$\chi_{\frac{m}{2}+2}(y) = \chi_{\frac{m}{2}+3}(y) = 1,$$

while if $m/2$ is even

$$\chi_{\frac{m}{2}+2}(y) = 2 \qquad \text{and} \qquad \chi_{\frac{m}{2}+3}(y) = 0.$$

Similarly, if $m/2$ is odd,

$$\chi_{\frac{m}{2}+2}(xy) = 0 \qquad \text{and} \qquad \chi_{\frac{m}{2}+3}(xy) = 1,$$

while if $m/2$ is even,

$$\chi_{\frac{m}{2}+2}(xy) = 0 \qquad \text{and} \qquad \chi_{\frac{m}{2}+3}(xy) = 2.$$

Now by computation with the irreducible characters of D_{2m} (which may be read off from Example 1.4 (7)) we see the following:

For $m/2$ odd,

$$\gamma_{\frac{m}{2}+2} = \gamma_{\frac{m}{2}+3} = \psi_{++} \oplus \gamma'.$$

For $m/2$ even,

$$\gamma_{\frac{m}{2}+2} = \psi_{++} \oplus \psi_{-+} \oplus \gamma'$$

and

$$\gamma_{\frac{m}{2}+3} = \psi_{++} \oplus \psi_{--} \oplus \gamma'$$

where

$$\gamma' = \phi_2 \oplus \phi_4 \oplus \phi_6 \oplus \cdots \oplus \phi_k,$$

with $k = (m-1)/2$ for m odd and $k = \frac{m}{2} - 1$ for m even.

(3) Consider A_4. We adopt the notation of Example 3.9. One may then check that the following table is correct, where an entry is the number of elements of the given conjugacy class fixed by the given element (and hence the trace of that element in the representation on that conjugacy class):

	1	I	T	T^2
$\{1\}$	1	1	1	1
$\{I, J, K\}$	3	3	0	0
$\{T, TI, TJ, TK\}$	4	0	1	1
$\{T^2, T^2I, T^2J, T^2K\}$	4	0	1	1

Then computation with the characters of A_4 (cf. Example 4.31) gives

$$\gamma_1 = \tau$$
$$\gamma_2 = \tau \oplus \pi^*(\theta_1) \oplus \pi^*(\theta_2)$$
$$\gamma_3 = \gamma_4 = \tau \oplus \alpha.$$

8.7 Concluding Remarks

In this section we make three remarks. The first is only a slight variation of what we have already done, while the last two are results that fit in quite naturally with our development, but whose proofs are beyond the scope of this book.

(7.1) Definition. *Let* **F** *be a field of characteristic zero or prime to the order of the finite group* G. *The* **F-representation ring** $R_{\mathbf{F}}(G)$ *of the group* G *is the free* **Z**-*module with basis* $\{\sigma_i\}_{i=1}^t$, *the irreducible* **F**-*representations of* G. *The elements* $\sigma = \sum_{i=1}^t m_i \sigma_i$ *of* $R_{\mathbf{F}}(G)$ *are called* **virtual representations**, *and those with* $m_i \geq 0$ *for all* i *are called* **proper representations**.

The formation of $R_{\mathbf{F}}(G)$ is a special case of a more general construction.

(7.2) Definition. *Let* $\mathcal{R}$ *be a ring. Let* F *be the free abelian group with basis*

$$\{P : P \text{ is a projective } \mathcal{R}\text{-module}\},$$

and let N *be the subgroup spanned by*

$$\{M - M_1 - M_2 : \text{ there is a short exact sequence of } \mathcal{R}\text{-modules:}$$
$$0 \longrightarrow M_1 \longrightarrow M \longrightarrow M_2 \longrightarrow 0\}.$$

Then let $K(\mathcal{R}) = F/N$.

Thus $R_{\mathbf{F}}(G) = K(\mathbf{F}(G))$. This equality is as **Z**-modules. But $R_{\mathbf{F}}(G)$, in fact, has more structure. Recall that we have defined the intertwining number of two representations V and W of G, which we shall here denote by i_G. Then i_G extends by linearity to $R_{\mathbf{F}}(G)$. It is easy to check that i_G is then a symmetric bilinear form on $R_{\mathbf{F}}(G)$. (If $\mathbf{F}$ is algebraically closed, this form is isometric to $t[1]$, and the irreducible $\mathbf{F}$-representations $\{\sigma_j\}_{j=1}^t$ of G form an orthonormal basis of $R_{\mathbf{F}}(G)$ with respect to i_G.)

Also, let H be a subgroup of G, W a virtual $\mathbf{F}$-representation of H, and V a virtual $\mathbf{F}$-representation of G. Then Frobenius reciprocity is the equality

$$i_G(\operatorname{Ind}_H^G(W),\, V) = i_H(W,\, \operatorname{Res}_H^G(V)),$$

which says that in a certain sense induction and restriction are adjoints of each other. This can be made precise: induction and restriction are an example of what is known as a pair of "adjoint functors."

Finally, note that the tensor product of representations extends by linearity to $R_{\mathbf{F}}(G)$, so $R_{\mathbf{F}}(G)$ has the structure of a commutative ring (or $\mathbf{Z}$-algebra) as well. Note too that we may form the character of a virtual representation—the character of $\sum_{i=1}^t m_i\sigma_i$ is $\sum_{i=1}^t m_i\chi\sigma_i$. In the special case where $\mathbf{F} = \mathbf{C}$, if $\{\sigma_i\}_{i=1}^t$ are the irreducible complex representations of G with characters $\{\chi_i\}_{i=1}^t$, then the map $\sigma \mapsto \chi_\sigma$ gives a ring isomorphism between $R_{\mathbf{C}}(G)$ (under direct sum and tensor product) and $\mathbf{Z}[\chi_1, \dots, \chi_t]$ (under sum and product).

(7.3) *Remark.* The reader will recall that we defined a good field $\mathbf{F}$ for G in Definition 1.3 and an excellent one in Definition 3.1. We often used the hypothesis of excellence, but in our examples goodness sufficed. This is no accident. If $\mathbf{F}$ is a good field and $\mathbf{F}'$ an excellent field containing $\mathbf{F}$, then all $\mathbf{F}'$-representations of G are in fact defined over $\mathbf{F}$, i.e., every $\mathbf{F}'$-representation is of the form $V = \mathbf{F}' \otimes_{\mathbf{F}} W$, where W is an $\mathbf{F}(G)$-module. (In other words, if V is defined by $\sigma : G \to \operatorname{Aut}(V)$, then V has a basis $\mathcal{B}$ such that for every $g \in G$, $[\sigma(g)]_{\mathcal{B}}$ is a matrix with coefficients in $\mathbf{F}$.) This was conjectured by Schur and proven by Brauer. We remark that there is no proof of this on "general principles"; Brauer actually proved more, showing how to write all representations as linear combinations of particular kinds of induced representations. (Of course, in the easy case where G is abelian, we showed this result in Corollary 2.3.)

(7.4) *Remark.* In Theorem 4.28, we proved that the degrees d_i of complex irreducible representations of G are divisors of n, the order of G. If $\mathbf{F}$ is an algebraically closed field of characteristic p prime to the order of G, all we showed was that the degrees of the irreducible $\mathbf{F}$-representations are prime to p (Corollary 4.6). In fact, it is the case that the degrees d_i are independent of the characteristic (assuming, of course, that $(p, n) = 1$), and, indeed, one can obtain the characteristic p representations from the complex ones.

8.8 Exercises

1. Verify the assertions of Example 1.8 directly (i.e., not as consequences of our general theory or by character computations). In particular, for (3) and (4), find explicit bases exhibiting the isomorphisms.

2. Let $\pi : G \to H$ be an epimorphism. Show that a representation σ of H is irreducible if and only if $\pi^*(\sigma)$ is an irreducible representation of G.

3. Let H be a subgroup of G. Show that if $\mathrm{Res}_H^G(\sigma)$ is irreducible, then so is σ, but not necessarily conversely.

4. Verify the last assertion of Example 1.7.

5. Show that τ and $\mathcal{R}_0$ are the only irreducible $\mathbf{Q}$-representations of $\mathbf{Z}_p$.

6. Find all irreducible and all indecomposable $\mathbf{F}_p$-representations of $\mathbf{Z}_p$. ($\mathbf{F}_p$ denotes the unique field with p elements.)

7. In Example 3.8, compute $\pi^*(\theta_i) \otimes \rho$ in two ways: by finding an explicit basis and by using characters.

8. Do the same for the representations $\pi^*(\theta_i) \otimes \alpha$ of Example 3.9.

9. In Example 3.8 (resp., 3.9) prove directly that ρ (resp., α) is irreducible.

10. In Example 3.10, verify that the characteristic polynomials of $\alpha(U)$ and $\alpha'(U)$ are as claimed. Also, verify that $\pi^*(\psi_-) \otimes \pi^*(\phi_1) = \pi^*(\phi_1)$ both directly and by using characters.

11. Verify that D_8 and Q_8 have the same character table (cf. Remark 4.32).

12. Show that the representation ρ of Q_8 constructed in Example 3.8 cannot be defined over $\mathbf{R}$, although its character is real valued, but that 2ρ can (cf. Remark 4.33).

13. Let $\mathbf{F}$ and G be arbitrary. If α is an irreducible $\mathbf{F}$-representation of G and β is a 1-dimensional $\mathbf{F}$-representation of G, show that $\alpha \otimes \beta$ is irreducible.

14. Find an example to illustrate Remark 4.9 (1). (Hint: Let $G = D_6$.)

15. (a) Find an example to illustrate both phenomena remarked on in Remark 4.9 (2). (Hint: Let $G = D_6$.)
    (b) Show that neither phenomenon remarked on in Remark 4.9 (2) can occur if h is in the center of the group G.

16. Prove Lemma 4.18.

17. Prove Proposition 4.24.

18. Show that the following is the expanded character table of S_4:

	C_1	C_2	C_3	C_4	C_5
	1	6	3	8	6
τ	1	1	1	1	1
α	3	1	-1	0	1
$\pi^*(\phi_1)$	2	0	2	-1	0
$\pi^*(\psi_-) \otimes \alpha$	3	-1	-1	0	1
$\pi^*(\psi_-)$	1	-1	1	1	-1

19. Show that the following is the expanded character table of S_5 in two ways—by the method of Example 4.30 and the method of Example 5.28.

	C_1	C_2	C_3	C_4	C_5	C_6	C_7
	1	10	15	20	20	30	24
τ	1	1	1	1	1	1	1
$\widetilde{\alpha}_4$	4	2	0	1	-1	0	1
$\widetilde{\alpha}_5$	5	1	1	-1	1	-1	0
$\widetilde{\alpha}_6$	6	0	-2	0	0	0	1
$\varepsilon \otimes \widetilde{\alpha}_5$	5	-1	1	-1	-1	1	0
$\varepsilon \otimes \widetilde{\alpha}_4$	4	-2	0	1	1	0	-1
ε	1	-1	1	1	-1	-1	1

20. Let $\mathbf{F}$, G_1, and G_2 be arbitrary. If σ_i is an irreducible $\mathbf{F}$-representation of G_i for $i = 1, 2$, show that $\sigma_1 \otimes \sigma_2$ is an irreducible $\mathbf{F}$-representation of $G_1 \times G_2$.

21. If G_1 and G_2 are finite and $\mathbf{F}$ is excellent for $G_1 \times G_2$, show that all irreducible $\mathbf{F}$-representations of G_1 and G_2 are as in the last problem. (Hint: Use Theorem 3.4.) Also, show this without any restrictions.

22. Compute the "multiplication table" for irreducible complex representations of S_4, i.e., for any two irreducible representations σ_1 and σ_2, decompose $\sigma_1 \otimes \sigma_2$ into a direct sum of irreducibles. (Hint: Use characters.)

23. Do the same for A_5.

24. Do the same for S_5.

25. Let P_1 and P_2 be disjoint sets and let σ_i be a permutation representation of G on P_i for $i = 1, 2$. Show that $\sigma_1 \oplus \sigma_2$ is a permutation representation of G on $P_1 \cup P_2$ and that $\sigma_1 \otimes \sigma_2$ is a permutation representation of G on $P_1 \times P_2$.

26. Verify all the group and character computations of Example 5.21.

27. Find all systems of domains of imprimitivity for the permutation action of D_{2m} on the vertices of a regular m-gon.

28. Determine the conjugation representation of S_4 on itself (cf. Example 6.19).

29. (a) Let G be the nonabelian group of order 21. Find all irreducible complex representations of G.

    (b) More generally, let p and q be primes with p dividing $q - 1$, and let G be the nonabelian group of order pq. Find all irreducible complex representations of G. (Hint: They are all either one-dimensional or induced from one-dimensional representations.)

30. Let $\sigma : G \to \mathrm{Aut}(V)$ be an irreducible representation of G. If $g \in Z(G)$, the center of G, show that $\sigma(g)$ is a homothety.

Appendix

A.1 Equivalence Relations and Zorn's Lemma

If X is a set, a **binary relation** on X is a subset R of $X \times X$. If $(x, y) \in X \times X$ it is traditional to write xRy or $x \overset{R}{\sim} y$ instead of $(x, y) \in R$. The second notation is frequently shortened to $x \sim y$ and the relation is denoted by $\sim$ rather than R.

(A.1) Definition. *A binary relation* $\sim$ *on X is an* **equivalence relation** *if the following conditions are satisfied.*

(1) $x \sim x$ *for all $x \in X$. (reflexivity)*
(2) $x \sim y$ *implies* $y \sim x$. *(symmetry)*
(3) $x \sim y$ *and* $y \sim z$ *implies* $x \sim z$. *(transitivity)*

If $x \in X$ then we let $[x] = \{y \in X : x \sim y\}$. $[x]$ is called the **equivalence class** *of the element $x \in X$.*

Let X be a nonempty set. A **partition** of X is a family of nonempty subsets $\{A_i\}_{i \in I}$ of X such that $A_i \cap A_j = \emptyset$ if $i \neq j$ and $X = \bigcup_{i \in I} A_i$. Thus a partition of X is a collection of subsets of X such that each element of X is in exactly one of the subsets.

Partitions and equivalence relations are essentially the same concept as we see from the following two propositions.

(A.2) Proposition. *If $\sim$ is an equivalence relation on X, then the family of all equivalence classes is a partition of X.*

Proof. By Definition A.1 (1), if $x \in X$, then $x \in [x]$. Thus the equivalence classes are nonempty subsets of X such that

$$X = \bigcup_{x \in X} [x].$$

Now suppose that $[x] \cap [y] \neq \emptyset$. Then there is an element $z \in [x] \cap [y]$. Then $x \sim z$ and $y \sim z$. Therefore, by symmetry, $z \sim y$ and then, by transitivity, we conclude that $x \sim y$. Thus $y \in [x]$ and another application of transitivity shows that $[y] \subseteq [x]$ and by symmetry we conclude that $[x] \subseteq [y]$ and hence $[x] = [y]$.   □

(A.3) Proposition. *If $\{A_i\}_{i \in I}$ is a partition of X, then there is an equivalence relation on X whose equivalence classes are precisely the subsets A_i for $i \in I$.*

Proof. Define the relation $\sim$ on X by the rule $x \sim y$ if and only if x and y both are in the same subset A_i. Properties (1) and (2) of the definition of an equivalence relation are clear. Now check transitivity. Suppose that $x \sim y$ and $y \sim z$. Then x and y are in the same A_i, while y and z are in the same subset, which must be the same A_i that contains x since the family $\{A_i\}$ is pairwise disjoint. Thus, $\sim$ is in fact an equivalence relation.

Now suppose that $x \in A_i$. Then from the definition of $\sim$ we have that $x \sim y$ if and only if $y \in A_i$. Thus $[x] = A_i$ and the proposition is proved.   □

Remark. The partition of a set X into equivalence classes following an equivalence relation is a concept that is used repeatedly in the construction of quotient objects in algebra.

Now we give a brief introduction to Zorn's lemma.

(A.4) Definition. *Let X be a nonempty set. A binary relation $\leq$ on X is said to be a* **partial order** *if it satisfies the following:*

(1) $x \leq x$ *for all* $x \in X$. *(reflexivity)*
(2) $x \leq y$ *and* $y \leq x$ *implies* $x = y$.
(3) $x \leq y$ *and* $y \leq z$ *implies* $x \leq z$. *(transitivity)*

The standard example of a partially ordered set is the power set $\mathcal{P}(Y)$ of a nonempty set Y, where $A \leq B$ means $A \subseteq B$.

If X is a partially ordered set, we say that X is **totally ordered** if whenever $x, y \in X$ then $x \leq y$ or $y \leq x$. A **chain** in a partially ordered set X is a subset $C \subset X$ such that C with the partial order inherited from X is a totally ordered set.

If $S \subseteq X$ is nonempty, then an **upper bound** for S is an element $x_0 \in X$ (not necessarily in S) such that

$$s \leq x_0 \qquad \text{for all} \quad s \in S.$$

A **maximal element** of X is an element $m \in X$ such that

$$\text{if} \quad m \leq x, \qquad \text{then} \quad m = x.$$

Thus, m is maximal means that, whenever m and x are comparable in the partial order on X, $x \leq m$. It is possible for a partially ordered set to have many maximal elements, and it is possible for a partially ordered set to have no maximal elements. The most important criterion for the existence of maximal elements in a partially ordered set is Zorn's lemma.

(A.5) Proposition. (Zorn's Lemma) *Let X be a partially ordered set and assume that every chain in X has an upper bound. Then X has a maximal element.*

It turns out that Zorn's lemma is equivalent to the axiom of choice. We refer the reader to Section 16 of *Naive set theory* by P.R. Halmos for a detailed discussion of the derivation of Zorn's lemma from the axiom of choice. It is worth pointing out that many of the results proved using the axiom of choice (or its equivalent, Zorn's lemma) turn out to be equivalent to the axiom of choice. We mention only that the existence of a maximal ideal in a ring with identity (Theorem 2.2.16) is one such result.

Bibliography

The following is a collection of standard works in algebra:

1. E. Artin, *Galois Theory, Second Edition*, University of Notre Dame Press, South Bend, IN, 1959.
2. N. Bourbaki, *Algebra I, Chapters 1 - 3*, Springer-Verlag, New York, 1989.
3. N. Bourbaki, *Algebra II, Chapters 4 - 7*, Springer-Verlag, New York, 1990.
4. P. Cohn, *Algebra, Second Edition*, John Wiley, New York, Vol. I, 1982; Vol. II, 1989; Vol. III, 1991.
5. C.W. Curtis and I. Reiner, *Representation Theory of Finite Groups and Associative Algebras*, Interscience Publishers, New York, 1962.
6. L.G. Grove, *Algebra*, Academic Press, New York, 1983.
7. T.W. Hungerford, *Algebra*, Graduate Texts in Math. 73, Springer-Verlag, New York, 1974.
8. N. Jacobson, *Lectures in Abstract Algebra*, Van Nostrand, Princeton, NJ, Vol. I, 1951; Vol. II, 1953; Vol. III, 1964.
9. N. Jacobson, *Basic Algebra I*, W. H. Freeman, New York, 1985.
10. N. Jacobson, *Basic Algebra II*, W. H. Freeman, New York, 1989.
11. S. Lang, *Algebra*, Addison-Wesley, Reading, MA, 1965.
12. J.-P. Serre, *Linear Representations of Finite Groups*, Graduate Texts in Math. 42, Springer-Verlag, New York, 1977.
13. B.L. van der Waerden, *Modern Algebra, Vol. I, II*, Frederick Unger Publishers, New York, 1955.

Index of Notation

This list consists of all the symbols used in the text. Those without a page reference are standard set theoretic symbols; they are presented to establish the notation that we use for set operations and functions. The rest of the list consists of symbols defined in the text. They appear with a very brief description and a reference to the first occurrence in the text.

Symbol	Description	Page
$\subseteq$	set inclusion	
$A \subsetneqq B$	$A \subseteq B$ but $A \neq B$	
$\cap$	set intersection	
$\cup$	set union	
$A \setminus B$	everything in A but not in B	
$A \times B$	cartesian product of A and B	
$\lvert A \rvert$	cardinality of $A \quad (\in \mathbf{N} \cup \{\infty\})$	
$f : X \to Y$	function from X to Y	
$a \mapsto f(a)$	a is sent to $f(a)$ by f	
$1_X : X \to X$	identity function from X to X	
$f\lvert_Z$	restriction of f to the subset Z	
$\mathbf{N}$	natural numbers $= \{1, 2, \dots\}$	
$e, 1$	group identity	1
$\mathbf{Z}$	integers	2
$\mathbf{Z}^+$	nonnegative integers	54
$\mathbf{Q}$	rational numbers	2
$\mathbf{R}$	real numbers	2
$\mathbf{C}$	complex numbers	2
$\mathbf{Q}^*$	nonzero rational numbers	2
$\mathbf{R}^*$	nonzero real numbers	2
$\mathbf{C}^*$	nonzero complex numbers	2
$\mathbf{Z}_n$	integers modulo n	2
$\mathbf{Z}_n^*$	integers relatively prime to n (multiplication mod n)	2

Index of Terminology

Graduate Texts in Mathematics

(continued from page ii)